MW00453530

Antenna Theory and Design

Antenna Theory and Design

SECOND EDITION

Warren L. Stutzman
Gary A. Thiele

JOHN WILEY & SONS, INC.

New York • Chichester • Weinheim • Brisbane • Toronto • Singapore

ACQUISITIONS EDITOR	Charity Robey
MARKETING MANAGER	Katherine Hepburn
PRODUCTION EDITOR	Ken Santor
ILLUSTRATION COORDINATOR	Jaime Perea
COVER DESIGN	Carol Grobe

The cover art is the radiation pattern of a six-wavelength travelling wave line source and was created by George Ruhlmann.

This book was set in 10/12 Times Ten by University Graphics, Inc. and printed and bound by Hamilton Printing Company. The cover was printed by Phoenix Color Corp.

Recognizing the importance of preserving what has been written, it is a policy of John Wiley & Sons, Inc. to have books of enduring value published in the United States printed on acid-free paper, and we exert our best efforts to that end.

The paper in this book was manufactured by a mill whose forest management programs include sustained yield harvesting of its timberlands. Sustained yield harvesting principles ensure that the number of trees cut each year does not exceed the amount of new growth.

Library of Congress Cataloging in Publication Data:

Stutzman, Warren L.

Antenna theory and design / Warren L. Stutzman, Gary A. Thiele.—2nd ed.

p. cm.

Includes bibliographical references and index.

ISBN 0-471-02590-9 (cloth : alk. paper)

1. Antennas (Electronics) I. Thiele, Gary A. II. Title.

TK7874.6.S79 1998 97-35498

621.382'4—dc21 CIP

Printed in the United States of America

10 9 8 7 6 5 4

This book is dedicated to our wives,
Claudia and Jo Ann
and to our children
Darren and Dana
Eric, Scott, and Brad

Preface

Since the first edition of *Antenna Theory and Design* was published in 1981, there have been major expansions of application areas for antennas, mainly in wireless communications. In addition, in recent years new areas important to antennas have emerged into prominence. This second edition has been expanded to include new areas in antennas. Coverage of microstrip antennas (Sec. 5.8) as well as the use of antennas in systems and measurements (Chapter 9) have been added. In addition, the treatments of array antennas (Chapter 3), broadband antennas (Chapter 6), and aperture antennas (Chapter 7) have been expanded. Also, since the first edition was written there have been major advances in Computational Electromagnetics (CEM), due in part to the use of more sophisticated antennas and antenna systems. The second edition expands on the Method of Moments in Chapter 10, introduces a succinct treatment of the Finite Difference-Time Domain (FD-TD) technique for antennas in Chapter 11, and adds the topic of the Physical Theory of Diffraction (PTD) to high frequency methods in Chapter 12. The objective in the second edition has been to preserve the simplicity of the first edition, while adding modern topics.

This book is a textbook and finds its widest use in the college classroom. Thus, the primary purpose is to emphasize the understanding of principles and the development of techniques for examining and designing antenna systems. Handbooks are available to supplement the fundamentals and antennas discussed here. We have found that the first edition is in wide use by practicing engineers as well as students. This is because of the applied nature of the material and the treatment of basic topics that are directly useable for analyzing practical antennas. This is illustrated by the material in Chapters 1 to 6 and 9, which do not rely heavily on mathematics and use calculus sparingly.

Antenna Theory and Design covers antennas from three perspectives: antenna fundamentals, measurement techniques, and the design of popular antennas. The first four chapters stress antenna fundamentals. Since the student has probably had little exposure to antennas, many fundamentals are presented in Chapter 1. The emergence of antenna theory from Maxwell's equations is developed, along with a physical explanation of how antennas radiate. The four types of antenna elements (electrically small, resonant, broadband, and aperture) are introduced. The discipline of antennas has its own terminology that is quite different from other areas of engineering, so Chapter 1 includes definitions of many antenna terms. Chapter 2 examines simple radiating systems, such as dipoles, in order to solidify the principles of Chapter 1 and to equip the reader to move forward with analysis of antenna systems, such as arrays, that are treated in Chapter 3. Arrays are covered early in the book to introduce the relationship between the current distribution on an antenna and its spatial radiation characteristics using elementary mathematics. In addition, arrays are widely used in practice today. The discrete approach to antennas (arrays) is followed in Chapter 4 with line source antennas, which introduce the continuous form of antennas.

Chapters 5 to 7 give details on commonly used antenna elements. Chapter 5 surveys the resonant antenna elements encountered in practice, including dipoles, yagis, and microstrip patches. Chapter 6 covers broadband antennas such as helix, spiral, and log-periodic antennas. Chapter 7 treats aperture antennas. Emphasis in these chapters is on the operating principles using the fundamentals introduced in the Chapters 1 to 4 and on design guidelines. As appropriate, data are presented using numerical or experimental models, or computations based on theoretical formulations. In addition, empirical formulas are often presented for easy evaluation of performance parameters.

The synthesis of arrays and continuous antennas is presented in Chapter 8 for shaped main-beam or low side-lobe applications. The use of antennas as devices in systems is covered in Chapter 9, along with antenna measurements.

Chapters 10 to 12, as noted above, introduce CEM techniques for evaluating simple antenna elements as well as large complex antenna systems. Here, as in all the book, actual code statements are not listed. The wide variety of computing environments and the availability of high-level mathematics applications packages makes this inappropriate and unnecessary. Instead, some key computational and visualization antenna software packages are made available on the World Wide Web (see Appendix G).

It is important to be aware of the background that is assumed for this book. It is not necessary that the reader have complete mastery of the following subjects, but exposure to these topics is very helpful. A basic course in electromagnetics, such as is commonly required in engineering and physics, is assumed. Mathematics used often includes complex numbers, trigonometry, vector algebra, and the major coordinate systems (rectangular, cylindrical, and spherical). Vector calculus is used at various points and scalar integration is frequently used.

This book can be readily adapted to various academic programs at both introductory and advanced levels. For a first course, the text is usually used in a senior elective or entry level graduate course. A one-semester introductory course usually covers Chapter 1 to 6. For a master's degree-level course, parts of Chapters 7, 8, and 9 can be added. In a one-quarter senior course, material in the latter parts of Chapters 3, 4, or 5 can be eliminated. A second course can focus on advanced design, synthesis, and systems using Chapters 7, 8, and 9. Alternatively, a second course can specialize on computational methods using Chapters 10 to 12.

Several features have been included to aid in learning and in preparation for further self study. Defined terms follow the IEEE standard definitions. Literature references found at the end of each chapter provide sources for further reading. In addition, the bibliography in Appendix H lists literature sources by technical topic. The appendices also include information on the radio spectrum, data on materials, and important mathematical relations.

The authors are indebted to the many individuals who provided invaluable technical assistance to this second edition. The reviewers of the entire manuscript (two of whom also reviewed the first edition) gave essential input on the organization of the book and on several technical issues. Many students offered critical remarks during classroom testing of the manuscript. In addition, special thanks are owed to those who gave detailed evaluations, including Keith Carver (Secs. 5.8 and 6.2), David Jackson (Sec. 5.8), Ahmad Safaai-Jazi (Secs. 6.2 and 8.4), Dave Olver (Chapter 11), Buck Walter (Secs. 4.4 and 10.12), Gerald Ricciardi (Sec. 5.8), Marco Terada (Sec. 7.6) and Krish Pasala (Sec. 12.15). One author (Gary Thiele) extends special thanks to his son, Eric T. Thiele, for many long, valuable discussions on FD-TD, for reviewing Chapter 11, and for generating the numerical data used in several illustrations in Chapter 11.

Finally, we recognize our wives, Claudia and Jo Ann, for enduring countless hours of neglect during the preparation for both editions. The same recognition goes to our children, Darren and Dana, and Eric, Scott, and Brad.

Warren L. Stutzman
Gary A. Thiele

Contents

Chapter 1

Antenna Fundamentals and Definitions

1.1 INTRODUCTION

Communication between humans was first by sound through voice. With the desire for long distance communication came devices such as drums. Then, visual methods such as signal flags and smoke signals were used. These optical communication devices, of course, utilized the light portion of the electromagnetic spectrum. It has been only very recent in human history that the electromagnetic spectrum, outside the visible region, has been employed for communication, through the use of radio.

The radio antenna is an essential component in any radio system. An antenna is a device that provides a means for radiating or receiving radio waves. In other words, it provides a transition from a guided wave on a transmission line to a "free-space" wave (and vice versa in the receiving case). Thus, information can be transferred between different locations without any intervening structures. The possible frequencies of the electromagnetic waves carrying this information form the electromagnetic spectrum (the radio frequency bands are given in Appendix A). One of humankind's greatest natural resources is the electromagnetic spectrum and the antenna has been instrumental in harnessing this resource. A brief history of antenna technology [1–4] and a discussion of the uses of antennas follow.

Perhaps the first radiation experiment was performed in 1842 by Joseph Henry of Princeton University, the inventor of wire telegraphy. He "threw a spark" in a circuit in an upper room and observed that needles were magnetized by the current in a receiving circuit located in the cellar. This experiment was extended to a distance of over a kilometer. Henry also detected lightning flashes with a vertical wire on the roof of his house. These experiments marked the beginning of wire antennas.

Based on his observations in 1875 that telegraph key closures radiate, Thomas Edison patented a communication system in 1885 that employed top-loaded, vertical antennas.

The theoretical foundations for antennas rest on Maxwell's equations, which James Clerk Maxwell (1831–1879) presented before the Royal Society in 1864, that unify electric and magnetic forces into a single theory of electromagnetism. Maxwell also predicted that light is explained by electromagnetics and that light and electromagnetic disturbances both travel at the same speed.

In 1887 the German physicist Heinrich Hertz (1857–1894) was able to verify experimentally the claim of Maxwell that electromagnetic actions propagate through air. Hertz discovered that electrical disturbances could be detected with a single loop of the proper dimensions for resonance that contains an air gap for sparks to occur. The primary source of electrical disturbances studied by Hertz consisted of two metal plates in the same plane, each with a wire connected to an induction coil; this early antenna is similar to the capacitor-plate dipole antenna described in Section 2.1 and was called a "Hertzian dipole." Hertz also constructed loop antennas. Motivated by the need for more directive radiation, he also invented reflector antennas. In 1888 he constructed a parabolic cylinder reflector antenna from a sheet of zinc; see Fig. 1-1*a*. It was fed with a dipole along the focal line and operated at 455 MHz.

Guglielmo Marconi (1874–1937), an Italian inventor, also built a microwave parabolic cylinder reflector in 1895 for his original code transmission at 1.2 GHz. But his subsequent work was at lower frequencies for improved communication range. The transmitting antenna for the first transatlantic radio communication in 1901 consisted of a 70-kHz spark transmitter connected between the ground and a system of 50 wires, forming a 48-m tall fan monopole; see Fig. 1-1*b*. The antenna resembles a variation of the discone antenna described in Sec. 6.3. The receiving antenna was supported by kites.

Although Marconi is credited as the pioneer of radio, Mahlon Loomis (1826–1886), a dentist and inventor in Washington, DC, received a U.S. patent in 1872 for an "Improvement in Telegraphying" in which he described the use of an "aerial" to radiate and recieve "pulsations." In October 1866, Loomis demonstrated his wireless signaling system to U.S. senators in the Blue Ridge Mountains of Virginia using wire supported by kites at both the transmitting and receiving antennas about twenty miles apart.

The Russian physicist Alexander Popov (1859–1905) also recognized the importance of Hertz's discovery of radio waves and began working on ways of receiving

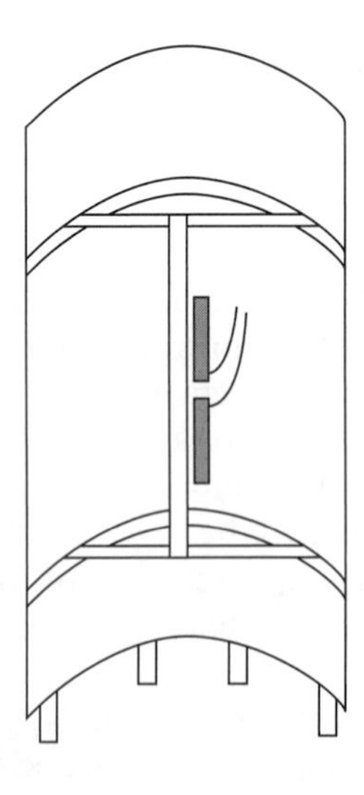

(*a*) The 455-MHz cylinder reflector antenna invented by Hertz in 1888

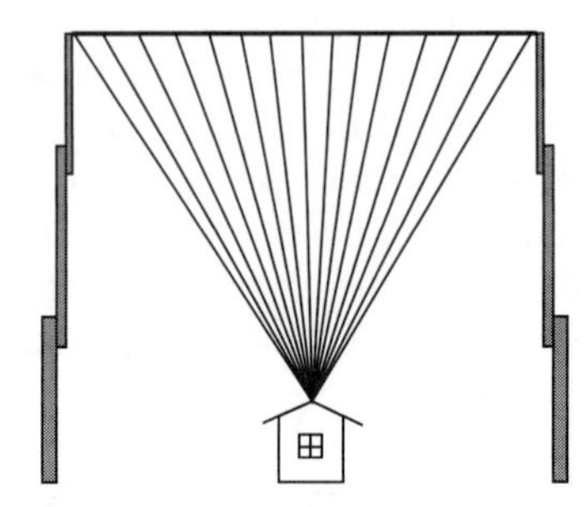

(*b*) The monopole transmitting antenna used by Marconi at 70 kHz for the first transatlantic radio communication

Figure 1-1 Examples of early antennas.

them a year before Marconi. He is sometimes credited with using the first antenna in the first radio system by sending a signal over a 3-mile ship-to-shore path in 1897. However, it was Marconi who developed radio commercially and also pioneered transoceanic radio communication. Marconi may be considered to be the father of what was then called wireless. Since then the term "radio" has been used, but "wireless" has also returned to popular use.

In 1912 the Institute of Radio Engineers was formed by the merger of the Wireless Institute and the Society of Radio Engineers. The importance of antennas is punctuated by the fact that the first article of the first issue of *Proceedings of the I.R.E.* was on antennas [5].

Antenna developments in the early years were limited by the availability of signal generators. Resonant length antennas (e.g., a half-wavelength dipole) of manageable physical size were possible about 1920 after the De Forest triode tube was used to produce continuous wave signals up to 1 MHz. Just before World War II, microwave (about 1 GHz) klystron and magnetron signal generators were developed along with hollow pipe waveguides. These led to the development of horn antennas, although Chunder Bose (1858–1937) in India produced the first electromagnetic horn antenna many years earlier. The first commercial microwave radiotelephone system in 1934 was operated between England and France at 1.8 GHz. The need for radar during the war spawned many "modern" antennas, such as large reflectors, lenses, and waveguide slot arrays [6].

Let us now direct our attention to the uses of antennas. Electromagnetic energy can be transported using a transmission line. Alternatively, no guiding structure is needed if antennas are used. For a transmitter-receiver spacing of R, the power loss of a transmission line is proportional to $(e^{-\alpha R})^2$, where α is the attenuation constant of the transmission line. If the antennas are used in a line of sight configuration, the power loss is proportional to $1/R^2$. Many factors enter into the decision of whether to use transmission lines or antennas. Generally speaking, at low frequencies and short distances transmission lines are practical. But high frequencies are attractive because of the available bandwidth. As distances become large and frequency increases, the signal losses and the costs of using transmission lines become large, and thus the use of antennas is favored. A notable exception to this is the fiber optic transmission line, which has very low loss. Transmission lines offer the advantages of not being subject to interference that is often encountered in radio systems and added bandwidth is achieved by laying new cable. However, there are significant costs and construction delays associated with cable.

In several applications, antennas *must* be used. For example, mobile communications involving aircraft, spacecraft, ships, or land vehicles require antennas. Antennas are also popular in broadcast situations where one transmit terminal can serve an unlimited number of receivers, which can be mobile (e.g., car radio). Nonbroadcast radio applications such as municipal radio (police, fire, rescue) and amateur radio also require antennas. Also, personal communication devices such as pagers and cellular telephones are commonplace.

There are also many noncommunication applications for antennas. These include remote sensing and industrial applications. Remote sensing systems are either active (e.g., radar) or passive (e.g., radiometry) and receive scattered energy or inherent emissions from objects, respectively. The received signals are processed to infer information about the objects or scenes. Industrial applications include cooking and drying with microwaves.

Other factors that influence the choice of the type of transmission system include historical reasons, security, and reliability. Telephone companies began interconnecting multiple transmit-receive terminals with transmission lines before radio technology was available. Currently, domestic telephone companies employ microwave radio and fiber optic transmission lines for long distance telephone calls. Satellite radio links are used heavily for international telephone calls. In addition, satellite-based communication systems are the primary means of distributing television program material to affiliate stations. Television programs distributed directly to consumers by satellite is increasing worldwide. Also, very small aperture terminals (VSAT) for satellite radio systems are widely used in private data networks to interconnect, for example, chain retail stores. Transmission lines inherently offer more security than radio. However, radio links using digital communications can be secured with coding techniques. Also, additional security is unnecessary for most communication systems. Another factor to be considered is reliability. For example, radio signals are affected by environmental conditions such as structures along the signal path, the ionosphere, and weather. Furthermore, interference is always a threat to radio systems. On the other hand, cables are vulnerable to being damaged by earthquakes or being accidentally dug up. All these factors must be examined together with the costs associated with using transmission lines and antennas. Cable systems often require expensive land purchase or lease. Every year radio equipment decreases in cost and increases in reliability. This tends to tip the scale in favor of radio systems. Cable and radio communication systems will continue to be used in the future with the choice depending on the specific application. For high reliability, both cable and radio are employed to provide diversity.

As we shall see, antennas cannot be miniaturized and replaced by a chip as often happens in electronics. Although long-used types of antennas will remain in use far into the future, new applications will require innovative antenna systems. For example, the demand for more communications is leading to personal communications systems (PCS) where each person will be freed of wire connections by carrying a small radio that can be used anywhere on the globe. The first of the seven IEEE New Technology Directions Committee grand challenges for electrotechnology is, "To make any person anywhere in the world reachable at his or her discretion, at any time by communication methods independent of connecting wires and cables." A "wireless" society is possible only through the use of antennas. Indeed, the future of antennas is very bright.

The next two sections of this first chapter provide a basic understanding of how antennas operate and an overview of the types of antennas encountered in practice. The remainder of Chap. 1 is devoted to developing the fundamental principles and terminology used throughout the book.

References to specific literature works are found at the end of the chapter where the citation occurs. In addition, a complete bibliography is found in Appendix H. About 150 books are listed by their topical coverage to aid the student in locating further details. The IEEE definitions of antenna terms [Ref. 1 in App. H] are followed very closely in this book.

1.2 HOW ANTENNAS RADIATE

Before we proceed with a mathematical development of antennas that is necessary for engineering design, it is instructive to explain the basic principles of radiation. Radiation is a disturbance in the electromagnetic fields that propagates away from the source of the disturbance so that the total power associated with the wave in a lossless medium is constant with radial distance. This disturbance is created by a time-varying current source that has an accelerated charge distribution associated with it. We, therefore, begin our discussion of radiation with a single accelerated charge.

Consider a single electric charge moving with constant velocity in the z-direction of Fig. 1-2. Prior to arrival at point A, the static electric field lines extend radially away from the charge to infinity and move with the charge. At point A, the charge begins to be accelerated (i.e., velocity is increased) until reaching point B, where it continues on at the acquired velocity. The static electric field (often called the Coulomb field) originates at the charge and is directed radially away from the charge. The radial field lines outside the circle of radius r_A in Fig. 1-2 originated when the charge was at point A. The circle of radius r_B is centered on point B, which is the charge position at the end of the acceleration period Δt. Interior to r_B, the electric field lines extend radially away from point B. The distance between the circles is that distance light would travel in time Δt, or $\Delta r = r_B - r_A = \Delta tc$. Since the charge moves slowly compared to the speed of light, $\Delta z \ll \Delta r$ and the circles are nearly concentric; the distance Δz in Fig. 1-2 is shown large relative to Δr for clarity. The electric field lines in the Δr region are joined together because of the required continuity of electric field lines in the absence of charges. This region is obviously one of disturbed field structure. This disturbance was caused by acceleration of the charge, which ended a time r_B/c earlier than the instant represented in Fig. 1-2. This

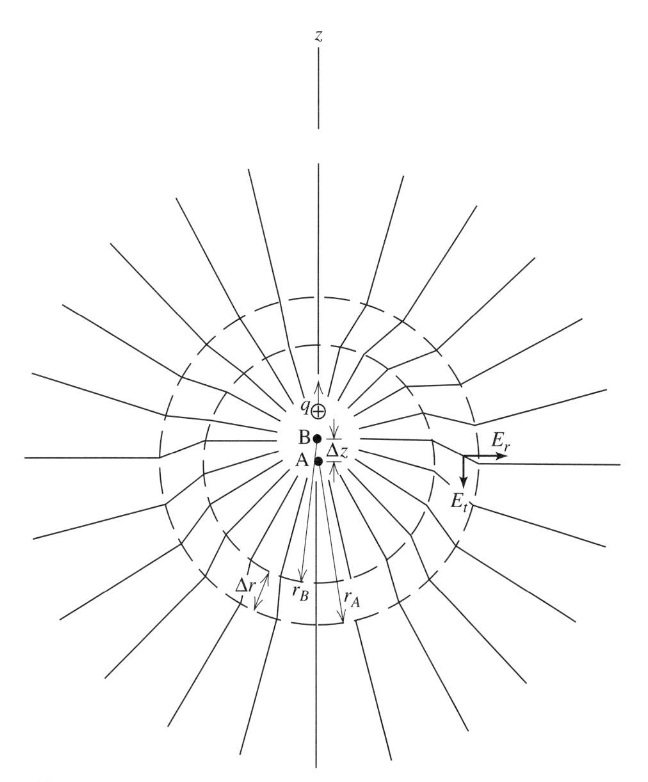

Figure 1-2 Illustration of how an accelerated charged particle radiates. Charge q moves with constant velocity in the $+z$-direction until it reaches point A (time $t = 0$), after which it accelerates to point B (time $t = \Delta t$) and then maintains its velocity. The electric field lines shown here are for a time r_B/c after the charge passed point B.

disturbance expands outward and has a transverse component E_t, which is the radiated field that persists as the disturbance propagates to infinity.

This example illustrates that radiation is a disturbance. It is directly analogous to a transient wave created by a stone dropped into a calm lake, where the disturbance of the lake surface continues to propagate radially away from the impact point long after the stone has disappeared. If charges are accelerated back and forth (i.e., oscillate), a regular disturbance is created and radiation is continuous. Antennas are designed to support charge oscillations.

The directional properties of radiation are evident in the accelerated charge example. The disturbance in Fig. 1-2 is maximum in a direction perpendicular to the charge acceleration direction, and we shall see in this chapter that maximum radiation occurs perpendicular to a straight wire antenna.

We can now explain how an actual antenna operates. To do this, we begin with the open-circuited transmission line of Fig. 1-3, which has a standing wave pattern with a zero current magnitude at the wire end and nulls every half wavelength from the end. The currents are in opposite directions on the wires, as indicated by arrows in Fig. 1-3*a*. In transmission lines, the conductors guide the waves and the power resides in the region surrounding the conductors as manifested by the electric and magnetic fields. The fields for the open-ended transmission line are shown in Fig. 1-3*a*. The electric fields originate from or terminate on charges on the wires and are

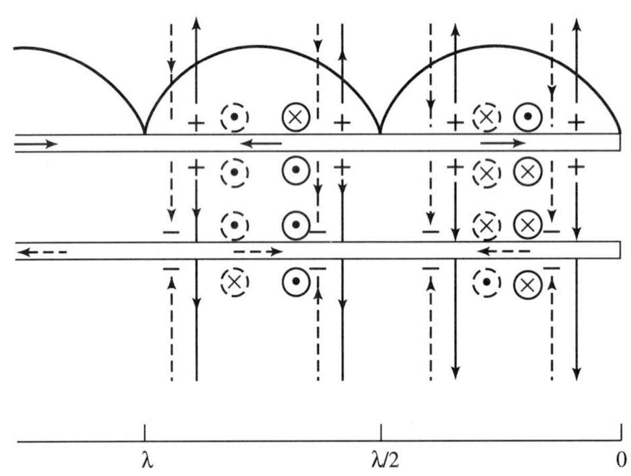

(*a*) Open-circuited transmission line showing currents, charges, and fields. The electric fields are indicated with lines and the magnetic fields with arrow heads and tails, solid (dashed) for those arising from the top (bottom) wire.

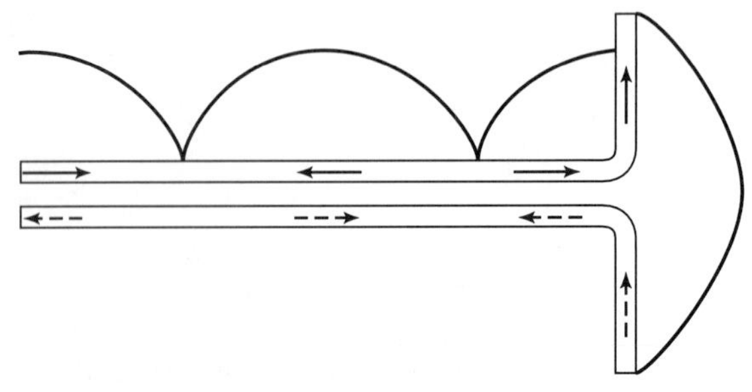

(*b*) Peak currents on a half-wavelength dipole created by bending out the ends of the transmission line.

Figure 1-3 Evolution of a dipole antenna from an open-circuited transmission line.

perpendicular to the wires. The magnetic fields encircle the wires. Note that all fields reinforce between the wires and cancel elsewhere. This is true for a wire spacing that is much smaller than a wavelength, as is usually the case. If the ends of the wires are bent outward as shown in Fig. 1-3*b*, the reinforced fields between the wires are exposed to space. Note that the currents on the vertical wire halves, which are each a quarter wavelength in this case, are no longer opposed as with the transmission line, but are both upwardly directed. In reality, the currents on the dipole are approximately sinusoidal as shown, but the transmission line currents are not pure standing waves due to the improved impedance match presented by the antenna compared to the open circuit. The situation of Fig. 1-3*b* is for a peak current condition. As time proceeds and current oscillations occur, disturbances are created that propagate away from the wire, much as the accelerated single charge.

The time dynamics of the fields associated with an oscillating dipole charge distribution are shown in Fig. 1-4 [7]. This is similar to the electrostatic dipole with equal, but opposite signed, separated charges. In this case, the charge distributions oscillate at frequency f. As the charge distributions at the ends oscillate, a current flows between them that is uniform with distance. This is the ideal dipole of Sec. 1.6. In Fig. 1-4, an oscillating current of frequency f (and period $T = 1/f$) was turned on a quarter period before $t = 0$. The upward-flowing current creates an excess of charges on the upper half of the dipole and a deficit of charges on the lower half. Peak charge buildup occurs at $t = 0$ as shown in Fig. 1-4*a* and produces a voltage between the dipole halves. The positive charges on the top are attracted to the negative charges on the bottom half of the dipole, creating a current. The current

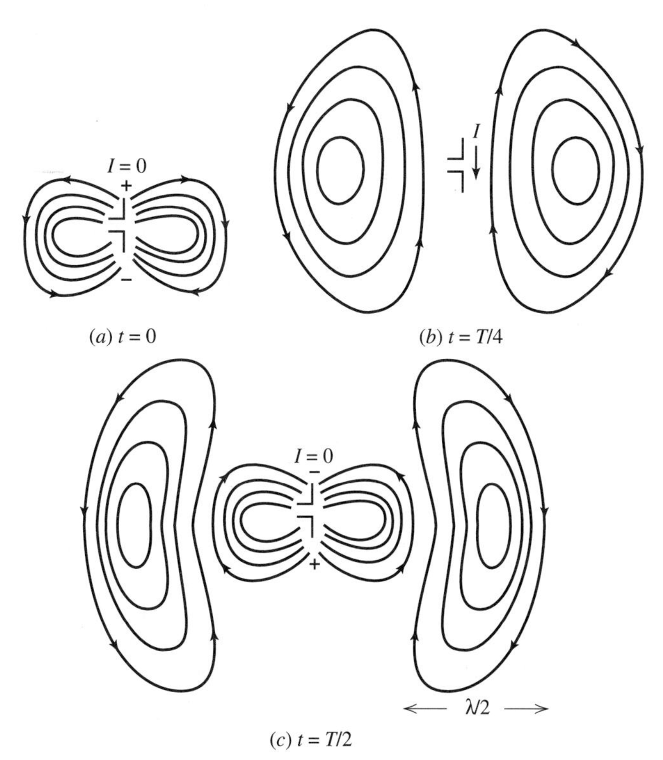

Figure 1-4 Electric fields of an oscillating dipole for various instants of time. The oscillations are of frequency f with a period of $T = 1/f$.

is maximum at $t = T/4$ as shown in Fig. 1-4*b*, at which time the charges have been neutralized and there are no longer charges for the termination of electric field lines, which form closed loops near the dipole. During the next quarter cycle, negative charges accumulate at the top end of the dipole as shown in Fig. 1-4*c*. Near the dipole, the fields are most intense normal to the oscillating charges on the dipole, just as we found with the single accelerated charge. As time progresses, the electric field lines detach from the dipole, forming closed loops in space. Viewed in terms of current, the conduction current on the antenna converts to a displacement current in space, consisting of the longitudinal fields near the antenna and solenoidal (loops) fields away from the antenna. Thus, current continuity is satisfied. This process continues, producing radiation via electric field components that are transverse to the radial direction and propagate to large distances from the antenna. We shall see in Sec. 1.6 that the mathematical solution of the (oscillating) dipole produces the property required for successful radiation: fields decrease with distance as $1/r$. In contrast, the electric field of an electrostatic dipole decreases as $1/r^3$. The time-space behavior of fields from an antenna is revisited in Sec. 11.8 and 11.9.

1.3 OVERVIEW OF ANTENNAS

An antenna acts to convert guided waves on a transmission structure into free space waves. Figure 1-3 illustrates a parallel wire transmission line feeding a half-wave dipole antenna. The official IEEE definition of an antenna follows this concept: "That part of a transmitting or receiving system that is designed to radiate or receive electromagnetic waves." Most antennas are reciprocal devices and behave the same on transmit as on receive. Antennas are treated as transmitting or receiving as appropriate for the particular situation. In the receiving mode, antennas act to collect incoming waves and direct them to a common feed point where a transmission line is attached. In some cases, antennas focus radio waves just as lenses focus optical waves. In all cases, antennas have directional characteristics; that is, electromagnetic power density is radiated from a transmitting antenna with intensity that varies with angle around the antenna.

In this section, we introduce the parameters used to evaluate antennas and then discuss the four types of antennas. The parameters are defined and developed in more detail in the remainder of this chapter after the brief overview given here.

Antenna performance parameters are listed in Table 1-1. The *radiation pattern* (or simply, *pattern*) gives the angular variation of radiation at a fixed distance from an antenna when the antenna is transmitting. Radiation is quantified by noting the value of power density S at a fixed distance r from the antenna. When receiving, the antenna responds to an incoming wave from a given direction according to the pattern value in that direction. The typical pattern in Fig. 1-5 shows the pattern main beam and side lobes. This directive antenna, with a single narrow main beam, is used in point-to-point communications. In some applications, the shape of the main beam is important. On the other hand, an omnidirectional antenna with constant radiation in one plane is used in broadcast situations.

An antenna is essentially a spatial amplifier and *directivity* expresses how much greater the peak radiated power density is for an antenna than it would be if all the radiated power were distributed uniformly around the antenna. Fig. 1-5 shows the radiation pattern of a real antenna compared to an *isotropic* spatial distribution; also see Fig. 1-20. The spatial enhancement that can be achieved by an antenna is evident. *Gain* G is directivity reduced by the losses on the antenna.

Table 1-1 Antenna Performance Parameters

- **Radiation Pattern F(θ, ϕ):** Angular variation of radiation around the antenna, including:
  Directive, single or multiple narrow beams
  Omnidirectional (uniform radiation in one plane)
  Shaped main beam
- **Directivity *D*:** Ratio of power density in the direction of the pattern maximum to the average power density at the same distance from the antenna.
- **Gain *G*:** Directivity reduced by the losses on the antenna.
- **Polarization:** The figure traced out with time by the instantaneous electric field vector associated with the radiation from an antenna when transmitting. Antenna polarizations: Linear, Circular, Elliptical
- **Impedance Z_A:** Input impedance at the antenna terminals.
- **Bandwidth:** Range of frequencies over which important performance parameters are acceptable.
- **Scanning:** Movement of the radiation pattern in space. Scanning is accomplished by mechanical movement or by electronic means such as adjustment of antenna current phase.
- **System Considerations:** Size, weight, power handling, radar cross section, environmental operating conditions, etc.

The third parameter, *polarization*, describes the vector nature of electric fields radiated by an antenna and is discussed in detail in Sec. 1.10. The figure traced out with time by the tip of the instantaneous electric field vector determines the polarization of the wave. A straight wire antenna radiates a wave with linear polarization parallel to the wire. Another popular polarization is circular. In general, polarization is elliptical. *Dual polarized* antennas enable the doubling of communication capacity by carrying separate information on orthogonal polarizations over the same physical link at the same frequency.

The *input impedance* of an antenna is the ratio of the voltage to current at the antenna terminals. The usual goal is to match antenna input impedance to the characteristic impedance of the connecting transmission line. *Bandwidth* is the range of frequencies with acceptable antenna performance as measured by one or more of the performance parameters; see Chap. 6 for commonly used definitions of bandwidth. Finally, it is often desired to scan the main beam of an antenna over a region of space. This can be accomplished by moving an antenna, by electronic scanning or by a combination of mechanical and electronic scanning.

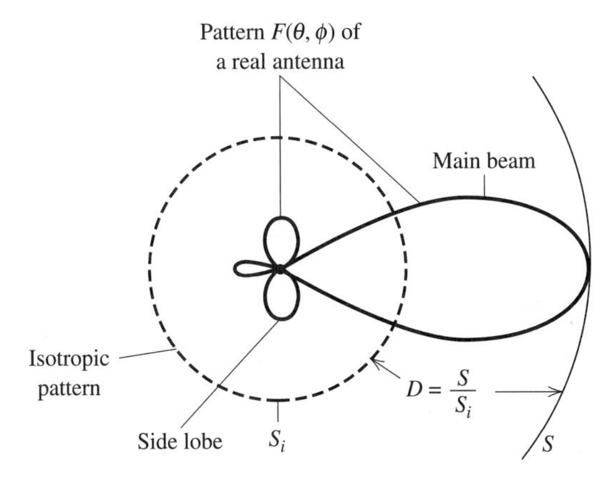

Figure 1-5 Illustration of radiation pattern $F(\theta, \phi)$ and directivity D. The power densities at the same distance are S and S_i for the real and isotropic antennas, respectively.

- **Electrically small antennas:** The extent of the antenna structure is much less than a wavelength λ.
  Properties:
    Very low directivity
    Low input resistance
    High input reactance
    Low radiation efficiency
  Examples:

- **Resonant antennas:** The antenna operates well at a single or selected narrow frequency bands.
  Properties:
    Low to moderate gain
    Real input impedance
    Narrow bandwidth

  Examples:

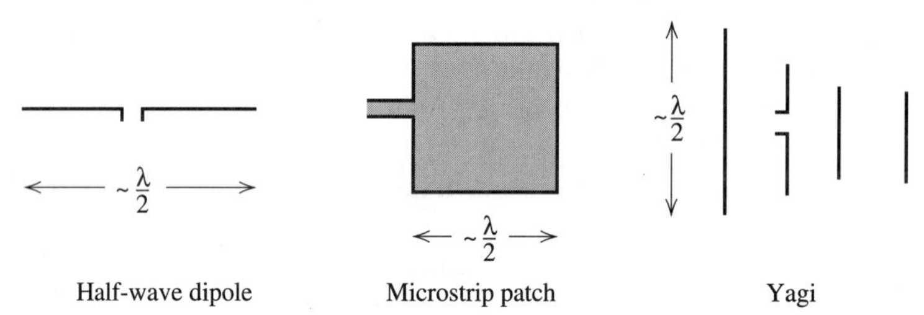

- **Broadband antennas:** The pattern, gain, and impedance remain acceptable and are nearly constant over a wide frequency range, and are characterized by an active region with a circumference of one wavelength or an extent of a half-wavelength, which relocates on the antenna as frequency changes.
  Properties:
    Low to moderate gain
    Constant gain
    Real input impedance
    Wide bandwidth

  Examples:

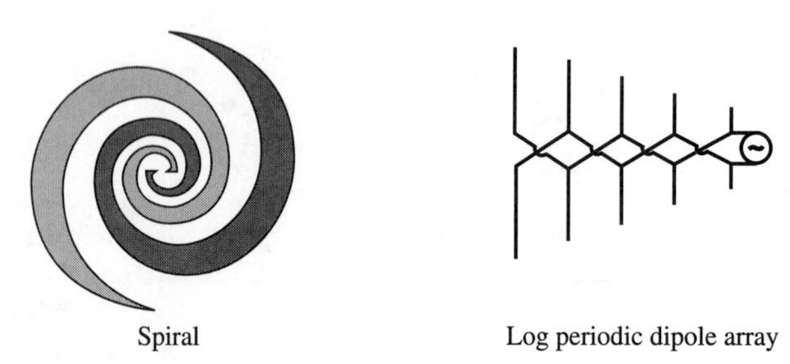

Figure 1-6 Types of antennas.

- **Aperture antennas:** Has a physical aperture (opening) through which waves flow.
  Properties:
    High gain
    Gain increases with frequency
    Moderate bandwidth

  Examples:

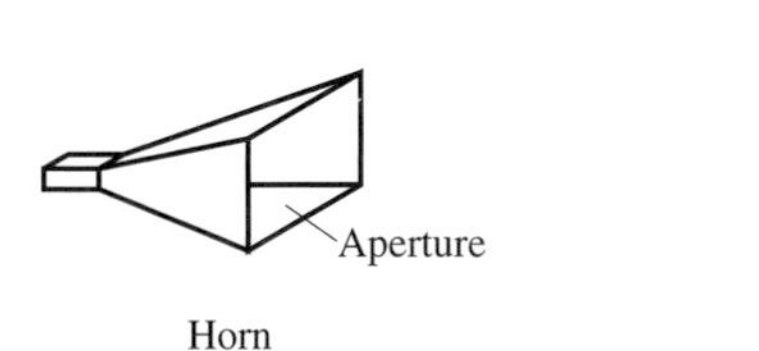

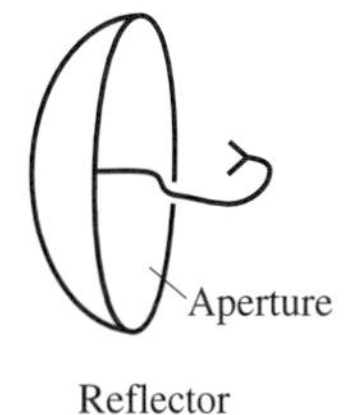

Figure 1-6 (continued).

There are trade-offs between parameter values. Usually, performance cannot be improved significantly for one parameter without sacrificing one or more of the other parameter levels. This is the antenna design challenge.

Antennas can be divided into four basic types by their performance as a function of frequency. These are introduced so that the common features can be grasped early in the study of antennas. When you encounter a new antenna, try to determine which type it is. The antenna types in Fig. 1-6 are listed in the order that they are commonly used across the radio spectrum; see Appendix A for lists of frequency bands. This discussion serves as an overview and should be referred to from time to time as your knowledge of antennas builds. Electrically small antennas are used at VHF frequencies and below. Resonant antennas are mainly used from HF to low GHz frequencies. Broadband antennas are mainly used from VHF to middle GHz frequencies. Aperture antennas are mainly used at UHF and above.

Electrically small (or simply, *small*) *antennas* are much less than a wavelength in extent. Recall from electromagnetics that wavelength λ is related to frequency f through the speed of light c as $\lambda = c/f$. Electrically small antennas are simple in structure and their properties are not sensitive to construction details. The vertical monopole used for AM reception on cars is a good example. It is about 0.003λ long and has a pattern that is nearly omnidirectional in the horizontal plane. This is often a desirable property, but its low input resistance and high input reactance are serious disadvantages. Also, small antennas are inefficient because of ohmic losses on the structure.

Resonant antennas are popular when a simple structure with good input impedance over a narrow band of frequencies is needed. It has a broad main beam and low or moderate (a few dB) gain. The half-wavelength long dipole is a prominent example.

Many applications require an antenna that operates over a wide frequency range. A *broadband antenna* has acceptable performance as measured with one or more parameters (pattern, gain, and/or impedance) over a 2:1 bandwidth ratio of upper to lower operating frequency. A broadband antenna is characterized by an active region. Propagating (or, traveling) waves originate at the feed point and travel without radiation to the active region where most of the power is radiated. A broadband antenna with circular geometry has an active region where the circumference is one wavelength and produces circular polarization. An example is the spiral antenna illustrated in Fig. 1-6 that can have a 20:1 bandwidth. A broadband antenna made up of linear elements or straight edges has an active region where these elements are about a half-wavelength in extent and produces linearly polarized radiation parallel to the linear elements. Since only a portion of a broadband antenna is responsible for radiation at a given frequency, the gain is low. But it may be an advantage to have gain that is nearly constant with frequency, although low. Also, the traveling wave nature of a broadband antenna means that it has a real-valued input impedance.

Aperture antennas have an opening through which propagating electromagnetic waves flow. A horn antenna is a good example; it acts as a "funnel," directing the waves into the connecting waveguide. The aperture is usually several wavelengths long in one or more dimensions. The pattern usually has a narrow main beam, leading to high gain. The pattern main beam narrows with increasing frequency for a fixed physical aperture size. Bandwidth is moderate (as much as 2:1).

1.4 ELECTROMAGNETIC FUNDAMENTALS

This and the next section present a brief review of electromagnetic field principles and the solution of Maxwell's equations for radiation problems. Any basic electromagnetic fields textbook can be consulted for the details on these topics.

The fundamental electromagnetic equations are[1]

$$\nabla \times \mathscr{E} = -\frac{\partial \mathscr{B}}{\partial t} \tag{1-1}$$

$$\nabla \times \mathscr{H} = \frac{\partial}{\partial t}\mathscr{D} + \mathscr{J}_{\mathbf{T}} \tag{1-2}$$

$$\nabla \cdot \mathscr{D} = \rho_T(t) \tag{1-3}$$

$$\nabla \cdot \mathscr{B} = 0 \tag{1-4}$$

$$\nabla \cdot \mathscr{J}_{\mathbf{T}} = -\frac{\partial}{\partial t}\rho_T(t) \tag{1-5}$$

The first four of these differential equations are frequently referred to as Maxwell's equations and the last as the continuity equation. The curl equations together with the continuity equation are equivalent to the curl and divergence equations. In time-varying field problems, the curl equations with the continuity equation is the most convenient formulation. Each of these differential equations has an integral counterpart.

If the sources $\rho_T(t)$ and $\mathscr{J}_{\mathbf{T}}(t)$ vary sinusoidally with time at radian frequency ω, the fields will also vary sinusoidally and are frequently called time-harmonic fields. The fundamental electromagnetic equations and their solutions are considerably simplified if phasor fields are introduced as follows:[2]

$$\mathscr{E} = \mathrm{Re}(\mathbf{E}e^{j\omega t}), \qquad \mathscr{H} = \mathrm{Re}(\mathbf{H}e^{j\omega t}), \qquad \text{etc.} \tag{1-6}$$

where phasor quantities $\mathbf{E}$, $\mathbf{H}$, $\mathbf{D}$, $\mathbf{B}$, ρ_T, and $\mathbf{J_T}$ are complex-valued functions of spatial coordinates only (i.e., time dependence is not shown). Using the phasor

[1]Time-varying quantities will be denoted with script quantities, for example, $\mathscr{E} = \mathscr{E}(x, y, z, t)$.

[2]The student is cautioned that some authors use $e^{-j\omega t}$, which leads to sign differences in subsequent developments.

definitions of the electromagnetic quantities from (1-6) in (1-1) to (1-5) and eliminating the $e^{j\omega t}$ factors that appear on both sides of the equations yields

$$\nabla \times \mathbf{E} = -j\omega\mathbf{B} \tag{1-7}$$

$$\nabla \times \mathbf{H} = j\omega\mathbf{D} + \mathbf{J_T} \tag{1-8}$$

$$\nabla \cdot \mathbf{D} = \rho_T \tag{1-9}$$

$$\nabla \cdot \mathbf{B} = 0 \tag{1-10}$$

$$\nabla \cdot \mathbf{J_T} = -j\omega\rho_T \tag{1-11}$$

The time derivatives in (1-1) to (1-5) have been replaced by a $j\omega$ factor in (1-7) to (1-11) and time-varying electromagnetic quantities have been replaced by their phasor counterpart. This process is similar to the solution of network equations where the time-dependent differential equations are Laplace-transformed and the time derivatives are thus replaced by $j\omega$ (or s). Equations (1-7) to (1-10) are often referred to as the time-harmonic form of Maxwell's equations because they apply to sinusoidally varying (i.e., time-harmonic) fields.

If more than one frequency is present, the time-varying forms of the electromagnetic quantities can be found by inverse transforms after (1-7) to (1-11) have been solved for the phasor quantities as a function of radian frequency ω. This is again analogous to the procedure used to solve network problems. Fortunately, this is not usually necessary in antenna problems since the bandwidth of the signals is usually very small. In the typical case, a carrier frequency is accompanied by some form of modulation giving a spread of frequencies around the carrier. For analysis purposes, we use a single frequency equal to the carrier frequency. Thus, all subsequent material in this book (except in Chap. 11) will assume time-harmonic fields.

The total current density $\mathbf{J_T}$ is composed of an impressed, or source, current $\mathbf{J}$ and a conduction current density term $\sigma\mathbf{E}$, which occurs in response to the impressed current:

$$\mathbf{J_T} = \sigma\mathbf{E} + \mathbf{J} \tag{1-12}$$

The role played by the impressed current density is that of a known quantity. It is quite frequently an assumed current density on an antenna, but as far as the field equations are concerned, it is a known function. The current density $\sigma\mathbf{E}$ is a current density flowing on a nearby conductor due to the fields created by source $\mathbf{J}$ and can be computed after the field equations are solved for $\mathbf{E}$. In addition to conductivity σ, a material is further characterized by permittivity ε and permeability μ, where[3]

$$\mathbf{D} = \varepsilon\mathbf{E} \tag{1-13}$$

and

$$\mathbf{B} = \mu\mathbf{H} \tag{1-14}$$

We now rewrite the field equations in preparation for their solution. Substituting (1-12) and (1-13) into (1-8) gives

$$\nabla \times \mathbf{H} = j\omega\left(\varepsilon + \frac{\sigma}{j\omega}\right)\mathbf{E} + \mathbf{J} = j\omega\varepsilon'\mathbf{E} + \mathbf{J} \tag{1-15}$$

[3]In general, ε and μ can be complex, but in most antenna problems they can be approximated as real constants.

where we have defined $\varepsilon' = \varepsilon - j(\sigma/\omega)$. For antenna problems, we are usually solving for the fields in air surrounding an antenna where $\sigma = 0$ and $\varepsilon' = \varepsilon$. We therefore use ε instead of ε' in subsequent developments. However, if the conductivity is nonzero, ε can be replaced by $\varepsilon' = \varepsilon - j(\sigma/\omega)$. Note also that **E** and **H** are the fields of primary interest in antennas. They are properly referred to as electric and magnetic field intensities and have units of V/m and A/m, respectively. For conciseness, it is common to refer to them simply as electric and magnetic fields.

Let ρ be the source charge corresponding to the source current density **J**. Then using (1-12) to (1-14) in (1-7) and (1-9) to (1-11), and repeating (1-15), we have [see Prob. 1.4-2 (for 1-18)]

$$\nabla \times \mathbf{E} = -j\omega\mu\mathbf{H} \tag{1-16}$$

$$\nabla \times \mathbf{H} = j\omega\varepsilon\mathbf{E} + \mathbf{J} \tag{1-17}$$

$$\nabla \cdot \mathbf{E} = \frac{\rho}{\varepsilon} \tag{1-18}$$

$$\nabla \cdot \mathbf{H} = 0 \tag{1-19}$$

$$\nabla \cdot \mathbf{J} = -j\omega\rho \tag{1-20}$$

These are the time-harmonic electromagnetic field equations with source current density **J** and source charge density ρ shown explicitly. Sometimes, it is convenient to introduce a fictitious magnetic current density **M**. Then (1-16) becomes

$$\nabla \times \mathbf{E} = -j\omega\mu\mathbf{H} - \mathbf{M} \tag{1-21}$$

Magnetic currents are useful as equivalent sources that replace complicated electric fields.

The solution of the fundamental electromagnetic equations is not complete until the boundary conditions are satisfied. A sufficient set of boundary conditions in the time-harmonic form is

$$\hat{\mathbf{n}} \times (\mathbf{H}_2 - \mathbf{H}_1) = \mathbf{J}_s \tag{1-22}$$

$$(\mathbf{E}_2 - \mathbf{E}_1) \times \hat{\mathbf{n}} = \mathbf{M}_s \tag{1-23}$$

where the electric and magnetic surface currents $\mathbf{J}_s$ and $\mathbf{M}_s$ flow on the boundary between two homogeneous media with constitutive parameters ε_1, μ_1, σ_1, and ε_2, μ_2, σ_2. $\mathbf{M}_s$ is zero unless an equivalent magnetic current sheet is used. The unit normal to the boundary surface $\hat{\mathbf{n}}$ is directed from medium 1 into medium 2. The cross products with the unit normal form the tangential components to the boundary, and these equations can be written as

$$H_{\tan 2} = H_{\tan 1} + J_s \tag{1-24}$$

$$E_{\tan 2} = E_{\tan 1} + M_s \tag{1-25}$$

These boundary conditions are derived from the integral form of (1-17) and (1-21). If one side is a perfect electrical conductor, the boundary conditions become

$$H_{\tan} = J_s \tag{1-26}$$

$$E_{\tan} = 0 \tag{1-27}$$

The tangential boundary conditions on the magnetic field intensity are illustrated in Fig. 1-7 for the general case and for the case where medium 1 is a perfect con-

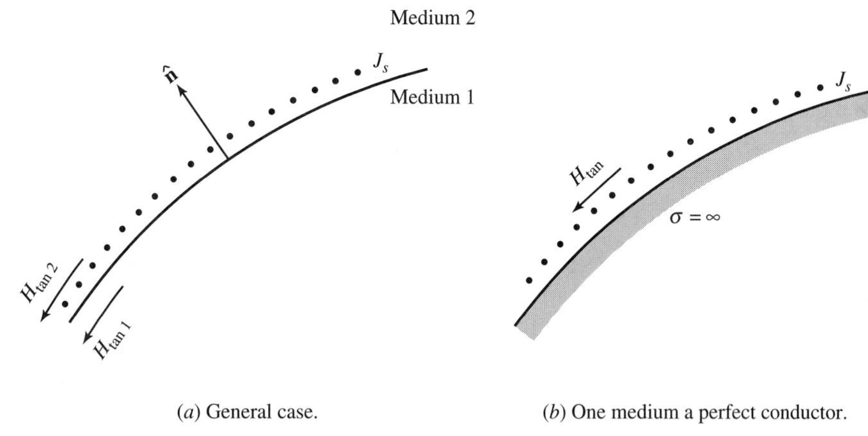

(*a*) General case. (*b*) One medium a perfect conductor.

Figure 1-7 Magnetic field intensity boundary condition.

ductor. It is important to note that all field quantities in the boundary condition equations are evaluated at the boundary and the equations apply to each point along the boundary.

Also derivable from Maxwell's curl equations is a conservation of power equation, or Poynting's theorem. Consider a volume v bounded by a closed surface s. The complex power P_s delivered by the sources in v equals the sum of the power P_f flowing out of s, the time-average power $P_{d_{av}}$ dissipated in v, plus the time-average stored power in v:

$$P_s = P_f + P_{d_{av}} + j2\omega(W_{m_{av}} - W_{e_{av}}) \tag{1-28}$$

The complex power flowing out through closed surface s is found from

$$P_f = \frac{1}{2} \oiint_s \mathbf{E} \times \mathbf{H}^* \cdot d\mathbf{s} \tag{1-29}$$

where $d\mathbf{s} = ds\hat{\mathbf{n}}$ and $\hat{\mathbf{n}}$ is the unit normal to the surface directed *out* from the surface. Note that $\mathbf{E}$ and $\mathbf{H}$ are peak phasors, not rms, leading to 1/2 in power expressions. The integrand inside this integral is defined as the *Poynting vector*:

$$\mathbf{S} = \tfrac{1}{2}\mathbf{E} \times \mathbf{H}^* \tag{1-30}$$

which is a power density with units of W/m^2. The time-average dissipated power in volume v bounded by closed surface s is

$$P_{d_{av}} = \frac{1}{2} \iiint_v \sigma |E|^2 \, dv \tag{1-31}$$

The time-average stored magnetic energy is

$$W_{m_{av}} = \frac{1}{2} \iiint_v \frac{1}{2} \mu |H|^2 \, dv \tag{1-32}$$

The time-average stored electric energy is

$$W_{e_{av}} = \frac{1}{2} \iiint_v \frac{1}{2} \varepsilon |E|^2 \, dv \tag{1-33}$$

If the source power is not known explicitly, it can be calculated from the volume current density as follows:

$$P_s = -\frac{1}{2}\iiint_v \mathbf{E} \cdot \mathbf{J}^* \, dv \tag{1-34}$$

If magnetic current density is used, the term $\mathbf{H}^* \cdot \mathbf{M}$ is added to the integrand in the preceding equation.

From (1-29), we see that the integral of the complex Poynting vector $\frac{1}{2}\mathbf{E} \times \mathbf{H}^*$ over a closed surface s gives the total complex power flowing out through the surface s. It is assumed that the complex Poynting vector represents the complex power density in watts per square meter at a point. Then the complex power through any surface s (not necessarily closed) can be found by integrating the complex Poynting vector over that surface. We are particularly interested in real power (the real component of the complex power that represents the electric and magnetic field intensities being in-phase). The real power flowing through surface s is

$$P = \mathrm{Re}\left(\iint_s \mathbf{S} \cdot d\mathbf{s}\right) = \frac{1}{2}\mathrm{Re}\left(\iint_s \mathbf{E} \times \mathbf{H}^* \cdot d\mathbf{s}\right) \tag{1-35}$$

The reference direction for this average power flow is that of the specified unit normal $\hat{\mathbf{n}}$ contained in $d\mathbf{s} = ds\hat{\mathbf{n}}$.

1.5 SOLUTION OF MAXWELL'S EQUATIONS FOR RADIATION PROBLEMS

This section develops procedures for finding fields radiated by an antenna based on Maxwell's equations. Subsequent antenna analysis in this book begins with these basic relations and it is not necessary to return to Maxwell's equations.

The antenna problem consists of solving for the fields that are created by an impressed current distribution $\mathbf{J}$. In the simplest approach, this current distribution is obtained during the solution process. How to obtain the current distribution will be discussed at various points in the book, but for the moment suppose we have the current distribution and wish to determine the fields $\mathbf{E}$ and $\mathbf{H}$. As mentioned in the previous section, we need only work with the two curl equations of Maxwell's equations as given by (1-16) and (1-17). These are two coupled, linear, first-order differential equations. They are coupled because the unknown functions $\mathbf{E}$ and $\mathbf{H}$ appear in both equations. Thus, these equations must be solved simultaneously. In order to simplify the solution for $\mathbf{E}$ and $\mathbf{H}$ with a given $\mathbf{J}$, we introduce the scalar and vector potential functions Φ and $\mathbf{A}$.

The vector potential is introduced by noting from (1-19) that the divergence of $\mathbf{H}$ is zero:

$$\nabla \cdot \mathbf{H} = 0 \tag{1-36}$$

Therefore, the vector field $\mathbf{H}$ has only circulation; for this reason, it is often called a solenoidal field. Because it possesses only a circulation, it can be represented by the curl of some other vector function as follows:

$$\boxed{\mathbf{H} = \frac{1}{\mu}\nabla \times \mathbf{A}} \tag{1-37}$$

where **A** is the (*magnetic*) *vector potential*. To be more precise, (1-37) is possible because it satisfies (1-36) identically; that is, from (C-9) $\nabla \cdot \nabla \times \mathbf{A} \equiv 0$ for any **A**. The curl of **A** is defined by (1-37), but its divergence is yet to be specified for a complete definition of **A**.

The scalar potential is introduced by substituting (1-37) into (1-16), which gives

$$\nabla \times (\mathbf{E} + j\omega\mathbf{A}) = 0 \tag{1-38}$$

The expression in parentheses is an electric field, and since its curl is zero, it is a conservative field and behaves as a static electric field. The (*electric*) *scalar potential* Φ is defined from

$$\mathbf{E} + j\omega\mathbf{A} = -\nabla\Phi \tag{1-39}$$

because this definition satisfies (1-38) identically, that is, from (C-10) $\nabla \times \nabla\Phi \equiv 0$ for any Φ. Solving (1-39) for the total electric field gives

$$\mathbf{E} = -j\omega\mathbf{A} - \nabla\Phi \tag{1-40}$$

which may be a familiar result.

The fields **E** and **H** are now expressed in terms of potential functions by (1-37) and (1-40). If we knew the potential functions, then the fields could be obtained. We now discuss the solution for the potential functions. Substituting (1-37) into (1-17) gives

$$\nabla \times \mathbf{H} = \frac{1}{\mu} \nabla \times \nabla \times \mathbf{A} = j\omega\varepsilon\mathbf{E} + \mathbf{J} \tag{1-41}$$

Using the following vector identity, from (C-17),

$$\nabla \times \nabla \times \mathbf{A} \equiv \nabla(\nabla \cdot \mathbf{A}) - \nabla^2\mathbf{A} \tag{1-42}$$

and (1-40) in (1-41) yields

$$\nabla(\nabla \cdot \mathbf{A}) - \nabla^2\mathbf{A} = j\omega\mu\varepsilon(-j\omega\mathbf{A} - \nabla\Phi) + \mu\mathbf{J} \tag{1-43}$$

or

$$\nabla^2\mathbf{A} + \omega^2\mu\varepsilon\mathbf{A} - \nabla(j\omega\mu\varepsilon\Phi + \nabla \cdot \mathbf{A}) = -\mu\mathbf{J} \tag{1-44}$$

As we mentioned previously, the divergence of **A** is yet to be specified. A convenient choice would be one that eliminates the third term of (1-44). It is the Lorentz condition (perhaps more properly attributed to L. Lorenz rather than H. Lorentz [8]):

$$\nabla \cdot \mathbf{A} = -j\omega\mu\varepsilon\Phi. \tag{1-45}$$

Then (1-44) reduces to

$$\boxed{\nabla^2\mathbf{A} + \omega^2\mu\varepsilon\mathbf{A} = -\mu\mathbf{J}} \tag{1-46}$$

The choice of (1-45) leads to a decoupling of variables: that is, (1-46) involves **A** and not Φ. This is the *vector wave equation*. It is a differential equation that can be solved for **A** after the impressed current **J** is specified. The fields are easily found then from (1-37) and

$$\boxed{\mathbf{E} = -j\omega\mathbf{A} - j\,\frac{\nabla(\nabla \cdot \mathbf{A})}{\omega\mu\varepsilon}} \tag{1-47}$$

where this equation was obtained from (1-40) and (1-45). Notice that only a knowledge of **A** is required. A more cumbersome approach would be to solve the scalar wave equation

$$\nabla^2\Phi + \omega^2\mu\varepsilon\Phi = -\frac{\rho}{\varepsilon} \tag{1-48}$$

in addition to the vector wave equation. It is left as a problem to derive (1-48). If this approach is used, **E** is found from (1-40). Note that ρ in (1-48) is related to **J** in (1-46) by the continuity equation of (1-20).

The vector wave equation (1-46) is solved by forming three scalar equations. This begins by decomposing **A** into rectangular components using (C-18):

$$\nabla^2\mathbf{A} = \hat{\mathbf{x}}\nabla^2 A_x + \hat{\mathbf{y}}\nabla^2 A_y + \hat{\mathbf{z}}\nabla^2 A_z \tag{1-49}$$

Rectangular components are used because the unit vectors in rectangular components can be factored out of the Laplacian since they are not themselves functions of coordinates. This feature is unique to the rectangular coordinate system. Although **A** is always decomposed into rectangular components, the Laplacian of each component of **A** is expressed in a coordinate system appropriate to the geometry of the problem. The solution proceeds by substituting (1-49) into (1-46) and equating rectangular components:

$$\begin{aligned} \nabla^2 A_x + \beta^2 A_x &= -\mu J_x \\ \nabla^2 A_y + \beta^2 A_y &= -\mu J_y \\ \nabla^2 A_z + \beta^2 A_z &= -\mu J_z \end{aligned} \tag{1-50}$$

where $\beta^2 = \omega^2\mu\varepsilon$. The real-valued constant

$$\beta = \omega\sqrt{\mu\varepsilon} \tag{1-51}$$

is recognized as the phase constant for a plane wave.

The three equations in (1-50) are identical in form. After solving one of these equations, the other two are easily solved. We first find the solution for a point source. This unit impulse response solution can then be used to form a general solution by viewing an arbitrary source as a collection of point sources. The differential equation for a point source is

$$\nabla^2\psi + \beta^2\psi = -\delta(x)\,\delta(y)\,\delta(z) \tag{1-52}$$

where ψ is the response to a point source at the origin, and $\delta(\ )$ is the unit impulse function, or Dirac delta function (see Appendix F.1). In spite of the fact that the point source is of infinitesimal extent, its associated current has a direction. This is because in solving practical problems, the point source represents a small subdivision of current that does have a direction. If the point source current is taken as z-directed, then

$$\psi = A_z \tag{1-53}$$

Since the point source is zero everywhere except at the origin, (1-52) becomes

$$\nabla^2\psi + \beta^2\psi = 0 \tag{1-54}$$

away from the origin.

This is the complex *scalar wave equation* or *Helmholtz equation*. Because of spherical symmetry, the Laplacian is written in spherical coordinates and ψ has only radial dependence. The two solutions to (1-54) are $e^{-j\beta r}/r$ and $e^{+j\beta r}/r$. These correspond to waves propagating radially outward and inward, respectively. The physically meaningful solution is the one for waves traveling away from the point source. Evaluating the constant of proportionality (see Prob. 1.5-2), we have for the point source solution:

$$\psi = \frac{e^{-j\beta r}}{4\pi r} \tag{1-55}$$

This is the solution to (1-52) and is the magnitude and phase variation with distance r away from a point source located at the origin. If the source were positioned at an arbitrary location, we must compute the distance R between the source location and observation point P (see Fig. 1-8). Then

$$\psi = \frac{e^{-j\beta R}}{4\pi R} \tag{1-56}$$

The point source serves as a starting point for the ideal dipole antenna solution, which is discussed in the next section.

For an arbitrary z-directed current density, the vector potential is also z-directed. If we consider the source to be a collection of point sources weighted by the distribution J_z, the response A_z is a sum of the point source responses of (1-56). This is expressed by the integral over the source volume v':

$$A_z = \iiint_{v'} \mu J_z \frac{e^{-j\beta R}}{4\pi R}\, dv' \tag{1-57}$$

Similar equations hold for the x- and y-components. The total solution is then the sum of all components, which is

$$\mathbf{A} = \iiint_{v'} \mu \mathbf{J} \frac{e^{-j\beta R}}{4\pi R}\, dv' \tag{1-58}$$

This is the solution to the vector wave equation (1-46). The geometry is shown in Fig. 1-8. The coordinate system shown is used to describe both the source point and field point. $\mathbf{r}'$ is the vector from the coordinate origin to the source point, and $\mathbf{r}_p$ is the vector from the coordinate origin to the field point P. The vector $\mathbf{R}$ is the vector

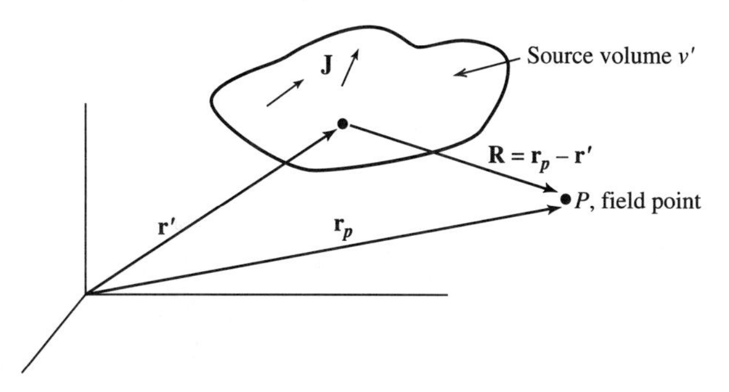

Figure 1-8 Vectors used to solve radiation problems.

from the source point to the field point and is given by $\mathbf{r}_p - \mathbf{r}'$. This geometry is standard and will be used here.

We can summarize rather simply the procedure for finding the fields generated by a current distribution **J**. First, **A** is found from (1-58). The **H** field is found from (1-37). The **E** field can be found from (1-47), but frequently it is simpler to find **E** from (1-17) as

$$\mathbf{E} = \frac{1}{j\omega\varepsilon}(\nabla \times \mathbf{H} - \mathbf{J}) \tag{1-59}$$

in the source region, or from

$$\mathbf{E} = \frac{1}{j\omega\varepsilon}\nabla \times \mathbf{H} \tag{1-60}$$

if the field point is removed in distance from the source; that is, if $\mathbf{J} = 0$ at point P.

1.6 THE IDEAL DIPOLE

The principles presented in the previous section are used in this section to find the fields of an infinitesimal element of current. We shall use the term **ideal dipole** for a uniform amplitude current that is electrically small with $\Delta z \ll \lambda$. It is ideal in the sense that the current is uniform in both magnitude and phase over the radiating element extent. Such a discontinuous current is difficult to realize in practice; practical realizations that approximate the ideal dipole are presented in Sec. 2.1. The term *current element* is often used for the ideal dipole to describe its application as a section of a larger current associated with an actual antenna. Thus, any practical antenna can first be decomposed into filaments of continuous current that are then subdivided into ideal dipoles. The fields from the antenna are then found by summing contributions from the ideal dipoles. Other terms used for the ideal dipole are *Hertzian electric dipole*, *electric dipole*, *infinitesimal dipole*, and *doublet*. An electrically small, center-fed wire antenna has a current distribution that tapers to zero from the center to the ends of the wire. This *short dipole* antenna has the same pattern as an ideal dipole and is discussed further in Secs. 1.9 and 2.1.

Consider an element of current of length Δz along the z-axis centered on the coordinate origin. It is of constant amplitude I. In this case, the volume integral of (1-58) for vector potential reduces to the one-dimensional integral[4]

$$\mathbf{A} = \hat{\mathbf{z}}\mu I \int_{-\Delta z/2}^{\Delta z/2} \frac{e^{-j\beta R}}{4\pi R}\, dz' \tag{1-61}$$

The length Δz is very small compared to a wavelength and to the distance R; see Fig. 1-9. Since Δz is very small, the distance R from points on the current element

[4] The result in (1-61) could also be obtained by representing the current density on the dipole as

$$\mathbf{J} = I\,\delta(x')\,\delta(y')\hat{\mathbf{z}} \qquad \text{for} \qquad -\frac{\Delta z}{2} < z' < \frac{\Delta z}{2}$$

Substituting this into (1-58) yields

$$\mathbf{A} = \hat{\mathbf{z}}\mu I \int_{-\infty}^{\infty} \delta(x')\, dx' \int_{-\infty}^{\infty} \delta(y')\, dy' \int_{-\Delta z/2}^{\Delta z/2} \frac{e^{-j\beta R}}{4\pi R}\, dz'$$

from which (1-61) follows.

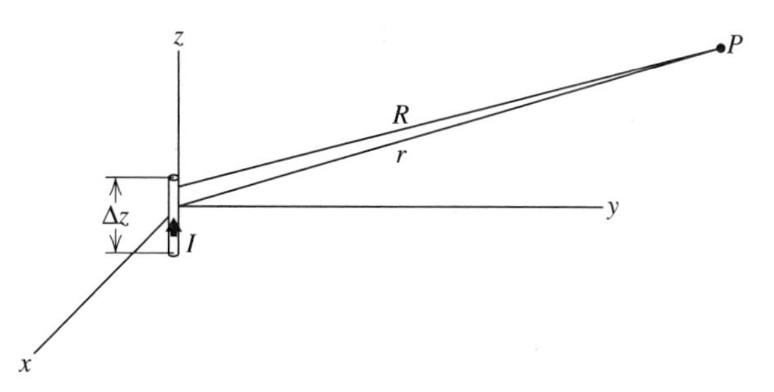

Figure 1-9 The ideal dipole. The current I is uniform, $\Delta z \ll \lambda$, and $R \approx r$.

to the field point approximately equals the distance r from the origin to the field point. Substituting r for R in (1-61) and integrating gives

$$\mathbf{A} = \frac{\mu I e^{-j\beta r}}{4\pi r} \Delta z \, \hat{\mathbf{z}} \tag{1-62}$$

This is exactly true for a point current element and is approximately true for a small ($\Delta z \ll \lambda$ and $\Delta z \ll R$) but finite uniform current element. The vector potential A_z for a point source was also derived in the previous section; see (1-55) in which $I\,\Delta z = 1$. For many current sources, we can readily make the substitution of r for R in the denominator of the integrand in (1-61), but usually cannot make the same substitution in the exponent. However, in the case of a very small source, we can use r for R in both the denominator and exponent.

We are now ready to calculate the electromagnetic fields created by the ideal dipole. The magnetic field is found from (1-37) as

$$\mathbf{H} = \frac{1}{\mu} \nabla \times \mathbf{A} = \frac{1}{\mu} \nabla \times (A_z \hat{\mathbf{z}}) \tag{1-63}$$

If we apply the vector identity (C-16), the preceding equation becomes

$$\mathbf{H} = \frac{1}{\mu} (\nabla A_z) \times \hat{\mathbf{z}} + \frac{1}{\mu} A_z (\nabla \times \hat{\mathbf{z}}) = \frac{1}{\mu} (\nabla A_z) \times \hat{\mathbf{z}} \tag{1-64}$$

The second term is zero because the curl of a constant vector is zero. Substituting (1-62) into (1-64), we have

$$\mathbf{H} = \nabla \left(\frac{I\,\Delta z e^{-j\beta r}}{4\pi r} \right) \times \hat{\mathbf{z}} \tag{1-65}$$

Applying the gradient in spherical coordinates from (C-33) gives

$$\begin{aligned} \mathbf{H} &= \frac{I\,\Delta z}{4\pi} \frac{\partial}{\partial r} \left(\frac{e^{-j\beta r}}{r} \right) \hat{\mathbf{r}} \times \hat{\mathbf{z}} \\ &= \frac{I\,\Delta z}{4\pi} \left[\frac{-j\beta e^{-j\beta r}}{r} - \frac{e^{-j\beta r}}{r^2} \right] \hat{\mathbf{r}} \times \hat{\mathbf{z}} \end{aligned} \tag{1-66}$$

From (C-3), we have

$$\hat{\mathbf{r}} \times \hat{\mathbf{z}} = \hat{\mathbf{r}} \times (\hat{\mathbf{r}} \cos\theta - \hat{\boldsymbol{\theta}} \sin\theta) = -\hat{\boldsymbol{\phi}} \sin\theta \tag{1-67}$$

Substituting (1-67) into (1-66) gives

$$\mathbf{H} = \frac{I\,\Delta z}{4\pi} \left[\frac{j\beta}{r} + \frac{1}{r^2} \right] e^{-j\beta r} \sin\theta \, \hat{\boldsymbol{\phi}} \tag{1-68}$$

The electric field can be obtained from (1-60) as

$$\mathbf{E} = \frac{I\,\Delta z}{4\pi}\left[\frac{j\omega\mu}{r} + \sqrt{\frac{\mu}{\varepsilon}}\frac{1}{r^2} + \frac{1}{j\omega\varepsilon r^3}\right]e^{-j\beta r}\sin\theta\,\hat{\boldsymbol{\theta}} + \frac{I\,\Delta z}{2\pi}\left[\sqrt{\frac{\mu}{\varepsilon}}\frac{1}{r^2} + \frac{1}{j\omega\varepsilon}\frac{1}{r^3}\right]e^{-j\beta r}\cos\theta\,\hat{\mathbf{r}} \tag{1-69}$$

Here, β is given by (1-51) and is related to wavelength as

$$\beta = \omega\sqrt{\mu\varepsilon} = \frac{2\pi}{\lambda} \tag{1-70}$$

Note that if the medium surrounding the dipole is air or free space, $\beta = \omega\sqrt{\mu_o\varepsilon_o}$, where μ_o and ε_o are the permeability and permittivity of free space.

Eqs. (1-68) and (1-69) can be written as

$$\mathbf{H} = \frac{I\,\Delta z}{4\pi}\,j\beta\left(1 + \frac{1}{j\beta r}\right)\frac{e^{-j\beta r}}{r}\sin\theta\,\hat{\boldsymbol{\phi}} \tag{1-71a}$$

$$\mathbf{E} = \frac{I\,\Delta z}{4\pi}\,j\omega\mu\left[1 + \frac{1}{j\beta r} - \frac{1}{(\beta r)^2}\right]\frac{e^{-j\beta r}}{r}\sin\theta\,\hat{\boldsymbol{\theta}} + \frac{I\,\Delta z}{2\pi}\,\eta\left[\frac{1}{r} - j\frac{1}{\beta r^2}\right]\frac{e^{-j\beta r}}{r}\cos\theta\,\hat{\mathbf{r}} \tag{1-71b}$$

If βr is large (i.e., $\beta r \gg 1$, or $r \gg \lambda$ since $\beta = 2\pi/\lambda$), then all terms having inverse powers of $j\beta r$ are small compared to unity, and (1-71) reduces to

$$\mathbf{E} = \frac{I\,\Delta z}{4\pi}\,j\omega\mu\,\frac{e^{-j\beta r}}{r}\sin\theta\,\hat{\boldsymbol{\theta}} \tag{1-72a}$$

$$\mathbf{H} = \frac{I\,\Delta z}{4\pi}\,j\beta\,\frac{e^{-j\beta r}}{r}\sin\theta\,\hat{\boldsymbol{\phi}} \tag{1-72b}$$

These are the fields of an ideal dipole at large distances from the dipole. The ratio of these electric and magnetic field components is

$$\frac{E_\theta}{H_\phi} = \frac{\omega\mu}{\beta} = \frac{\omega\mu}{\omega\sqrt{\mu\varepsilon}} = \sqrt{\frac{\mu}{\varepsilon}} = \eta \tag{1-73}$$

where $\eta = \sqrt{\mu/\varepsilon}$ is the intrinsic impedance of the medium (for free space $\eta_o = 376.7\ \Omega \approx 120\pi\ \Omega$). This is a property of plane waves. Also, as we shall see, at large distances from any antenna the fields are related in this manner.

Using the fields of (1-72) in (1-30) gives an expression for the complex power flowing density out of a sphere of radius r surrounding the ideal dipole:

$$\begin{aligned}\mathbf{S} &= \frac{1}{2}\mathbf{E} \times \mathbf{H}^* \\ &= \frac{1}{2}\left(\frac{I\,\Delta z}{4\pi}\right)^2 j\omega\mu\,\frac{e^{-j\beta r}}{r}\sin\theta\,\hat{\boldsymbol{\theta}} \times (-j\beta)\,\frac{e^{+j\beta r}}{r}\sin\theta\,\hat{\boldsymbol{\phi}} \\ &= \frac{1}{2}\left(\frac{I\,\Delta z}{4\pi}\right)^2 \omega\mu\beta\,\frac{\sin^2\theta}{r^2}\,\hat{\mathbf{r}}\end{aligned} \tag{1-74}$$

which is real-valued and radially directed, both characteristics of radiation. The total power flowing out through a sphere of radius r surrounding the ideal dipole using (1-29) is

$$P_f = \iint \mathbf{S} \cdot d\mathbf{s} = \frac{1}{2}\left(\frac{I\,\Delta z}{4\pi}\right)^2 \omega\mu\beta \int_0^{2\pi} d\phi \int_0^{\pi} \sin^3\theta\, d\theta$$

$$= \frac{1}{2}\left(\frac{I\,\Delta z}{4\pi}\right)^2 \omega\mu\beta 2\pi \frac{4}{3}$$

$$= \frac{\omega\mu\beta}{12\pi}(I\,\Delta z)^2 \tag{1-75}$$

This is a real quantity, and real power indicates dissipated power. It is dissipated in the sense that it travels away from the source. In fact, the average power going out through a sphere of radius r can be found as indicated in (1-35) by taking the real part of (1-75), which leaves it unchanged. This power expression is independent of r, and thus if we integrate over a sphere of larger radius, we still have the same total power streaming through it. We refer to this type of power as *radiated power*. The fields in (1-72) are called *radiation fields*.

The general field expressions of (1-71) are valid at any distance from an ideal dipole and are important in some applications and for understanding the input impedance of a dipole. For distances so close to the dipole that $\beta r \ll 1$, or $r \ll \lambda$, only the dominant terms with the largest inverse powers of r need be retained in each component of (1-71):

$$\mathbf{H}^{\mathbf{nf}} = \frac{I\,\Delta z e^{-j\beta r}}{4\pi r^2} \sin\theta\,\hat{\boldsymbol{\phi}} \tag{1-76a}$$

$$\mathbf{E}^{\mathbf{nf}} = -j\eta \frac{I\,\Delta z}{4\pi\beta}\frac{e^{-j\beta r}}{r^3} \sin\theta\,\hat{\boldsymbol{\theta}} - j\eta \frac{I\,\Delta z}{2\pi\beta}\frac{e^{-j\beta r}}{r^3} \cos\theta\,\hat{\mathbf{r}} \tag{1-76b}$$

These are referred to as the *near fields* of the antenna. Actually, the magnetic field of (1-76a) which varies as $1/r^2$ is that of a short, steady or slowly oscillating current, that is, an induction field. The electric fields of (1-76b) vary as $1/r^3$ and are those of an electrostatic or quasi-static dipole with charges of $+q$ and $-q$ spaced Δz apart. Note that the electric field components E_θ^{nf} and E_r^{nf} are in-phase, but are in phase-quadrature with the magnetic field H_ϕ^{nf}, indicating reactive power. This can be shown directly using these near fields in the complex Poynting vector expression of (1-30):

$$\mathbf{S}^{\mathbf{nf}} = \frac{1}{2}\left[E_\theta^{nf} H_\phi^{nf*}\hat{\mathbf{r}} - E_r^{nf} H_\phi^{nf*}\hat{\boldsymbol{\theta}}\right]$$

$$= -\frac{j\eta}{2\beta}\left(\frac{I\,\Delta z}{4\pi}\right)^2 \frac{1}{r^5}\left(\sin^2\theta\hat{\mathbf{r}} - \sin\theta\,\hat{\boldsymbol{\theta}}\right) \tag{1-77}$$

Note that this power density vector is imaginary and therefore has no time-average radial power flow. The radiation fields, in contrast, are in-phase giving a real-valued Poynting vector that is radially directed; see (1-72) and (1-74). The imaginary power density corresponds to standing waves, rather than traveling waves associated with radiation, and indicate stored energy as in any reactive device. The quadrature phase relationship between the electric and magnetic field components of (1-76) indicates that energy is interchanged between these fields with time. That is, at one instant of time electric fields near the dipole are strong close to maximum charge regions and a quarter-period later energy is stored in the magnetic field, primarily close to the center of the dipole where the current is maximum.

The imaginary power density in the near field is manifested by a reactive component of the antenna input impedance. The real part of the input impedance represents radiation if ohmic losses on the antenna structure can be neglected. Antenna impedance will be discussed further in Sec. 1.9. The power density associated with radiation exists everywhere and passes through the near field. The radiated power density from (1-74) and the near-field power density from (1-77) are both maximum for $\theta = 90°$. The distance for which the maximum radiated and reactive powers are equal for the ideal dipole is $r = \lambda/2\pi$. That is, interior to this radius reactive power dominates. This region is sometimes referred to as the *radiansphere*.

At large distances from an antenna, called the *far-field region*, all power is radiated power. The far field is further characterized by the fact that the angular distribution around the antenna (i.e., the radiation pattern) is independent of distance from the antenna. Field regions and the distance away from an antenna where the far field begins are discussed further in Sec. 1.7.3.

1.7 RADIATION PATTERNS

We briefly introduce the radiation pattern in Sec. 1.3 as a description of the angular variation of radiation level around an antenna. This is perhaps the most important characteristic of an antenna. In this section, we present several definitions associated with patterns and develop the general procedures for calculating radiation patterns.

1.7.1 Radiation Pattern Basics

A *radiation pattern* (*antenna pattern*) is a graphical representation of the radiation (far-field) properties of an antenna. We have seen that the radiation fields from a transmitting antenna vary inversely with distance, e.g., $1/r$. The variation with observation angles (θ, ϕ), however, depends on the antenna.

Radiation patterns can be understood by examining the ideal dipole. The fields radiated from an ideal dipole are shown in Fig. 1-10*a* over the surface of a sphere of radius r that is in the far field. The length and orientation of the field vectors follow from (1-72); they are shown for an instant of time for which the fields are peak. The angular variation of E_θ and H_ϕ over the sphere is $\sin\theta$. An electric field probe antenna moved over the sphere surface and oriented parallel to E_θ will have an output proportional to $\sin\theta$; see Fig. 9-7. Any plane containing the z-axis has the same radiation pattern since there is no ϕ variation in the fields. A pattern taken in one of these planes is called an *E-plane pattern* because it contains the electric vector. A pattern taken in a plane perpendicular to an E-plane and cutting through the test antenna (the xy-plane in this case) is called an *H-plane pattern* because it contains the magnetic field H_ϕ. The E- and H-plane patterns, in general, are referred to as *principal plane patterns*. The E- and H-plane patterns for the ideal dipole are shown in Figs. 1-10*b* and 1-10*c*. These are polar plots in which the distance from the origin to the curve is proportional to the field intensity; they are often called polar patterns or polar diagrams.

The complete pattern for the ideal dipole is shown in isometric view with a slice removed in Fig. 1-10*d*. This solid polar radiation pattern resembles a "doughnut"

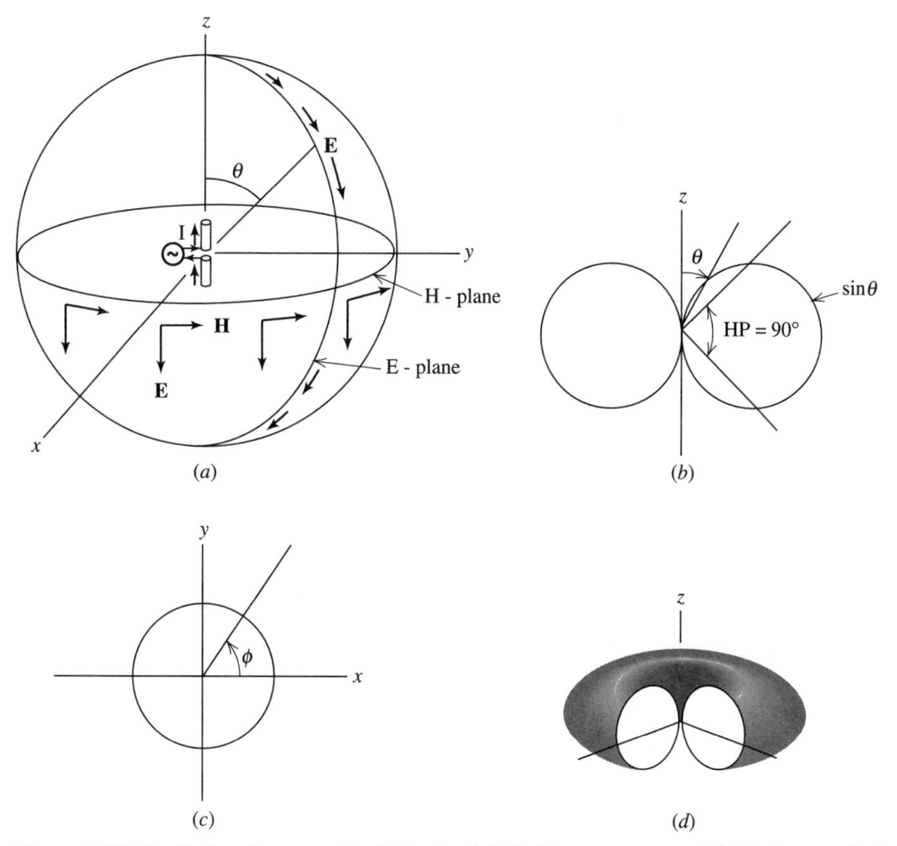

Figure 1-10 Radiation from an ideal dipole. (*a*) Field components. (*b*) *E*-plane radiation pattern polar plot of $|E_\theta|$ or $|H_\phi|$. (*c*) *H*-plane radiation pattern polar plot of $|E_\theta|$ or $|H_\phi|$. (*d*) Three-dimensional plot of radiation pattern.

with no hole. It is referred to as an *omnidirectional pattern* since it is uniform in the *xy*-plane. Omnidirectional antennas are very popular in ground-based applications with the omnidirectional plane horizontal. When encountering new antennas, the reader should attempt to visualize the complete pattern in three dimensions.

1.7.2 Radiation from Line Currents

Radiation patterns in general can be calculated in a manner similar to that used for the ideal dipole if the current distribution on the antenna is known. This is done by first finding the vector potential given in (1-58). As a simple example, consider a filament of current along the z-axis and located near the origin. Many antennas can be modeled by this **line source**; straight wire antennas are good examples. In this case, the vector potential has only a z-component and the vector potential integral is one-dimensional[5]:

$$A_z = \mu \int I(z') \frac{e^{-j\beta R}}{4\pi R} dz' \tag{1-78}$$

[5]This result could also be obtained by using $J_z(r') = I(z')\,\delta(x')\,\delta(y')$ in (1-57), where $dv' = dx'\,dy'\,dz'$.

Due to the symmetry of the source, we expect that the radiation fields will not vary with ϕ. This is because as the observer moves around the source such that r and z are constant, the appearance of the source remains the same; thus, its radiation fields are also unchanged. Therefore, for simplicity we will confine the observation point to a fixed ϕ in the yz-plane ($\phi = 90°$) as shown in Fig. 1-11. Then from Fig. 1-11, we see that

$$r^2 = y^2 + z^2 \tag{1-79}$$

$$z = r \cos \theta \tag{1-80}$$

$$y = r \sin \theta \tag{1-81}$$

If we apply the general geometry of Fig. 1-8 to this case, $\mathbf{r}_p = \mathbf{r} = y\hat{\mathbf{y}} + z\hat{\mathbf{z}}$ and $\mathbf{r}' = z'\hat{\mathbf{z}}$ lead to $\mathbf{R} = \mathbf{r}_p - \mathbf{r}' = y\hat{\mathbf{y}} + (z - z')\hat{\mathbf{z}}$, and then

$$R = \sqrt{y^2 + (z - z')^2} = \sqrt{y^2 + z^2 - 2zz' + (z')^2} \tag{1-82}$$

Substituting (1-79) and (1-80) into (1-81), to put all field point coordinates into the spherical coordinate system, gives

$$R = \{r^2 + [-2r \cos \theta \, z' + (z')^2]\}^{1/2} \tag{1-83}$$

In order to develop approximate expressions for R, we expand (1-83) using the binomial theorem (F-4):

$$\begin{aligned} R &= (r^2)^{1/2} + \frac{1}{2}(r^2)^{-1/2}[-2r \cos \theta \, z' + (z')^2] + \frac{\frac{1}{2}(-\frac{1}{2})}{2}(r^2)^{-3/2} \\ &\quad \cdot [-2r \cos \theta \, z' + (z')^2]^2 + \cdots \\ &= r - z' \cos \theta + \frac{(z')^2 \sin^2 \theta}{2r} + \frac{(z')^3 \sin^2 \theta \cos \theta}{2r^2} + \cdots \end{aligned} \tag{1-84}$$

The terms in this series decrease as the power of z' increases if z' is small compared to r. This expression for R is used in the radiation integral (1-78) to different degrees of approximation. In the denominator (which affects only the amplitude), we let

$$R \approx r \tag{1-85}$$

We can do this because in the far field r is very large compared to the antenna size, so $r \gg z' \geq z' \cos \theta$. In the phase term $-\beta R$, we must be more accurate when computing the distance from points along the line source to the observation point.

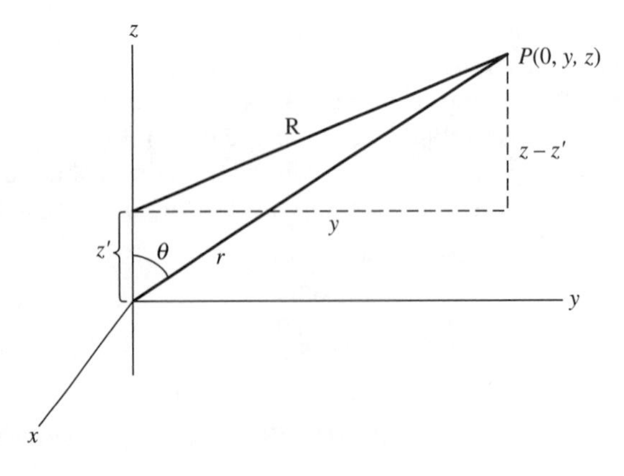

Figure 1-11 Geometry used for field calculations of a line source along the z-axis.

The integral (1-78) sums the contributions from all the points along the line source. Although the amplitude of waves due to each source point is essentially the same, the phase can be different if the path length differences are a sizable fraction of a wavelength. We, therefore, include the first two terms of the series in (1-84) for the R in the numerator of (1-78), giving

$$R \approx r - z' \cos \theta \tag{1-86}$$

Using the far-field approximations (1-85) and (1-86) in (1-78) yields

$$A_z = \mu \int I(z') \frac{e^{-j\beta(r - z' \cos \theta)}}{4\pi r} dz' = \mu \frac{e^{-j\beta r}}{4\pi r} \int I(z') e^{j\beta z' \cos \theta} dz' \tag{1-87}$$

where the integral is over the extent of the line source. This may be recognized as a Fourier-transform-type integral; see Sec. 4.3. Next, the magnetic field is found using (1-37):

$$\begin{aligned} \mathbf{H} &= \frac{1}{\mu} \nabla \times \mathbf{A} = \frac{1}{\mu} \nabla \times (A_z \hat{\mathbf{z}}) \\ &= \frac{1}{\mu} \nabla \times (-A_z \sin \theta \hat{\boldsymbol{\theta}} + A_z \cos \theta \hat{\mathbf{r}}) \end{aligned} \tag{1-88}$$

where (C-3) was used. Since A_z is a function of r and θ, the curl in spherical coordinates, as given by (C-35), leads to

$$\mathbf{H} = \hat{\boldsymbol{\phi}} \frac{1}{\mu} \frac{1}{r} \left[\frac{\partial}{\partial r} (-r A_z \sin \theta) - \frac{\partial}{\partial \theta} (A_z \cos \theta) \right] \tag{1-89}$$

Substitution of (1-87) into the above gives

$$\begin{aligned} \mathbf{H} &= \hat{\boldsymbol{\phi}} \frac{1}{\mu} \left\{ \mu \frac{-\sin \theta}{4\pi r} \int I(z') e^{j\beta z' \cos \theta} dz' \frac{\partial}{\partial r} e^{-j\beta r} - \frac{e^{-j\beta r}}{4\pi r^2} \frac{\partial}{\partial \theta} \right. \\ &\quad \left. \cdot \left[\mu \cos \theta \int I(z') e^{j\beta z' \cos \theta} dz' \right] \right\} \\ &= \hat{\boldsymbol{\phi}} \frac{1}{\mu} \frac{e^{-j\beta r}}{4\pi r} \left\{ j\beta \, \mu \sin \theta \int I(z') e^{j\beta z' \cos \theta} dz' \right. \\ &\quad \left. - \frac{1}{r} \frac{\partial}{\partial \theta} \left[\mu \cos \theta \int I(z') e^{j\beta z' \cos \theta} dz' \right] \right\} \end{aligned} \tag{1-90}$$

The ratio of the first term to the second term above is of the order βr. If $\beta r \gg 1$, the second term is small compared to the first and can be neglected, as we did for the far-field approximation of the ideal dipole in Sec. 1.6. Thus, (1-90) becomes

$$\mathbf{H} = \hat{\boldsymbol{\phi}} \frac{j\beta}{\mu} \sin \theta \, \mu \frac{e^{-j\beta r}}{4\pi r} \int I(z') e^{j\beta z' \cos \theta} dz' = \frac{j\beta}{\mu} \sin \theta \, A_z \, \hat{\boldsymbol{\phi}} \tag{1-91}$$

The electric field is found from (1-47), which is

$$\mathbf{E} = -j\omega \mathbf{A} - j \frac{\nabla(\nabla \cdot \mathbf{A})}{\omega \mu \epsilon} \tag{1-92}$$

Using (1-78) in (1-79) and retaining only the r^{-1} term (and assuming $\beta r \gg 1$) lead to the far-field approximation

$$\mathbf{E} = -j\omega A_\theta \hat{\boldsymbol{\theta}} = j\omega \sin \theta \, A_z \hat{\boldsymbol{\theta}} \tag{1-93}$$

Note that this result is the portion of the first term of (1-92) that is transverse to $\hat{\mathbf{r}}$ because $-j\omega\mathbf{A} = -j\omega(-A_z \sin\theta\hat{\boldsymbol{\theta}} + A_z \cos\theta\hat{\mathbf{r}})$. This is an important general result for z-directed sources that is not restricted to line sources.

The radiation fields from a z-directed line source (any z-directed current source in general) are H_ϕ and E_θ, and are found from (1-91) and (1-93). The only remaining problem is to calculate A_z, which is given by (1-57) in general and by (1-87) for z-directed line sources. The calculation of A_z is the focal point of antenna analysis. We will return to this topic after pausing to further examine the characteristics of the far-field region.

The ratio of the radiation field components as given by (1-91) and (1-93) yields

$$E_\theta = \frac{\omega\mu}{\beta} H_\phi = \frac{\omega\mu}{\omega\sqrt{\mu\varepsilon}} H_\phi = \eta H_\phi \tag{1-94}$$

where $\eta = \sqrt{\mu/\varepsilon}$ is the intrinsic impedance of the medium. Thus, the radiation fields are perpendicular to each other and to the direction of propagation $\hat{\mathbf{r}}$ and their magnitudes are related by (1-94). These are the familiar properties of a plane wave. They also hold for the general form of a "transverse electromagnetic (TEM) wave" that has both the electric and magnetic fields transverse to the direction of propagation. In general, radiation from a finite antenna is a special case of a TEM wave, called a "spherical wave," that propagates radially outward from the antenna and the radiation fields have no radial components. Spherical wave behavior is also characterized by the $e^{-j\beta r}/4\pi r$ factor in the field expressions; see (1-91). The $e^{-j\beta r}$ phase factor indicates a traveling wave propagating radially outward from the origin and the $1/r$ magnitude dependence leads to constant power flow just as with the infinitesimal dipole. In fact, the radiation fields of all antennas of finite extent display this dependence with distance from the antenna.

Another way to view radiation field behavior is to note that spherical waves appear to an observer in the far field to be a plane wave. This "local plane wave behavior" occurs because the radius of curvature of the spherical wave is so large that the phase front is nearly planar over a local region.

1.7.3 Far-Field Conditions and Field Regions

The results for the line current from the previous section are easily generalized to an arbitrary, but finite-sized, antenna. In the far field of an antenna, the fields exhibit local plane wave behavior and have $1/r$ magnitude dependence. In this section, we develop the conditions for determining the minimum distance from an antenna for far-field behavior. This begins with a geometric interpretation for far-field approximations.

If parallel lines (or rays) are drawn from each point on a line current as shown in Fig. 1-12, the distance R to the far field is geometrically related to r by (1-86), which was derived by neglecting high-order terms in the expression for R in (1-84). The parallel ray assumption is exact only when the observation point is at infinity, but it is a good approximation in the far field. Radiation calculations often start by assuming parallel rays and then determining R for the phase by geometrical techniques. From the general source shown in Fig. 1-13, we see that

$$R = r - r' \cos\alpha \tag{1-95}$$

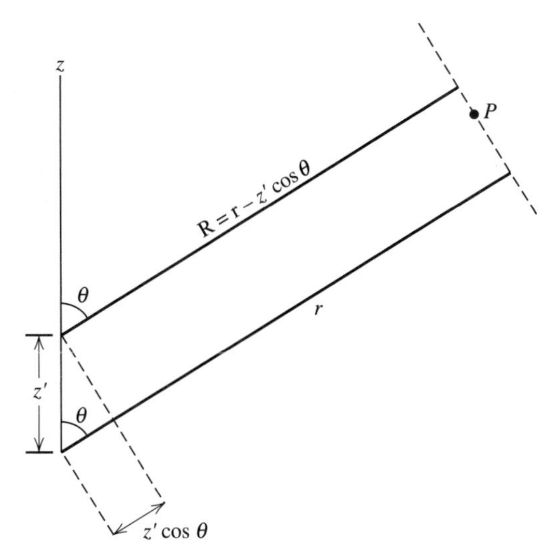

Figure 1-12 Parallel ray approximation for far-field calculations of a line source.

Using the definition of dot product, we have

$$R = r - r' \frac{\mathbf{r} \cdot \mathbf{r}'}{rr'}$$

= Def. of dot Product.

or

$$\boxed{R = r - \hat{\mathbf{r}} \cdot \mathbf{r}'} \tag{1-96}$$

This is a general approximation to R for the phase factor in the radiation integral. Notice that if $\mathbf{r}' = z'\hat{\mathbf{z}}$, as for line sources along the z-axis, (1-96) reduces to (1-86).

The definition of the distance from the source where the far field begins is where errors due to the parallel ray approximation become insignificant. The distance

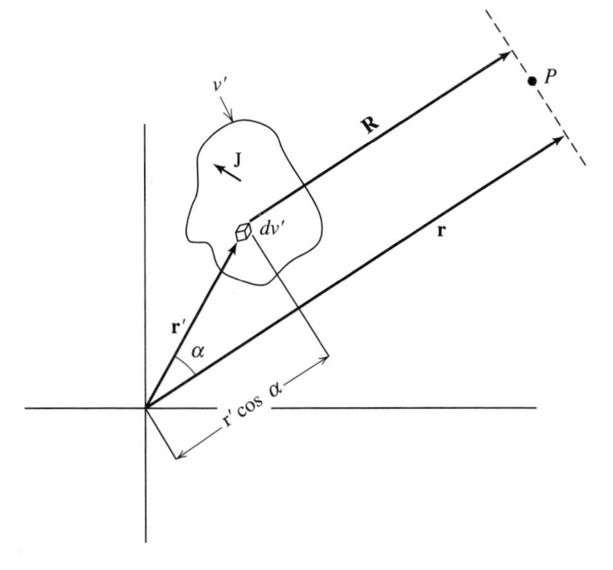

Figure 1-13 Parallel ray approximation for far-field calculations of a general source.

where the far field begins, r_{ff}, is taken to be that value of r for which the path length deviation due to neglecting the third term of (1-84) is a sixteenth of a wavelength. This corresponds to a phase error (by neglecting the third term) of $2\pi/\lambda \cdot \lambda/16 = \pi/8$ rad $= 22.5°$.

If D is the length of the line source, r_{ff} is found by equating the maximum value of the third term of (1-84), which occurs for $z' = D/2$ and $\theta = 90°$, to a sixteenth of a wavelength:

$$\frac{(D/2)^2}{2r_{ff}} = \frac{\lambda}{16} \tag{1-97}$$

Solving for r_{ff} gives

$$\boxed{r_{ff} = \frac{2D^2}{\lambda}} \tag{1-98}$$

The far-field region is $r \geq r_{ff}$ and r_{ff} is called the *far-field distance*, or Rayleigh distance.

The far-field conditions are summarized as follows:

$$r > \frac{2D^2}{\lambda} \tag{1-99a}$$

$$r \gg D \qquad \textit{far-field conditions} \tag{1-99b}$$

$$r \gg \lambda \tag{1-99c}$$

The condition $r \gg D$ was mentioned in association with the approximation $R \approx r$ of (1-85) for use in the magnitude dependence. The condition $r \gg \lambda$ follows from $\beta r = (2\pi r/\lambda) \gg 1$ that was used to reduce (1-90) to (1-91). Usually, the far field is taken to begin at a distance given by (1-98), where D is the maximum dimension of the antenna. This is usually a sufficient condition for antennas operating in the UHF region and above. At lower frequencies, where the antenna can be small compared to the wavelength, the far-field distance may have to be greater than $2D^2/\lambda$ in order that all conditions in (1-99) are satisfied. See Prob. 1.7-4.

The concept of field regions was introduced in Sec. 1.6 and illustrated with the fields of an ideal dipole. We can now generalize that discussion to any finite antenna of maximum extent D. The distance to the far field is $2D^2/\lambda$. This zone was historically called the *Fraunhofer region* if the antenna is focused at infinity; that is, if the rays at large distances from the antenna when transmitting are parallel. In the far-field region, the radiation pattern is independent of distance. For example, the $\sin\theta$ pattern of an ideal dipole is valid anywhere in its far field. The zone interior to r_{ff}, called the near field, is divided into two subregions. The *reactive near-field region* is closest to the antenna and is that region for which the reactive field dominates over the radiative fields. This region extends to a distance $0.62\sqrt{D^3/\lambda}$ from the antenna (see Prob. 1.7-7), as long as $D \gg \lambda$. We mentioned in Sec. 1.6 that for an ideal dipole, for which $D = \Delta z \ll \lambda$, this distance is $\lambda/2\pi$. Between the reactive near-field and far-field regions is the *radiating near-field region* in which the radiation fields dominate and where the angular field distribution depends on distance from the antenna. For an antenna focused at infinity, the region is sometimes referred to as the *Fresnel region*. We can summarize the field region distances for cases where $D \gg \lambda$ as follows:

Region	Distance from antenna (r)	
Reactive near field	0 to $0.62\sqrt{D^3/\lambda}$	(1-100a)
Radiating near field	$0.62\sqrt{D^3/\lambda}$ to $2D^2/\lambda$	(1-100b)
Far field	$2D^2/\lambda$ to ∞	(1-100c)

1.7.4 Steps in the Evaluation of Radiation Fields

The derivation for the fields radiated by a line source in Sec. 1.7.2 can be generalized for application to any antenna. Fortunately, the derivation itself need not be repeated each time an antenna is analyzed. That is, it is not necessary to return to Maxwell's equation with each new antenna system. Instead, we work from the results of the line source and its generalizations, which can be reduced to the three-step procedure detailed below:

1. **Find A.** Select a coordinate system most compatible with the geometry of the antenna, using the notation of Fig. 1-8. In general, use (1-58) with $R \approx r$ in the magnitude factor and the parallel ray approximation of (1-96) for determining phase differences over the antenna. These yield

$$\mathbf{A} = \mu \frac{e^{-j\beta r}}{4\pi r} \iiint_{v'} \mathbf{J} e^{j\beta \hat{\mathbf{r}} \cdot \mathbf{r}'}\, dv' \tag{1-101}$$

For z-directed sources,

$$\mathbf{A} = \hat{\mathbf{z}} \mu \frac{e^{-j\beta r}}{4\pi r} \iiint_{v'} J_z e^{j\beta \hat{\mathbf{r}} \cdot \mathbf{r}'}\, dv' \tag{1-102}$$

For z-directed line sources on the z-axis,

$$\mathbf{A} = \hat{\mathbf{z}} \mu \frac{e^{-j\beta r}}{4\pi r} \int I(z') e^{j\beta z' \cos\theta}\, dz' \tag{1-103}$$

which is (1-87).

2. **Find E.** In general, use the component of

$$\mathbf{E} = -j\omega \mathbf{A} \tag{1-104}$$

which is transverse to the direction of propagation $\hat{\mathbf{r}}$. This is expressed formally as

$$\mathbf{E} = -j\omega \mathbf{A} - (-j\omega \mathbf{A} \cdot \hat{\mathbf{r}})\hat{\mathbf{r}} = -j\omega (A_\theta \hat{\boldsymbol{\theta}} + A_\phi \hat{\boldsymbol{\phi}}) \tag{1-105}$$

which arises from the component of $\mathbf{A}$ tangent to the far-field sphere. For z-directed sources,

$$\mathbf{E} = j\omega \sin\theta\, A_z \hat{\boldsymbol{\theta}} \tag{1-106}$$

which is (1-93)

3. **Find H.** In general, use the plane wave relation.

$$\mathbf{H} = \frac{1}{\eta}\hat{\mathbf{r}} \times \mathbf{E} \tag{1-107}$$

This equation expresses the fact that in the far field, the directions of **E** and **H** are perpendicular to each other and to the direction of propagation, and also that their magnitudes are related by η. For z-directed sources,

$$H_\phi = \frac{E_\theta}{\eta} \tag{1-108}$$

which is (1-73).

The most difficult step is the first, calculating the radiation integral. This topic will be discussed many times throughout the book, but to immediately develop an appreciation for the process, we will present an example. This uniform line source example also provides a specific setting for introducing general radiation pattern concepts and definitions.

EXAMPLE 1-1 *The Uniform Line Source*

The **uniform line source** is a line source for which the current is constant along its extent. If we use a z-directed uniform line source centered on the origin and along the z-axis, the current is

$$I(z') = \begin{cases} I_o & x' = 0, \quad y' = 0, \quad |z'| \leq \dfrac{L}{2} \\ 0 & \textit{elsewhere} \end{cases} \tag{1-109}$$

where L is the length of the line source; see Fig. 1-14. We first find A_z from (1-103) as follows:

$$\begin{aligned} A_z &= \mu \frac{e^{-j\beta r}}{4\pi r} \int_{-L/2}^{L/2} I_o e^{j\beta z' \cos\theta}\, dz' \\ &= \mu \frac{e^{-j\beta r}}{4\pi r} I_o \left[\frac{e^{j\beta(L/2)\cos\theta} - e^{-j\beta(L/2)\cos\theta}}{j\beta\cos\theta} \right] \\ &= \mu \frac{I_o L e^{-j\beta r}}{4\pi r} \frac{\sin[(\beta L/2\cos\theta]}{(\beta L/2)\cos\theta}. \end{aligned} \tag{1-110}$$

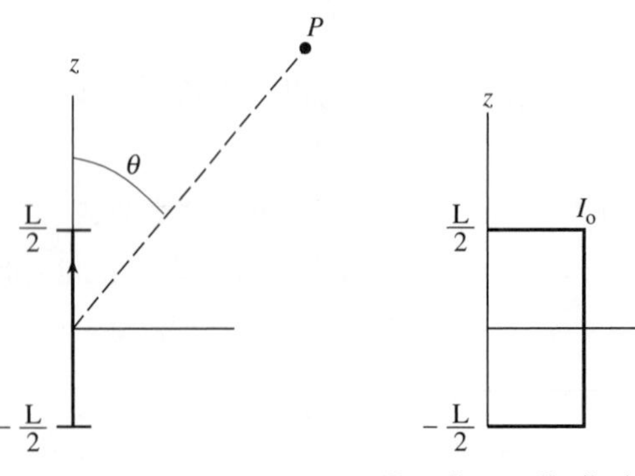

Figure 1-14 The uniform line source (Example 1-1).

The electric field from (1-106) is then

$$\mathbf{E} = j\omega \sin\theta\, A_z \hat{\boldsymbol{\theta}} = \frac{j\omega\mu I_o L e^{-j\beta r}}{4\pi r} \sin\theta \frac{\sin[(\beta L/2)\cos\theta]}{(\beta L/2)\cos\theta}\hat{\boldsymbol{\theta}} \tag{1-111}$$

The magnetic field is simply found from this using $H_\phi = E_\theta/\eta$.

1.7.5 Radiation Pattern Definitions

Since the radiation pattern is the variation over a sphere centered on the antenna, r is constant and we have only θ and ϕ variation of the field. It is convenient to normalize the field expression such that its maximum value is unity. This is accomplished as follows for a z-directed source that has only a θ-component of $\mathbf{E}$:

$$F(\theta, \phi) = \frac{E_\theta}{E_\theta(\max)} \tag{1-112}$$

where $F(\theta, \phi)$ is the *normalized field pattern* and $E_\theta(\max)$ is the maximum value of the magnitude of E_θ over a sphere of radius r.

In general, E_θ can be complex-valued and, therefore, so can $F(\theta, \phi)$. In this case, the phase is usually set to zero at the same point the magnitude is normalized to unity. This is appropriate since we are only interested in relative phase behavior. The pattern magnitude is obtained by taking the absolute value of (1-112).

An element of current on the z-axis has a normalized field pattern from (1-72a) of

$$F(\theta) = \frac{(I\,\Delta z/4\pi)j\omega\mu(e^{-j\beta r}/r)\sin\theta}{(I\,\Delta z/4\pi)j\omega\mu(e^{-j\beta r}/r)} = \sin\theta \tag{1-113}$$

and there is no ϕ variation. The normalized field pattern for the uniform line source is from (1-111) in (1-112)

$$F(\theta) = \sin\theta \frac{\sin[(\beta L/2)\cos\theta]}{(\beta L/2)\cos\theta} \tag{1-114}$$

and again there is no ϕ variation. The second factor of this expression is the function $\sin(u)/u$ and we will encounter it frequently. It has a maximum value of unity at $u = 0$; this corresponds to $\theta = 90°$, where $u = (\beta L/2)\cos\theta$. Substituting $\theta = 90°$ in (1-114) gives unity and we see that $F(\theta)$ is properly normalized.

In general, a normalized field pattern can be written as the product

$$\boxed{F(\theta, \phi) = g(\theta, \phi) f(\theta, \phi)} \tag{1-115}$$

where $g(\theta, \phi)$ is the *element factor* and $f(\theta, \phi)$ the *pattern factor*. The pattern factor comes from the integral over the current and is strictly due to the distribution of current in space. The element factor is the pattern of an infinitesimal current element in the current distribution. For example, we found for a z-directed current element that $F(\theta) = \sin\theta$. This is, obviously, also the element factor, so

$$g(\theta) = \sin\theta \tag{1-116}$$

for a z-directed current element. Actually, this factor originates from (1-93) and can be interpreted as the projection of the current element in the θ-direction. In other words, at $\theta = 90°$ we see the maximum length of the current, whereas at $\theta = 0°$ or

180° we see the endview of an infinitesimal current that yields no radiation. The $\sin\,\theta$ factor expresses the fraction of the size of the current as seen from the observation angle θ. On the other hand, the pattern factor $f(\theta,\,\phi)$ represents the integrated effect of radiation contributions from the current distribution, which can be treated as being made up of many current elements. The pattern value in a specific direction is then found by summing the parallel rays from each current element to the far field with the magnitude and phase of each included. The radiation integral of (1-101) sums the far-field contributions from the current elements and when normalized yields the pattern factor. Antenna analysis is usually easier to understand by considering the antenna to be transmitting as we have here. However, most antennas are reciprocal and thus their radiation properties are identical when used for reception; see Sec. 9.4.

A typical power pattern is shown in Fig. 1-15 as a polar plot. The rays from various parts of an antenna arrive in the far field with different magnitude and phase due to variations in the current where the ray originated on the antenna and due to phase changes arising from path length differences to the far field. These rays interfere, as computed through the radiation integral, and produce a "lobing" effect.

The radiation lobe containing the direction of maximum radiation is the *major lobe*, *main lobe*, or *main beam*. It is the most intense portion of the radiation pattern and is caused by the fact that the rays from various parts of the antenna arrive in the far field more nearly in-phase than they do for other directions. For a source with constant phase, all rays arrive in-phase in the direction normal to the antenna and the pattern is maximum there. For the ideal dipole, we have said that the source is so small that there are essentially no phase differences for rays along the source and thus the pattern factor is unity.

For the z-directed uniform line source pattern of (1-114), we can identify the factors as

$$g(\theta) = \sin\,\theta \tag{1-117}$$

and

$$f(\theta) = \frac{\sin[(\beta L/2)\cos\,\theta]}{(\beta L/2)\cos\,\theta} \tag{1-118}$$

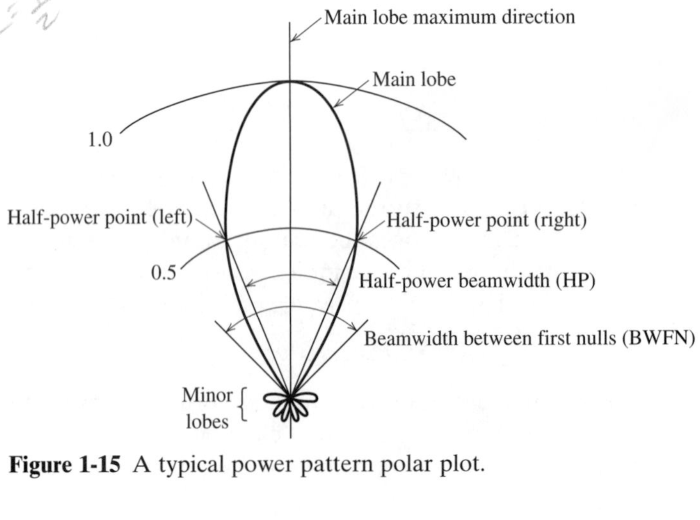

Main lobe maximum direction
Main lobe
1.0
Half-power point (left)
Half-power point (right)
0.5
Half-power beamwidth (HP)
Beamwidth between first nulls (BWFN)
Minor lobes

Figure 1-15 A typical power pattern polar plot.

For long line sources ($L \gg \lambda$), the pattern factor of (1-118) is much sharper than the element factor $\sin\theta$, and the total pattern is approximately that of (1-118); that is, $F(\theta) \approx f(\theta)$. Hence, in many cases we need only work with $f(\theta)$, which is obtained from (1-103). If we allow the beam to be scanned (this will be discussed in Sec. 1.7.6), the element factor becomes important as the pattern maximum approaches the z-axis.

Frequently, the directional properties of the radiation from an antenna are described by another form of radiation pattern, the power pattern. The power pattern gives angular dependence of the power density and is found from the θ, ϕ variation of the r-component of the Poynting vector. For z-directed sources, $H_\phi = E_\theta/\eta$ so the r-component of the Poynting vector is $(1/2)E_\theta H_\phi^* = |E_\theta|^2/2\eta$ and the normalized power pattern is simply the square of its field pattern magnitude $P(\theta) = |F(\theta)|^2$. The general normalized *power pattern* is

$$\boxed{P(\theta, \phi) = |F(\theta, \phi)|^2} \tag{1-119}$$

The normalized power pattern for a z-directed current element is

$$P(\theta) = \sin^2\theta \tag{1-120}$$

and for a z-directed uniform line source is

$$P(\theta) = \left\{\sin\theta \, \frac{\sin[(\beta L/2)\cos\theta]}{(\beta L/2)\cos\theta}\right\}^2 \tag{1-121}$$

Frequently, patterns are plotted in decibels. *It is important to recognize that the field (magnitude) pattern and power pattern are the same in decibels.* This follows directly from the definitions. For field intensity in decibels,

$$|F(\theta, \phi)|_{\text{dB}} = 20 \log|F(\theta, \phi)| \tag{1-122}$$

and for power in decibels,

$$P(\theta, \phi)_{\text{dB}} = 10 \log P(\theta, \phi) = 10 \log|F(\theta, \phi)|^2 = 20 \log|F(\theta, \phi)| \tag{1-123}$$

and we see that

$$\boxed{P(\theta, \phi)_{\text{dB}} = |F(\theta, \phi)|_{\text{dB}}} \tag{1-124}$$

1.7.6 Radiation Pattern Parameters

A typical antenna power pattern is shown in Fig. 1-15 as a polar plot in linear units (rather than decibels). It consists of several *lobes*. The *main lobe* (or main beam or major lobe) is the lobe containing the direction of maximum radiation. There is also usually a series of lobes smaller than the main lobe. Any lobe other than the main lobe is called a *minor lobe*. Minor lobes are composed of *side lobes* and *back lobes*. Back lobes are directly opposite the main lobe, or sometimes they are taken to be the lobes in the half-space opposite the main lobe. The term side lobe is sometimes reserved for those minor lobes near the main lobe, but is most often taken to be synonymous with minor lobe; we will use the latter convention.

The radiation from an antenna is represented mathematically through the radiation pattern function, $F(\theta, \phi)$ for field and $P(\theta, \phi)$ for power. This angular distribution of radiation is visualized through various graphical representations of the pattern, which we discuss in this section. Graphical representations also are used to introduce definitions of pattern parameters that are commonly used to quantify radiation pattern characteristics.

A three-dimensional plot as in Fig. 1-10*d* gives a good overall impression of the entire radiation pattern, but cannot convey accurate quantitative information. Cuts through this pattern in various planes are the most popular pattern plots. They usually include the *E*- and *H*-plane patterns; see Figs. 1-10*b* and 1-10*c*. Pattern cuts are often given various fixed ϕ values, leaving the pattern a function of θ alone; we will assume that is the case here. Typically, the side lobes are alternately positive- and negative-valued; see Fig. 4-1*a*. In fact, a pattern in its most general form can be complex-valued. Then we use the magnitude of the field pattern $|F(\theta)|$ or the power pattern $P(\theta)$.

A measure of how well the power is concentrated into the main lobe is the (relative) *side lobe level*, which is the ratio of the pattern value of a side lobe peak to the pattern value of the main lobe. The largest side lobe level for the whole pattern is the *maximum (relative) side lobe level*, frequently abbreviated as SLL. In decibels, it is given by

$$\mathrm{SLL_{dB}} = 20 \log \frac{|F(\mathrm{SLL})|}{|F(\max)|} \tag{1-125}$$

where $|F(\max)|$ is the maximum value of the pattern magnitude and $|F(\mathrm{SLL})|$ is the pattern value of the maximum of the highest side lobe magnitude. For a normalized pattern, $F(\max) = 1$.

The width of the main beam is quantified through *half-power beamwidth* HP, which is the angular separation of the points where the main beam of the power pattern equals one-half the maximum value:

$$\mathrm{HP} = |\theta_{\mathrm{HP\,left}} - \theta_{\mathrm{HP\,right}}| \tag{1-126}$$

where $\theta_{\mathrm{HP\,left}}$ and $\theta_{\mathrm{HP\,right}}$ are points to the "left" and "right" of the main beam maximum for which the normalized power pattern has a value of one-half (see Fig. 1-15). On the field pattern $|F(\theta)|$, these points correspond to the value $1/\sqrt{2}$. For example, the $\sin\theta$ pattern of an ideal dipole has a value of $1/\sqrt{2}$ for θ values of $\theta_{\mathrm{HP\,left}} = 135°$ and $\theta_{\mathrm{HP\,right}} = 45°$. Then $\mathrm{HP} = |135° - 45°| = 90°$. This is shown in Fig. 1-10*b*. Note that the definition of HP is the magnitude of the difference of the half-power points and the assignment of left and right can be interchanged without changing HP. In three dimensions, the radiation pattern major lobe becomes a solid object and the half-power contour is a continuous curve. If this curve is essentially elliptical, the pattern cuts that contain the major and minor axes of the ellipse determine what the IEEE defines as the *principal half-power beamwidths.*

Antennas are often referred to by the type of pattern they produce. An *isotropic antenna*, which is hypothetical, radiates equally in all directions giving a constant radiation pattern. An *omnidirectional antenna* produces a pattern that is constant in one plane; the ideal dipole of Fig. 1-10 is an example. The pattern shape resembles a "doughnut." We often refer to antennas as being broadside or endfire. A *broadside antenna* is one for which the main beam maximum is in a direction normal to the plane containing the antenna. An *endfire antenna* is one for which the main beam is in the plane containing the antenna. For a linear current on the *z*-axis, the broadside direction is $\theta = 90°$ and the endfire directions are 0° and 180°. For example, an ideal dipole is a broadside antenna. For *z*-directed line sources, several patterns are possible. Figure 1-16 illustrates a few $|f(\theta)|$ patterns. The entire pattern

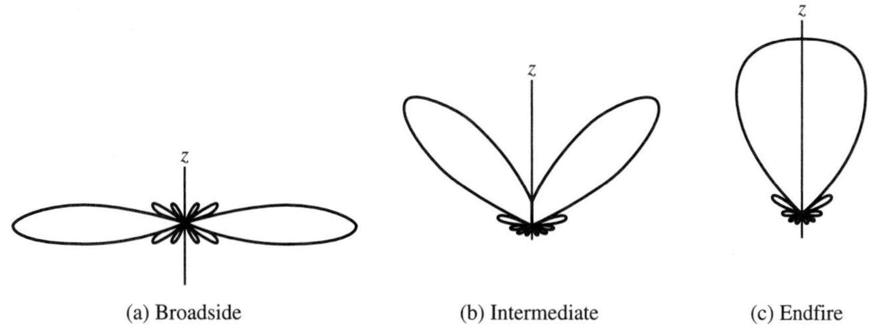

Figure 1-16 Polar plots of uniform line source patterns $|f(\theta)|$.

in three dimensions is imagined by rotating the pattern about the z-axis. The full pattern can then be generated from the E-plane patterns shown. The broadside pattern of Fig. 1-16*a* is called a *fan beam*. The full three-dimensional endfire pattern for Fig. 1-16*b* has a single lobe in the endfire direction. This single lobe is referred to as a *pencil beam*. Note that the $\sin\theta$ element factor, which must multiply these patterns to obtain the total pattern, will have a significant effect on the endfire pattern.

1.8 DIRECTIVITY AND GAIN

One very important description of an antenna is how much it concentrates energy in one direction in preference to radiation in other directions. This characteristic of an antenna is called its directivity and is equal to its power gain if the antenna is 100% efficient. Usually, power gain is expressed relative to a reference such as an isotropic radiator or half-wavelength dipole.

Toward the definition of directivity, let us begin by recalling that the power radiated by an antenna from (1-29) is

$$P = \iint \mathbf{S} \cdot d\mathbf{s} = \frac{1}{2} \operatorname{Re} \iint (\mathbf{E} \times \mathbf{H}^*) \cdot d\mathbf{s} \tag{1-127}$$

$$= \frac{1}{2} \operatorname{Re} \int_0^{2\pi} \int_0^{\pi} (E_\theta H_\phi^* - E_\phi H_\theta^*) r^2 \sin\theta \, d\theta \, d\phi \tag{1-128}$$

In general, there will be both θ- and ϕ-components of the radiation fields. From (1-107), we find that

$$H_\phi = \frac{E_\theta}{\eta} \quad \text{and} \quad H_\theta = -\frac{E_\phi}{\eta} \tag{1-129}$$

Using these in (1-128) gives

$$P = \frac{1}{2\eta} \iint (|E_\theta|^2 + |E_\phi|^2) r^2 \, d\Omega \tag{1-130}$$

where $d\Omega$ = element of solid angle = $\sin\theta \, d\theta \, d\phi$, which is shown in Fig. 1-17. The integral can be evaluated over any surface enclosing the antenna; however, for simplicity a spherical surface centered on the origin is usually used. Since the mag-

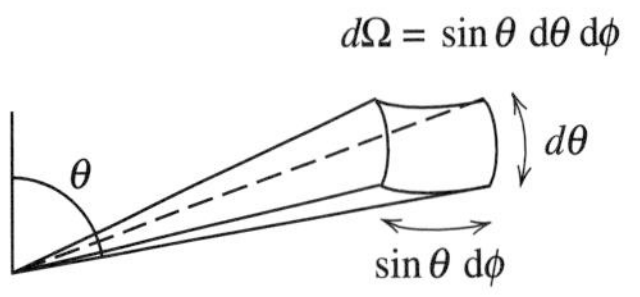

$d\Omega = \sin\theta \, d\theta \, d\phi$

$d\theta$

θ

$\sin\theta \, d\phi$

Figure 1-17 Element of solid angle $d\Omega$.

nitude variations of the radiation fields are $1/r$, we find it convenient to introduce *radiation intensity*, which is defined from

$$U(\theta, \phi) = \tfrac{1}{2} \operatorname{Re}(\mathbf{E} \times \mathbf{H}^*) \cdot r^2\hat{\mathbf{r}} = S(\theta, \phi) r^2 \qquad (1\text{-}131)$$

Radiation intensity is the power radiated in a given direction per unit solid angle and has units of watts per square radian (or steradian, sr). The advantage to using radiation intensity is that it is independent of distance r. Radiation intensity can be expressed as

$$U(\theta, \phi) = U_m |F(\theta, \phi)|^2 \qquad (1\text{-}132)$$

where U_m is the maximum radiation intensity, and $|F(\theta, \phi)|^2$ is the power pattern normalized to a maximum value of unity in the direction $(\theta_{\max}, \phi_{\max})$, and

$$U_m = U(\theta_{\max}, \phi_{\max}) \qquad (1\text{-}133)$$

The total power radiated is obtained by integrating the radiation intensity over all angles around the antenna:

$$P = \iint U(\theta, \phi) \, d\Omega = U_m \iint |F(\theta, \phi)|^2 \, d\Omega \qquad (1\text{-}134)$$

An isotropic source with uniform radiation in all directions is only hypothetical but is sometimes a useful concept. The radiation intensity of an isotropic source is constant over all space, at a value of U_{ave}. Then $P = \iint U_{\text{ave}} \, d\Omega = U_{\text{ave}} \iint d\Omega = 4\pi U_{\text{ave}}$ since there are 4π sr in all space (see Prob. 1.8-1). For nonisotropic sources, the radiation intensity is not constant throughout space, but an average power per steradian can be defined as

$$U_{\text{ave}} = \frac{1}{4\pi} \iint U(\theta, \phi) \, d\Omega = \frac{P}{4\pi} \qquad (1\text{-}135)$$

The average radiation intensity U_{ave} equals the radiation intensity $U(\theta, \phi)$ that an isotropic source with the same input power P would radiate.

As an example, consider the ideal dipole again; we find from (1-72) and (1-131) that

$$U(\theta, \phi) = \frac{1}{2} \left(\frac{I \, \Delta z}{4\pi} \right)^2 \beta\omega\mu \sin^2\theta \qquad (1\text{-}136)$$

so

$$U_m = \frac{1}{2} \left(\frac{I \, \Delta z}{4\pi} \right)^2 \beta\omega\mu \qquad (1\text{-}137)$$

and

$$F(\theta, \phi) = \sin\theta \qquad (1\text{-}138)$$

The average radiation intensity follows from the total radiated power expression (1-75) for an ideal dipole as

$$U_{\text{ave}} = \frac{P}{4\pi} = \frac{(\beta\omega\mu/12\pi)(I\,\Delta z)^2}{4\pi} = \frac{1}{3}\left(\frac{I\,\Delta z}{4\pi}\right)^2 \beta\omega\mu$$

$$= \frac{2}{3}\,U_m \qquad \textit{ideal dipole} \tag{1-139}$$

Thus, $U_m = 1.5U_{\text{ave}}$ for the ideal dipole, which means that in the direction of maximum radiation, the radiation intensity is 50% more than that which would occur from an isotropic source radiating the same total power.

Directivity. *Directivity* is defined as the ratio of the radiation intensity in a certain direction to the average radiation intensity, or

$$D(\theta, \phi) = \frac{U(\theta, \phi)}{U_{\text{ave}}} \tag{1-140}$$

If we divide the numerator and denominator by r^2, then we have power densities. So, directivity is also the ratio of the power density in a certain direction at a given range r to the average power density at that range, or

$$D(\theta, \phi) = \frac{U(\theta, \phi)/r^2}{U_{\text{ave}}/r^2} = \frac{\frac{1}{2}\,\text{Re}(\mathbf{E} \times \mathbf{H}^*) \cdot \hat{\mathbf{r}}}{P/4\pi r^2} \tag{1-141}$$

Substitution of (1-135) for U_{ave} in (1-140) yields

$$D(\theta, \phi) = \frac{U(\theta, \phi)}{\dfrac{1}{4\pi}\displaystyle\iint U(\theta, \phi)\, d\Omega} = \frac{|F(\theta, \phi)|^2}{\dfrac{1}{4\pi}\displaystyle\iint |F(\theta, \phi)|^2\, d\Omega}$$

$$= \frac{4\pi}{\Omega_A}\,|F(\theta, \phi)|^2 \tag{1-142}$$

where Ω_A is the *beam solid angle* defined by

$$\boxed{\Omega_A = \iint |F(\theta, \phi)|^2\, d\Omega} \tag{1-143}$$

This result shows that directivity is entirely determined by pattern shape. Beam solid angle is the solid angle through which all the power would be radiated if the power per unit solid angle (radiation intensity) equaled the maximum value over the beam area. This is illustrated in Fig. 1-18. From (1-134) and (1-143), we see that

$$P = U_m \Omega_A \tag{1-144}$$

This can also be inferred from Fig. 1-18*b*.

When directivity is quoted as a single number without reference to a direction, maximum (peak) directivity is usually intended. Maximum directivity follows from (1-140) as

$$D = \frac{U_m}{U_{\text{ave}}} \tag{1-145}$$

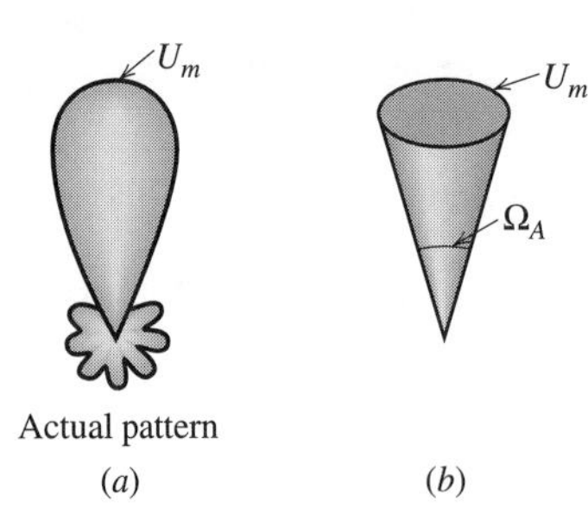

Figure 1-18 Antenna beam solid angle Ω_A. (*a*) Plot of radiation intensity $U(\theta, \phi)$ from an actual antenna. (*b*) Plot of radiation intensity with all radiation from the actual antenna concentrated into a cone of solid angle Ω_A with constant radiation intensity equal to the maximum of the actual pattern.

Using (1-135) and (1-144) in (1-145) gives

$$D = \frac{U_m}{P/4\pi} = \frac{4\pi U_m}{P} = \frac{4\pi U_m}{U_m \Omega_A} \tag{1-146a}$$

or

$$\boxed{D = \frac{4\pi}{\Omega_A}} \tag{1-146b}$$

Also from (1-132) in (1-140) we see that

$$D(\theta, \phi) = \frac{U_m |F(\theta, \phi)^2|}{U_{\text{ave}}} = D|F(\theta, \phi)|^2 \tag{1-147}$$

and since $|F(\theta, \phi)|^2$ has a maximum value of unity, the maximum value of directivity is D.

The concept of directivity is illustrated in Fig. 1-19. If the radiated power were distributed isotropically over all of space, the radiation intensity would have a maximum value equal to its average value as shown in Fig. 1-19*a*; that is, $U_m = U_{\text{ave}}$ or $\Omega_A = 4\pi$. Thus, the directivity of this isotropic pattern is unity. The distribution of radiation intensity $U(\theta, \phi)$ for an actual antenna is shown in Fig. 1-19b. It has a maximum radiation intensity in the direction $(\theta_{\max}, \phi_{\max})$ of $U_m = DU_{\text{ave}}$ and an average radiation intensity of $U_{\text{ave}} = P/4\pi$. By directing the radiated power P in a preferred direction, we can increase the radiation intensity in that direction by a factor of D over what it would be if the same radiated power had been isotropically radiated.

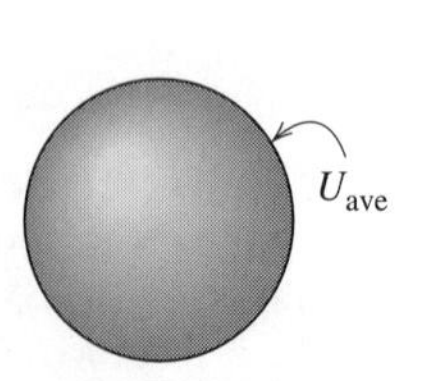

(*a*) Radiation intensity distributed isotropically

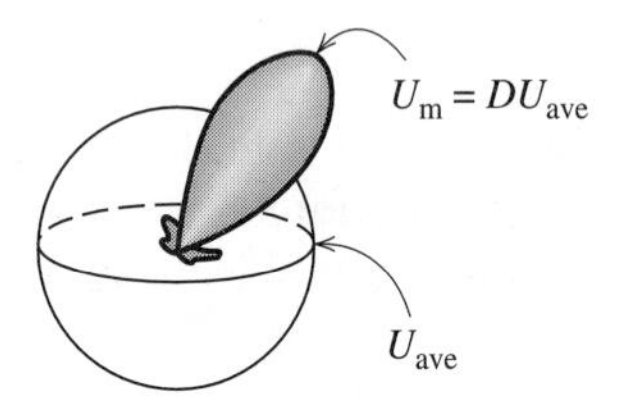

(*b*) Radiation intensity from an actual antenna

Figure 1-19 Illustration of directivity.

EXAMPLE 1-2 *Directivity of an Ideal Dipole*

The directivity of an ideal dipole can now be easily calculated using (1-139) in (1-145):

$$D = \frac{U_m}{U_{\text{ave}}} = \frac{U_m}{\frac{2}{3}U_m} = \frac{3}{2} \qquad \textit{ideal dipole} \tag{1-148}$$

Usually directivity is calculated directly from (1-146b), and the directivity calculation reduces to one of finding the beam solid angle. To illustrate, we use the ideal dipole. Substituting (1-138) in (1-143) leads to

$$\Omega_A = \int_0^{2\pi}\int_0^{\pi} |\sin\theta|^2 \sin\theta \, d\theta \, d\phi = 2\pi\frac{4}{3} = \frac{8\pi}{3} \tag{1-149}$$

and we obtain the same value of directivity from

$$D = \frac{4\pi}{\Omega_A} = \frac{4\pi}{8\pi/3} = \frac{3}{2} \tag{1-150}$$

Thus, the directivity of an ideal dipole is 50% greater than that of an isotropic source, which has a directivity of 1.

EXAMPLE 1-3 *Directivity of a Sector Omnidirectional Pattern*

An ideal omnidirectional antenna would have constant radiation in the horizontal plane ($\theta = 90°$) and would fall rapidly to zero outside that plane. Suppose that the pattern in the vertical plane is constant out to $\pm\frac{\pi}{6}$ ($\pm 30°$) from horizontal. The pattern expression is then written as

$$F(\theta) = \begin{cases} 1 & \frac{1}{3}\pi < \theta < \frac{2}{3}\pi \\ 0 & \text{elsewhere} \end{cases} \tag{1-151}$$

The solid angle of the pattern from (1-143) is

$$\begin{aligned}\Omega_A &= \iint |F(\theta, \phi)|^2 \, d\Omega = \int_0^{2\pi}\int_{\pi/3}^{2\pi/3} 1^2 \sin\theta \, d\theta \, d\phi \\ &= (2\pi)[-\cos\theta]_{\pi/3}^{2\pi/3} = (2\pi)(0.5 + 0.5) = 2\pi \end{aligned} \tag{1-152}$$

The directivity from (1-146b) is

$$D = \frac{4\pi}{\Omega_A} = \frac{4\pi}{2\pi} = 2 \tag{1-153}$$

Gain. As noted above, directivity is solely determined by the radiation pattern of an antenna. When an antenna is used in a system (say, as a transmitting antenna), we are actually interested in how efficiently the antenna transforms available power at its input terminals to radiated power, together with its directive properties. *Power gain* (or simply *gain*) is used to quantify this and is defined as 4π times the ratio of the radiation intensity in a given direction to the net power accepted by the antenna from the connected transmitter, or

$$G(\theta, \phi) = \frac{4\pi U(\theta, \phi)}{P_{\text{in}}} \tag{1-154}$$

where $G(\theta, \phi)$ is the gain and $U(\theta, \phi)$ is the radiation intensity of the antenna in the direction (θ, ϕ) including the effect of any losses on the antenna, and P_{in} is the input power accepted by the antenna. This definition does not include losses due to mismatches of impedance or polarization, which are discussed in Sec. 9.1. The maximum value of gain is the maximum of (1-154), so

$$G = \frac{4\pi U_m}{P_{in}} \tag{1-155}$$

Thus, gain can be expressed as a function of θ and ϕ and can also be given as a value in a specific direction. If no direction is specified and the gain value is not given as a function of θ and ϕ, it is assumed to be the maximum gain.

Directivity can be written from (1-146a) as $D = 4\pi U_m/P$. Comparing this with (1-155), we see that the only difference between maximum gain and directivity is the power value used. Directivity can be viewed as the gain an antenna would have if all input power appeared as radiated power; that is, $P_{in} = P$. Gain reflects the fact that real antennas do not behave in this fashion and some of the input power is lost on the antenna. The portion of input power P_{in} that does not appear as radiated power is absorbed on the antenna and nearby structures. This prompts us to define *radiation efficiency* e_r as

$$e_r = \frac{P}{P_{in}} \tag{1-156}$$

Note that

$$0 \le e_r \le 1 \tag{1-157}$$

Using (1-156) in (1-54) gives

$$G(\theta, \phi) = e_r \frac{4\pi U(\theta, \phi)}{P} = e_r \frac{U(\theta, \phi)}{U_{ave}} = e_r D(\theta, \phi) \tag{1-158}$$

Similarly, for maximum gain

$$\boxed{G = e_r D} \tag{1-159}$$

Thus, the maximum gain of an antenna is equal to its purely directional characteristic of maximum directivity multiplied by radiation efficiency.

The terminology found in the literature is inconsistent and often incorrect on the topics of directivity and gain. Directivity and gain can be functions of angle or be maximum values, that is, $D(\theta, \phi)$ or D, and $G(\theta, \phi)$ or G. Formerly, the term "directive gain" was used for directivity as a function of angle, but its use is no longer recommended by the IEEE. If no other information is given during a discussion of directivity or gain, it can safely be assumed that the maximum value is intended.

Units for Directivity and Gain. Since gain is a power ratio it can be calculated in decibels as follows:

$$G_{dB} = 10 \log G \tag{1-160}$$

Similarly for directivity,

$$D_{\text{dB}} = 10 \log D \tag{1-161}$$

For example, the directivity in decibels of an ideal dipole is

$$D_{\text{dB}} = 10 \log 1.5 = 1.76 \text{ dB} \qquad \textit{ideal dipole} \tag{1-162}$$

Frequently, gain is used to describe the performance of the antenna relative to some standard reference antenna. This *relative gain* is defined as the ratio of the maximum radiation intensity from the antenna U_m to the maximum radiation intensity from a reference antenna $U_{m,\text{ref}}$ with the same input power, or

$$G_{\text{ref}} = \frac{U_m}{U_{m,\text{ref}}} \tag{1-163}$$

This is a convenient definition from a measurement standpoint. The formal definition of gain employs a hypothetical lossless isotropic antenna as a reference antenna. This can be shown by noting that the lossless isotropic reference antenna has a maximum radiation intensity of $P_{\text{in}}/4\pi$ since all its input power is radiated, and substituting this into (1-163) for $U_{m,\text{ref}}$ leads to (1-155).

It is common at frequencies below 1 GHz to quote gain values relative to that of a half-wave dipole. The directivity of a half-wave dipole is 1.64, or 2.15 dB; see Sec. 5.1. Gain relative to a half-wave dipole carries the units of dBd. The unit dBi is often used instead of dB to emphasize that an isotropic antenna is the reference. In addition, the term *absolute gain*, which is synonymous with gain, is sometimes used. As a numerical example, consider an antenna with a gain of 6.1 dB; its gain can be written in the following ways:

$$G = 6.1 \text{ dB} = 6.1 \text{ dBi} = 3.95 \text{ dBd} \tag{1-164}$$

1.9 ANTENNA IMPEDANCE, RADIATION EFFICIENCY, AND THE SHORT DIPOLE

The input impedance of an antenna is the impedance presented by the antenna at its terminals. Thus, suitable terminals must be defined for an antenna. The input impedance will be affected by other antennas or objects that are nearby, but for this discussion we assume that the antenna is isolated. Input impedance is composed of real and imaginary parts:

$$Z_A = R_A + jX_A \tag{1-165}$$

The input resistance R_A represents dissipation, which occurs in two ways. Power that leaves the antenna and never returns (i.e., radiation) is a form of dissipation. There are also ohmic losses associated with heating on the antenna structure, but on many antennas ohmic losses are small compared to radiation losses. However, ohmic losses are usually significant on electrically small antennas, which have dimensions much less than a wavelength. The input reactance X_A represents power stored in the near field of the antenna. As a consequence of reciprocity, the impedance of an antenna is identical during reception and transmission.

First, we discuss the input resistance. The average power dissipated in an antenna is

$$P_{\text{in}} = \frac{1}{2} R_A |I_A|^2 \tag{1-166}$$

where I_A is the current at the input terminals. Note that a factor of $\frac{1}{2}$ is present because current I_A is the peak value in the time waveform. Separating the dissipated power into radiative and ohmic losses gives

$$\begin{aligned} P_{\text{in}} &= P + P_{\text{ohmic}} \\ \frac{1}{2} R_A |I_A|^2 &= \frac{1}{2} R_r |I_A|^2 + \frac{1}{2} R_{\text{ohmic}} |I_A|^2 \end{aligned} \tag{1-167}$$

where we define the radiation resistance of an antenna referred to the input terminals as

$$R_r = \frac{2P}{|I_A|^2} \tag{1-168}$$

It follows from (1-167) that

$$R_A = R_r + R_{\text{ohmic}} \tag{1-169}$$

where R_{ohmic} is resistance associated with ohmic losses that include the directly driven part of the antenna as well as losses in other portions of the antenna structure such as a ground plane. Ohmic resistance of an antenna is defined as

$$R_{\text{ohmic}} = \frac{2P_{\text{ohmic}}}{|I_A|^2} = \frac{2(P_{\text{in}} - P)}{|I_A|^2} \tag{1-170}$$

The radiated power is found using (1-35):

$$P = \frac{1}{2} \iint_{s_{\text{ff}}} (\mathbf{E} \times \mathbf{H}^*) \cdot d\mathbf{s} \tag{1-171}$$

where s_{ff} is a surface in the far field, usually spherical. P is real-valued because the power density $\mathbf{S} = \frac{1}{2}\mathbf{E} \times \mathbf{H}^*$ is real-valued in the far field.

Radiation resistance can be defined relative to the current at any point on the antenna, but we reserve R_r for radiation resistance referred to the input terminals. Radiation resistance relative to the maximum current I_m that occurs on the antenna R_{rm} is obtained by using I_m in place of I_A in (1-168). In this section, we discuss center-fed electrically short antennas, which always have a current maximum at the input, so $R_r = R_{rm}$. We discuss this topic again in Sec. 5.1.

The power radiated from an ideal dipole of length $\Delta z \ll \lambda$ and input current $I_A = I$ is given by (1-75), which together with (1-168) gives the radiation resistance:

$$\begin{aligned} R_r &= \frac{2P}{|I_A|^2} = \frac{2}{I^2} \frac{\omega\mu\beta}{12\pi} (I\,\Delta z)^2 = \frac{\sqrt{\mu}\omega\sqrt{\mu}\sqrt{\varepsilon}}{\sqrt{\varepsilon}\, 6\pi} \beta(\Delta z)^2 \\ &= \eta \frac{\beta^2}{6\pi} (\Delta z)^2 = \eta \frac{2}{3} \pi \left(\frac{\Delta z}{\lambda}\right)^2 \end{aligned}$$

$$R_r = 80\pi^2 \left(\frac{\Delta z}{\lambda}\right)^2 \ \Omega \qquad \textit{ideal dipole} \tag{1-172}$$

For ideal dipoles, R_r is very small since $\Delta z \ll \lambda$.

The relative amounts of input power dissipated by radiation and ohmic losses determine the efficiency of the antenna. This is expressed by the radiation efficiency

e_r that was introduced in Sec. 1.8 and defined in (1-156) as the ratio of total radiated power to the net power accepted by the antenna, so

$$e_r = \frac{P}{P_{\text{in}}} = \frac{P}{P + P_{\text{ohmic}}} \tag{1-173}$$

Substituting (1-167) and (1-168) into (1-173) yields

$$e_r = \frac{\frac{1}{2} R_r |I_A|^2}{\frac{1}{2} R_r |I_A|^2 + \frac{1}{2} R_{\text{ohmic}} |I_A|^2} = \frac{R_r}{R_r + R_{\text{ohmic}}} = \frac{R_r}{R_A} \tag{1-174}$$

where (1-169) was used. Except for low frequencies, the skin depth $\delta = \sqrt{2/\omega\mu\sigma}$ is much smaller than the conductor radius and then the ohmic resistance for an antenna of length L that carries an axially uniform current is

$$R_{\text{ohmic}} \approx \frac{L}{2\pi a} R_s \tag{1-175}$$

where L is the length of the wire, a the wire radius, and R_s the surface resistance:

$$R_s = \sqrt{\frac{\omega\mu}{2\sigma}} \tag{1-176}$$

For many antennas, radiation efficiency is nearly 100%. For electrically small antennas, however, the radiation efficiency can be low. We, therefore, take a closer look at them.

The ideal dipole has a uniform current as shown in Fig. 1-20*a*. In reality, the current on a straight wire antenna must smoothly go to zero at the wire ends. The current distribution on a center-fed wire dipole of length $\Delta z \ll \lambda$, called a **short dipole**, is approximately triangular in shape as illustrated in Fig. 1-20*b*. If end loading such as with metal plates (see Fig. 2-3) is added to the short dipole, the radial current reduces to zero at the edge of the plates, giving a nearly uniform current on the vertical portion of the dipole, which permits use of the ideal dipole model. More will be said about short dipoles in Sec. 2.1.

Pattern calculations for the ideal dipole were performed in Sec. 1.6 assuming that the magnitude and phase differences of rays coming from points on the wire due to different path lengths were negligible. Since the short dipole also satisfies $\Delta z \ll \lambda$, the pattern will also be the same $\sin\theta$ radiation pattern as the ideal dipole. In

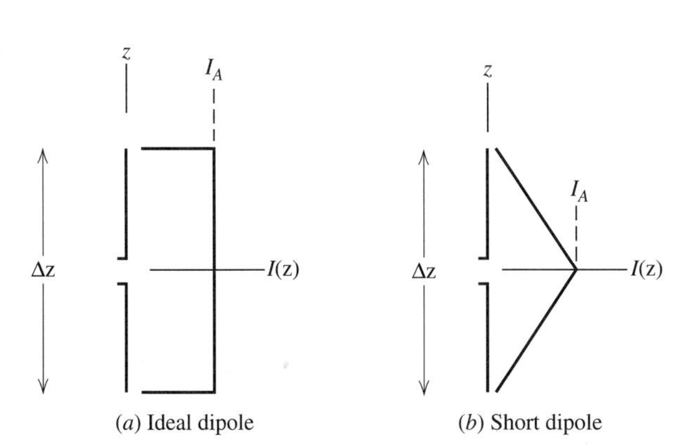

Figure 1-20 The ideal dipole and short dipole with current distributions; $\Delta z \ll \lambda$. I_A is the value of the input current at the terminals in the center of each antenna. The short dipole of (*b*) is that which is encountered in practice.

addition, the ideal dipole and short dipole will have the same directivity value of 1.5 because pattern shape completely determines directivity.

The triangular current distribution of the short dipole leads to an equivalent length that is one-half that of its physical length. This is because the equivalent length is proportional to the area under the current versus distance curves shown in Fig. 1-20, which follows from the radiation integral of (1-103) with $\exp(j\beta z' \cos\theta) \approx 1$ for short dipoles. The radiated fields are, in turn, proportional to this equivalent length. Since the radiation resistance is proportional to the integral of the far-zone electric field squared and the patterns of the ideal and short dipoles are the same, the radiation resistances are proportional to the equivalent lengths squared. The area of the triangle shape current on the short dipole is one-half that of the uniform current shape, so the radiation resistance is one fourth that of the ideal dipole. Dividing (1-171) by 4 gives

$$R_r = 20\pi^2\left(\frac{\Delta z}{\lambda}\right)^2 \ \Omega \qquad \textit{short dipole} \tag{1-177}$$

The ohmic resistance for the short dipole is also reduced from that of the ideal dipole. The ohmic resistance of a short dipole is found by first determining the power dissipation from ohmic losses, which at any point along the antenna is proportional to the current squared. In fact, in general the total power dissipated is evaluated by integrating the current squared over the wire antenna, which together with (1-170) yields

$$R_{\text{ohmic}} = \frac{2P_{\text{ohmic}}}{|I_A|^2} = \frac{1}{|I_A|^2}\frac{R_s}{2\pi a}\int_{-L/2}^{L/2} |I(z)|^2\, dz \tag{1-178}$$

It is easy to show that this reduces to (1-175) for a uniform current of length $L = \Delta z$. The short dipole triangular current of Fig. 1-20*b* can be written as a function of position along the wire as

$$I(z) = I_A\left(1 - \frac{2|z|}{\Delta z}\right), \qquad |z| \le \frac{\Delta z}{2} \tag{1-179}$$

Using this in (1-178) yields

$$R_{\text{ohmic}} = \frac{\Delta z}{2\pi a}\frac{R_s}{3} \qquad \textit{short dipole} \tag{1-180}$$

Notice that this is one-third that for an ideal dipole of the same length Δz. Since the radiation resistance for the short dipole is one-fourth that of an ideal dipole, the radiation resistance is decreased more relative to the ohmic resistance, and thus the efficiency is lower for the short dipole than it is for an ideal dipole of the same length.

EXAMPLE 1-4 *Radiation Efficiency of an AM Car Radio Antenna*

Most fender-mount car radio antennas are 31 in. long. We assume that the fender images the monopole antenna, forming a dipole 62 in. long ($L = 1.575$ m) and 1/8 in. in diameter

(a = 0.159 cm). Monopoles are treated in Sec. 2.3, but here we use a dipole model. For an operating frequency of 1 MHz (λ = 300 m), the electrical length is 0.00525λ, indicating an electrically small antenna. If a short dipole model is assumed, the radiation resistance follows from (1-177) as

$$R_r = 20\pi^2\left(\frac{1.575}{300}\right)^2 = 0.00545\ \Omega \qquad (1\text{-}181)$$

Using the conductivity of steel (see App. B.1) in (1-176) gives

$$R_s = \sqrt{\frac{4\pi \times 10^{-7} \cdot 2\pi \times 10^6}{2 \cdot 2 \times 10^6}} = 1.40 \times 10^{-3}\ \Omega \qquad (1\text{-}182)$$

The ohmic resistance from (1-180) is

$$R_{\text{ohmic}} = \frac{L}{2\pi a}\frac{R_s}{3} = \frac{1.575}{2\pi \cdot 1.59 \times 10^{-3}}\frac{1.40 \times 10^{-3}}{3} = 0.0736\ \Omega \qquad (1\text{-}183)$$

The radiation efficiency from (1-174), (1-181), and (1-183) is

$$e_r = \frac{R_r}{R_r + R_{\text{ohmic}}} = \frac{0.00545}{0.00545 + 0.0736} = 6.7\% \qquad (1\text{-}184)$$

The 6.7% radiation efficiency of the above example is low. Low efficiency in receiving antennas for broadcast applications is overcome by using high-power transmitters operating into tall antennas that are efficient. Thus, cost and complexity are concentrated into a few transmitting stations, allowing inexpensive and simple receiving antennas.

In addition to loss of efficiency, ohmic losses on antennas have another undesirable effect. As with any resistive element in an electrical system, ohmic losses on antennas are noise sources. This can be a problem for receiving applications when the signal is low. For frequencies around 1 MHz and below, external noise, mainly due to lightning, is significant and always present. The external noise picked up by the antenna is proportional to the antenna radiation resistance and is usually larger than the noise arising from internal ohmic resistance. Antenna noise is discussed further in Sec. 9.2.

We now return to the reactive part of the input impedance. In contrast to radiated power that contributes to the real part of the input impedance, the reactive part of the input impedance represents power stored in the near field. This behavior is very similar to a complex load impedance in circuit theory. Antennas that are electrically small (i.e., much smaller than a wavelength) have a large input reactance, in addition to a small radiation resistance. For example, the short dipole has a capacitive reactance, whereas an electrically small loop antenna has an inductive reactance. This is an expected result from low-frequency circuit theory. The reactance of a short dipole is approximated by [9]:

$$X_A = -\frac{120}{\pi\dfrac{\Delta z}{\lambda}}\left[\ln\left(\frac{\Delta z}{2a}\right) - 1\right]\ \Omega \qquad \textit{short dipole} \qquad (1\text{-}185)$$

This gives a large capacitive reactance for very short dipoles. The total input impedance of the short dipole is $R_r + R_{\text{ohmic}} + jX_A$, where X_A is given above, R_r is given by (1-177), and R_{ohmic} is given by (1-180).

EXAMPLE 1-5 ***Input Reactance of an AM Car Radio Antenna***

We now return to Example 1-4 and calculate the reactive portion of the input impedance. Using $\Delta z/\lambda = 0.00525$ and $\Delta z/2a = 157.5/2 \cdot 0.159 = 495.3$ in (1-185) gives $X_A = -37{,}870\ \Omega$. This is a very large capacitive reactance, leading to a severe impedance mismatch. Also see Fig. 5-6.

Antenna impedance is important to the transfer of power from a transmitter to an antenna or from an antenna to a receiver. For example, to maximize the power transferred from a receiving antenna, the receiver impedance should be a conjugate match to the antenna impedance (equal resistances, equal magnitude and opposite sign reactances). Receivers have real-valued impedance, typically 50 Ω, so it is necessary to "tune out" the antenna reactance with a matching network. There are two disadvantages to using matching networks: Ohmic losses in the network components such as tuning coils reduce efficiency, and second, a matching network provides a match only over a narrow band of frequencies, which reduces the operational bandwidth. Impedance-matching techniques are discussed in Section 5.3.

1.10 ANTENNA POLARIZATION

A monochromatic electromagnetic wave, which varies sinusoidally with time, is characterized at an observation point by its frequency, magnitude, phase, and polarization. The first three of these are familiar parameters, but polarization is often not well understood by students and practicing engineers. The polarization of an antenna is the polarization of the wave radiated in a given direction by the antenna when transmitting. In this section, we first discuss the possible polarizations of an electromagnetic wave, and then antenna polarization will follow directly from wave polarization. A complete discussion of wave and antenna polarization is found in [10].

The phase front (surface of constant phase) of a wave radiated by a finite-sized radiator becomes nearly planar over small observation regions. This wave is referred to as a *plane wave* and its electric and magnetic fields lie in a plane. The *polarization* of a plane wave is the figure the instantaneous electric field traces out with time at a fixed observation point. An example is the vertical, linearly polarized wave in Fig. 1-21, which shows the spatial variation of the electric field at a fixed instant of time.

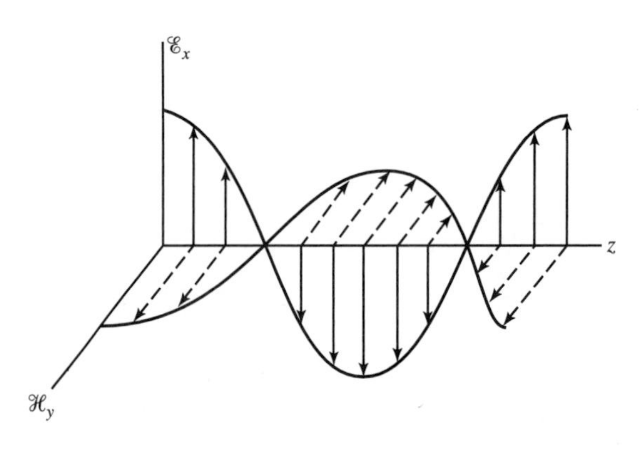

Figure 1-21 The spatial behavior of the electric (solid) and magnetic (dashed) fields of a linearly (vertical) polarized wave for a fixed instant of time. (From [10]. Reprinted by permission of Artech House, Inc., Boston, MA.)

As time progresses, the electric field ($\mathscr{E}_x$) at a fixed point oscillates back and forth along a vertical line. For a completely polarized wave, the figure traced out is, in general, an ellipse. As indicated in Fig. 1-21, the temporal and spatial variations of the magnetic field are similar to those for the electric field, except that the magnetic field is perpendicular to the electric field. Waves can have a nonperiodic behavior, but we will not consider such randomly polarized wave components because antennas cannot generate them.

There are some important special cases of the polarization ellipse. If the electric field vector moves back and forth along a line, it is said to be *linearly polarized*; see Figs. 1-22*a* and 1-22*b*. An example is the electric field from an ideal dipole or any linear current. If the electric field vector remains constant in length but rotates around in a circular path, it is *circularly polarized*. Rotation at radian frequency ω is in one of two directions, referred to as the sense of rotation. If the wave is traveling toward the observer and the vector rotates clockwise, it is *left-hand polarized*. The left-hand rule applies here: With the thumb of the left hand in the direction of propagation, the fingers will curl in the direction of rotation of the instantaneous electric field $\mathscr{E}$. If it rotates counterclockwise, it is *right-hand polarized*. Left- and right-hand circularly polarized waves are shown in Figs. 1-22*c* and 1-22*d*. A helix antenna produces circularly polarized waves and the sense of rotation of the wave is the same as that of the helix windings; for example, a right-hand wound helix produces a right-hand circularly polarized wave; see Sec. 6.2. Finally, Figs. 1-22*e* and 1-22*f* show the most general cases of left-hand and right-hand sensed *elliptical polarizations*.

The time-space behavior of the important special case of circular polarization is

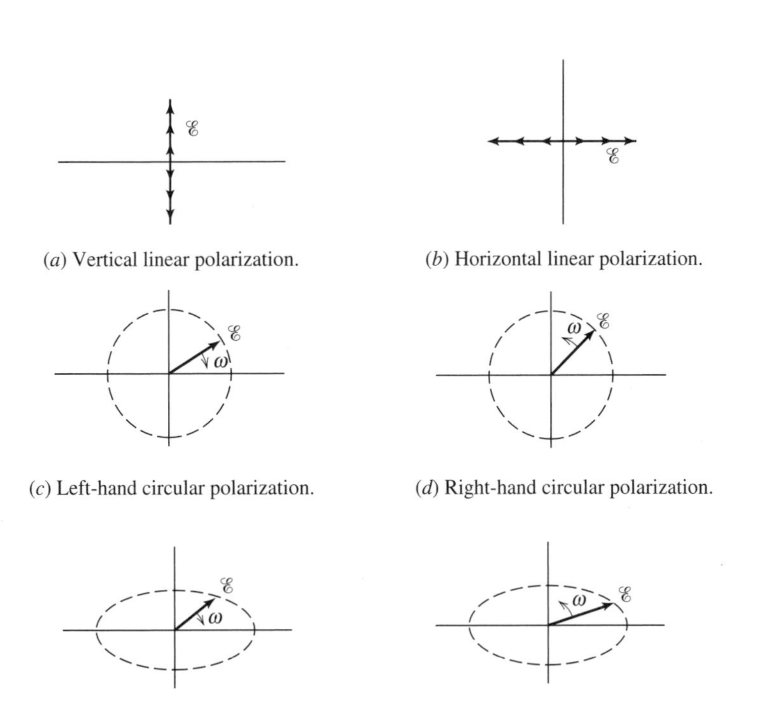

(*a*) Vertical linear polarization. (*b*) Horizontal linear polarization.

(*c*) Left-hand circular polarization. (*d*) Right-hand circular polarization.

(*e*) Left-hand elliptical polarization. (*f*) Right-hand elliptical polarization.

Figure 1-22 Some wave polarization states. The wave is approaching.

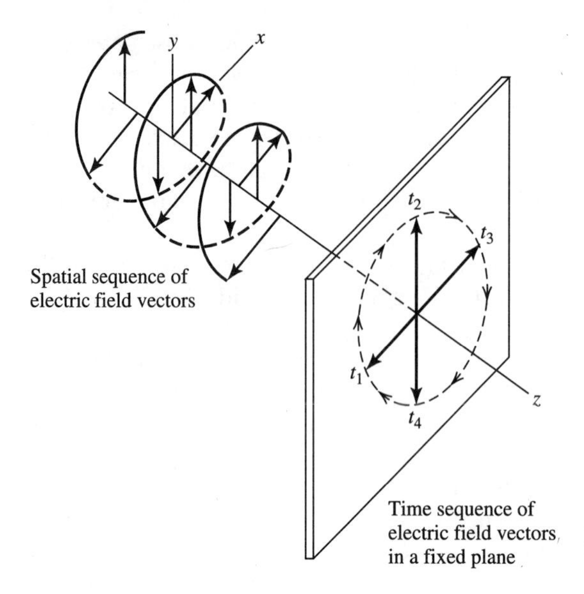

Figure 1-23 Perspective view of a left-hand circularly polarized wave shown at a fixed instant of time and the time sequence of electric field vectors as the wave passes through a fixed plane in the $+z$-direction. (From [10]. Reprinted by permission of Artech House, Inc., Boston, MA.)

difficult to visualize. Figure 1-23 provides a space perspective view of a left-hand circularly polarized wave. As the vector pattern translates along the $+z$-axis, the electric field at a fixed point appears to rotate clockwise in the xy-plane (yielding a left-hand circularly polarized wave). This is illustrated with the time sequence of vectors in the fixed plane shown in Fig. 1-23.

A general polarization ellipse is shown in Fig. 1-24 with a reference axis system. The wave associated with this polarization ellipse is traveling in the $+z$-direction. The sense of rotation can be either left or right. The instantaneous electric field vector $\mathscr{E}$ has components $\mathscr{E}_x$ and $\mathscr{E}_y$ along the x- and y-axes. The peak values of these components are E_1 and E_2. The angle γ describes the relative values of E_1 and E_2 from

$$\gamma = \tan^{-1}\frac{E_2}{E_1}, \qquad 0° \leq \gamma \leq 90° \tag{1-186}$$

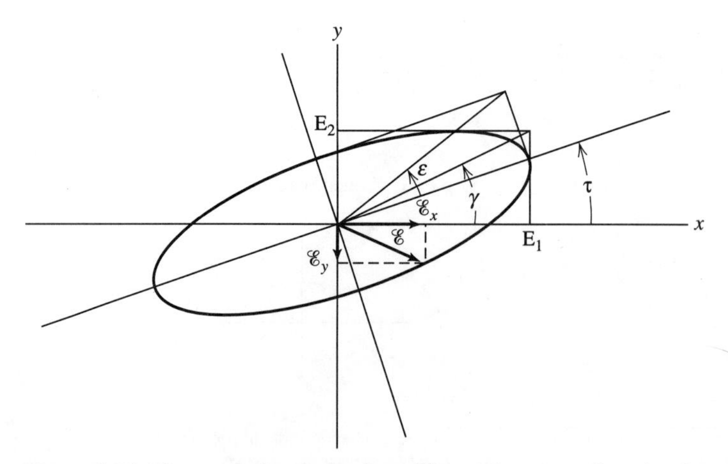

Figure 1-24 The general polarization ellipse. The wave direction is out of the page in the $+z$-direction. The tip of the instantaneous electric field vector $\mathscr{E}$ traces out the ellipse.

The tilt angle of the ellipse τ is the angle between the x-axis (horizontal) and the major axis of the ellipse. The angle ε is seen from Fig. 1-24 to be

$$\varepsilon = \cot^{-1}(-AR), \qquad 1 \leq |AR| \leq \infty, \qquad -45° \leq \varepsilon \leq 45° \tag{1-187}$$

where the axial ratio of the ellipse $|AR|$ is the ratio of the major axis electric field component to that along the minor axis. The sign of AR is positive for right-hand sense and negative for left-hand sense. Axial ratio is often expressed in dB as 20 log $|AR|$.

The instantaneous electric field for the wave of Fig. 1-24 can be written as (with $z = 0$ for simplicity)

$$\mathscr{E} = \mathscr{E}_x\hat{\mathbf{x}} + \mathscr{E}_y\hat{\mathbf{y}} = E_1 \cos \omega t\,\hat{\mathbf{x}} + E_2 \cos(\omega t + \delta)\hat{\mathbf{y}} \tag{1-188}$$

where δ is the phase by which the y-component leads the x-component. This representation describes the ellipse shape as time t progresses. If the components are in-phase ($\delta = 0$), the net vector is linearly polarized. The orientation of the linear polarization depends on the relative values of E_1 and E_2. For example, if $E_1 = 0$, vertical linear polarization results; if $E_2 = 0$, horizontal linear results; if $E_1 = E_2$, the polarization is linear at 45° with respect to the axes. Linear polarization is a collapsed ellipse with infinite axial ratio. If δ is nonzero, the axial ratio is finite. When $\delta > 0$, $\mathscr{E}_y$ leads $\mathscr{E}_x$ in-phase and the sense of rotation is left-hand. For $\delta < 0$, the sense is right-hand. If $E_1 = E_2$ (thus, $\gamma = 45°$) and $\delta = \pm 90°$, the polarization is circular (+90° is left-hand and −90° is right-hand). The axial ratio magnitude of a circularly polarized wave is unity.

The phasor form of (1-188) is

$$\mathbf{E} = E_1\hat{\mathbf{x}} + E_2 e^{j\delta}\hat{\mathbf{y}} \tag{1-189}$$

which can be written as (see Prob. 1.10-3)

$$\mathbf{E} = \sqrt{E_1^2 + E_2^2}(\cos \gamma\,\hat{\mathbf{x}} + \sin \gamma\, e^{j\delta}\hat{\mathbf{y}}) = |E|\hat{\mathbf{e}} \tag{1-190}$$

The factor $|E|$ is the field magnitude and $\hat{\mathbf{e}}$ is the complex vector representation for the field and is normalized to unity magnitude. Thus, γ and δ completely specify the polarization state of the wave. In fact, either pair of angles (ε, τ) or (γ, δ) uniquely define the polarization state of a wave. The transformations between these angles are

$$\gamma = \frac{1}{2} \cos^{-1}(\cos 2\varepsilon \cos 2\tau) \tag{1-191}$$

$$\delta = \tan^{-1}\left(\frac{\tan 2\varepsilon}{\sin 2\tau}\right) \tag{1-192}$$

The *polarization of an antenna* is the polarization of the wave radiated by the antenna when transmitting. Therefore, all the discussions on wave polarization apply directly to antenna polarization. The polarization of waves radiated by an antenna will vary with direction. Usually, the polarization characteristics of an antenna remain relatively constant over its main beam and the polarization on the main beam peak is used to describe the antenna polarization. However, the radiation from side lobes can differ greatly in polarization from that of the main beam. When measuring the radiation from an antenna, both E_θ and E_ϕ should be measured to be complete. The principal plane patterns of a perfectly linearly polarized antenna, such as a line source on the z-axis, are completely specified when a linearly polarized probe antenna is oriented to respond to E_θ.

Reciprocal antennas have identical radiation patterns on transmit and receive. This extends to the vector nature of the radiation that includes polarization. A transmit antenna is polarization matched to a receive antenna if its polarization ellipse axial ratio, sense, and major axis orientation are the same as those of the receive antenna (in the direction of the transmit antenna). For example, a right-hand circularly polarized receiving antenna is polarized matched to a right circularly polarized wave. As a mechanical analogy, consider a right-hand threaded rod corresponding to a right-hand circularly polarized (RHCP) wave and a right-hand tapped hole representing a RHCP antenna. The rod and hole are matched when screwed in or out, corresponding to reception or transmission.

It is interesting to examine the polarizations used in the U.S. broadcast FM radio and TV industry. Historically, TV and FM broadcast transmitting antennas were horizontally polarized. In recent years, CP has been used for TV and FM since the FCC now allows transmitters to have the maximum EIRP in both horizontal polarization and vertical polarization. This was permitted because nearly all vehicle receiving antennas are VP.

REFERENCES

1. Jack Ramsay, "Highlights of Antenna History," *IEEE Ant. & Prop. Soc. Newsletter*, pp. 8–20, Dec. 1981.
2. Sir Edmund Whittaker, *A History of the Theories of Aether and Electricity, Vol. 1: The Classical Theories*, Harper Torchbooks, New York, 1960.
3. R. C. Hansen, Ed. *Microwave Scanning Antenna, Vol. I: Apertures*, Academic Press, New York, 1964, Chap. 2.
4. E. Larsen, *Telecommunications—A History*, Frederick Muller Ltd., London, 1977.
5. M. I. Pupin, "A Discussion on Experimental Tests of the Radiation Law for Radio Oscillators," *Proc. Inst. Radio Engineers*, Vol. 1, pp. 3–13, 1913.
6. S. Silver, Ed., *Microwave Antenna Theory and Design*, M.I.T. Radiation Laboratory Series, Vol. 12, McGraw-Hill, New York, 1949.
7. S. A. Schelkunoff and H. T. Friis, *ANTENNAS: Theory and Practice*, John Wiley & Sons, New York, 1952, p. 127.
8. J. Van Bladel, "Lorenz or Lorentz?," *IEEE Ant. & Prop. Magazine*, Vol. 33, p. 69, April 1991.
9. R. C. Johnson, Ed., *Antenna Engineering Handbook*, 3rd ed., McGraw-Hill, New York, 1993, Chap. 4.
10. W. L. Stutzman, *Polarization in Electromagnetic Systems*, Artech House, Boston, 1992.

PROBLEMS

1.3-1 *What is this antenna?* (a) Locate an antenna that you would like to know more about. It can be one you see in your community or one you find in a catalog or magazine. Investigate and learn more about the antenna as you move through this book. For now make a sketch of the antenna and describe its location and surroundings. (b) After covering a substantial part of this book, provide an explanation of the antenna in (a) including the type of antenna it is, its operating frequency, how it is being used, and performance parameter values that you can estimate such as beamwidth and gain.

1.4-1 Use (1-6) in (1-1) to derive (1-7).

1.4-2 Use (1-13) in (1-19) together with (1-11), (1-12), and (1-20) to derive (1-18).

1.4-3 Assuming ε and μ are real and $\mathbf{M} = 0$, derive (1-28) through (1-34) using the identity (C-19).

1.4-4 Write the complex power equation for a series *RLC* network driven by a voltage generator in a form analogous to the Poynting theorem.

1.5-1 Derive (1-48) starting with (1-18).

1.5-2 a. Show that $\psi = Ce^{-j\beta r}/r$ satisfies (1-54) at all points except the origin.

b. By integrating (1-52) over a small volume containing the origin, substituting $\psi = Ce^{-j\beta r}/r$, and letting r approach zero, show that $C = (4\pi)^{-1}$, thus proving (1-55).

1.6-1 Show that (1-71b) follows from (1-69).

1.6-2 The expression for the electric field intensity of an ideal dipole can be derived in two ways.

a. Derive (1-71b) using the magnetic field intensity expression (1-71a) in (1-60).

b. Derive (1-71b) using the vector potential expression (1-62) in (1-45).

1.6-3 For a z-directed current element $I\,\Delta z$ in free space and located at the origin of a spherical coordinate system:

a. Calculate the complex Poynting vector in the general case, where r can be in the near-field region. Use the fields of (1-71).

b. Then find the expression for the time-average power flowing out through a sphere of radius r enclosing the current element. Your answer will be that of (1-75). Why?

1.6-4 Show that the electric field for the ideal dipole in (1-71b) satisfies Maxwell's equation $\nabla \cdot \mathbf{E} = 0$.

1.7-1 Prove (1-93) by using (1-87) in (1-92) and retaining only $1/r$ terms; that is, using $\beta r \gg 1$.

1.7-2 *Uniform line source.*

a. Find the half-power beamwidth of the uniform line source pattern factor $|f(\theta)|$ of (1-118). Your answer should be in the form

$$\mathrm{HP} = K\lambda/L \quad \text{for} \quad L \gg \lambda.$$

Determine the constant K. *Hint:* First find the values u_{HP} of $u = (\beta L/2)\cos\theta$ for which $|f(u_{\mathrm{HP}})| = 1/\sqrt{2}$. Then use the approximation $\cos^{-1}(\pm x) \approx \pi/2 \mp x$ for x small.

b. Calculate the maximum side lobe level for the pattern in decibels relative to the main beam maximum. The side lobe maximum can be located by differentiating $f(u)$ with respect to u, setting equal to zero, and solving for u.

c. Suppose now that the current has a linear phase taper across it so that

$$I(z') = I_o e^{j\beta_o z'}$$

What is $f(\theta)$ now? If we let $\beta_o = -\beta\cos\theta_o$ where is the pattern maximum (main beam pointing direction)? This is how the scanned beams of Fig. 1-16 are generated.

1.7-3 Equation (1-96) can be derived without initially assuming that the rays are parallel. Derive (1-96) by writing $R = [(\mathbf{r} - \mathbf{r}')\cdot(\mathbf{r} - \mathbf{r}')]^{1/2}$, expanding, factoring out an r, neglecting the smallest term, and using the first two terms of the binomial expansion.

1.7-4 Using the inner boundary of the far-field to be $r_{\mathrm{ff}} = 2L^2/\lambda$ for a linear antenna of length L, find r_{ff} for the following three antennas: $L = 5\lambda$, a half-wave dipole ($L = \lambda/2$), and a short dipole ($L = 0.01\lambda$). Is the far-field boundary you have computed valid for each of these; if not, why not?

1.7-5 It can be shown that criteria for the far-field distance corresponding to (1-99b) and (1-99c) are more accurately given by $r > 5D$ and $r > 1.6\lambda$. Using these together with (1-99a), plot a single graph of r/λ (vertical axis) versus D/λ for the far-field boundary. Indicate which region of the graph corresponds to the far field.

1.7-6 A car radio antenna is 1 m long and operates at a frequency of 1 MHz. Use the graph of Prob. 1.7-5 to find the far-field distance.

1.7-7 Derive the near-field region boundary expression in (1-100) of $0.62\sqrt{D^3/\lambda}$. Do this by finding the maximum error of the fourth term in (1-84) with $z' = D/2$ and equating to $\pi/8$.

1.8-1 Show that there is 4π *sr* in all space by integrating $d\Omega$ over a sphere.
1.8-2 A *power* pattern is given by $|\cos^n \theta|$ for $0 \leq \theta \leq \pi/2$ and is zero for $\pi/2 \leq \theta \leq \pi$. (a) Calculate the directivity for $n = 1, 2,$ and 3, (b) Find the HP values in degrees for each n, (c) Sketch the patterns for the n values on one polar plot and comment on them, (d) Explain the directivity value for the case of $n = 0$.
1.8-3 An antenna has a far-field pattern which is independent of ϕ but which varies with θ as follows:

$$F = 1 \quad \text{for} \quad 0° \leq \theta \leq 30°.$$
$$F = 0.5 \quad \text{for} \quad 60° \leq \theta \leq 120°.$$
$$F = 0.707 \quad \text{for} \quad 150° \leq \theta \leq 180°.$$
$$F = 0 \quad \text{for} \quad 30° < \theta < 60° \quad \text{and} \quad 120° < \theta < 150°.$$

Find the directivity. Also find the directivity in the direction $\theta = 90°$.
1.8-4 For a single-lobed pattern the beam solid angle is approximately given by

$$\Omega_A \approx \text{HP}_E\text{HP}_H$$

where HP_E and HP_H are the half-power beamwidths in radians of the main beam in the E and H planes. Show that

$$D \approx \frac{41{,}253}{\text{HP}_{E°}\text{HP}_{H°}}$$

where $\text{HP}_{E°}$ and $\text{HP}_{H°}$ are the E and H plane half-power beamwidths in degrees.
1.8-5 A horn antenna with low side lobes has half-power beamwidths of 29° in both principal planes. Use the approximate expression in Prob. 1.8-4 to compute the directivity of the horn in decibels.
1.8-6 A sector pattern has uniform radiation intensity over a specified angular region and is zero elsewhere. An example is

$$F(\theta) = \begin{cases} 1 & \dfrac{\pi}{2} - \alpha < \theta < \dfrac{\pi}{2} + \alpha \\ 0 & \text{elsewhere} \end{cases}$$

Derive an expression for the directivity corresponding to this pattern.
1.8-7 An airplane is flying parallel to the ground (in the z-direction). For a surface search radar an antenna is required that uniformly illuminates the ground over some region. The so-called cosecant pattern will do this. From the figure we see that $h = r\cos\left(\dfrac{\pi}{2} - \theta\right)$ or

$$r = \frac{h}{\sin\theta} = h \csc\theta.$$

This expresses how much farther the radiation must travel to reach the ground as θ is deceased. The radiation pattern

$$F(\theta) = \csc\theta,$$

will just compensate for the $1/r$ field variation with distance. If, in addition, the ϕ variation is a sector pattern of small angular extent ϕ_0, then

$$F(\theta, \phi) = \begin{cases} \csc\theta & \theta_1 < \theta < \dfrac{\pi}{2},\ 0 < \phi < \phi_0 \\ 0 & \text{elsewhere.} \end{cases}$$

Derive an expression for the directivity.

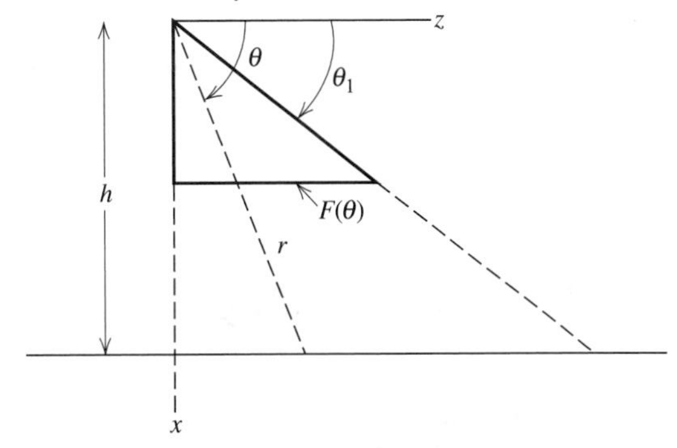

1.8-8 *Gaussian Pattern.* A circularly symmetric, narrow beam antenna pattern is frequently modeled by a Gaussian shape given by

$$F(\theta) = e^{-4 \ln\sqrt{2}\,(\theta/HP)^2}$$

Derive expressions for the directivity associated with this pattern in terms of the half-power beamwidth HP in radians and in degrees. Do this by approximating $\sin\theta$ by θ in the integration for Ω_A and extending the integration limits to infinity.

1.8-9 An antenna has a directivity of 20 and a radiation efficiency of 90%. Compute the gain in dB.

1.8-10 Compute the gain of an antenna which has a radiation efficiency of 95% and the following radiation pattern

$$F(\theta) = \begin{cases} 1 & 0 \le \theta < 20^\circ \\ 0.707 & 20^\circ \le \theta < 120^\circ \\ 0 & 120^\circ \le \theta < 180^\circ \end{cases}$$

1.9-1 A 2-m-long dipole made of 6.35-mm (0.25-in.) diameter aluminum is operated at 500 kHz. Compute its radiation efficiency, assuming

a. The current is uniform
b. The current is triangular.

1.9-2 A citizen's band radio at 27 MHz uses a half-wavelength long antenna that has a radiation resistance of 70 Ω. Compute the radiation efficiency if the antenna is made with 6.35-mm-diameter aluminum. As a rough approximation assume that the current is triangular.

1.9-3 Use the ohmic resistance formula of (1-178) to verify the expression for R_{ohmic} for

a. A uniform current given by (1-175)
b. A triangular current given by (1-180).

1.9-4 A cordless telephone operating at 50 MHz has a 38-cm long monopole antenna made of 4-mm diameter aluminum tubing. Compute the radiation efficiency.

1.10-1 The instantaneous electric field components of an elliptically polarized wave are $\mathscr{E}_x = E_1 \cos(\omega t - \beta z)$ and $\mathscr{E}_y = E_2 \cos(\omega t - \beta z + \delta)$. Specify E_1, E_2, and δ for the following polarizations:

a. Linear with $E_1 \neq 0$ and $E_2 \neq 0$.
b. Right circular.
c. Left circular.
d. Elliptical with $E_1 = E_2$.
e. Elliptical with $\delta = 90°$.

1.10-2 Write the frequency domain form of the total vector electric fields given in Prob. 1.10-1.

1.10-3 Start with (1-189) and prove (1-190). Use the fact that the magnitude of **E** follows from $|\mathbf{E}|^2 = \mathbf{E} \cdot \mathbf{E}^*$. Also note that γ in Fig. 1-24 is in a triangle with sides E_1 and E_2 and hypotenuse $|\mathbf{E}|$.

1.10-4 Prove that a RHCP wave normally incident on a plane perfect conductor changes to LHCP upon reflection.

Chapter 2

Some Simple Radiating Systems and Antenna Practice

In this chapter, we introduce the simple antenna forms of the electrically small dipole, half-wave dipole, and electrically small loop. These antennas are fundamental to antenna practice, and they are used in the discussion of arrays in the next chapter. We return to a more in-depth treatment of wire antennas in Chap. 5. Image theory is also presented in this chapter for use in examining antennas operated in the presence of a perfect ground plane. Wireless communication systems are also discussed in this chapter to provide a motivation for further study in antennas by showing how antennas are frequently applied in practice.

2.1 ELECTRICALLY SMALL DIPOLES

An antenna whose dimensions are small compared to the wavelength at the frequency of operation is an *electrically small antenna* and is the most basic antenna; see Fig. 1-6. How much smaller than a wavelength an electrically small antenna must be depends on the application, but generally is taken to be on the order of a tenth of a wavelength in extent or less. We have already encountered two electrically small antennas in Chap. 1, the ideal dipole and the short dipole. We revisit these dipoles in their practical forms in this section.

Electrical size and physical size can be quite different. An antenna operating at low frequencies can be physically large but electrically small, that is, a small fraction of a wavelength in extent. This is especially true for frequencies in the low MHz range and below. Electrically small antennas are inherently inefficient. However, this often is not a serious problem in receiving systems and physically small antennas offer the advantages in size, weight, cost, and portability.

The simplest electrically small antenna is the **short dipole** shown in Fig. 2-1*a* as a wire with a feed point in the center. It has been suggested that the resemblance of the arms of the dipole to the feelers, or antennae, of an insect leads to the use of the name *antenna* [1]. The current distribution of the short dipole is nearly triangular in shape as modeled in Fig. 1-20*b*. This is because the current distribution on thin

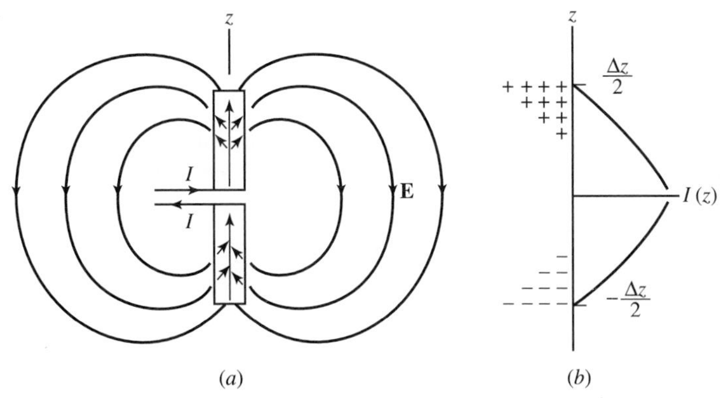

Figure 2-1 Short dipole, $\Delta z \ll \lambda$.
(*a*) Current on the antenna and the electric fields surrounding it.
(*b*) Current and charge distributions.

wire antennas (diameter $\ll \lambda$) is approximately sinusoidal and also must be zero at the wire ends. Since the arms of the short dipole are a fraction of a wavelength long, only a small portion of the sine wave current appears on the arm and is therefore nearly linear.

The decreasing current toward the wire ends requires that charges peel off and appear on the wire surface as shown in Fig. 2-1*a*. The current and charge distributions shown in Fig. 2-1*b* are for an instant of time when the input current at the dipole terminals is maximum. Since the input current is changing sinusoidally with time, the current and charge distributions on the dipole do also. This charge accumulation leads to a displacement current density $j\omega\varepsilon\mathbf{E}$ in the space surrounding the dipole. The displacement current density, in turn, gives rise to an electromagnetic wave that propagates outward from the source as illustrated in Fig. 1-4. Displacement current in space couples a transmitting antenna to a receiving antenna, much as a conduction current provides coupling between lumped elements in a circuit. The radiation pattern of all forms of the electrically small dipole (with its radiating portion along the z-axis) is $\sin\theta$ as shown in Fig. 1-10.

In the ideal dipole, all charge accumulates at the ends of the antenna. In fact, the ideal dipole can be analyzed as either a uniform current or two point charges oscillating at radian frequency ω (see Prob. 2.1-1) as shown in Fig. 2-2. The charge dipole model shows that charge accumulates at the ends of the antenna, leading to a higher radiation resistance. In fact, the ideal dipole radiation resistance of (1-172) is four times that of the short dipole given by (1-177).

The input reactance of a short dipole is capacitive. This can be seen by visualizing the antenna as an open-circuited transmission line as in Fig. 1-3. When the distance

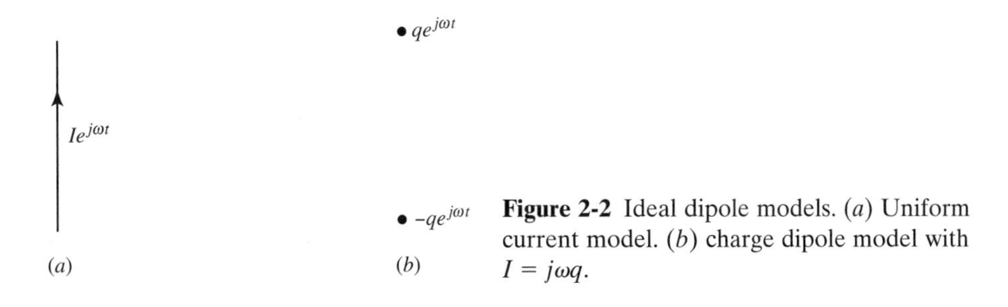

Figure 2-2 Ideal dipole models. (*a*) Uniform current model. (*b*) charge dipole model with $I = j\omega q$.

from the end of the antenna to the feed point is much less than a quarter wavelength, the input impedance is capacitive, since from transmission line theory the impedance a distance s from an open-circuit termination is $-jZ_o \cos(\beta s)$. Simple transmission line theory only gives qualitative results when radiation is present. An approximate expression for the capacitive reactance of the short dipole is given in (1-185). Moment method computation techniques are used for accurate impedance evaluation; see Sec. 10.5. Loading coils are frequently used to tune out this capacitance.

The larger radiation resistance associated with the uniform current of the ideal dipole can be realized in practice by providing a mechanism for charge accumulation at the wire ends. One method of accomplishing this is to place metal plates at the ends of the wire. This is called a **capacitor-plate antenna**, or top-hat-loaded dipole antenna. Figure 2-3 shows the construction of the antenna and the current and charges on it. If $\Delta z \ll \lambda$, the radial currents on the plates will produce fields that almost cancel in the far field, since the currents are opposite-directed and the phase difference due to separation is small ($\beta \, \Delta z \ll 2\pi$). If, in addition, $\Delta z \ll \Delta r$, the plates will provide for charge storage such that the current on the wire is constant. The capacitor-plate antenna then closely approximates the uniform current ideal dipole model. Frequently in practice, radial wires are used for the top loading in place of the solid plates.

Another small antenna used to approximate the ideal dipole is the **transmission line loaded antenna** as shown in Fig. 2-4*a*. The results of transmission line theory can be borrowed to determine the current distribution. The current is essentially sinusoidal along the wire with a zero at the ends. This current distribution is sketched in Fig. 2-4*b* for $L < \lambda/4$. If $\Delta z \ll \lambda$, the fields from the currents on the horizontal wires essentially cancel in the far field. If also $\Delta z \ll L$, the horizontal wires provide an effective place for the charge to be stored and the current on the vertical section is nearly constant as illustrated in Fig. 2-4*b*. Then radiation comes from a short section over which the current is nearly constant and the antenna approximates an ideal dipole.

The monopole form of the transmission line loaded dipole shown in Fig. 2-4*c* is called the **inverted-L antenna**. The inverted-L and variations of it such as the **inverted-F antenna** are popular in small handheld radio units [2].

Transmission line loading ideas can be extended by attaching several horizontal wires to the ends of the short vertical section. If the transmission lines of Fig. 2-4*a*

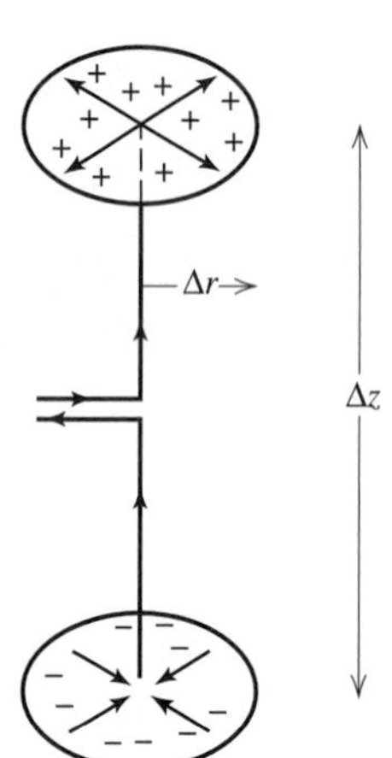

Figure 2-3 Capacitor-plate antenna. The arrows on the antenna indicate current. The charges on the plates are also shown.

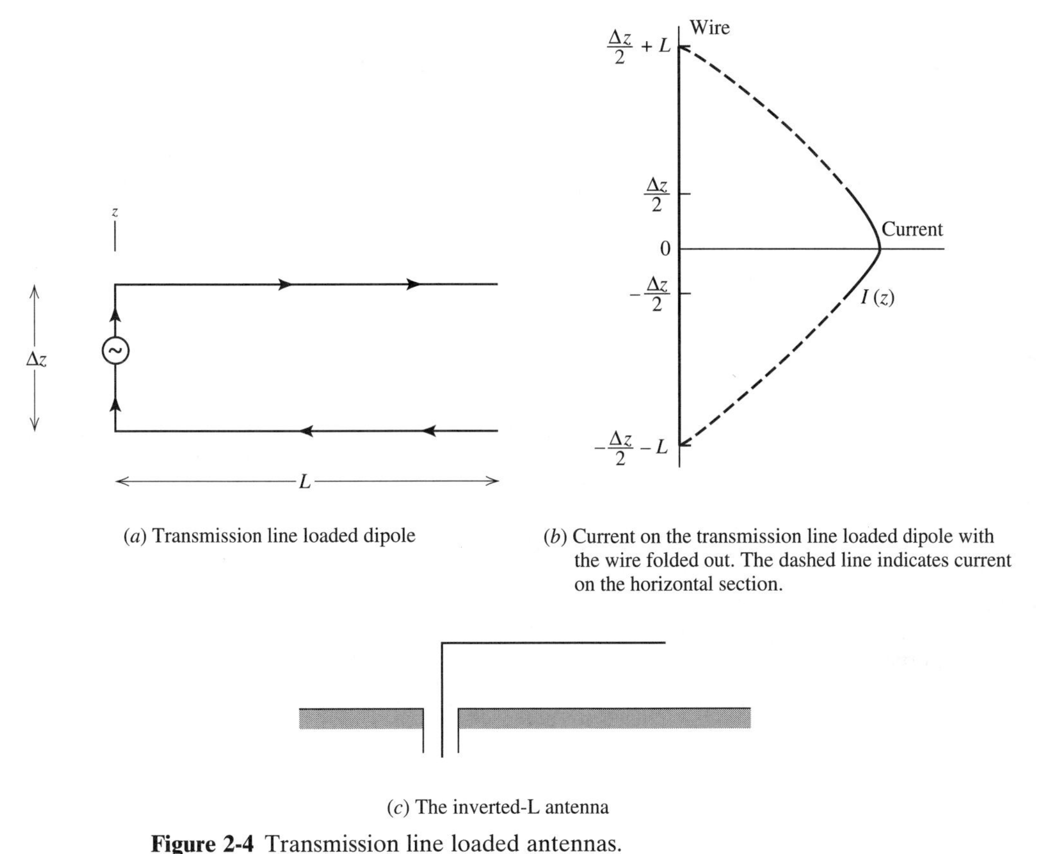

(*a*) Transmission line loaded dipole

(*b*) Current on the transmission line loaded dipole with the wire folded out. The dashed line indicates current on the horizontal section.

(*c*) The inverted-L antenna

Figure 2-4 Transmission line loaded antennas.

are extended in opposite directions, the reactance is one-half its former value (by paralleling identical capacitive elements). As more wires are added, the reactance is further reduced and the structure approaches that of a capacitor-plate antenna.

At different portions of the frequency spectrum, electrically small antennas are used for different reasons. For instance, in the VLF region where wavelength is very large, an electrically short vertical radiator is used with a large-top hat load. The top-hat loading makes the antenna appear like the capacitor-plate antenna of Fig. 2-3. Further up the spectrum, such as in the AM broadcast band, receiving antennas are usually small electrically, as we saw in Section 1.9. AM transmitting antennas are not small, but are of resonant size as discussed in the next section. At VHF frequencies and above, electrically small antennas are only used in special situations.

2.2 DIPOLES

A very widely used antenna is the **half-wave dipole** antenna. It is a linear current whose amplitude varies as one-half of a sine wave with a maximum at the center. For simplicity, we will assume this to be a filament of current. Also, it could be imagined to flow on an infinitely thin, perfectly conducting, half-wavelength long wire. This is a good approximation of a wire half-wave dipole that has a diameter much smaller than its length. The advantage of a half-wave dipole is that it can be made to resonate and present a zero input reactance, thus eliminating the need for tuning to achieve a conjugate impedance match. Input impedance of dipole antennas is discussed in detail in Section 5.1, but for now we point out that to obtain a resonant condition for a half-wave dipole, the physical length must be somewhat shorter than a free space half-wavelength, and as the antenna wire thickness is increased, the length must be reduced more to achieve resonance.

As usual, the current distribution is placed along the z-axis and for the half-sine wave current on the half-wave dipole, the current distribution is written as

$$I(z) = I_m \sin\left[\beta\left(\frac{\lambda}{4} - |z|\right)\right], \qquad |z| \leq \frac{\lambda}{4} \tag{2-1}$$

where $\beta = 2\pi/\lambda$. This current goes to zero at the ends (for $z = \pm\lambda/4$) and its maximum value I_m occurs at the center ($z = 0$) as shown in Fig. 2-5*a*. From this current, we can calculate the radiation pattern. Since it is a z-directed line source, we can use (1-103) in (1-106) to find the electric field as

$$E_\theta = j\omega\mu \sin\theta \frac{e^{-j\beta r}}{4\pi r} \int I(z')e^{j\beta z' \cos\theta}\, dz' \tag{2-2}$$

Substituting (2-1) into the integral of (2-2) and evaluating gives

$$\begin{aligned} f_{\text{un}} &= \int I(z')e^{j\beta z'\cos\theta}\, dz' = \int_{-\lambda/4}^{\lambda/4} I_m \sin\left(\frac{\pi}{2} - \beta|z'|\right)e^{j\beta z'\cos\theta}\, dz' \\ &= I_m \int_{-\lambda/4}^{0} \sin\left(\frac{\pi}{2} + \beta z'\right)e^{j\beta z'\cos\theta}\, dz' \\ &\quad + I_m \int_{0}^{\lambda/4} \sin\left(\frac{\pi}{2} - \beta z'\right)e^{j\beta z'\cos\theta}\, dz' \end{aligned} \tag{2-3}$$

where f_{un} is the unnormalized pattern factor. Using the integral (F-11)

$$\int \sin(a + bx)e^{cx}\, dx = \frac{e^{cx}}{b^2 + c^2}\left[c \sin(a + bx) - b\cos(a + bx)\right] \tag{2-4}$$

in (2-3), we have

$$\begin{aligned} f_{\text{un}} &= I_m \frac{e^{j\beta z'\cos\theta}}{\beta^2 - \beta^2\cos^2\theta}\left[j\beta\cos\theta \sin\left(\frac{\pi}{2} + \beta z'\right) - \beta\cos\left(\frac{\pi}{2} + \beta z'\right)\right]_{-\lambda/4}^{0} \\ &\quad + I_m \frac{e^{j\beta z'\cos\theta}}{\beta^2 - \beta^2\cos^2\theta}\left[j\beta\cos\theta \sin\left(\frac{\pi}{2} - \beta z'\right) + \beta\cos\left(\frac{\pi}{2} - \beta z'\right)\right]_{0}^{\lambda/4} \\ &= \frac{I_m}{\beta^2 \sin^2\theta}\left[j\beta\cos\theta - e^{-j(\pi/2)\cos\theta}(-\beta) + e^{j(\pi/2)\cos\theta}(\beta) - j\beta\cos\theta\right] \\ &= \frac{I_m}{\beta\sin^2\theta}\, 2\cos\left(\frac{\pi}{2}\cos\theta\right) \end{aligned} \tag{2-5}$$

Substituting this into (2-2) gives

$$E_\theta = j\omega\mu \frac{2I_m}{\beta}\frac{e^{-j\beta r}}{4\pi r}\sin\theta \frac{\cos[(\pi/2)\cos\theta]}{\sin^2\theta} \tag{2-6}$$

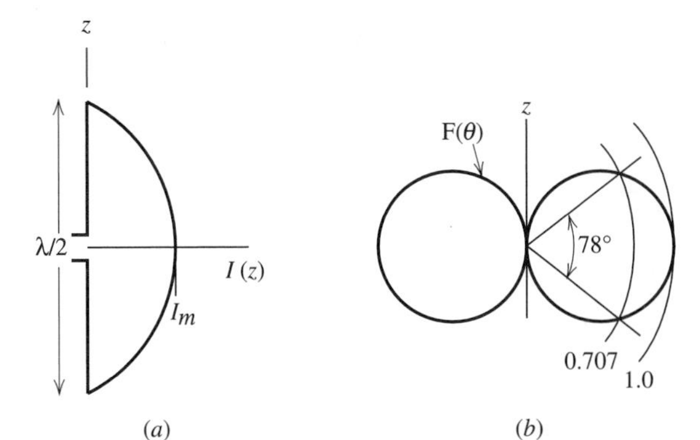

Figure 2-5 The half-wave dipole. (a) Current distribution $I(z)$. (b) Radiation pattern $F(\theta)$.

In this expression, we can identify the element factor $g(\theta) = \sin\theta$ and the normalized pattern factor

$$f(\theta) = \frac{\cos[(\pi/2)\cos\theta]}{\sin^2\theta} \tag{2-7}$$

Both $g(\theta)$ and $f(\theta)$ are maximum for $\theta = \pi/2$ and have a value of unity there. The complete (normalized) far-field pattern is then [see (1-115)]

$$F(\theta) = g(\theta)f(\theta) = \frac{\cos[(\pi/2)\cos\theta]}{\sin\theta} \qquad \textit{half-wave dipole} \tag{2-8}$$

This pattern is plotted in Fig. 2-5b in linear, polar form. The input impedance of an infinitely thin half-wavelength dipole is $73 + j42.5\ \Omega$. If it is slightly reduced in length to achieve resonance, the input impedance is about $70 + j0\ \Omega$.

So far, we have introduced three dipole antennas: the ideal, short, and half-wave dipoles. The characteristics and performance of these dipoles are listed in Fig. 2-6. The ideal and short dipoles, with uniform and triangular current distributions, respectively, have identical patterns. Both have a half-power beamwidth of 90° and a directivity of 1.5; see Fig. 1-10. The half-wave dipole has a narrower beamwidth of 78°, and, thus, a higher directivity value of 1.64, which will be derived in Sec. 5.1.

In Section 2.1, we briefly discussed one viewpoint on the phenomena of radiation. We are now ready to discuss another viewpoint in which the fields in space are considered to be produced by currents and charges on the antenna. We know that in a complete system at any instant of time there must be equal numbers of positive and negative charges, and if these were static fields (i.e., zero frequency and infinite wavelength), the fields at a great distance from the positive and negative sources would cancel. However, when the distance between positive and negative oscillating sources becomes a significant fraction of a wavelength, the phase shift (or retardation) due to different path lengths from positive and negative sources to an observation point prevents cancellation. In the case of the half-wave dipole, the current is essentially in phase and radiation will be strongest in a direction normal to the dipole and weakest along the axis of the dipole. For longer dipoles, the current on some sections of the dipole will be out of phase with others, leading to partial or total cancellation in the far field in the broadside direction and reinforcement in off-broadside directions; see Figs. 5-3 and 5-4. This explains why most practical wire antennas are on the order of a wavelength or less in size.

Dipole Type	Length	Current	Pattern	HP	D	D (dB)	R_r (Ω)	R_{ohmic} (Ω)	Current Distribution
Ideal	$L \ll \lambda$	Uniform	$\sin\theta$	90°	1.5	1.76	$80\pi^2\left(\frac{L}{\lambda}\right)^2$	$\frac{R_s}{2\pi a}L$	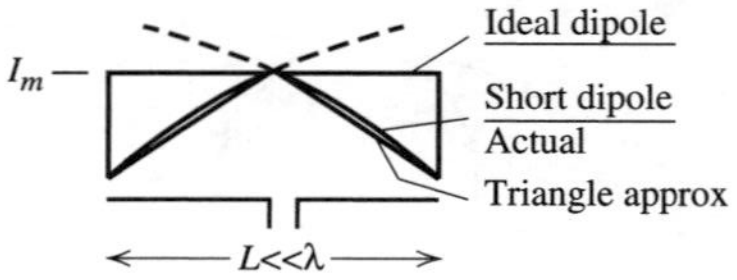 I_m; Ideal dipole
Short	$L \ll \lambda$	Triangle	$\sin\theta$	90°	1.5	1.76	$20\pi^2\left(\frac{L}{\lambda}\right)^2$	$\frac{R_s}{2\pi a}\frac{L}{3}$	Short dipole: Actual, Triangle approx; $L \ll \lambda$
Half-wave	$L = 0.5\,\lambda$	Sinusoid	$\dfrac{\cos\left(\frac{\pi}{2}\cos\theta\right)}{\sin\theta}$	78°	1.64	2.15	~70	$\frac{R_s}{2\pi a}\frac{\lambda}{4}$	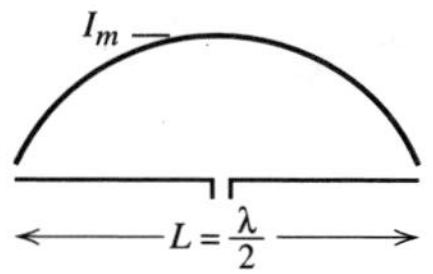 I_m; $L = \frac{\lambda}{2}$

Figure 2-6 Characteristics and performance of some dipole antennas.

2.3 ANTENNAS ABOVE A PERFECT GROUND PLANE

Our treatment of antennas thus far has been for a free space environment. In practice, environmental effects are small for elevated high gain antennas. However, the radiation properties of antennas with broad beams are affected by their surrounding environment. Both the pattern and impedance are influenced by the presence of nearby objects. The most commonly encountered object is a *ground plane*. The real earth is a ground plane and is discussed in Sec. 5.6. The ideal form of a ground plane is planar, infinite in extent, and perfectly conducting, and is referred to as a *perfect ground plane*. The perfectly conducting assumption is very mild and any good conductor such as aluminum or copper is very accurately modeled as a perfect conductor. The infinite extent assumption is more severe. Accurate evaluation of a finite-sized ground plane can be found from the moment method or geometrical theory of diffraction techniques, which are discussed in Chaps. 10 and 12. In most cases, a perfect ground plane is well approximated by a solid metal plate or a planar wire grid system that is large compared to the size of the antenna, if the antenna is not far from the conducting plane. In this section, we present image theory to model an antenna operating in the presence of a perfect ground plane and apply the theory to monopole antennas.

2.3.1 Image Theory

An antenna operating in the presence of a perfect ground plane produces two rays at each observation angle, a direct ray from the antenna and a second ray due to reflection from the ground plane such that Snell's law of reflection is satisfied. This is the approach used in Sec. 5.6 to analyze antennas above perfect and imperfect ground planes. Here, we develop the solution from first principles and it will be seen that the image antenna acts as an equivalent source for the reflected ray.

Consider first an ideal dipole near a perfect ground plane and oriented perpendicular to the ground plane as shown in Fig. 2-7*a*. Ground planes are usually horizontal, so this situation is referred to as a vertical ideal dipole above a perfect ground plane. We wish to find the fields **E** and **H** above the plane PP'. The uniqueness of the solution to a differential equation (the wave equation) plus its boundary conditions permits introduction of an equivalent system that is different below PP' but satisfies the same boundary conditions on PP' and has the same sources above PP'. Such an equivalent system, which produces the same fields above PP' as the original system, has an image source the same distance below the plane PP' and similarly

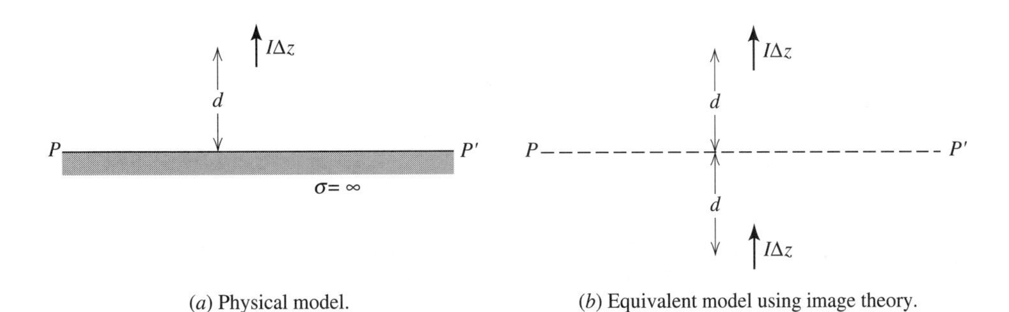

(*a*) Physical model. (*b*) Equivalent model using image theory.

Figure 2-7 Ideal dipole above and perpendicular to a perfectly conducting ground plane.

directed. In this case, the image source is the virtual ideal dipole as shown in Fig. 2-7*b*.

It is a simple matter to prove that the boundary condition of zero tangential electric field along plane PP' is satisfied by this source configuration. To do this, we examine the electric field expression for an ideal dipole given by (1-71b). The complete expression must be used because the ground plane can be, and usually is, in the near field of the antenna. The radial component varies as $\cos\theta$ and the θ-component varies as $\sin\theta$, where θ is the angle from the axis along the direction of the current element. Let θ_1 and θ_2 be the angles from the line of the current elements to a point on the plane PP' for the primary source and its image, respectively. The radial components from the sources are then

$$E_{r1} = C\cos\theta_1 \tag{2-9}$$

$$E_{r2} = C\cos\theta_2 \tag{2-10}$$

The constant C is the same for each field component since the amplitude of the sources is the same and points on the boundary are equidistant from the current elements. From Fig. 2-8*a* we see that

$$\theta_1 + \theta_2 = 180° \tag{2-11}$$

so,

$$E_{r1} = C\cos(180° - \theta_2) = -C\cos\theta_2 \tag{2-12}$$

Comparing this to (2-10), we see that

$$E_{r1} = -E_{r2} \qquad \textit{along boundary} \tag{2-13}$$

Thus along the plane PP', the radial components are equal in magnitude and opposite in phase. E_{r2} is directed radially out from the image source since θ_2 is less than 90°, and then $\cos\theta_2$ is positive. On the other hand, E_{r1} is radially inward toward the primary source since (2-12) is negative. Figure 2-8*a* illustrates this and shows that the projections of each along PP' will cancel. A similar line of reasoning for the θ-components leads to

$$E_{\theta 1} = D\sin\theta_1 = D\sin\theta_2 \tag{2-14}$$

$$E_{\theta 2} = D\sin\theta_2 \tag{2-15}$$

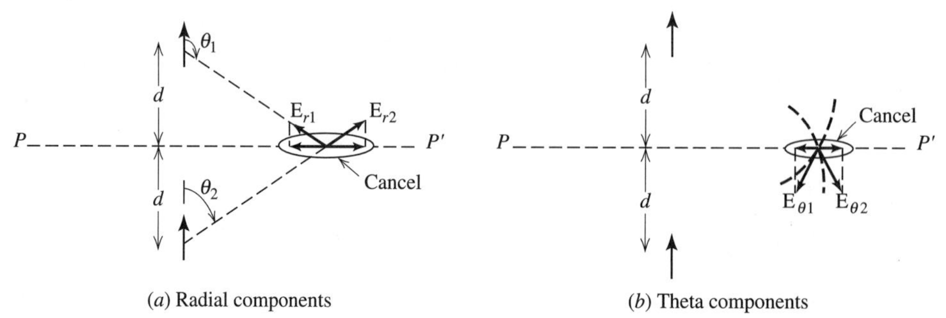

(*a*) Radial components (*b*) Theta components

Figure 2-8 The ideal dipole and its image in a ground plane of Fig. 2-7. The source and its image acting together give zero tangential electric field intensity along the plane PP' where the original perfect ground plane was located.

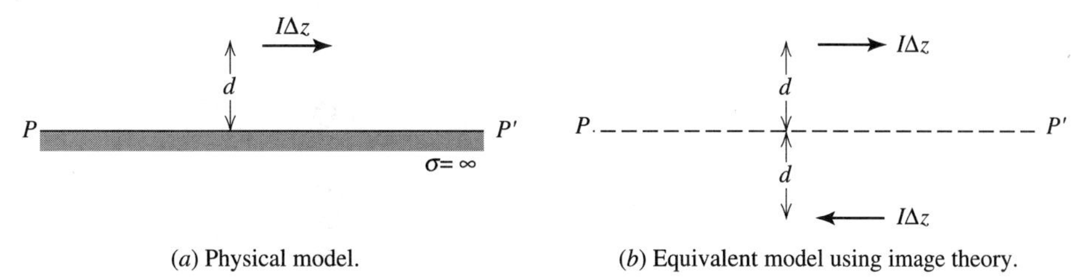

(*a*) Physical model. (*b*) Equivalent model using image theory.

Figure 2-9 Ideal dipole above and parallel to a perfect ground plane.

where D is a constant and thus

$$E_{\theta 1} = E_{\theta 2} \qquad \textit{along boundary} \tag{2-16}$$

Figure 2-8*b* demonstrates that the net projection of these θ-components along plane PP' is zero.

We have shown that the total tangential electric field intensity is zero along the image plane PP' for an ideal dipole perpendicular to the plane and its image was acting together. Therefore, since the source configuration above the plane and the boundary conditions were not altered, the system of Fig. 2-7*b* is equivalent to the original problem of Fig. 2-7*a*. The systems are equivalent in the sense that the fields above the plane PP' are identical. The above derivation can be reversed by starting with the two sources of Fig. 2-7*b* and then introducing a perfect ground plane with its surface along plane PP', thus arriving at Fig. 2-7*a*. The essential feature to remember is that *the fields above a perfect ground plane from a primary source acting in the presence of the perfect ground plane are found by summing the contributions of the primary source and its image, each acting in free space.*

An ideal dipole oriented parallel to a perfect ground plane (i.e., horizontal) has an image that again is equidistant below the image plane, but in this case the image is oppositely directed as shown in Fig. 2-9. The equivalent model of Fig. 2-9*b*, which gives the same fields above plane PP' as the physical model of Fig. 2-9*a*, can be proven by simple sketches similar to those of Fig. 2-8.

The image of a current element oriented in any direction with respect to a perfect ground plane can be found by decomposing it into perpendicular and parallel components, forming the images of the components, and constructing the image from these image components. An example is shown in Fig. 2-10. The image of an arbitrary current distribution is obtained in a similar fashion. The current is decomposed into perpendicular and parallel current elements whose images are readily found. The image current distribution is then the vector sum of these image current elements.

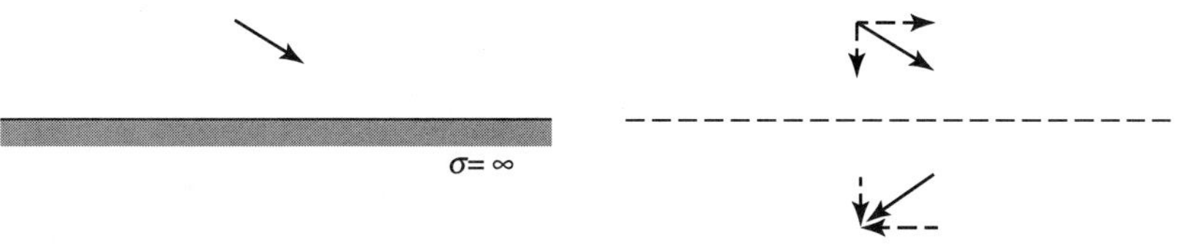

(*a*) Physical model. (*b*) Equivalent model using image theory.

Figure 2-10 Ideal dipole above and obliquely oriented relative to a perfect ground plane.

2.3.2 Monopoles

The principles of image theory are illustrated in this section with several forms of the monopole antenna. A **monopole** is a dipole that has been divided in half at its center feed point and fed against a ground plane. Three monopoles and their images in a perfect ground plane are shown in Fig. 2-11. High-frequency monopoles are often fed from coaxial cables behind the ground plane as shown in Fig. 2-12*a*.

The currents and charges on a monopole are the same as on the upper half of its dipole counterpart, but the terminal voltage is only half that of the dipole. The voltage is half because the gap width of the input terminals is half that of the dipole, and the same electric field over half the distance gives half the voltage. The input impedance for a monopole is therefore half that of its dipole counterpart, or

$$Z_{A,\text{mono}} = \frac{V_{A,\text{mono}}}{I_{A,\text{mono}}} = \frac{\frac{1}{2}V_{A,\text{dipole}}}{I_{A,\text{dipole}}} = \frac{1}{2}Z_{A,\text{diople}} \tag{2-17}$$

This is easily demonstrated for the radiation resistance. Since the fields only extend over a hemisphere, the power radiated is only half that of a dipole with the same current. Therefore, the radiation resistance of a monopole is given by

$$R_{r,\text{mono}} = \frac{P_{\text{mono}}}{\frac{1}{2}|I_{A,\text{mono}}|^2} = \frac{\frac{1}{2}P_{\text{dipole}}}{\frac{1}{2}|I_{A,\text{dipole}}|^2} = \frac{1}{2}R_{r,\text{dipole}} \tag{2-18}$$

For example, the radiation resistance of a short monopole is from (1-177)

$$R_{r,\text{mono}} = 40\pi^2\left(\frac{h}{\lambda}\right)^2 \qquad \textit{for } h \ll \lambda \tag{2-19}$$

where h is the length of the monopole and $\Delta z = 2h$.

The radiation pattern of a monopole above a perfect ground plane, as in Fig. 2-12, is the same as that of a dipole similarly positioned in free space since the fields above the image plane are the same. Therefore, a monopole fed against a perfect ground plane radiates one-half the total power of a similar dipole in free space because the power is distributed in the same fashion but only over half as much

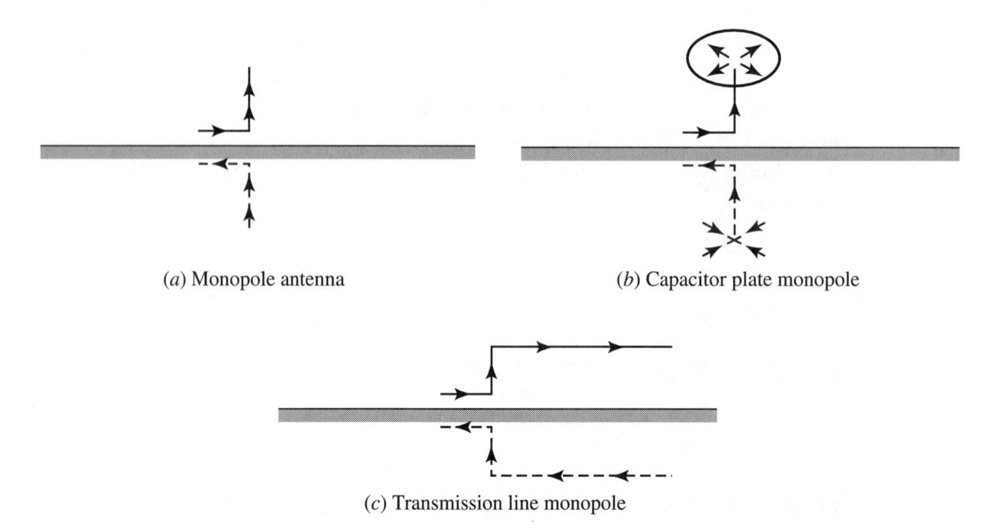

Figure 2-11 Monopole antennas over perfect ground planes with their images (dashed).

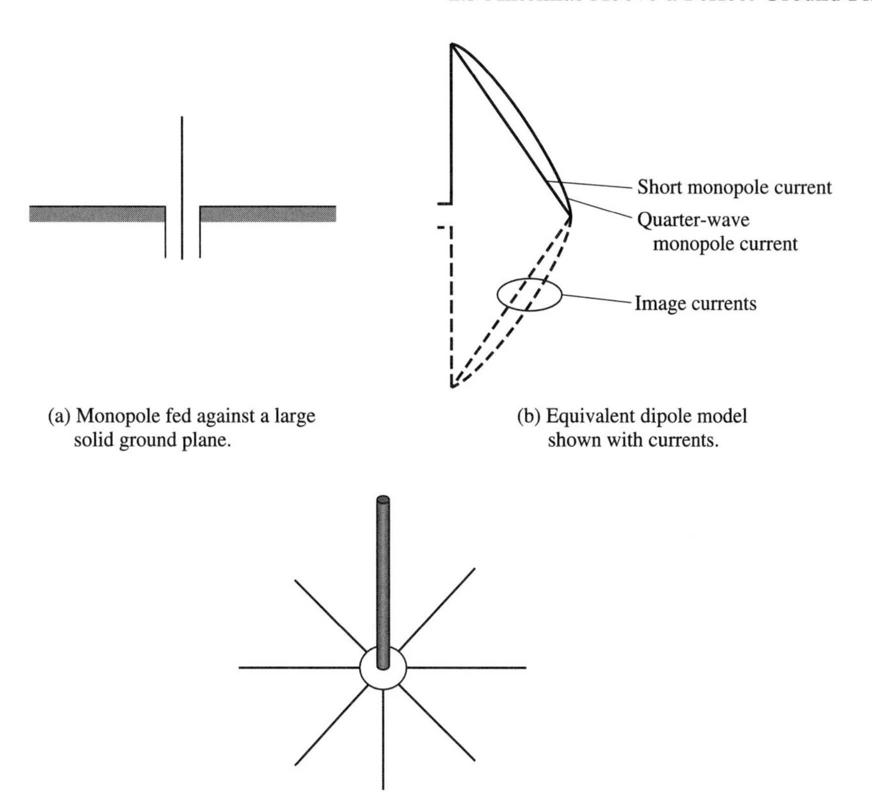

(a) Monopole fed against a large solid ground plane.

(b) Equivalent dipole model shown with currents.

(c) Practical monopole antenna with radial wires to simulate a ground plane.

Figure 2-12 Monopole antennas fed against a ground plane with a coaxial cable.

space. As a result, the beam solid angle of a monopole above a perfect ground plane is one-half that of a similar dipole in free space, leading to a doubling of the directivity:

$$D_{\text{mono}} = \frac{4\pi}{\Omega_{A,\text{mono}}} = \frac{4\pi}{\frac{1}{2}\Omega_{A,\text{dipole}}} = 2D_{\text{dipole}} \tag{2-20}$$

This can be shown in another way. If a dipole in free space has a maximum radiation intensity of U_m, a monopole of half the length above a perfect ground plane with the same current will have same value of U_m since the fields are the same. The total radiated power for the dipole is P, so the power radiated from the monopole is $\frac{1}{2}P$. The directivity from (1-145) for the two antennas is

$$D_{\text{dipole}} = \frac{U_m}{U_{\text{ave}}} = \frac{U_m}{P/4\pi} \tag{2-21}$$

and

$$D_{\text{mono}} = \frac{U_m}{\frac{1}{2}P/4\pi} = 2D_{\text{dipole}} \tag{2-22}$$

The directivity increase does not come from an increase in the radiation intensity (and, hence, field intensity) but rather from a decrease in average radiation intensity. This, in turn, comes about because only half the power radiated by a dipole

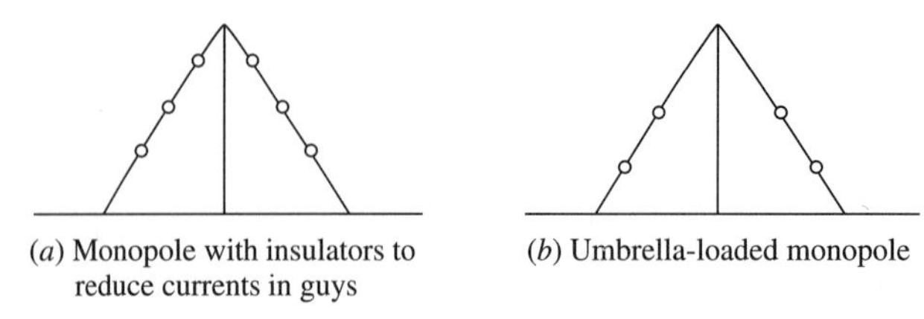

Figure 2-13 Monopoles with supporting guy wires.

is radiated by a monopole. The directivity of a short monopole, for example, is 2(1.5) = 3.

The directivity of a quarter-wave monopole is twice that of a half-wave dipole in free space; that is, from Fig. 2-6 and (2-22)

$$D = 2(1.64) = 3.28 = 5.16 \text{ dB} \qquad \lambda/4\text{-}monopole \tag{2-23}$$

The input impedance of an infinitesimally thin quarter-wave monopole from Fig. 2-6 and (2-17) is

$$Z_A = \frac{1}{2}(72 + j42.5) = 36 + j21.3\ \Omega \qquad \lambda/4\text{-}monopole \tag{2-24}$$

At low frequencies, a monopole that is a quarter wavelength long or less can be rather large physically. For example, in the standard AM broadcast band at 1 MHz, the wavelength is 300 m, and a quarter-wave monopole is 75 m tall. Such a large structure is usually not self-supporting, and guy wires are employed for support. Currents can exist in these guy wires in a downward direction tending to cancel the effect of the vertical element. Insulators are added to break up these currents, as in Fig. 2-13.

If currents are allowed to continue from the monopole out onto the guys, there is a partial top-loading effect for towers shorter than a quarter wavelength, thereby increasing the radiation resistance. See Fig. 2-13*b*. The loading is usually not enough to give uniform current on the vertical member. Also, the downward angle of the guys gives a slight canceling of the fields from the vertical current. For a comparable-length monopole, the umbrella-loaded version has a lower radiation resistance than the capacitor-plate monopole. Experimental data are available in the literature for umbrella-loaded monopoles [3].

2.4 SMALL LOOP ANTENNAS

A closed loop current whose maximum dimension is less than about a tenth of a wavelength is called a **small loop antenna**. Again, small is interpreted as meaning electrically small, or small compared to a wavelength. In this section, we use two methods to solve for the radiation properties of small loop antennas. First, we show that the small loop is the dual of an ideal dipole, and by observing the duality contained in Maxwell's equations, we use the results previously derived for the ideal dipole to write the fields of a small loop. Next, we derive the fields of a small loop directly and show that the results are the same as those obtained using duality.

2.4.1 Duality

Frequently, an antenna problem arises for which the structure is the dual of an antenna whose solution is known. If antenna structures are duals, it is possible to

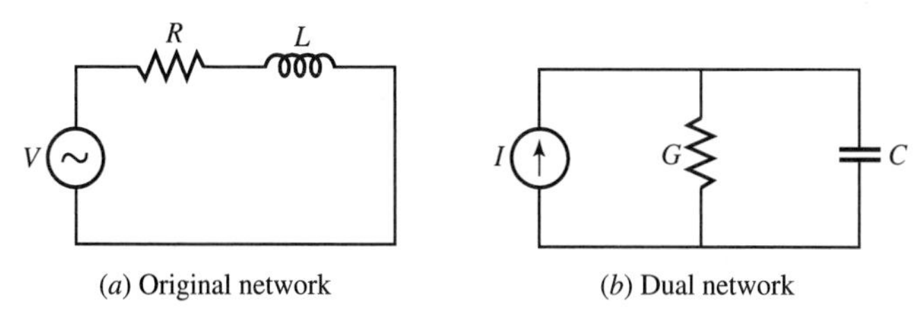

(a) Original network (b) Dual network

Figure 2-14 Dual networks: I (=) V, G (=) R, C (=) L.

write the fields for one antenna from the field expressions of the other by interchanging parameters using the principle of duality. Before examining the small loop, we discuss the general principle of duality as applied to antennas.

Dual antenna structures are similar to dual networks. For example, consider a simple network of a voltage source applied to a series connection of a resistor R and an inductor L as in Fig. 2-14a. The dual network of Fig. 2-14b is a current source I (=) V applied to the parallel combination of conductance G (=) R and capacitance C (=) L. The symbol "(=)" means replace the quantity on the left with the quantity on the right, much as the equal sign in a computer program statement. Since the networks are duals, the solutions are duals. In this example, the original series network can be described by the mesh equation

$$V = RI + j\omega LI \tag{2-25}$$

The dual of this mesh equation is a node equation obtained by replacing V by I, R by G, and L by C. The node equation for the dual parallel network is then

$$I = GV + j\omega CV \tag{2-26}$$

Returning to the antenna problem, suppose we have an electric current source with current density $\mathbf{J}_1$ and boundary conditions on materials present $(\varepsilon_1, \mu_1, \sigma_1)$. Maxwell's equations for this system from (1-16) and (1-15) are

$$\nabla \times \mathbf{E}_1 = -j\omega\mu_1\mathbf{H}_1 \tag{2-27}$$

$$\nabla \times \mathbf{H}_1 = j\omega\varepsilon_1'\mathbf{E}_1 + \mathbf{J}_1 \tag{2-28}$$

where $\mathbf{E}_1$ and $\mathbf{H}_1$ are the fields generated by $\mathbf{J}_1$ with materials $(\varepsilon_1, \mu_1, \sigma_1)$ present. Now suppose a fictitious magnetic current source with magnetic current density $\mathbf{M}_2$ exists with materials $(\varepsilon_2, \mu_2, \sigma_2)$ present. Maxwell's equations for this system from (1-15) and (1-21) are

$$\nabla \times \mathbf{H}_2 = j\omega\varepsilon_2'\mathbf{E}_2 \tag{2-29}$$

$$\nabla \times \mathbf{E}_2 = -j\omega\mu_2\mathbf{H}_2 - \mathbf{M}_2 \tag{2-30}$$

where $\mathbf{E}_2$ and $\mathbf{H}_2$ are the fields arising from $\mathbf{M}_2$.

The electric and magnetic systems are duals if the procedure in Table 2-1 can be performed. This is easy to demonstrate. To see if (2-29) and (2-30) are the duals of (2-27) and (2-28), we substitute the quantities in the left-hand column of Table 2-1 into (2-29) and (2-30) for the corresponding quantities of the right-hand column. This yields

$$\nabla \times \mathbf{E}_1 = j\omega\mu_1(-\mathbf{H}_1) \tag{2-31}$$

$$\nabla \times (-\mathbf{H}_1) = -j\omega\varepsilon_1'\mathbf{E}_1 - \mathbf{J}_1 \tag{2-32}$$

Table 2-1 Dual Radiating Systems. Radiating System #1 with Electric Currents and System #2 with Magnetic Currents Are Duals If One Can:

Replace the Following in System #2	By the Following in System #1
$\mathbf{M}_2$	$\mathbf{J}_1$
ε_2'	μ_1
μ_2	ε_1'
$\mathbf{E}_2$	$-\mathbf{H}_1$
$\mathbf{H}_2$	$\mathbf{E}_1$

Thus, the equations of the electric system, (2-27) and (2-28), are dual to the equations of the magnetic system, (2-29) and (2-30), just as (2-25) and (2-26) are dual equations. Since the equations of the systems are dual, the solutions will be also. Before illustrating this, we summarize the principle of duality:

If the sources of two systems are duals, that is,

$$\mathbf{M}_2 \ (=) \ \mathbf{J}_1 \tag{2-33}$$

and **if** the boundary conditions are also dual,[1] that is,

$$\mu_2 \ (=) \ \varepsilon_1', \qquad \varepsilon_2' \ (=) \ \mu_1 \tag{2-34}$$

then the fields of system #2 can be found from the solution of system #1 by the substitutions

$$\mathbf{E}_2 \ (=) \ -\mathbf{H}_1, \qquad \mathbf{H}_2 \ (=) \ \mathbf{E}_1 \tag{2-35}$$

in the field expressions for system #1 along with the substitutions in (2-34).

Now we use duality to find the fields of a small current loop from a knowledge of the fields of an ideal electric dipole. A current loop can be represented as a fictitious (ideal) magnetic dipole with uniform magnetic current I^m and length Δz. The sources are duals as required by (2-33) if we let

$$I^m \ (=) \ I^e \tag{2-36}$$

where I^e is the current of an ideal electric dipole of length Δz. Since no materials are present, there are no boundary conditions. The ideal electric dipole has field solutions of the form

$$\mathbf{E}_1 = E_{\theta 1}\hat{\boldsymbol{\theta}} + E_{r1}\hat{\mathbf{r}} \tag{2-37}$$

$$\mathbf{H}_1 = H_{\phi 1}\hat{\boldsymbol{\phi}} \tag{2-38}$$

The fields of the dual magnetic dipole are then found from (2-35) as

$$\mathbf{E}_2 \ (=) \ -\mathbf{H}_1 = -H_{\phi 1}\hat{\boldsymbol{\phi}} \tag{2-39}$$

$$\mathbf{H}_2 \ (=) \ \mathbf{E}_1 = E_{\theta 1}\hat{\boldsymbol{\theta}} + E_{r1}\hat{\mathbf{r}} \tag{2-40}$$

[1]Note that $\varepsilon_1' = \varepsilon_1 - j(\sigma_1/\omega)$. If magnetic conductors of magnetic conductivity σ_2^m were assumed to exist in system #2, then μ_2 would become $\mu_2' = \mu_2 - j(\sigma_2^m/\omega)$ and ε_1' would be replaced by μ_2', or equivalently σ_1 replaced by σ_2^m.

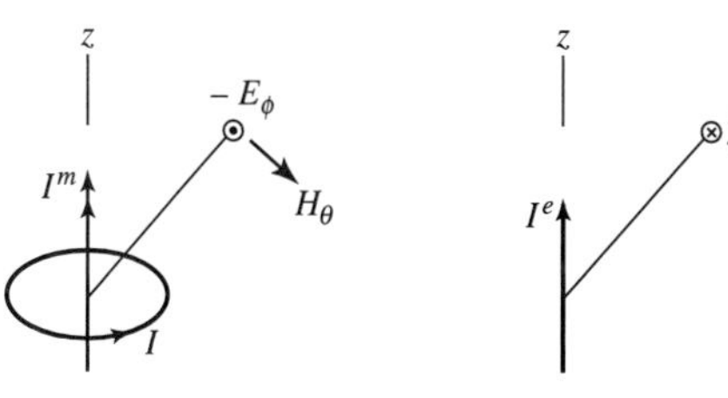

(*a*) Small current loop and equivalent magnetic dipole.

(*b*) Ideal electric dipole.

Figure 2-15 Radiation field components of ideal magnetic and electric dipoles.

if we make the substitutions

$$\mu_2 = \mu \; (=) \; \varepsilon_1' = \varepsilon \quad \text{and} \quad \varepsilon_2' = \varepsilon \; (=) \; \mu_1 = \mu \tag{2-41}$$

which follow from (2-34) and the fact that in both systems the surrounding medium is a homogeneous material of μ and ε. Note that β remains the same since replacing μ by ε and ε by μ in $\omega\sqrt{\mu\varepsilon}$ yields $\omega\sqrt{\varepsilon\mu}$. Now, using (2-36) and (2-41) in the ideal electric dipole field expressions of (1-71) together with (2-39) and (2-40) gives

$$\mathbf{E}_2 = -\frac{I^m\,\Delta z}{4\pi}\, j\beta\left(1 + \frac{1}{j\beta r}\right)\frac{e^{-j\beta r}}{r}\sin\theta\,\hat{\boldsymbol{\phi}} \tag{2-42}$$

$$\begin{aligned}\mathbf{H}_2 = {} & \frac{I^m\,\Delta z}{4\pi}\, j\omega\varepsilon\left[1 + \frac{1}{j\beta r} + \frac{1}{(j\beta r)^2}\right]\frac{e^{-j\beta r}}{r}\sin\theta\,\hat{\boldsymbol{\theta}} \\ & + \frac{I^m\,\Delta z}{2\pi}\, j\omega\varepsilon\left[\frac{1}{j\beta r} + \frac{1}{(j\beta r)^2}\right]\frac{e^{-j\beta r}}{r}\cos\theta\,\hat{\mathbf{r}}\end{aligned} \tag{2-43}$$

These are the complete field expressions (valid in the near-field region) for a small loop of electric current. The far-field components are obtained by retaining only those terms that vary as r^{-1}, giving

$$\mathbf{E}_2 = -I^m\,\Delta z\; j\beta\,\frac{e^{-j\beta r}}{4\pi r}\sin\theta\,\hat{\boldsymbol{\phi}} \tag{2-44}$$

$$\mathbf{H}_2 = I^m\,\Delta z\; j\omega\varepsilon\,\frac{e^{-j\beta r}}{4\pi r}\sin\theta\,\hat{\boldsymbol{\theta}} \tag{2-45}$$

These radiation fields as well as those of the ideal electric dipole are shown in Fig. 2-15. Both antennas have the same radiation pattern, sin θ. The magnetic field component H_ϕ of the ideal electric dipole is easily remembered by use of the right-hand rule. Place the thumb of your right hand along the current of the dipole and pointing in the direction of current flow. Your fingers will then curl in the direction of the magnetic field. This statement is implicit in Ampere's law of (2-28). A similar relationship holds for the magnetic dipole, except the left-hand rule is used and the field obtained is the electric field component $-E_\phi$. This follows from (2-30). Another realization for a magnetic dipole in addition to the small loop is a narrow slot in a ground plane, whose fields can be found from an equivalent magnetic current along the long axis of the slot.

2.4.2 The Small Loop Antenna

Using duality, we found the field expressions for a small loop of uniform current. However, these expressions contain the equivalent magnetic dipole current amplitude I^m. By solving the small loop problem directly, we can establish the relationship between the current I in the loop and I^m. This can be accomplished by dealing only with the far-field region.

It turns out that the radiation fields of small loops are independent of the shape of the loop and depend only on the area of the loop. Therefore, we will select a square loop as shown in Fig. 2-16*a* to simplify the mathematics. The current has constant amplitude I and zero phase around the loop. Each side of the square loop is a short uniform electric current segment that is modeled as an ideal dipole. The two sides parallel to the x-axis have a total vector potential that is x-directed and is given by

$$A_x = \frac{\mu I \ell}{4\pi}\left(\frac{e^{-j\beta R_1}}{R_1} - \frac{e^{-j\beta R_3}}{R_3}\right) \tag{2-46}$$

which follows from (1-62). The minus sign in the second term arises because the current in side 3 is negative x-directed. Similarly for sides 2 and 4, we find

$$A_y = \frac{\mu I \ell}{4\pi}\left(\frac{e^{-j\beta R_2}}{R_2} - \frac{e^{-j\beta R_4}}{R_4}\right) \tag{2-47}$$

The far-field approximation is that the distances used for amplitude variations are nearly equal (i.e., $R_1 \approx R_2 \approx R_3 \approx R_4 \approx r$) and the phase differences are found from assuming parallel rays emanating from each side. By comparing the parallel path lengths, we find from geometrical considerations that

$$\begin{aligned} R_1 &= r + \frac{\ell}{2}\sin\theta\sin\phi, \qquad & R_2 &= r - \frac{\ell}{2}\sin\theta\cos\phi \\ R_3 &= r - \frac{\ell}{2}\sin\theta\sin\phi, \qquad & R_4 &= r + \frac{\ell}{2}\sin\theta\cos\phi \end{aligned} \tag{2-48}$$

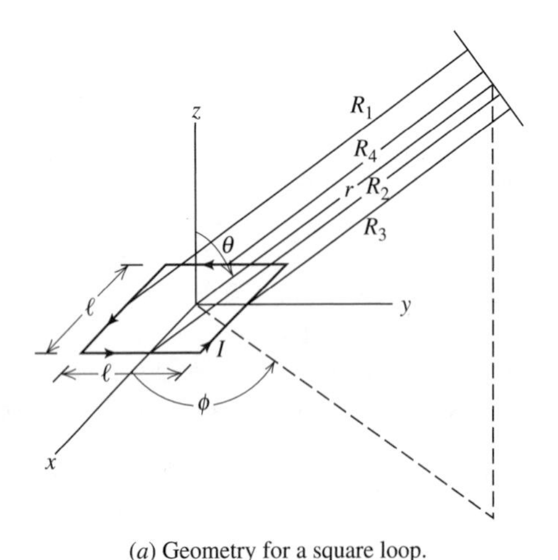

(*a*) Geometry for a square loop.

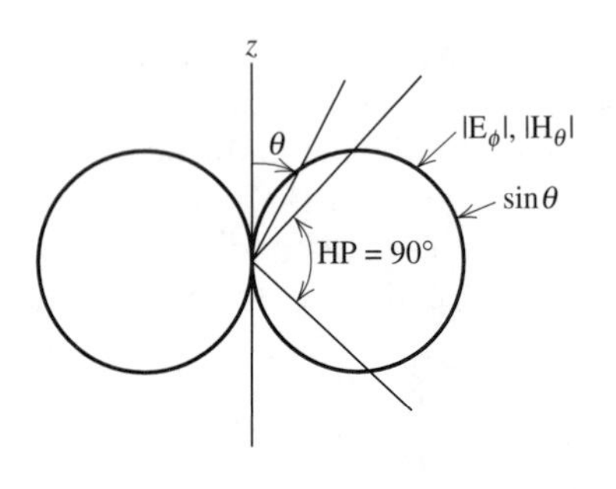

(*b*) Small loop radiation pattern.

Figure 2-16 The small loop antenna.

Substituting these into the exponents and r into the denominators of (2-46) and (2-47), we have

$$A_x = \frac{\mu I \ell e^{-j\beta r}}{4\pi r} \left(e^{-j\beta(\ell/2)\sin\theta\sin\phi} - e^{+j\beta(\ell/2)\sin\theta\sin\phi}\right)$$

$$A_y = \frac{\mu I \ell e^{-j\beta r}}{4\pi r} \left(e^{+j\beta(\ell/2)\sin\theta\cos\phi} - e^{-j\beta(\ell/2)\sin\theta\cos\phi}\right)$$

or

$$\begin{aligned} A_x &= -2j \frac{\mu I \ell e^{-j\beta r}}{4\pi r} \sin\left(\frac{\beta\ell}{2}\sin\theta\sin\phi\right) \\ A_y &= 2j \frac{\mu I \ell e^{-j\beta r}}{4\pi r} \sin\left(\frac{\beta\ell}{2}\sin\theta\cos\phi\right) \end{aligned} \tag{2-49}$$

Since the loop is small compared to a wavelength, $\beta\ell = 2\pi\ell/\lambda$ is also small and the sine functions in (2-49) can be replaced by their arguments, giving

$$\begin{aligned} A_x &\approx -j \frac{\mu I e^{-j\beta r}}{4\pi r} \beta\ell^2 \sin\theta\sin\phi \\ A_y &\approx j \frac{\mu I e^{-j\beta r}}{4\pi r} \beta\ell^2 \sin\theta\cos\phi \end{aligned} \tag{2-50}$$

Combining components to form the total vector potential gives

$$\mathbf{A} = A_x\hat{\mathbf{x}} + A_y\hat{\mathbf{y}} = j\beta\ell^2 \frac{\mu I e^{-j\beta r}}{4\pi r} \sin\theta\,(-\sin\phi\,\hat{\mathbf{x}} + \cos\phi\,\hat{\mathbf{y}}) \tag{2-51}$$

The term in parentheses is the unit vector $\hat{\boldsymbol{\phi}}$ in (C-6), so

$$\mathbf{A} = j\beta S \frac{\mu I e^{-j\beta r}}{4\pi r} \sin\theta\,\hat{\boldsymbol{\phi}} \tag{2-52}$$

where S is the area of the loop. All of $\mathbf{A}$ is transverse to the direction of propagation, so the radiation electric field from (1-104) is $-j\omega\mathbf{A}$, giving

$$\mathbf{E} = \eta\beta^2 S \frac{I e^{-j\beta r}}{4\pi r} \sin\theta\,\hat{\boldsymbol{\phi}} \tag{2-53}$$

since $\omega\mu\beta = \omega^2\mu\sqrt{\mu\varepsilon} = \sqrt{\mu/\varepsilon}\,\omega^2\mu\varepsilon = \eta\beta^2$. The radiation magnetic field is

$$\mathbf{H} = \frac{1}{\eta}\hat{\mathbf{r}} \times \mathbf{E} = -\beta^2 S \frac{I e^{-j\beta r}}{4\pi r} \sin\theta\,\hat{\boldsymbol{\theta}} \tag{2-54}$$

Comparing (2-53) or (2-54) to the magnetic dipole radiation fields of (2-44) or (2-45), we find that

$$I^m\,\Delta z = j\omega\mu I S \tag{2-55}$$

This completes the relationship between the small current loop and its equivalent magnetic dipole. The complete field expressions for a small loop of *magnetic moment IS* are found from (2-42) and (2-43) using (2-55). The fields depend only on the magnetic moment (current and area) and not the loop shape. And the radiation pattern for a small loop, independent of its shape, equals that of an ideal electric dipole; see Fig. 2-16*b*. The radiation fields from a large loop are derived in Sec. 5.7.

The loop antenna has been used since Hertz first used it as a receiver in his experiments in 1888. It has an omnidirectional doughnut radiation pattern that is needed in many applications. The horizontal small loop (in the xy-plane) and vertical (z-directed) short dipole both have uniform radiation in the horizontal (xy) plane, but the loop provides horizontal polarization (E_ϕ), whereas the dipole is vertically polarized (E_θ). We now discuss the impedance properties of the small loop and introduce the multiturn loop and ferrite core loop.

The impedance of a small loop antenna is quite different from its ideal dipole dual. Whereas the ideal dipole is capacitive, the small loop is inductive. We discuss the input resistance first. The radiation resistance is found by calculating the power radiated using the small loop radiation fields with (1-128), which yields

$$P = 10I^2(\beta^2 S)^2 \tag{2-56}$$

The radiation resistance is then

$$R_r = \frac{2P}{I^2} = 20(\beta^2 S)^2 \approx 31{,}200\left(\frac{S}{\lambda^2}\right)^2 \ \Omega \tag{2-57}$$

This result provides a reasonable approximation to the radiation resistance of an actual small loop antenna for a loop perimeter less than about three-tenths of a wavelength.

The radiation resistance of a loop antenna can be increased significantly by using multiple turns. The magnetic moment of an N turn loop is NIS, where S is the area of a single turn. The radiation resistance is then

$$R_r = 20(\beta^2 NS)^2 \approx 31{,}200\left(\frac{NS}{\lambda^2}\right)^2 \ \Omega \tag{2-58}$$

The radiation resistance thus goes up as N^2. Another way to enhance the radiation resistance is to wind the loop turns around a ferrite core. A ferrite core of effective relative permeability μ_{eff} has a phase constant of $\beta = \omega\sqrt{\mu\varepsilon} = \omega\sqrt{\mu_o\varepsilon_o}\sqrt{\mu_{\text{eff}}} = (2\pi/\lambda)\sqrt{\mu_{\text{eff}}}$, where λ is the free-space wavelength. The relative effective permeability depends on the core size and shape and is usually less than the relative permeablity of the core material [4]. The radiation resistance of a coil of N turns wound on a ferrite core is then

$$R_r \approx 31{,}200\left(N\mu_{\text{eff}}\frac{S}{\lambda^2}\right)^2 \ \Omega \tag{2-59}$$

A multiturn loop wound on a linear ferrite core is referred to as a **loop-stick antenna**. It is a commonly used low-frequency receiving antenna. For example, it is used with most AM broadcast receivers. At frequencies around 1 MHz (e.g., the AM broadcast band), the recommended ferrite material with $\mu_r = 100$ has an effective relative permeability of 40.

Small loop antennas also have considerable ohmic resistance. For a rectangular loop of wire ℓ_1 by ℓ_2, the ohmic resistance of the wire is given approximately by

$$R_w = \frac{2\ell_1\ell_2}{\pi d^2} R_s\left\{\frac{1}{[(\ell_1/2a)^2 - 1]^{1/2}} + \frac{1}{[(\ell_2/2a)^2 - 1]^{1/2}}\right\} \tag{2-60}$$

where a is the wire radius and R_s is the surface resistance of (1-176). If ℓ_1 and ℓ_2 are much larger than a (i.e., the wire is thin), then (2-60) reduces to

$$R_w = \frac{2(\ell_1 + \ell_2)}{2\pi a} R_s \tag{2-61}$$

This formula can be generalized to loops of arbitrary shape as follows:

$$R_w = \frac{L_m}{2\pi a} R_s \tag{2-62}$$

where L_m is the mean length of the wire loop. For a circular loop, this becomes

$$R_w = \frac{2\pi b}{2\pi a} R_s = \frac{b}{a} R_s \qquad \textit{circular loop} \tag{2-63}$$

where b is the mean loop radius and a is the wire radius; see (1-175).

As mentioned previously, the small loop antenna is inherently inductive. The inductance of a small ℓ_1 by ℓ_2 rectangular loop is given by

$$L = \frac{\mu}{\pi}\left(\ell_2 \cosh^{-1}\frac{\ell_1}{2a} + \ell_1 \cosh^{-1}\frac{\ell_2}{2a}\right) \qquad \textit{rectangular loop} \tag{2-64}$$

For a small circular loop of radius b, the inductance for $a \ll b$ is [5]

$$L = \mu b\left[\ln\left(\frac{8b}{a}\right) - 1.75\right] \qquad \textit{circular loop} \tag{2-65}$$

Small loop antennas have several applications. The small loop is very popular as a receiving antenna. For example, single turn small loop antennas are used in pagers. Multiturn small loops are popular in AM broadcast receivers. Small loop antennas are also used in direction-finding receivers and for field strength probes.

Radiation resistance decreases much faster with decreasing frequency for a small loop (f^{-4}) than for a short dipole (f^{-2}). Multiturn loops are used to increase radiation resistance; see (2-59). However, loss and inductance of an N turn loop must be multiplied by N. But wire losses can be reduced by decreasing the number of turns in a loop and using a ferrite core to maintain radiation resistance. In practice, a variable capacitor is used to tune out inductance by placing it in parallel with the loop. This section is concluded with an example to illustrate numerical results.

EXAMPLE 2-1 A Small Circular Loop Antenna

To illustrate the impedance calculations for small loop antennas, consider a circular loop with a mean loop circumference of 0.2λ and a wire radius of 0.001λ. Then $b = 0.1\lambda/\pi$ and $a = 0.001\lambda$ in (2-57) yields the radiation resistance as

$$R_r = 31{,}200\left(\frac{\pi b^2}{\lambda^2}\right)^2 = 31{,}200\left(\frac{0.01}{\pi}\right)^2 = 0.316\ \Omega \tag{2-66}$$

The reactance from (2-65) is

$$X_{\text{in}} = \omega L = 2\pi\frac{c}{\lambda} b\mu\left[\ln\left(\frac{8b}{a}\right) - 1.75\right] \tag{2-67}$$

where c is the velocity of light. For an air-filled loop, $\mu = \mu_o$ and then

$$X_A = 2\pi \frac{3 \times 10^8}{\lambda} \frac{0.1\lambda}{\pi} 4\pi \times 10^{-7}\left[\ln \frac{0.8}{0.001\pi} - 1.75\right] = 285.8\ \Omega \tag{2-68}$$

To determine the ohmic resistance a frequency must be specified, say 1 MHz. (Note that the loop is physically large at 1 MHz, having a circumference of 60 m and a wire diameter of 0.3 m.) Further suppose the wire is copper, and then $\mu = \mu_o$ and $\sigma = 5.7 \times 10^7$ S/m in (1-176) gives

$$R_s = \sqrt{\frac{\mu_o 2\pi f}{2\sigma}} = \sqrt{\frac{4\pi \times 10^{-7} \cdot 2\pi \times 10^6}{2 \cdot 5.7 \times 10^7}} = 2.63 \times 10^{-4}\ \Omega \tag{2-69}$$

And from (2-63),

$$R_w = \frac{b}{a} R_s = \frac{0.1}{0.001\pi} 2.63 \times 10^{-4} = 8.38 \times 10^{-3}\ \Omega \tag{2-70}$$

Adding this to (2-66) gives the total input resistance

$$R_A = R_r + R_w = 0.324\ \Omega \tag{2-71}$$

The input impedance is thus

$$Z_A = R_A + jX_A = 0.324 + j\,285.8\ \Omega \tag{2-72}$$

The radiation efficiency of this loop is

$$e_r = \frac{R_r}{R_A} = \frac{0.316}{0.324} = 97.5\% \tag{2-73}$$

2.5 ANTENNAS IN COMMUNICATION SYSTEMS

It is important to have an appreciation for the role played by antennas when they are employed in their primary application area of communication links. A simple communication link is shown in Fig. 2-17. We first discuss the basic properties of a receiving antenna. The receiving antenna with impedance Z_A and terminated in load impedance Z_L is modeled as shown in Fig. 2-18. The total power incident on the receiving antenna is found by summing up the incident power density over the "area" of the receive antenna, called effective aperture. How an antenna converts this incident power into available power at its terminals depends on the type of antenna used, its pointing direction, and polarization. In this section, we discuss the basic relationships for power calculations and illustrate their use in communication links. Additional details on communication links as well as the application areas of radiometry and radar are treated in Secs. 9.1 to 9.3.

Directivity and Gain. For system calculations, it is usually easier to work with directivity rather than its equivalent, maximum effective aperture. The relation can be established by examining an infinitesimal dipole and generalizing. The maximum effective aperture of an ideal, lossless dipole of length Δz is found by orienting the dipole for maximum response, which is parallel to the incoming linearly polarized electric field E^i. Then the open circuit voltage is found from

$$V = E^i\,\Delta z \qquad \textit{ideal dipole receiving antenna} \tag{2-78}$$

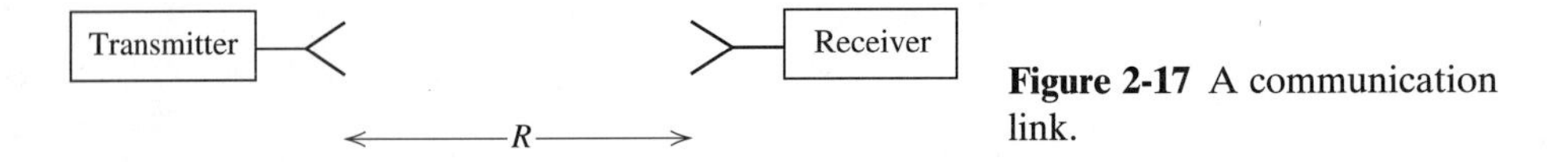

Figure 2-17 A communication link.

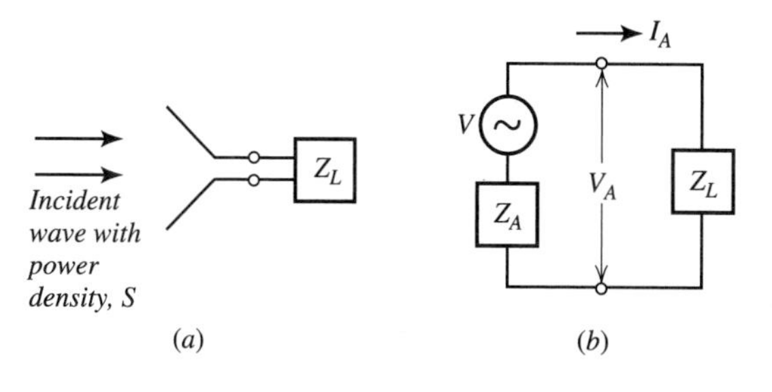

Figure 2-18 Equivalent circuit for a receiving antenna. (*a*) Receive antenna connected to a receiver with load impedance Z_L. (*b*) Equivalent circuit.

The power available from the antenna is realized when the antenna impedance is matched by a load impedance of $Z_L = R_r - jX_A$ if we assume $R_{\text{ohmic}} = 0$. The maximum available power is then (see Sec. 9.1)

$$P_{Am} = \frac{1}{8}\frac{|V|^2}{R_r} = \frac{1}{8}\frac{|E^i|^2}{R_r}(\Delta z)^2 \tag{2-79}$$

where (2-78) was used. The available power can also be calculated by examining the incident wave. The power density (Poynting vector magnitude) in the incoming wave is

$$S = \frac{1}{2}|\mathbf{E} \times \mathbf{H}^*| = \frac{1}{2}\frac{|E^i|^2}{\eta} \tag{2-80}$$

The available power is found using the *maximum effective aperture* A_{em}, which is the collecting area of the antenna. The receiving antenna collects power from the incident wave in proportion to its maximum effective apertures:

$$P_{Am} = S\,A_{em} \tag{2-81}$$

The maximum available power P_{Am} will be realized if the antenna is directed for maximum response, is polarization-matched to the wave, and is impedance-matched to its load. The "maximum" refers to the assumption that there are no ohmic losses on the antenna.

Maximum effective aperture for the ideal dipole is found using (2-79) and (2-80) with (2-81):

$$A_{em} = \frac{P_{Am}}{S} = \frac{\dfrac{1}{8}\dfrac{|V|^2}{R_r}}{\dfrac{1}{2}\dfrac{|E^i|^2}{\eta}} = \frac{1}{4}\frac{\eta}{R_r}\frac{|E^i|^2(\Delta z)^2}{|E^i|^2} = \frac{1}{4}\frac{\eta}{R_r}(\Delta z)^2 = \frac{1}{4}\frac{\eta(\Delta z)^2}{\eta\dfrac{2}{3}\pi\left(\dfrac{\Delta z}{\lambda}\right)^2}$$

$$= \frac{3}{8\pi}\lambda^2 = 0.119\lambda^2 \tag{2-82}$$

where the ideal dipole radiation resistance value from (1-172) was used. The maximum effective aperture of an ideal dipole is independent of its length Δz (as long as $\Delta z \ll \lambda$). However, it is important to note that R_r is proportional to $(\Delta z/\lambda)^2$ so that even though A_{em} remains constant as the dipole is shortened, its radiation resistance decreases rapidly and it is more difficult to realize this maximum effective aperture because of the required conjugate impedance match of the receiver to the antenna.

The directivity of the ideal dipole can be written in the following manner:

$$D = \frac{3}{2} = \frac{4\pi}{\lambda^2}\frac{3}{8\pi}\lambda^2 \qquad \textit{ideal dipole} \tag{2-83}$$

Grouping factors this way permits identification of A_{em} from (2-82). Thus,

$$\boxed{D = \frac{4\pi}{\lambda^2}A_{em}} \tag{2-84}$$

Although we derived this for an ideal dipole, *this relationship is true for any antenna.* For an isotropic antenna, the directivity by definition is unity, so from (2-84) with $D = 1$:

$$A_{em} = \frac{\lambda^2}{4\pi} \qquad \textit{isotropic antenna} \tag{2-85}$$

Comparing this to $D = 4\pi/\Omega_A$, we see that

$$\boxed{\lambda^2 = A_{em}\Omega_A} \tag{2-86}$$

which is also a general relationship. We can extract some interesting concepts from this relation. For a fixed wavelength, A_{em} and Ω_A are inversely proportional; that is, as the maximum effective aperture increases (as a result of increasing its physical size), the beam solid angle decreases, which means power is more concentrated in angular space (i.e., directivity goes up, which also follows from $D = 4\pi/\Omega_A$). For a fixed maximum effective aperture (i.e., antenna size), as wavelength decreases (frequency increases), the beam solid angle also decreases, leading to increased directivity.

In practice, antennas are not completely lossless. In Sec. 1.9, we saw that power available at the terminals of a transmitting antenna was not all transformed into radiated power. The power received by a receiving antenna is reduced to the fraction e_r (radiation efficiency) of what it would be if the antenna were lossless. This is represented by defining *effective aperture*:

$$A_e = e_r A_{em} \tag{2-87}$$

and the available power with antenna losses included, analogous to (2-81), is

$$P_A = S\,A_e \tag{2-88}$$

This simple equation is very intuitive and indicates that a receiving antenna acts to convert incident power (flux) density in W/m^2 to power delivered to the load in W. Losses associated with a mismatch between the polarization of the incident wave and receiving antenna as well as impedance mismatch between the antenna and load are not included in A_e. These losses are not inherent to the antenna, but depend on how it is used in the system. The concept of gain was introduced to account for losses on an antenna, that is, $G = e_r D$; see (1-159). We can form a gain expression from the directivity expression by multiplying both sides of (2-84) by e_r and using (2-87):

$$G = e_r D = \frac{4\pi}{\lambda^2}e_r A_{em} = \frac{4\pi}{\lambda^2}A_e$$

or

$$G = \frac{4\pi}{\lambda^2} A_e \tag{2-89}$$

We will show in Sec. 7.3 that for electrically large antennas, effective aperture is equal to or less than the physical aperture area of the antenna A_p, which is expressed using *aperture efficiency* ε_{ap}:

$$A_e = \varepsilon_{ap} A_p \tag{2-90}$$

It is important to note that although we developed the general relationships of (2-84), (2-86), and (2-89) for receiving antennas, they apply to transmitting antennas as well. The relationships are essential for communication system computations that we consider next.

Communication Links. We are now ready to completely describe the power transfer in the communication link of Fig. 2-17. If the transmitting antenna were isotropic, it would have power density at distance R of

$$S = \frac{U_{\text{ave}}}{R^2} = \frac{P_t}{4\pi R^2} \tag{2-91}$$

where P_t is the time-average input power accepted by the transmitting antenna and (1-131) and (1-135) have been used. For a transmitting antenna that is not isotropic but has gain G_t and is pointed for maximum power density in the direction of the receiver, we have for the power density incident on the receiving antenna:

$$S = \frac{G_t U_{\text{ave}}}{R^2} = \frac{G_t P_t}{4\pi R^2} \tag{2-92}$$

Using this in (2-88) gives the available received power as

$$P_r = S A_{er} = \frac{G_t P_t A_{er}}{4\pi R^2} \tag{2-93}$$

where A_{er} is the effective aperture of the receiving antenna and we assume it to be pointed and polarized for maximum response. Now from (2-89), $A_{er} = G_r \lambda^2/4\pi$, so (2-93) becomes

$$P_r = P_t \frac{G_t G_r \lambda^2}{(4\pi R)^2} \tag{2-94}$$

which gives the available power in terms of the transmitted power, antenna gains, and wavelength. Or, we could use $G_t = 4\pi A_{et}/\lambda^2$ in (2-93), giving

$$P_r = P_t \frac{A_{et} A_{er}}{R^2 \lambda^2} \tag{2-95}$$

which is called the *Friis transmission formula.*

The power transmission formula (2-94) is very useful for calculating signal power levels in communication links. It assumes that the transmitting and receiving antennas are matched in impedance to their connecting transmission lines, have identical polarizations, and are aligned for polarization match. It also assumes the antennas are pointed toward each other for maximum gain. If any of the above conditions are not met, it is a simple matter to correct for the loss introduced by polarization mismatch, impedance mismatch, or antenna misalignment. The antenna misalignment effect is easily included by using the power gain value in the appropriate direction.

The effect and evaluation of polarization and impedance mismatch are discussed in Sec. 9.1, but here we discuss how they are included in systems. Figure 2-18 shows the network model for a receiving antenna with input antenna impedance Z_A and an attached load impedance Z_L that can be a transmission line connected to a distant receiver. The power delivered to the terminating impedance is given by

$$\boxed{P_D = pqP_r} \tag{2-96}$$

where

P_D = power delivered from the antenna
P_r = power available from the receiving antenna
p = polarization efficiency (or polarization mismatch factor), $0 \leq p \leq 1$
q = impedance mismatch factor, $0 \leq q \leq 1$
A_e = effective aperture (area)

An overall efficiency, or *total efficiency* $\varepsilon_{\text{total}}$, can be defined that includes the effects of polarization and impedance mismatch:

$$\varepsilon_{\text{total}} = pq\varepsilon_{ap} \tag{2-97}$$

Then $P_D = \varepsilon_{\text{total}} P_r$. It is convenient to express (2-96) in dB form:

$$P_D(\text{dBm}) = 10 \log p + 10 \log q + P_r(\text{dBm}) \tag{2-98}$$

where the unit dBm is power in decibels above a milliwatt; for example, 30 dBm is 1 W. Both powers could also be expressed in units of decibels above a watt, dBW. The power transmission formula (2-94) can also be expressed in dB form as

$$\begin{aligned} P_r(\text{dBm}) = P_t(\text{dBm}) + G_t(\text{dB}) + G_r(\text{dB}) - 20 \log R(\text{km}) \\ - 20 \log f(\text{MHz}) - 32.44 \end{aligned} \tag{2-99}$$

where $G_t(\text{dB})$ and $G_r(\text{dB})$ are the transmit and receive antenna gains in decibels, $R(\text{km})$ is the distance between the transmitter and receiver in kilometers, and $f(\text{MHz})$ is the frequency in megahertz.

EIRP. A frequently used concept in communication systems is that of *effective (or equivalent) isotropically radiated power, EIRP*. It is formally defined as the power gain of a transmitting antenna in a given direction multiplied by the net power accepted by the antenna from the connected transmitter. ERP, effective radiated power, is similar to EIRP but with antenna gain relative to that of a half-wave dipole instead of relative to an isotropic antenna. As an example of EIRP, suppose an observer is located in the direction of maximum radiation from a transmitting antenna with input power P_t; then

$$\text{EIRP} = P_t G_t \tag{2-100}$$

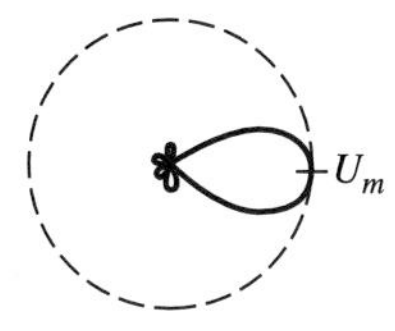

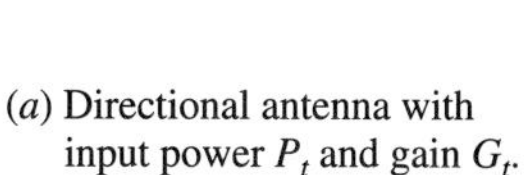

(*a*) Directional antenna with input power P_t and gain G_t.

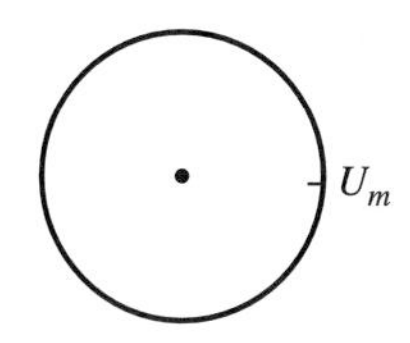

(*b*) Isotropic antenna with input power $P_t G_t$ and unity gain.

Figure 2-19 Illustration of effective isotropically radiated power, EIRP. In both (*a*) and (*b*), EIRP $= 4\pi U_m$.

The radiation intensity there is U_m as illustrated in Fig. 2-19*a* and $G_t = 4\pi U_m / P_t$, so

$$\text{EIRP} = P_t \frac{4\pi U_m}{P_t} = 4\pi U_m \tag{2-101}$$

The same radiation intensity could be obtained from a lossless isotropic antenna (with power gain $G_i = 1$) if it had an input power P_{in} equal to $P_t G_t$ as illustrated in Fig. 2-19*b*. In other words, to obtain the same radiation intensity produced by the directional antenna in its pattern maximum direction, an isotropic antenna would have to have an input power G_t times greater. Effective isotropically radiated power is a parameter used in the broadcast industry. FM radio stations often mention their effective radiated power when they sign off at night.

EXAMPLE 2-3 *Direct Broadcast Satellite Reception*

Reception of high-quality television channels at home with an inexpensive, small terminal is the result of technology development, including new antenna designs. The typical system transmits from 12.2 to 12.7 GHz with 120 W of power and an EIRP of about 55 dBW in each 24-MHz transponder that handles several compressed digital video channels. The receiving system uses a 0.46-m (18-in.) diameter offset fed reflector antenna with $\varepsilon_{ap} = 0.7$. In this example, we perform the system calculations using the following link parameter values:

$$f = 12.45 \text{ GHz} \quad \text{(midband)}$$

$$P_t(\text{dBW}) = 20.8 \text{ dBW} \quad (120 \text{ W})$$

$$G_t(\text{dB}) = \text{EIRP(dBW)} - P_t(\text{dBW}) = 55 - 20.8 = 34.2 \text{ dB}$$

$$R = 38{,}000 \text{ km} \quad \text{(typical slant path length)}$$

$$G_r = \frac{4\pi}{\lambda^2} \varepsilon_{ap} A_p = \frac{4\pi}{(0.024)^2} 0.7 \left(\pi \frac{(0.46)^2}{4} \right) = 2538$$

$$= 34 \text{ dB} \quad (70\% \text{ aperture efficiency})$$

The received power from (2-99) is

$$\begin{aligned} P_r(\text{dBW}) &= P_t(\text{dBW}) + G_t(\text{dB}) + G_r(\text{dB}) - 20 \log R(\text{km}) \\ &\quad - 20 \log f(\text{MHz}) - 32.44 \\ &= 20.8 + 34.2 + 34 - 20 \log(38{,}000) - 20 \log(12{,}450) - 32.44 \\ &= 20.8 + 34.2 + 34 - 91.6 - 81.9 - 32.4 \\ &= -116.9 \text{ dBW} \end{aligned} \tag{2-102}$$

This is 2×10^{-12} W! Without the high gains of the antennas (68 dB combined), this signal would be hopelessly lost in noise. This example is revisited in Sec. 9.2 for noise calculations.

2.6 PRACTICAL CONSIDERATIONS FOR ELECTRICALLY SMALL ANTENNAS

In this chapter, we examined several simple but basic radiators. Some of these were electrically small radiators, whereas one (the half-wave dipole) was of resonant size. We will say more about resonant antennas in Chap. 5, but not much more will be said about electrically small antennas. Thus, it is appropriate at this point to consider the practical limitations of electrically small antennas.

An electrically small antenna is one that is smaller than a radiansphere (see Sec. 1.6). It is characterized by a radiation resistance that is much less than its reactance and by a far-field pattern that is independent of the antenna size. An electrically small antenna behaves like a simple electric and/or magnetic dipole. The electric dipole is physically realizable, whereas the magnetic dipole is simulated by a current loop. Although the radiation pattern and the directivity of an electrically small antenna are independent of size or frequency, the radiation resistance and especially the reactance are not. This makes it difficult to transfer power from the antenna to a load or from a generator to the antenna as the frequency changes. An antenna with this characteristic has a high Q, where Q is defined as $2\pi f$ times the peak energy stored/average power radiated. Practically speaking, high Q means that the input impedance is very sensitive to small changes in frequency. An electrically small antenna is well approximated by a lumped resonant circuit where impedance bandwidth $\approx 1/Q$.

An analysis by McLean [6], based on the fields surrounding a small radiator, predicts the minimum Q that can be achieved for electrically small antennas in relation to the volume of the smallest sphere that can enclose the antenna. The fields of an ideal dipole (electric or magnetic) have the smallest Q of all possible antenna types. For either type of dipole just enclosed by a sphere of radius a, the Q is given by

$$Q = \left[\frac{1}{\beta^3 a^3} + \frac{1}{\beta a}\right] e_r \qquad \text{(2-103)}$$

This was obtained from McLean's lossless result by multiplying Q by the radiation efficiency e_r. Equation (2-103) is plotted in Fig. 2-20 for several values of the radiation efficiency. It is interesting to note that the curves vary as $1/(\beta a)^3$ for small a and the near fields of an ideal dipole vary as $1/(\beta r)^3$. A practical electrically small antenna will have a Q greater than that in Fig. 2-20 since the curves represent a fundamental limit that can be approached but not exceeded. The increasing Q with diminishing size implies a fundamental limitation on the usable bandwidth of an electrically small antenna. The concept of bandwidth will be considered more formally in Chap. 5 and 6, but here it is the frequency range over which the antenna is usable without retuning to a resonant condition (i.e., tuning out the reactance). Thus, high Q and small bandwidth are characteristic limitations of small antennas.

In addition to having high Q and narrow bandwidth, electrically small antennas tend to be superdirective. By *superdirectivity*, we mean a directivity that is greater than normal for an antenna of a given electrical size. For antennas greater in size than a wavelength, it will be shown in later material that directivity is directly proportional to L/λ for a linear antenna of length L and to A_p/λ^2 for an aperture antenna of area A_p. This proportionality breaks down for electrically small antennas since the directivity cannot go to zero (in violation of the directivity definition) as

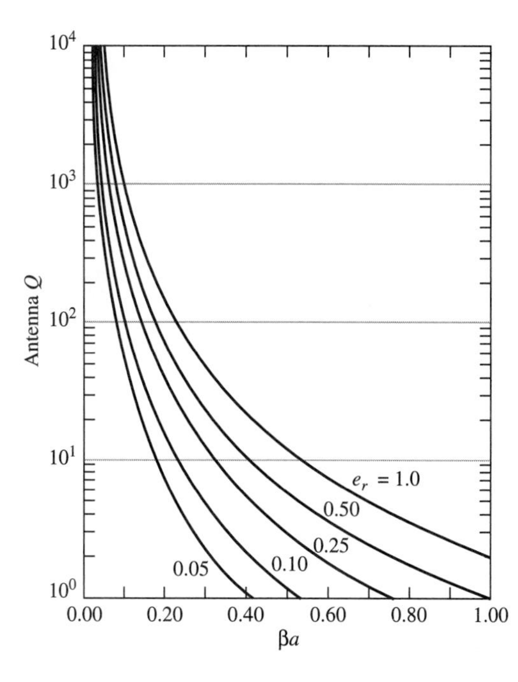

Figure 2-20 Q of an ideal antenna as a function of βa with radiation efficiency e_r as a parameter, where a is the minimum radius of a sphere enclosing the antenna.

the antenna size approaches zero, but instead is constant at 1.5 independent of the actual (small) size.

What is taking place as the antenna becomes electrically smaller is a sharp increase in the stored energy quite close to the antenna, while the directivity of the antenna remains constant. This can be interpreted as radiating energy into what is termed the invisible region where, for the linear radiator, for example, θ can be outside the range of 0 to π in the directivity calculation of (1-142). The more the pattern extends outside this range, the more superdirective the antenna becomes. Superdirectivity and Q are closely related. Supedirectivity, however, does not imply supergain. For example, an electrically small antenna with a radiation efficiency less than 0.667 will have a gain of less than unity, that is, less than 0 dB. Thus, another characteristic of electrically small antennas is that they have gains that are sensitive to size and frequency and that are less than unity. More will be said about superdirectivity in Chap. 4 and the concept of an invisible region will be illustrated further in Chap. 3.

REFERENCES

1. S. A. Schelkunoff and H. T. Friis, *ANTENNAS: Theory and Practice*, Wiley, New York, 1952, p. 5.
2. A. Fujimoto, A. Henderson, K. Hirasawa, and J. James, *Small Antennas*, Wiley, New York, 1987, Sec. 2.4.
3. A. F. Gangi, S. Sensiper, and G. R. Dunn, "The characteristics of electrically short, umbrella top-loaded antennas," *IEEE Trans. Ant. Prop.*, Vol. AP-13, pp. 864–871, Nov. 1965. Also see W. Weeks, *Antenna Engineering*, McGraw-Hill, New York, 1968, pp. 44–46.
4. W. J. Polydoroff, *High-Frequency Magnetic Materials*, Wiley, New York, 1960, Chap. 5.
5. R. C. Johnson, Ed., *Antenna Engineering Handbook*, 3rd ed., McGraw-Hill, New York, 1993, Ch. 5.
6. J. S. McLean, "A re-examination of the fundamental limits on the radiation Q of electrically small antennas," *IEEE Trans. Ant. Prop.*, Vol. AP-44, pp. 672–676, May 1996.

PROBLEMS

2.1-1 Use the oscillating charge model for an ideal dipole as shown in Fig. 2-2*b* to derive the electric field expressions of (1-71b). *Hints:* The far-field scalar potential function for this problem is

$$\Phi = \frac{q}{4\pi\varepsilon_o}\left[\frac{e^{-j\beta(r-(\Delta z/2)\cos\theta]}}{r-(\Delta z/2)\cos\theta} - \frac{e^{-j\beta(r+(\Delta z/2)\cos\theta]}}{r+(\Delta z/2)\cos\theta}\right]$$

where the parallel ray approximation was used and the $e^{j\omega t}$ time dependence was suppressed. Use $r \gg \Delta z$, $\lambda \gg \Delta z$, and $I = j\omega q$ to show that

$$\Phi \approx \frac{e^{-j\beta r}}{4\pi r^2}\frac{I\,\Delta z}{j\omega\varepsilon_o}(1 + j\beta r)\cos\theta$$

Then make use of (1-40).

2.1-2 The current density on an actual short dipole antenna of Fig. 2-1b can be written as

$$\mathbf{J} = \hat{\mathbf{z}}J_o \sin\left[\beta\left(\frac{\Delta z}{2} - |z|\right)\right]$$

Find an expression for the associated charge density.

2.1-3 Show that the capacitance of the capacitor of the capacitor-plate antenna of Fig. 2-3 is given by

$$C = \frac{\pi(\Delta r)^2\varepsilon_o}{\Delta z}$$

Assume that capacitance is entirely due to the end plates and neglect fringing.

2.1-4 a. Using the capacitance formula in Prob. 2.1-3, calculate the capacitive reactance of a capacitor-plate dipole for which $\Delta r = 0.01\lambda$ and $\Delta z = 0.02\lambda$.

b. Calculate the radiation resistance of this antenna.

2.2-1 Sketch the current distribution on a half-wave dipole for various instants during the time cycle of the current oscillation.

2.2-2 Show that the pattern factor for half-wave dipole in (2-7) is normalized to unity at $\theta = \pi/2$.

2.2-3 Calculate and plot the radiation pattern $F(\theta)$ for a half-wave dipole in (2-8) for $0 \le \theta \le 180°$. Plot in linear, polar form as shown in Fig. 2-5*b*.

2.2-4 Show that the ohmic resistance of a half-wave dipole from (1-178) is given by

$$R_{\text{ohmic}} = \frac{R_s}{2\pi a}\frac{\lambda}{4}$$

2.2-5 Use the results of Prob. 2.2-4 to calculate the radiation efficiency of a half-wave dipole at 100 MHz if it is made of aluminum wire 6.35 mm (0.25 in.) in diameter. Assume the radiation resistance to be 70 Ω.

2.3-1 Show that the image theory model of Fig. 2-9*b* for an ideal dipole parallel to a perfect ground plane yields zero tangential electric field along plane PP'.

2.3-2 For a thin monopole as shown in Fig. 2-10*a* that is a quarter wavelength long:

a. Rough sketch the radiation pattern in polar form as a function of θ, if the monopole is along the z-axis.

b. What is the directivity?

c. What is the input impedance?

2.4-1 Use (1-96) to derive the far-field distance expressions (2-48) for the small square loop.

2.4-2 Verify that the power radiated from a small loop is given by (2-56).

2.4-3 Show that (2-61) follows from (2-60).

2.4-4 Compute the radiation efficiency of a small single turn loop antenna at 1 MHz if it is made of No. 20 AWG copper wire and has a loop radius of 0.2 m.

2.4-5 Compute the inductance of the loop antenna in Prob. 2.4-4.

2.4-6 A single turn circular loop 15 cm in radius is made of 3-mm-diameter copper wire.

Calculate the radiation resistance, ohmic resistance, input impedance, and radiation efficiency at 1 MHz.

2.4-7 An AM broadcast receiver operating at 1 MHz uses a loop stick antenna with 500 turns of No. 30 copper wire wound on a core of ferrite with $\mu_{\text{eff}} = 38$ and a cross section that is 1 cm $\times$ 3 mm. Find the radiation resistance and the radiation efficiency, neglecting ferrite core losses.

2.4-8 A single turn square loop antenna that is 0.5 m on each side operates at 30 MHz. The wire is aluminum with a diameter of 2 cm. Compute (a) the radiation resistance, (b) the input impedance, and (c) the radiation efficiency.

2.5-1 Calculate the beam solid angle Ω_A for an ideal dipole in steradians (square radians) and square degrees. Use the fact that $A_{em} = 0.119\lambda^2$ for an ideal dipole.

2.5-2 A half-wave dipole has a directivity of 2.15 dB. Derive an expression for its maximum effective aperture in terms of wavelength squared.

2.5-3 Suppose a transmitting antenna produces a maximum far-zone electric field in a certain direction given by

$$E = 90I\frac{e^{-j\beta r}}{r}$$

where I is the peak value of the terminal current. The input resistance of the lossless antenna is 50 Ω. Find the maximum effective aperture of the antenna A_{em}. Your answer will be a number times wavelength squared.

2.5-4 A parabolic reflector antenna with a circular aperture of 3.66-m diameter has a 6.30 m^2 effective aperture area. Compute the gain in dB at 11.7 GHz.

2.5-5 The effective aperture of a 1.22-m-diameter parabolic reflector antenna is 55% of the physical aperture area. Compute the gain in dB at 20 GHz.

2.5-6 Compute the gain in dB of a 0.3-m circular diameter aperture antenna with 70% aperture efficiency at 5, 10, and 20 GHz. This problem approximates the performance of a small satellite earth terminal antenna over the range of commonly used frequencies and illustrates the frequency dependence of gain for a fixed aperture.

2.5-7 Derive the dB form of the power transmission equation (2-99) from (2-94).

2.5-8 Write a power transfer equation similar to (2-99) but with distance R in units of miles.

2.5-9 Calculate the received power in watts for the DBS system of Example 2-3 using (2-94).

2.5-10 A 150-MHz VHF transmitter delivers 20 W into an antenna with 10-dB gain. Compute the power in W available from a 3-dB gain receiving antenna 50 km away.

2.5-11 A low earth orbit (LEO) satellite system transmits 1 W at 1.62 GHz using a 29-dB gain antenna with spot beams directed toward users on the earth that are a maximum of 1500 km away. Find the required satellite transmit power in order for the power received by a user at the maximum distance to be at least -100 dBm if the user has a 1-dB gain antenna directed toward the satellite.

2.5-12 A cellular telephone base station transmitter at 850 MHz delivers 20 W into a 10-dB gain antenna. Compute the power in W available from a 3-dB gain mobile receiving antenna 20 km away.

2.5-13 This problem compares the performance of the wireless system of Prob. 2.5-10 to wire line systems using coaxial and fiber optic cables.

a. Compute the coaxial cable loss in dB for the 50 km distance using an attenuation of 0.1 dB/m.

b. What is the net loss for the wireless system of Prob. 2.5-10? That is, find the net loss between the transmit and receive antenna input ports.

c. Would repeater amplifiers be necessary in the cable system?

d. Repeat (a) and (b) for the case of a 500-m path length.

e. Repeat (a) and (b) for a 500-m path at 300 MHz for a cable attenuation of 0.14 dB/m. The antenna gains are the same.

f. Compute the loss in dB for distances of 50 km and 500 m of a fiber optic cable with an attenuation of 1 dB/km.

g. Tabulate numerical results.

2.5-14 The factor $(\lambda/4\pi R)^2$ in the communication link formula of (2-94) is often referred to as "free-space loss." It is the loss encountered in the free-space region between isotropic transmit and receive antennas. The frequency dependence of f^{-2} in this loss is not easily understood until the case of an isotropic transmit antenna and a receiving antenna with a fixed effective aperture is considered. Determine the frequency dependence for this link.
2.5-15 It is often stated that operating a communication link at a higher frequency permits the use of smaller-diameter antennas. To address this question in a specific way, suppose the operating frequency of a link is doubled. The transmit and receive antennas are of the same diameter and there are no changes in efficiencies, the propagation medium, or the transmit power. How much does received power increase or decrease after the frequency is doubled?
2.5-16 Derive a power transfer equation in a form involving the effective isotropically radiated power of the transmitter, the effective aperture of the receiving antenna, and free-space spreading loss $1/4\pi R^2$. Start with (2-95).
2.5-17 An FM broadcast radio station has a 2-dB gain antenna system and 100 kW of transmit power. Calculate the effective isotropically radiated power in kW.
2.6-1 (a) An electrically small antenna has a Q of 60. The smallest sphere that can enclose the antenna has a radius of 0.0159λ. Find the antenna gain in dB. (b) Repeat (a) if the antenna Q is 40 and the radius is 0.0318λ. (c)What is the directivity of these antennas in dB? (d) Explain what negative dB gain means.
2.6-2 Show that the radius of a radian sphere $\lambda/2\pi$ corresponds to the distance from an ideal dipole where the power density contained in the near field equals that in the far field in the direction of maximum radiation, $\theta_o = 90°$.

Chapter 3

Arrays

Several antennas can be arranged in space and interconnected to produce a directional radiation pattern. Such a configuration of multiple radiating elements is referred to as an **array antenna**, or simply, an **array**. Many small antennas can be used in an array to obtain a level of performance similar to that of a single large antenna. The mechanical problems associated with a single large antenna are traded for the electrical problems of feeding several small antennas. However, with the advancements in solid-state technology, it is possible to realize the feed network required for excitation with reasonable cost. Arrays offer the unique capability of electronic scanning of the main beam. By changing the phase of the exciting currents in each element antenna of the array, the radiation pattern can be scanned through space. The array is then called a **phased array**. Phased arrays have many applications, particularly in radar.

The concept of phased arrays was proposed in 1889, but the first successful array (a two-element receiving array) did not appear until about 1906. The introduction of shortwave radio equipment in the 1920s made possible the use of reasonably sized antenna arrays, providing a convenient way to achieve a directive radiation pattern for radio communications. Around the time of World War II, array antennas operating at VHF, UHF, and, later, microwave frequencies were introduced for use in radar systems. Today, arrays are used extensively.

Arrays are found in many geometrical configurations. The most elementary is that of a **linear array** in which the array element centers lie along a straight line. The elements in an array can form a **planar array**. A popular planar array is the rectangular array in which the element centers are contained within a rectangular area. A class of arrays that is still emerging is that of **conformal arrays**, where the array element locations conform to a nonplanar surface. This is a great advantage for arrays on the skin of a vehicle such as an airplane.

Arrays offer many advantages over aperture antennas. For example, the narrow main beam of a parabolic reflector antenna is scanned by slewing the entire structure, whereas arrays can be phase-scanned at the speed of the control electronics without moving the antenna. In addition, it is possible to track multiple targets with a phased array. As already mentioned, arrays can be conformed to surfaces. But, arrays also present challenges. The advantage of avoiding the mechanical difficulties associated with slewing a large aperture antenna is balanced by the complexity of the network required to feed the elements of an array. Additional concerns are bandwidth limitations and mutual coupling between the elements.

The field of antennas, similar to circuits, can be divided into digital and analog implementations. Continuous, electrically large antennas are the analog portion and must be analyzed using integrals that are often difficult to evaluate. Arrays form the digital portion and can be analyzed using simple summations. For this reason, we present arrays early in our antenna studies. But arrays are also an important area within the field of antennas. They also offer an opportunity to understand the relationship between the distribution of current in space and the resulting radiation using simple mathematics.

The radiation pattern of an array is determined by the type of individual elements used, their orientations, their positions in space, and the amplitude and phase of the currents feeding them. To simplify our discussion of arrays, we will begin by letting each element of the array be an isotropic point source. The resulting radiation pattern is called the *array factor*. In this chapter, the array factors for several simple arrays will be examined before considering general uniformly excited linear arrays. The principle of pattern multiplication introduced in Sec. 3.3 permits inclusion of the array element effect. Array directivity is discussed in Sec. 3.4. By controlling the current amplitudes in an array, the pattern can be shaped for special applications. The relationship between the radiation pattern of an array and its element current amplitudes is illustrated by several linear array examples in Sec. 3.5. The effects of mutual coupling between elements of a real array on impedance are detailed in Sec. 3.6. Multidimensional arrays are introduced in Sec. 3.7. Scanning of the array pattern by element phase control is discussed in Sec. 3.8.

3.1 THE ARRAY FACTOR FOR LINEAR ARRAYS

The fundamental configuration for elements in an array is the linear array shown in Fig. 3-1. Linear arrays are used widely in practice and their operating principles can be used to understand more complex array geometries. The array of Fig. 3-1 has identical elements and is operated as a receiving antenna. The pattern characteristics of an array can be explained for operation as a transmitter or receiver, whichever is more convenient, since antennas usually satisfy the conditions for reciprocity. The output of each element can be controlled in amplitude and phase as indicated by the attenuators and phase shifters in Fig. 3-1. As we shall see, amplitude and phase control provide for custom shaping of the radiation pattern and for scanning of the pattern in space.

The basic array antenna model consists of two parts, the pattern of one of the elements by itself, the *element pattern*, and the pattern of the array with the actual

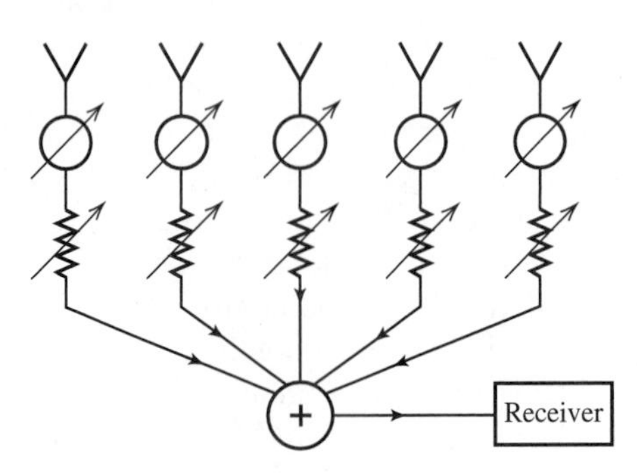

Figure 3-1 A typical linear array. The symbols ⌀ and (attenuator symbol) indicate variable phase shifters and attenuators. The output currents are summed before entering the receiver.

elements replaced by isotropic point sources, the array factor. The total pattern of the array is then the product of the element pattern and array factor; this will be discussed in detail in Sec. 3.3. We treat the array factor first.

The array factor corresponding to the linear array of Fig. 3-1 is found by replacing each element by isotropic radiators, but retaining the element locations and excitations as shown in Fig. 3-2. The array is receiving a plane wave arriving at an angle θ from the line of the elements and the planes of equal phase (i.e., wavefronts) are shown. Rays perpendicular to the wavefronts indicate the direction of travel of the wave. With the reference wavefront taken to be of zero phase, the distance to the nth element has a corresponding phase delay (found by multiplying by β) of ξ_n. That is, each element is excited with phase ξ_n, due to the spatial phase delay effect of the incoming plane wave. The amplitudes of excitation are constant, taken to be unity, because a plane wave has uniform amplitude. The resulting excitations of $1e^{j\xi_0}$, $1e^{j\xi_1}$, . . . are shown for each element in Fig. 3-2. The elements themselves do not weight the outputs since they are isotropic radiators that respond equally to all incoming wave directions.

An *isotropic radiator* is a hypothetical, lossless antenna occupying a point in space and when transmitting radiates uniformly in all directions; see Fig. 1-19*a*. It is sometimes referred to as a *point source*. The radiation fields of an isotropic radiator at the origin of a spherical coordinate system are proportional to

$$I_o = \frac{e^{-j\beta r}}{4\pi r} \tag{3-1}$$

where I_o is the current of the point source. This can be seen by examining the radiation field expressions (1-72) for an ideal dipole and dropping the angular dependence. The far-field pattern is obtained from the angular dependence (i.e., at

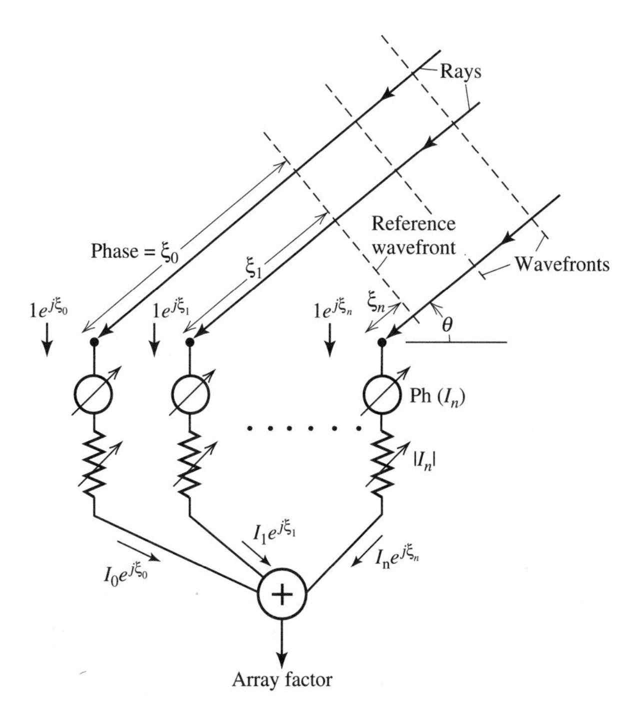

Figure 3-2 Equivalent configuration of the array in Fig. 3-1 for determining the array factor. The elements of the array are replaced by isotropic point sources.

constant r) of the radiation fields; thus, the pattern of a point source, from (3-1), is constant and is given by

$$\text{AF} = I_o \tag{3-2}$$

where AF is the array factor for this "array" of only one point source. Since I_o is constant, the array factor in (3-2) could have been written as unity, but as elements are added to the array, each with a different current, it is necessary to account for their relative field strengths as determined by their element currents.

The array factor for the array of Fig. 3-1 is found from the array of Fig. 3-2 that has isotropic radiators for array elements in place of the actual elements. The array factor for this receiving array is then the sum of the isotropic radiator receiving antenna responses $\{e^{j\xi_0}, e^{j\xi_1}, \ldots\}$ weighted by the amplitude and phase shift $\{I_0, I_1, \ldots\}$ introduced in the transmission path connected to each element. The array factor of the array shown in Fig. 3-2 is thus

$$\text{AF} = I_0 e^{j\xi_0} + I_1 e^{j\xi_1} + I_2 e^{j\xi_2} + \cdots \tag{3-3}$$

where $\xi_0, \xi_1, \ldots$ are the phases of an incoming plane wave at the element locations designated 0, 1, ... For convenience, these phases are usually relative to the coordinate origin; that is, the phase of the wave arriving at the nth element leads the phase of the wave arriving at the origin by ξ_n.

The expression in (3-3) is very general and can be applied to any geometry. However, instead of proceeding with a general form for the array factor, it is much more instructive to consider simple geometries in order to understand the basic behavior of arrays. This is accomplished through examples of two-element arrays of various spacings and phasings. Treating these examples from the transmitting point of view permits inference of the general pattern features by inspection.

EXAMPLE 3-1 *Two Isotropic Point Sources with Identical Amplitude and Phase Currents, and Spaced One-Half Wavelength Apart (Fig. 3-3)*

Figure 3-3*a* shows how the pattern of this example can be approximated by inspection. At points in the far field along the perpendicular bisector of the line joining the point sources (x-axis), path lengths from each point source are equal. Since the amplitudes and phases of each source are also equal, the waves arrive in phase and equal in amplitude in the far field along the x-axis. Thus, the total field is double that for one source. The situation is different along the axis of the array (z-axis). If we look to the right along the $+z$-axis, waves coming from the left source must travel one-half wavelength before reaching the source on the right. This amounts to a 180° phase lag. The waves then continue traveling to the right along the $+z$-axis and maintain this same phase relationship on out to the far field. Thus, in the far field, waves from the two sources traveling in the $+z$-direction arrive 180° out-of-phase (due to the one-half wavelength separation of the sources) and are equal in amplitude (since the sources are). Therefore, there is a perfect cancellation and the total field is zero. The same reasoning can be used to see the effect in the $-z$-direction. The total pattern has a relative value of 2 in the $\pm x$-directions, 0 in the $\pm z$-directions, and a smooth variation in between (because the phase difference between waves from two sources changes smoothly from 0 to 180° as the observer moves from the broadside direction to the axial direction along a constant radius from the array center). This pattern is sketched in Fig. 3-3*b*. The pattern in three dimensions can be imagined by holding the z-axis in your fingertips and spinning the pattern shown to sweep out the total pattern. The three-dimensional pattern in Fig. 3-3*c* is a doughnut-type pattern similar to that for an ideal dipole.

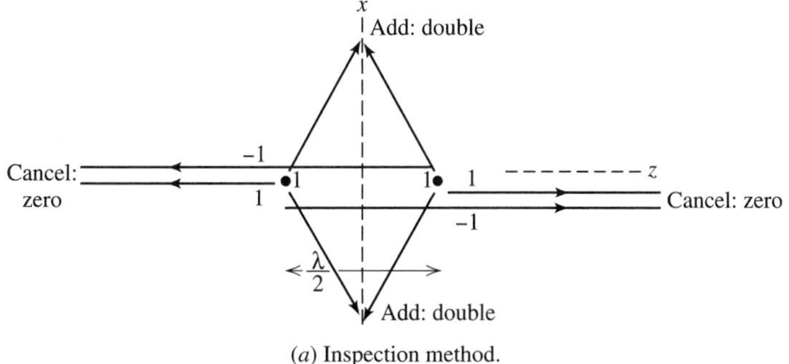

(*a*) Inspection method.

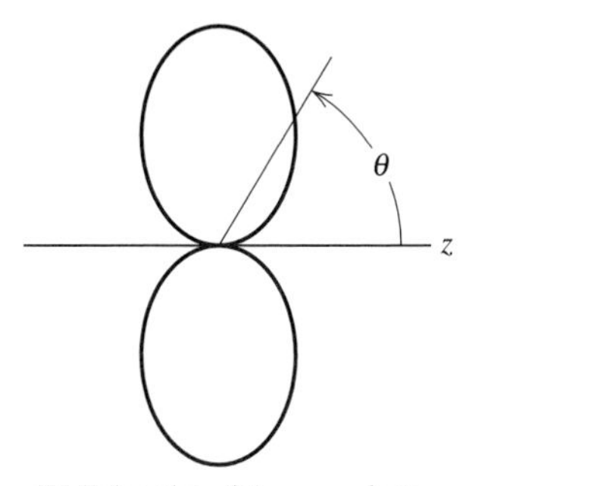

(*b*) Polar plot of the array factor $f(\theta) = \cos[(\pi/2)\cos\theta]$.

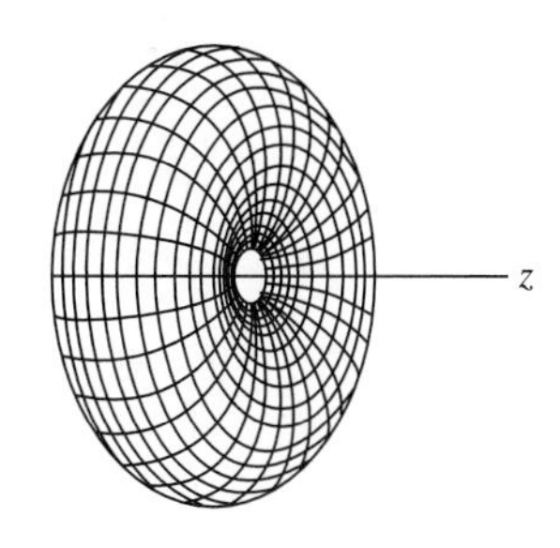

(*c*) 3D polar pattern.

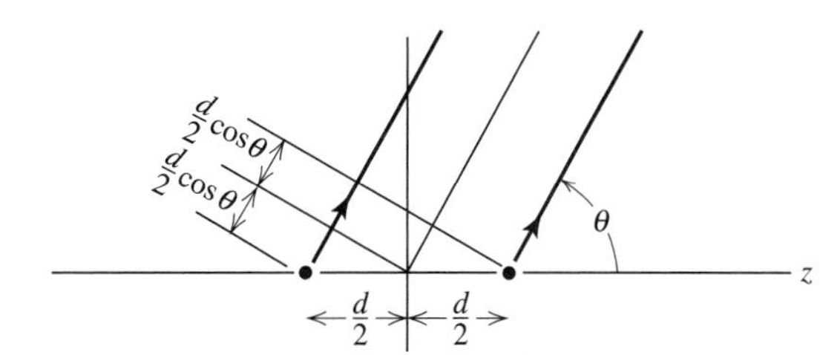

(*d*) Geometry for pattern calculation using rays.

Figure 3-3 Two isotropic point sources with identical amplitude and phase currents, and spaced one-half wavelength apart (Example 3-1).

We can also calculate the array factor exactly. If we use phases corresponding to the path length differences shown in Fig. 3-3*d* in (3-3), the array factor is

$$\mathrm{AF} = 1e^{-j\beta(d/2)\cos\theta} + 1e^{j\beta(d/2)\cos\theta} = 2\cos\left(\beta\frac{d}{2}\cos\theta\right) \tag{3-4}$$

The distance between the elements is $d = \lambda/2$, so $\beta d/2 = \pi/2$ and (3-4) becomes

$$\mathrm{AF} = 2\cos\left(\frac{\pi}{2}\cos\theta\right) \tag{3-5}$$

Normalizing the array factor for a maximum value of unity gives

$$f(\theta) = \cos\left(\frac{\pi}{2}\cos\theta\right) \tag{3-6}$$

This is maximum for $\theta = \pi/2$ since $\cos[(\pi/2)\cdot 0] = 1$ and zero for $\theta = 0$ since $\cos[(\pi/2)\cdot 1] = 0$. This result agrees with the inspection method that leads to Fig. 3-3*b*.

EXAMPLE 3-2 ***Two Isotropic Point Sources with Identical Amplitudes and Opposite Phases, and Spaced One-Half Wavelength Apart (Fig. 3-4)***

If we consider the array to be transmitting, the gross features of the pattern can be determined by inspection as shown in Fig. 3-4*a*. The path lengths from each point source to a point on the x-axis are the same. But the left source is 180° out-of-phase with respect to the right source; thus, waves arriving at points on the x-axis are 180° out-of-phase and equal in amplitude, giving a zero field. Along the z-axis (in both directions), the 180° phase difference in the currents is compensated for by the half-wavelength path difference between waves from the two sources. For example, in the $+z$-direction the waves from the left source arrive at the location of the right source, lagging the phase of waves from the right source by 360° (180° from the distance traveled and 180° from the excitation lag). This is an in-phase condition and thus the waves add in the far field, giving a relative maximum. From these few pattern values, the entire pattern can be sketched, yielding a plot similar to that of Fig. 3-4*b*. The three-dimensional polar plot of the pattern shown in Fig. 3-4*c* has the shape of a dumbbell.

We calculate the array factor exactly using (3-3) and Fig. 3-3*d* as

$$\mathrm{AF} = -1e^{-j\beta(d/2)\cos\theta} + 1e^{j\beta(d/2)\cos\theta} = 2j\sin\left(\beta\frac{d}{2}\cos\theta\right) \tag{3-7}$$

Using $d = \lambda/2$ and normalizing, we have

$$f(\theta) = \sin\left(\frac{\pi}{2}\cos\theta\right) \tag{3-8}$$

Plotting this pattern, we obtain the same result as with the inspection method (see Fig. 3-4*b*).

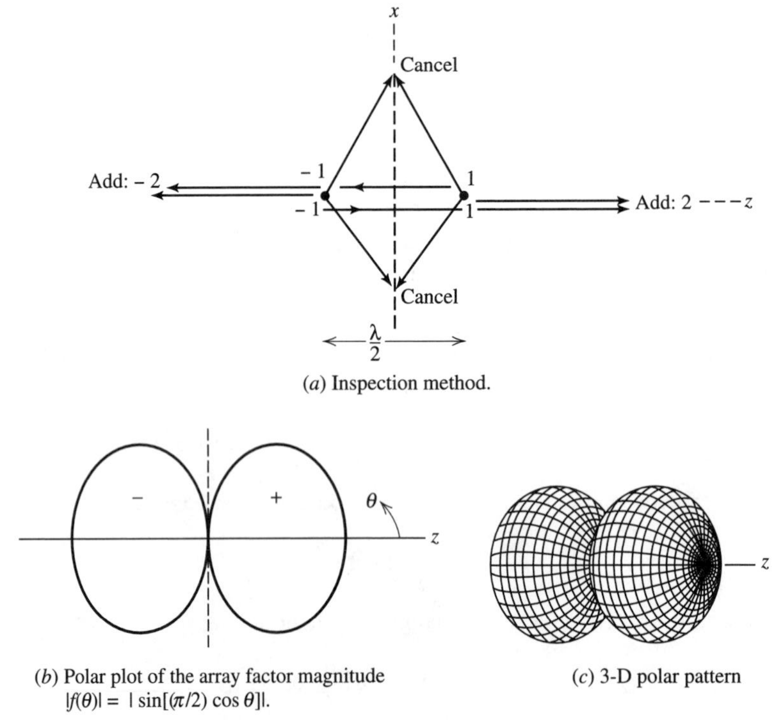

(*a*) Inspection method.

(*b*) Polar plot of the array factor magnitude $|f(\theta)| = |\sin[(\pi/2)\cos\theta]|$.

(*c*) 3-D polar pattern

Figure 3-4 Two isotropic point sources with identical amplitudes and opposite phases, and spaced one-half wavelength apart (Example 3-2).

EXAMPLE 3-3 *Two Isotropic Point Sources with Identical Amplitudes and 90° Out-of-Phase, and Spaced a Quarter-Wavelength Apart (Fig. 3-5)*

Waves leaving the left source of the transmitting array in Fig. 3-5 and traveling in the $+z$-direction arrive at the right source delayed by 90° due to the quarter-wavelength path. But the excitation of the right source lags the left source by 90° so waves in the $+z$-direction are in step and add in the far field. For waves leaving the right-hand source and traveling in the $-z$-direction, the phase at the location of the left source is 180° with respect to the wave from the left source (90° from the path difference and 90° from the excitation). See Fig. 3-5*b*. At angles between $\theta = 0°$ ($+z$-direction) and 180° ($-z$-direction), there is a smooth pattern variation from 2 (perfect addition) to 0 (perfect cancellation). This pattern is shown in Fig. 3-5*c* and is the so-called *cardioid pattern*. It is used frequently in the area of acoustics for microphone patterns. The response is strong in the direction of the microphone input and weak in the direction where the speakers are aimed to reduce feedback.

Using these excitations in (3-3) and Fig. 3-3*d*, we can calculate the array factor expression as follows:

$$\begin{aligned}\text{AF} &= 1e^{-j\beta(d/2)\cos\theta} + 1e^{-j(\pi/2)}e^{j\beta(d/2)\cos\theta}\\ &= e^{-j(\pi/4)}[e^{-j[\beta(d/2)\cos\theta-\pi/4]} + e^{j[\beta(d/2)\cos\theta-\pi/4)}]\\ &= e^{-j(\pi/4)}\, 2\cos\left(\frac{\beta d}{2}\cos\theta - \frac{\pi}{4}\right)\end{aligned} \tag{3-9}$$

Substituting $d = \lambda/4$ and normalizing give

$$f(\theta) = \cos\left[\frac{\pi}{4}(\cos\theta - 1)\right] \tag{3-10}$$

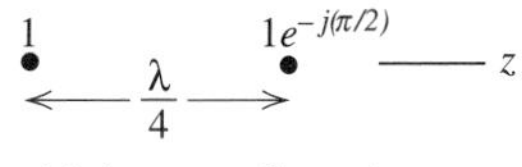

(*a*) Array configuration.

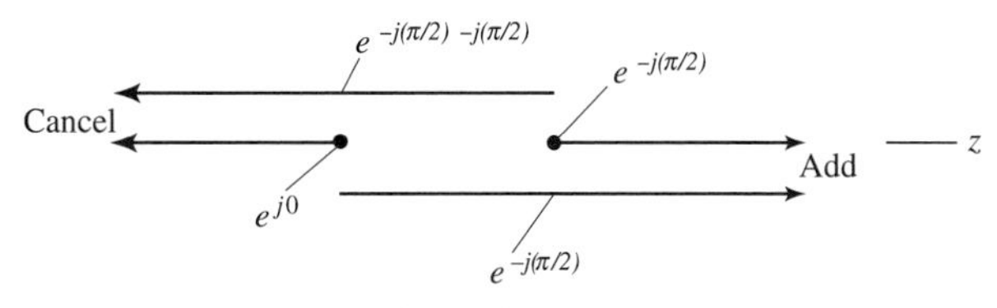

(*b*) Inspection method.

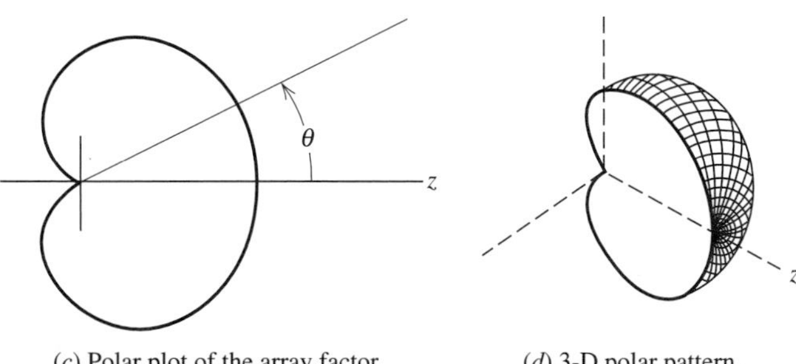

(*c*) Polar plot of the array factor $f(\theta) = \cos[(\pi/4)(\cos\theta - 1)]$.

(*d*) 3-D polar pattern.

Figure 3-5 Two isotropic point sources with identical amplitudes and the right element lagging the left by 90°, and spaced a quarter-wavelength apart (Example 3-3). This pattern shape is called a cardioid pattern.

This function has a maximum value of unity for $\theta = 0°$, $1/\sqrt{2}$ for $\theta = 90°$, and 0 for $\theta = 180°$. This agrees with the pattern of Fig. 3-5*c* obtained by inspection.

EXAMPLE 3-4 *Two Identical Isotropic Point Sources Spaced One Wavelength Apart (Fig. 3-6)*

Since the currents are in-phase, the fields of each element add perfectly (i.e., double) in the $\pm x$-directions. Also, since the phase lag of the field from one element is 360° (one wavelength additional path length) with respect to the other, their effects add perfectly in the far field in the $\pm z$-directions. However, with the one-wavelength spacing there are directions of perfect cancellation as indicated in Fig. 3-6*b*. To determine these directions, we reason as follows. For perfect cancellation, the waves from the two sources must be 180° out-of-phase. This means a path length difference of one-half wavelength. Since the path length difference as a function of θ is $\lambda \cos \theta$ (see Fig. 3-3*d*), we solve for the values of θ such that

$$\lambda \cos \theta = \pm\frac{\lambda}{2} \quad \text{or} \quad \cos \theta = \pm\frac{1}{2} \tag{3-11}$$

The solutions are 60° and 120°. By filling in smooth variations between the maxima and zeros indicated in Fig. 3-6*b*, the pattern of Fig. 3-6*c* results.

The exact array factor calculation parallels that of Example 3-1 except that with $d = \lambda$ in (3-4),

$$\text{AF} = 2 \cos\left(\beta \frac{d}{2} \cos \theta\right) = 2 \cos(\pi \cos \theta) \tag{3-12}$$

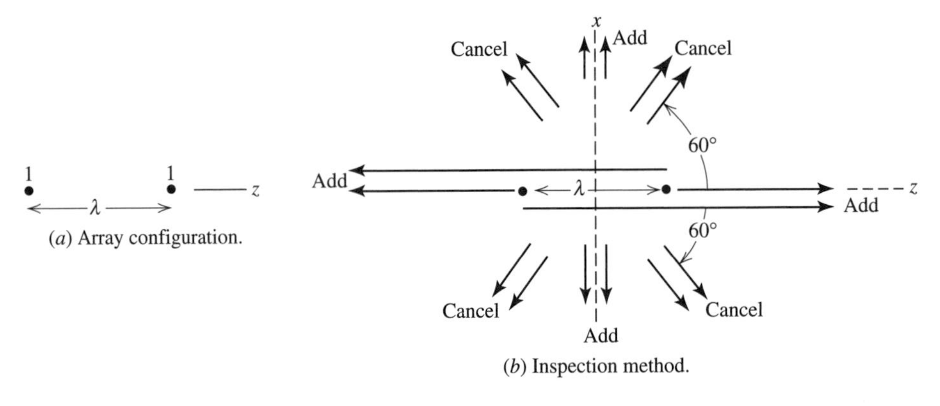

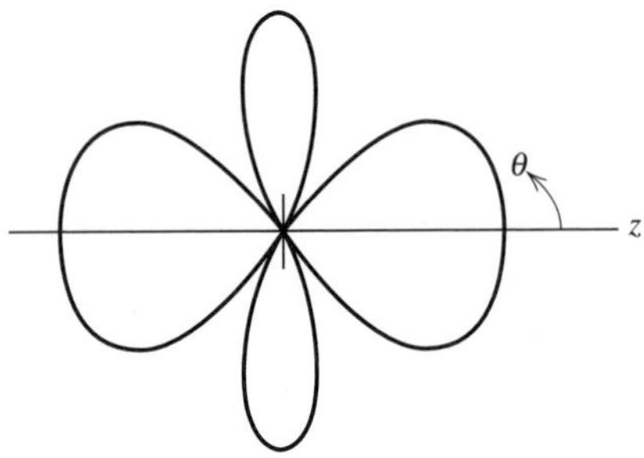

(*c*) Polar plot of array factor magnitude $|f(\theta)| = |\cos(\pi \cos \theta)|$.

Figure 3-6 Two isotropic point sources with identical amplitude and phase currents, and spaced one wavelength apart (Example 3-4).

The normalized array factor is

$$f(\theta) = \cos(\pi \cos \theta) \tag{3-13}$$

Note that $|f(\theta)|$ has a maximum value of unity for $\theta = 0°$, $90°$, and $180°$ and is zero for $\theta = 60°$ and $120°$. These are the same results we obtained by inspection in Fig. 3-6*c*. This example illustrates the fact that multiple lobes will appear for spacings greater than a half-wavelength.

The inspection method is difficult to use in all but the simplest arrays. Therefore, we examine the general array factor given by (3-3) for the case of equally spaced arrays. We will study its properties and develop a method to obtain a quick sketch of the radiation pattern.

Suppose we have a linear array of several elements. If the elements are equally spaced as shown in Fig. 3-7, the array factor expression (3-3) can be simplified. The angle θ is that of an incoming plane wave relative to the axis of the receiving array. The isotropic sources respond equally in all directions, but when their outputs are added together (each weighted according to I_n), a directional response is obtained. The phase of the wave arriving at the origin is set arbitrarily to zero, so $\xi_0 = 0$. The incoming waves at element 1 arrive before those at the origin since the distance is shorter by an amount $d \cos \theta$. The corresponding phase lead of waves at element 1 relative to those at 0 is $\xi_1 = \beta d \cos \theta$. This process continues and (3-3) becomes

$$\mathrm{AF} = I_0 + I_1 e^{j\beta d \cos \theta} + I_2 e^{j\beta 2d \cos \theta} + \cdots = \sum_{n=0}^{N-1} I_n e^{j\beta n d \cos \theta} \tag{3-14}$$

Now consider the array to be transmitting. If the current has a linear phase progression (i.e., relative phase between adjacent elements is the same), we can separate the phase explicitly as

$$I_n = A_n e^{jn\alpha} \tag{3-15}$$

where the $n + 1^{\text{th}}$ element leads the nth element in phase by α. Then (3-14) becomes

$$\mathrm{AF} = \sum_{n=0}^{N-1} A_n e^{jn(\beta d \cos \theta + \alpha)} \tag{3-16}$$

Define

$$\psi = \beta\, d \cos \theta + \alpha \tag{3-17}$$

Then

$$\mathrm{AF} = \sum_{n=0}^{N-1} A_n e^{jn\psi} \tag{3-18}$$

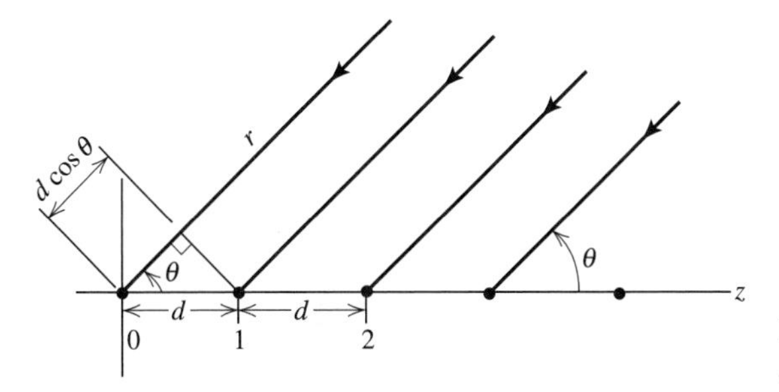

Figure 3-7 Equally spaced linear array of isotropic point sources.

This array factor is a function of ψ and is a Fourier series. This form is convenient for calculations, but we usually want field plots in terms of the polar angle θ.

The nonlinear transformation from ψ to θ given by (3-17) can be accomplished graphically. For example, consider two elements spaced one-half wavelength apart and with identical currents as in Example 3-1. We found the normalized array factor in (3-6) to be $f(\theta) = \cos[(\pi/2)\cos\theta]$. In this case, ψ from (3-17) is

$$\psi = \beta d \cos\theta + \alpha = \pi \cos\theta \tag{3-19}$$

since $d = \lambda/2$ and $\alpha = 0$. Now f is expressed in terms of ψ as

$$f(\psi) = \cos\frac{\psi}{2} \tag{3-20}$$

This is a rather simple function to plot. To obtain a plot of $|f|$ as a function of θ, first plot $|f(\psi)|$ from (3-20) as shown in Fig. 3-8. Then draw a circle of radius $\psi = \pi$ below it as shown, since (3-19) is a polar equation of a circle. For an arbitrary value of ψ, say, ψ_1, drop a line straight down until it intersects the circle. The values of $\theta = \theta_1$ and $|f| = f_1$ corresponding to $\psi = \psi_1$ are indicated in the figure. Locating several points taken in this fashion produces the desired sketch. Note that as θ ranges from 0 to π, ψ goes from π to $-\pi$ in this case. The resulting polar plot is shown in Fig. 3-10b. It is the same as the result obtained using inspection in Fig. 3-3.

Before proceeding with more specific examples, we consider a general array factor and how a polar pattern is obtained from it. The magnitude of a typical array factor is plotted as a function of ψ in Fig. 3-9. Below it a circle is constructed with a radius equal to βd and its center located at $\psi = \alpha$. The angle θ is as shown. It is very simple

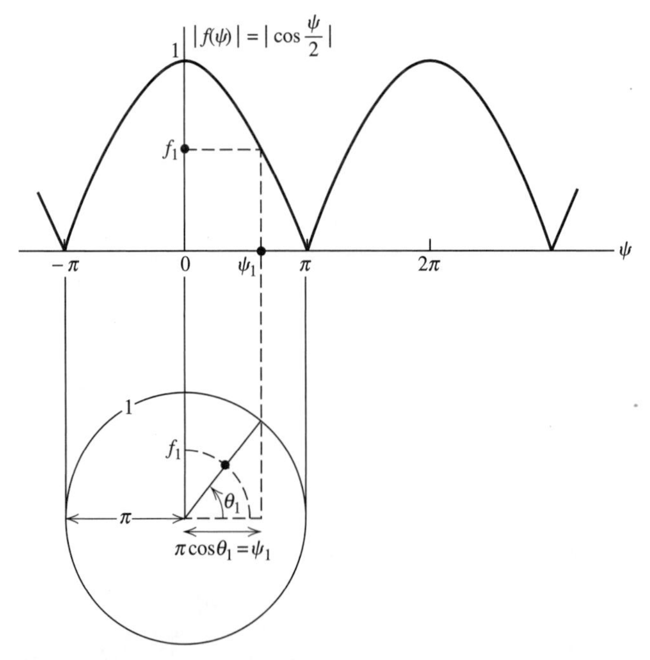

Figure 3-8 Procedure for obtaining the polar plot of the array factor of two elements spaced one-half wavelength apart with identical currents.

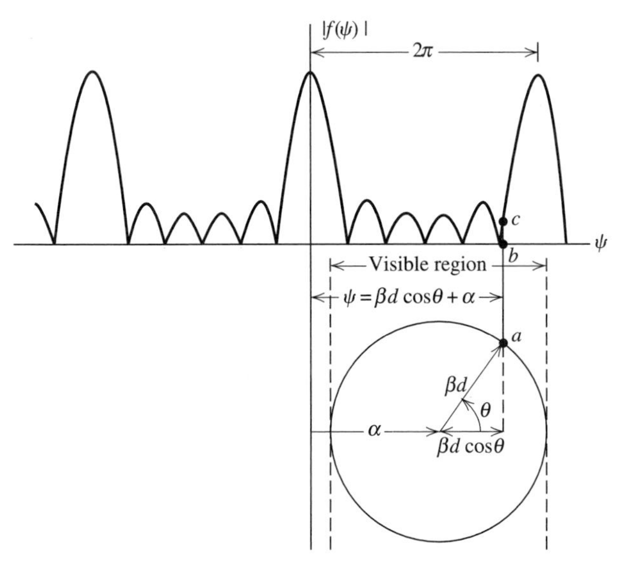

Figure 3-9 Construction technique for finding the array factor as a function of polar angle θ.

to use this plot. For a given value of θ, locate the intersection of a radial line from the origin of the circle and the perimeter, point a. The corresponding value of ψ, at point b, is on a vertical line from a. The array factor value corresponding to these values of ψ and θ is then point c, also on the vertical line from a. Notice that the distance from the $\psi = 0$ axis to a point, say, at a, can be written as $\psi = \alpha + \beta d \cos \theta$, which is (3-17).

To illustrate the procedure further, we will find the polar plots of the array factors for some two-element arrays with uniform current amplitudes that were discussed earlier in this section. The array factor as a function of ψ, from (3-18) with $N = 2$, is

$$\text{AF} = 1 + e^{j\psi} = e^{j(\psi/2)}(e^{-j(\psi/2)} + e^{j(\psi/2)}) = 2e^{j(\psi/2)} \cos \frac{\psi}{2} \tag{3-21}$$

where $A_0 = A_1 = 1$. Taking the magnitude eliminates the exponential factor and normalization removes the factor of 2, giving

$$|f(\psi)| = \left| \cos \frac{\psi}{2} \right| \tag{3-22}$$

which also follows from (3-20). The array factor $|f(\psi)|$ is the same for all two-element arrays with the same current amplitudes and is plotted in Fig. 3-10a. Of course, ψ changes with element spacing and phasing. For example, if the spacing is a half-wavelength and the phases of each element are zero ($\alpha = 0$), the pattern is obtained as shown in Fig. 3-8 with the resulting pattern plotted in Fig. 3-10b. This is Example 3-1 discussed earlier. For Example 3-2, $d = \lambda/2$ and $\alpha = \pi$. The resulting polar plot of the array factor using the procedures of Fig. 3-9 is shown in Fig. 3-10c. The array factor for Example 3-3 with $d = \lambda/4$ and $\alpha = -\pi/2$ is shown in Fig. 3-10d.

By examining the general array factor expression in (3-18), some general prop-

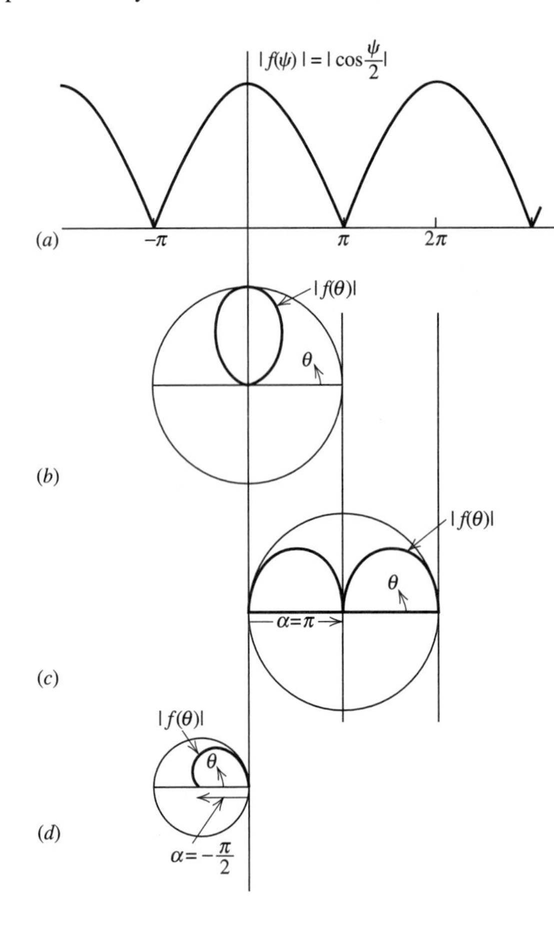

Figure 3-10 Array factors for two-element arrays with equal amplitude currents. (*a*) Universal array factor. (*b*) Polar plot for $d = \lambda/2$, $\beta d = \pi$, $\alpha = 0$ (Example 3-1). (*c*) Polar plot for $d = \lambda/2$, $\beta d = \pi$, $\alpha = \pi$ (Example 3-2). (*d*) Polar plot $d = \lambda/4$, $\beta d = \pi/2$, $\alpha = -\pi/2$ (Example 3-3).

erties can be derived that aid in the construction of pattern plots. First, the array factor is periodic in the variable ψ with period 2π. This is easily shown as follows:

$$\text{AF}(\psi + 2\pi) = \sum A_n e^{jn(\psi+2\pi)} = \sum A_n e^{jn\psi} e^{jn2\pi} = \sum A_n e^{jn\psi} = \text{AF}(\psi) \quad (3\text{-}23)$$

The array factor of a linear array along the z-axis is a function of θ but not of ϕ (the element pattern may be though). In other words, the array factor is a pattern that has rotational symmetry about the line of the array. Therefore, its complete structure is determined by its values for

$$0 < \theta < \pi \quad (3\text{-}24)$$

This is called the *visible region*. This corresponds to $-1 < \cos\theta < 1$ or $-\beta d < \beta d \cos\theta < \beta d$ or

$$\alpha - \beta d < \psi < \alpha + \beta d \quad (3\text{-}25)$$

Hence, the visible region in terms of θ and ψ is given by (3-24) and (3-25), respectively. The element spacing of the array in terms of a wavelength, d/λ, determines the size of the circle in Fig. 3-9 and thus how much of the array factor appears in the visible region. The visible region in the variable ψ is of length $2\beta d$, as seen from (3-25). This is the diameter of the circle in Fig. 3-9. Suppose that exactly one period appears in the visible region. Since the period is 2π, we have $2\pi = 2\beta d = 2(2\pi/\lambda)\,d$ or $d/\lambda = \frac{1}{2}$. Thus, *exactly one period of the array factor appears in the visible region when the element spacing is one-half wavelength.* Less than one period is visible if $2\beta d < 2\pi$, which corresponds to $d/\lambda < \frac{1}{2}$, that is, for spacings less than one-half wavelength. For spacings greater than one-half wavelength, more than one period will be visible. For one-wavelength spacings, two periods will be visible. For spacings larger than a half-wavelength, there may be more than one major lobe in the visible region, depending on the element phasings. Additional major lobes that rise to an intensity equal to that of the main lobe are called *grating lobes.* In the one-wavelength spaced, two-element array factor of Fig. 3-6*c*, there are grating lobes at $\theta =$ 0 and 180°, in addition to the desired lobe in the $\theta = 90°$ direction. In most situations, it is undesirable to have grating lobes. As a result, most arrays are designed so the element spacings are less than one wavelength.

3.2 UNIFORMLY EXCITED, EQUALLY SPACED LINEAR ARRAYS

An array is usually comprised of identical elements positioned in a regular geometrical arrangement. In fact, this is the definition adopted by the IEEE. However, arrays are encountered in practice with unequal interelement spacings. Usually, a modifier (e.g., equally or unequally spaced) is included to be completely clear about the array geometry. The examples presented in this chapter are for equally spaced arrays, and unequally spaced arrays are treated using the theory of Sec. 3.7.

3.2.1 The Array Factor Expression

A very important special case of equally spaced linear arrays is that of the uniformly excited array. This is an array whose element current amplitudes are identical, so

$$A_0 = A_1 = A_2 = \cdots \tag{3-26}$$

In this section, we consider only element phasings of a linear form accounted for by interelement phase shift α. The array factor from (3-18) is then

$$\text{AF} = A_0 \sum_{n=0}^{N-1} e^{jn\psi} = A_0(1 + e^{j\psi} + \cdots + e^{j(N-1)\psi}) \tag{3-27}$$

Only a few short steps are required to sum this geometric series. First, multiply (3-27) by $e^{j\psi}$ to obtain

$$\text{AF}\, e^{j\psi} = A_0(e^{j\psi} + e^{j2\psi} + \cdots + e^{jN\psi}) \tag{3-28}$$

Subtracting this from (3-27) gives

$$\text{AF}(1 - e^{j\psi}) = A_0(1 - e^{jN\psi})$$

or

$$\text{AF} = \frac{1 - e^{jN\psi}}{1 - e^{j\psi}} A_0 \tag{3-29}$$

This is rewritten in a more convenient form as follows:

$$\begin{aligned} \text{AF} &= A_0 \frac{e^{jN\psi} - 1}{e^{j\psi} - 1} = A_0 \frac{e^{jN\psi/2}}{e^{j\psi/2}} \frac{e^{jN\psi/2} - e^{-jN\psi/2}}{e^{j\psi/2} - e^{-j\psi/2}} \\ &= A_0 e^{j(N-1)\psi/2} \frac{\sin(N\psi/2)}{\sin(\psi/2)} \end{aligned} \tag{3-30}$$

The phase factor $e^{j(N-1)\psi/2}$ is not important unless the array output signal is further combined with the output from another antenna. In fact, if the array were centered about the origin, the phase factor would not be present since it represents the phase shift of the array phase center relative to the origin. Neglecting the phase factor in (3-30) gives

$$\mathrm{AF} = A_0 \frac{\sin(N\psi/2)}{\sin(\psi/2)} \tag{3-31}$$

This expression is maximum for $\psi = 0$ and the maximum value from (3-27) is

$$\mathrm{AF}(\psi = 0) = A_0(1 + 1 + \cdots + 1) = A_0 N \tag{3-32}$$

Dividing this into (3-31) gives the normalized array factor

$$\boxed{f(\psi) = \frac{\sin(N\psi/2)}{N \sin(\psi/2)}} \qquad \textit{UE, ESLA} \tag{3-33}$$

This is the normalized array factor for an N element, uniformly excited, equally spaced linear array (UE, ESLA) that is centered about the coordinate origin. This function is similar to a $(\sin u)/u$ function, with the major difference that the side lobes do not die off without limit for increasing argument. In fact, the function (3-33) is periodic in 2π, which is true in general as we showed in (3-23).

A number of trends can be seen by examining array factor plots for various values of N as shown in Fig. 3-11:

1. As N increases, the main lobe narrows.
2. As N increases, there are more side lobes in one period of $f(\psi)$. In fact, the number of full lobes (one main lobe and the side lobes) in one period of $f(\psi)$ equals $N - 1$. Thus, there will be $N - 2$ side lobes and one main lobe in each period.
3. The minor lobes are of width $2\pi/N$ in the variable ψ and the major lobes (main and grating) are twice this width.
4. The side lobe peaks decrease with increasing N. A measure of the side lobe peaks is the *side lobe level* that we have defined as

$$\mathrm{SLL} = \frac{|\text{maximum value of largest side lobe}|}{|\text{maximum value of main lobe}|} \tag{3-34}$$

   and it is often expressed in decibels. The side lobe level of the array factor for $N = 5$ is -12 dB and it is -13 dB for $N = 20$. SLL approaches the value of a uniform line source, -13.3 dB, as N is increased.
5. $|f(\psi)|$ is symmetric about π. It is left as an exercise to show this.

The radiation field polar plots in the variable θ can be obtained from $f(\psi)$ as discussed in Sec. 3.1. For example, consider the two-element case. Then (3-33) becomes

$$f(\psi) = \frac{\sin \psi}{2 \sin(\psi/2)} \tag{3-35}$$

This is a universal pattern function for all equal amplitude two-element arrays and is plotted in Fig. 3-10*a*. Note that by the techniques used in Sec. 3.1, we found that the array factor for a two-element array was $\cos(\psi/2)$; see (3-20). It can be shown that this is identical to (3-35).

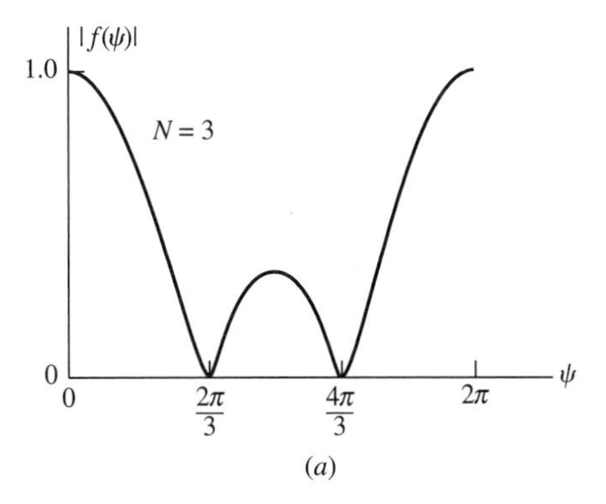

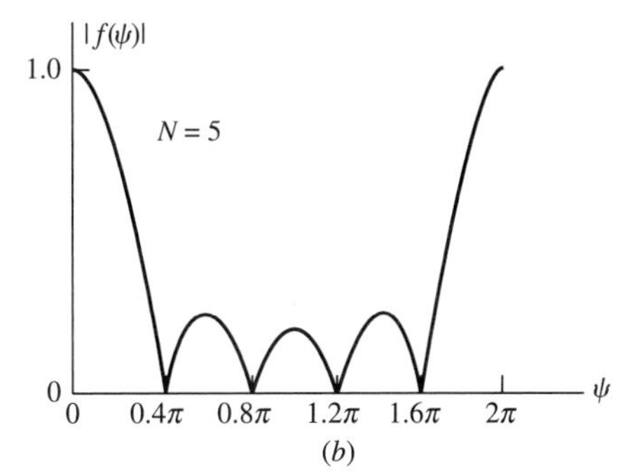

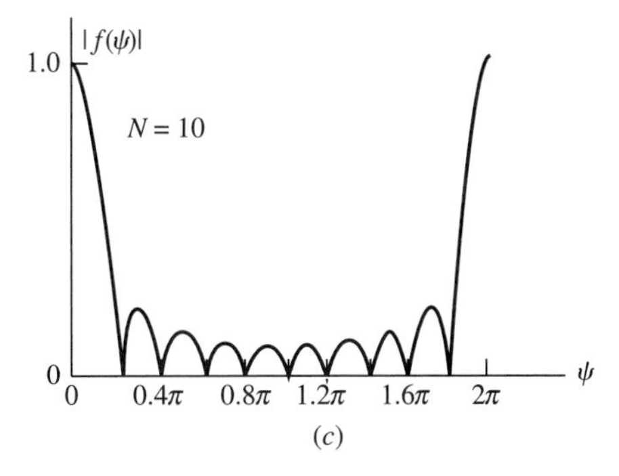

Figure 3-11 Array factor of an equally spaced, uniformly excited linear array for a few array numbers. (*a*) Three elements. (*b*) Five elements. (*c*) Ten elements.

EXAMPLE 3-5 *Four-Element Linear Array (Fig. 3-12)*

The universal array factor for a four-element, uniformly excited, equally spaced array is plotted in Fig. 3-12*b*. Let us find the array factor plot for the special case of half-wavelength spacing and 90° interelement phasing (i.e., $\alpha = \pi/2$). The array excitations are shown in Fig. 3-12*a*. The pattern plot can be sketched quickly by locating prominent features such as maxima and zeros. Then vertical lines are dropped down from these points to the circle below. From the intersection points with the circle, straight lines are drawn in to the center of the circle. The perimeter of the circle has a pattern value of unity and the center a value of 0. For linear polar plots such as this one, the magnitude of the pattern factor is linearly proportional to the distance from the origin. For example, if the circle radius is 4 cm and the pattern value to be plotted is 0.25, the pattern point is 1 cm from the origin along a radial line at the appropriate angle θ. After locating the relative maxima and the zeros, a smooth curve is drawn, joining these points. The complete polar plot is shown in Fig. 3-12*c*. Note that a polar plot can be made larger or smaller by expanding or contracting the construction circle.

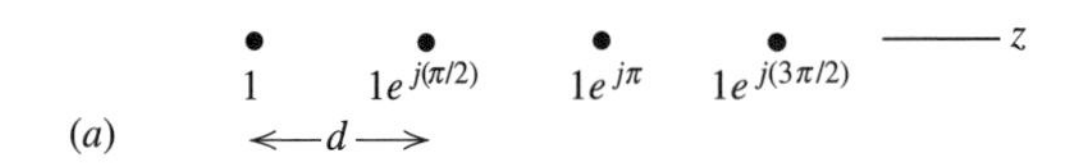

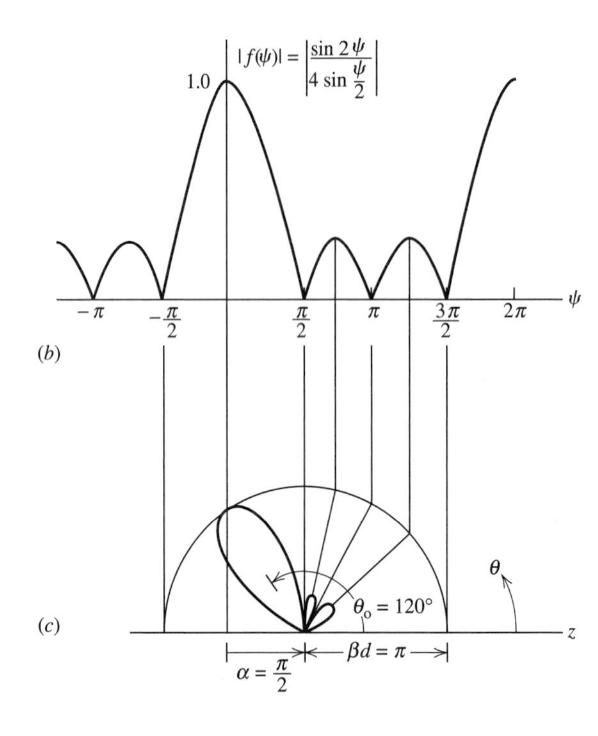

Figure 3-12 Array factor for a four-element, uniformly excited, equally spaced phased array (Example 3-5). (*a*) The array excitations. (*b*) Universal pattern for $N = 4$. (*c*) Polar plot for $d = \lambda/2$ and $\alpha = \pi/2$.

3.2.2 Main Beam Scanning and Beamwidth

A maximum of an array factor occurs for $\psi = 0$. Let θ_o be the corresponding value of θ for which the array factor is maximum. Then from (3-17), we have $0 = \beta d \cos \theta_o + \alpha$, or

$$\alpha = -\beta d \cos \theta_o \tag{3-36}$$

This is the element-to-element phase shift in the excitation currents required to produce an array factor main beam maximum in a direction θ_o relative to the line along which the array elements are disposed. Thus, if we want an array factor maximum in the $\theta = \theta_o$ direction, the required element currents from (3-15) with (3-36) are

$$I_n = e^{jn\alpha} = e^{-jn\beta d \cos \theta_o} \tag{3-37}$$

for a uniformly excited, equally spaced linear array. For the broadside case ($\theta_o = 90°$), $\alpha = 0$. For the endfire case ($\theta_o = 0°$ or $180°$), $\alpha = -\beta d$ or βd. In the example illustrated in Fig. 3-12, $\alpha = \pi/2$ and $d = \lambda/2$ so $\theta_o = \cos^{-1}(-\alpha/\beta d) = \cos^{-1}(-\frac{1}{2}) = 120°$. This main beam scanning by phase control feature can be explicitly incorporated into ψ by substituting (3-36) into (3-17), giving

$$\psi = \beta d(\cos \theta - \cos \theta_o) \tag{3-38}$$

Scanning is discussed further in Sec. 3.8.

A measure of the width of the main beam of a uniformly excited, equally spaced linear array is given by the *beamwidth between first nulls*, BWFN, which is illustrated in Fig. 1-15 for a general pattern. The main beam nulls are where the array factor (3-33) first goes to zero in a plane containing the linear array. The zeros of the numerator of (3-33) occur for $N\psi_{FN}/2 = \pm n\pi$. When the denominator also goes to zero ($\frac{1}{2}\psi_{FN} = \pm n\pi$), the pattern factor is unity, corresponding to the main beam ($n = 0$) and grating lobes. The first nulls associated with the main beam occur for $N\psi_{FN}/2 = \pm\pi$. For a broadside array ($\alpha = 0°$), $\psi = \beta d \cos\theta$, so the angles θ for the first nulls are found from

$$\pm\pi = \frac{N}{2}\frac{2\pi}{\lambda} d \cos\theta_{FN} \tag{3-39}$$

or

$$\theta_{FN} = \cos^{-1}\left(\pm\frac{\lambda}{Nd}\right) \tag{3-40}$$

The BWFN is then

$$\text{BWFN} = |\theta_{FN\,\text{left}} - \theta_{FN\,\text{right}}| \tag{3-41}$$

$$= \left|\cos^{-1}\left(-\frac{\lambda}{Nd}\right) - \cos^{-1}\left(+\frac{\lambda}{Nd}\right)\right| \tag{3-42}$$

For long arrays (length $L = Nd \gg \lambda$), we can approximate (3-42) as follows:

$$\text{BWFN} \approx \left|\frac{\pi}{2} + \frac{\lambda}{Nd} - \left(\frac{\pi}{2} - \frac{\lambda}{Nd}\right)\right| = \frac{2\lambda}{Nd} \qquad \textit{near broadside} \tag{3-43}$$

For an endfire array (see Fig. 1-16*c*), the beamwidth between first nulls is twice that from the main beam maximum to the first null. For long arrays it is approximately

$$\text{BWFN} \approx 2\sqrt{\frac{2\lambda}{Nd}} \qquad \textit{endfire} \tag{3-44}$$

Half-power beamwidth (HP) is a more popular measure of the main beam size than BWFN. Both depend on the array length Nd and main beam pointing angle θ_o. For a long ($Nd \gg \lambda$) uniformly excited linear array, the HP is approximately [2]

$$\text{HP} \approx 0.886 \frac{\lambda}{Nd} \csc\theta_o \qquad \textit{near broadside} \tag{3-45}$$

and

$$\text{HP} \approx 2\sqrt{0.886 \frac{\lambda}{Nd}} \qquad \textit{endfire} \tag{3-46}$$

Comparing the formulas for HP and BWFN, we can see that HP is roughly one-half of the corresponding BWFN value for long, uniformly excited linear arrays.

3.2.3 The Ordinary Endfire Array

In many applications, antennas are required to produce a single pencil beam. The array factor for a broadside array produces a fan beam, although the proper selection of array elements may yield a total pattern that has a single pencil beam. Another way to achieve a single pencil beam is by the proper design of an endfire array. We have said that an endfire condition results when $\theta_o = 0°$ or $180°$, which corresponds to $\alpha = -\beta d$ or $+\beta d$. Such arrays for which $\alpha = \pm\beta d$ are referred to as *ordinary endfire* arrays. If the spacing d is a half-wavelength, there will be two identical endfire lobes (see Fig. 3-10*c*). There are several ways to eliminate one of these lobes, thus leaving a single pencil beam. The most obvious way is to reduce the spacing below a half-wavelength. The visible region is $2\beta d$ wide in the variable ψ, and to eliminate the unwanted major lobe (grating lobe), we should reduce the visible region (and thus the spacing d) below the half-wavelength spacing value of 2π. Since the grating lobe half-width (maximum to null) is $2\pi/N$, we can eliminate most of it by reducing the visible region by at least π/N, that is,

$$2\beta d \leq 2\pi - \frac{\pi}{N} \qquad \textit{ordinary endfire} \tag{3-47}$$

Dividing this by 2β gives the condition on the spacings as

$$d \leq \frac{\lambda}{2}\left(1 - \frac{1}{2N}\right) \qquad \textit{ordinary endfire} \tag{3-48}$$

An ordinary endfire array with spacing d satisfying (3-48) produces a single endfire beam at $\theta = 0°$ for $\alpha = -\beta d$ or at $\theta = 180°$ for $\alpha = \beta d$.

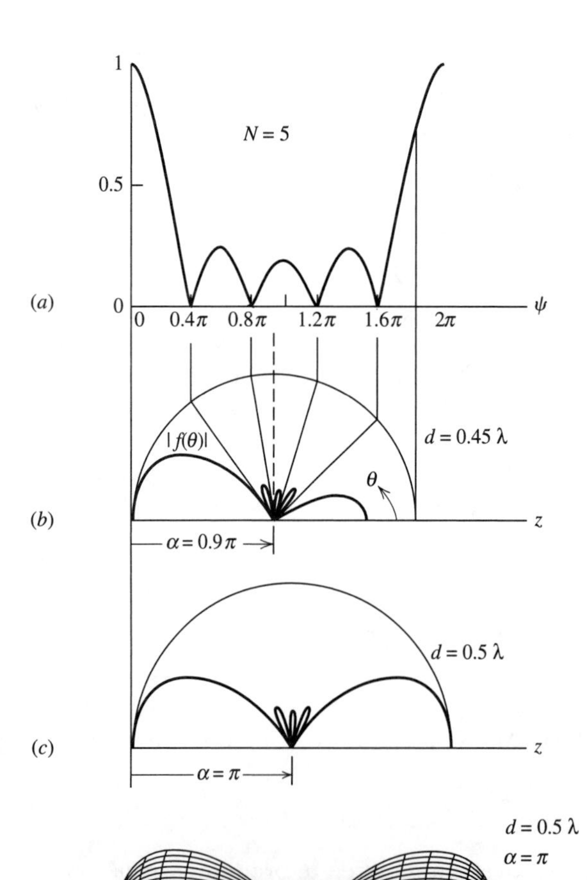

Figure 3-13 Five-element uniformly excited, equally spaced linear array (Example 3-6). (*a*) Universal pattern plot. (*b*) Polar plot for ordinary endfire case with $d = 0.45\lambda$ and $\alpha = 0.9\pi$. (*c*) and (*d*) Plots for endfire case with $d = 0.5\lambda$.

EXAMPLE 3-6 ***Five-Element Ordinary Endfire Linear Array (Fig. 3-13)***

From (3-48) for a five-element array, we must have $d \leq (\lambda/2)(1 - 1/10) = 0.45\lambda$. If we select $d = 0.45\lambda$ with a main beam direction $\theta_o = 180°$, the required element-to-element phase shift is $\alpha = -\beta d \cos \theta_o = \beta d = (2\pi/\lambda)(0.45\lambda) = 0.9\pi$. The pattern construction process is shown in Figs. 3-13*a* and 3-13*b*. Note the single endfire main lobe. If spacing is not reduced in accordance with (3-48), two main lobes appear as shown in Fig. 3-13*c* for $d = 0.5\lambda$. The corresponding three-dimensional polar plot is shown in Fig. 3-13*d*.

3.2.4 The Hansen–Woodyard Endfire Array

In the ordinary endfire case, the interelement excitation phase, $\alpha = \pm\beta d$, exactly equals the spatial phase delay of waves in the endfire direction. It is possible to make the main beam narrower and thus increase directivity by increasing the interelement phase shift, thereby moving some of the main beam outside of the visible region. If the phase shift is increased over the ordinary endfire case such that

$$\alpha = \pm(\beta d + \delta) \tag{3-49}$$

it is called the *Hansen–Woodyard condition* for increased directivity [3]. This condition was obtained by studying several long line sources, but also applies to long arrays.

To illustrate the Hansen–Woodyard condition, return to Fig. 3-13 and notice that as α is increased, the circle moves to the right but the radius of the circle remains the same for βd unchanged. This causes the main beam to narrow since part of the main lobe of the $|f(\psi)|$ plot does not appear in the visible region. However, the side lobes become larger relative to the main beam and the back lobe increases in magnitude. To prevent the back lobe from becoming equal to or greater than the main beam, the phase $|\alpha|$ must be less than π. Using this in (3-49) gives

$$\alpha = \beta d + \delta < \pi \tag{3-50}$$

Hansen and Woodyard [3] found that maximum directivity is obtained when $\delta \approx 2.94/(N - 1)$ when the array is long and much greater than a wavelength; also see (3-82). The simpler form of $\delta \approx \pi/N$ is usually used. Then (3-49) and (3-50) yield the phasing and spacing for a Hansen–Woodyard endfire array:

$$\alpha = \pm\left(\beta d + \frac{\pi}{N}\right) \tag{3-51a}$$

Hansen–Woodyard

$$d < \frac{\lambda}{2}\left(1 - \frac{1}{N}\right) \tag{3-51b}$$

EXAMPLE 3-7 ***Five-Element Hansen–Woodyard Endfire Linear Array (Fig. 3-14)***

The array in this example has five elements, so from (3-51) we must have $d < (\lambda/2)(1 - 1/5) = 0.4\lambda$. Choosing $d = 0.37\lambda$ leads to $\alpha = \beta d + \pi/N = 0.74\pi + 0.2\pi = 0.94\pi$. The pattern is shown in Fig. 3-14. The main beam is narrower than when the ordinary endfire condition is used (see Fig. 3-13), but the side lobes are higher. Nevertheless, the array exhibits increased directivity. The directivity as a function of spacing is compared to that of an ordinary endfire five-element array in Fig. 3-15. Note that directivity peaks near the spacing limit of 0.45λ from (3-48) for ordinary endfire arrays and near 0.4λ for Hansen–Woodyard endfire arrays.

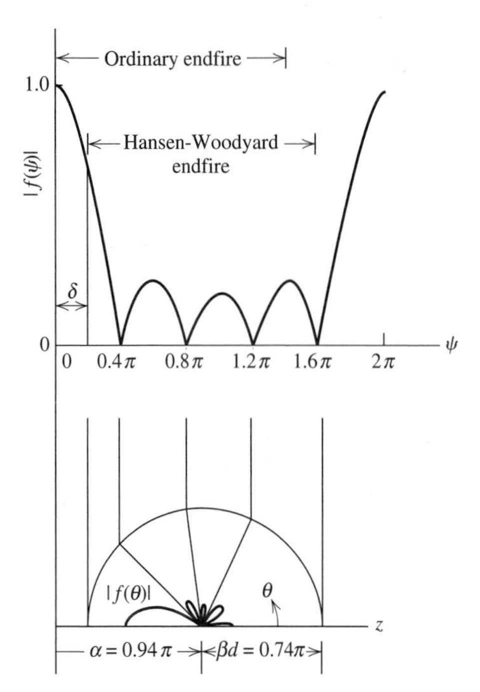

Figure 3-14 Single endfire beam for a five-element Hansen–Woodyard increased directivity array with $\alpha = 0.94\pi$ and $d = 0.37\lambda$ (Example 3-7).

More directivity is achieved with the Hansen–Woodyard array than with an ordinary endfire array of the same number of elements. This effect is referred to as superdirectivity and, in general, is accomplished with small spacings and phase control; see Secs. 2.6. This causes the peak of the main beam to move into the invisible region, narrowing the visible main beam, reducing beam solid angle, and increasing directivity. Superdirectivity is discussed further in Sec. 4.3.

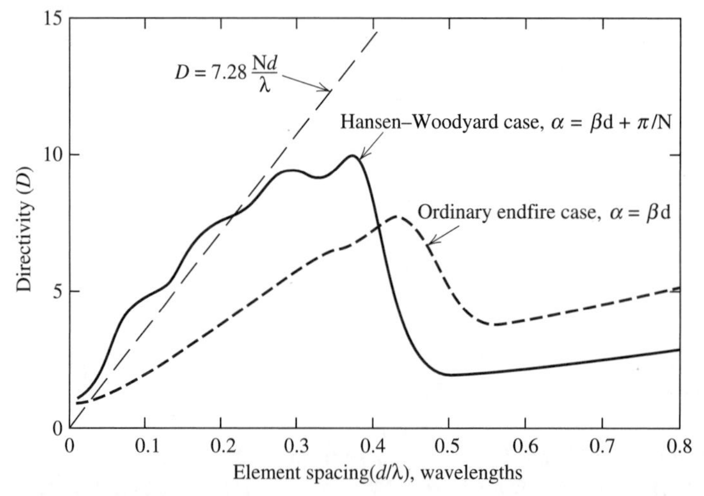

Figure 3-15 Comparison of directivities for two five-element equally spaced, uniformly excited endfire arrays: ordinary endfire (dotted curve) and Hansen–Woodyard endfire (solid curve). Also shown is the directivity approximation (3-82) for the Hansen–Woodyard case.

3.3 PATTERN MULTIPLICATION

So far in our study of arrays, we have discussed only arrays of isotropic point sources. Actual arrays have element antennas that, of course, are not isotropic. In this section, we discuss how to compute the radiation pattern of actual arrays. We will find that the array factor plays a major role in these pattern calculations.

When the elements of an array are placed along a line and the currents in each element also flow in the direction of that line, the array is said to be *collinear*. As a simple example of a collinear array, suppose we have N short dipoles as shown in Fig. 3-16. The elements are equally spaced a distance d apart and have currents I_0, $I_1, I_2, \ldots, I_{N-1}$. The total current is the sum of the z-directed short dipole currents and thus is z-directed and the vector potential is also. The vector potential integral in (1-103) reduces to a sum over the element currents (modeled as ideal dipoles) as[1]

$$\begin{aligned} A_z = \mu \frac{e^{-j\beta r}}{4\pi r} \Delta z [I_0 + I_1 e^{j\beta d \cos\theta} + I_2 e^{j\beta 2d \cos\theta} + \cdots \\ + I_{N-1} e^{j\beta(N-1)\, d \cos\theta}] = \mu \frac{e^{-j\beta r}}{4\pi r} \Delta z \sum_{n=0}^{N-1} I_n e^{j\beta n d \cos\theta} \end{aligned} \tag{3-52}$$

in the far field. Then from (1-106),

$$E_\theta = j\omega\mu \frac{e^{-j\beta r}}{4\pi r} \Delta z \sin\theta \sum_{n=0}^{N-1} I_n e^{j\beta n d \cos\theta} \tag{3-53}$$

From this expression, we can identify $\sin\theta$ as the pattern of a single element by itself, called the *element pattern*. The remaining factor

$$\text{AF} = \sum_{n=0}^{N-1} I_n e^{j\beta n d \cos\theta} \tag{3-54}$$

is the array factor of (3-14). The array factor is a sum of fields from isotropic point sources located at the center of each array element and is found from the element currents (amplitudes and phases) and their locations. On the other hand, the ele-

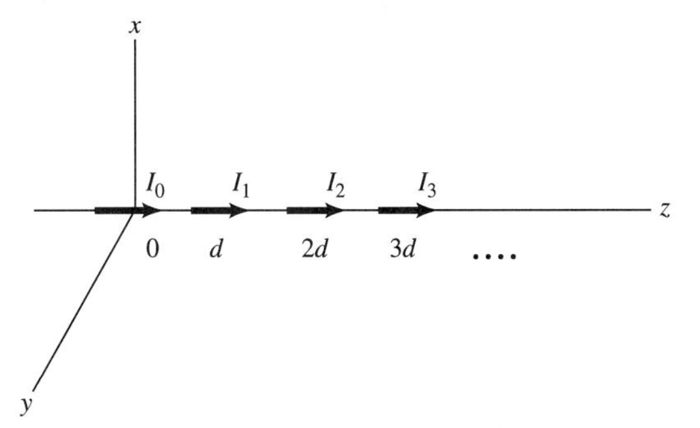

Figure 3-16 A collinear array of short dipoles.

[1]This result could also be obtained by writing the z-directed current density as

$$J_z = \delta(x')\,\delta(y')[I_0\,\delta(z') + I_1\,\delta(z'-d) + I_2\,\delta(z'-2d) + \cdots I_{N-1}\,\delta(z'-(N-1)\,d)]\,\Delta z$$

and substituting this into (1-102), giving

$$A_z = \mu \frac{e^{-j\beta r}}{4\pi r} \Delta z \int_{-\infty}^{\infty} [I_0\,\delta(z') + I_1\,\delta(z'-d) + \cdots] e^{j\beta z' \cos\theta}\, dz'$$

from which (3-52) follows.

ment pattern is that factor of the radiation pattern determined by the individual properties of an element based on its current distribution and orientation in space. We shall see that this factoring process holds in general if the elements have the same pattern and are similarly oriented.

We now consider a slightly more complicated case. Suppose for the sake of explanation, we have N identical element antennas forming a collinear array along the z-axis. The nth element is centered at $z = z_n$ and has a current distribution $i_n(z')$. We are now relaxing the equal spacing constraint. The total current along the z-axis is

$$I(z') = \sum_{n=0}^{N-1} i_n(z') \tag{3-55}$$

The vector potential is then

$$A_z = \mu \frac{e^{-j\beta r}}{4\pi r} \int_{-\infty}^{\infty} \sum_{n=0}^{N-1} i_n(z') e^{j\beta z' \cos\theta} \, dz' \tag{3-56}$$

The far-field elecric field from this and (1-106) is

$$E_\theta = j\omega\mu \frac{e^{-j\beta r}}{4\pi r} \sum_{n=0}^{N-1} E_n(\theta) \tag{3-57}$$

where

$$E_n(\theta) = \sin\theta \int_{-\infty}^{\infty} i_n(z') e^{j\beta z' \cos\theta} \, dz' \tag{3-58}$$

is the pattern of the nth element.

If the array possesses no symmetry, (3-57) cannot be simplified. But if the array elements are similar, a great deal of simplification is possible. By *similar* we mean that the currents of each antenna element are in the same direction, of the same length, and have the same distribution (although there may be different current amplitudes and phases for each element). Then the patterns of (3-58) will be similar; that is, they will have the same spatial variation but may have different amplitudes and phases. In the example at hand, the currents are all z-directed. Now assume that each element is of length ℓ, has a normalized current distribution over its length of $i(z')$, and an input current of I_n. Then

$$i_n(z') = I_n i(z' - z_n) \tag{3-59}$$

where z_n is the position of the nth element center along the z-axis. Substituting this into (3-58) gives

$$E_n(\theta) = \sin\theta \, I_n \int_{z_n-\ell/2}^{z_n+\ell/2} i(\xi - z_n) e^{j\beta\xi \cos\theta} \, d\xi \tag{3-60}$$

where ξ replaced z'. Let $\tau = \xi - z_n$; then (3-60) becomes

$$\begin{aligned} E_n(\theta) &= \sin\theta \, I_n \int_{-\ell/2}^{\ell/2} i(\tau) e^{j\beta(\tau + z_n)\cos\theta} \, d\tau \\ &= \sin\theta \left[\int_{-\ell/2}^{\ell/2} i(\tau) e^{j\beta\tau\cos\theta} \, d\tau \right] I_n e^{j\beta z_n \cos\theta} \end{aligned} \tag{3-61}$$

To maintain consistent notation, we replace τ by z', yielding

$$E_n(\theta) = \sin\theta \left[\int_{-\ell/2}^{\ell/2} i(z') e^{j\beta z' \cos\theta} \, dz' \right] I_n e^{j\beta z_n \cos\theta} \tag{3-62}$$

The pattern for each element of an array of similar elements given by (3-62) is a product of the pattern of the current distribution, and the amplitude and phase of excitation I_n, and the last factor represents the spatial phase due to the displacement from the origin. Substituting (3-62) into (3-57) gives

$$E_\theta = j\omega\mu \frac{e^{-j\beta r}}{4\pi r}\left[\sin\theta \int_{-\ell/2}^{\ell/2} i(z')e^{j\beta z'\cos\theta}\,dz'\right]\sum_{n=0}^{N-1} I_n e^{j\beta z_n \cos\theta} \tag{3-63}$$

The factor

$$\sin\theta \int_{-\ell/2}^{\ell/2} i(z')e^{j\beta z'\cos\theta}\,dz' \tag{3-64}$$

when normalized is the element pattern $g_a(\theta)$ of any element in the array of similar elements. The sum

$$\text{AF} = \sum_{n=0}^{N-1} I_n e^{j\beta z_n \cos\theta} \tag{3-65}$$

is the unnormalized array factor.

In going from (3-57) to (3-63), it was necessary to assume that the elements of the array were similar. When this is true, the electric field can be written as a product of an element pattern, as in (3-64), and an array factor, as in (3-65). Note that the array factor of (3-65) is the pattern of a linear array of N point sources located at positions $\{z_n\}$ on the z-axis. If the elements are equally spaced, (3-14) results. If further, they are uniformly excited, the array factor reduces to (3-31). This result is not restricted to collinear elements but can be applied to any array of similar elements. This is discussed below.

The process of factoring the pattern of an array into an element pattern and an array factor is referred to as the *principle of pattern multiplication.* It is stated as follows: The electric field pattern of an array consisting of similar elements is the product of the pattern of one of the elements (the element pattern) and the pattern of an array of isotropic point sources with the same locations, relative amplitudes, and phases as the original array (the array factor).

In Section 1.7, we wrote the normalized electric field pattern of a single antenna as a product of a normalized element factor g and a normalized pattern factor f. For array antennas, we expand this concept and call the pattern of one-element antenna in the array an element pattern g_a. It, in turn, is composed of an element factor that is the pattern of an infinitesimal piece of current on the array element (i.e., an ideal dipole) and a pattern factor that is the pattern due to its current distribution. The complete (normalized) pattern of an array antenna is

$$F(\theta, \phi) = g_a(\theta, \phi)f(\theta, \phi) \tag{3-66}$$

where $g_a(\theta, \phi)$ is the normalized pattern of a single element antenna of the array (the element pattern) and $f(\theta, \phi)$ is the normalized array factor.

EXAMPLE 3-8 *Two Collinear, Half-Wavelength Spaced Short Dipoles (Fig. 3-17)*

To illustrate pattern multiplication, consider two collinear short dipoles spaced a half-wavelength apart and equally excited. The element pattern is $\sin\theta$ for an element along the z-axis

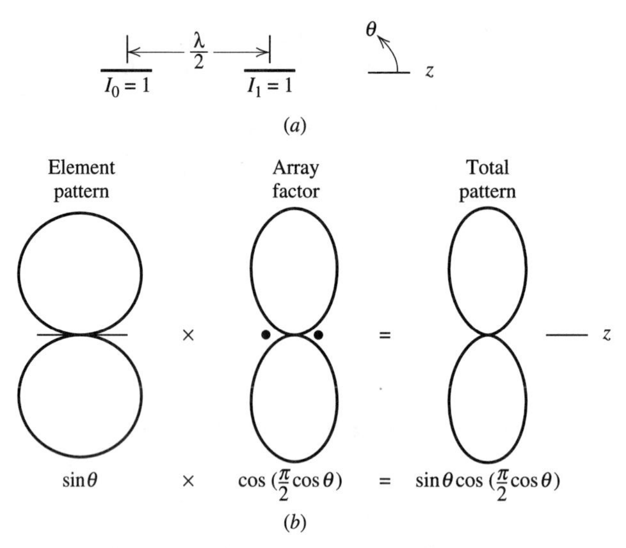

Figure 3-17 Array of two half-wavelength spaced, equal amplitude, equal phase, collinear short dipoles (Example 3-8). (*a*) The array. (*b*) The pattern.

and the array factor was found in (3-6) to be $\cos[(\pi/2)\cos\theta]$. The total pattern is then $\sin\theta \cos[(\pi/2)\cos\theta]$. The patterns are illustrated in Fig. 3-17.

Collinear arrays are in widespread use in base stations for land mobile communications. Half-wave dipoles spaced more than a half-wavelength apart are popular. The array axis is oriented vertically, producing an omnidirectional pattern in the horizontal plane as required for point-to-multipoint communications. Lengthening the array by adding elements narrows the beamwidth in the elevation plane, increasing the directivity and extending the usable range to a mobile unit.

The principle of pattern multiplication can be used directly for many different geometries. For example, suppose line sources, positioned along the z-axis are not z-directed, as in a collinear array, but are parallel as shown in Fig. 3-18. Let γ be the spherical polar angle from the x-axis; note that $0° \le \gamma \le 180°$ and $\cos\gamma = \sin\theta \cos\phi$. The element pattern is then found from the following expression that is analogous to (3-64):

$$\sin\gamma \int_{-\ell/2}^{\ell/2} i(x')e^{j\beta x' \cos\gamma}\, dx' \tag{3-67}$$

The array factor of (3-65) is unchanged.

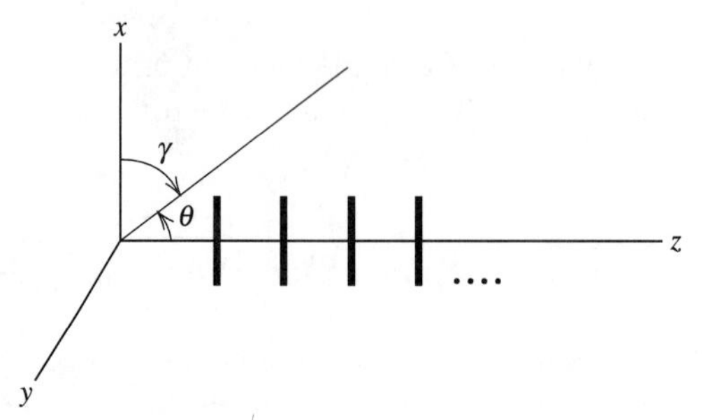

Figure 3-18 A linear array of parallel line sources.

EXAMPLE 3-9 ***Two Parallel, Half-Wavelength Spaced Short Dipoles (Fig. 3-19)***

The complete pattern for the array of two parallel short dipoles in Fig. 3-19*a* is found by pattern multiplication as indicated in Figs. 3-19*b* and 3-19*c*. It is difficult to visualize the pattern in three dimensions from the principal plane patterns. Figure 3-19*d* is a three-dimensional polar pattern plot tilted to show the broad null along the z-axis and a narrow null along the x-axis. It is obvious from the pattern that this array has few applications. It is

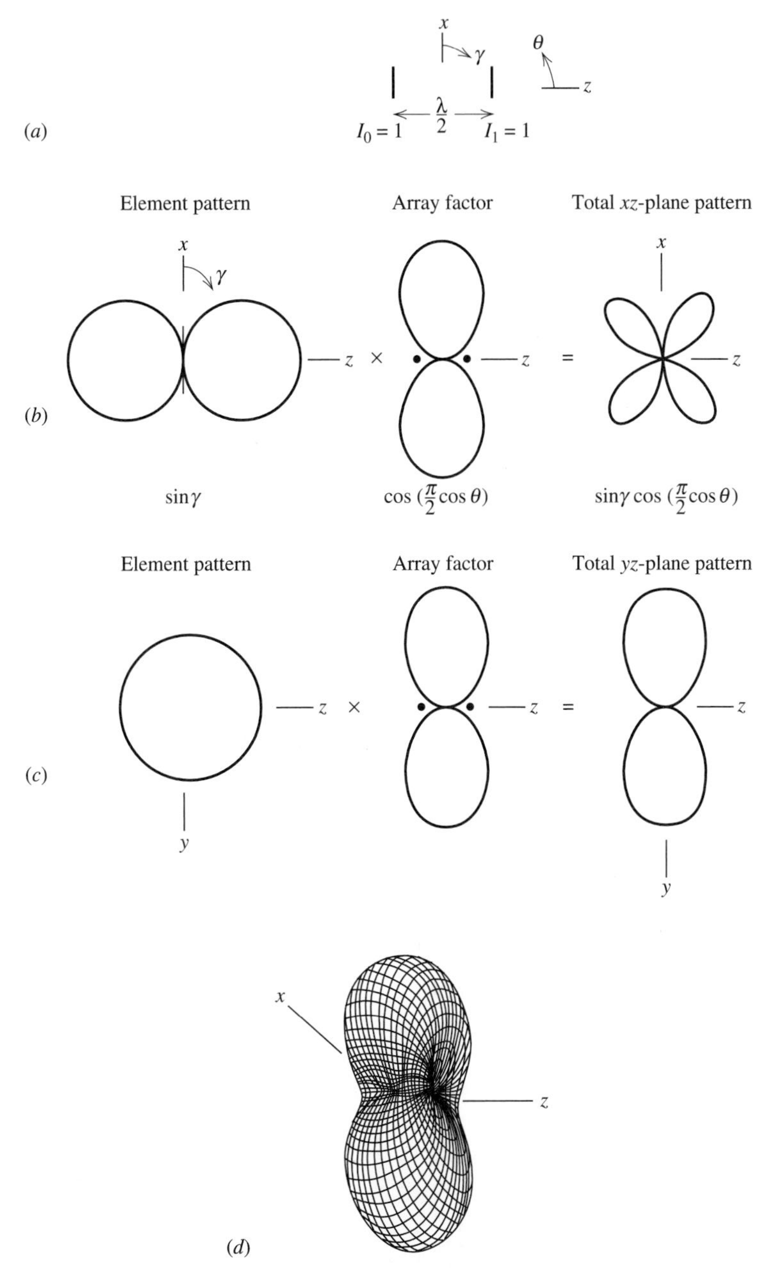

Figure 3-19 Array of two half-wavelength spaced, equal amplitude, equal phase parallel short dipoles. (*a*) The array. (*b*) The xz-plane pattern. (*c*) The yz-plane pattern. (*d*) Three-dimensional pattern.

presented here to illustrate how the element can significantly affect the pattern and to aid in understanding the pattern multiplication process.

EXAMPLE 3-10 ***Five-Element Endfire Array of Parallel Half-Wave Dipoles***

To illustrate parallel element arrays further, suppose the element antennas of Fig. 3-18 are half-wave dipoles. Also suppose there are five elements arranged and excited for ordinary endfire as in Example 3-6. The complete pattern is the product of the single half-wave dipole element pattern and the array factor found from five isotropic sources. The element pattern for a half-wave dipole element along the x-axis is

$$g_a(\gamma) = \frac{\cos[(\pi/2)\cos\gamma]}{\sin\gamma} \tag{3-68}$$

which is (2-10) with θ replaced by γ. Since $\cos\gamma = \sin\theta\cos\phi$, then

$$\sin\gamma = \sqrt{1 - \sin^2\theta\cos^2\phi}$$

and (3-68) becomes

$$g_a(\theta, \phi) = \frac{\cos[(\pi/2)\sin\theta\cos\phi]}{\sqrt{1 - \sin^2\theta\cos^2\phi}} \tag{3-69}$$

The array factor is (3-33) with $N = 5$, or

$$f(\psi) = \frac{\sin(\frac{5}{2}\psi)}{5\sin(\frac{1}{2}\psi)} \tag{3-70}$$

For this example, $\alpha = 0.9\pi$ and $d = 0.45\lambda$ so $\psi = \beta d\cos\theta + \alpha = 0.9\pi\cos\theta + 0.9\pi$, and (3-70) is

$$f(\theta) = \frac{\sin(2.25\pi\cos\theta + 2.25\pi)}{5\sin(0.45\pi\cos\theta + 0.45\pi)} \tag{3-71}$$

The total pattern of the array in terms of θ and ϕ is then the product of (3-69) and (3-71):

$$F(\theta, \phi) = \frac{\cos[(\pi/2)\sin\theta\cos\phi]}{\sqrt{1 - \sin^2\theta\cos^2\phi}} \frac{\sin(2.25\pi\cos\theta + 2.25\pi)}{5\sin(0.45\pi\cos\theta + 0.45\pi)} \tag{3-72}$$

The polar plot of this pattern is easily obtained by multiplying the plot in Fig. 2-5*b*, where the axis of symmetry is now the x-axis instead of the z-axis, times the polar plot of Fig. 3-13. This is a polar plot similar to the array factor plot except that the endfire lobes are slightly narrower, and there is a pattern zero in the $\gamma = 0°$ direction caused by the element pattern.

3.4 DIRECTIVITY OF UNIFORMLY EXCITED, EQUALLY SPACED LINEAR ARRAYS

Now that we have developed a method for obtaining the entire pattern expression for an antenna array, we can discuss the directivity of various arrays. Directivity is determined entirely from the radiation pattern. The array gain can be found by multiplying the array directivity by the radiation efficiency of one element if all elements are alike.

To derive directivity expressions, we use $D = 4\pi/\Omega_A$, first finding the beam solid angle as

$$\Omega_A = \iint |F(\theta, \phi)|^2\, d\Omega = \iint |g_a(\theta, \phi)|^2 |f(\theta)|^2\, d\Omega \tag{3-73}$$

where $g_a(\theta, \phi)$ and $f(\theta)$ are the normalized element pattern and linear array factor and $d\Omega = \sin\theta\, d\theta\, d\phi$.

We begin by assuming the elements are equally spaced, uniformly excited, and isotropic. This assumption leads to approximate results for situations where the element pattern is much broader than the array factor and the main beams of both are aligned. The appropriate array factor from (3-33) is

$$|f|^2 = \left|\frac{\sin(N\psi/2)}{N\sin(\psi/2)}\right|^2 \tag{3-74}$$

$$= \frac{1}{N} + \frac{2}{N^2}\sum_{m=1}^{N-1}(N-m)\cos m\psi \tag{3-75}$$

where (3-75) is another form for (3-74). This identity can be shown to be true for $N = 2$ since from (3-75), $|f(\psi)|^2 = \frac{1}{2} + \frac{1}{2}\cos\psi \equiv \cos^2(\psi/2)$ as in (3-20). With the simple expression in (3-75), it is easier to perform the integration in (3-73) in terms of the variable ψ. Using $g_a(\theta, \phi) = 1$, $\psi = \beta d\cos\theta + \alpha$, and $\sin\theta\, d\theta = -(1/\beta d)\, d\psi$ in (3-73) gives

$$\Omega_A = \int_0^{2\pi} d\phi \int_0^{\pi} |f(\theta)|^2 \sin\theta\, d\theta = 2\pi \int_{\beta d+\alpha}^{-\beta d+\alpha} |f(\psi)|^2 \left(-\frac{1}{\beta d}\right) d\psi$$

$$= \frac{2\pi}{\beta d}\int_{-\beta d+\alpha}^{\beta d+\alpha} |f(\psi)|^2\, d\psi \tag{3-76}$$

Substituting (3-75) in the above yields

$$\Omega_A = \frac{2\pi}{\beta d}\left[\frac{1}{N}\int_{-\beta d+\alpha}^{\beta d+\alpha} d\psi + \frac{2}{N^2}\sum_{m=1}^{N-1}(N-m)\int_{-\beta d+\alpha}^{\beta d+\alpha}\cos m\psi\, d\psi\right]$$

$$= \frac{2\pi}{\beta d}\left[\frac{1}{N}\psi\bigg|_{-\beta d+\alpha}^{\beta d+\alpha} + \frac{2}{N^2}\sum_{m=1}^{N-1}(N-m)\frac{\sin m\psi}{m}\bigg|_{-\beta d+\alpha}^{\beta d+\alpha}\right]$$

$$= \frac{2\pi}{\beta d}\left[\frac{1}{N}(2\beta d) + \frac{2}{N^2}\sum_{m=1}^{N-1}\frac{N-m}{m}[\sin m(\beta d+\alpha) - \sin m(-\beta d+\alpha)]\right]$$

$$= \frac{4\pi}{N} + \frac{4\pi}{N^2}\sum_{m=1}^{N-1}\frac{N-m}{m\beta d}\, 2\cos m\alpha \sin m\beta d \tag{3-77}$$

where (D-6) was used in the last step.

The foregoing assumes that the array factor reaches its peak of unity in the visible region. If this is not the case, renormalization is required. This possibility was included explicitly in the Hansen–Woodyard array development with δ in (3-49). Normalizing the array factor to unity maximum in the visible region leads to the following general form for the directivity of a linear array of N isotropic, uniformly excited elements spaced distance d apart with interelement phase shift α [4]:

$$D = \frac{4\pi}{\Omega_A} = \frac{\left|\dfrac{\sin(N\,\delta/2)}{N\sin(\delta/2)}\right|^2}{\dfrac{1}{N} + \dfrac{2}{N^2}\displaystyle\sum_{m=1}^{N-1}\frac{N-m}{m\beta d}\sin m\beta d\cos m\alpha} \tag{3-78}$$

where δ is zero and the numerator is unity except for endfire operation; see (3-49).

The directivity expression of (3-78) reduces to a simple result for a broadside array of half-wavelength spaced elements. For multiple half-wavelength spacings, $d = n\lambda/2$ with n an integer, $\beta d = n\pi$ and $\sin m\beta d = 0$. The array gives a broadside pattern when all elements are in-phase, or $\alpha = 0$. These conditions simplify (3-78) to

$$D = N \qquad d = n\frac{\lambda}{2}, \quad \alpha = 0 \tag{3-79}$$

The directivity of a broadside array of isotropic elements as a function of the spacing in terms of a wavelength, d/λ, is plotted in Fig. 3-20 for several element numbers N. Notice that the directivity equals N at integer multiples of a half-wavelength. Also, the directivity curves for each N take a sharp dip for spacings near one and two wavelengths. This is caused by the emergence of grating lobes into the visible region. For example, see Fig. 3-6 where full grating lobes appear for one-wavelength spacing.

The directivity of a broadside array of isotropic elements is approximated by

$$D \approx 2\frac{L}{\lambda} = 2\frac{Nd}{\lambda} \qquad \textit{broadside} \tag{3-80}$$

where $L = Nd$ is the array length. This is a straight-line approximation to the curves in Fig. 3-20, being very accurate for d up to almost one wavelength. This approximation is shown in Fig. 3-20 for $N = 10$. Note that (3-80) is exact for $d = \lambda/2$, since then (3-80) equals N as in (3-79).

The directivities for arrays with ordinary and Hansen–Woodyard endfire phase conditions are plotted in Fig. 3-15 as a function of spacing. Note that $D = N$ for ordinary endfire when $d = n\lambda/2$. The directivity increase when both the Hansen–Woodyard endfire phasing and spacing conditions are satisfied is evident in Fig. 3-15. The directivity values for Examples 3-6 and 3-7 of 7.4 and 10.0 can be found in Fig. 3-15. Approximations exist for endfire arrays of isotropic elements. For an ordinary endfire array with $\alpha = \pm\beta d$ and the spacing satisfying (3-48), the directivity is

$$D \approx 4\frac{L}{\lambda} \qquad \textit{ordinary endfire} \tag{3-81}$$

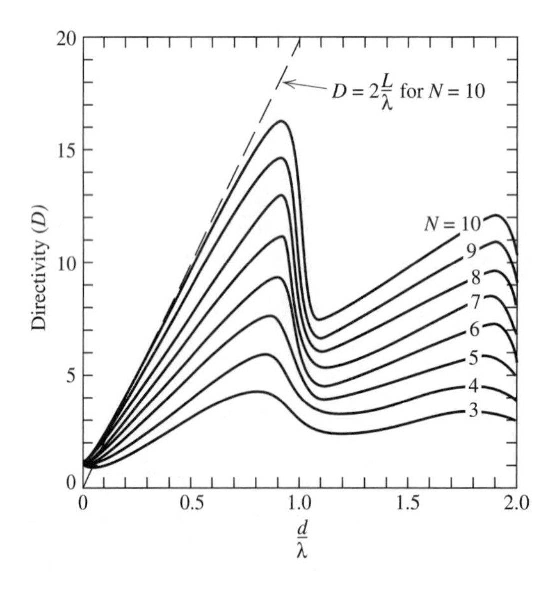

Figure 3-20 Directivity as a function of element spacing for a broadside array of isotropic elements for several element numbers N.

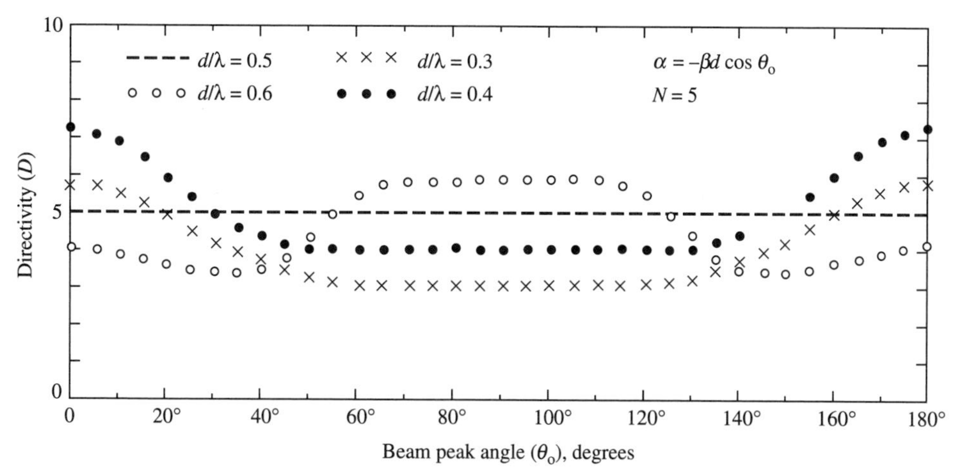

Figure 3-21 Variation of directivity with scan angle for five-element uniformly excited arrays of various element spacings. The elements are isotropic.

For an endfire array of the Hansen–Woodyard type, it is given by

$$D \approx 7.28 \frac{L}{\lambda} \qquad \textit{Hansen–Woodyard} \tag{3-82}$$

This approximation is shown in Fig. 3-15.

It is also interesting to examine the directivity expression of (3-78) for various scan angles. A few cases are given in Fig. 3-21. It is apparent from the figure that for $d = n\lambda/2$, directivity is independent of scan angle. This will be shown mathematically in Sec. 3.5. For the four cases shown, the greatest directivity in the broadside direction ($\theta_o = 90°$) is for the largest spacing. In fact, we obtain even higher directivities for spacings up to $d = 0.8\lambda$ in the $N = 5$ case (see Fig. 3-20). As can be seen from this example, the directivity of linear arrays remains constant over a wide range of scan angles near broadside; this will be explained in Sec. 3.8. The greatest directivity in the endfire direction ($\theta_o = 0$ or 180°) is for the largest spacing that satisfies the single main beam criterion of (3-48), which is $d \leq 0.45\lambda$ for $N = 5$. For the four spacings shown, $d = 0.4\lambda$ is the largest spacing satisfying this single endfire beam condition and, thus, displays the largest endfire directivity.

Inclusion of element pattern effects increases the difficulty of array directivity computations. In general, directivity is evaluated by integrating to obtain Ω_A and then using $D = 4\pi/\Omega_A$. The following formula is one of the few available for the directivity of linear arrays:

$$D = \frac{1}{\dfrac{a_0}{N} + \dfrac{2}{N^2} \displaystyle\sum_{m=1}^{N-1} \frac{N-m}{m\beta d} (a_1 \sin m\beta d + a_2 \cos m\beta d) \cos m\alpha} \tag{3-83}$$

where a_0, a_1, and a_2 are given in Table 3-1 for various element patterns [5, 6]. The directivity of long arrays ($L \gg \lambda$) is primarily controlled by the array factor if the element pattern is of low directivity and its major lobe is aligned with that of the array factor. In cases such as these, the approximate formulas of (3-80) to (3-82) can be used.

Table 3-1 Parameters for Use in Computing the Directivity of Uniform Current Amplitude, Equally Spaced Linear Arrays; see (3-83).

Element Type	$\lvert g_a(\theta, \phi)\rvert^2$	a_0	a_1	a_2
Isotropic	1	1	1	0
Collinear short dipoles (Fig. 3-16)	$\sin^2 \theta$	$\frac{2}{3}$	$\frac{2}{(m\beta d)^2}$	$\frac{-2}{m\beta d}$
Parallel short dipoles (Fig. 3-18)	$1 - \sin^2 \theta \cos^2 \phi = \sin^2 \gamma$	$\frac{2}{3}$	$1 - \frac{1}{(m\beta d)^2}$	$\frac{1}{m\beta d}$

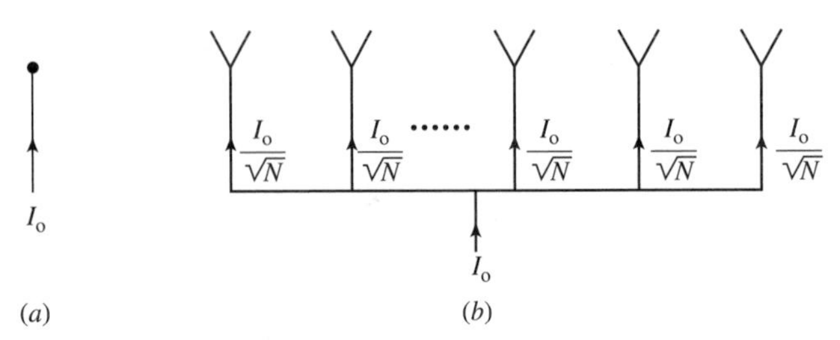

Figure 3-22 Array directivity is the ratio of the maximum radiation intensity of the array compared to that for an isotropic element with the same input power. (*a*) Reference isotropic antenna. (*b*) Array with the same total input power divided equally among the elements.

It is important to note that array directivity represents the increase in the radiation intensity in the direction of maximum radiation over that of a single element. Consider a single isotropic element and an array of N equally excited isotopic elements as shown in Fig. 3-22. The input power to the array is assumed to divide equally among the array elements, so the element powers are $1/N$ of the input power and the element currents are $1/\sqrt{N}$ of the input current. The radiation intensity U_o for the isotropic element is proportional to its input power, which in turn is proportional to the input current squared I_o^2. The maximum radiation intensity $U_{\max}$ of the array in Fig. 3-22*b* is a factor of D greater than that for a single isotropic element with the same input power.

3.5 NONUNIFORMLY EXCITED, EQUALLY SPACED LINEAR ARRAYS

We have seen that the main beam of an endfire array can be narrowed by changing the phase from that which is required for the ordinary endfire case. We can also shape the beam and control the level of the side lobes by adjusting the current amplitudes in an array. General synthesis procedures for achieving a specified pattern are presented in Chap. 8. In this section, a few simple techniques for controlling side lobe levels and beamwidth are introduced. Several examples are given that reveal the relationship between the array current distribution and the radiation pattern. The directivity for arrays with nonuniform excitation are also examined.

The array factor of (3-18) can be written as a polynomial in terms of $Z = e^{j\psi}$ as follows:

$$\mathrm{AF} = \sum_{n=0}^{N-1} A_n e^{jn\psi} = \sum_{n=0}^{N-1} A_n Z^n \tag{3-84}$$

where the current amplitudes A_n are real and can be different for each n. S. A. Schelkunoff [7] applied the algebra of polynomials to array factors. He showed the connection between placement in the complex plane of the $N - 1$ zeros (roots) of the array polynomial in (3-84) and the radiation pattern and element currents. However, we examine the relationship between the element excitation and the array factor in a direct fashion. It is a simple matter to investigate element current distributions using a computer to perform the array factor summation. We present the results of several such calculations. The influence of the element current amplitudes is apparent since we use the same five-element, broadside linear array with a half-wavelength element spacing throughout this section.

The pattern of a uniform array with all current amplitudes equal is plotted in linear, polar form in Fig. 3-23*a* and the element currents are shown in Fig. 3-24*a*. If the element current amplitudes form a triangle as shown in Fig. 3-24*b*, the radiation pattern of Fig. 3-23*b* results. Notice that the side lobes are considerably smaller than those of the uniformly illuminated array, but at the expense of increased beamwidth. This increased beamwidth (from 20.8 to 26.0°) is responsible for reduced directivity (from 5 to 4.26).

The side lobe reduction introduced by the triangular amplitude taper suggests that perhaps an amplitude distribution exists such that all side lobes are completely eliminated. Indeed, this is possible if the ratios of the currents are equal to the coefficients of the binomial series. To see how this comes about, first consider a two-element array with equal amplitudes and spacing d. The array factor from (3-84) is $\mathrm{AF} = 1 + e^{j\psi}$ and can be written in terms of $Z = e^{j\psi}$ as

$$\mathrm{AF} = 1 + Z \tag{3-85}$$

If the spacing for this broadside array is less than, or at most equal to, a half-wavelength, the array factor will have no side lobes (see Fig. 3-3). Now consider an array formed by taking the product of two array factors of this type:

$$\mathrm{AF} = (1 + Z)(1 + Z) = 1 + 2Z + Z^2 \tag{3-86}$$

This corresponds to a three-element array with the current amplitudes in the ratio 1:2:1. Since this array is simply the square of one that had no side lobes, the three-element array also has no side lobes. This process can also be viewed as arraying of two of the two-element arrays such that the centers of each subarray are spaced d apart. This leads to a coincidence of two elements in the middle of the total array, thus giving a current of 2. The total array factor is the product of the "element pattern," which is a two-element subarray pattern, and the array factor that is again a two-element, equal amplitude array. Thus, the total array factor is the square of one subarray pattern. Continuing this process for an N element array, we obtain

$$\mathrm{AF} = (1 + Z)^{N-1} \tag{3-87}$$

which is a binomial series; see (F-4). For $N = 5$,

$$\mathrm{AF} = (1 + Z)^4 = 1 + 4Z + 6Z^2 + 4Z^3 + Z^4 \tag{3-88}$$

Therefore, the ratios of the current amplitudes are 1:4:6:4:1. This current distribution is shown in Fig. 3-24*c* and the resulting pattern is shown in Fig. 3-23*c*. This

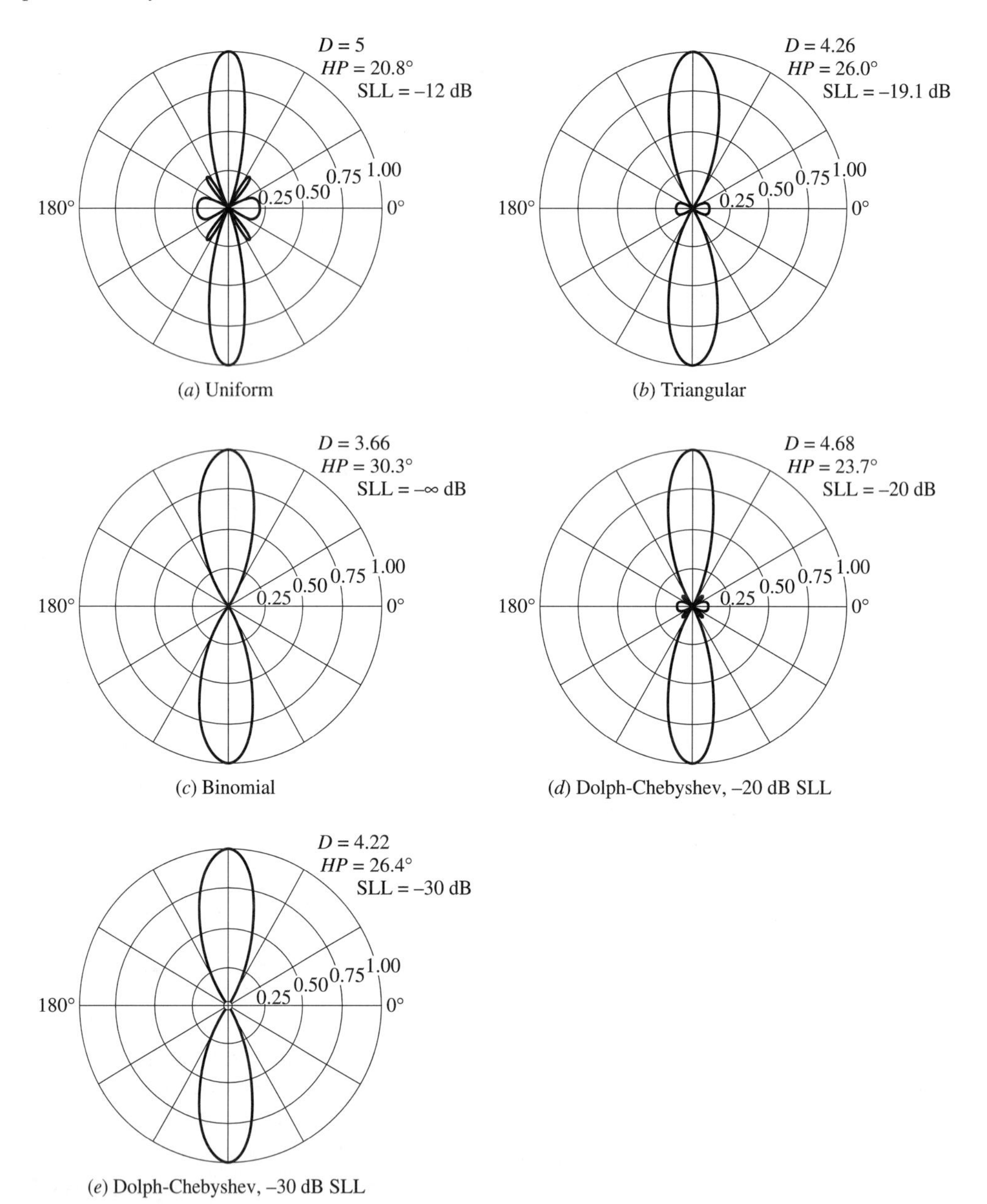

Figure 3-23 Patterns of several uniform phase ($\theta_o = 90°$), equally spaced ($d = \lambda/2$) five-element linear arrays with various amplitude distributions. The currents are plotted in Fig. 3-24. (*a*) Uniform currents, 1:1:1:1:1. (*b*) Triangular current amplitude distribution, 1:2:3:2:1. (*c*) Binomial current amplitude distribution, 1:4:6:4:1. (*d*) Dolph–Chebyshev current amplitude distribution, 1:1.61:1.94:1.61:1, for a side lobe level of −20 dB. See Example 8-5. (*e*) Dolph–Chebyshev current amplitude distribution, 1:2.41:3.14:2.41:1, with a side lobe level of −30 dB.

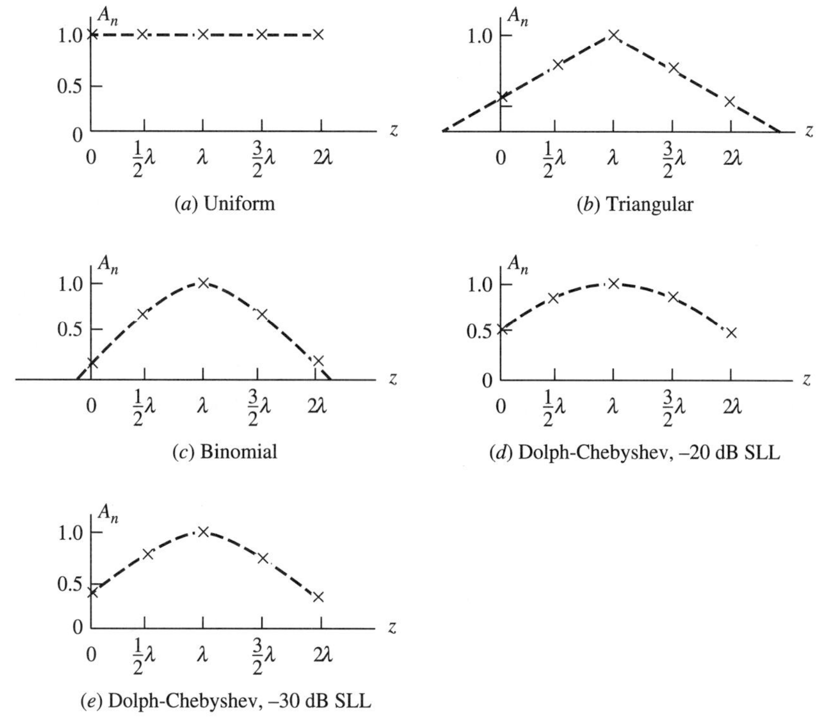

Figure 3-24 Current distributions corresponding to the patterns of Fig. 3-23. The current phases are zero ($\alpha = 0$). Currents are normalized to unity at the array center.

pattern is broader than either the uniform or triangular distribution cases and has a lower directivity, but it has no side lobes.

From these three five-element array examples, a trend has emerged: *As the current amplitude is tapered more toward the edges of the array, the side lobes tend to decrease and the beamwidth increases.* This beamwidth/side lobe level tradeoff can be optimized. In other words, it is possible to determine the element current amplitudes such that the beamwidth is minimum for a specified side lobe level, or conversely to specify the beamwidth and obtain the lowest possible side lobe level. This array is referred to as a Dolph–Chebyshev array and it provides a pattern with all side lobes of the same level. The Dolph–Chebyshev array synthesis procedure is explained in detail in Sec. 8.4.1. For a five-element array with an element spacing of a half-wavelength and a specified side lobe level of −20 dB, the Dolph–Chebyshev current distribution is plotted in Fig. 3-24*d* and the corresponding pattern is shown in Fig. 3-23*d*. If the side lobe level for the Dolph–Chebyshev array is specified to be −30 dB, the distribution is that of Fig. 3-24*e* and the corresponding pattern is shown in Fig. 3-23*e*. Note that the main beam is slightly broader than in the previous case where the side lobe level was 10 dB higher.

The discussion of nonuniformly excited arrays thus far was for amplitudes that are tapered toward the ends of the linear array. If the amplitude distribution becomes larger at the ends of the array (called an inverse taper), we expect the opposite effect; that is, the side lobe level increases and the beamwidth decreases. Suppose, for example, that we invert the triangular distribution such that the amplitudes are 3:2:1:2:3. The resulting pattern shown in Fig. 3-25 demonstrates the expected decrease in beamwidth and increase in side lobe level. Although the di-

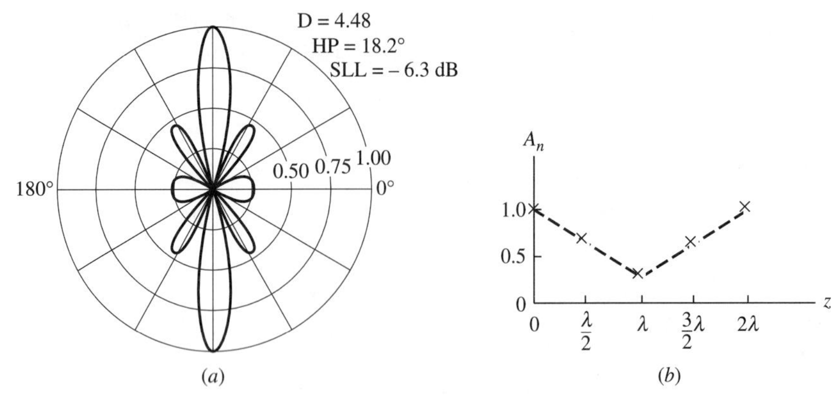

Figure 3-25 The inverse triangular tapered, five-element linear array with $d = \lambda/2$ and $\theta_o = 90°$.
(a) The array factor.
(b) The current distribution, 3:2:1:2:3.

rectivity for the inverse triangular tapered current is greater than that for the triangular taper of Fib. 3-23b, it is still not as large as that produced by the uniform distribution.

The directivity values were given for each of the examples in this section. We close this section by developing the directivity expression. With little additional complexity, the treatment can be expanded to include unequal element spacings as well as nonuniform excitation. The element positions along the z-axis are z_n and the element current amplitudes are A_n. If the element phases are linear with distance, then $\alpha_n = -\beta z_n \cos\theta_o$, where θ_o is the angle of the pattern maximum; the applications of this type of phasing are discussed in Sec. 3.8. The array factor of (3-65) is then appropriate and when normalized is

$$f(\theta) = \frac{\sum_{n=0}^{N-1} A_n e^{j\alpha_n} e^{j\beta z_n \cos\theta}}{\sum_{n=0}^{N-1} A_n} \tag{3-89}$$

The appropriate beam solid angle expression is

$$\begin{aligned} \Omega_A &= 2\pi \int_0^{\pi} |f(\theta)|^2 \sin\theta \, d\theta \\ &= \frac{2\pi}{\left(\sum_{k=0}^{N-1} A_k\right)^2} \sum_{m=0}^{N-1} \sum_{p=0}^{N-1} A_m A_p e^{j(\alpha_m - \alpha_p)} \int_0^{\pi} e^{j\beta(z_m - z_p)\cos\theta} \sin\theta \, d\theta \end{aligned} \tag{3-90}$$

Evaluating the integral in the above expression and applying the result to $D = 4\pi/\Omega_A$ yields

$$D = \frac{\left(\sum_{k=0}^{N-1} A_k\right)^2}{\sum_{m=0}^{N-1} \sum_{p=0}^{N-1} A_m A_p e^{j(\alpha_m - \alpha_p)} \dfrac{\sin[\beta(z_m - z_p)]}{\beta(z_m - z_p)}} \tag{3-91}$$

where $\alpha_n = -\beta z_n \cos\theta_o$ and the elements can have any positions z_n and current amplitudes A_n. This general result simplifies for a broadside, equally spaced array to

$$D = \frac{\left(\sum_{k=0}^{N-1} A_k\right)^2}{\sum_{m=0}^{N-1}\sum_{p=0}^{N-1} A_m A_p \dfrac{\sin[(m-p)\beta d]}{(m-p)\beta d}} \qquad \alpha_n = 0, \quad z_n = nd \tag{3-92}$$

As another special case, when the spacings are a multiple of a half-wavelength, (3-91) reduces to

$$D = \frac{\left(\sum_{n=0}^{N-1} A_n\right)^2}{\sum_{n=0}^{N-1} (A_n)^2} \qquad d = \frac{\lambda}{2}, \lambda, \ldots \tag{3-93}$$

Note that this is independent of scan angle θ_o, as indicated in Fig. 3-21 for $d = \lambda/2$. Also, if the amplitudes are uniform, (3-93) yields $D = N$ as given by (3-79). For a further example, consider the triangular excitation with the pattern of Fig. 3-23*b*. The directivity value from (3-93) is $[2(1) + 2(2) + 3]^2/[2(1)^2 + 2(2)^2 + (3)^2] = 4.26$. Equation (3-93) is a very instructive formula because it shows that directivity is a measure of the coherent radiation from the linear array. The numerator is proportional to the square of the *total* coherent field, whereas the denominator is proportional to the *sum* of the squares of the field from each of the elements.

There is no closed-form expression for directivity that includes element pattern effects in an array with weighted excitations that is analogous to (3-83), which is for uniformly illuminated linear arrays of simple element types. Instead directivity is found by integrating the pattern to find Ω_A and using $D = 4\pi/\Omega_A$. This is a relatively easy task with a math applications computer package.

An approximate formula for array directivity often appears in the literature [8]. It is the product of the directivity of a single element and the directivity of the array with isotropic elements, $D = D_e D_i$. This formula is for large arrays without grating lobe effects, but the reader is cautioned that its accuracy varies greatly with the array geometry [9]. See Prob. 3.5-7. Generally applicable approximate directivity techniques are discussed in Sec. 7.3.3.

In this section, we showed that the side lobe levels of a linear array can be controlled with the current amplitude distribution. To achieve low side lobe levels, it is necessary to taper the element current amplitudes from the center to the edges of the array. However, low side lobe levels are obtained at the cost of reduced directivity due to an increase in beamwidth.

3.6 MUTUAL COUPLING

To this point in our treatment of array antennas, the following assumptions have been used: The element terminal currents are proportional to their excitations, the current distributions on elements in the array are identical, and pattern multiplication is valid. As might be expected, in a real array the elements interact with each other and alter the currents (and thus impedances) from that which would exist if the elements were isolated. This interaction, called *mutual coupling*, changes the current magnitude, phase, and distribution on each element. This is manifested by a total array pattern that is different from the no-coupling case. In addition, these effects depend on frequency and scan direction. In this section, we discuss the effects of mutual coupling on impedance and pattern and present techniques for determining the impedance and pattern of an array in the presence of mutual coupling.

3.6.1 Impedance Effects of Mutual Coupling

The three mechanisms responsible for mutual coupling are illustrated in Fig. 3-26*a*. First is direct space coupling between array elements. Second, indirect coupling can occur by scattering from nearby objects such as a support tower. Third, the feed network that interconnects elements in the array provides a path for coupling. In many practical arrays, feed network coupling can be minimized through proper impedance matching at each element. This permits each element in the array to be modeled with independent generators as in Fig. 3-26*b*, where the mth element has an applied generator voltage and terminal impedance given by V_m^g and Z_m^g. The voltage and current at the element terminals, V_m and I_m, include all coupling effects. An array of N elements is then treated as an N port network using conventional circuit analysis, giving

$$\begin{aligned} V_1 &= Z_{11}I_1 + Z_{12}I_2 + \cdots + Z_{1N}I_N \\ V_2 &= Z_{12}I_1 + Z_{22}I_2 + \cdots + Z_{2N}I_N \\ &\vdots \\ V_N &= Z_{1N}I_1 + Z_{2N}I_2 + \cdots + Z_{NN}I_N \end{aligned} \tag{3-94}$$

where V_n and I_n are the impressed current and voltage in the nth element, and Z_{nn} is the *self-impedance* of the nth element when all other elements are open-circuited. The *mutual impedance* Z_{mn} ($= Z_{nm}$ by reciprocity) between the two terminal pairs

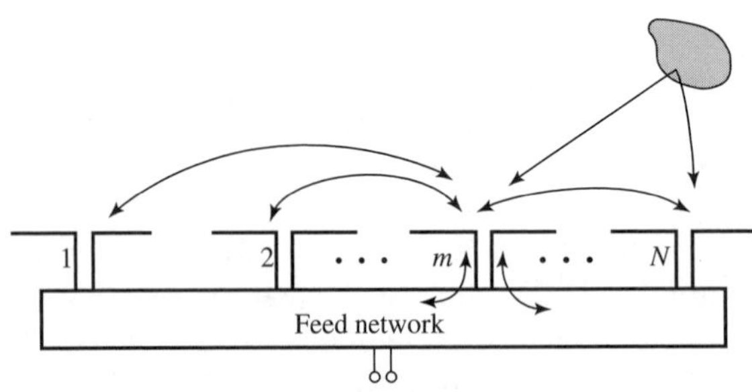

(*a*) Mechanisms for coupling between elements of an array.

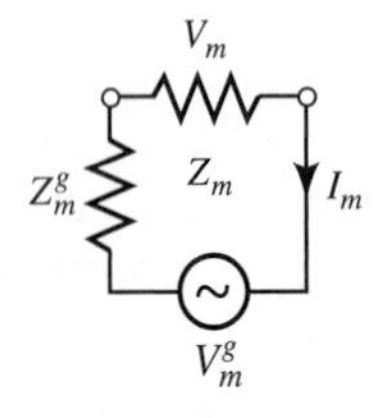

(*b*) Model for m^{th} element in an array.

Figure 3-26 Mutual coupling in a fully excited array antenna.

of elements m and n is the open circuit voltage produced at the first terminal pair divided by the current supplied to the second when all other terminals are open-circuited; that is,

$$Z_{mn} = \left.\frac{V_m}{I_n}\right|_{I_i=0} \qquad \textit{for all } i \textit{ except } i = n \tag{3-95}$$

Mutual impedance is, in general, difficult to compute or measure. However, numerical methods such as the method of moments discussed in Chap. 10 can be used to compute mutual impedance. We now discuss how the mutual impedance between two antennas is measured; this approach is easily generalized for the determination of the mutual impedance between any two elements of an arbitrary array Z_{mn}. Suppose an antenna when isolated in free space has a voltage V_1 and a current I_1, so the input impedance is

$$Z_{11} = \frac{V_1}{I_1} \qquad \textit{single isolated element} \tag{3-96}$$

If a second antenna is brought into proximity with the first, then radiation from the first antenna induces currents on the second, which in turn radiate and also influence the current on the first antenna. The second antenna can either be excited or unexcited (parasitic), but in any case it has terminal current I_2. Then the total voltage at the first antenna is

$$V_1 = Z_{11}I_1 + Z_{12}I_2 \tag{3-97}$$

Similarly, the voltage at the terminals of the second antenna is expressed by

$$V_2 = Z_{21}I_1 + Z_{22}I_2. \tag{3-98}$$

Note that (3-94) is a generalization of (3-97) and (3-98).

Now suppose the second antenna has a load impedance Z_2 across its terminals ($V_2^g = 0$) such that $V_2 = -Z_2^g I_2$. We can write (3-98) as

$$-Z_2^g I_2 = Z_{21}I_1 + Z_{22}I_2 \tag{3-99}$$

Solving for I_2, we obtain

$$I_2 = \frac{-Z_{21}I_2}{Z_{22} + Z_2^g} = \frac{-Z_{12}I_1}{Z_{22} + Z_2^g} \tag{3-100}$$

Substituting this into (3-97) and dividing by I_1, we find that

$$\frac{V_1}{I_1} = Z_1 = Z_{11} - \frac{(Z_{12})^2}{Z_{22} + Z_2^g} \tag{3-101}$$

This expresses the input impedance in terms of the two self-impedances (Z_{11} and Z_{22}), the mutual impedance Z_{12}, and the load Z_2^g at the unexcited terminals of antenna 2.

The above discussion suggests the equivalent circuit of Fig. 3-27 for the coupling between two resonant antennas (see Prob. 1.7-4). For a single isolated antenna (i.e., antenna 2 very far away), $Z_{12} = 0$, and (3-101) gives the input impedance equal to the self-impedance, $Z_1 = Z_{11}$. If antenna 2 is open-circuited, then $Z_2^g = \infty$ and (3-101) gives $Z_1 = Z_{oc} = Z_{11}$. Open circuiting implies that the current all along antenna 2 is reduced to zero. This occurs for antennas such as half-wave dipoles, where resonant behavior is eliminated by open circuiting. In other antennas (such

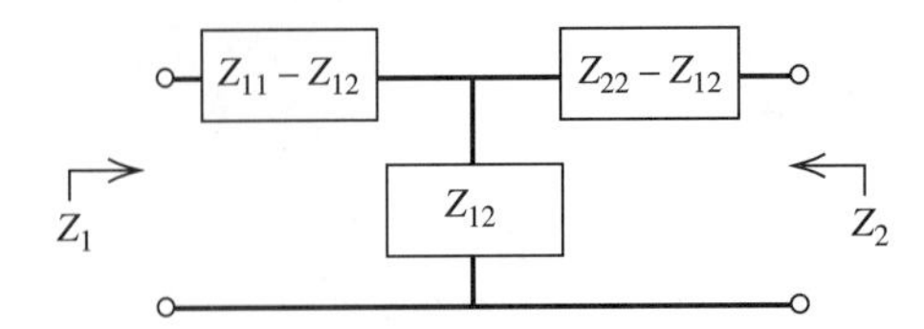

Figure 3-27 Network representation of the coupling between two antennas.

as full-wave dipoles), even with an open circuit there will be current induced on the antenna. In this case, the second antenna should be removed.

The general procedure for determining mutual impedance from open-circuit and short-circuit measurements involves the following steps [2, pp. 157–160]:

1. Open circuit (or remove) antenna 2. Measure $Z_{oc} = Z_{11}$ at the terminals to antenna 1. For identical antennas, $Z_{22} = Z_{11}$.
2. Short circuit antenna 2. Measure Z_{sc} at the terminals to antenna 1.
3. Compute Z_{12}, using

$$Z_{12} = \sqrt{Z_{oc}(Z_{oc} - Z_{sc})} \tag{3-102}$$

This follows from (3-101) with $Z_2^g = 0$.

The proper sign must be chosen with (3-102). This is aided by examining variations with spacing in the limit of small spacing and maintaining continuity through zero crossings [10].

In order to illustrate mutual impedance behavior, consider two resonant half-wave dipoles. Figure 3-28 shows the mutual impedance between two parallel half-wave dipoles as a function of the spacing between their centers calculated using the method of moments in Chap. 10. Figure 3-29 presents the same results for two collinear half-wave dipoles. Examining these figures reveals the following trends, which are general guidelines [11]:

1. The strength of the coupling decreases (but not smoothly) as spacing increases, roughly as $1/d^2$.
2. The far-field pattern of each element predicts coupling strength. If the elements are oriented such that they are illuminated by a pattern maximum, then

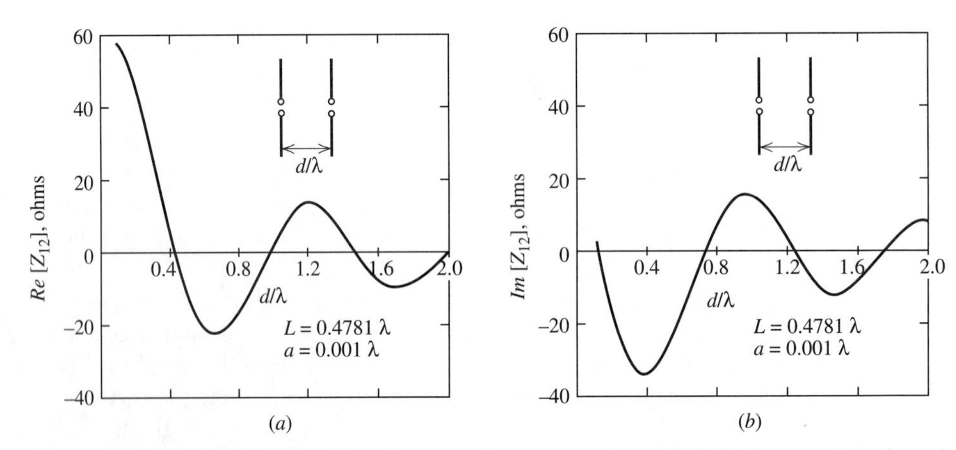

Figure 3-28 The mutual impedance between two resonant parallel dipoles as a function of their spacing relative to a wavelength. (*a*) The real part. (*b*) The imaginary part.

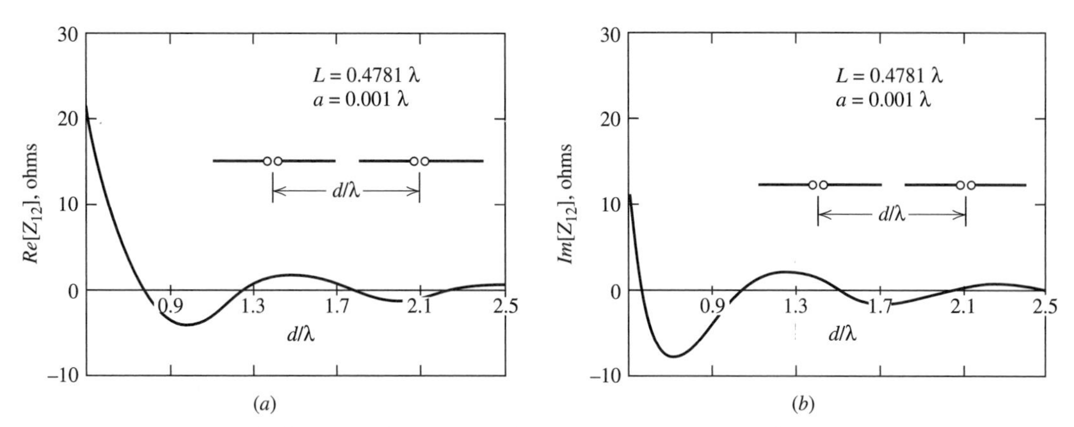

Figure 3-29 The mutual impedance between two resonant collinear dipoles as a function of spacing relative to a wavelength. (*a*) The real part. (*b*) The imaginary part.

the coupling will be appreciable. If, on the other hand, the individual patterns exhibit a null in the direction of the coupled antennas, the coupling will be small.

3. Elements with electric field orientations (i.e., polarizations) that are parallel will couple more than when collinear.
4. Larger antenna elements with broadside patterns have lower coupling to neighboring elements.

For example, two similar parallel elements such as dipoles will couple much more strongly than two collinear dipoles, as dictated by rules 2 and 3 and as seen from Figs. 3-28 and 3-29.

The input impedance of the mth element in the presence of all elements and with all mutual coupling included is found from (3-94) as

$$Z_m = \frac{V_m}{I_m} = Z_{m1}\frac{I_1}{I_m} + Z_{m2}\frac{I_2}{I_m} + \cdots + Z_{mN}\frac{I_N}{I_m} \tag{3-103}$$

This is referred to as the *active impedance*, or driving point impedance. This equation clearly shows the dependence of element input impedance on the mutual impedances and the terminal currents. This include the phase of the currents, which are varied in phase-scanned arrays. Active impedance can be found from (3-103) using mutual impedance values calculated as in Fig. 3-28 or from measurements using (3-102).

3.6.2 Array Pattern Evaluation Including Mutual Coupling

As we have mentioned, mutual coupling not only affects impedance, but also radiation properties such as far-field pattern and polarization. For complete accuracy, the pattern of an array antenna must include the variations in the excitation currents as well as the patterns of each element acting under the influence of all coupling effects. This is a difficult task; however, simplifying approximations are usually possible. Here, we use the model of Fig. 3-26*a* that represents the usual application, often referred to as "free excitation," which models the constant incident power situation of Fig. 3-22. It is well suited to the scattering matrix formulation discussed in [2, Chap. 3] and [8, Sec. 2.1]. Two common approaches to the evaluation of array patterns that use the element currents and element patterns are presented here [10].

In the *isolated element pattern* approach, all coupling effects in the total array pattern are accounted for in the excitations:

$$F_{un}(\theta, \phi) = g_i(\theta, \phi) \sum_{m=1}^{N} I_m e^{j\xi_m} \tag{3-104}$$

where ξ_m is the total phase contribution (usually referenced to the center of the array) due to spatial phase delay. It is the classical array pattern approach of Sec. 3.3 consisting of the product of an isolated element pattern $g_i(\theta, \phi)$ and an array factor. Without coupling effects, the currents $\{I_m\}$ vary in proportion to the excitation voltages. Coupling effects are included using the simple circuit model of the mth element in Fig. 3-26b:

$$I_m = \frac{V_m^g}{Z_m^g + Z_m} \tag{3-105}$$

In the ideal case of an infinite array of identical elements with equal spacings, each element sees the same operating environment and active impedances are identical so that all Z_m equal Z_A. Then the currents are proportional to voltages across the element terminals:

$$I_m = \frac{V_m^g}{Z^g + Z_A} \propto V_m \tag{3-106}$$

This situation applies to large, equally spaced arrays. It must be pointed out that the common implementation of (3-104) uses the approximation of (3-106) for finite arrays, thereby ignoring terminal current variations due to mutual coupling and only including generator voltage variations. It is difficult to obtain accurate current information so that (3-104) can be evaluated, and the active-element pattern method is usually employed; this method is discussed next.

In the *active-element pattern* approach, all coupling effects are accounted for through the active element. The active-element pattern $g_{ae}^n(\theta, \phi)$ is obtained by exciting only the nth element and loading all other elements with the generator impedance Z^g. The active-element pattern arises from direct radiation from the nth element combined with fields reradiated from the other elements, which in turn receive their power through spatial coupling from the element n. The array pattern in this formulation is

$$F_{un}(\theta, \phi) = \sum_{n=1}^{N} g_{ae}^n(\theta, \phi) I_n e^{j\xi_n} \tag{3-107}$$

Here, the currents $\{I_n\}$ are proportional to excitation voltages $\{V_n\}$ as in (3-106). All mutual coupling effects are incorporated into the active-element patterns $g_{ae}^n(\theta, \phi)$, which depend on the element characteristics and the array geometry. To represent the possibility of gain variations, the active-element pattern levels are relative to a reference element near the center of the array.

It would be tedious to measure active-element patterns for each element in an array. Fortunately, this is usually not necessary. For a large array of identical elements in an equally spaced array, each element sees the same environment of nearest neighbors, except for the edge elements. The appropriate approximation is to factor (3-107) using an *average active-element pattern* $g_{ae}(\theta, \phi)$, which is the normalized pattern for a typical central element in the array:

$$F_{un}(\theta, \phi) = g_{ae}(\theta, \phi) \sum_{n=1}^{N} I_n e^{j\xi_n} \tag{3-108}$$

The advantage of this approach is that the summation in the above equation is the array factor based on simple theory without mutual coupling. All coupling effects are contained in the average active-element pattern, which is found by a single pattern measurement of a central element in a large array.

Figure 3-30 shows measured active-element patterns for a linear array of eight elements spaced 0.57λ apart. The polar-dB patterns shown above each element correspond to the case where that element is active and the remaining elements are connected to loads of Z^g. The pattern of the interior element (#5) is very symmetric. The active-element patterns for the remaining interior elements (#2, 3, 4, 6, 7) are nearly identical to that for #5. The patterns of the edge elements (#1 and #8) are slightly asymmetric, but are nearly mirror images of each other. If the pattern asymmetries for the edge elements are not severe, the active-element pattern for a central element can be used for $g_{ae}(\theta, \phi)$ to find the total array pattern with little error. This is found by writing (3-108) in normalized form giving

$$\boxed{F(\theta, \phi) = g_{ae}(\theta, \phi) f(\theta, \phi)} \tag{3-109}$$

where

$$\begin{aligned} g_{ae}(\theta, \phi) &= \text{average active-element pattern} \\ f(\theta, \phi) &= \text{array factor} \\ F(\theta, \phi) &= \text{array pattern} \end{aligned}$$

This is the modeling approach usually used in practice.

The use of "active" is different for active impedance and for active-element pattern. Active impedance for the nth element in an array Z_n is defined with all elements in the array excited. In contrast, the active-element pattern $g_{ae}^{n}(\theta, \phi)$ is defined for the nth element excited and all other elements loaded in their generator impedance. This distinction should be clear; if not, the student should review this section.

The isolated- and active-element patterns can be quite different when mutual coupling is strong. Similarly, the total array pattern computed with and without mutual coupling effects included can be significantly different. Array patterns computed with and without mutual coupling are compared in Sec. 10.11 for dipole ar-

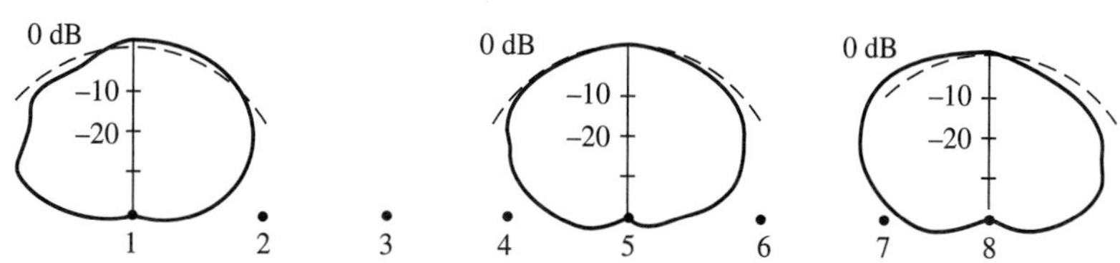

Figure 3-30 Measured active-element patterns for three elements, an interior element (#5) and the end elements (#1 and #8), of a linear array of eight microstrip patch elements spaced 0.57λ apart.

rays; also see [12]. In addition to pattern, the gain and polarization of the total array are affected by mutual coupling. The same guidelines discussed in Sec. 3.6.1 for mutual impedance also apply to patterns. Mutual coupling is especially important in phased arrays and is revisited in Sec. 3.8.

3.7 MULTIDIMENSIONAL ARRAYS

Linear arrays have a number of limitations. For instance, they can be phase-scanned in only a plane containing the line of the elements' centers. The beamwidth in a plane perpendicular to the line of element centers is determined by the element beamwidth in that plane. This usually limits the realizable gain. Thus, multidimensional arrays are used for applications requiring a pencil beam, high gain, or main beam scanning in any direction. With advances in fabrication and integrated feed electronics, the costs of large multidimensional arrays are affordable in many situations. Multidimensional arrays are classified by three characteristics: The geometric shape of the surface on which the element centers are located, the perimeter of the array, and the grid geometry of the element centers. The surface on which elements are placed can be linear, circular, planar, etc. The perimeter of planar arrays, for example, is usually circular, rectangular, or square in shape. Figure 3-31 illustrates a planar array with a rectangular perimeter. The array grid can have equal or unequal row and column spacings. A planar array with equal element spacings of d_x and d_y in the principal planes such as in Fig. 3-31 is referred to as having a rectangular grid. If $d_x = d_y$, the grid is said to be square. A triangular grid is also widely used. When the array conforms to a complicated surface such as the fuselage of an aircraft, the array is said to be conformal. In this section, we present techniques for analyzing arrays of arbitrary geometry as well as a few important special case geometries.

The pattern multiplication principles developed in Sec. 3.3 for linear arrays apply to arrays of any geometry as long as the elements are similar. That is, if the elements are identical and oriented in the same direction, the total array pattern is factorable as in (3-109), which includes mutual coupling effects. This is the usual situation and permits us to confine attention to the array factor $f(\theta, \varphi)$ when studying multidimensional arrays. In this section, we develop the array factor for an arbitrary geometry.

The elements for an arbitrary three-dimensional array are located with position vectors from the origin to the mnth element:

$$\mathbf{r}'_{\mathbf{mn}} = x'_{mn}\hat{\mathbf{x}} + y'_{mn}\hat{\mathbf{y}} + z'_{mn}\hat{\mathbf{z}} \tag{3-110}$$

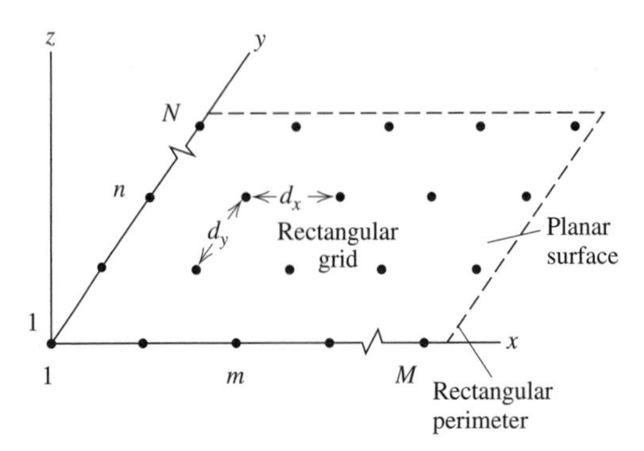

Figure 3-31 Geometry of a planar array.

The array factor is then

$$\mathrm{AF}(\theta, \phi) = \sum_{n=1}^{N} \sum_{m=1}^{M} I_{mn} e^{j(\beta \hat{\mathbf{r}} \cdot \mathbf{r}'_{mn} + \alpha_{mn})} \tag{3-111}$$

which, when normalized, is $f(\theta, \phi)$. This equation is general but is directly applicable to the common situation of an array on a surface. The double summation is useful in geometries that employ "rows" and "columns." The phase term α_{mn} is that portion of the excitation current phase used to scan the main beam and is shown explicitly. A common geometry for phased arrays is planar. The array factor for a planar array in the xy-plane, as in Fig. 3-31, follows from (3-111) as

$$\mathrm{AF}(\theta, \phi) = \sum_{n=1}^{N} \sum_{m=1}^{M} I_{mn} e^{j\alpha_{mn}} e^{j\xi_{mn}} \tag{3-112}$$

where

$$\begin{aligned}
\xi_{mn} &= \beta \hat{\mathbf{r}} \cdot \hat{\mathbf{r}}'_{mn} = \beta[x'_{mn} \sin\theta \cos\phi + y'_{mn} \sin\theta \sin\phi] \\
\alpha_{mn} &= -\beta[x'_{mn} \sin\theta_o \cos\phi_o + y'_{mn} \sin\theta_o \sin\phi_o] \\
\theta_o, \phi_o &= \text{main beam pointing direction}
\end{aligned}$$

This formulation is a generalization of that given for a linear array in Sec. 3.2.2; note that the z-axis is normal to the plane of the array, whereas in our treatment of linear arrays the z-axis is along the array. If all rows parallel to the x-axis have the same current distribution and if all columns have identical current distributions, then the current is separable (e.g., $I_{mn} = I_{xm} I_{yn}$) and (3-112) reduces to

$$\mathrm{AF}(\theta, \phi) = \sum_{m=1}^{M} I_{xm} e^{j\xi_{xm}} \cdot \sum_{n=1}^{N} I_{yn} e^{j\xi_{yn}} \tag{3-113}$$

where the phase of the current for beam steering is not shown explicitly and

$$\xi_{xm} = \beta x'_m \sin\theta \cos\phi \qquad \text{and} \qquad \xi_{yn} = \beta y'_n \sin\theta \sin\phi$$

This is a product of two linear array factors associated with the row and column current distributions. The patterns in the principal planes ($\phi = 0°, 90°$) are those of the corresponding linear arrays (row, column). Planar arrays normally have separable current distributions, so linear array analysis can be applied directly to find the principal plane patterns.

For planes off the principal planes, called intercardinal planes, the pattern is the product of the row and column linear array patterns if the current distribution is separable. This means that in an off-principal plane away from the main beam, the side lobes will be very low if the principal plane patterns have low side lobes, since they are a product of low side lobe levels; see Fig. 7-7 for the analogous situation in aperture antennas. Another powerful equivalent linear array technique can be used for off-principal planes, whether the current distribution is separable or not. The technique works by projecting each element onto a line in the pattern plane. The projection lines locate elements in the equivalent array. If multiple array elements lie on the same projection line, the currents are summed to determine the equivalent array element current.

Arrays are often operated with a conducting ground plane parallel to the plane of the array to greatly reduce radiation in the hemisphere behind the ground plane. This increases the array directivity. The array is usually placed a quarter-wavelength in front of the array. This gives a total of a half-wavelength path length for the reflected waves from each element. But an additional 180° phase shift is encountered during reflection, so the reflected wave reinforces the direct wave for each element. The image theory of Sec. 2.3.1 is used to find the pattern. See Probs. 3.7-7 and 3.7-8. A second way to find the pattern of an array backed by a ground plane is to consider each element and its image as a pair and form a new element pattern in place of each element in the planar array. See Prob. 3.7-9 for an example.

If sufficient symmetry exists, the patterns of multidimensional planar arrays can often be found using sequential linear array analysis along each principal direction; see Prob. 3.7-10. Multidimensional arrays will be treated at other points in the book. Circular and planar arrays are discussed in Sec. 10.11. The directivity of multidimensional arrays can be found using the techniques in Sec. 7.3.

3.8 PHASED ARRAYS AND ARRAY FEEDING TECHNIQUES

Many antenna system applications require that the main beam pointing direction be changed with time, or scanned. This can be done by mechanically slewing a single antenna or an array with fixed phase to the elements. However, mechanical slewing requires a positioning system that can be costly and scan too slowly. Phased arrays, on the other hand, scan the main beam at electronic speeds. In general, a **phased array** is defined as an array antenna whose main beam maximum direction is controlled by varying the phase or time delay to the elements. Phased arrays are used in radar where rapid target tracking is required, in direction finding, and in communications applications where the radiation pattern must be adjusted to accommodate varying traffic conditions. Array technology is moving toward the integration of transmit/receive electronics and associated controllers. The antenna is then a subsystem rather than a separate device. The term *smart antenna* has been coined for antennas that include control functions such as beam scanning. Smart antennas are finding wide use in several commercial and military applications. A unique advantage offered by phased arrays is the ability to form multiple main beams pointing in different directions simultaneously. In situations that require two-dimensional scanning, hybrid approaches are used that combine mechanical pointing in one plane and electronic scanning in the orthogonal plane. An example is AWACS (Airborne Warning and Control System), which is flown on top of U.S. military aircraft. The array of 4,000 waveguide slots is scanned electronically in elevation and rotated mechanically inside a radome for azimuth scan. This airborne surveillance radar achieves ultra-low side lobes that are below −40 dB. In this section, we explain the methods of scanning and the feed networks required to support scanning functions.

3.8.1 Scan Principles

The principles of phase scanning have been discussed throughout this chapter. The goal is to form a smooth phase front (i.e., a wave focused at infinity) that is tipped to change the main beam direction, which is normal to the phase front. We illustrate this with a linear array along the z-axis that can have unequally spaced elements by generalizing (3-65):

$$\mathrm{AF}(\theta) = \sum_{n=0}^{N-1} I_n e^{j\xi_n} = \sum_{n=0}^{N-1} A_n e^{j(\alpha_n+\delta_n)} e^{j\xi_n} \tag{3-114}$$

The spatial phase for elements with arbitrary positions $\{z_n\}$ along the z-axis is

$$\xi_n = \beta z_n \cos \theta \tag{3-115}$$

The current for the nth element has magnitude A_n and phase $\alpha_n + \delta_n$, where α_n is the portion of the phase that varies linearly with distance along the z-axis and is responsible for steering the main beam peak to angle θ_o:

$$\alpha_n = -\beta z_n \cos \theta_o \tag{3-116}$$

Since α_n varies linearly with position z_n, it is referred to as *linear phase*, or *uniform progressive phase*. The slope, or constant of proportionality, is $-\beta \cos \theta_o$. The remaining part of the excitation phase δ_n is nonlinear with distance and is useful in beam shaping as discussed in Sec. 8.3. For an equally spaced array, $z_n = nd$ and $\alpha_n = n\alpha$, where $\alpha = -\beta d \cos \theta_o$ as in (3-36).

As the pattern of an array is scanned off broadside, the main beam widens. This effect is called *beam broadening*. We illustrate this for a linear array of five isotropic elements spaced 0.4λ apart. Figure 3-32 shows a series of patterns for increasing off-broadside scan angles. Notice the increase in the beamwidth of the main beam with scan off broadside. The full pattern for this array is obtained by rotating the pattern about the z-axis. Two examples of three-dimensional patterns are shown in Figs. 3-32*b* and 3-32*f*. As the main beam is scanned away from broadside, the increase in beam solid angle of the main beam is just about compensated for by the reduced solid angle of the total pattern (formed by rotation of the pattern about the array axis). Thus, directivity of the array factor remains relatively constant with scan for spacings less than a half-wavelength and for scan angles not close to endfire; see Fig. 3-21. For spacings slightly greater than a half-wavelength, a grating lobe begins to appear for scan angles near endfire and the directivity decreases; again, refer to Fig. 3-21. Since isotropic elements were assumed, these remarks apply to array factors. When the element pattern effects are included for the case of a directive, broadside element pattern, directivity will decrease with scan angle.

As we noted in Sec. 3.1, for half-wavelength spacings there is exactly one period of the array factor in the visible region and no grating lobe will be visible, except for endfire operation that produces two endfire beams. For spacings larger than a half-wavelength, part or all of a grating lobe may be visible depending on scan angle. For one wavelength spacing or more, there will be visible grating lobes. When spacings of several wavelengths are used, many grating lobes are visible and the array is called an **interferometer**. Each major lobe has a narrow beamwidth but there are many of them. Large element spacings, however, permit electrically large elements with relatively narrow beamwidth patterns that act to decrease the size of the grating lobes. In normal applications, grating lobes limit the performance of phased arrays and are to be avoided. Grating lobe peaks will be not appear in the visible region if element spacings are restricted as follows:

$$d < \frac{\lambda}{1 + |\cos \theta_{o\,\max}|} \qquad \textit{to avoid grating lobes} \tag{3-117}$$

where $\theta_{o\,\max}$ is the maximum main beam scan angle with respect to the line of the array. This relation is derived by solving (3-38) for the first grating lobe at $\psi = 2\pi$, where $\theta = 0°$. For broadside operation, $\theta_{o\,\max} = 90°$ and (3-117) gives $d < \lambda$. If scanning to endfire ($\theta_{o\,\max} = 0, 180°$) is desired, then $d < \lambda/2$. This result is based

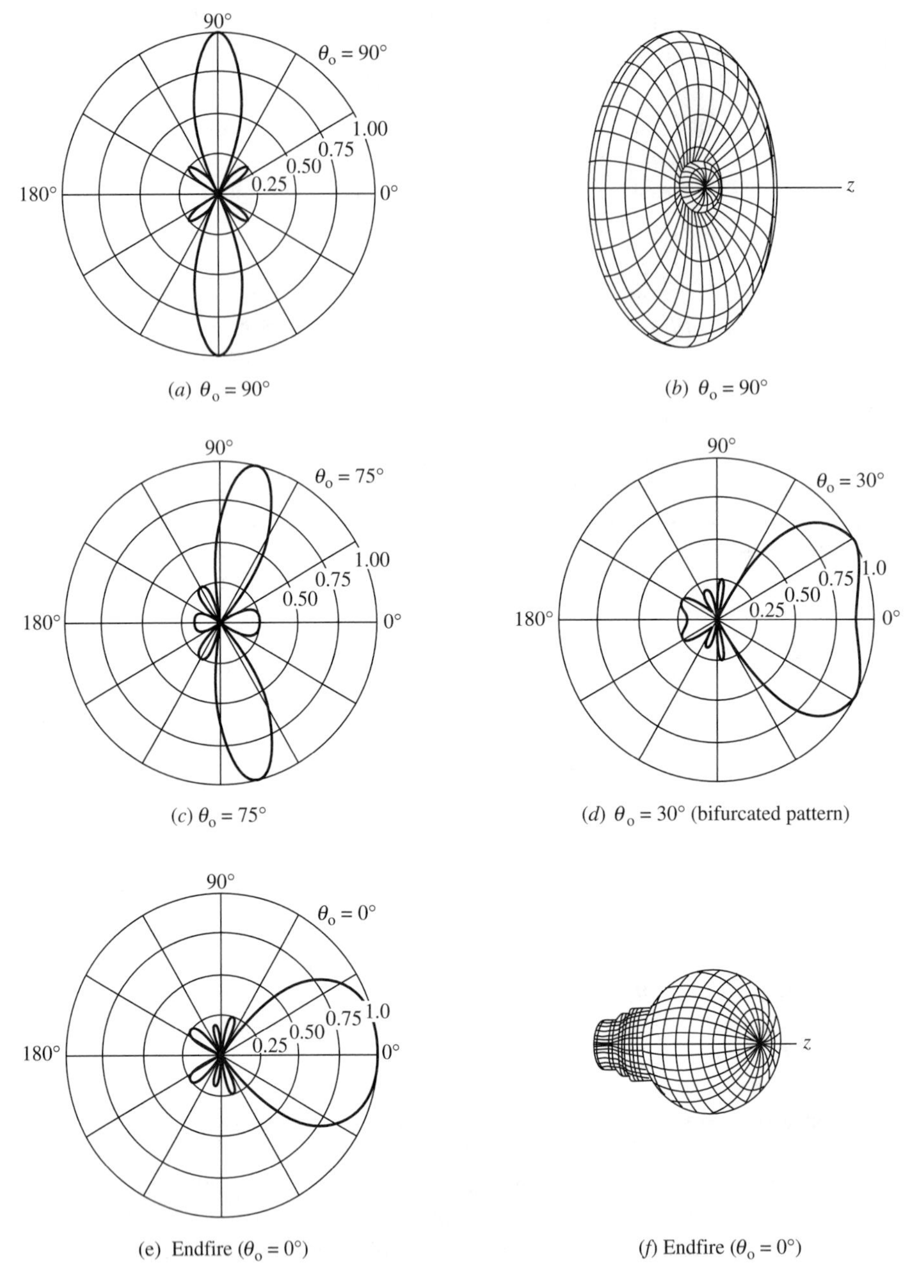

Figure 3-32 Example of phase-scanned patterns for a five-element linear array along the z-axis with elements equally spaced at $d = 0.4\lambda$ and with uniform current magnitudes for various main beam pointing angles θ_o.

on an omnidirectional element pattern in the plane of scan. Larger spacings are permitted for directive element patterns since they diminish the effect of grating lobes.

As we saw in the previous section, the above linear array analysis can also be applied to find the patterns of a planar array. The array factor and associated phases of the individual elements α_{mn} needed to steer the main beam peak to the direction (θ_o, ϕ_o) for a multidimensional array are given in (3-112).

3.8.2 Feed Networks for Beam Scanning

The many attractive features of arrays come with the penalty that each element of the array must have a transmission path to the receiver (or transmitter). In addition, phased arrays must control the phase of the elements. The hardware-connecting elements of an array are called a *feed network* or *beam-forming network* (BFN). Feed networks can take one of three basic geometric forms: parallel, series, or space, as illustrated in Fig. 3-33. The *parallel feed* of Fig. 3-33*a* is also called a *corporate feed* because of its similarity to an organization diagram of a corporation. The path length to each element from the feed point is equal; thus, the phase of the excitations will also be equal. Note that impedance effects must be included in feed design. For example, parallel combining of two elements will half the impedance. Techniques for impedance matching are discussed with microstrip arrays in Sec. 5.7.

The *series feed* of Fig. 3-33*b* is easy to construct and requires little feed network hardware. As the wave travels down the transmission line, it is attenuated because of power radiated from the elements and this power loss must be accounted for when determinating the element excitations. The phase of each element is determined by the connecting transmission line lengths and mutual coupling effects. Series-fed arrays can be scanned by changing frequency, which changes the electrical length between elements. This is referred to as *frequency scanning*. However, except for some radar applications, system constraints usually require the frequency to be fixed. An example of a series-fed array is a waveguide with slots milled in one wall that act as radiating elements.

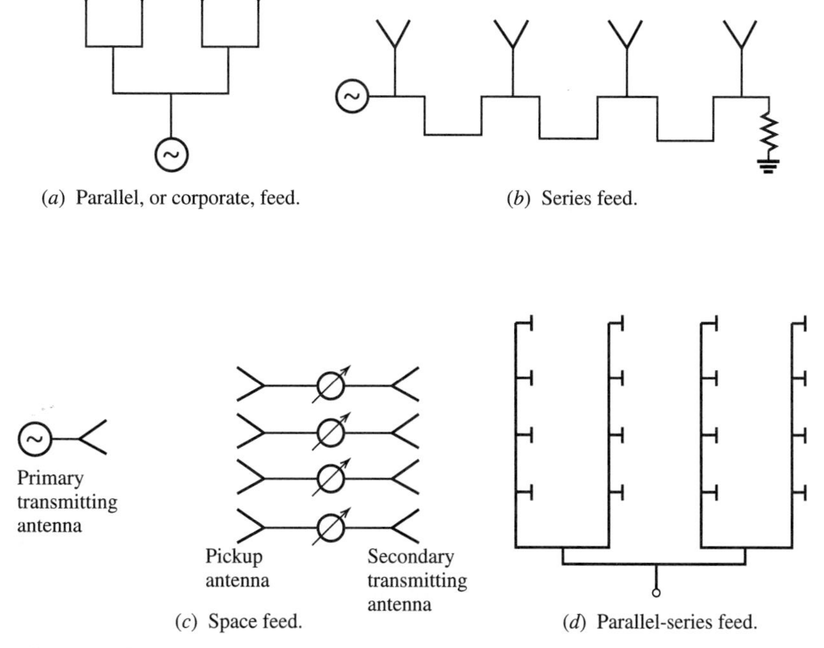

Figure 3-33 Types of array feed networks.

The losses, cost, weight, and size of a BFN can be avoided by using a *space feed* as illustrated in Fig. 3-33*c*. In the space feed, a primary antenna is connected to the transmitter and illuminates several pickup antennas that are connected to secondary radiating elements that form the array. The amplitude distribution of the array is determined by the primary antenna pattern and pickup antenna placement. The phase introduced by different path lengths to each pickup antenna can be compensated for with phase shifters on the radiating elements, which can also used for beam scanning.

The three types of feed geometries were illustrated for linear arrays, but they also apply to multidimensional arrays. The parallel feed network (realized with radio frequency interconnections) for arrays with many elements will become large. To avoid this, multidimensional arrays often employ a hybrid combination of feed types. A parallel-series *hybrid feed* for a planar array is shown in Fig. 3-33*d*. The rows are fed in parallel and the columns are series-fed. The parallel-parallel hybrid feed facilitates the use of subarrays that have the same amplitude and/or phase in each element of the subarray. The feed types are illustrated here as operating at the same frequency as that of the antenna. However, down conversion to a lower frequency before beam forming is often employed. This permits the use of less expensive electronic components to form the beam and often includes digital processing. This is a popular approach in modern adaptive arrays for use in both military and commercial wireless applications. Another non-RF realization is the *optical feed*, which is a recent innovation that converts radio waves to optical frequencies and back again in the feed network to take advantage of the small size of optical devices.

The physical construction of the feed network and transmit/receive electronics behind the array face takes one of two forms, brick or tile. In the *brick* construction, the complete feed hardware modules for one (or a few elements) is placed perpendicular to the array face behind it. The *tile* construction consists of several parallel layers with each layer containing the same components; that is, one layer might have all the phase shifters and another might have all the low-noise amplifiers.

Beam steering is accomplished in the feed network using *electronic scanning*. There are four methods of electronic scanning: frequency scanning, phase scanning, time-delay scanning, and beam switching [13]. *Frequency scanning* was discussed above in association with series feeds. *Phase scanning* is electronic beam pointing by variation of the element current phases and it also was discussed above. *Time-delay scanning* is a form of phase scanning in which phase change is achieved by switching in different lengths of transmission paths. It overcomes the instantaneous bandwidth limitations of phase shifters. *Beam switching* involves the use of different transmission paths in the beam-forming network for each beam and has one input port for each beam. A beam-switched BFN, such as the *Butler matrix*, avoids the use of variable phase shifters [14, 15]. This *multiple beam feed network* has beam ports that can be used to form simultaneous beams, or the port outputs can be weighted and combined for pattern control.

Electronic beam scanning is commonly realized with a phased array using phase shifters that are either analog or digital. Analog phase shifters can be set to any phase over its range. Digital phase shifters introduce 2^M discrete multiples of $360°/2^M$ phase, where M is the number of bits of the phase shifter. For example, a 3-bit phase shifter can be set to one of the following $2^3 = 8$ phases; 0, 45, 90, 135, 180, 225, 270, or 315°. Typically, phased arrays require 4- to 7-bit phase shifters, which have 22.5 to 2.8° phase increments. Quantization of the phase with digital phase shifters increases the side lobe levels [8, Sec. 7.3]. There are many forms of phase shifters that can be broadly classified as either ferrite or semiconductor-diode [15]. Ferrites in the transmission path alter the phase in proportion to an applied static magnetic field. Ferrite phase shifters can be either digital for fast switching or analog for more accuracy, lower loss, and lower cost. Diode phase shifters offer fabrication advantages and can be integrated with other feed network electronics.

The simplest BFN is completely passive and consists of power dividers and fixed phase shifters. Placement of active devices at the subarray level gives a semidistributed approach that reduces the effects of active device failure. The most flexible design employs fully distributed BFNs with active devices at the element level.

Another feed configuration used in tracking systems is the *sum and difference feed*. This feed network combines the left and right halves of an array both in-phase and out-of-phase, creating a *sum pattern* and a *difference pattern*. The sum pattern is that of a conventional array with a broad main beam for coarse angular tracking of distant sources in communications or targets in radar. The difference pattern has a sharp on-axis null that is used for fine angle tracking. See Prob. 3.8-7.

3.8.3 Scan Blindness

Phased array performance varies during beam scanning mostly due to the changes introduced by mutual coupling. The most serious of these effects is *scan blindness*, which is manifested by a dramatic reduction of radiated power for certain "blind" scan angles. In transmitting applications such as radar, when the feed network is configured to steer to a blind scan angle, generator power is reflected rather than radiated, which can damage the transmitter electronics. If $\Gamma_m(\theta_o, \phi_o)$ is the voltage reflection coefficient of the mth element in a fully excited phased array, the power delivered to that element is

$$P_m = P_{\text{inc}}[1 - |\Gamma_m(\theta_o, \phi_o)|^2] \tag{3-118}$$

where P_{inc} is the incident power; see Prob. 3.8-6. Based on this, we can express the active-element pattern as [16]

$$g_{ae}^m(\theta_o, \phi_o) = g_i(\theta_o, \phi_o)[1 - |\Gamma_m(\theta_o, \phi_o)|^2] \tag{3-119}$$

where $g_i(\theta_o, \phi_o)$ is the isolated element pattern. If we factor the total array pattern as in Sec. 3.6, the average active-element pattern for large arrays with equal element spacings is

$$g_{ae}(\theta_o, \phi_o) \approx g_i(\theta_o, \phi_o)[1 - |\Gamma(\theta_o, \phi_o)|^2] \tag{3-120}$$

As the array is scanned by changing (θ_o, ϕ_o), the peak of the total array pattern follows the active element pattern.

Usually, an array is matched at broadside so that $\Gamma(0, 0) = 0$ and impedance mismatch increases with scan angle away from broadside [8, p. 66]. If scanned far enough, a complete mismatch may occur and $|\Gamma(\theta_o, \phi_o)| = 1$. Then (3-120) shows that the active-element pattern is zero and no radiation occurs, giving scan blindness. Scan blindness is associated with surface wave-like phenomena and usually occurs for spacings less than those for which grating lobes occur; see (3-117) [8, p. 343]. It usually can be avoided by using spacings of a half-wavelength or less. Scan blindness is illustrated with a planar array of dipoles in Sec. 10.11.3.

3.9 PERSPECTIVE ON ARRAYS

One of the more difficult choices the antenna designer has to make is between an array and a single large antenna such as a reflector. The decision is based on available space, power handling, cost, and requirements on scanning. Large printed arrays can be constructed at reasonable cost, but feed network losses reduce overall efficiency. The tradeoff between arrays and aperture antennas is discussed further in Sec. 5.7.

Electronic scanning arrays will find many applications in the future as fabrication costs decrease. Smart antennas make use of electronic scanned arrays. Arrays also offer the opportunity for combining several functions in a single antenna using a *multifunctional array*. For example, several communication functions can be handled along with radar through the same array. The many antennas protruding from a vehicle can be replaced by one multifunctional array on the surface of the vehicle. Arrays can also be used with aperture antennas. One such use is an array placed at the focal point of a reflector antenna to achieve rapid electronic scanning of the reflector beam over a limited angular sector while the whole reflector is moved mechanically.

In this chapter, the relationship between a current distribution in space and the resulting radiation pattern was established. We found that the array factor for an equally spaced array has the form of a Fourier series; see (3-18). In the next chapter, we will see that the pattern factor for a continuous linear current distribution has the form of a Fourier transform. The same general principles apply to both discrete and continuous current distributions. For example, as the current magnitude taper increases toward the ends of the source, the side lobes decrease and the main beam widens. The operation of continuous aperture antennas discussed in subsequent chapters can be understood using the principles introduced in this chapter for array antennas.

REFERENCES

1. J. Ramsay, "Highlights of Antenna History," *IEEE Ant. Prop. Society Newsletter*, Vol. 6, pp. 8–20, Dec. 1981.
2. R. C. Hansen, Ed., *Microwave Scanning Antennas*, Vol. II, *Array Theory and Practice*, Academic Press, New York, 1966, pp. 23–29.
3. W. W. Hansen and J. R. Woodyard, "A New Principle in Directional Antenna Design," *Proc. IRE*, Vol. 26, pp. 333–345, March 1938.
4. M. T. Ma, *Theory and Application of Antenna Arrays*, Wiley, New York, 1974, Sec. 1.3.
5. H. Bach, "Directivity Diagrams for Uniform Linear Arrays," *Microwave Journal*, Vol. 15, pp. 41–44, Dec. 1972.
6. H. Bach, "Directivity of Basic Linear Arrays," *IEEE Trans. Ant. Prop.*, Vol. AP-18, pp. 107–110, Jan. 1970.
7. S. A. Schelkunoff, "A Mathematical Theory of Linear Arrays," *Bell Syst. Tech. J.*, Vol. 22, pp. 80–107, Jan. 1943.
8. R. J. Mailloux, *Phased Array Antenna Handbook*, Artech House, Boston, MA, 1994, Chap. 2.
9. B. J. Forman, "Directivity of Scannable Planar Arrays," *IEEE Trans. Ant. Prop.*, Vol. AP-20, pp. 245–252, May 1972.
10. A Parfitt, D. Griffin, and P. Cole, "Mutual Coupling Between Metal Strip Antennas on Finite Size, Electrically Thick Dielectric Substrates," *IEEE Trans. Ant. Prop.*, Vol. 41, pp. 108–115, Jan. 1993.
11. A. Rudge, K. Milne, A. Olver, and P. Knight, Eds., *The Handbook of Antenna Design*, Vol. II, Peregrinus, London, 1982, Sec. 10.3.
12. D. Kelley and W. Stutzman, "Array Antenna Pattern Modeling Methods That Include Mutual Coupling Effects," *IEEE Trans. Ant. Prop.*, Vol. 41, pp. 1625–1632, Dec. 1993.
13. E. Brookner, Ed., *Practical Phased-Array Antenna Systems*, Artech House, Boston, 1991, Chap. 1.

14. J. Butler and R. Lowe, "Beam-Forming Matrix Simplifies Design of Electronically Scanned Antennas," *Electronic Design*, Vol. 7, pp. 170–173, April 12, 1961.
15. R. C. Johnson, Ed., *Antenna Engineering Handbook*, 3rd ed., McGraw-Hill, New York, 1993, Chap. 20.
16. D. Pozar, "The Active Element Pattern," *IEEE Trans. Ant. Prop.*, Vol. 42, pp. 1176–1178, Aug. 1994.

PROBLEMS

3.1-1 Consider an array of two elements spaced one wavelength apart with currents that are equal in amplitude and 180° out-of-phase.
a. Use the inspection method to rough sketch the polar plot of the array factor.
b. Derive the exact array factor as a function of θ if the elements are on the z-axis.
c. For what angles of θ is this array factor maximum?
d. What is the expression for the normalized array factor $|f(\theta)|$?
e. Show that (3-20) reduces to your answer in (d).

3.1-2 Use the techniques of Fig. 3-9 to obtain a polar plot of the array factor of the array given in Prob. 3.1-1.

3.1-3 Use the techniques of Fig. 3-9 to obtain a polar plot of the array factor of a two-element, one-wavelength spaced array with equal amplitude and equal phase currents (Example 3-4).

3.1-4 Usually, the interelement spacing of an array is about one-half wavelength. Spacings much greater than this produce major lobes in undesired directions. To illustrate this point, use the techniques of Fig. 3-9 to sketch the array factor for a two-element array with equal amplitude, in-phase elements in polar form for the following spacings: (a) $d = 3\lambda/4$ and (b) $d = 2\lambda$. Examples 3-1 and 3-4 and this problem show the effects of spacing on an array of fixed excitation.

3.1-5 Using the array factor for a two-element broadside array ($\alpha = 0$) with equal current amplitude point source elements, show that the directivity expression is

$$D = \frac{2}{1 + (\sin \beta d)/\beta d}$$

Hint: Change from variable θ to $\psi = \beta d \cos \theta$.

3.1-6. Plot the directivity expression of Prob. 3.1-5 as a function of d from zero to two wavelengths.

3.2-1 Prove that the array factor magnitude $|f(\psi)|$ for a uniformly excited, equally spaced linear array is symmetric about $\psi = \pi$.

3.2-2 Show that the array factor expressions (3-20) and (3-35) for a two-element uniformly excited array are identical.

3.2-3 Drive (3-44).

3.2-4 The expression for the half-power beamwidth of the array factor for a broadside, uniformly excited, equally spaced, linear array may be approximated as

$$\text{HP} \approx K \frac{\lambda}{Nd}$$

for $Nd \gg \lambda$. Determine K for $N = 10$ and 20, and compare to (3-45).

3.2-5 In this problem, the effects of phasings and spacings on a simple array are illustrated. Consider an equally spaced five-element array with uniform current amplitudes. Sketch the array factors for:
a. $d = \lambda/2$, broadside case ($\theta_o = 90°$).
b. $d = \lambda$, broadside case.
c. $d = 2\lambda$, broadside case.
d. $d = \lambda/2$, $\theta_o = 45°$.
e. $d = \lambda/2$, $\theta_o = 0°$.

These five plots can be obtained from one universal pattern plot as discussed in Secs. 3.1 and 3.2. For the last two cases, determine the interelement phase shift α required to steer the main beam as specified.

3.2-6 Repeat Prob. 3.2-5 using a computer code (see Appendix G).

3.2-7 Design a five-element uniformly excited, equally spaced linear array for:

a. Main beam maximum at broadside.

b. Main beam maximum at 45° from broadside ($\theta_o = 45°$).

In each case, select the element spacing and linear phasing such that the beamwidth is as small as possible and also so that no part of a grating lobe appears in the visible region. Sketch the polar plots of the patterns.

3.2-8 Design and plot the array factor for an ordinary endfire, five-element, uniformly excited linear array with spacings $d = 0.35\lambda$. Use $\theta_o = 180°$ and find α.

3.2-9 Design a linear array of five isotropic elements for ordinary endfire spacing and phasing, and a single main beam using the maximum spacing. Plot the polar pattern.

3.2-10 Design a linear array of five isotropic elements for Hansen–Woodyard increased directivity with $d = 0.35\lambda$. Plot the polar pattern. (Note the differences in the results for Prob. 3.2-8 and Example 3-7.)

3.2-11 Design a linear array of 10 isotropic elements for Hansen–Woodyard increased directivity with $d = 0.4\lambda$. Plot the polar pattern.

3.2-12 Show that the array factor for a uniformly excited, equally spaced linear array approaches the pattern factor of a uniform line source (i.e., neglect the element factor) in the limit of small array element spacings.

3.3-1 Two collinear half-wave dipoles are spaced a half-wavelength apart (but not quite touching) with equal amplitude and equal phase terminal currents. What is the expression for the far-field pattern $F(\theta)$ if the element centers are along the z-axis? Use pattern multiplication ideas to rough sketch the pattern.

3.3-2 a. Repeat Prob. 3.3-1 for one-wavelength spacing.

b. Plot $F(\theta)$ directly to check your pattern multiplication result.

3.3-3 Two parallel half-wave dipoles are spaced one wavelength apart with equal amplitude and equal phase terminal currents. The element centers are along the z-axis and the dipoles are parallel to the x-axis. Write the expressions for the far-field pattern $F(\theta, \phi)$. Rough sketch this pattern, using pattern multiplication ideas, in both the xz-plane and yz-plane.

3.3-4 A linear array of three, quarter-wavelength long, vertical monopoles is operated against an infinite, perfectly conducting ground plane. Let the element feeds be along the z-axis, the ground plane in the yz-plane, and the monopoles in the x-direction.

a. Design the array as a Hansen–Woodyard increased directivity endfire array, that is, determine the element spacings and phasings (choose $d = 0.3\lambda$).

b. Use the universal array factor plot for three uniformly excited elements to obtain a polar plot of the array factor for this problem.

c. Write the expression for the complete pattern.

d. Using pattern multiplication ideas, rough sketch the complete far-field patterns in the xz-plane and yz-plane.

3.3-5 Three collinear half-wave dipoles are spaced $d = 3/4\lambda$ apart and are excited with uniform magnitude and phase. Use simple array theory to obtain the polar-linear pattern, starting with the universal pattern plot followed by pattern multiplication.

3.3-6 An array of four identical *small* circular loop antennas oriented with the loops in the xy-plane are spaced $3\lambda/4$ apart along the x-axis. For equal amplitude and phase excitation, use simple array theory and pattern multiplication to obtain the polar-linear patterns in the xy-, xz-, and yz-planes.

3.3-7 Suppose a truck uses a Citizens Band radio to communicate at 27 MHz. The antenna system is two quarter-wave monopoles parallel to the x-axis (assumed to operate above a perfect ground plane) mounted on mirrors 2.78 m apart along the z-axis and fed with equal amplitude and phase. Use simple array theory to obtain sketches of the array patterns in the three principal planes.

3.3-8 *Design problem.* Design a broadside linear array of four parallel half-wave dipole antennas for as narrow a beamwidth as possible and with no level outside the main beam above -8 dB relative to the main beam peak. The excitations are uniform. (a) Find the spacing d. (b) Sketch the polar patterns in the E- and H-planes.

3.4-1 Calculate the directivities in decibels for the following broadside arrays of point sources:

a. $N = 2$, $d = \lambda/2$.

b. $N = 10$, $d = \lambda/2$.

c. $N = 15$, $d = \lambda$.

3.4-2 Evaluate (3-78) for $d = 3\lambda/8$ and $N = 10$ for:

a. Broadside, and compare the results to that of (3-80),

b. Ordinary endfire, and compare the result to that of (3-81).

3.4-3 Evaluate (3-78) and plot D as a function of d/λ for $N = 5$ and endfire operation. Compare to Fig. 3-21.

3.4-4 The approximate directivity formula of (3-80) for long, broadside linear arrays of isotropic elements can be checked in the following two ways using HP $\approx 0.886\ \lambda/L$ from (3-45):

a. Use $D = 4\pi/\Omega_A$ to find D in terms of λ/L by approximating Ω_A as 2π HP.

b. It has been shown [17] that the following formula gives good results for broadside collinear arrays:

$$D = \frac{101}{\text{HP}_d - 0.0027(\text{HP}_d)^2}$$

where HP_d is the half-power beamwidth of the array pattern in degrees. Use HP $\approx 0.886\ \lambda/L$ and $L \gg \lambda$ to find $D \approx KL/\lambda$; give the value of K.

3.4-5 Use a computer to calculate the directivity for the arrays given in Table 3-1. Treat N, D/λ, and α input variables. Use it to (a) plot the curve in Fig. 3-20 for $N = 5$, and (b) plot a directivity versus d/λ curve for an array of eight collinear short dipoles.

3.4-6 Show that $D = N$ for an ordinary endfire linear array of uniformly excited isotropic elements with spacing $d = \lambda/4$.

3.4-7 Evaluate the directivity in decibels of a uniformly excited, broadside linear array of eight isotropic elements spaced 0.7λ apart in two ways: (a) use Fig. 3-20 and (b) use (3-80).

3.4-8 Compute the exact directivity of a broadside linear array of four isotropic elements with uniform currents and spaced 0.8λ apart. Compare to the approximate result based on $2L/\lambda$.

3.4-9 Derive the normalization factor in the numerator of the directivity expression in (3-78). Evaluate for a Hansen–Woodyard array of five isotropic elements.

3.4-10 *Design problem.* Base station communication antennas are often constructed using a collinear array of half-wave dipoles oriented vertically to produce an omnidirectional pattern in the horizontal plane. The objective in this problem is to maximize gain by selecting the proper element spacing. Assuming uniform amplitude and phase excitation, design a maximum-directivity four-element array for the middle of the cellular telephone band (824 to 894 MHz). Show solution details. Give the values of spacing d in wavelengths and directivity D in decibels at midband and at the band edges. Plot the vertical pattern in polar-linear form. Sketch the array, showing its physical length along with a feed network.

3.5-1 Use an array pattern plotting computer code (see Appendix G) to plot the array factor for the following arrays: (a) Fig. 3-24*a*; (b) Fig. 3-24*b*; (c) Fig. 3-24*c*; (d) Fig. 3-24*d*; (e) Fig. 3-24*e*; (f) Fig. 3-25. Give the HP value and maximum side lobe level and its location.

3.5-2 (a) Show that (3-91) follows from (3-90). (b) Show that (3-93) follows from (3-91).

3.5-3 Verify the directivity values associated with Figs. 3-23*c* through 3-23*e*, and Fig. 3-25.

3.5-4 *Shaped beam patterns.* Array analysis techniques can be used for shaped beam patterns. In this problem, we use the currents derived in Examples 8-3 and 8-4 for producing a sector radiation pattern. The arrays are 20 half-wave spaced isotropic elements. Use an array computer code (see Appendix G) to generate the pattern for (a) the Fourier series synthesized array of Example 8-3 and (b) the Woodward–Lawson synthesized array of Example 8-4.

3.5-5 *Binomial array.* Consider a linear array of elements spaced $d = \lambda/2$ apart with binomial current weightings.

a. Derive the normalized array factor expression in θ.

b. Derive an expression for the directivity.

c. Evaluate directivity for $N = 5$.

3.5-6 An often quoted approximate directivity formula is $D \approx D_e D_i$, where D_e and D_i are the directivities of one element in the array and of the array with isotropic elements, respectively. Evaluate this approximation for the following arrays of short dipoles and compare to the exact values found from (3-83):

a. $N = 4$, broadside, collinear, $d = \lambda/2$.
b. $N = 4$, endfire, parallel, $d = \lambda/2$.
c. $N = 4$, broadside, parallel, $d = \lambda/2$.
d. $N = 3$, broadside, collinear, $d = \lambda/2$.
e. $N = 3$, broadside, parallel, $d = \lambda/2$.

3.5-7 *Array directivity computation.* Consider a collinear array of dipoles.

a. Write a computer code to evaluate the array directivity by direct integration. Verify its accuracy by comparing to values obtained using (3-83) for four collinear short dipoles.

b. Use your computer code to find the directivity of collinear arrays of 2, 3, and 4 half-wave dipoles spaced a half-wavelength apart and excited with uniform amplitude and phase. Tabulate directivities as ratios and compare to the directivities for isotropic-element arrays D_i. Include in the table the directivity found using the approximation $D \approx D_e D_i$; see Prob. 3.5-6.

c. Repeat (b) for the $N = 4$ case with a binomial current amplitude distribution.

3.6-1 Two antennas have the following self- and mutual impedances:

$$Z_{11} = 70\angle 0°, \qquad Z_{22} = 100\angle 45°, \qquad Z_{12} = 60\angle -10°$$

a. Find the input impedance to antenna 1, if antenna 2 is short-circuited.

b. Find the voltage induced at the open-circuited terminals of antenna 2 when the voltage applied to antenna 1 is $1\angle 0°$ V.

3.6-2 Derive (3-102), making use of Fig. 3-27.

3.7-1 Compute and tabulate the element current phases for the four scan cases in Fig. 3-32. The center element has zero phase.

3.7-2 Use an array pattern plotting code (see Appendix G) to plot the array factor for the following arrays. Give the HP value and the maximum side lobe level and its location. Compute the excitation phases for each element: (a) Fig. 3-32*a*; (b) Fig. 3-32*c*; (c) Fig. 3-32*e*.

3.7-3 A linear array of three parallel half-wave dipoles along the z-axis is series fed with a transmission line of electrical length equal to the spacing, $d = \lambda/3$. Assume equal amplitude currents.

a. Sketch the array, showing the feed line.

b. Use simple array theory to obtain polar-linear pattern sketches in the xz-, yz-, and xy-planes.

c. Use a computer code to plot the same patterns as in (b).

3.7-4 Show how the general phase term $\beta\hat{\mathbf{r}} \cdot \mathbf{r}_n'$ of (3-111) reduces to that of (3-65), $\beta z_n \cos\theta$, for a linear array.

3.7-5 A uniformly excited, unequally spaced linear array of four isotropic sources is shown below. Using the principle of pattern multiplication, sketch the array pattern, showing your intermediate steps. Can (3-33) and the graphical procedure be used to verify your final result? Why?

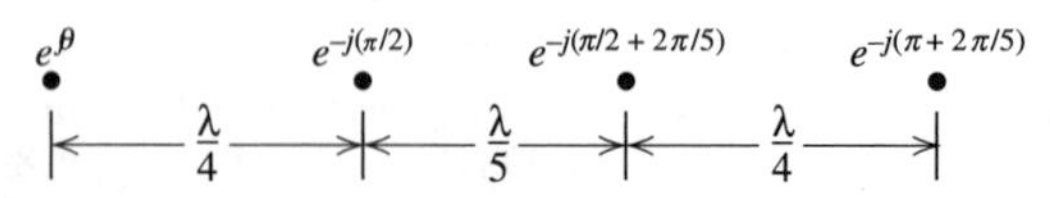

3.7-6 A planar array of four isotropic elements is arranged in the xy-plane with the following positions and currents: $(\lambda/4, \lambda/4)$, $+1$; $(-\lambda/4, \lambda/4)$, $+1$; $(-\lambda/4, -\lambda/4)$, -1; $(\lambda/4, -\lambda/4)$, -1. Use simple array modeling techniques to obtain sketches of the xz- and yz-plane patterns.

3.7-7 A half-wave dipole along the x-axis is backed by a perfect ground plane that is parallel to it and one quarter wavelength away from it (in the z-direction). Sketch the antenna system including coordinates. Use appropriate techniques to obtain polar pattern sketches in the xz- and yz-planes.

3.7-8 A four-element linear array of parallel, in-phase, half-wave dipoles is located $\lambda/4$ in front of a large planar reflector located in the xy-plane. Assume the reflector to be a perfect ground plane. If the dipoles are parallel to the x-axis and spaced $\lambda/2$ apart, sketch the complete pattern in the xy- and yz planes. Show your reasoning.

3.7-9 A two-element array of vertical short dipoles is operated a quarter-wavelength above a perfect ground plane as shown. The elements are a half-wavelength apart and are excited with equal amplitude and opposite phase. Obtain polar plots for the radiation pattern of this radiating system in the xz- and yz-planes. Carefully explain how you obtain these plots.

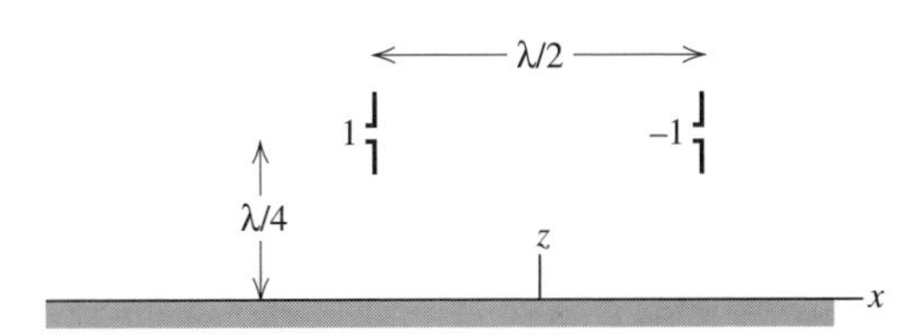

3.7-10 A two-dimensional, uniformly excited array of isotropic elements as shown below is to be analyzed. Use the principle of pattern multiplication with a pair of elements oriented vertically as the "element" and the four pairs as the "array." Give the pattern expression $F(\theta, \phi)$ and sketch the patterns in the xz-, yz-, and xy-planes.

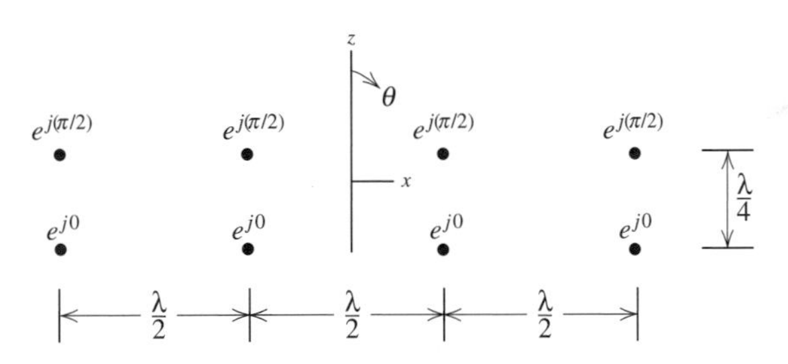

3.7-11 An 18-element planar array has three rows of six isotropic elements. The rows have current amplitudes in the ratios of 1:2:3:3:2:1 and the columns are weighted uniformly. The row and column interelement spacings are both equal to d.

a. Using the projection technique, sketch the equivalent linear array for determining the pattern in the 45° plane. Show current amplitudes and spacings.

b. Evaluate and plot the patterns in decibels in the 45° plane for the full planar array and for the equivalent linear array found in (a) for $d = 0.7\lambda$.

3.8-1 An interferometer is constructed from five collinear half-wave dipoles spaced two wavelengths apart. Sketch the polar plot of the complete array pattern.

3.8-2 An array of four collinear, 0.4-λ spaced short dipoles is fed with uniform current amplitude and phases such that the main beam peak is 30° off broadside.

a. Use simple array techniques to obtain a sketch of the array pattern.

b. Give the phases of each element.

c. Draw the array, showing a parallel feed network with parallel wire feed lines that will produce the proper phase.

3.8-3 Derive (3-117).

3.8-4 A parallel fed, uniform array of five half-wave dipole elements has half-wavelength spacing at 300 MHz. The five dipoles are located along the z-axis as shown in Fig. 3-18, and they all have the same phase. If the array is operated at 360 MHz, compare the pattern in the yz-plane at 360 MHz with that at 300 MHz.

3.8-5 Repeat Prob. 3.8-4 if the array is series fed, starting with the element closest to the coordinate origin. Assume all elements have the same amplitudes and that there is 1 m of transmission line between adjacent elements that are a half-wavelength apart at 300 MHz.

3.8-6 Derive (3-118) by expressing V_m as $V_{inc}(1 + \Gamma_m)$ and I_m as $I_{inc}(1 - \Gamma_m)$.

3.8-7 *Sum and difference patterns.* Consider a linear array of eight half-wave dipole elements with equal excitation amplitudes that are parallel to the x-axis with their centers along the z-axis and spaced 0.6λ apart. Plot the polar-dB pattern in the yz-plane for the following cases:

a. All elements in-phase, forming the sum pattern. Use simple array theory.

b. The left four elements at 0° phase and the right four elements at 180° phase, forming the difference pattern. Use simple array theory.

c. Repeat (b) but use a moment method code (see Chap. 10) to include mutual coupling effects. For the elements, use $L = 0.47\lambda$ and $a = 0.005\lambda$.

Chapter 4

Line Sources

In Chapter 1, we found that far-zone fields are obtained by a radiation integral over the current distribution. For a line source along the z-axis, the far-zone electric field intensity from (1-103) and (1-106) is

$$\mathbf{E} = \hat{\boldsymbol{\theta}} j\omega\mu \frac{e^{-j\beta r}}{4\pi r} \sin\theta \int_{-L/2}^{L/2} I(z')e^{j\beta z' \cos\theta}\, dz' \tag{4-1}$$

where the line source current distribution $I(z')$ is of length L centered symmetrically about the origin as shown in Fig. 1-14. The far-zone magnetic field intensity is then simply $H_\phi = E_\theta/\eta$. The element factor is $\sin\theta$. The pattern factor is $f(\theta)$ and is obtained by normalizing the integral in (4-1). This pattern factor is solely determined by the current distribution $I(z')$.

In Chapter 3, we found that the far-zone fields of an array are obtained by summing over the individual element currents. For an array of collinear ideal dipoles, the far-zone electric field intensity from (3-53) is

$$\mathbf{E} = \hat{\boldsymbol{\theta}} j\omega\mu \frac{e^{-j\beta r}}{4\pi r} \Delta z \sin\theta \sum_{n=0}^{N-1} I_n e^{j\beta nd \cos\theta} \tag{4-2}$$

The factor $\sin\theta$ is, in this case, the element pattern and the summation is the array factor. Note the similarities between (4-1) and (4-2); the integral in (4-1) is replaced by a summation in (4-2), z' is replaced by nd, and $I(z')$ is replaced by I_n. The line source is, in a sense, a continuous array. It will become apparent to us in the discussion of the line source which follows that much of what we know about the pattern characteristics of discrete arrays is also true of line sources. Line sources are important because many antennas can be modeled as a line source or combination of line sources.

4.1 THE UNIFORM LINE SOURCE

We begin our discussion of line sources by considering an important special case, that of the uniform line source. A uniform line source has a current distribution with uniform amplitude and linear phase progression given by

$$I(z') = \begin{cases} I_o e^{j\beta_o z'} & -\dfrac{L}{2} < z' < \dfrac{L}{2} \\ 0 & \text{elsewhere} \end{cases} \tag{4-3}$$

where β_o is the phase shift per unit length along the line source. The unnormalized pattern factor of the uniform line source is

$$f_{\text{un}}(u) = \int_{-L/2}^{L/2} I(z')e^{j\beta z' \cos\theta}\, dz' = I_o L \frac{\sin u}{u} \tag{4-4}$$

where

$$u = (\beta \cos\theta + \beta_o)\frac{L}{2} \tag{4-5}$$

The evaluation of (4-4) is similar to that given in (1-110) for a broadside uniform line source.

It is convenient to introduce an angle θ_o such that

$$\beta_o = -\beta \cos\theta_o \tag{4-6}$$

Then (4-5) becomes

$$u = \frac{\beta L}{2}(\cos\theta - \cos\theta_o) \tag{4-7}$$

The far-zone electric field from (4-1) and (4-4) is

$$E_\theta = \frac{j\omega\mu e^{-j\beta r}}{4\pi r} I_o L \sin\theta \frac{\sin u}{u} \tag{4-8}$$

The pattern factor of this uniform line source field expression is

$$f(u) = \frac{\sin u}{u} \tag{4-9}$$

The pattern factor is shown in Fig. 4-1 without using absolute values. The maximum occurs for $u = 0$ and is unity (0 dB) there. The nulls occur at multiples of π and are separated by π, except for the beamwidth between first nulls, which is 2π.

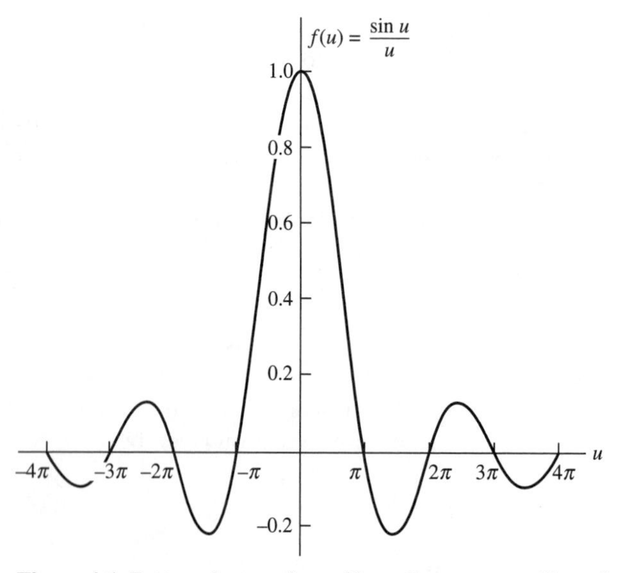

Figure 4-1 Pattern factor of a uniform line source of length L and $u = (\beta L/2)\cos\theta$.

The half-power beamwidth of the uniform line source pattern factor is found from solving

$$\frac{1}{\sqrt{2}} = \frac{\sin u_{\text{HP}}}{u_{\text{HP}}} \tag{4-10}$$

The solutions to this are $u_{\text{HP}} = \pm 1.39$. Then from (4-7),

$$\theta_{\text{HP}} = \cos^{-1}\left(\frac{2}{\beta L}\, u_{\text{HP}} + \cos\theta_o\right) = \cos^{-1}\left(\pm 0.443\,\frac{\lambda}{L} + \cos\theta_o\right) \tag{4-11}$$

The plus sign corresponds to the half-power point on the right of the main beam maximum and the minus sign to the left half-power point. So from (1-126),

$$\begin{aligned} \text{HP} &= \left|\theta_{\text{HP left}} - \theta_{\text{HP right}}\right| \\ &= \left|\cos^{-1}\left(-0.443\,\frac{\lambda}{L} + \cos\theta_o\right) - \cos^{-1}\left(0.443\,\frac{\lambda}{L} + \cos\theta_o\right)\right| \end{aligned} \tag{4-12}$$

This formula is general but useful only when both half-power points appear in the visible region ($0 \leq \theta \leq 180°$), which in turn requires that the arguments of the arccosines in (4-12) are between -1 and $+1$. For a broadside uniform line source, $\theta_o = 90°$ and (4-12) reduces to (see Prob. 4.1-1)

$$\text{HP} = 2\sin^{-1}\left(0.443\,\frac{\lambda}{L}\right) \qquad (\theta_o = 90°) \tag{4-13}$$

For long ($L \gg \lambda$) line sources, this is approximately

$$\text{HP} \approx 0.886\,\frac{\lambda}{L}\ \text{rad} = 50.8\,\frac{\lambda}{L}\ \text{degrees} \qquad (\theta_o = 90°) \tag{4-14}$$

since $\sin^{-1}(x) \approx x$ for $x \ll 1$. For an endfire uniform line source, only one half-power point appears in the visible region and then

$$\text{HP} = 2\cos^{-1}\left(1 - 0.443\,\frac{\lambda}{L}\right) \qquad (\theta_o = 0° \text{ or } 180°) \tag{4-15}$$

For long ($L \gg \lambda$) line sources, this may be approximated as (see Prob. 4.1-2)

$$\text{HP} \approx 2\sqrt{0.886\,\frac{\lambda}{L}}\ \text{rad} \qquad (\theta_o = 0° \text{ or } 180°) \tag{4-16}$$

Since (4-16) leads to wider beamwidths than does (4-14), we conclude that beamwidth increases as the pattern is scanned away from broadside (see Fig. 4-3).

The half-power beamwidth expression $\text{HP} = 0.886(\lambda/L)$ for the broadside uniform line source was developed using two approximations. The effect of the element factor $\sin\theta$ was neglected and also it was assumed that the line source was long. With a few examples, we can see how these approximations affect the beamwidth. In Table 4-1, half-power beamwidth values for three uniform line sources are presented for various levels of approximation. The first column is the HP found from the complete pattern expression

$$F(\theta) = \sin\theta\,\frac{\sin[(\beta L/2)\cos\theta]}{(\beta L/2)\cos\theta} \qquad (\theta_o = 90°) \tag{4-17}$$

Table 4-1 Half-Power Beamwidth Evaluation for Broadside Uniform Line Sources

Length L	Exact Value from Complete Pattern F of (4-17)	Value from Pattern Factor $f = \frac{\sin u}{u}$	Value from $\mathrm{HP} = 0.886 \frac{\lambda}{L}$
2λ	24.766°	25.591°	25.382°
5λ	10.112°	10.166°	10.153°
10λ	5.071°	5.080°	5.076°

The third column is the HP obtained from only the pattern factor of (4-9). The last column is that of (4-14). Note that even for five wavelengths, all values are in very close agreement. We can also see that as the length increases, the approximations improve.

The largest side lobe is the first one (i.e., the one closest to the main beam). The side lobe maxima locations are found by differentiating (4-9) and setting it equal to zero. This leads to

$$u_{\mathrm{SL}} = \tan u_{\mathrm{SL}} \tag{4-18}$$

The intersections of the straight-line curve u_{SL} with the curve $\tan u_{\mathrm{SL}}$ give the side lobe maximum locations (the main beam maximum is at $u_{\mathrm{SL}_0} = 0$). The first side lobe maximum occurs for $u_{\mathrm{SL}_1} = \pm 1.43\pi$. This is not precisely midway between the pattern nulls at π and 2π. The side lobe maxima are slightly closer to the main beam than midway between their nulls. Evaluating (4-9) at the first side lobe maximum location gives 0.217 or -13.3 dB.

The polar plot of the pattern factor of a uniform line source can be obtained from a universal pattern factor in a manner very similar to that used for linear arrays. The uniform line source universal pattern factor is shown in Fig. 4-2*a*. It is used for *all* source lengths L and scan angles θ_o. A typical case is shown in Fig. 4-2*b*. The transformation (4-7) between u and θ is illustrated graphically by the dashed lines. Pattern values for a given value of θ can be found from the universal pattern factor using this graphical transformation. The radius of the circle used in the transformation is $\beta L/2$ and its origin is at the value of u equal to $-(\beta L/2)\cos\theta_o$.

As an example, consider a three-wavelength uniform line source. The universal pattern factor is shown in Fig. 4-3*a*. The polar plot for the broadside case is illustrated in Fig. 4-3*b*. The pattern factor for a main beam maximum angle of 45° is polar-plotted in Fig. 4-3*c*. The endfire case is shown in Fig. 4-3*d*. Notice that the main beam (and also the side lobes) widen near endfire, as pointed out earlier. The current distributions required to produce those patterns are shown in Fig. 4-4. The amplitudes are constant in all cases, as illustrated in Fig. 4-4*a*. The required linear phase distributions for main beam scanning are depicted in Fig. 4-4*b*.

The effects of the element factor on the total pattern are shown in Fig. 4-5 for the three-wavelength uniform line source. In the broadside case of Fig. 4-5*a*, the element factor has a relatively minor effect. However, in the endfire case of Fig. 4-5*b* where the pattern factor alone produces a single endfire beam, the element factor effect on the total pattern produces a null in the endfire direction, thus bifurcating the main beam.

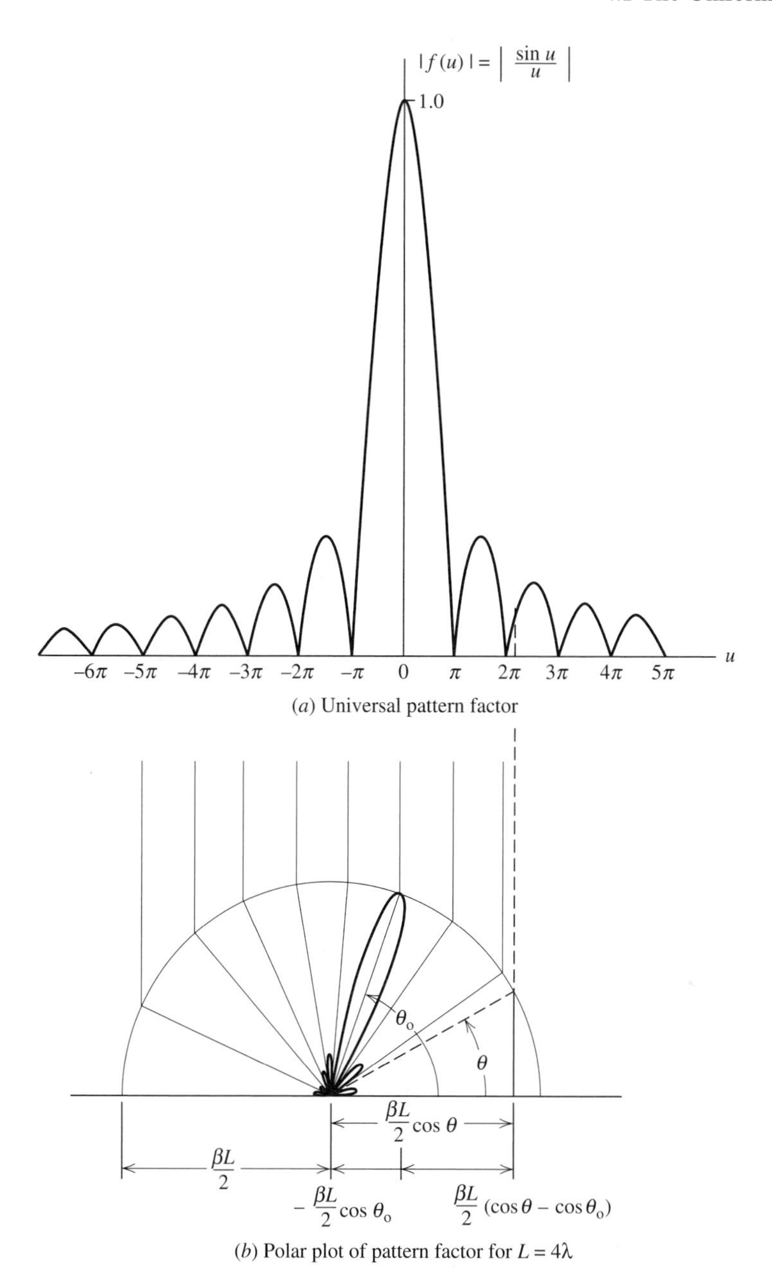

Figure 4-2 Illustration of obtaining a polar plot from the universal pattern factor of a uniform line source.

Next, we consider the directivity of the uniform line source. The directivity can be found easily if the element factor is assumed to have a negligible effect on the pattern. Then, we can work with the pattern factor f alone. First, the beam solid angle is from (1-143) and (4-9):

$$\Omega_A = \int_0^{2\pi} \int_0^{\pi} \left| \frac{\sin u}{u} \right|^2 \sin \theta \, d\theta \, d\phi \tag{4-19}$$

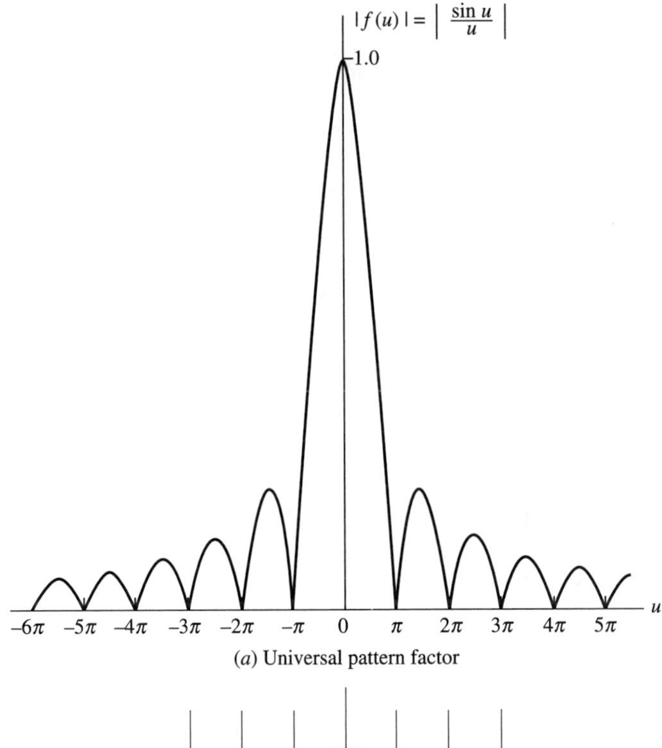

(*a*) Universal pattern factor

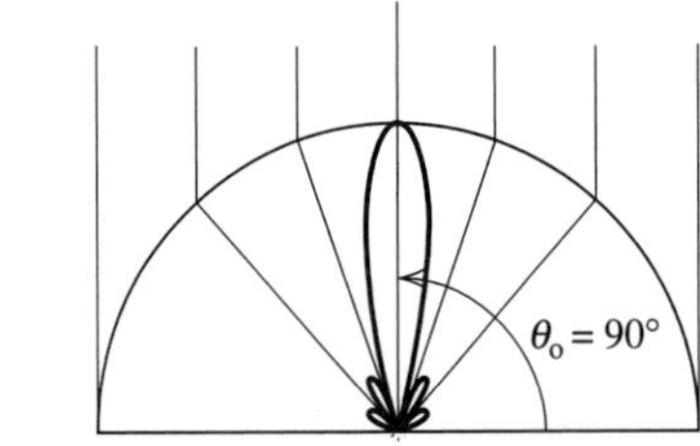

(*b*) Polar plot of pattern factor for $\beta_o L/2 = 0$, ($\theta_o = 90°$).

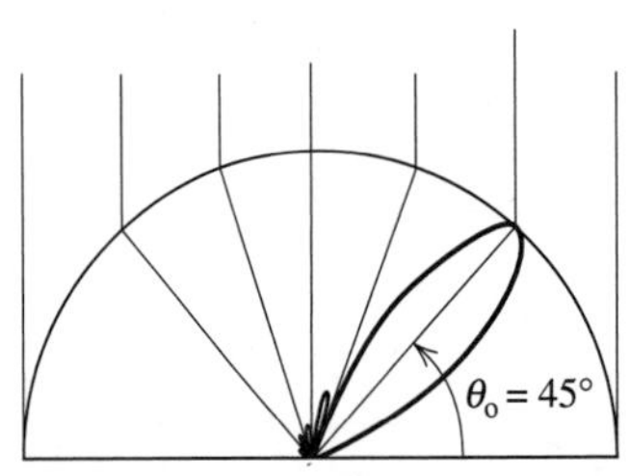

(*c*) Polar plot of pattern factor for $\beta_o L/2 = -2.12\pi$, ($\theta_o = 45°$).

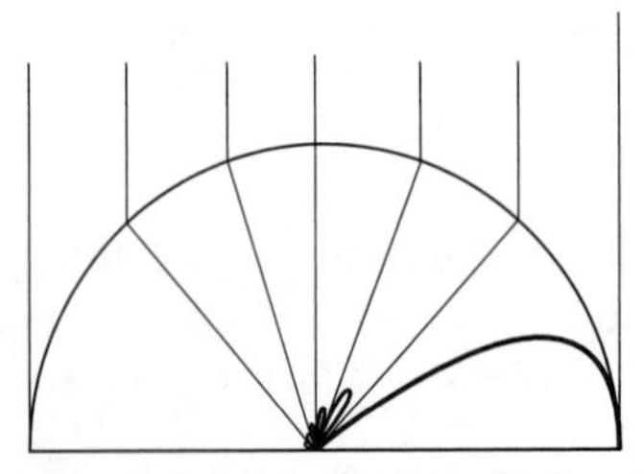

(*d*) Polar plot of pattern factor for $\beta_o L/2 = -\beta L/2 = -3\pi$, ($\theta_o = 0°$).

Figure 4-3 Pattern factors for a three-wavelength long ($L = 3\lambda$) uniform line source for various scan conditions.

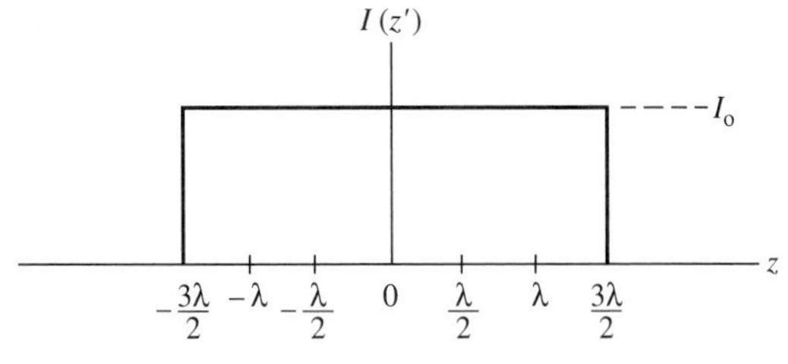

(*a*) Current amplitude distribution

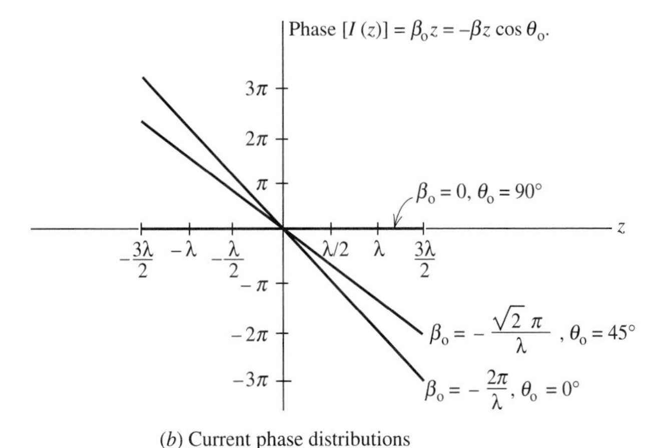

(*b*) Current phase distributions

Figure 4-4 Current distributions for the three-wavelength uniform line source patterns of Fig. 4-3.

with the element factor g set to unity. If we change the θ integration variable to u as given by (4-5), then $du = -(\beta L/2)\sin\theta\, d\theta$ and (4-19) becomes

$$\Omega_A = \int_0^{2\pi} d\phi \int_{(\beta+\beta_o)L/2}^{(-\beta+\beta_o)L/2} \frac{\sin^2 u}{u^2} \frac{du}{-(\beta L/2)}$$

$$= 2\frac{\lambda}{L}\int_{(\beta_o-\beta)L/2}^{(\beta_o+\beta)L/2} \frac{\sin^2 u}{u^2}\, du \tag{4-20}$$

The evaluation of this expression for the general case is discussed in Prob. 4.1-7. For the broadside case ($\beta_o = 0$), the limits on the integral are $-\beta L/2$ to $\beta L/2$. If further $L \gg \lambda$, then $\beta L/2 \gg 1$ and we approximate the limits as $-\infty$ to $+\infty$, and if we use (F-12), the definite integral has a value of π. Thus, $\Omega_A \approx 2\lambda\pi/L$ and $D = 4\pi/\Omega_A$ yields

$$D_u = 2\frac{L}{\lambda} \qquad \text{(broadside, } L \gg \lambda\text{)} \tag{4-21}$$

where the subscript u indicates a uniform line source. For the endfire case ($\beta_o = \pm\beta$), the integral limits are 0 and $\beta L/2$ that are approximated as 0 and ∞ when $L \gg \lambda$; this yields a value of $\pi/2$ for the integral. So, $\Omega_A \approx \lambda\pi/L$ and

$$D_u = 4\frac{L}{\lambda} \qquad \text{(endfire, } L \gg \lambda\text{)} \tag{4-22}$$

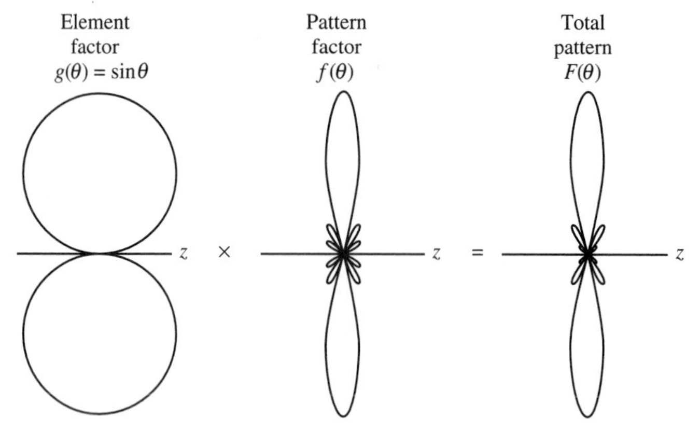

(*a*) Broadside case ($\theta_o = 90°$, $\beta_o = 0$). Pattern factor is from Fig. 4-3*b*.

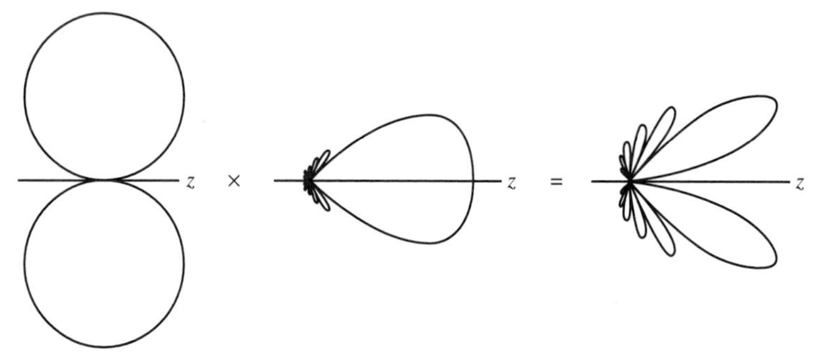

(*b*) Endfire case ($\theta_o = 0°$, $\beta_o L/2 = -3\pi$). Pattern factor is from Fig. 4-3*d*.

Figure 4-5 Total patterns for a three-wavelength uniform line source.

These are the same directivity results we obtained for linear arrays; see (3-80) and (3-81). The uniform line source exhibits the most directivity that can be obtained from a linear phase source of fixed length. Other current distributions will yield lower directivities. We found this principle to hold true in Section 3.5 for discrete current distributions (arrays).

From the beamwidth and directivity relationships presented here for the uniform line source, we can begin to get a feel for the pattern changes as a function of source length and scan angle. First, consider the pattern factor alone. As the length increases, the beamwidth decreases and the directivity increases. The side lobe level (if the line source is long enough for the first side lobe maximum to be visible) remains constant with length variations; it is always -13.3 dB for a uniform line source. For a scanned line source, the beamwidth increases as the main beam is scanned away from broadside. However, the total main beam volume (obtained by rotating the E-plane pattern about the z-axis) decreases and, consequently, Ω_A decreases, which in turn leads to an increase in directivity. The beamwidth and directivity change slowly for scan angles near broadside but change rapidly near endfire. The complete pattern must include the element factor effects. For long sources ($L \gg \lambda$), the pattern factor $f(\theta)$ has a much narrower pattern than the element factor $g(\theta) = \sin\theta$ and the total pattern obtained from $g(\theta)\, f(\theta)$ is closely approximated by $f(\theta)$. The side lobe level, beamwidth, and directivity values are then accurately determined from the pattern factor $f(\theta)$ alone, except near endfire where the element factor becomes significant since it forces the total pattern to zero in the $\theta = 0$ and 180° directions, as illustrated in Fig. 4-5*b*.

EXAMPLE 4-1 *Plane Wave Incident on a Slit*

A simple physical example of a uniform line source is a long narrow slit in a good conductor that has a uniform plane wave incident on it, as illustrated in Fig. 4-6. Phase fronts (planes

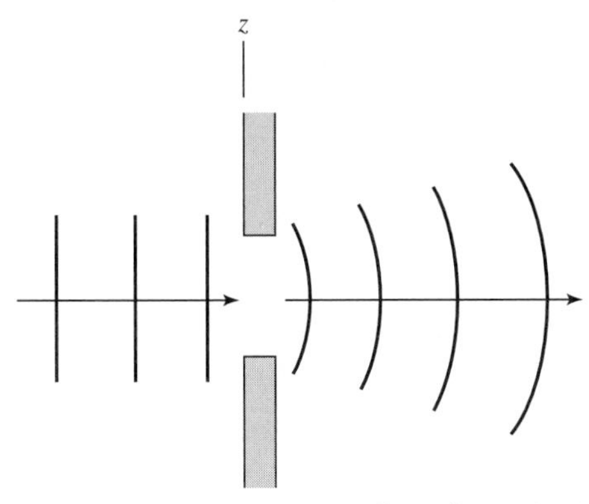

(a) Broadside case: $\theta_o = 90°$ and $\beta_o = -\beta \cos 90° = 0$.

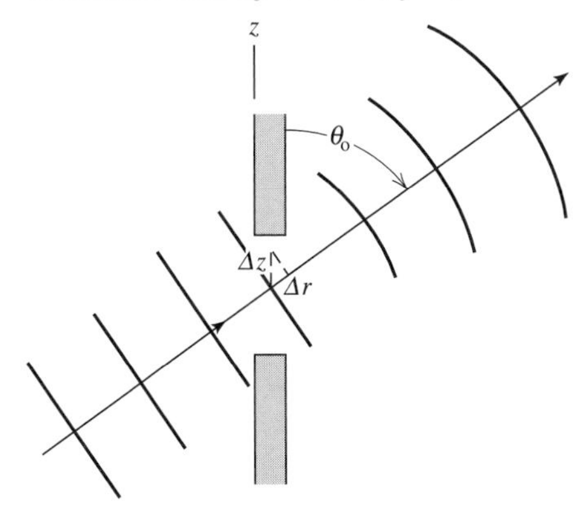

(b) Intermediate case: $\beta_o = -\beta \cos\ \theta_o$.

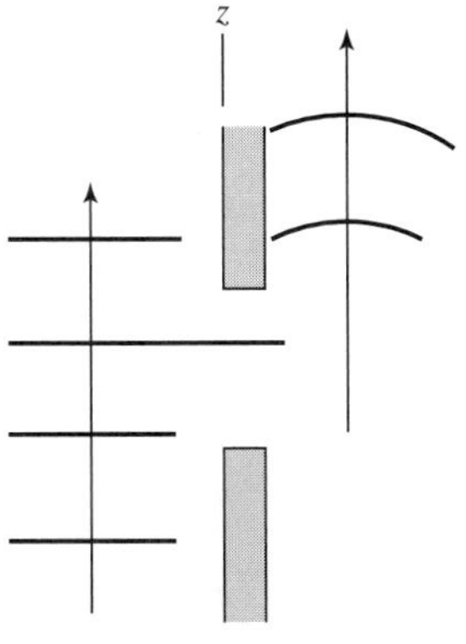

(c) Endfire case: $\theta_o = 0$ and $\beta_o = -\beta$.

Figure 4-6 Example of a uniform line source: an infinitely long slit of width L in a good conductor illuminated by a uniform plane wave from the left.

of constant phase) are indicated by the parallel lines. In Fig. 4-6*a*, the wave is normally incident on the slit. Thus, the slit has a uniform amplitude excitation and also has uniform phase since the phase fronts are parallel to the slit. The slit then behaves as a uniform line source with uniform phase across it. This equivalence of a field distribution to a current distribution will be discussed further in Section 7.1. It is obvious, however, that the maximum radiation on the right-hand side of the slit will be primarily in the direction of propagation of the incident wave coming from the left, that is, $\theta_o = 90°$. For a line source, the phase shift along the source is related to the direction of maximum radiation θ_o by $\beta_o = -\beta \cos \theta_o$, where in this case β is the phase constant of the incident plane wave. Since $\theta_o = 90°$, $\beta_o = -\beta \cos 90° = 0$. This says that there is no phase shift along the slit. We already observed that this must be true for a plane wave normally incident on the slit.

If the direction of propagation of the plane wave makes an angle θ_o with the slit plane, there will be a phase shift along the slit due to different arrival times of the wave. In fact, this phase shift is given by $\beta_o z'$, where β_o is the phase shift per meter along the slit and we have assumed zero phase at $z' = 0$. But the phase shifts β rad/m in the direction of propagation, so the phase shift for distance Δr along the direction of propagation is $\Delta\phi = -\beta\,\Delta r$ (since the wave propagates as $e^{-j\beta r}$). The same phase shift is encountered in the distance Δz along the slit, or $\Delta\phi = \beta_o\,\Delta z$ (see Fig. 4-6*b*). But $\Delta r = \Delta z \cos \theta_o$, and since the phase shifts are equal, we have $\Delta\phi = -\beta\,\Delta z \cos \theta_o = \beta_o\,\Delta z$. Thus, $\beta_o = -\beta \cos \theta_o$ as given by (4-6), which was then a convenient definition. It is obvious from Fig. 4-6*b* that the maximum radiation from the slit or its equivalent line source will occur in the direction of propagation of the wave $\theta = \theta_o$.

In Fig. 4-6*c*, the incident wave is traveling parallel to the slit. The phase shift per meter along the slit is obviously equal to the negative of the wave phase constant. This also follows from $\beta_o = -\beta \cos \theta_o = -\beta$ for $\theta_o = 0°$. The radiated wave on the right side is endfire in this case.

4.2 TAPERED LINE SOURCES

Many antennas that can be modeled by line sources are designed to have tapered distributions. This is because if the current amplitude decreases toward the ends of a line source, the pattern side lobes are lowered and the main beam widens. In many applications, low side lobes are necessary and a wider main beam is accepted as a consequence. This tradeoff between side lobe level and half-power beamwidth is a major consideration to the antenna engineer.

An an example, consider a current distribution with the so-called cosine taper, where

$$I(z') = \begin{cases} I_o \cos\left(\dfrac{\pi}{L} z'\right) e^{j\beta_o z'} & -\dfrac{L}{2} < z' < \dfrac{L}{2} \\ 0 & \text{elsewhere} \end{cases} \tag{4-23}$$

The shape of this current distribution is plotted in Fig. 4-7*a*. The unnormalized pattern factor is then found as follows:

$$\begin{aligned} f_{\text{un}}(\theta) &= I_o \int_{-L/2}^{L/2} \cos\left(\frac{\pi}{L} z'\right) e^{j(\beta \cos \theta + \beta_o)z'}\, dz' \\ &= \frac{I_o}{2} \int_{-L/2}^{L/2} [e^{j(\pi/L + \beta \cos \theta + \beta_o)z'} + e^{-j(\pi/L - \beta \cos \theta - \beta_o)z'}]\, dz' \\ &= \frac{I_o}{2} \left[\frac{e^{j(\pi/L + \beta \cos \theta + \beta_o)z'}}{j(\pi/L + \beta \cos \theta + \beta_o)} + \frac{e^{-j(\pi/L - \beta \cos \theta - \beta_o)z'}}{-j(\pi/L - \beta \cos \theta - \beta_o)} \right]_{-L/2}^{L/2} \end{aligned} \tag{4-24}$$

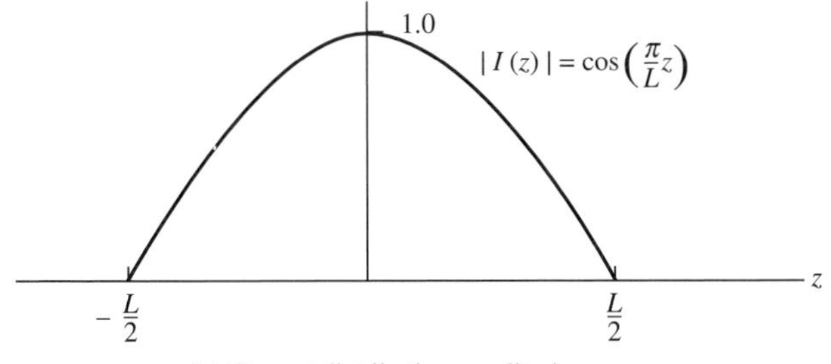

(*a*) Current distribution amplitude

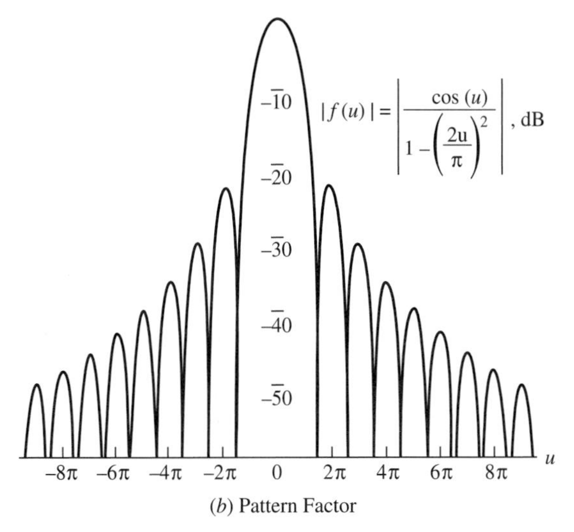

(*b*) Pattern Factor

Figure 4-7 Current distribution and pattern factor for a cosine-tapered line source.

Evaluating the above expression leads to

$$f_{\text{un}}(\theta) = I_o \frac{2L}{\pi} \frac{\cos[(\beta \cos \theta + \beta_o)L/2]}{1 - [(\beta \cos \theta + \beta_o)L/\pi]^2} \tag{4-25}$$

Using $\beta_o = -\beta \cos \theta_o$ as in (4-6) and normalizing such that the pattern factor is unity for $\theta = \theta_o$ give

$$f(\theta) = \frac{\cos[(\beta L/2)(\cos \theta - \cos \theta_o)]}{1 - [(\beta L/\pi)(\cos \theta - \cos \theta_o)]^2} \tag{4-26}$$

This pattern can be written in terms of u using (4-7) as

$$f(u) = \frac{\cos u}{1 - (2u/\pi)^2} \tag{4-27}$$

This pattern is plotted in Fig. 4-7*b*. Compare the side lobe level to that of Fig. 4-10a for the uniform line source.

The side lobe level for the cosine-tapered line source is −23.0 dB and the beamwidth is given by

$$\text{HP} \approx 1.19 \frac{\lambda}{L} \text{ rad} = 68.2 \frac{\lambda}{L} \text{ degrees} \tag{4-28}$$

Table 4-2 Characteristics of Tapered Line Source Distributions

a. Triangular taper

$$I(z) = 1 - \frac{2}{L}|z| \qquad |z| \leq \frac{L}{2}$$

$$f(u) = \left[\frac{\sin(u/2)}{u/2}\right]^2$$

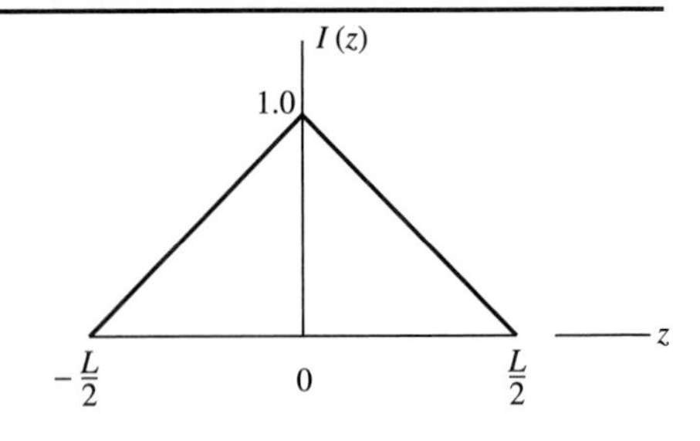

HP (rad)	Side Lobe Level (dB)	D/D_u
$1.28(\lambda/L)$	−26.6	0.75

b. Cosine tapers

$$I(z) = \cos^n\left(\frac{\pi z}{L}\right) \qquad |z| \leq \frac{L}{2}$$

$$f(u) = \frac{\sin u}{u} \qquad n = 0$$

$$f(u) = \frac{\cos u}{1 - (2u/\pi)^2} \qquad n = 1$$

$$f(u) = \frac{1}{1 - (u/\pi)^2}\frac{\sin u}{u} \qquad n = 2$$

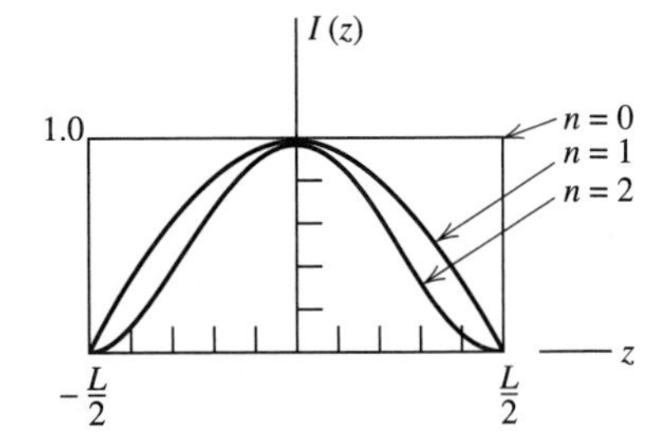

n	HP (rad)	Side Lobe Level (dB)	D/D_u	Type
0	$0.886\frac{\lambda}{L}$	−13.3	1.00	Uniform line source
1	$1.19\frac{\lambda}{L}$	−23.0	0.810	Cosine taper
2	$1.44\frac{\lambda}{L}$	−31.7	0.667	Cosine-squared taper

c. Cosine on a pedestal

$$I(z) = C + (1 - C)\cos\frac{\pi z}{L}$$

$$f(u) = \frac{C\dfrac{\sin u}{u} + (1 - C)\dfrac{2}{\pi}\dfrac{\cos u}{1 - (2u/\pi)^2}}{C + (1 - C)\dfrac{2}{\pi}}$$

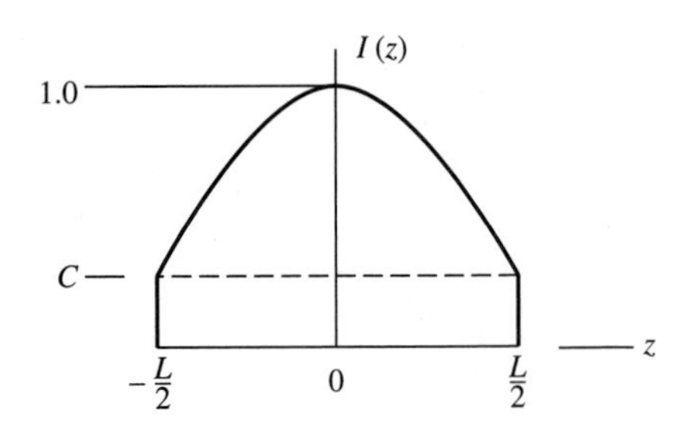

Table 4-2 (continued)

Edge Illumination				
C	$-20 \log C$ (dB)	HP (rad)	Side Lobe Level (dB)	D/D_u
0.3162	−10	$1.03 \frac{\lambda}{L}$	−20	0.92
0.1778	−15	$1.08 \frac{\lambda}{L}$	−22	0.88
0	$-\infty$	$1.19 \frac{\lambda}{L}$	−23	0.81

Note 1: The pattern expressions are valid for any value of $u = (\beta L/2)(\cos\theta - \cos\theta_o)$. However, the half-power beamwidth values and directivities are approximations for broadside line sources ($\theta_o = 90°$ and $u = (\beta L/2)\cos\theta$) and $L \gg \lambda$.
Note 2: The directivity for each line source is found from the ratio D/D_u as

$$D = \frac{D}{D_u} D_u = \frac{D}{D_u} 2 \frac{L}{\lambda}$$

for broadside line sources with $L \gg \lambda$.
Note 3: The element factor $\sin\theta$ has been neglected in the calculations leading to the values in this table. For long, broadside line sources, its effect is minimal.

for the broadside case. The side lobe level is 10 dB lower and the beamwidth is 38% greater than a uniform line source of the same length. Although the side lobes are reduced from those of the uniform line source, the main beam widening leads to smaller directivity than obtained from a uniform line source. The ratio D/D_u is used to compare the directivity of a tapered line source to that of a uniform line source of the same length. For the cosine taper, $D/D_u = 0.810$. The actual directivity D from (4-21) is then

$$D = 0.810 D_u = 1.620 \frac{L}{\lambda} \qquad \text{(broadside, } L \gg \lambda) \tag{4-29}$$

If the current amplitude taper is increased as in the case of a cosine-squared taper, the side lobes are reduced even more and the beamwidth is further widened. The pattern parameters of the cosine-squared case, as well as many other important cases, are summarized in Table 4-2 [1–3].

As a further example, consider the triangular current taper given in Table 4-2*a*. The pattern $(\sin u/u)^2$ is the square of the uniform line source pattern. This property is apparent when the pattern of Fig. 4-8 for the triangular line source is compared to that of the uniform line source in Fig. 4-1. The first nulls of the triangular line source are twice as far out as for the uniform line source pattern. Thus, the beamwidth between first nulls is twice as large. The half-power beamwidth is 44% larger (from $0.886\lambda/L$ to $1.28\lambda/L$). Also, the side lobes of the triangular line source are twice as wide in the variable u and the side lobe level in decibels is twice as small, −13.3 dB for the uniform line source and −26.6 dB for the triangular line source. The directivity (from Table 4-2*a*) is 75% of the uniform line source value.

From Table 4-2, we can generalize and make some statements about current

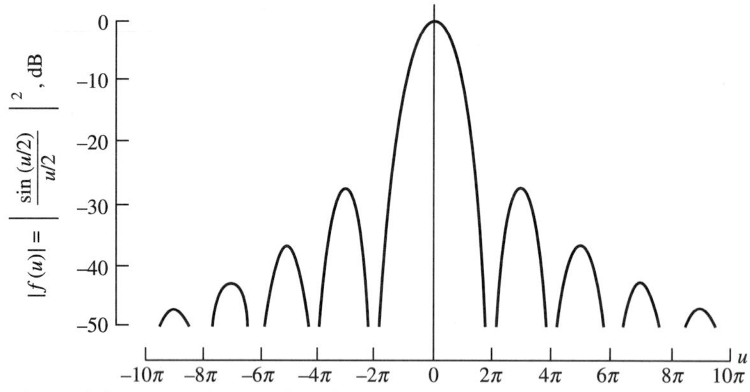

Figure 4-8 Pattern factor of a triangular tapered line source.

amplitude distributions and their influence on the far-field pattern. We assume that the current is of constant phase. *As the taper of the current amplitude from the center to the edges of a line source becomes more severe, the side lobes decrease and the beamwidth increases.* Consequently, the directivity decreases. There is then a trade-off between the side lobe level and the beamwidth for the continuous source just as there was for discrete sources (see Section 3.5). The antenna engineer must decide on a compromise between beamwidth and side lobe level for each specific design problem.

EXAMPLE 4-2 *A Cylindrical Parabolic Reflector Antenna*

A cylindrical parabolic reflector antenna (see Fig. 4-9) can be modeled by line sources. Suppose the parabolic surface is 10λ across at the edges of the reflector (i.e., the aperture) in Fig. 4-9 and that the field distribution in the aperture in the y-direction is that of a cosine on a pedestal with -15 dB edge illumination. Then from Table 4-2c, the half-power beamwidth is

$$\text{HP} = 1.08\lambda/L = 0.108 \text{ rad} = 6.2°$$

and the side lobe level is -22 dB. Figure 12-21 shows a pattern calculated via a one-dimensional aperture integration for an aperture distribution that is nearly a cosine on a pedestal (Fig. 12-22). It has -22 dB side lobes and a half-power beamwidth of 6.2° in the H-plane. There is no E-plane data because the antenna in Sec. 12.6 is infinite in the x-direction (i.e.,

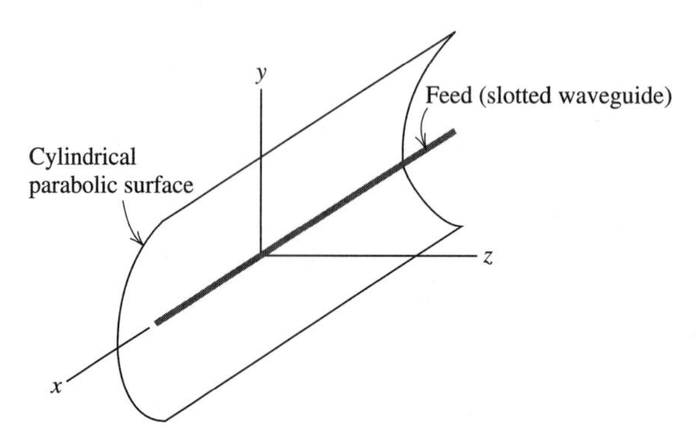

Figure 4-9 Parabolic cylindrical reflector with x-polarized feed along the axis of the cylinder.

a two-dimensional rather than a three-dimensional problem). Suppose, however, that the antenna is also 10λ in the x-direction (E-plane), as implied by Fig. 4-9, and the aperture distribution is uniform in the E-plane. The E-plane pattern would then be modeled by a uniform line source. Aperture directivity can be calculated from line source formulas if the principal plane distributions are separable; see (7-86). Much more will be said about aperture antennas in Chap. 7, where, as in this example, line source results will play an important role.

4.3 FOURIER TRANSFORM RELATIONS BETWEEN THE FAR-FIELD PATTERN AND THE SOURCE DISTRIBUTION

The far-field pattern and its (nonperiodic) source distribution form a Fourier transform pair. To see this, consider (4-1) where $I(z') = |I(z')|e^{j\beta_o z'}$ and write

$$F_{\text{un}}(\theta) = \sin\theta \int_{-L/2}^{L/2} |I(z')| e^{j(\beta\cos\theta + \beta_o)z'}\, dz' \tag{4-30}$$

or

$$f_{\text{un}}(\theta) = \frac{F_{\text{un}}(\theta)}{\sin\theta} = \int_{-L/2}^{L/2} |I(z')| e^{j(\beta\cos\theta + \beta_o)z'}\, dz' \tag{4-31}$$

where once again, the element pattern $\sin\theta$ has been absorbed into the far field of the line source. Thus, $f_{\text{un}}(\theta)$ can be viewed as the far field of a line source in which the element pattern is isotropic. Since $I(z')$ is zero for $z' > L/2$ and $z' < -L/2$, the limits on (4-31) may be extended to infinity. Thus,

$$f_{\text{un}}(\theta) = \int_{-\infty}^{\infty} |I(z')| e^{j(\beta\cos\theta + \beta_o)z'}\, dz' \tag{4-32}$$

which is recognized as one-half of a Fourier transform pair. The other half of the (antenna) pair is

$$I(z') = \frac{1}{2\pi}\int_{-\infty}^{\infty} f_{\text{un}}(\theta) e^{-jz'\beta\cos\theta}\, d(\beta\cos\theta) \tag{4-33}$$

From circuit theory, the Fourier transform (circuit) pair can be written as

$$f(t) = \frac{1}{2\pi}\int_{-\infty}^{\infty} g(\omega) e^{j\omega t}\, d\omega \tag{4-34}$$

and

$$g(\omega) = \int_{-\infty}^{\infty} f(t) e^{-j\omega t}\, dt \tag{4-35}$$

If we let $\cos\theta$ and $\beta z'$ correspond to t and ω, respectively, then z'/λ corresponds to frequency f. The quantity z'/λ is called spatial frequency with units of hertz per radian. For real values of θ and $|\cos\theta| \le 1$, the field distribution associated with $f_{\text{un}}(\theta)$ represents radiated power, whereas for $|\cos\theta| > 1$, it represents reactive or stored power (e.g., see Secs. 2.6 or 4.4). The pattern $f_{\text{un}}(\theta)$, or angular spectrum, represents an angular distribution of plane waves. For $|\cos\theta| \le 1$, the angular spectrum is the same as the far-field pattern $f_{\text{un}}(\theta)$.

In circuit theory, a very narrow pulse (in time) has a large or wide frequency

Table 4-3 Some Common Fourier Transform Pairs

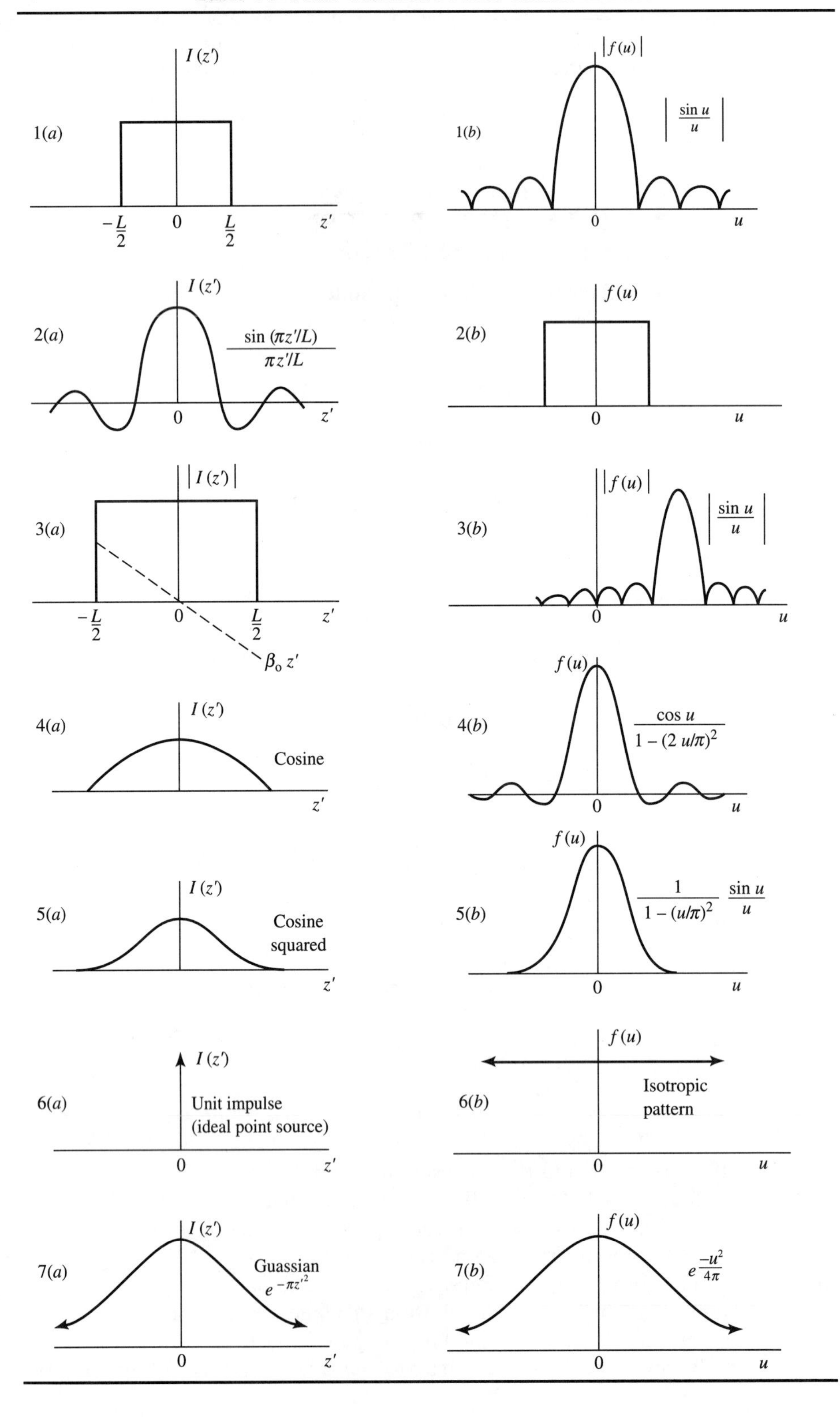

spectrum. To pass such a pulse through a filter requires that the filter have a wide passband. Similarly, an antenna with a very narrow far-field pattern must pass a wide band of spatial frequencies. That is, the antenna must be electrically large. Thus, the antenna may be viewed as a spatial filter, a concept widely used in radiometry and radio astronomy.

Probably the greatest value in recognizing that the source and far-field function form a Fourier transform pair is that one can utilize the vast amount of information available on Fourier transform theory, particularly in circuit theory. For example, Table 4-3 shows some common Fourier transform pairs found in antenna theory. Some of these also appear in Table 4-2 that was constructed without reliance on a knowledge of Fourier transforms. The pairs in Table 4-3 work in either direction. That is, the source distributions and far-field patterns in Table 4-3 may be interchanged, at least theoretically. In some cases, however, the resulting distributions are not practical as far-field patterns or as source distributions. And finally, although our familiarity with (4-32) may imply that the application of (4-33) is equally straightforward, it is not. Application of (4-33) leads to techniques in antenna synthesis. Antenna synthesis is discussed in Chap. 8.

4.4 SUPERDIRECTIVE LINE SOURCES

In Sec. 2.6, it was pointed out that electrically small antennas have a directivity larger than warranted by their electrical size. That is, they are superdirective. This section examines the superdirectivity of line sources that are not electrically small. In general, linear sources with $L > \lambda$ are superdirective if the directivity is higher than that obtained using a phase distribution $e^{\pm j\beta_o z'}$ with $|\cos\theta_o| \le 1$. Thus, the Hansen–Woodyard endfire arrays in Chap. 3 exhibited superdirective properties, since $|\alpha| > \beta d$ implies $|\cos\theta_o| > 1$.

Superdirectivity is produced by an interference process whereby the main beam is scanned into the invisible region (see Fig. 4-10), where $|u| > \pi L/\lambda$ or $|\cos\theta_o| > 1$. This causes energy to be stored in the near field, resulting in a large antenna Q. The reactive power is found approximately by integrating over the invisible region and the radiated power is found, of course, by integrating the pattern factor over the visible region (where $|u| \le \pi L/\lambda$). To quantify superdirectivity, a superdirective ratio R_{SD} may be defined as the ratio of radiated power plus reactive power to radiated power, which for a broadside line source is [2]

$$R_{SD} = \frac{\displaystyle\int_{-\infty}^{\infty} |f(u)|^2\, du}{\displaystyle\int_{-\pi L/\lambda}^{\pi L/\lambda} |f(u)|^2\, du} \tag{4-36}$$

For other than broadside, the limits in the denominator change (see Probs. 4.4-1 and 4.4-2).

Since the Q may be expressed as the ratio of reactive power to radiated power,

$$Q = \frac{\displaystyle\int_{-\infty}^{-\pi L/\lambda} |f(u)|^2\, du + \int_{\pi L/\lambda}^{\infty} |f(u)|^2\, du}{\displaystyle\int_{-\pi L/\lambda}^{\pi L/\lambda} |f(u)|^2\, du} \tag{4-37}$$

Comparing the previous two equations gives $R_{SD} = 1 + Q$.

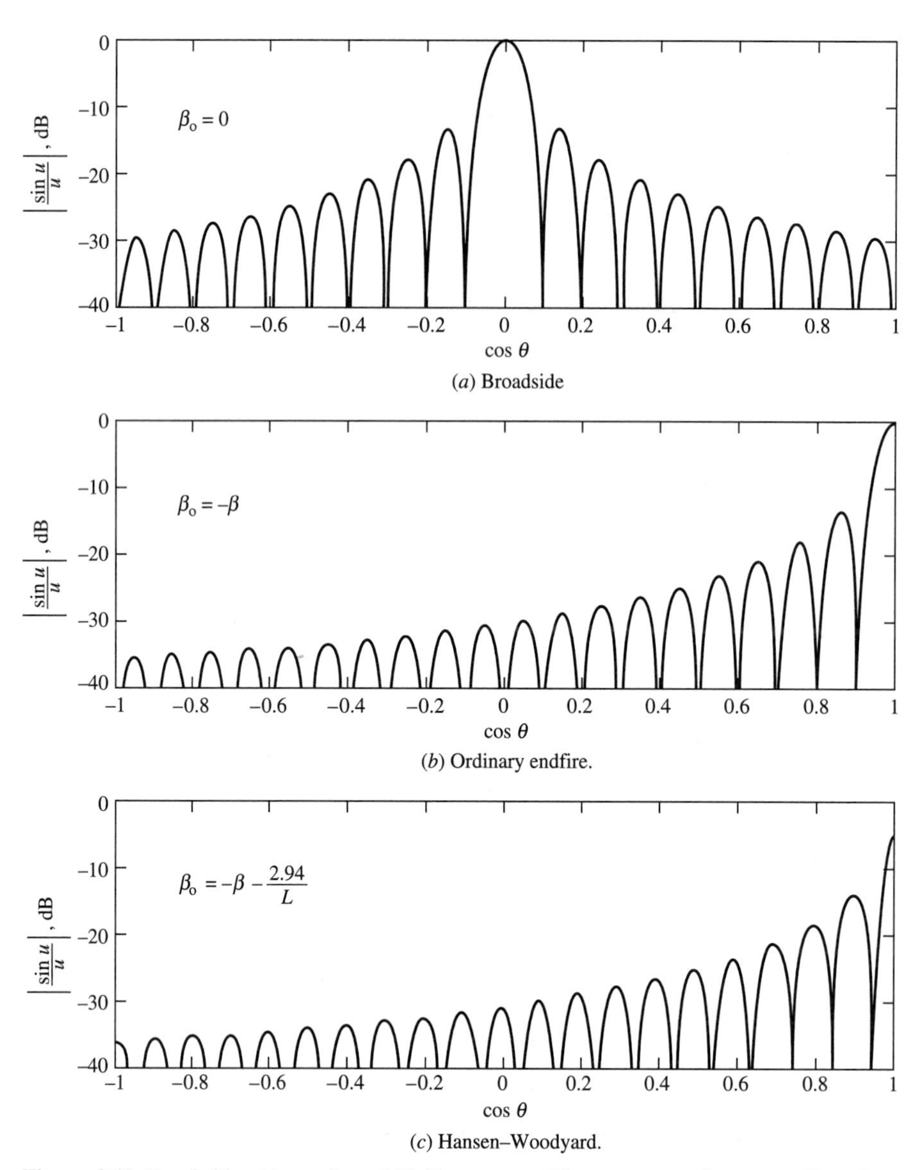

Figure 4-10 Far-field patterns for a 10λ line source: The corresponding superdirective ratios are found in Table 4-4.

To investigate superdirectivity for a uniform line source, the directivity can be written as

$$D_u \simeq R_{\text{SD}} \frac{2L}{\lambda} \tag{4-38}$$

Comparing (4-21) and (4-22) for long line sources, we conclude $R_{\text{SD}} = 1$ for the broadside line source and $R_{\text{SD}} = 2$ for the ordinary endfire line source. For a 10λ line source, this is approximately true as Table 4-4 shows. The exact values for R_{SD} in the table were obtained using (4-36). Table 4-4 indicates that for the Hansen–

Table 4-4 Superdirectivity for a 10λ Line Source

Case	R_{SD}
Broadside	1.01
Ordinary endfire	2.01
Hansen–Woodyard	8.03

Woodyard case, moderate levels of superdirectivity are achievable. The value of $R_{SD} = 2.01$ for the ordinary endfire case does not indicate a superdirective condition since R_{SD} was achieved with the linear phase distribution $e^{\pm j\beta_o z'}$ and $|\cos\theta_o| \le 1$. In the broadside case, superdirectivity ratios even modestly greater than unity are not practical since rapid precisely controlled variations of phase are required. Even if such rapid phase variations could be achieved in practice, the resulting superdirectivity would not result in supergain because of a decrease in e_r, the radiation efficiency, due to the ohmic losses that would inevitably occur.

EXAMPLE 4-3 Superdirectivity of a 10λ Broadside Line Source

It is desired to calculate R_{SD} for a broadside line source when $L = 10\lambda$:

$$f(u) = \frac{\sin u}{u} \quad \text{where } u = \frac{\beta L}{2}\cos\theta \quad \text{since } \theta_o = 90°$$

The superdirective ratio is found using (4-36). The numerator of (4-36) has a value of π from (F-12). The denominator of (4-36) is evaluated using integration by parts. Let

$$dy = \frac{1}{u^2}\,du \qquad \text{and} \qquad x = \sin^2 u$$

so that

$$y = -\frac{1}{u} \qquad \text{and} \qquad dx = 2\sin u\cos u\,du$$

Now

$$\int x\,dy = xy - \int y\,dx$$

or in this case,

$$\int_{-10\pi}^{10\pi} \frac{\sin^2 u}{u^2}\,du = -\frac{\sin^2 u}{u}\bigg|_{-10\pi}^{10\pi} - \int_{-10\pi}^{10\pi}\left(\frac{-1}{u}\right) 2\sin u\cos u\,du$$

$$= 0 + \int_{-10\pi}^{10\pi} \frac{\sin 2u}{2u}\,d(2u) = 2\,\text{Si}(20\pi)$$

where the sine integral of (F-13) has been used. Thus,

$$R_{SD} = \frac{\pi}{2\,\text{Si}(20\pi)} = \frac{3.14159}{3.10976} = 1.01$$

and the broadside entry in Table 4-4 has been confirmed.

REFERENCES

1. J. F. Ramsay, "Lambda Functions Describe Antenna/Diffraction Patterns," *Microwave Journal*, Vol. 6, pp. 69–107, June 1967.
2. R. C. Hansen, Ed., *Microwave Scanning Antennas*, Vol. I, Academic Press, New York, 1964, Chap. 1.
3. S. Silver, Ed., *Microwave Antenna Theory and Design*, Vol. 12, MIT Radiation Laboratory Series, Boston Technical Publishers, Inc., Lexington, MA, 1964, p. 187.

PROBLEMS

4.1-1 Show that

$$\cos^{-1}(-x) - \cos^{-1}(x) = 2 \sin^{-1}(x)$$

and thereby proving (4-13). To do this, introduce α such that $x = \sin \alpha$ and use $\cos(\alpha \pm \pi/2) = \mp\sin \alpha$.

4.1-2 Prove the half-power beamwidth expression for an endfire, uniform line source. Start with (4-15) and derive (4-16). *Hint:* Let $\alpha = \cos^{-1}(1 - y)$ where $y = 0.443(\lambda/L)$, then form $\cos^2 \alpha$, neglect y^2, expand 1 as $\cos^2 \alpha + \sin^2 \alpha$, and use $\sin \alpha \approx \alpha$.

4.1-3 Show that the far-zone electric field expression E_θ for a broadside, uniform line source approximates that of an ideal dipole for short line sources ($L \ll \lambda$).

4.1-4 Compute the half-power beamwidths (in degrees) and the directivities (in decibels) for the following uniform line sources:

a. Eight-wavelength broadside, uniform line source.
b. Eight-wavelength endfire, uniform line source.
c. Sixteen-wavelength broadside, uniform line source.
d. Sixteen-wavelength endfire, uniform line source.

4.1-5 a. Use the universal pattern factor for a uniform line source to obtain polar plots of a four-wavelength uniform line source for two cases: broadside and endfire ($\theta_o = 0°$).

b. Measure the half-power beamwidths from the polar plots obtained in part (a).

c. Calculate the half-power beamwidths in degrees using (4-14) and (4-16). The agreement between these results and those of (b) depends mainly on how accurately you constructed the polar plot.

4.1-6 Verify the half-power beamwidth values in Table 4-1 for the three levels of approximation for uniform line sources with the following lengths: (a) 2λ, (b) 5λ, and (c) 10λ.

4.1-7 *Uniform line source directivity.*

a. Show that (4-20) leads to the expression

$$\frac{\beta L}{D_u} = \frac{\cos a - 1}{a} + \frac{\cos b - 1}{b} + \mathrm{Si}(a) + \mathrm{Si}(b)$$

where D_u is the directivity of a uniform line source with excitation phase shift per unit length of β_o, $a = (\beta - \beta_o)L$, $b = (\beta + \beta_o)L$, and Si is the sine integral function defined in (F-13).

b. Plot the directivity relative to that of the broadside, very long, uniform line source case, that is, $D_u/(2L/\lambda)$, for $\beta L = 10$ and $\beta L = 100$ as a function of θ_o from 90° to 0°.

c. What does the expression in part (a) reduce to for the broadside case ($\theta_o = 90°$)?

d. As L becomes much larger than a wavelength, show that your result in part (c) gives (4-21).

e. Use the result from part (c) and plot the directivity relative to that of a broadside, very long, uniform line source (i.e., $\lambda D_u/2L$) for βL from 1 to 10. This result shows how well the long line-source directivity approximation behaves.

4.2-1 Verify for the cosine-tapered line source pattern of (4-27) that (a) HP $= 1.19(\lambda/L)$ in the broadside case for $L \gg \lambda$, and (b) the side lobe level is -23.0 dB.

4.2-2 Construct the linear, polar plot of the pattern factor for a broadside cosine-tapered line source that is three wavelengths long. Proceed as in Fig. 4-3.

4.2-3 A 3-m long, broadside line source operating at 1 GHz has a cosine-squared tapered current distribution.

a. Compute the half-power beamwidth in degrees.

b. Compute the directivity in decibels.

4.2-4 Evaluate the half-power beamwidths in degrees and the directivities in decibels of 10λ long line sources with the following current distributions: (a) uniform, (b) triangular, (c) cosine, (d) cosine-squared, and (e) cosine on a -10-dB pedestal.

4.2-5 *Triangular current-tapered line source.*

a. From the current distribution in Table 4-2*a*, derive the normalized pattern factor $f(u)$.

b. Verify that the half-power beamwidth is given by $1.28(\lambda/L)$ for $L \gg \lambda$ and the side lobe level is -26.6 dB. You may do this by substitution, and also you may find some of the results from the uniform line source helpful.

4.2-6 The pattern from a triangular-tapered current distribution is the square of that of the uniform current distribution. From Fourier transform theory, how are the current distributions related?

4.2-7 Dipole antennas with lengths less than a half-wavelength have current distributions that are nearly triangular (see Fig. 1-20*b*).

a. Write the complete electric field expression E_θ in the far field for a broadside line source with a triangular current distribution.

b. Approximate the expression of part (a) for short dipoles ($L \ll \lambda$).

c. Compare this to the far-field expression for E_θ of an ideal dipole. Discuss.

4.2-8 Derive the pattern factor expression in Table 4-2 for a cosine-squared line source current distribution. Also verify the half-power beamwidth expression.

4.2-9 A broadside line source has a cosine on a -10-dB pedestal current distribution. It operates at 200 MHz and has a length of 20 m. Compute (a) the half-power beamwidth in degrees and (b) the directivity in decibels.

4.2-10 Derive the pattern factor expression in Table 4-2 for a cosine on a pedestal current distribution for a line source.

4.2-11 The directivity of a line source can be calculated from

$$D = \frac{2}{\lambda} \frac{\left| \int_{-L/2}^{L/2} I(z)\, dz \right|^2}{\int_{-L/2}^{L/2} |I(z)|^2\, dz}$$

This is the one-dimensional analogy of (7-66). Use this formula to:

a. Derive $D_u = 2L/\lambda$, the directivity of a uniform line source.

b. Derive an expression for D/D_u of a cosine on a pedestal current distribution. Evaluate for $C = 1$, 0.3162, 0.1778, and 0.

4.4-1 Verify the ordinary endfire value for R_{SD} in Table 4-4. *Note:* $u = (\beta L/2)(\cos\theta - 1)$.

4.4-2 Verify the Hansen–Woodyard value for R_{SD} in Table 4-4. *Note:* $u = 0.5[\beta L(\cos\theta - 1) - 2.94]$.

4.4-3 Consider a line source to have a wave traveling on it with a phase velocity v. Hence, $\beta_o = \beta(c/v)$. Determine the ratio of c/v for radiation in these three cases: (a) broadside; (b) ordinary endfire; (c) supergain (endfire). For these three cases, identify whether the phase velocity is that of a fast wave (i.e., $v > c$), a slow wave (i.e., $v < c$), or neither.

Chapter 5

Resonant Antennas: Wires and Patches

In this chapter, we discuss the important topics of wire antennas and patches. Wire antennas are the oldest and still the most prevalent of all antenna forms. Just about every imaginable shape and configuration of wires has a useful antenna application. Wire antennas can be made from either solid wire or tubular conductors. They are relatively simple in concept, easy to construct, and very inexpensive.

To obtain completely accurate solutions for wire antennas, the current on the wire must be solved for, subject to the boundary condition that the tangential electric field is zero along the wire. This approach gives rise to an integral equation, for which many approximate solutions have been reported over the last several decades [1]. These classical solutions are rather tedious and limited to a few simple wire shapes. On the other hand, modern numerical methods implemented on the digital computer are rather simple in concept and applicable to many wire antenna configurations. These numerical (moment method) techniques are discussed in Chapter 10. In this chapter, we adopt a simple approach to solving for the properties of wire antennas. This affords a conceptual understanding of how wire antennas operate, as well as yielding surprisingly accurate engineering results. For example, during the discussion of the loop antenna in Section 5.7 a detailed comparison of results from simple theory and the more exact numerical methods demonstrates the accuracy of simple theory.

In this chapter, we discuss several resonant wire antennas such as straight wire dipoles, vee dipoles, folded dipoles, Yagi–Uda arrays, and loops. Microstrip patch antennas, which are resonant antennas, are also described. A resonant antenna is a standing wave antenna (e.g., a dipole) with zero input reactance at resonance.

Other wire antennas that are broadband, such as traveling-wave antennas, the helix, and log-periodic, are presented in the next chapter. Methods of feeding wire antennas and their performance in the presence of an imperfect ground plane are included here. Most of the developments in this chapter utilize the principles set forth thus far. Design data and guidelines for the construction and use of wire antennas are emphasized.

5.1 DIPOLE ANTENNAS

We have discussed short dipoles in Sec. 1.6 and 2.1 and the half-wave dipole in Sec. 2.2. In this section, dipoles of arbitrary length are examined. The dipole antenna has received intensive study [1–3]. We will use a simple but effective approach that involves an assumed form for the current distribution. The radiation integral may then be evaluated and thus also the pattern parameters. For dipoles, we assume that the current distribution is sinusoidal. This is a good approximation verified by measurements. The current must, of course, be zero at the ends. We are, in effect, using the current distribution that is found on an open-circuited parallel wire transmission line. It is assumed that if the end of such a transmission line is bent out to form a wire antenna, the current distribution along the bent portion is essentially unchanged. Although this is not strictly true, it is a good approximation for thin antennas, for which the conductor diameter is on the order of 0.01λ or smaller [4].

5.1.1 Straight Wire Dipoles

A straight dipole antenna is shown in Fig. 5-1 oriented along the z-axis. It is fed at the center from a balanced two-wire transmission line, that is, the currents on each wire are equal in magnitude and opposite in direction. The current distribution along the antenna is assumed to be sinusoidal and can be written as

$$I(z) = I_m \sin\left[\beta\left(\frac{L}{2} - |z|\right)\right], \qquad |z| < \frac{L}{2} \tag{5-1}$$

The dipole is surrounded by free space, thus, the phase constant is that of free space, β.

It is helpful to visualize the current distribution on an antenna. Figure 5-2 shows the current on a dipole for $L < \lambda/2$. The solid lines indicate actual currents on the antenna and the dotted lines indicate extensions of the sine wave function. As a note of caution with this visualization, the dotted portion of the current distribution does not appear on the transmission line [5]. For this case, I_m in (5-1) is not the maximum current attained on the antenna. The maximum current on the antenna shown in Fig. 5-2 is at the input terminals where $z = 0$ and is of a value $I_m \sin(\beta L/2)$. The arrows in Fig. 5-2 show the current direction. The currents on the top and bottom halves of the antenna are in the same direction at any instant of time, and thus the radiation effects from each half reinforce. The transmission line, however,

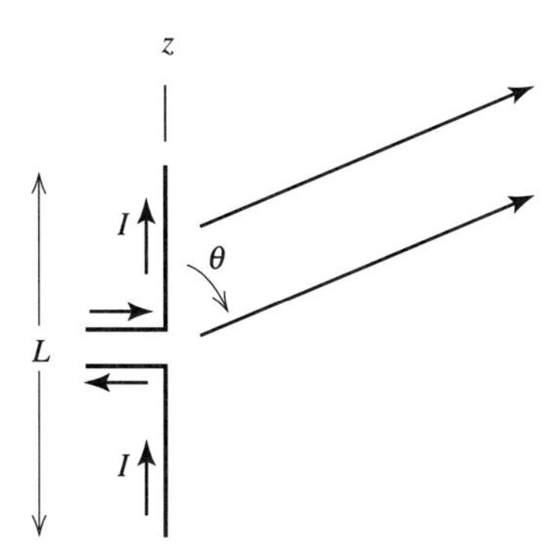

Figure 5-1 The dipole antenna.

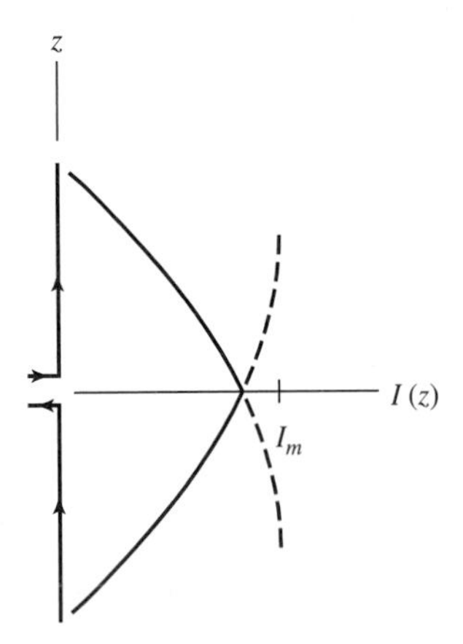

Figure 5-2 Current distribution on a dipole of length $L < \lambda/2$.

has oppositely directed currents that have canceling radiation effects for typical close conductor spacings. (See Fig. 2-4.)

In Fig. 5-3, current distributions on various dipoles are plotted together with the antennas used to generate them. The sinusoidal curves superimposed on the antennas indicate the intensity of the current on the wire, that is, the value of the curve at point z is the current value *on the wire* at the same point z. Again, the arrows indicate current directions. To construct plots such as these, begin on the z-axis at one end of the wire where the current is zero and draw a sine wave while moving toward the feed point. The current on the other half is then the mirror image. For dipoles longer than one wavelength, the currents on the antenna are not all in the same direction. Over a half-wave section, the current is in-phase and adjacent half-wave sections are of opposite phase. We would then expect to see some large canceling effects in the radiation pattern. This will be shown later to be precisely what happens. For all the current distributions presented, the plots represent the

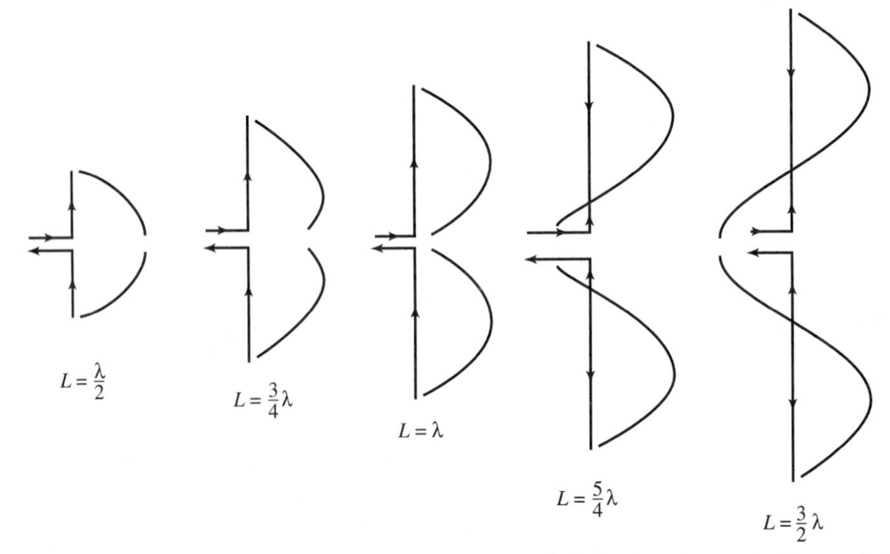

Figure 5-3 Current distributions for various center-fed dipoles. Arrows indicate relative current directions for these maximum current conditions.

maximum excitation state. It is assumed that a sinusoidal waveform generator of radian frequency $\omega = 2\pi c/\lambda$ is connected to the input transmission line. The standing wave pattern of the current at any instant of time is obtained by multiplying (5-1) by cos ωt, which follows from (1-6).

To obtain the dipole radiation pattern, we first evaluate the radiation integral

$$f_{\text{un}} = \int_{-L/2}^{L/2} I(z')e^{j\beta z' \cos\theta}\, dz' \tag{5-2}$$

Substituting the current expression from (5-1) gives

$$f_{\text{un}} = \int_{-L/2}^{0} I_m \sin\left[\beta\left(\frac{L}{2} + z'\right)\right] e^{j\beta z' \cos\theta}\, dz' + \int_{0}^{L/2} I_m \sin\left[\beta\left(\frac{L}{2} - z'\right)\right] e^{j\beta z' \cos\theta}\, dz' \tag{5-3}$$

Evaluating these integrals (see Prob. 5.1-1) gives the unnormalized pattern

$$f_{\text{un}} = \frac{2I_m}{\beta} \frac{\cos[(\beta L/2)\cos\theta] - \cos(\beta L/2)}{\sin^2\theta} \tag{5-4}$$

Using this in (4-1) leads to the complete far-zone electric field

$$E_\theta = j\omega\mu \sin\theta \frac{e^{-j\beta r}}{4\pi r} \frac{2I_m}{\beta} \frac{\cos[(\beta L/2)\cos\theta] - \cos(\beta L/2)}{\sin^2\theta} \tag{5-5}$$

Noting that $\omega\mu/\beta = \eta$, we see that this expression simplifies to

$$E_\theta = j\eta \frac{e^{-j\beta r}}{2\pi r} I_m \frac{\cos[(\beta L/2)\cos\theta] - \cos(\beta L/2)}{\sin\theta} \tag{5-6}$$

The θ-variation of this function determines the far-field pattern. For $L = \lambda/2$, it is

$$F(\theta) = \frac{\cos[(\pi/2)\cos\theta]}{\sin\theta} \qquad (L = \lambda/2) \tag{5-7}$$

This expression was also derived in Sec. 2.2; see (2-8). This is the normalized electric field pattern of a **half-wave dipole**. The half-power beamwidth is 78° and its pattern plot is shown in Fig. 5-4.

For a center-fed dipole with $L = \lambda$, the normalized electric field pattern from (5-6) is

$$F(\theta) = \frac{\cos(\pi\cos\theta) + 1}{2\sin\theta} \qquad (L = \lambda) \tag{5-8}$$

The half-power beamwidth for this **full-wave dipole** is 47°. Its pattern is also shown in Fig. 5-4. If $L = \frac{3}{2}\lambda$, the pattern function is

$$F(\theta) = 0.7148 \frac{\cos(\frac{3}{2}\pi\cos\theta)}{\sin\theta} \qquad (L = \tfrac{3}{2}\lambda) \tag{5-9}$$

The factor 0.7148 is the normalization constant. As predicted earlier, for dipoles of length greater than one wavelength, the pattern of the three-halves wavelength dipole shown in Fig. 5-4 has a multiple lobe structure due to the canceling effect of oppositely directed currents on the antenna. This effect is also visible in the $\frac{5}{4}$ wavelength case.

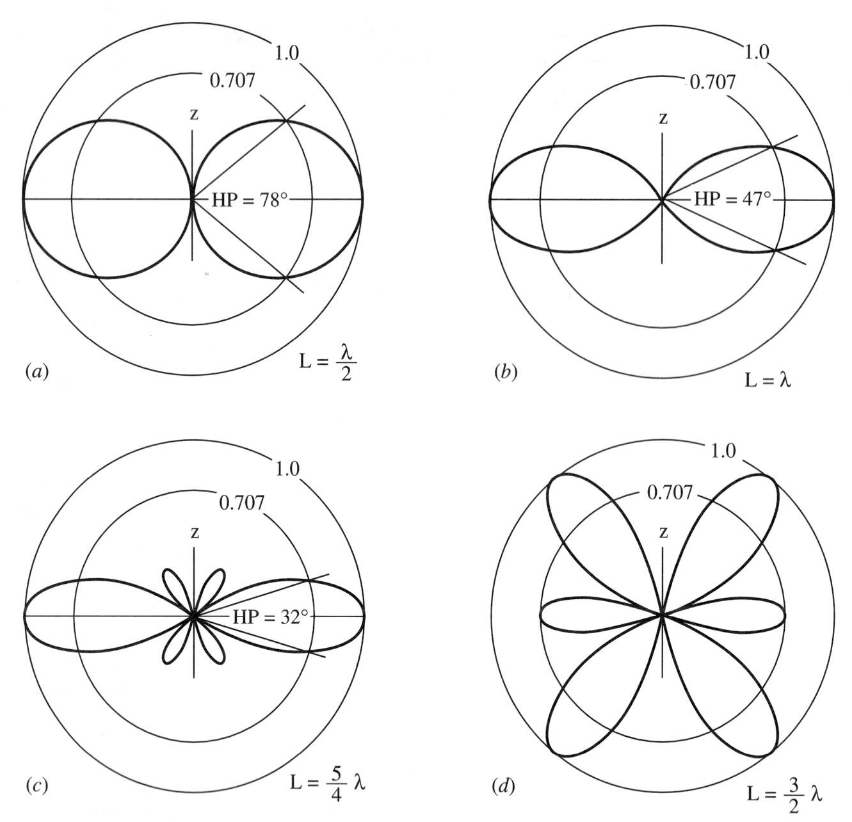

Figure 5-4 Radiation patterns of center-fed straight dipole antennas of length L.

As L/λ becomes very small, the dipole pattern variation in (5-6) approaches $\sin\theta$. Thus, we see again that the pattern of a short dipole along the z-axis is $\sin\theta$. Recall that the short dipole pattern has a 90° half-power beamwidth; see Fig. 1-10*b*.

To obtain the radiation resistance, first the radiated power must be found. Substituting (5-6) into (1-130) gives

$$P = \frac{1}{2\eta}\int_0^{2\pi}\int_0^{\pi} \eta^2 \frac{I_m^2}{(2\pi r)^2}\left\{\frac{\cos[(\beta L/2)\cos\theta] - \cos(\beta L/2)}{\sin\theta}\right\}^2 r^2 \sin\theta\, d\theta\, d\phi$$
$$= \frac{\eta}{8\pi^2} I_m^2 \int_0^{2\pi} d\phi\, 2\int_0^{\pi/2} \frac{\{\cos[(\beta L/2)\cos\theta] - \cos(\beta L/2)\}^2}{\sin\theta}\, d\theta \tag{5-10}$$

Changing the integration variable to $\tau = \cos\theta$, so $d\tau = -\sin\theta\, d\theta$, gives

$$P = \frac{\eta}{2\pi} I_m^2 \int_1^0 \frac{\{\cos[(\beta L/2)\tau] - \cos(\beta L/2)\}^2}{1-\tau^2}(-d\tau)$$
$$= \frac{\eta}{4\pi} I_m^2 \int_0^1 \left(\frac{\{\cos[(\beta L/2)\tau] - \cos(\beta L/2)\}^2}{1+\tau} + \frac{\{\cos[(\beta L/2)\tau] - \cos(\beta L/2)\}^2}{1-\tau}\right) d\tau \tag{5-11}$$

where in the last expression the identity

$$\frac{1}{1-u^2} = \frac{1}{2}\left(\frac{1}{1+u} + \frac{1}{1-u}\right) \tag{5-12}$$

was used. Equation (5-11) can be evaluated in terms of sine and cosine integral functions; see (F-13) and (F-14). A simpler expression for the special case of the half-wave dipole is obtainable in terms of a single cosine integral function. Thus, when $\beta L/2 = \pi/2$, (5-11) becomes

$$P = \frac{\eta}{4\pi} I_m^2 \int_0^1 \left[\frac{\cos^2(\pi\tau/2)}{1+\tau} + \frac{\cos^2(\pi\tau/2)}{1-\tau}\right] d\tau \tag{5-13}$$

Changing variables again as $v = 1 - \tau$ and $w = 1 + \tau$ and substituting into (5-13) give

$$\begin{aligned} P &= \frac{\eta}{4\pi} I_m^2 \left[\int_1^0 \frac{-\sin^2(\pi v/2)}{v}\, dv + \int_1^2 \frac{\sin^2(\pi w/2)}{w}\, dw\right] \\ &= \frac{\eta}{4\pi} I_m^2 \int_0^2 \frac{\sin^2(\pi v/2)}{v}\, dv = \frac{\eta}{4\pi} I_m^2 \int_0^2 \frac{1 - \cos \pi v}{2v}\, dv \end{aligned} \tag{5-14}$$

Changing the variable of integration to $t = \pi v$ leads to

$$P = \frac{\eta}{8\pi} I_m^2 \int_0^{2\pi} \frac{1 - \cos t}{t}\, dt = \frac{\eta}{8\pi} I_m^2 \,\mathrm{Cin}(2\pi) = \frac{\eta}{8\pi} I_m^2 (2.44) \tag{5-15}$$

where $\mathrm{Cin}(x)$ is related to the cosine integral function by (F-16) and is tabulated in [6]. In this case, $\mathrm{Cin}(2\pi) = 2.44$. Using this and $\eta = 120\pi$ in (5-15) leads to the radiation resistance for a half-wave dipole as

$$R_r = \frac{2P}{I_m^2} = \frac{2(15\, I_m^2\, 2.44)}{I_m^2} = 73\ \Omega \quad \left(L = \frac{\lambda}{2}\right) \tag{5-16}$$

The infinitely thin dipole antenna also has a reactive impedance component. For the half-wave dipole, the reactance is inductive, and the complete input impedance is

$$Z_A = 73 + j42.5\ \Omega \qquad \left(L = \frac{\lambda}{2}\right) \tag{5-17}$$

This can be calculated for an infinitely thin dipole by a classical procedure known as the induced emf method [7]. However, the input impedance of dipoles with finite wire diameter can be calculated using the numerical methods of Chap. 10, where the form of the current is not assumed. The results of such a calculation for the input resistance and reactance of a small-diameter, center-fed dipole are given in Figs. 5-5 and 5-6. The resonance effects are evident in these plots. Note that the input reactance is capacitive for small lengths, as we pointed out in Sec. 2.1.

The dotted curve in Fig. 5-5 is the input resistance from (1-172) for an ideal dipole with uniform current. It does not give good results for an actual wire dipole as shown by the solid curve of Fig. 5-5. However, the triangular current approximation with $R_{ri} = 20\pi^2(L/\lambda)^2$ from (1-177) does give a good approximation to the input radiation resistance for short dipoles as demonstrated by the dashed curve of Fig. 5-5. Some

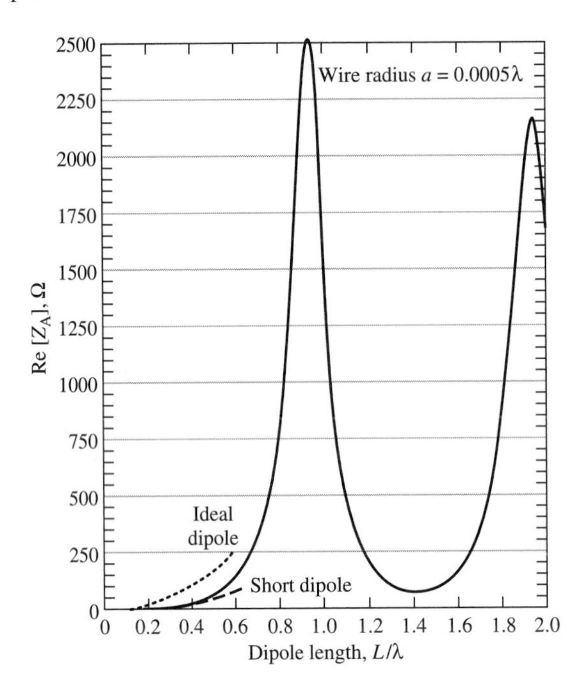

Figure 5-5 Calculated input resistance of a center-fed wire dipole of 0.0005λ radius as a function of length L (solid curve). Also shown is the input resistance $R_{ri} = 80\pi^2(L/\lambda)^2$ of an ideal dipole with a uniform current distribution (dotted curve) and the input resistance $R_{ri} = 20\pi^2(L/\lambda)^2$ of a short dipole with a triangular current distribution approximation (dashed curve).

simple formulas that approximate the input resistance of wire dipoles are given in Table 5-1 [8]. For example, using the second formula for $L = \lambda/2$ gives $R_{ri} = 24.7(\pi/2)^{2.4} = 73.0\ \Omega$, which agrees with (5-17). The values obtained from Table 5-1 also agree closely with those of Fig. 5-5.

Input resistance can be related to radiation resistance. There are several ways to define radiation resistance by using different current reference points. Usually, ra-

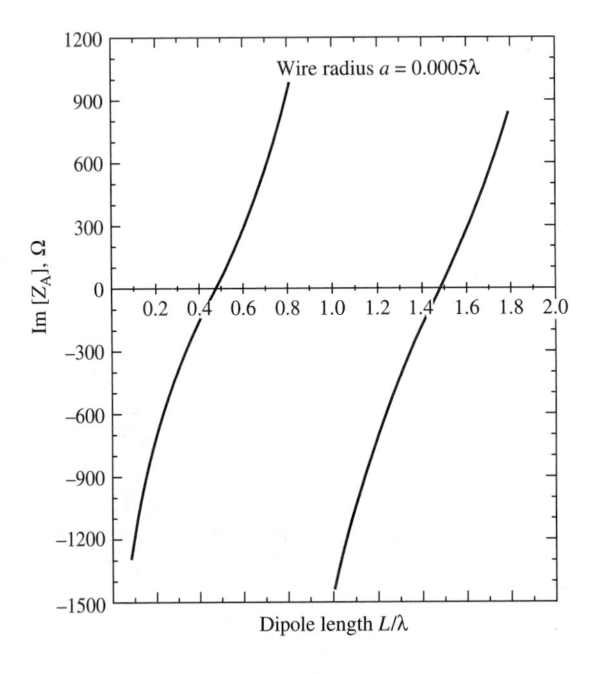

Figure 5-6 Calculated input reactance of center-fed wire dipole of radius 0.0005λ as a function of length L.

Table 5-1 Simple Formulas for the Input Resistance of Dipoles

Length L	Input Resistance (R_{ri}), Ω
$0 < L < \dfrac{\lambda}{4}$	$20\pi^2\left(\dfrac{L}{\lambda}\right)^2$
$\dfrac{\lambda}{4} < L < \dfrac{\lambda}{2}$	$24.7\left(\pi\dfrac{L}{\lambda}\right)^{2.4}$
$\dfrac{\lambda}{2} < L < 0.637\lambda$	$11.14\left(\pi\dfrac{L}{\lambda}\right)^{4.17}$

diation resistance is defined using the current distribution maximum I_m, whether or not it actually occurs on the antenna. We shall use the symbol R_{rm} for this definition. It is also useful to refer the radiation resistance to the input terminal point. In this case, the symbol R_{ri} is used. These definitions can be related by writing the radiated power as

$$P = \tfrac{1}{2}I_m^2 R_{rm} = \tfrac{1}{2}I_A^2 R_{ri} \tag{5-18}$$

For dipoles that are odd integer multiples of a half-wavelength long, $I_m = I_A$ and $R_{rm} = R_{ri}$. A third radiation resistance, denoted by R_r, is often used; it is the radiation resistance relative to the maximum current that occurs on the antenna. For dipoles less than a half-wavelength long, the current maximum on the antenna always occurs at the center, and then $R_{ri} = R_r$ for center-fed dipoles; this was discussed in Sec. 1.9. In practice, we are interested in input resistance, so R_{ri} is of primary importance. It is related to R_{rm} for center-fed dipoles by setting $z = 0$ in (5-1), giving

$$I_A = I_m \sin\frac{\beta L}{2} \tag{5-19}$$

and substituting into (5-18), which yields

$$R_{ri} = \frac{I_m^2}{I_A^2} R_{rm} = \frac{R_{rm}}{\sin^2(\beta L/2)} \tag{5-20}$$

R_{ri} is the component of input resistance due to radiation and equals the total input resistance R_A if ohmic losses are neglected, which we shall do unless otherwise indicated.

For dipole lengths, $L = \lambda, 2\lambda, 3\lambda, \ldots, \beta L/2 = \pi, 2\pi, 3\pi, \ldots$, and R_{ri} from (5-20) is infinite. For example, the one-wavelength dipole of Fig. 5-3*c* has a current zero at its feed point and thus an infinite input impedance. This, of course, is based on the perfect sine wave current distribution. Dipoles of finite thickness have large but finite values of input impedance for lengths near integer multiples of wavelength, as seen in Fig. 5-5. This effect arises from the deviation of the current distribution from that of (5-1) for dipole lengths near integer multiples of the wavelength: there is always a finite input current on an actual dipole. For other-length dipoles, the sinusoidal current distribution is a good approximation for thin wire dipole antennas.

Table 5-2 Wire Lengths Required to Produce a Resonant Half-Wave Dipole for a Wire Diameter of $2a$ and Length L

Length to Diameter Ratio, $L/2a$	Percent Shortening Required	Resonant Length L	Dipole Thickness Class
5000	2	0.49λ	Very thin
50	5	0.475λ	Thin
10	9	0.455λ	Thick

By reducing the length of the half-wave dipole slightly, the antenna can be made to resonate ($X_A = 0$). The input impedance of the infinitely thin half-wave dipole is then about $70 + j0\ \Omega$. In Fig. 5-6, the dipole of radius 0.0005λ resonates for lengths corresponding to the intersections with the horizontal ($X_A = 0$) axis. The first intersection is the half-wave dipole case and the resonant length is slightly less than $\lambda/2$. It turns out that as the wire thickness increases, the dipole must be shortened more to obtain resonance. Approximate length values for resonance are given in Table 5-2. For the dipole of 0.0005λ wire radius, the length-to-diameter ratio, $L/2a$, is 500 for the half-wave case. From Table 5-2, we see that about 4% shortening ($L = 0.48\lambda$) would be required to produce resonance. This agrees closely with the resonance point from Fig. 5-6. In practice, wire antennas are constructed slightly longer than required. Then a transmitter is connected to the antenna and the standing wave ratio (or reflected power) is monitored on the feed transmission line. The ends of the antenna are trimmed until a low value of standing wave ratio is obtained. Note that as the length is reduced to obtain resonance, the input resistance also decreases. For example, for a thick dipole with $L/2a = 50$ and $L = 0.475\lambda$, the second formula of Table 5-1 gives $R_A = 64.5\ \Omega$; the reactance is, of course, zero.

Since dipoles are resonant-type structures, their bandwidth is low. The VSWR as a function of frequency for a half-wave dipole is shown in Fig. 5-7. In general, bandwidth is defined as "the range of frequencies within which the performance of

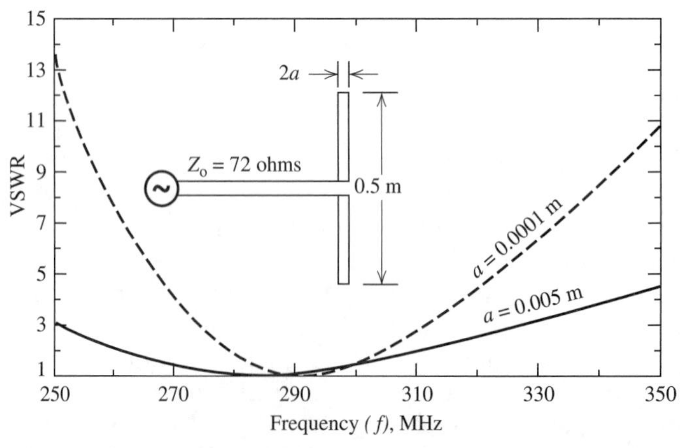

Figure 5-7 Calculated VSWR as a function of frequency for dipoles of different wire diameters.

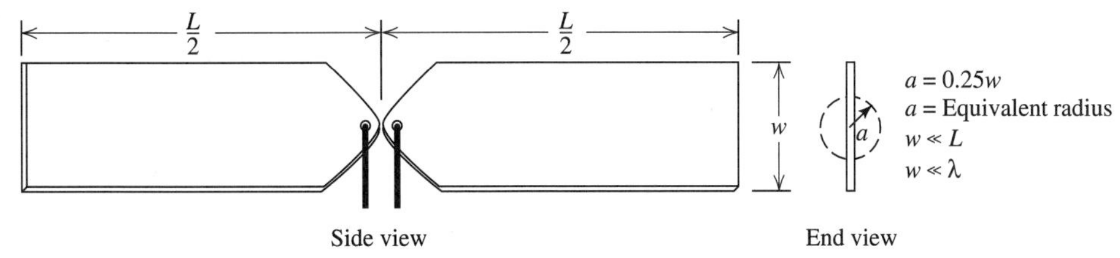

Figure 5-8 Thin metal strip dipole.

the antenna, with respect to some characteristic, conforms to a specified standard" [9]. In this case, let the specified standard be a VSWR less than 2.0:1. From Fig. 5-7, we see that the bandwidths are $310 - 262 = 48$ MHz and $304 - 280 = 24$ MHz, respectively, for $a = 0.005$ m ($L/2a = 50$) and $a = 0.0001$ m ($L/2a = 2500$). In terms of percent relative to the design frequency (300 MHz), the bandwidths are 16 and 8%. It is an important general principle that the thicker the dipole, the wider is its bandwidth. Also, note that the minimum VSWR for the thicker dipole occurs at a lower frequency than for the thinner one. In fact, if we use the rules in Table 5-2, the resonant frequencies are calculated to be 285 and 294 MHz for wire radii of 0.005 and 0.0001 m. These values agree well with the minimum points of the curves in Fig. 5-7.

The improved bandwidth offered by the thick circularly cylindrical dipole in Fig. 5-7 can also be achieved with a flat metallic strip as Fig. 5-8 indicates. The relationship between the circularly cylindrical dipole radius and the width of the metallic strip for equivalent performance under certain conditions is $a = 0.25w$ [10]. The advantage of the flat strip dipole is primarily an economic one.

Finally, we compute the directivity of a half-wave dipole. It is found from $D = 4\pi U_m/P$. The radiated power P was evaluated in (5-15). Using the far-zone electric field of (5-6) leads to the maximum radiation intensity as

$$U_m = \frac{r^2}{2\eta}|E_\theta|^2_{\max} = \frac{1}{2\eta}\frac{\eta^2 I_m^2}{(2\pi)^2} = \frac{\eta}{8\pi^2} I_m^2 \tag{5-21}$$

So,

$$D_{(\lambda/2)\text{dipole}} = \frac{4\pi U_m}{P} = \frac{4\pi(\eta/8\pi^2)I_m^2}{(\eta/8\pi)I_m^2(2.44)} = 1.64 = 2.15 \text{ dB} \tag{5-22}$$

This is only slightly greater than the directivity value of 1.5 for an ideal dipole with uniform current. So for very short dipoles, the directivity is 1.5 and increases to 1.64 as the length is increased to a half-wavelength. As length is increased further, directivity also increases. A full-wave dipole has a directivity of 2.41. Even more directivity is obtained for a length of about 1.25λ. As the length is increased further, the pattern begins to break up (see Fig. 5-4) and directivity drops sharply. See Prob. 5.1-12.

5.1.2 The Vee Dipole

Wire dipole antennas that are not straight also appear in practice. One such antenna is the **vee dipole** shown in Fig. 5-9. This antenna may be visualized as an open-circuited transmission line that has been bent so that ends of length h have an

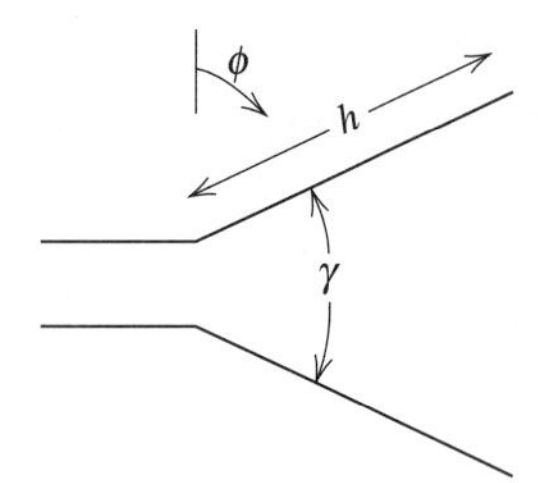

Figure 5-9 The vee dipole antenna.

included angle of γ. The angle γ for which the directivity is greatest in the direction of the bisector of γ is given by

$$\begin{aligned} \gamma &= 152\left(\frac{h}{\lambda}\right)^2 - 388\left(\frac{h}{\lambda}\right) + 324, \qquad 0.5 \leq \frac{h}{\lambda} < 1.5 \\ \gamma &= 11.5\left(\frac{h}{\lambda}\right)^2 - 70.5\left(\frac{h}{\lambda}\right) + 162, \qquad 1.5 \leq \frac{h}{\lambda} \leq 3.0 \end{aligned} \tag{5-23}$$

where the resulting angle γ is in degrees. The corresponding directivity is

$$D = 2.95\left(\frac{h}{\lambda}\right) + 1.15 \tag{5-24}$$

These equations were empirically derived for antennas with $0.5 \leq h/\lambda \leq 3.0$ using the computational methods (MoM) of Chap. 10.

The directivity of a vee dipole can be greater than that of a straight dipole. This can be seen from the pattern in Fig. 5-10 where $h = 0.75\lambda$ and γ from (5-23) is 118.5°. Notice that the direction of maximum radiation is $\phi = 90°$ while radiation in the $\phi = 270°$ direction is about 2 dB less. Even more significant is the low level of the side lobes. For the most part, it is the reduced side lobe levels of the vee dipole that give it a greater directivity than the straight dipole version (see Fig. 5-4*d*). The directivity for the vee dipole of Fig. 5-10 from (5-24) is $D = 2.94(0.75)$

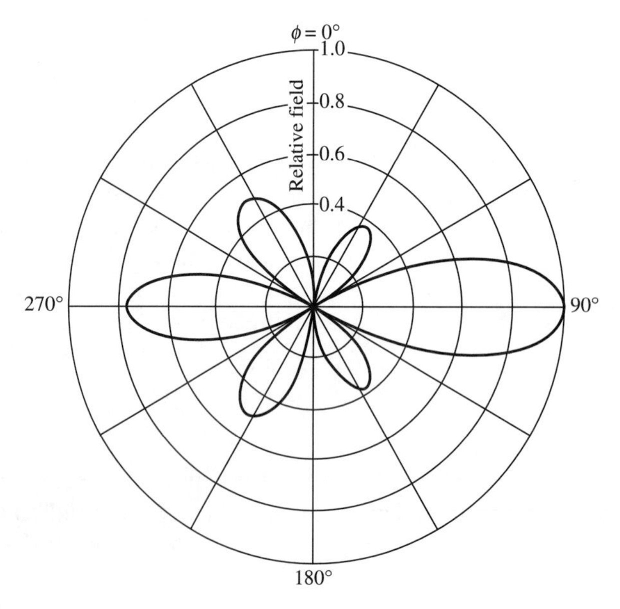

Figure 5-10 Far-field pattern of a vee dipole shown in Fig. 5-9 with arm length $h = 0.75\lambda$, $\gamma = 118.5°$, and $a = 0.0005\lambda$.

+ 1.15 = 3.355 = 5.26 dB. The directivity of a 1.5λ long straight wire dipole is about 2.2, or 3.4 dB.

The input impedance of a vee dipole antenna is generally less than that of a straight dipole of the same length. For example, the input impedance of the vee dipole in Fig. 5-10 is $106 + j17\ \Omega$, which is less than for the straight dipole version ($L = 1.5\lambda$) as found from Figs. 5-5 and 5-6.

5.2 FOLDED DIPOLE ANTENNAS

An extremely practical wire antenna is the **folded dipole**. It consists of two parallel dipoles connected at the ends forming a narrow wire loop, as shown in Fig. 5-11, with dimension d much smaller than L and much smaller than a wavelength. The feed point is at the center of one side. The folded dipole is essentially an unbalanced transmission line with unequal currents. Its operation is analyzed by considering the current to be composed of two modes: the transmission line mode and the antenna mode. The currents for these modes are illustrated in Fig. 5-12.

The currents in the transmission line mode have fields that tend to cancel in the far field since d is small. The input impedance Z_t for this mode is given by the equation for a transmission line with a short circuit load

$$Z_t = jZ_o \tan \beta \frac{L}{2} \tag{5-25}$$

where Z_o is the characteristic impedance of the transmission line.

In the antenna mode, the fields from the currents in each vertical section reinforce in the far field since they are similarly directed. In this mode the charges "go around the corner" at the end, instead of being reflected back toward the input as in an ordinary dipole, which leads to a doubling of the input current for resonant lengths. The result of this is that the antenna mode has an input current that is half that of a dipole of resonant length.

Suppose a voltage V is applied across the input terminals of a folded dipole. The total behavior is determined by the superposition of the equivalent circuits for each mode in Fig. 5-13. Note that if the figures for each mode are superimposed and the voltages are added, the total on the left is V and on the right is zero as it should be. The transmission line mode current is

$$I_t = \frac{V}{2Z_t} \tag{5-26}$$

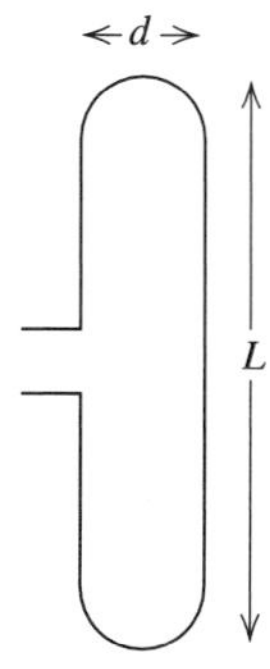

Figure 5-11 The folded dipole antenna.

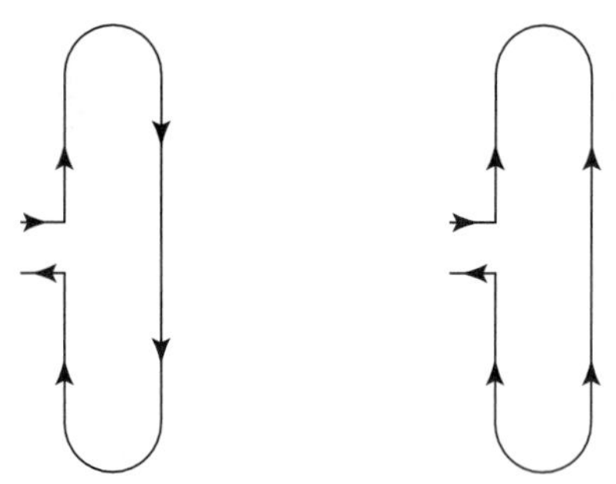

Figure 5-12 The current modes on a folded dipole antenna.

For the antenna mode, the total current is the sum of each side, or I_a. The excitation for this current is $V/2$; thus, the antenna current is

$$I_a = \frac{V}{2Z_d} \tag{5-27}$$

where to a first-order approximation Z_d is the input impedance for an ordinary dipole of the same wire size [11]. The total current on the left is $I_t + \frac{1}{2}I_a$ and the total voltage is V, so the input impedance of the folded dipole is

$$Z_A = \frac{V}{I_t + \frac{1}{2}I_a} \tag{5-28}$$

Substituting (5-26) and (5-27) in (5-28) yields

$$Z_A = \frac{4Z_t Z_d}{Z_t + 2Z_d} \tag{5-29}$$

As an example, consider the popular **half-wave folded dipole**. From (5-25) with $L = \lambda/2$, $Z_t = jZ_o \tan[(2\pi/\lambda)(\lambda/4)] = jZ_o \tan(\pi/2) = \infty$. Then (5-29) gives

$$Z_A = 4Z_d \qquad \left(L = \frac{\lambda}{2}\right) \tag{5-30}$$

Thus, the half-wave folded dipole provides a four-fold increase in impedance over its dipole version. Since the half-wave dipole (at resonance) has a real input impedance, the half-wave folded dipole has also.

The current on the half-wave folded dipole is particularly easy to visualize. We will discuss this current and also rederive the impedance. If the vertical wire section on the right in Fig. 5-11 were cut directly across from the feed point and the wire

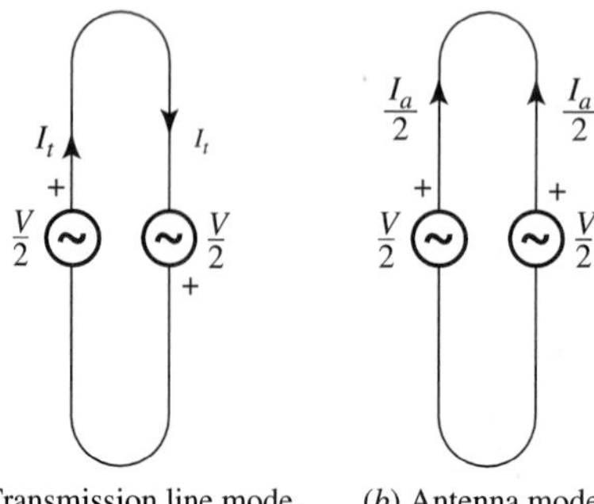

Figure 5-13 Mode excitation and current for a voltage V applied to the terminals of a folded dipole. Superposition of these modes gives the complete folded dipole model.

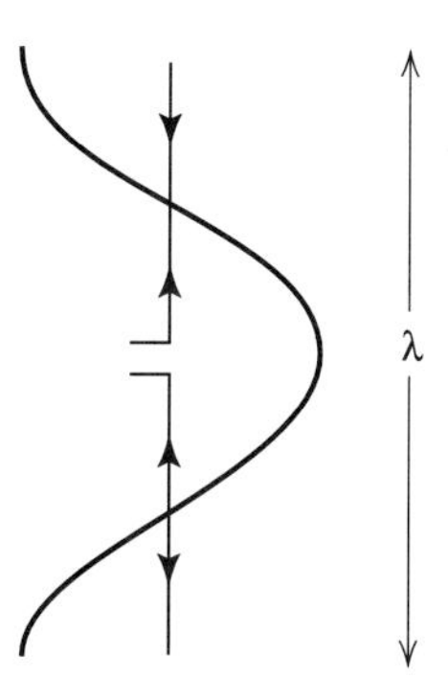

Figure 5-14 Current for the antenna mode of a half-wave folded dipole that has been folded out without disturbing the current.

folded out without disturbing the current, it would appear as shown in Fig. 5-14. The current is not zero at the ends because they are actually connected. Perhaps a better way to view this is to fold the current back down and note that currents on the folded part are now upside down as shown in Fig. 5-15*a*. The same total current (and thus the same pattern) is obtained with both the folded and the ordinary dipoles in Fig. 5-15. The difference is that the folded dipole has two closely spaced currents equal in value, whereas in the ordinary dipole they are combined on one wire. From this, it is easy to see that the input currents in the two cases are related as

$$I_f = \tfrac{1}{2} I_d \qquad \left(L = \frac{\lambda}{2}\right) \tag{5-31}$$

The input powers are

$$P_f = \tfrac{1}{2} Z_A I_f^2 \qquad \text{(folded dipole)} \tag{5-32}$$

and

$$P_d = \tfrac{1}{2} Z_d I_d^2 \qquad \text{(dipole)} \tag{5-33}$$

Since the total currents are the same in the half-wave case, the radiated powers are also. Equating (5-32) and (5-33) and using (5-31) give

$$\tfrac{1}{2} Z_d I_d^2 = \tfrac{1}{2} Z_A \tfrac{1}{4} I_d^2$$

or

$$Z_A = 4 Z_d \qquad \left(L = \frac{\lambda}{2}\right) \tag{5-34}$$

This result is an independent confirmation of the result in (5-30).

The input impedance of a half-wave folded dipole (at resonance) is four times

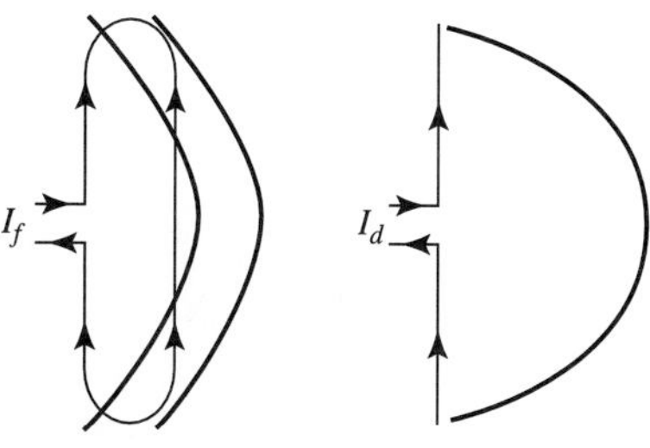

(*a*) Folded dipole. (*b*) Dipole. **Figure 5-15** Currents on half-wave dipoles.

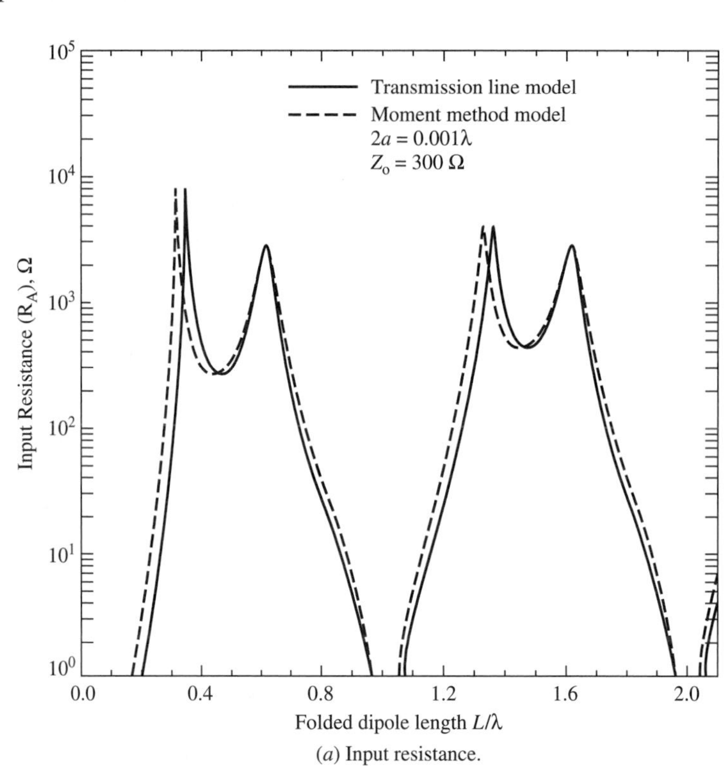

(*a*) Input resistance.

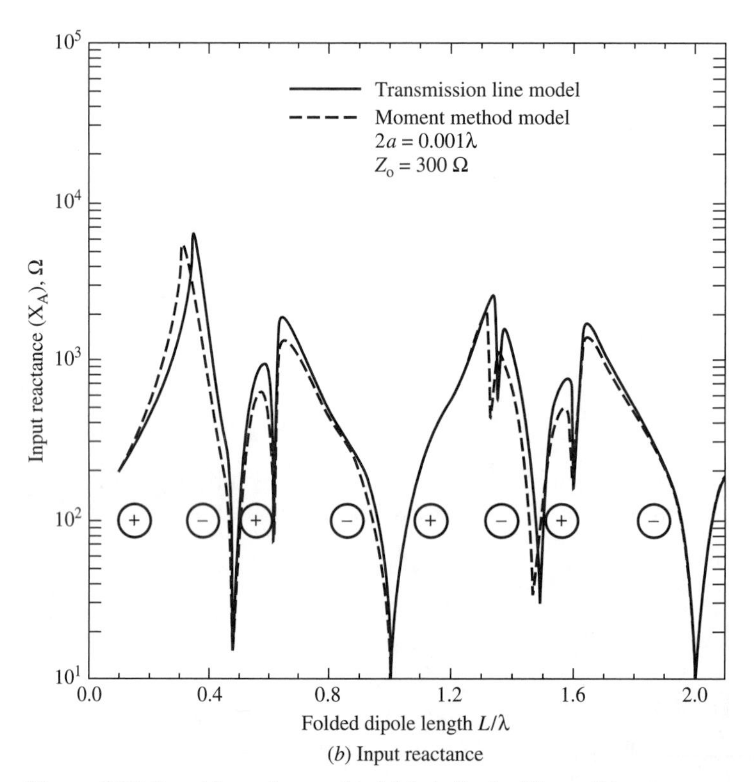

(*b*) Input reactance

Figure 5-16 Input impedance of a folded dipole. The solid curves are calculated from the transmission line model. The dashed curves are calculated from more accurate numerical methods. The wire radius a is 0.0005λ and wire spacing d is $12.5a$.

that of an ordinary dipole. A resonant half-wave dipole has about 70 Ω of input resistance, so a half-wave folded dipole then has an input impedance of

$$Z_f = 4(70) = 280\ \Omega \qquad \text{(half-wave)} \tag{5-35}$$

This impedance is very close to the 300 Ω of common twin-lead transmission line.

The input impedance curves for a folded dipole of finite wire thickness are given in Fig. 5-16 as a function of folded dipole length. The solid curves were obtained from the transmission line model. The wire spacing $d = 12.5a$ is such that the characteristic impedance corresponds to a 300 Ω transmission line [$Z_o = (\eta/\pi)\ln(d/a) = 120\ln(12.5a/a) \approx 300\ \Omega$]. The folded dipole input impedance is then found from (5-25) and (5-29). As an example, consider a folded dipole of length $L = 0.8\lambda$, spacing $d = 12.5a$, and radius $a = 0.0005\lambda$. From (5-25),

$$Z_t = j300 \tan 0.8\pi = -j218\ \Omega \tag{5-36}$$

From Figs. 5-5 and 5-6,

$$Z_d = 950 + j950 \tag{5-37}$$

Using these in (5-29) yields

$$Z_A = 28 - j461\ \Omega \qquad (L = 0.8\lambda) \tag{5-38}$$

This result agrees with the values shown in Fig. 5-16.

Also shown in Fig. 5-16 as dashed curves is the input impedance calculated using the more exact methods of Chap. 10. The agreement between the simple transmission line model and the numerical method results is quite good. Both methods show that the real part of the input impedance is slightly less than 300 Ω at the first resonance ($L \approx 0.48\lambda$) and slightly larger than 300 Ω at the second resonance ($L \approx 1.47\lambda$). It is this characteristic of the folded dipole that makes it useful at harmonically related frequencies. Note too the very low value of Z_A when $L \approx \lambda$, $2\lambda, \ldots$. This can easily be explained from the transmission line model, since then $\tan(\beta L/2) \approx \tan \pi = 0$ and thus $Z_t = 0$ and Z_A from (5-29) is zero.

The folded dipole is used as an FM broadcast band receiving antenna, and it can be simply constructed by cutting a piece of 300-Ω twin-lead transmission line about a half-wavelength long (1.5 m at 100 MHz). The ends are soldered together such that the overall length L is slightly less than a half-wavelength at the desired frequency (usually 100 MHz). One wire is then cut in the middle and connected to the twin-lead transmission line feeding the receiver.

Occasionally, two different wire sizes are used for a folded dipole as shown in Fig. 5-17. The input impedance for the half-wave case is given by

$$Z_A = (1 + c)^2 Z_d \qquad \left(L = \frac{\lambda}{2}\right) \tag{5-39}$$

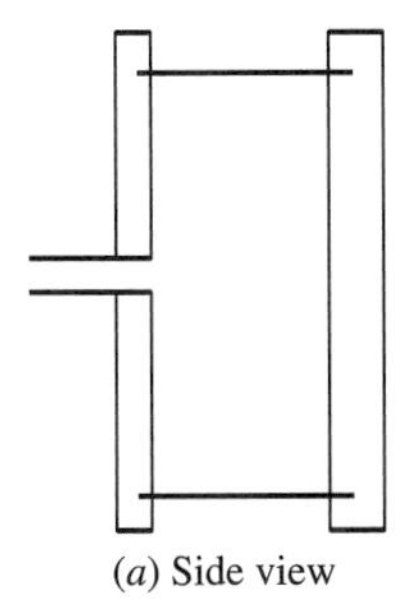

(*a*) Side view

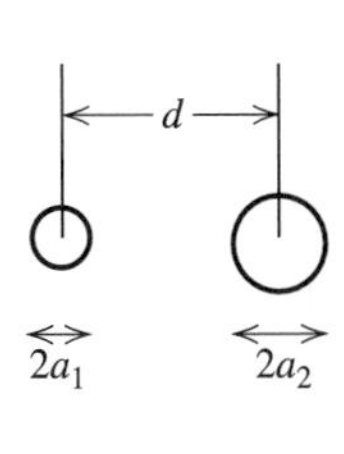

(*b*) End view

Figure 5-17 Folded dipole antenna constructed from two different size conductors.

For given values of d, a_1, and a_2, the value of c can be found [11, 12]. As is frequently the case, if a_1 and a_2 are much less than d, c is approximately given by

$$c \approx \frac{\ln(d/a_1)}{\ln(d/a_2)} \tag{5-40}$$

The folded dipole antenna is a very popular wire antenna. The reasons for this are its impedance properties, ease of construction, and structural rigidity. The equal-size conductor half-wave folded dipole has an input impedance very close to that of a 300-ohm twin-lead-type transmission line as seen from Fig. 5-16. Also, by changing the conductor radii, the input impedance can be changed. In addition to having desirable impedance properties, the half-wave folded dipole has a wider bandwidth than an ordinary half-wave dipole [13]. In part for these reasons, a folded dipole is frequently used as a feed antenna for Yagi–Uda arrays and other popular antennas.

5.3 FEEDING WIRE ANTENNAS

When connecting an antenna to a transmission line, it is important to make effective use of all available power from the transmitter in the transmit case and from the antenna in the receive case. There are two primary considerations: the impedance match between the antenna and transmission line, and the excitation of the current distribution on the antenna. In this section, these general topics are discussed along with specific applications to wire antennas.

First, consider impedance matching. A typical transmitter or receiver circuit is shown in Fig. 5-18. Usually, the transmitter or receiver has an impedance equal to that of the transmission line Z_o. However, the antenna impedance Z_A is frequently quite different from Z_o. The question is whether or not this is a problem. The answer depends on the application. In some cases, corrective measures such as a matching network are necessary. Let us examine the effects created by a mismatch. It is well known that maximum power is transferred when there is a conjugate impedance match. Also, if the system were operated with a poor match at the antenna, there would be reflections set up along the transmission line; that is, the voltage standing wave ratio (VSWR) is much greater than one. If the transmission line is of high quality (low loss), these reflections represent low-dissipative losses. For many applications, an extremely low VSWR is a luxury and not a necessity. This is demonstrated in Table 5-3, which follows from (9-9). For example, a VSWR of 2:1 leads to 89% power transmission. On the other hand, if the VSWR is very high, power is traveling back and forth along the transmission line, and if the line is lossy and/or of long length, dissipative losses may be significant.

High VSWR has other undesirable effects on a system. In high-power applications, very high voltages will be developed between the conductors at certain points

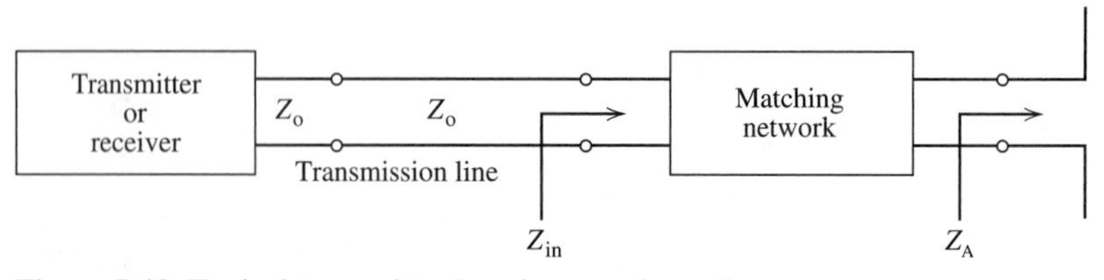

Figure 5-18 Typical transmitter/receiver configuration.

Table 5-3 VSWR and Transmitted Power for a Mismatched Antenna

VSWR	Percent Reflected Power $= \lvert\Gamma\rvert^2 \times 100 = \left(\frac{\text{VSWR} - 1}{\text{VSWR} + 1}\right)^2 \times 100$	Percent Transmited Power $= q \times 100 = (1 - \lvert\Gamma\rvert^2) \times 100$
1.0	0.0	100.0
1.1	0.2	99.8
1.2	0.8	99.2
1.5	4.0	96.0
2.0	11.1	88.9
3.0	25.0	75.0
4.0	36.0	64.0
5.0	44.4	55.6
5.83	50.0	50.0
10.0	66.9	33.1

along a transmission line. These are called "hot spots" and may cause arcing. Also, a high VSWR means that the impedance varies along the transmission line and further that the impedance at any point varies as the frequency is changed. This may affect transmitter operation. For example, the frequency of the transmitter can be changed by severe input impedance mismatch; this is called "frequency pulling."

If the impedance mismatch is unacceptable, there are several methods for improving the performance. Usually, the characteristic impedance Z_o is nearly real since low-loss transmission lines are used. For a match then, the antenna should have an input impedance equal to $Z_o + j0$. Sometimes, it is possible to select an antenna that achieves this. If this is not possible, a matching network can be employed as shown in Fig. 5-18. Such matching networks take many forms. One example is the quarter-wave transformer, which is a quarter-wave-length long transmission line with characteristic impedance $\sqrt{Z_o R_A}$, where R_A is the antenna input resistance. If the antenna impedance has a reactive component, other devices may be used. At UHF and microwave frequencies, tuning devices such as stub tuners and irises are introduced to transform the real part of the impedance to that of the transmission line as well as tuning out the reactive component. At low frequencies, reactive tuning is accomplished with variable capacitors and coils because the electrical dimensions of these lumped elements are small with respect to the wavelength.

There are disadvantages to using matching networks. For example, if a matching network is designed to obtain a near perfect match, it will usually be narrow-banded. If the matching network is designed to be broadbanded, it will usually not yield a near perfect match at all frequencies over the band, or perhaps at any frequencies over the band. A discussion of matching techniques may be found in [14] and [15].

On the other hand, there are several ways to change the input impedance of an antenna without using a matching network. For example, the input resistance of a dipole can be changed by displacing the feed point off center. If the feed point is a distance z_f from the center of the dipole, the current at the input terminals is

$$I_A = I_m \sin\left[\beta\left(\frac{L}{2} - |z_f|\right)\right] \tag{5-41}$$

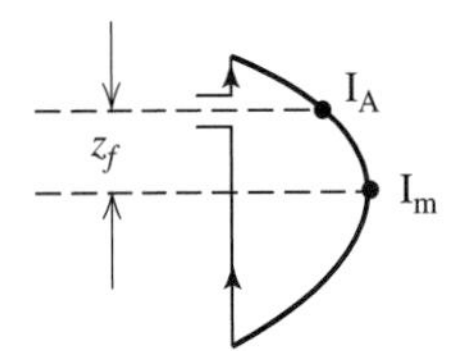

Figure 5-19 Half-wave dipole with displaced feed.

In the case of a half-wave dipole as shown in Fig. 5-19, $\beta L/2 = \pi/2$ and this reduces to

$$I_A = I_m \cos \beta z_f \tag{5-42}$$

The input resistance (not including ohmic losses) is found from (5-42) in (5-20), giving

$$R_A = \frac{I_m^2}{I_A^2} R_{rm} = \frac{R_{rm}}{\cos^2 \beta z_f} \qquad \left(L = \frac{\lambda}{2}\right) \tag{5-43}$$

As the feed point approaches the end of the wire, this result indicates that the input resistance increases toward infinity. In practice, the input resistance becomes very large as the feed point moves out. The pattern is essentially unchanged as the feed point shifts. For longer dipoles, the pattern and impedance differ significantly from the center-fed case as the feed point is displaced. For example, a full-wave dipole fed a quarter-wavelength from one end, as shown in Fig. 5-20, will have a current distribution that is significantly different from the center-fed full-wave dipole of Fig. 5-3*c* and that has a broadside null in the pattern.

The off-center feed arrangement is unsymmetrical and can lead to undesirable phase reversals in the antenna, as shown in Fig. 5-20. A symmetrical feed that increases the input resistance with increasing distance from the center point of the wire antenna is the *shunt feed.* A few forms of shunt matching are shown in Fig. 5-21. We will discuss the operation of the tee match; the remaining shunt matches behave in a similar fashion. The center section of the tee match may be viewed as being a shorted transmission line in parallel with a dipole of wide feed gap spacing. The shorted transmission line is less than a quarter-wavelength long and thus its impedance is inductive. Capacitance can be introduced to tune out this inductance by either shortening the dipole length or placing variable capacitors in the shunt legs. As the distance D is increased, the input impedance increases and peaks for a D of about half of the dipole length. As D is increased further, the impedance decreases and finally equals the folded dipole value when D equals the dipole length. The exact impedance value depends on the distances C and D, and the ratio of the

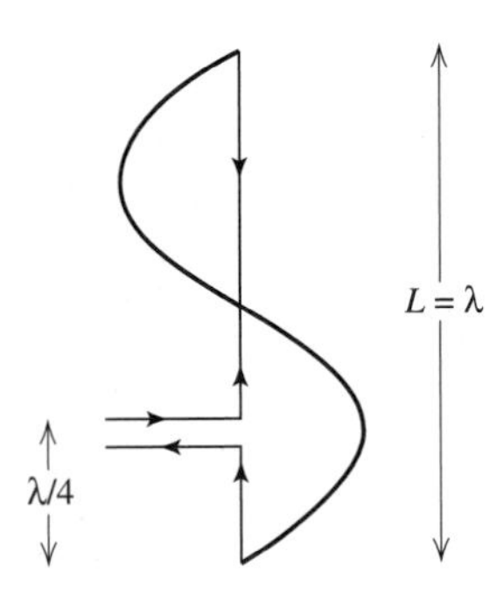

Figure 5-20 Current distribution on a full wave dipole for an off-center feed.

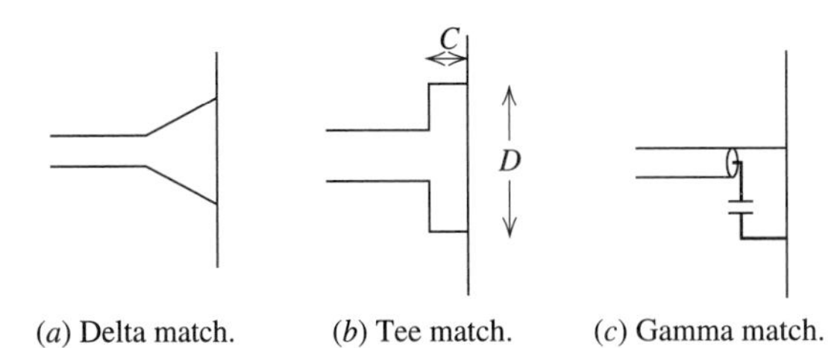

Figure 5-21 Shunt matching configurations.

dipole wire diameter to the shunt arm wire diameter (similar to the folded dipole behavior). In practice, sliding contacts are made between the shunt arms and the dipole for impedance adjustment. Shunt matches will radiate and do so in an undesirable fashion.

We now turn our attention to a separate but related problem of balancing currents on wire antennas. Many wire antennas are symmetrical in nature and, thus, the currents should also be symmetrical (or balanced). An example of balanced and unbalanced operation of a half-wave dipole is shown in Fig. 5-22. In the balanced case, the currents on the transmission line are equal in magnitude and opposite in direction, which yields very small radiation from the transmission line for closely spaced conductors. For unbalanced operation, as illustrated in Fig. 5-22*b*, the current I_1 is greater than I_2 and there is a net current flow on the transmission line leading to uncontrolled radiation that is not in the desired direction or of the desired polarization. Also, the unbalanced current on the antenna will change the radiation pattern from the balanced case. Thus, it is clear that balanced operation is desirable.

Transmission lines are referred to as balanced and unbalanced. Parallel wire lines are inherently balanced in that if an incident wave (with balanced currents) is launched down the line, it will excite balanced currents on a symmetrical antenna. On the other hand, a coaxial transmission line is not balanced. A wave traveling down the coax may have a balanced current mode, that is, the currents on the inner conductor and the inside of the outer conductor are equal in magnitude and opposite in direction. However, when this wave reaches a symmetrical antenna, a current may flow back on the outside of the outer conductor, which unbalances the antenna and transmission line. This is illustrated in Fig. 5-23. Note that the currents on the two halves of the dipole are unbalanced. The current I_3 flowing on the outside of the coax will radiate. The currents I_1 and I_2 in the coax are shielded from the external world by the thickness of the outer conductor. They could actually be unbalanced with no resulting radiation; it is the current on the outside surface of the outer conductor that must be suppressed. To suppress this outside surface current, a *balun* (contraction for "balanced to unbalanced transformer") is used.

The situation in Fig. 5-23 may be understood by examining the voltages that exist

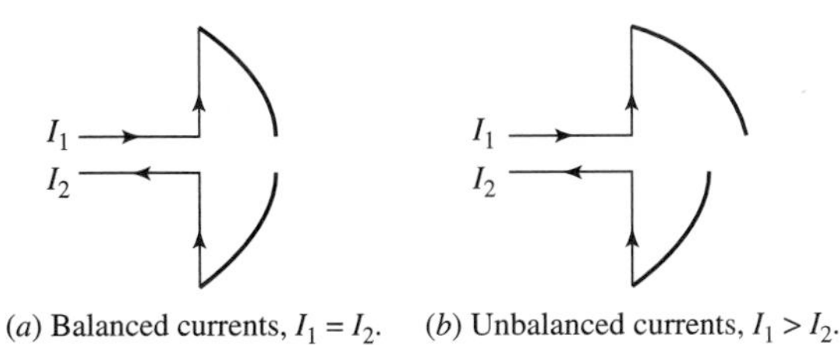

Figure 5-22 Balanced and unbalanced operation of a center-fed half-wave dipole.

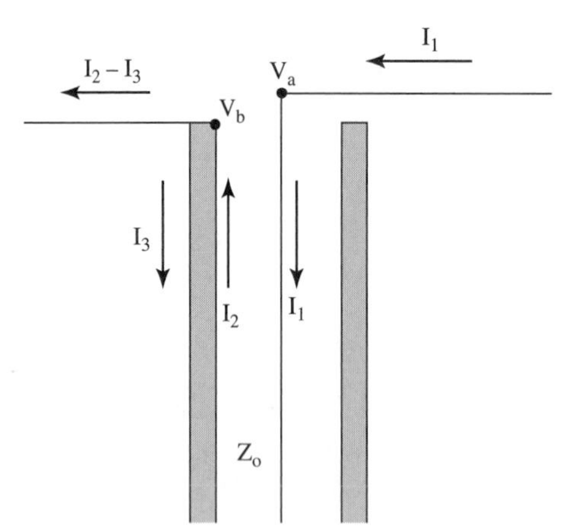

Figure 5-23 Cross section of a coaxial transmission line feeding a dipole antenna at its center.

at the terminals of the antenna. These voltages are equal in magnitude but opposite in phase (i.e., $V_a = -V_b$). Both voltages act to cause a current to flow on the outside of the coaxial line. If the magnitude of the currents on the outside of the coax produced by both voltages are equal, the net current would be zero. However, since one antenna terminal is directly connected to the outer conductor, its voltage V_b produces a much stronger current than the other voltage V_a. A balun is used to transform the balanced input impedance of the dipole to the unbalanced coaxial line such that there is no net current on the outer conductor of the coax.

To illustrate how a balun works, consider the *sleeve* (or *bazooka*) *balun* in Fig. 5-24. The sleeve and outer conductor of the coaxial line form another coaxial line of characteristic impedance Z_o' that is shorted a quarter-wavelength away from its input at the antenna terminals. The equivalent circuit for Fig. 5-23 is that of Fig. 5-25*a*. The equivalent circuit of Fig. 5-24 is that of Fig. 5-25*b*, which shows that both terminals see a very high impedance to ground. Thus, the situation in Fig. 5-25*b* is equivalent to the balanced condition of Fig. 5-25*c* wherein the currents I_1 and I_2 are equal.

An easily constructed balun form is the *folded balun* shown in Fig. 5-26. The quarter-wavelength of coax from the *a* terminal to the outer conductor of the transmission line does not affect the antenna impedance Z_A. The extra quarter-wave-

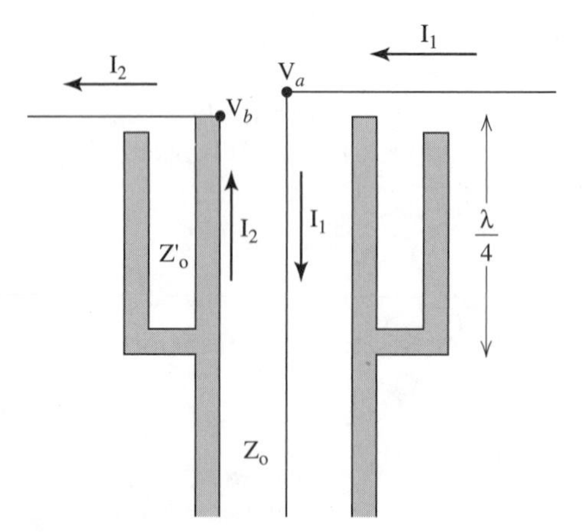

Figure 5-24 Cross section of a sleeve balun feeding a dipole at its center.

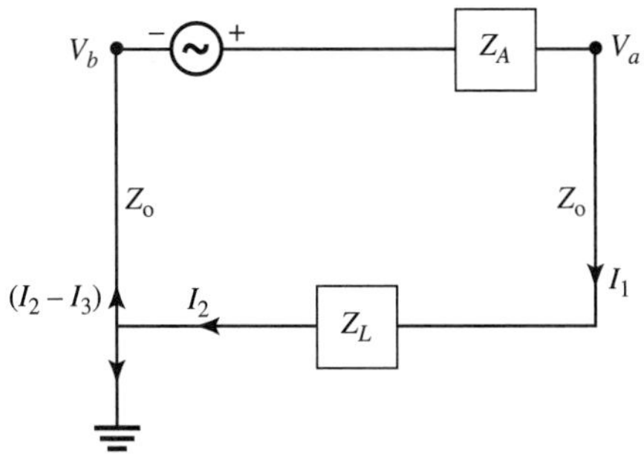

(*a*) Equivalent circuit of coax-fed dipole in Fig. 5-23.

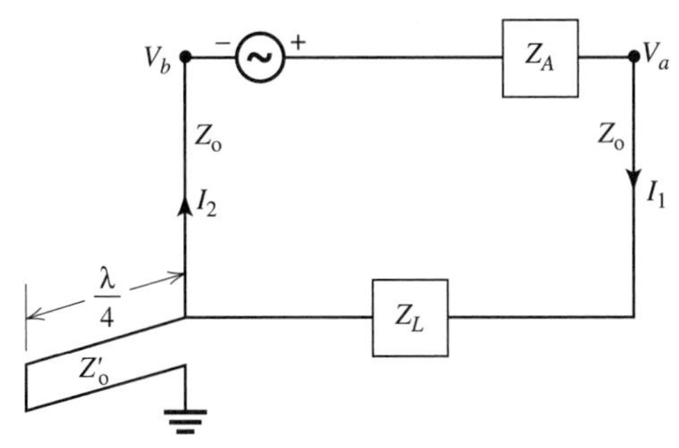

(*b*) Equivalent circuit of sleeve balun-fed dipole in Fig. 5-24.

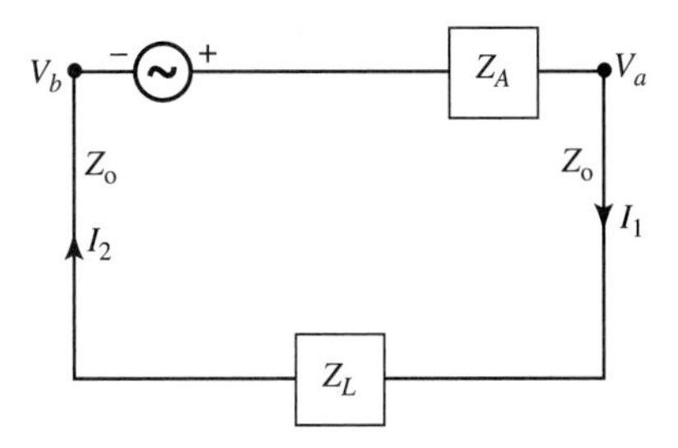

(*c*) Final equivalent circuit for Fig. 5-24 with quarter-wave transmission line removed, $I_1 = I_2$.

Figure 5-25 Equivalent circuits for a dipole fed from a coaxial transmission line of characteristic impedance Z_o and load impedance Z_L.

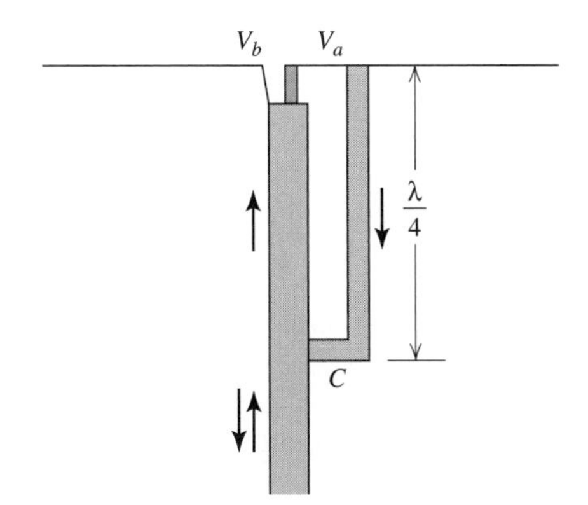

Figure 5-26 The folded balun.

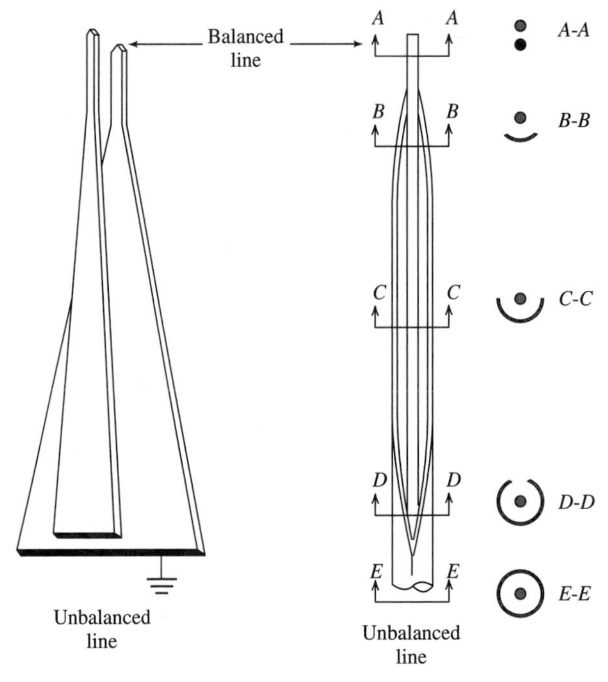

(*a*) Tapered microstrip balun. (*b*) Tapered coaxial balun.

Figure 5-27 Broadband baluns.

length of coax together with the outer conductor of the main transmission line forms another equivalent transmission line, which is a quarter-wavelength long and is shorted at *C*. Therefore, the short circuit at *C* is (ideally) transformed to an infinite impedance at the antenna terminals, which is in parallel with Z_A, leaving the input impedance unchanged. The quarter-wavelength line induces a cancelling current on the outside of the coaxial transmission line, so that the net current on the outside of the main coax below point *C* is zero as shown in Fig. 5-26. The folded and sleeve baluns are, of course, not broadband because of the quarter-wavelength involved in its construction.

Broadband baluns can be constructed by tapering a balanced transmission line to an unbalanced one over at least several wavelengths of transmission line length

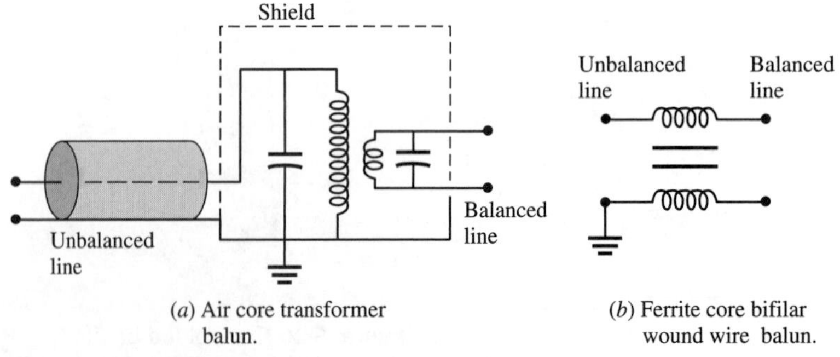

(*a*) Air core transformer balun. (*b*) Ferrite core bifilar wound wire balun.

Figure 5-28 Baluns used at lower frequencies.

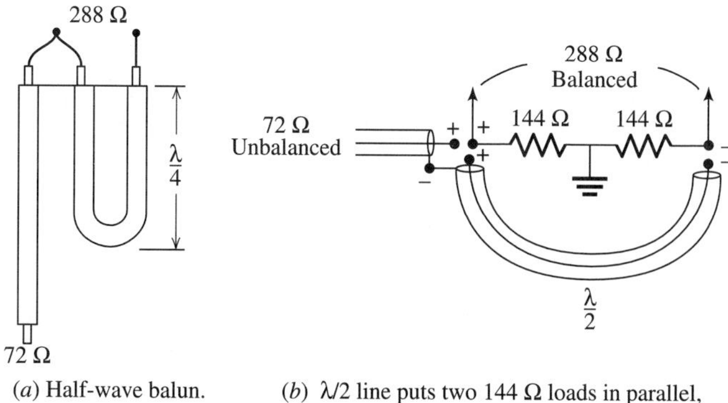

(*a*) Half-wave balun.

(*b*) λ/2 line puts two 144 Ω loads in parallel, transforming 288 Ω balanced to 72 Ω unbalanced.

Figure 5-29 A half-wave balun that provides an impedance setup ratio of 4:1.

as indicated in Fig. 5-27. Figure 5-27*a* shows a balanced transmission line tapering to an unbalanced microstrip line and Fig. 5-27*b* illustrates a balanced line tapering to an unbalanced coaxial line.

The baluns we have considered thus far are useful from microwave frequencies down to VHF. From VHF down to lower frequencies, it is impractical in many cases to employ these configurations and transformers are used as Fig. 5-28 indicates. Figure 5-28*a* is an air core transformer arrangement useful at lower frequencies. Figure 5-28*b* is bifilar wound ferrite core balun that can be used from VLF through UHF.

Impedance transformation may also be included in a balun for matching purposes. For example, the "four-to-one" balun in Fig. 5-29*a* will transform an unbalanced 72-Ω impedance to one that is 288-Ω balanced. Such a balun is useful with a folded dipole. To understand how the four-to-one balun works, consider Fig. 5-29*b* that shows the 288-Ω balanced impedance split into two 144-Ω parts and the connection between the 144-Ω impedances grounded. The 288-Ω impedance is still balanced (with respect to ground). Next, the negative terminal is connected via a half-wave-length section of transmission line to the positive terminal as shown in Fig. 5-29*b*. Thus, the unbalanced terminals present a 72-Ω impedance, whereas the balanced terminals present a 288-Ω impedance and the four-to-one balanced transformation is complete. A balun that leaves the impedance unchanged is often referred to as a "one-to-one" balun.

5.4 YAGI–UDA ANTENNAS

We saw in Chapter 3 that array antennas can be used to increase directivity. The arrays we examined had all elements active, requiring a direct connection to each element by a feed network. Array feed networks are considerably simplified if only a few elements are fed directly. Such an array is referred to as a *parasitic array*. The elements that are not directly driven (called parasites) receive their excitation by near-field coupling from the driven elements. A parasitic linear array of parallel dipoles is called a **Yagi–Uda antenna**, a Yagi–Uda array, or simply "Yagi." Yagi–Uda antennas are very popular because of their simplicity and relatively high gain. In this section, the principles of operation and design data for Yagis will be presented [16].

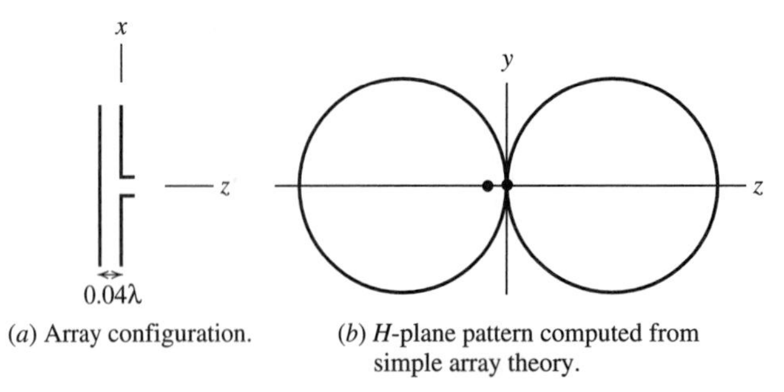

(*a*) Array configuration. (*b*) *H*-plane pattern computed from simple array theory.

Figure 5-30 A two-element array of half-wave resonant dipoles, one a driver and the other a parasite. The currents on both are equal in amplitude and opposite in phase.

The first research done on the Yagi–Uda antenna was performed by Shintaro Uda at Tohoku University in Sendai, Japan, in 1926 and was published in Japanese in 1926 and 1927. The work of Uda was reviewed in an article written in English by Uda's professor, H. Yagi, in 1928 [17].

The basic unit of a Yagi consists of three elements. To understand the principles of operation for a three-element Yagi, we begin with a driven element (or "driver") and add parasites to the array. Consider a driven element that is a resonant half-wave dipole. If a parasitic element is positioned very close to it, it is excited by the driven element with roughly equal amplitude, so the field incident on the parasite is

$$E_{\text{incident}} = E_{\text{driver}} \tag{5-44}$$

A current is excited on the parasite and the resulting radiated electric field, also tangent to the wire, is equal in amplitude and opposite in phase to the incident wave. This is because the electric field arriving at the parasite from the driver is tangential to it and the total electric field tangential to a good conductor is zero. Thus, the field radiated by the parasite is such that the total tangential field on the parasite is zero, or $0 = E_{\text{incident}} + E_{\text{parasite}}$. Combining this fact with (5-44) gives

$$E_{\text{parasite}} = -E_{\text{incident}} = -E_{\text{driver}} \tag{5-45}$$

From array theory, we know that two closely spaced, equal amplitude, opposite phase elements will have an endfire pattern; for example, see Fig. 3-4. The pattern of this simple two-element parasitic array for 0.04λ spacing is shown in Fig. 5-30*b*.

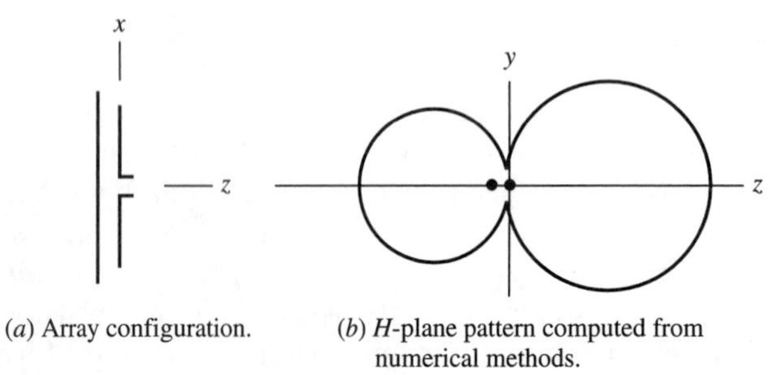

(*a*) Array configuration. (*b*) *H*-plane pattern computed from numerical methods.

Figure 5-31 Two-element Yagi–Uda antenna consisting of a driver of length 0.4781λ and a reflector of length 0.49λ spaced 0.04λ away. The wire radius for both is 0.001λ.

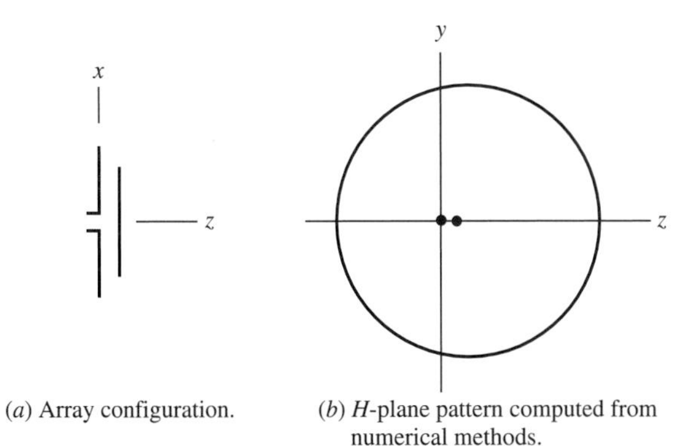

Figure 5-32 Two-element Yagi–Uda antenna consisting of a driver of length 0.4781λ and a director of length 0.45λ spaced 0.04λ away. The wire radius for both is 0.001λ.

The simplistic beauty of the Yagi is revealed by lengthening the parasite. The dual endfire beam is changed to a more desirable single endfire beam. This effect is illustrated for the two-element parasitic array of Fig. 5-31. The driver is a dipole of length 0.4781λ, which is a half-wave resonant length when operated in free space. The parasite is a straight wire of length 0.49λ and spaced a distance 0.04λ away from the driver. The H-plane pattern in Fig. 5-31*b* obtained from the numerical methods of Chapter 10 demonstrates the general trend of a parasite that is longer than the driver: A single main beam occurs in the endfire direction from the parasite to the driver along the line of array. Such a parasite is called a *reflector* because it appears to reflect radiation from the driver.

If the parasite is shorter than the driver, but now placed on the other side of the driver, the pattern effect is similar to that when using a reflector in the sense that main beam enhancement is in the same direction. The parasite is then referred to as a *director* since it appears to direct radiation in the direction from the driver toward the director. The parasitic array in Fig. 5-32*a* consisting of a driver and a director has the pattern shown in Fig. 5-32*b*.

The single endfire beam created by the use of a reflector or a director alone with a driver suggests that even further enhancement could be achieved with a reflector and a director on opposite sides of a driver. This is indeed the case. An example of a three-element Yagi is shown in Fig. 5-33*a*, which is a combination of the geometries of Figs. 5-31*a* and 5-32*a*. The pattern of Fig. 5-33*b* is improved over that of either two-element array. The E-plane pattern for the three-element Yagi is shown

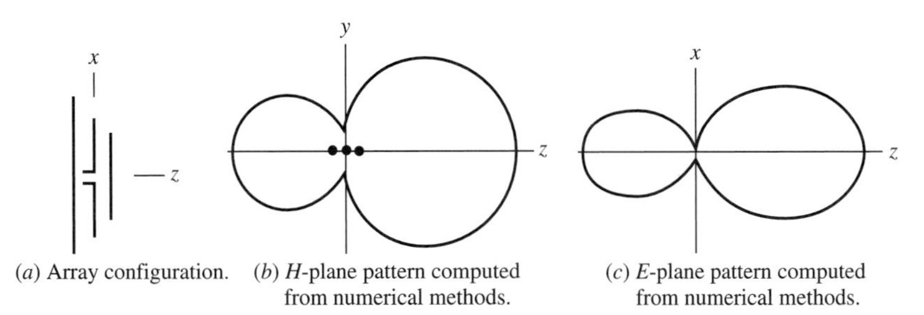

Figure 5-33 Three-element Yagi–Uda antenna consisting of a driver of length 0.4781λ, a reflector of 0.49λ, and a director of length 0.45λ, each spaced only 0.04λ apart. The wire radius for each is 0.001λ.

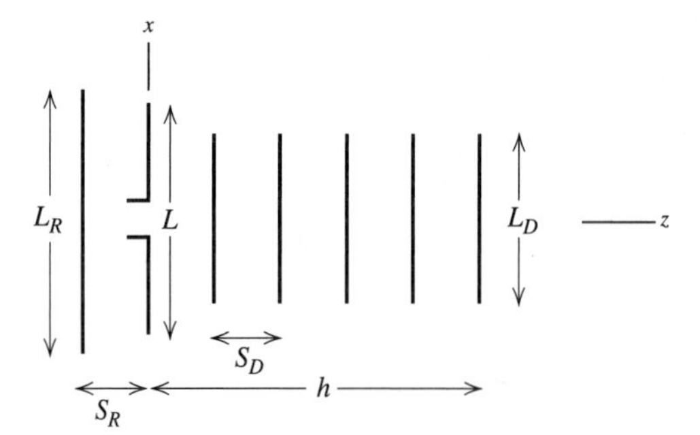

Figure 5-34 Configuration for a general Yagi–Uda antenna.

in Fig. 5-33*c*. It is essentially equal to the *H*-plane pattern multiplied by the element factor for the array, which is that of a half-wave dipole. Again, these patterns were obtained by numerical solution for exceptionally small element spacing (0.04λ).

The general Yagi configuration is shown in Fig. 5-34. The maximum directivity obtainable from a three-element Yagi is about 9 dBi or 7 dBd [18]. Optimum reflector spacing S_R (for maximum directivity) is between 0.15 and 0.25 wavelengths as Fig. 5-35 shows. Note that the gain above an isolated dipole is more than 2.5 dBd, whereas if a flat plate were used, instead of a simple wire-like element, the gain would be 3 dBd. Thus, a single wire-like reflector element is almost as effective as a flat plate in enhancing the gain of a dipole.

Director-to-director spacings are typically 0.2 to 0.35 wavelengths, with the larger spacings being more common for long arrays and closer spacings for shorter arrays. Typically, the reflector length is 0.5λ and the driver is of resonant length when no parasitic elements are present [19]. The director lengths are typically 10 to 20% shorter than their resonant length, the exact length being rather sensitive to the number of directors N_D and the interdirector spacing S_D.

The gain of the Yagi is related to its boom length as our study of uniform line sources in the previous chapter suggests, but for a parasitic array such as the Yagi, there is a smaller increase in gain per element as directors are added to the array

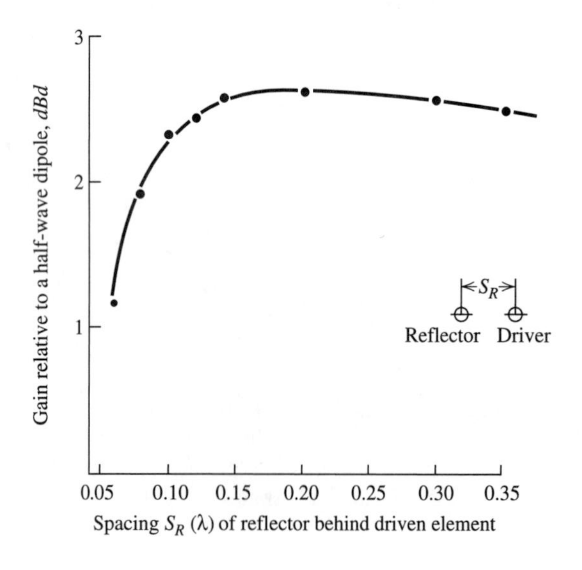

Figure 5-35 Measured gain [21] in dBd of a dipole and reflector element for different spacings S_R.

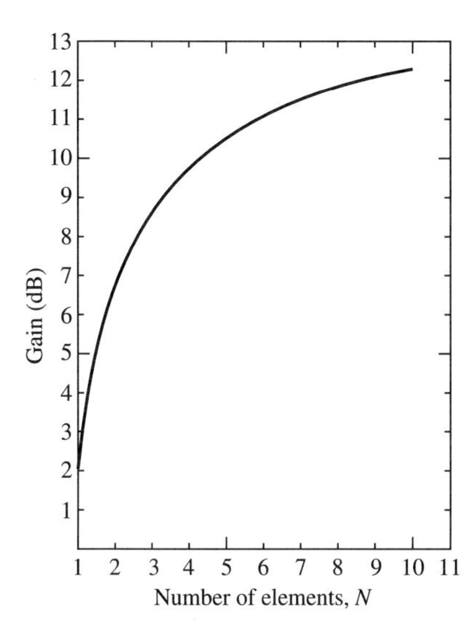

Figure 5-36 Gain of a typical Yagi–Uda antenna versus the total number of elements. The element spacings $S_R = S_D = 0.15\lambda$. The conductor diameters are 0.0025λ. (From Green [20].)

(if we assume S_D is fixed) since the Yagi is not uniformly excited (see Fig. 5-39). In fact, the addition of directors up to about 5 or 6 provides a significant increase in gain expressed in dB, whereas the addition of more directors is beyond the "point of diminishing returns" as Fig. 5-36 shows. Figure 5-36 plots the gain versus the number of elements N in the array (including one reflector and one driver) [20] for an interelement spacing for all elements of $S_R = S_D = 0.15\lambda$. Note that adding one director to increase N from 3 to 4 gives about a 1-dB gain increase, whereas adding one director to increase N from 9 to 10 yields only about an additional 0.2-dB gain.

The addition of more reflector elements results in a fractional dB increase in gain and is usually not done. The main effects of the reflector are on the driving point impedance at the feed point and on the back lobe of the array. Pattern shape, and therefore gain, are mostly controlled by the director elements. The director spacing and director length are interrelated, but the more sensitive parameter is the director length, which becomes more critical as the boom length increases.

An extensive decade-long experimental investigation by Viezbicke [21] at the National Bureau of Standards (later known as NIST) has produced a wealth of information on Yagi–Uda antenna design. An objective of the experimental investigation was to determine optimum designs for a specified boom length, $L_B = S_R + h$. Boom lengths from 0.2 to 4.2λ were included in the study. Some of Viezbicke's work is summarized in Table 5-4, which can be used for design purposes. Viezbicke's work and its summaries in [22] and [23] show how to correct the free-space parasitic element lengths for both the diameter of the conductors used (see Fig. 5-37) and for the diameter of a metal boom (see Fig. 5-38), if a metal boom is used. A metal boom may be used because the voltage distribution on the parasitic elements goes through a zero at the element center. Ideally, an infinitely thin metallic boom down the center of the array would not change the voltage distribution. However, metallic booms of practical size do have an effect that must be compensated for by increasing the parasitic element lengths. Alternatively, the parasitic elements may be insulated from the boom, in which case no compensation is required.

The Yagi–Uda antenna with at least several directors is an end-fire traveling-

Table 5-4 Optimized Lengths of Parasitic Dipoles for Yagi–Uda Array Antennas of Six Different Boom Lengths, L_B

$d/\lambda = 0.0085$	Boom length of Yagi–Uda Array, λ					
$S_R = 0.2\lambda$	0.4	0.8	1.20	2.2	3.2	4.2
Length of reflector, L_R/λ	0.482	0.482	0.482	0.482	0.482	0.475
Length of director D_n, (L_{D_n}/λ): D_1	0.442	0.428	0.428	0.432	0.428	0.424
D_2		0.424	0.420	0.415	0.420	0.424
D_3		0.428	0.420	0.407	0.407	0.420
D_4			0.428	0.398	0.398	0.407
D_5				0.390	0.394	0.403
D_6				0.390	0.390	0.398
D_7				0.390	0.386	0.394
D_8				0.390	0.386	0.390
D_9				0.398	0.386	0.390
D_{10}				0.407	0.386	0.390
D_{11}					0.386	0.390
D_{12}					0.386	0.390
D_{13}					0.386	0.390
D_{14}					0.386	
D_{15}					0.386	
Spacing between directors (S_D/λ)	0.20	0.20	0.25	0.20	0.20	0.308
Gain relative to half-wave dipole, dBd	7.1	9.2	10.2	12.25	13.4	14.2
Design curve (Fig. 5-37)	(*A*)	(*C*)	(*C*)	(*B*)	(*C*)	(*D*)
Front-to-back ratio, dB	8	15	19	23	22	20

Source: P. P. Viezbicke, "Yagi Antenna Design," NBS Tech. Note 688, National Bureau of Standards, Washington, DC, Dec. 1968.

wave antenna that supports a surface wave of the slow wave type (i.e., $c/v > 1$) (see Prob. 4.4-3). That is, the driver-reflector pair launches a wave onto the directors that slows the wave down such that the wave phase velocity v is less than that of the velocity of light c in free space. In other words, the phase delay per unit distance along the axis of the array in the forward direction is greater than that of the ordinary endfire condition. One might suspect that the additional phase delay beyond ordinary endfire required for maximum gain is that of the Hansen–Woodyard condition for uniform arrays (e.g., Sec. 3.2.4). If the boom length is quite long, this is approximately true.

The director currents on a well-designed Yagi are nearly equal as Fig. 5-39 indicates [16]. If the partial boom length h, as measured from the driver to the furthest director (see Fig. 5-34), is long ($h \gg \lambda$), the Hansen–Woodyard condition requires that the phase difference between the surface wave and a free-space wave at the director furthest from the driver (the terminal director) be approximately 180°. Thus,

$$h\beta_g - h\beta = \pi \tag{5-46}$$

or,

$$\lambda/\lambda_g = c/v = 1 + \lambda/2h \tag{5-47}$$

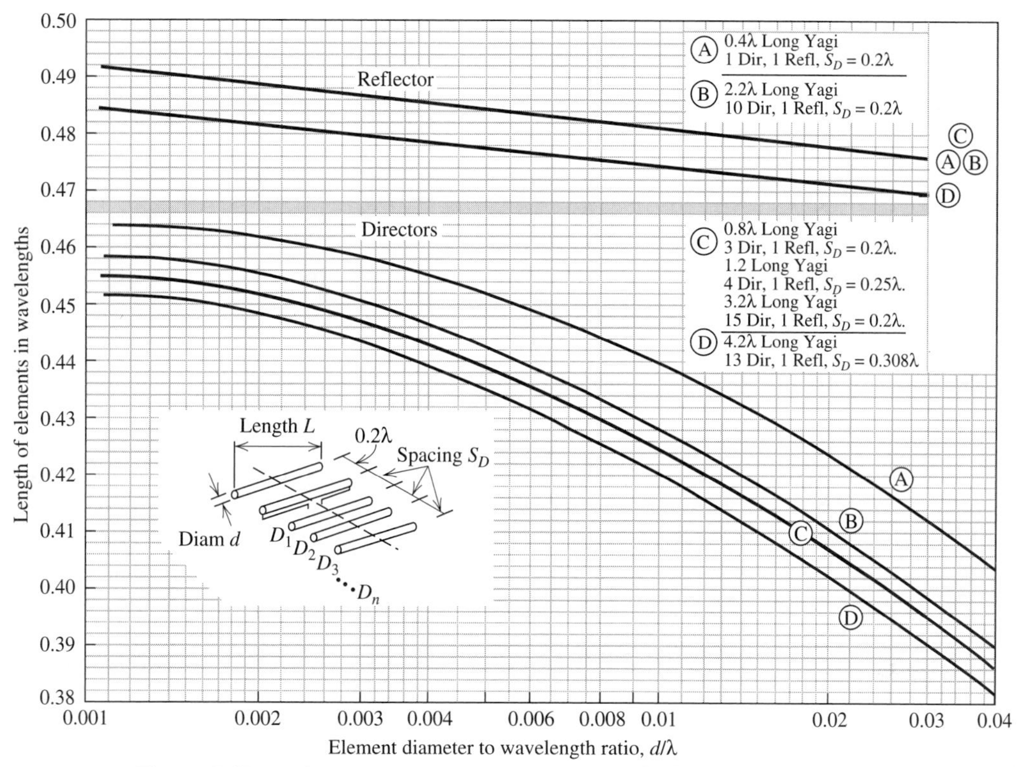

Figure 5-37 Design curves for Yagis in Table 5-4 [21].

where here β_g is a guided phase constant along the forward axis of the array, λ_g represents the corresponding guided wavelength, and λ is the unguided (free-space) wavelength. Note that $\beta_g = -\beta \cos \theta_o \approx -\beta(c/v)$ and recall that $c/v > 1$ implies $|\cos \theta_o| > 1$. Equation (5-47) is plotted as the upper dashed line in Fig. 5-40.

Experimental work by Ehrenspeck and Poehler [19] showed that the optimum

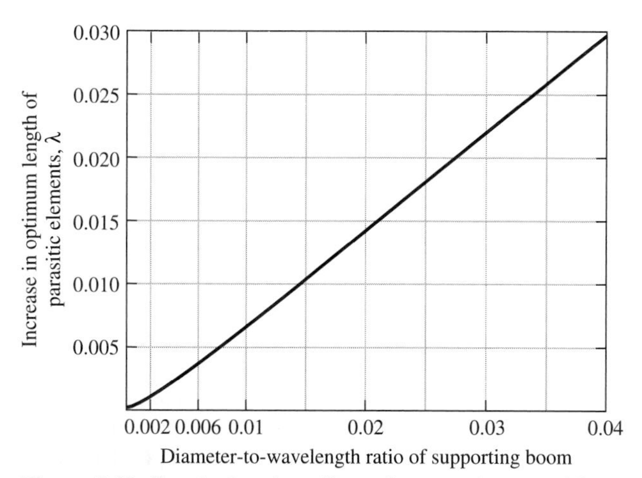

Figure 5-38 Graph showing effect of supporting metal boom on the length of Yagi parasitic elements [21].

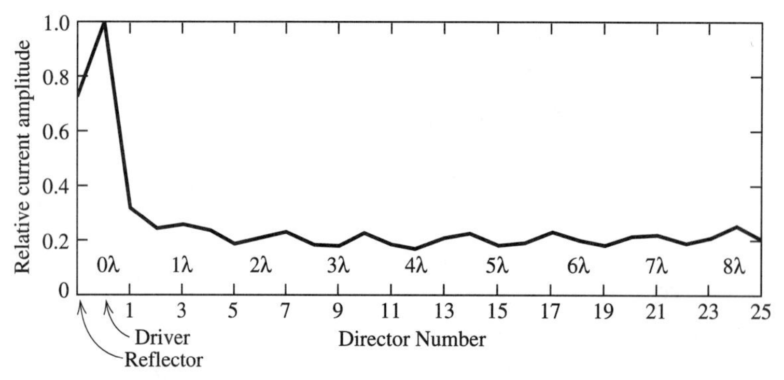

Figure 5-39 Relative current amplitudes for a 27-element Yagi array [16].

terminal phase difference is about 60° for short Yagis, rising to about 120° for $4\lambda < h < 8\lambda$, and then approaching 180° for $h > 20\lambda$. This is the solid curve in Fig. 5-40. Various data for Yagis and other endfire structures indicate that the optimum c/v values lie on or just below the solid curve in the shaded area. Other surface wave structures, with more efficient surface wave excitation than the Yagi, can have optimum c/v values that lie in the shaded region nearer to the dashed lower bound, but all of these surface wave structures approach the Hansen–Woodyard condition if they are very long.

Viewing the directors as a reactive surface over which the surface wave travels, and noting from Table 5-4 that director lengths tend to be shorter for longer boom lengths, lead us to surmise that the surface wave couples less to the reactive surface for long boom lengths so that the total phase delay is not excessive and falls on the solid curve in Fig. 5-40. Examining Table 5-4 for those arrays with a director spacing $S_D = 0.2\lambda$ shows that indeed the directors tend to be shorter for longer boom lengths.

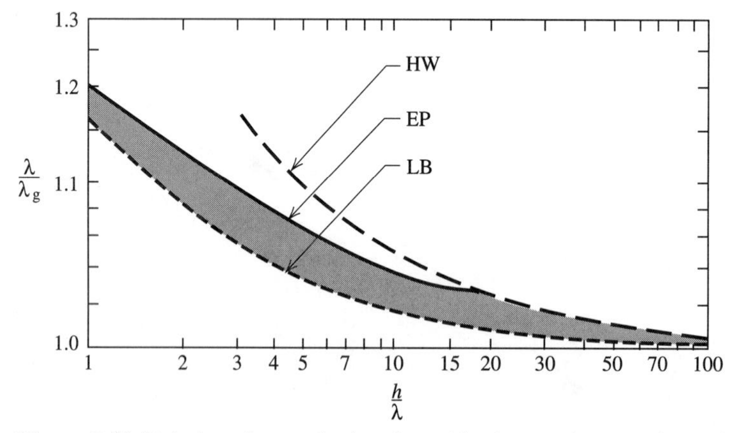

Figure 5-40 Relative phase velocity $c/v = \lambda/\lambda_g$ for maximum gain surface wave antennas as a function of antenna length h/λ. HW = Hansen–Woodyard condition (5-46); EP = Ehrenspeck and Poehler experimental values; LB = lower bound (for idealized surface wave excitation). (from *Antenna Engineering Handbook*, 3rd Ed., Richard C. Johnson, Ed. New York: McGraw-Hill, Inc., 1993. Used by permission.)

The Yagi is one of the more popular antennas used in the HF-VHF-UHF frequency range. It provides moderately high gain while offering low weight and low cost. It has a relatively narrow bandwidth (e.g., a few percent), which may be improved somewhat by using feeds other than a dipole, such as a folded dipole. The folded dipole also provides a higher input impedance than a dipole even though the driving point impedance of both are usually reduced considerably from their self-impedances by mutual coupling effects. Further, increased gain can be obtained by arraying or "stacking" Yagi antennas. Maximum gain results for a separation of almost one wavelength (see Fig. 3-20). Thus, for a given application, if a somewhat narrow bandwidth can be tolerated, the Yagi–Uda antenna can provide good gain (e. g., 9–12 dB) at low cost.

EXAMPLE 5-1 *TV Channel 12 Yagi Antenna Design*

A 12-element Yagi for TV channel 12 at 205.25 Mhz is to be designed using 1-cm-diameter elements insulated from a metallic boom [23]. The boom length is to be 2.2λ. Table 5-4 indicates that 0.2λ spacing is required. The wavelength at 205.25 MHz is 1.46 m. Thus, the spacing between all elements is 29.2 cm. To obtain the element lengths, the following four steps are followed:

1. Plot the element lengths from Table 5-4 on the design curves "B" in Fig. 5-37. The design curves are for conductor diameters of 0.0085λ.
2. Since the 1-cm conductor to be used is 0.0065λ in diameter, the element lengths in the table must be increased slightly. This is accomplished by drawing a vertical line at 0.0065λ on the horizontal axis. This line intersects the two applicable design curves, which gives the compensated lengths of the reflector and first director:

$$L_R = 0.483\lambda, \qquad L_{D_1} = 0.4375\lambda$$

   Notice the distance along the director "B" curve between the intersection of the vertical line at 0.0065λ and the location of the first director length from step 1 above. All the directors must be increased in length by an amount that is determined by this distance.
3. With a pair of dividers, measure the distance on the B director curve between the initial length and corrected length of the first director. Slide each of the other director lengths to the left by this amount to determine their compensated lengths:

$$L_{D_2} = 0.421\lambda$$
$$L_{D_3} = L_{D_{10}} = 0.414\lambda$$
$$L_{D_4} = L_{D_9} = 0.405\lambda$$
$$L_{D_5} = L_{D_6} = L_{D_7} = L_{D_8} = 0.398\lambda$$

4. The fourth step generally is to correct the lengths for the metallic boom using Fig. 5-38, if one is used. In this case, the boom is metallic, but the elements are insulated from it and no correction factor is required. (See Prob. 5.4-5.)

Calculated patterns for Example 5-1 using numerical methods (e.g., Chap. 10) are shown in Fig. 5-41 and the calculated directivity is 11.82 dBd, which agrees well with the gain value in Table 5-4. The calculated input impedance for a dipole driver is $26.5 + j23.7\ \Omega$. The calculated front-to-back ratio is 38.5 dB, owing to the almost total absence of a back lobe in the calculated pattern.

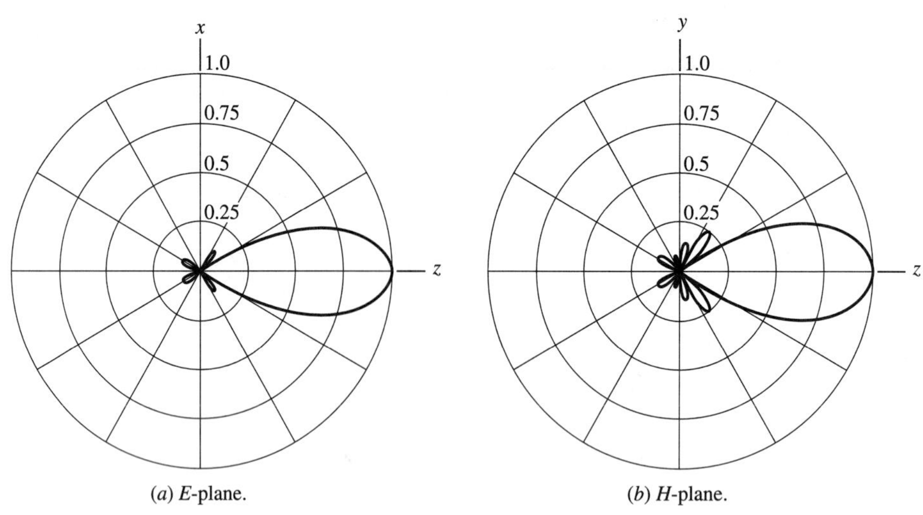

Figure 5-41 Calculated patterns for Example 5-1.

5.5 CORNER REFLECTOR ANTENNAS

Another practical antenna that produces a gain of 10 to 12 dB over a half-wave dipole is the corner reflector antenna invented by J. D. Kraus in 1938 [4]. His first experimental model was a 90° corner reflector. Although other corner angles can be used, the 90° corner illustrated in Fig. 5-42 is the most practical and the one that will be discussed here. The corner reflector is a gain standard at UHF frequencies.

The corner reflector antenna can be easily analyzed using the method of images and array theory. Consider Fig. 5-43 that shows the source and its three images. The array factor contribution from the feed element [#1] and image #4 from (3-4) is

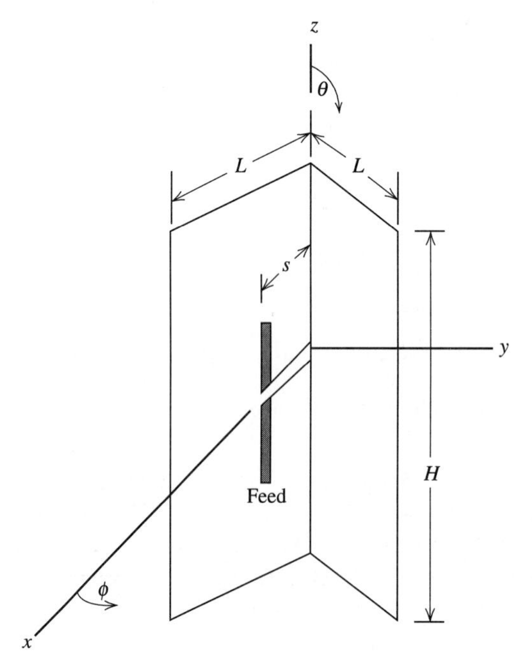

Figure 5-42 Right angle corner reflector with metal plates in the $\phi = \pm 45°$ planes.

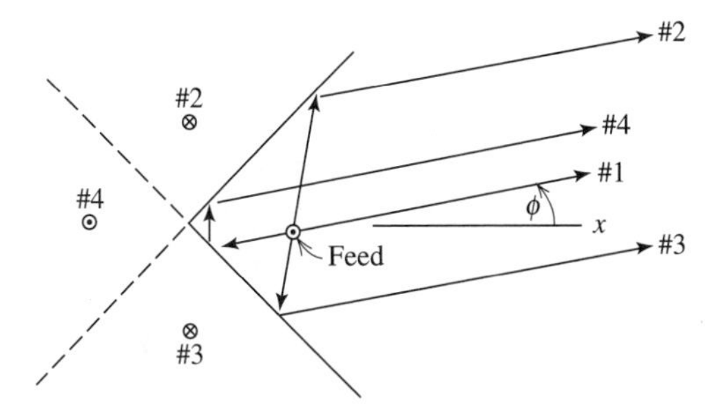

Figure 5-43 Right angle corner reflector with images shown and how they account for reflections.

$2 \cos[\beta s \cos(\phi)]$. The contribution from images #2 and #3 will be the same only rotated 90° and of opposite phase, or $-2 \cos[\beta s \cos(90° - \phi)]$. Thus, if we assume that the conducting reflecting sheets are infinite in extent, the array factor in the xy-plane (H-plane) valid in the region $-45° \leq \phi \leq 45°$ is

$$\text{AF}(\theta = 90°, \phi) = 2 \cos(\beta s \cos \phi) - 2 \cos(\beta s \sin \phi) \tag{5-48}$$

It follows that in the xz-plane or E-plane, the array factor may be constructed by a somewhat similar reasoning process:

$$\text{AF}(\theta, \phi = 0°) = \{-2 + 2 \cos[\beta s \cos(90° - \theta)]\}g(\theta) \tag{5-49}$$

where the element factor $g(\theta)$ is usually that of a half-wave dipole.

The pattern shape, gain, and feed point impedance will all be a function of the feed-to-corner spacing s (see Fig 5-44*a*). For the 90° corner reflector, the pattern will have no minor lobes within $-45° \leq \phi \leq 45°$ and good directivity if $0.25\lambda \leq s \leq 0.7\lambda$. The directivity will be greatest at $s = 0.5\lambda$ [24] when the conducting plates are of infinite extent, but the input impedance of a dipole feed will be high (i.e., around 125 Ω). Adjusting the spacing downward to 0.35λ will in theory produce a 70-Ω input impedance with a negligible decrease in gain. Often, a bow

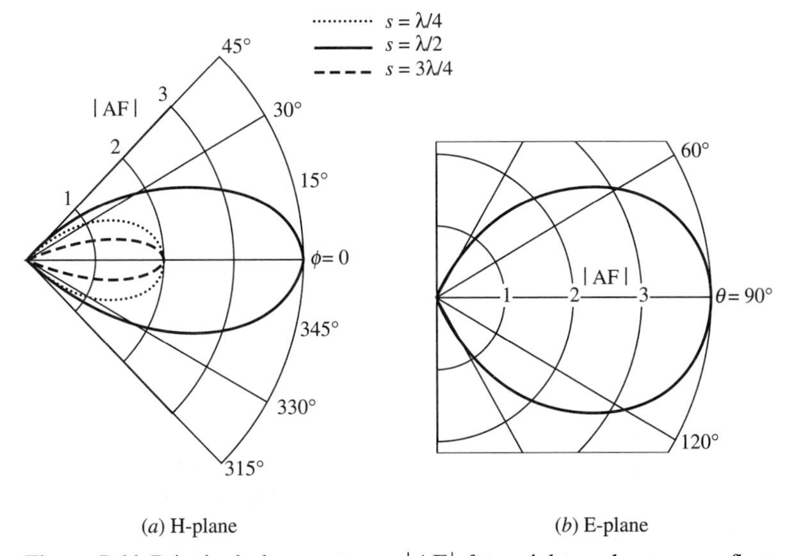

(*a*) H-plane (*b*) E-plane

Figure 5-44 Principal plane patterns, |AF|, for a right angle corner reflector composed of two (semi-infinite) half-planes and a $\lambda/2$ dipole feed. (Only the best directivity spacing is used in the E-plane.)

tie (see Fig. 6-32) is used for the feed because it has superior impedance bandwidth properties compared to an ordinary linear dipole.

Making the conducting plates of finite extent is, of course, necessary for a practical design. It can be shown by ray tracing that a length value of $L = 2s$ is a reasonable minimum length so that the main beam is not degraded by the finite extent of the conducting plates. The dimension H is usually chosen to be from 1.2 to 1.5 times the length of the feed so as to minimize the direct radiation by the dipole feed into the back region. The finite extent of the plates will result in a pattern broader than that predicted for infinite plates as in, for example, Fig. 5-44. The effect of the finite plate size on the feed driving point impedance is usually negligible.

5.6 WIRE ANTENNAS ABOVE AN IMPERFECT GROUND PLANE

The operation of low-frequency (roughly VHF and below) antennas is affected significantly by the presence of typical environmental surroundings, such as the earth, buildings, and so forth. In Sec. 2.3, we discussed the principles for analyzing antennas above a perfect ground plane. A perfect ground plane in its ideal form is an infinite, plane, perfect conductor. It is well approximated in practice by a planar good conductor that is large relative to the antenna extent. Image theory from Sec. 2.3.1 reveals that an antenna above a perfect ground plane, or an approximation of it, has an equivalent form that is an array. Array theory can then be used to obtain the radiation pattern above the ground plane.

In this section, we consider ground planes that are not well approximated by a perfect ground plane. Since low-frequency antennas are most affected by their surroundings and low-frequency antennas are usually wire antennas, the illustrations will be for wire antennas above a ground plane. The general principles can, however, be applied to many antenna types.

A ground plane can take many forms, such as radial wires around a monopole, the roof of a car, or the real earth. In many situations, the earth is well approximated as being infinite and planar, but it is a poor conductor. Good conductors have conductivities on the order of 10^7 S/m. Earth conductivity varies greatly, but is typically 10^{-1} to 10^{-3} S/m with rich soil at the high end and rocky or sandy soil at the low end. With these low conductivities, electric fields generated by a nearby antenna penetrate into the earth and excite currents that, in turn, give rise to $\sigma|E|^2$ ohmic losses. This loss appears as an increase in the input ohmic resistance and thus lowers the radiation efficiency of the antenna.

5.6.1 Pattern Effects of a Real Earth Ground Plane

The pattern of an antenna over a real earth is different from the pattern when the antenna is operated over a perfect ground plane. Approximate patterns can be obtained by using image theory. The same principles discussed in Sec. 2.3.1 for images in perfect ground planes apply, except that the strength of the image in a real ground will be reduced from that of the perfect ground plane case (equal amplitude and equal phase for vertical elements, and opposite phase for horizontal elements). The strength of the image can be approximated by weighting it with the plane wave reflection coefficient for the appropriate polarization of the field arriving at the ground plane. To illustrate, consider a short vertical dipole a distance h above a ground plane, shown in Fig. 5-45 together with its image. There is a direct and a reflected ray arriving in the far field. As can be seen, the reflected ray appears to

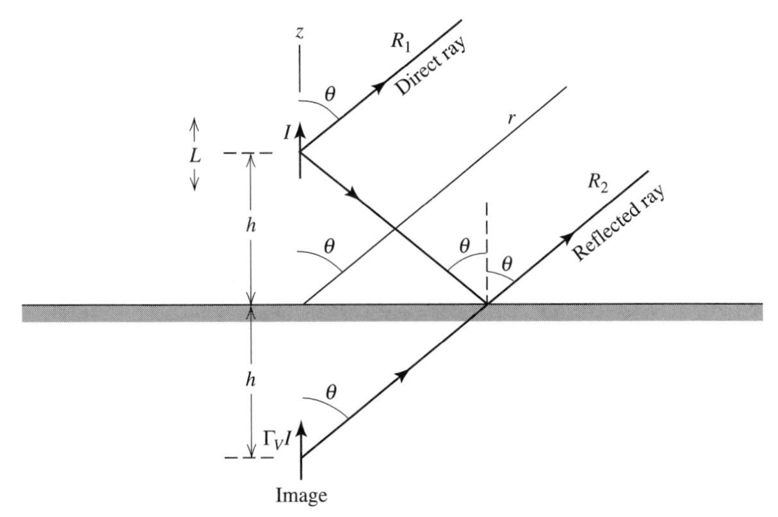

Figure 5-45 A short vertical dipole of current I above a real earth ground plane, together with its image of current $\Gamma_V I$.

be coming from the image antenna. The primary source and its image form an array. The electric field above the ground plane for this example, from (1-78) in (1-93), is

$$E_\theta = j\omega\mu \sin\theta\left(IL\,\frac{e^{-j\beta R_1}}{4\pi R_1} + \Gamma_V IL\,\frac{e^{-j\beta R_2}}{4\pi R_2}\right) \tag{5-50}$$

where L is the length of the short dipole and $\Gamma_V I$ is the current for the image dipole. Γ_V is the plane wave reflection coefficient for a planar earth and vertical incident polarization, when E is in the plane of incidence defined by a normal to the earth and the ray from the source to the normal at the earth. Using parallel rays for far-field calculations gives the far-field distance expressions

$$R_1 = r - h\cos\theta \quad \text{and} \quad R_2 = r + h\cos\theta \tag{5-51}$$

Then (5-50) reduces to

$$E_\theta = j\omega\mu\,\frac{IL}{4\pi}\,\frac{e^{-j\beta r}}{r}\,\sin\theta\left(e^{j\beta h\cos\theta} + \Gamma_V e^{-j\beta h\cos\theta}\right) \tag{5-52}$$

where $R_1 \approx R_2 \approx r$ was used in the denominator. This expression is valid above the ground plane. It contains an element pattern $\sin\theta$ and an array factor, in the brackets, for a two-element array with elements spaced $2h$ apart.

Similarly for a horizontally oriented short dipole as shown in Fig. 5-46, we have (in the xz-plane)

$$E_\theta = j\omega\mu\,\frac{IL}{4\pi}\,\frac{e^{-j\beta r}}{r}\,\cos\theta\left(e^{j\beta h\cos\theta} - \Gamma_V e^{-j\beta h\cos\theta}\right) \tag{5-53}$$

where the minus sign appears because the image current is in the opposite direction. This expression is valid only in the xz-plane. Γ_V is used because E is in the plane of incidence. The field in the yz-plane is given by

$$E_\phi = j\omega\mu\,\frac{IL}{4\pi}\,\frac{e^{-j\beta r}}{r}\left(e^{j\beta h\cos\theta} + \Gamma_H e^{-j\beta h\cos\theta}\right) \tag{5-54}$$

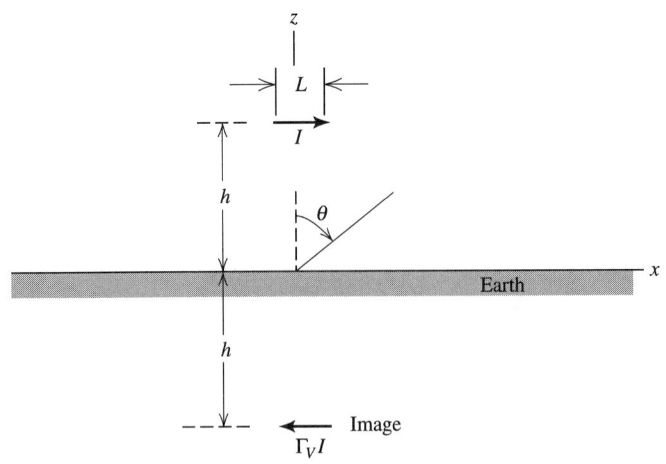

Figure 5-46 A short horizontal dipole of current I above a real earth ground plane together with its image of current $\Gamma_V I$ for the xz-plane. The image current in the yz-plane is $\Gamma_H I$.

The element pattern is unity because a dipole has an omnidirectional pattern in the plane normal to the dipole axis. The horizontal reflection coefficient Γ_H is used in this case because the electric field is perpendicular to the plane of incidence.

There is no minus sign in the second term of (5-54) because of the definition of Γ_H, which is [25]

$$\Gamma_H = \frac{\cos\theta - \sqrt{\varepsilon_r' - \sin^2\theta}}{\cos\theta + \sqrt{\varepsilon_r' - \sin^2\theta}} \tag{5-55}$$

This is the plane wave reflection coefficient for an incident electric field perpendicular to the plane of polarization (i.e., the plane formed by the surface normal and the direction of propagation). Further, for Γ_V we have [25]

$$\Gamma_V = \frac{\varepsilon_r' \cos\theta - \sqrt{\varepsilon_r' - \sin^2\theta}}{\varepsilon_r' \cos\theta + \sqrt{\varepsilon_r' - \sin^2\theta}} \tag{5-56}$$

This is the plane wave reflection coefficient for an incident electric field parallel to the plane of polarization. ε_r' is the relative complex effective dielectric constant (see Sec. 1.4) for the ground and is given by

$$\varepsilon_r' = \frac{\varepsilon'}{\varepsilon_o} = \varepsilon_r - j\,\frac{\sigma}{\omega\varepsilon_o} \tag{5-57}$$

ε_r and σ are the relative dielectric constant and conductivity of the ground plane. The earth has an average value of $\varepsilon_r = 15$. Ground conductivities across the United States vary from 10^{-3} to 3×10^{-2} S/m [25].

It is convenient to express the imaginary part of ε_r' as

$$\frac{\sigma}{\omega\varepsilon_o} = 18 \times 10^3\,\frac{\sigma}{f_{\text{MHz}}} \tag{5-58}$$

At low frequencies (e.g., 1 MHz and below), the imaginary part or loss-producing part of the complex permitivity dominates. At high frequencies (e.g., 100 MHz and above), the real part dominates.

The reflection coefficients are shown in Fig. 5-47 for a typical ground con-

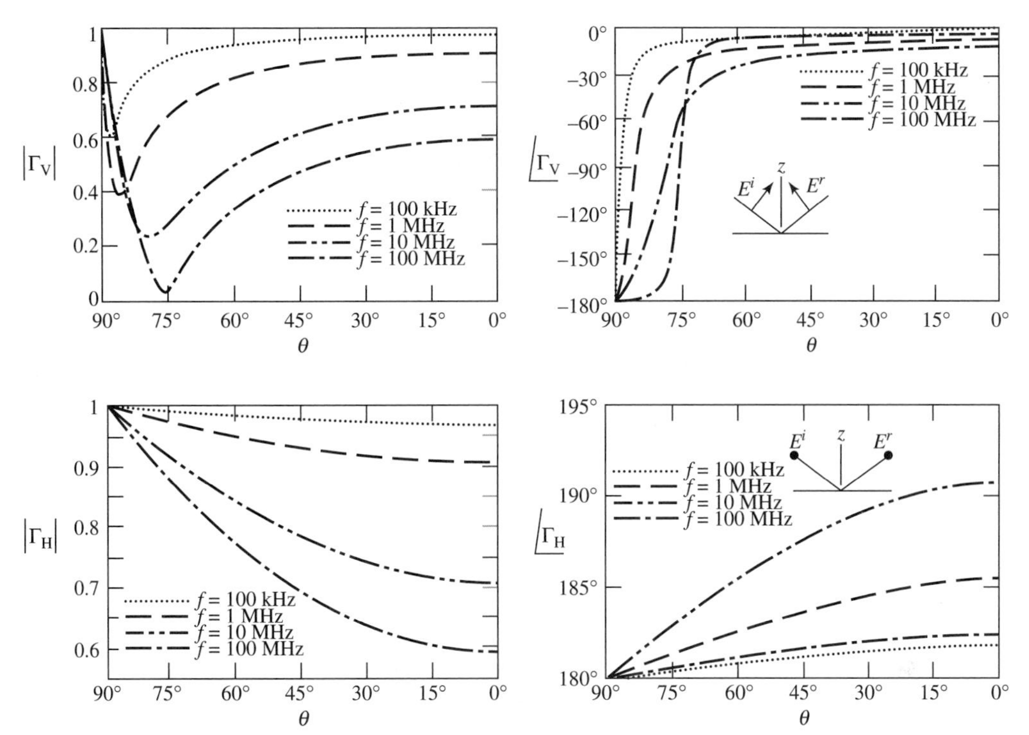

Figure 5-47 Magnitude and phase of Γ_V and Γ_H at four frequencies. Note that the horizon is at $\theta = 90°$. $\sigma = 12 \times 10^{-3}$ s/m and $\varepsilon_r = 15$.

ductivity $\sigma = 12 \times 10^{-3}$. There is a great deal of information in these curves. First, we note that Γ_H is close to -1 at low elevation angles ($\theta \sim 90°$) for all frequencies and ground conductivities. The situation is much different for Γ_V where both the magnitude and phase usually vary rapidly at low elevation angles due to the presence of the pseudo-Brewster angle when $\angle\Gamma_V = -90°$. This behavior does not occur when either $\sigma \rightarrow \infty$ or $\omega \rightarrow 0$ since $|\Gamma_V| \rightarrow 1$ as can be shown from (5-56) and (5-57).

The use of plane wave reflection coefficients to obtain the image strength is only an approximation since antennas near a ground plane do not form plane waves incident on the ground plane. In addition to the radiation we have described above, there is a surface wave that propagates along the ground plane surface. For HF and VHF frequencies, the surface wave attenuates very rapidly. For grazing angles (θ near 90°), $\Gamma_V \approx -1$ and vertical antennas close to a real earth have zero radiation for $\theta = 90°$; see (5-52). In this case, the surface wave accounts for all propagation, as in daylight standard broadcast AM. The effect of neglecting the surface wave, and using the procedure given above, has been found not to be critical for vertical antennas [26]. For horizontal antennas, the antenna should be at least 0.2λ above the earth for the plane wave reflection coefficient method to be valid [27].

The elevation pattern for a short vertical dipole at the surface of various ground planes is shown in Fig. 5-48. When the ground plane is perfect ($\sigma = \infty$), the pattern above the ground plane is the same as that of a short dipole in free space, sin θ. Thus, in the perfect ground plane case, radiation is maximum along the ground plane, whereas for a real earth ground plane, the radiation maximum is tilted up away from the ground plane and is reduced in intensity, for the same input power, due to reduced efficiency. This is a general trend. The effect of a lossy earth on vertical antennas is to tilt the radiation pattern upward. A good radial ground system

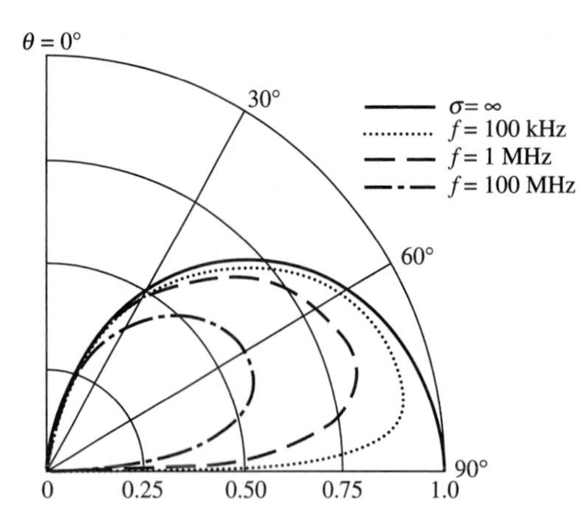

Figure 5-48 Elevation plane patterns of a vertical short dipole a distance h above a flat earth with $\varepsilon_r = 15$ and $\sigma = 1.2 \times 10^{-2}$ s/m for three frequencies compared to the perfect ground plane ($\sigma = \infty$) case.

(to be discussed in Sec. 5.6.2) makes the pattern behave more nearly like that for a perfect ground plane, that is, increase the low angle radiation (along the ground plane). Low angle radiation is particularly important for long-distance communication links that rely on ionospheric reflection (skip).

A short vertical dipole that is $\lambda/4$ above the ground plane forms a $\lambda/2$ spaced array with its image. For the perfect ground plane, $\varepsilon_r' = \infty$ and Γ_V from (5-56) is

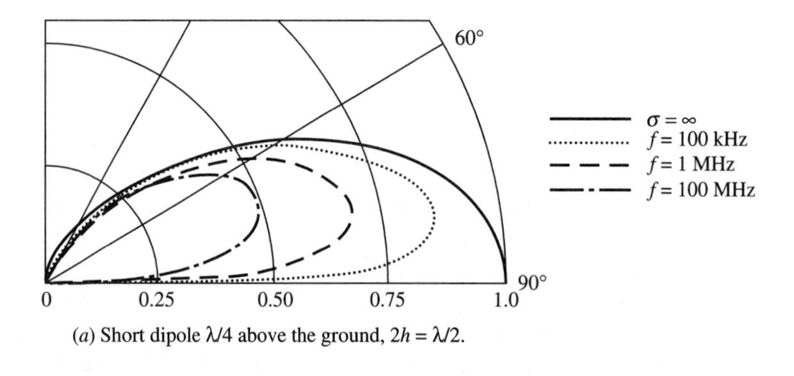

(*a*) Short dipole $\lambda/4$ above the ground, $2h = \lambda/2$.

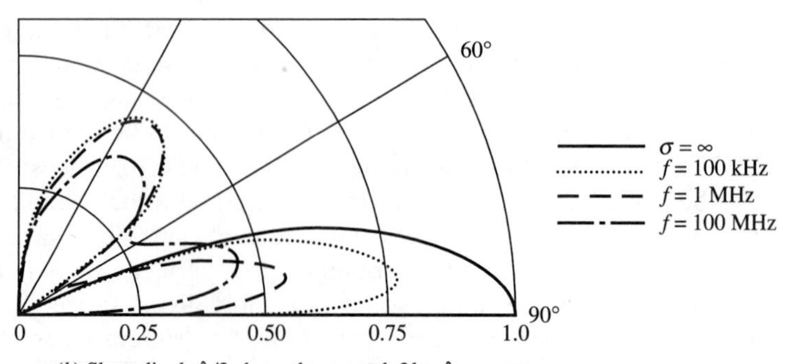

(*b*) Short dipole $\lambda/2$ above the ground, $2h = \lambda$.

Figure 5-49 Elevation plane patterns of a vertical short dipole at three frequencies compared to the perfect ground plane ($\sigma = \infty$) case.

+1. The array is then a $\lambda/2$ spaced, equally excited, in-phase collinear array. The pattern for this is given in Fig. 3-17 and is also plotted in Fig. 5-49*a* ($\sigma = \infty$). For a real earth ground plane, $\Gamma_V \approx -1$ at grazing angles (θ near 90°). The array contributions thus cancel, giving a null along the ground plane as shown in Fig. 5-49*a*. As the height h is increased to $\lambda/2$, the equivalent array of Fig. 5-45 has a λ spacing and multiple lobes appear in addition to the effects described for $h = \lambda/4$. The elevation patterns for $h = \lambda/2$ are plotted in Fig. 5-49*b*.

For a horizontal short dipole as shown in Fig. 5-46, the radiation is not the same for all planes through the z-axis, as for vertical antennas. In the yz-plane (perpendicular to the axis of the dipole), the radiation electric field is given by (5-54). The reflection coefficient Γ_H is exactly -1 for a perfect ground plane and approximately -1 for real earth ground planes at all angles θ if the frequency is low. The element pattern is isotropic since the elements are seen in end view in the H-plane (yz-plane). Thus, the array factor completely determines the pattern. The low elevation pattern effects of finite conductivity are much less pronounced for horizontal antennas than for vertical antennas.

The field expressions of (5-52) to (5-54) for short dipoles above a ground plane can be used for other antenna types by using the appropriate element pattern. In particular, $\sin\theta$ in (5-52) and $\cos\theta$ in (5-54) are replaced by the free-space pattern of the antenna considered.

5.6.2 Ground Plane Construction

An excellent ground plane can be constructed by using a metallic sheet that is much larger than the antenna extent. Such ground planes become impractical at low frequencies because of the size required. In this section, various techniques are discussed for increasing the apparent conductivity of a real earth ground.

Consider a vertical monopole antenna with its base at ground level. (See Sec. 2.3.2 for a discussion of monopoles over a perfect ground plane.) Currents flowing up the antenna leave the antenna and form displacement currents in air. Upon entering the earth, conduction currents are formed that converge toward the base of the antenna. Losses in an earth ground can be reduced by providing a highly conductive return path. This is commonly achieved with a *radial ground system*. The size of the wires used is not critical and is determined by the mechanical strength required. Number 8 AWG wire is typical. They need not be buried, but it is usually convenient to do so. However, they should not be buried too deep in order to minimize the extent of earth through which the fields must pass. Sometimes, the radial wires are linked together at the base of the monopole by a ring-shaped ground strap. Occasionally, one or more stakes are driven into the ground near the base of the monopole.

The ohmic resistance of the radial system and earth ground adds to the ohmic resistance of the monopole structure to determine the total ohmic resistance components of the input impedance. The efficiency of the antenna system depends on the proportion of radiation resistance and ohmic resistance; see (1-173). For high-power transmitting antennas, it is important to have a well-designed radial system to achieve high efficiency. On the other hand, for simple monopole structures, three equally spaced radial wires form the radial system.

The most sophisticated ground system such as used with a standard broadcast AM-transmitting antenna is 120 radial wires spaced equally, 3° apart, around the tower out to a distance of about a quarter-wavelength from the tower. In general, the length of the radials is roughly equal to the height of the monopole antenna. The value of the total ohmic resistance of a ground system with 120 radials for typical soils is plotted in Fig. 5-50 for a few frequencies as a function of radial length [28]. Note that at 3 MHz a ground system with 120 radials that are about a quarter-wavelength long (25 m) gives a ground system resistance of 1 Ω. Since the surface resistance of the earth varies as the square root of frequency [see (1-176)], the ground system resistance will be constant for lower frequencies if the length of the radials is increased in proportion to the square root of wavelength. For frequencies above 3 MHz, the curve for radial length in Fig. 5-50 is only slightly to the right of the 3-MHz curve. This is because after the radials reach a length of about a quarter-wavelength, most of the large current densities occur within the region of the radials and further length increase is of no major consequence.

The construction principles for a radial wire ground system on top of or in the earth can be summarized rather simply. The function of a radial system is to prevent the electromagnetic fields from the antenna from penetrating into the ground and exciting currents that, in turn, lead to $\sigma|E|^2$ ohmic loss. As can be seen from the above discussion, if 120 quarter-wavelength long radials are employed, the ohmic resistance introduced by the ground system will be at most a few ohms and usually well under an ohm. In most applications, it is impractical to install as many as 120 radials. Generally speaking, 50 radials about a quarter-wavelength long will reduce earth losses to a few ohms. When only a few radials are used, the added resistance of the ground can be several ohms. Also if the radial lengths (almost independent of the number used) are reduced below a tenth of a wavelength, the ground system

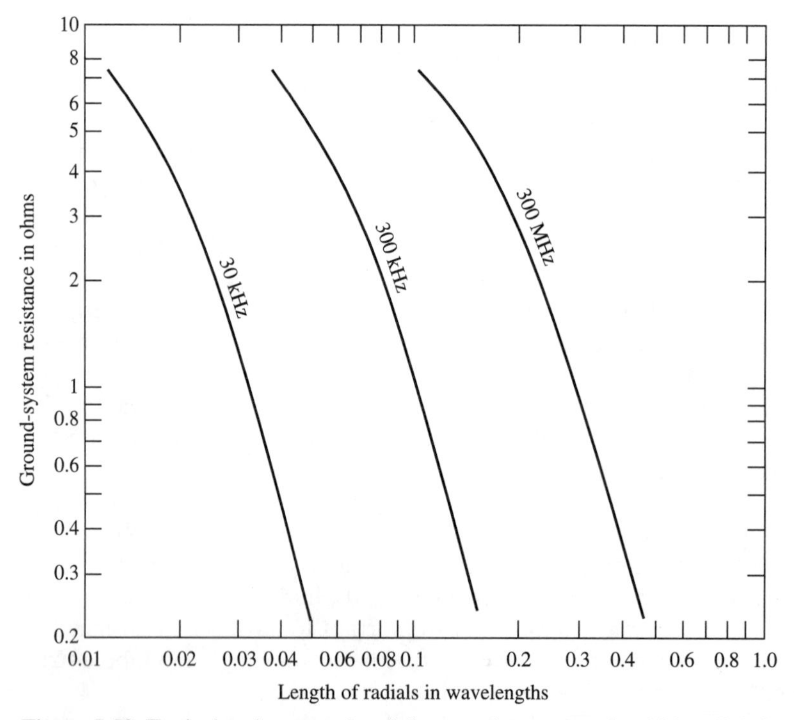

Figure 5-50 Typical resistance of radial ground systems using 120 radials in average soil. (From Griffith [28]. © 1962 McGraw-Hill. Used with permission of McGraw-Hill Book Company.)

resistance will increase significantly.[1] The radial wires can be laid on top of the ground or buried slightly (but never deeply buried). Wire selection is largely determined by mechanical considerations. As the number of radials is increased, the less current each one will have to carry and thus the smaller the wire diameter required. At the base of the antenna, the radials should be connected together and to one or more ground stakes.

At high frequencies (VHF and above), antennas are often mounted over metallic (solid, mesh, or radial wire) ground planes of relatively small extent. Then the dimensions and shape of the ground plane are important. In general, the radiation is greatest in the direction of the largest portion of the ground plane. For example, consider a monopole antenna mounted on an automobile. If it is placed on the right rear bumper, a pattern maximum occurs off of the left front of the car. When the antenna is mounted in the center of the car roof, there is some slight pattern enhancement in the forward and rear directions.

5.7 LARGE LOOP ANTENNAS

We found in Sec. 2.4 that the pattern and radiation resistance of electrically small loop antennas, which have a perimeter much less than a wavelength, are insensitive to loop shape and depend only on the loop area. Also, the radiation from a small loop is maximum in the plane of the loop and is zero along the axis normal to the loop. These facts are a consequence of the current amplitude and phase being constant around the loop, which holds if the loop perimeter length L is electrically small. For loop perimeters that are a sizable fraction of a wavelength or greater, the current amplitude and phase vary with position around the loop, causing performance variations with changing size. Equivalently, a fixed physical size large loop displays performance changes with varying frequency, which is characteristic of a resonant antenna.

Although commonly employed to avoid mathematical difficulties, the analysis of large loop antennas by assuming uniform current amplitude and phase (as in Prob. 5.7-1) yields inaccurate results in practice. As we show in this section, the current distribution is close to sinusoidal for resonant loops. For construction reasons, large loops usually have either a circular or square perimeter. Circular and square loops are usually operated near the first resonance point, which occurs for a perimeter length of slightly greater than one wavelength. The circular loop has been studied more extensively. Rigorous analysis techniques giving the fields from circular loop antennas [30] and approximate formulas for the directivity and radiation resistance of circular loop antennas [31] are available. A good summary of analysis and measurements of circular loops is found in [32]. In this section, we treat the large square loop in detail. Its performance is very similar to that of the circular loop; see Prob. 5.7-4. The one-wavelength square loop can be analyzed using the same techniques that we used for other resonant wire antennas. The approach, based on an assumed sinusoidal current distribution, is presented first. Next, accurate numerical method results are presented to support the approximate analysis. Finally, computed performance is given for the square loop with varying perimeter length.

The **one-wavelength square loop antenna** as shown in Fig. 5-51 has one-quarter wavelength sides. For a one-wavelength perimeter, it is reasonable to assume that

---

[1]More details and references for ground system design are available in [29].

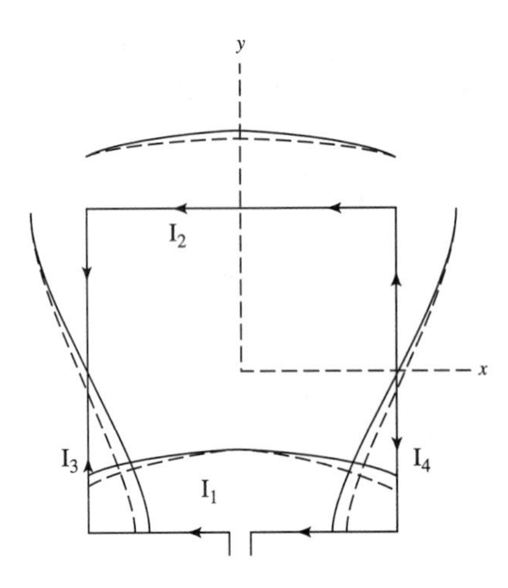

Figure 5-51 The one-wavelength square loop antenna. Each side is of length $\lambda/4$. The solid curve is the sinusoidal current distribution of (5-59). The dashed curve is the current magnitude obtained from more exact numerical methods.

the current distribution is sinusoidal. Then the current distribution is continuous around the loop as shown in Fig. 5-51 (solid curve). With the feed point in the center of a side parallel to the x-axis, this sinusoidal current is expressed as

$$\begin{aligned} \mathbf{I}_1 = \mathbf{I}_2 &= -\hat{\mathbf{x}} I_o \cos(\beta x'), \qquad |x'| \le \frac{\lambda}{8} \\ \mathbf{I}_4 = -\mathbf{I}_3 &= \hat{\mathbf{y}} I_o \sin(\beta y'), \qquad |y'| \le \frac{\lambda}{8} \end{aligned} \tag{5-59}$$

The solution for the radiation properties proceeds in the usual manner. First, the vector potential from (1-101) is

$$\mathbf{A} = \mu \frac{e^{-j\beta r}}{4\pi r} \int_{\text{loop}} \mathbf{I} e^{j\beta \hat{\mathbf{r}} \cdot \mathbf{r}'} \, dl \tag{5-60}$$

To find the phase function, the expressions for vectors from the origin to arbitrary positions on each side must be written. They are

$$\begin{aligned} \mathbf{r}_1' &= x'\hat{\mathbf{x}} - \frac{\lambda}{8}\hat{\mathbf{y}} \qquad & \mathbf{r}_2' &= x'\hat{\mathbf{x}} + \frac{\lambda}{8}\hat{\mathbf{y}} \\ \mathbf{r}_3' &= -\frac{\lambda}{8}\hat{\mathbf{x}} + y'\hat{\mathbf{y}} \qquad & \mathbf{r}_4' &= \frac{\lambda}{8}\hat{\mathbf{x}} + y'\hat{\mathbf{y}} \end{aligned} \tag{5-61}$$

where the numbered subscripts indicate the corresponding loop side. Using the expansion of $\hat{\mathbf{r}}$ from (C-4) and (5-61) in (5-60) with the loop integral broken into integrals over each side gives

$$\begin{aligned} \mathbf{A} = \mu \frac{e^{-j\beta r}}{4\pi r} I_o \Bigg[&-\hat{\mathbf{x}} \int_{-\lambda/8}^{\lambda/8} \cos(\beta x') e^{j\beta x' \sin\theta\cos\phi} \left(e^{-j(\pi/4)\sin\theta\sin\phi} + e^{j(\pi/4)\sin\theta\sin\phi}\right) dx' \\ &+ \hat{\mathbf{y}} \int_{-\lambda/8}^{\lambda/8} \sin(\beta y') e^{j\beta y' \sin\theta\sin\phi} \left(-e^{-j(\pi/4)\sin\theta\cos\phi} + e^{j(\pi/4)\sin\theta\cos\phi}\right) dy' \Bigg] \\ = \mu \frac{e^{-j\beta r}}{4\pi r} I_o \Bigg[&-\hat{\mathbf{x}} 2 \cos\left(\frac{\pi}{4}\sin\theta\sin\phi\right) \int_{-\lambda/8}^{\lambda/8} \cos(\beta x') e^{j\beta x' \sin\theta\cos\phi} \, dx' \\ &+ \hat{\mathbf{y}} 2j \sin\left(\frac{\pi}{4}\sin\theta\cos\phi\right) \int_{-\lambda/8}^{\lambda/8} \sin(\beta y') e^{j\beta x' \sin\theta\sin\phi} \, dy' \Bigg] \end{aligned} \tag{5-62}$$

The first factors in each of the above two terms in brackets are the array factors for the pairs of sides 1, 2 and 3, 4, respectively. Evaluation of the integrals and subsequent simplification lead to

$$\mathbf{A} = \mu \frac{e^{-j\beta r}}{4\pi r} \frac{2\sqrt{2}I_o}{\beta} \left\{ \hat{\mathbf{x}} \frac{\cos[(\pi/4)\cos\Omega]}{\sin^2\gamma} \left[\cos\gamma \sin\left(\frac{\pi}{4}\cos\gamma\right) - \cos\left(\frac{\pi}{4}\cos\gamma\right) \right] \right.$$
$$\left. - \hat{\mathbf{y}} \frac{\sin[(\pi/4)\cos\gamma]}{\sin^2\Omega} \left[\cos\Omega \cos\left(\frac{\pi}{4}\cos\Omega\right) - \sin\left(\frac{\pi}{4}\cos\Omega\right) \right] \right\} \tag{5-63}$$

where

$$\cos\gamma = \sin\theta\cos\phi \quad \text{and} \quad \cos\Omega = \sin\theta\sin\phi \tag{5-64}$$

The angles γ and Ω have a geometrical interpretation; they are the spherical polar angles (similar to θ) for the x- and y-axes; see (C-4).

The far-zone electric field components are

$$E_\theta = -j\omega A_\theta = -j\omega \mathbf{A} \cdot \hat{\boldsymbol{\theta}} = -j\omega(A_x \hat{\mathbf{x}} \cdot \hat{\boldsymbol{\theta}} + A_y \hat{\mathbf{y}} \cdot \hat{\boldsymbol{\theta}})$$
$$= -j\omega(A_x \cos\theta\cos\phi + A_y \cos\theta\sin\phi) \tag{5-65a}$$

$$E_\phi = -j\omega \mathbf{A} \cdot \hat{\boldsymbol{\phi}} = -j\omega(-A_x \sin\phi + A_y \cos\phi) \tag{5-65b}$$

Substituting A_x and A_y from (5-63) gives

$$E_\theta = \frac{jI_o \eta e^{-j\beta r}}{\sqrt{2}\pi r} \cos\theta \left\{ \frac{\sin\phi \sin[(\pi/4)\sin\theta\cos\phi]}{1 - \sin^2\theta \sin^2\phi} \right.$$
$$\cdot \left[\sin\theta\sin\phi\cos\left(\frac{\pi}{4}\sin\theta\sin\phi\right) - \sin\left(\frac{\pi}{4}\sin\theta\sin\phi\right) \right]$$
$$- \frac{\cos\phi\cos[(\pi/4)\sin\theta\sin\phi]}{1 - \sin^2\theta\cos^2\phi} \tag{5-66a}$$
$$\left. \cdot \left[\sin\theta\cos\phi\sin\left(\frac{\pi}{4}\sin\theta\cos\phi\right) - \cos\left(\frac{\pi}{4}\sin\theta\cos\phi\right) \right] \right\}$$

$$E_\phi = \frac{jI_o \eta e^{-j\beta r}}{\sqrt{2}\pi r} \left\{ \frac{\cos\phi \sin[(\pi/4)\sin\theta\cos\phi]}{1 - \sin^2\theta \sin^2\phi} \right.$$
$$\cdot \left[\sin\theta\sin\phi\cos\left(\frac{\pi}{4}\sin\theta\sin\phi\right) - \sin\left(\frac{\pi}{4}\sin\theta\sin\phi\right) \right]$$
$$+ \frac{\sin\phi\cos[(\pi/4)\sin\theta\sin\phi]}{1 - \sin^2\theta\cos^2\phi} \tag{5-66b}$$
$$\left. \cdot \left[\sin\theta\cos\phi\sin\left(\frac{\pi}{4}\sin\theta\cos\phi\right) - \cos\left(\frac{\pi}{4}\sin\theta\cos\phi\right) \right] \right\}$$

These expressions are rather involved but were derived in a straightforward fashion using the principles set forth in Sec. 1.7.

The far-field expressions simplify somewhat in the principal planes. In the xy-plane, which is the plane of the loop (an E-plane), $\theta = 90°$ and then (5-66) reduces to

$$E_\theta\left(\theta = \frac{\pi}{2}\right) = 0 \tag{5-67a}$$

$$E_\phi\left(\theta = \frac{\pi}{2}\right) = \frac{jI_o\eta e^{-j\beta r}}{\sqrt{2}\pi r}\frac{\pi}{4}\left\{\frac{\sin[(\pi/4)\cos\phi]}{(\pi/4)\cos\phi}\left[\sin\phi\cos\left(\frac{\pi}{4}\sin\phi\right) - \sin\left(\frac{\pi}{4}\sin\phi\right)\right]\right.$$
$$\left. + \frac{\cos[(\pi/4)\sin\phi]}{(\pi/4)\sin\phi}\left[\cos\phi\sin\left(\frac{\pi}{4}\cos\phi\right) - \cos\left(\frac{\pi}{4}\cos\phi\right)\right]\right\} \tag{5-67b}$$

The E_ϕ expression is plotted in Fig. 5-52*a* (solid curve) in normalized form. Along the x-axis ($\phi = 0°$ and $180°$), $E_\phi = 0$. This is true because the sides 3 and 4 alone each have patterns that are zero in the broadside direction since the current distributions on these sides are odd about their midpoints. Along the y-axis, (5-67b) reduces to

$$E_\phi\left(\theta = \frac{\pi}{2}, \phi = \frac{\pi}{2}\right) = -\frac{jI_o\eta e^{-j\beta r}}{\sqrt{2}\pi r}\frac{1}{\sqrt{2}} \tag{5-68}$$

In the xz-plane, which is an E-plane, (5-66) yields

$$E_\phi(\phi = 0) = 0 \tag{5-69a}$$

$$E_\theta(\phi = 0) = \frac{jI_o\eta e^{-j\beta r}}{\sqrt{2}\pi r}\frac{\sin\theta\sin[(\pi/4)\sin\theta] - \cos[(\pi/4)\sin\theta]}{\cos\theta} \tag{5-69b}$$

The normalized form of this E_θ expression is plotted in Fig. 5-52*b* (solid curve). It can be shown that (5-69b) goes to zero for $\theta = 90°$, as it should by (5-67a).

In the yz-plane, which is the H-plane, (5-66) reduces to

$$E_\theta\left(\phi = \frac{\pi}{2}\right) = 0 \tag{5-70a}$$

$$E_\phi\left(\phi = \frac{\pi}{2}\right) = -\frac{jI_o\eta e^{-j\beta r}}{\sqrt{2}\pi r}\cos\left(\frac{\pi}{4}\sin\theta\right) \tag{5-70b}$$

Figure 5-52*c* (solid curve) gives the plot of the normalized form of this E_ϕ expression. The $\cos[(\pi/4)\sin\theta]$ pattern is the array factor for two point sources at the midpoints of sides 1 and 2. Note that in the z-direction, (5-69) and (5-70) give the same result (for $\theta = 0°$): an electric field parallel to the x-axis given by

$$E_x = -\frac{jI_o\eta e^{-j\beta r}}{\sqrt{2}\pi r} \tag{5-71}$$

which is a factor of $\sqrt{2}$ greater than E_x in the y-direction given in (5-68). This is also seen in Fig. 5-52*c*. Comparing (5-71) to (5-68) shows that the level of E_x in the y-axis direction is 0.707 of that along the z-axis. Thus, the peak of the normalized xy-plane pattern in Fig. 5-52*a* along the y-axis multiplied by 0.707 matches up with the y-axis value in Fig. 5-52*c*.

From the patterns in Fig. 5-52, we can make some general conclusions about the radiation properties of the one-wavelength square loop antenna. Radiation is maximum normal to the plane of the loop (along the z-axis) and in that direction is polarized parallel to the loop side containing the feed. In the plane of the loop,

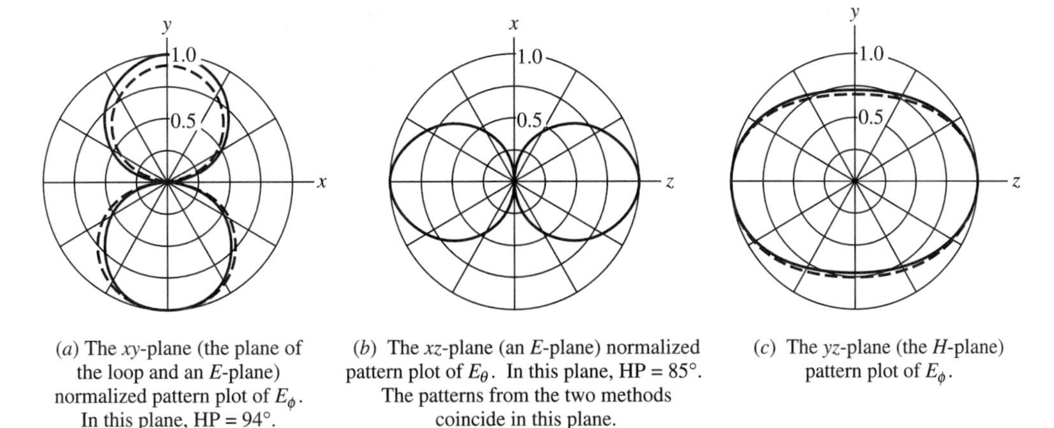

(a) The xy-plane (the plane of the loop and an E-plane) normalized pattern plot of E_ϕ. In this plane, HP = 94°.

(b) The xz-plane (an E-plane) normalized pattern plot of E_θ. In this plane, HP = 85°. The patterns from the two methods coincide in this plane.

(c) The yz-plane (the H-plane) pattern plot of E_ϕ.

Figure 5-52 Principal plane patterns for one-wavelength square loop antenna. The solid curves are the patterns based on a sinusoidal current distribution of Fig. 5-51. The dashed curves are the patterns arising from the current distribution obtained by the more exact numerical methods.

there is a null in the direction parallel to the side containing the feed point (along the x-axis), and there is a lobe in a direction perpendicular to the side containing the feed (along the y-axis). These results are quite different from the small loop antenna, which has a null on-axis and maximum (uniform) radiation in the plane of the loop.

The accuracy of our results can be investigated by solving the square loop problem without assuming the current distribution to be sinusoidal. The numerical methods of Chap. 10 applied to the one-wavelength loop antenna for a wire radius of 0.001λ renders the current magnitude shown in Fig. 5-51 (dashed curve). Note that the agreement is actually very good. The impact of the slight differences in these current distributions is revealed in Fig. 5-52. The dashed curves are the patterns corresponding to the exact current distribution and calculated by a radiation integral procedure similar to that detailed above for the assumed current. The agreement between the patterns arising from the simple current assumption and that of more exact methods is very good. In fact, in the xz-plane the agreement is nearly exact. This detailed comparison of the approximate antenna analysis methods employed thus far in the book to that of more exact (but more difficult) numerical methods serves to provide confidence that good engineering results can be obtained from reasonable assumptions about the operation of antennas.

The impedance of a square loop antenna with a wire radius of 0.001λ is plotted in Fig. 5-53 as a function of the perimeter. Note that for a one-wavelength perimeter, the input reactance is relatively small, and also note that resonance occurs for a 1.09λ perimeter. The input resistance for a one-wavelength perimeter is about 100 Ω. Other perimeter values give rather awkward input impedances. Similar results are obtained for circular loops.

The gain of the one-wavelength square loop is 3.09 dB, which is less than the 3.82-dB gain of a straight wire one-wavelength dipole. This is to be expected from the obviously less directive pattern of the loop in Fig. 5-52 compared to that of the one-wavelength dipole in Fig. 5-4.

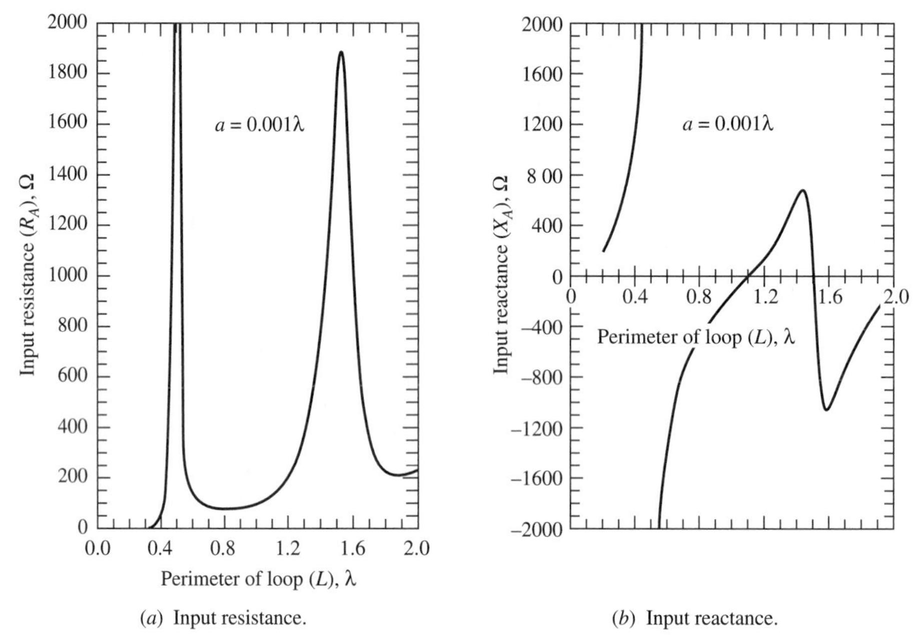

(a) Input resistance. (b) Input reactance.

Figure 5-53 Input impedance of a square loop antenna as a function of the loop perimeter in wavelengths. The loop is fed in the center of one side and has a wire radius of $a = 0.001\lambda$. Numerical calculation methods were used.

5.8 MICROSTRIP ANTENNAS

Printed antennas are constructed using printed circuit fabrication techniques such that a portion of the metallization layer is responsible for radiation. **Microstrip antenna** patch elements, and arrays of patches, are the most common form of printed antenna and were conceived in the 1950s. Extensive investigations of patch antennas began in the 1970s [33, 34] and resulted in many useful design configurations [35]. Printed antennas are popular with antenna engineers for their low profile, for the ease with which they can be configured to specialized geometries, and because of their low cost when produced in large quantities. This section explains the basic operating principles of microstrip elements and arrays. Simple formulas are given that produce approximate results. Lengthy formulas that are more general and more accurate are available [36, 37].

5.8.1 Microstrip Patch Antennas

A microstrip device in its simplest form is a layered structure with two parallel conductors separated by a thin dielectric substrate and the lower conductor acting as a ground plane. If the upper metallization is a long narrow strip, a microstrip transmission line is formed. If the upper conductor is a *patch* that is an appreciable fraction of a wavelength in extent, the device becomes a microstrip antenna, as illustrated in Fig. 5-54. The patch antenna belongs to the class of resonant antennas and its resonant behavior is responsible for the main challenge in microstrip antenna design—achieving adequate bandwidth. Conventional patch designs yield bandwidths as low as a few percent. The resonant nature of microstrip antennas also means that at frequencies below UHF they become excessively large. They are

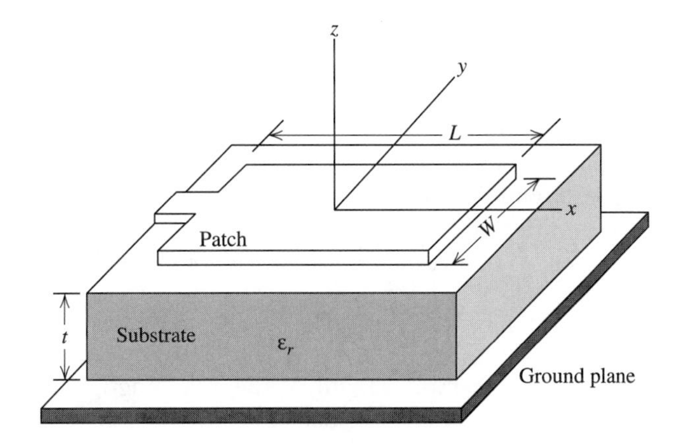

(a) Geometry for analyzing the edge-fed microstrip patch antenna.

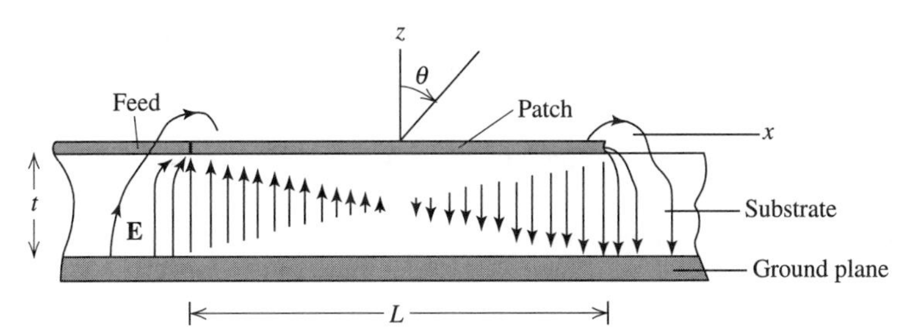

(b) Side view showing the electric fields.

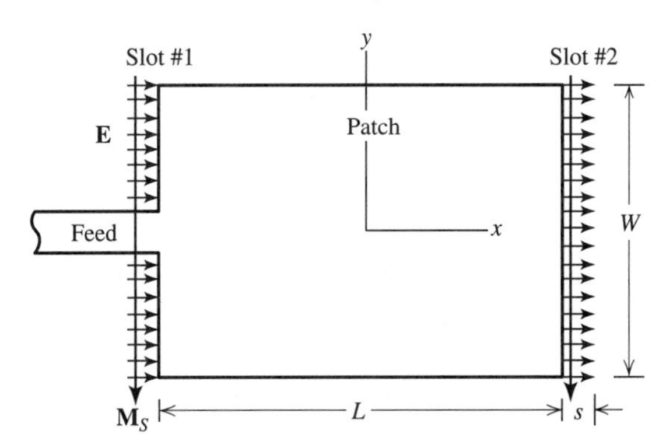

(c) Top view showing the fringing electric fields that are responsible for radiation. The equivalent magnetic surface $\mathbf{M}_S$ currents are also shown.

Figure 5-54 The half-wavelength rectangular patch microstrip antenna; $L \approx 0.49\lambda_d$.

typically used at frequencies from 1 to 100 GHz. The tradeoff in microstrip antennas is to design a patch with loosely bound fields extending into space while keeping the fields tightly bound to the feeding circuitry. This is to be accomplished with high radiation efficiency and with the desired polarization, impedance, and bandwidth.

The Rectangular Patch Antenna. Figure 5-54 shows the most commonly used microstrip antenna, a *rectangular patch* being fed from a microstrip transmission line. The substrate thickness t is much less than a wavelength. The rectangular patch is usually operated near resonance in order to obtain a real-valued input impedance.

Models are available for determining the resonant frequency, with the cavity model usually yielding accurate results; see [33]. The fringing fields act to extend the effective length of the patch. Thus, the length of a *half-wave* patch is slightly less than a half wavelength in the dielectric substrate material. This is similar to foreshortening a half-wave dipole to achieve resonance. The amount of length reduction depends on ε_r, t, and W. Formulas are available to estimate the resonant length [33, 36, 37], but empirical adjustments are often necessary in practice. An approximate value for the length of a resonant half-wavelength patch is [34]

$$L \approx 0.49\lambda_d = 0.49 \frac{\lambda}{\sqrt{\varepsilon_r}} \qquad \textit{half-wave patch} \tag{5-72}$$

where λ is the free-space wavelength, λ_d the wavelength in the dielectric, and ε_r the substrate dielectric constant. We focus our attention here on the half-wave patch antenna.

The region between the conductors acts as a half-wavelength transmission-line cavity that is open-circuited at its ends. Figure 5-54*b* shows the electric fields associated with the standing wave mode in the dielectric. The electric field lines are perpendicular to the conductors as required by boundary conditions and look much like those in a parallel plate capacitor. The fringing fields at the ends are exposed to the upper half-space ($z > 0$) and are responsible for the radiation. The standing wave mode with a half-wavelength separation between ends leads to electric fields that are of opposite phase on the left and right halves (i.e., positive and negative x). Therefore, the total fringing fields at the edges are 180° out of phase and equal in magnitude. Viewed from the top (see Fig. 5-54*c*), the x-components of the fringing fields are actually in-phase, leading to a broadside radiation pattern; that is, the peak radiation is in the $+z$-direction. This model suggests an "aperture field" analysis approach where the patch has two radiating slot apertures with electric fields in the plane of the patch. For the half-wave patch case, the slots are equal in magnitude and phase. The fields along the edges associated with slots 1 and 2 are constant, whereas those along the other edges, seen in side view in Fig. 5-54*b*, have odd symmetry and their radiation cancels in the broadside direction and is usually neglected. The width of the slots is often taken to be equal to the substrate thickness, that is, $s \approx t$. The patch radiation is linearly polarized in the xz-plane, that is, parallel to the electric fields in the slots.

The pattern of a rectangular patch antenna is rather broad with a maximum direction normal to the plane of the antenna. Pattern computation for the rectangular patch is easily performed by first creating equivalent magnetic surface currents, as shown in Fig. 5-54*c*, from the fringe electric fields using $\mathbf{M}_s = 2\mathbf{E}_a \times \hat{\mathbf{n}}$, where $\mathbf{E}_a$ is the fringe electric field in each of the edge slots; this follows from (1-23) or (7-2). The factor of 2 comes from the image of the magnetic current in the electric ground plane (see Fig. 7-4*c*) if we assume t is small. The far-field components follow from (7-26) (see Prob. 7.1-7) as

$$E_\theta = E_o \cos\phi\, f(\theta, \phi) \tag{5-73a}$$

$$E_\phi = -E_o \cos\theta \sin\phi\, f(\theta, \phi) \tag{5-73b}$$

where

$$f(\theta, \phi) = \frac{\sin\left[\dfrac{\beta W}{2} \sin\theta \sin\phi\right]}{\dfrac{\beta W}{2} \sin\theta \sin\phi} \cos\left(\frac{\beta L}{2} \sin\theta \cos\phi\right) \tag{5-73c}$$

and β is the usual free-space phase constant. The first factor is the pattern factor for a uniform line source of width W in the y-direction. The second factor is the array factor for a two-element array along the x-axis corresponding to the edge slots; see (3-8). The patch length L for resonance is given by (5-72). The patch width W is selected to give the proper radiation resistance at the input, often 50 Ω. The principal plane patterns follow from (5-73) as

$$F_E(\theta) = \cos\left(\frac{\beta L}{2}\sin\theta\right) \qquad E\text{-plane, } \phi = 0° \tag{5-74a}$$

$$F_H(\theta) = \cos\theta \frac{\sin\left[\frac{\beta W}{2}\sin\theta\right]}{\frac{\beta W}{2}\sin\theta} \qquad H\text{-plane, } \phi = 90° \tag{5-74b}$$

This simple pattern expression neglects substrate effects and slot width (i.e., fringing).

Typical input impedances at the edge of a resonant rectangular patch range from 100 to 400 Ω. An approximate expression for the input impedance (reactance is zero at resonance) of a resonant edge-fed patch is [36]

$$Z_A = 90\frac{\varepsilon_r^2}{\varepsilon_r - 1}\left(\frac{L}{W}\right)^2 \Omega \qquad \textit{half-wave patch} \tag{5-75}$$

Thus, the input impedance (resistance) is reduced by widening the patch. For example, for a dielectric of $\varepsilon_r = 2.2$, a width-to-length ratio of $W/L = 2.7$ gives a 50-Ω input impedance.

Techniques for feeding patches are summarized in Fig. 5-55. They can be classified into three groups: directly coupled, electromagnetically coupled, or aperture coupled. Direct coupling methods are the oldest and most popular, but only provide one degree of freedom to adjust impedance. The microstrip feed line exciting the patch edge and the coaxial probe are examples of direct feeds. The rectangular patch is normally fed along a patch centerline in the E-plane as shown in Fig. 5-55. This avoids excitation of a second resonant mode orthogonal to the desired mode, which would lead to excessive cross polarization.

The direct coaxial probe feed illustrated in Fig. 5-55*a* is simple to implement by extending the center conductor of the connector attached to the ground plane up to the patch. Impedance can be adjusted by proper placement of the probe feed. As the probe distance from the patch edge, Δx_p in Fig. 5-55*a*, is increased, the input resistance of (5-75) is reduced by the factor $\cos^2 \dfrac{\pi\,\Delta x_p}{L}$ [36]. A disadvantage of the probe feed is that it introduces an inductance that prevents the patch from being resonant if t is 0.1λ or greater. Also, probe radiation can be a source of cross polarization.

The microstrip feed of Fig. 5-54 is planar, permitting the patch and feed to be printed on a single metallization layer. This feed approach is especially well suited to arrays where the feed network can be printed with the elements. Changing the patch width, as determined with (5-75), to control impedance is often not convenient. However, the impedance of the edge-fed patch can be transformed by using a quarter-wave matching section of microstrip transmission line as shown in Fig. 5-55*b*. That is, the antenna input impedance Z_A can be matched to a transmission line of characteristic impedance Z_o (often 50 Ω) with a section of transmission line

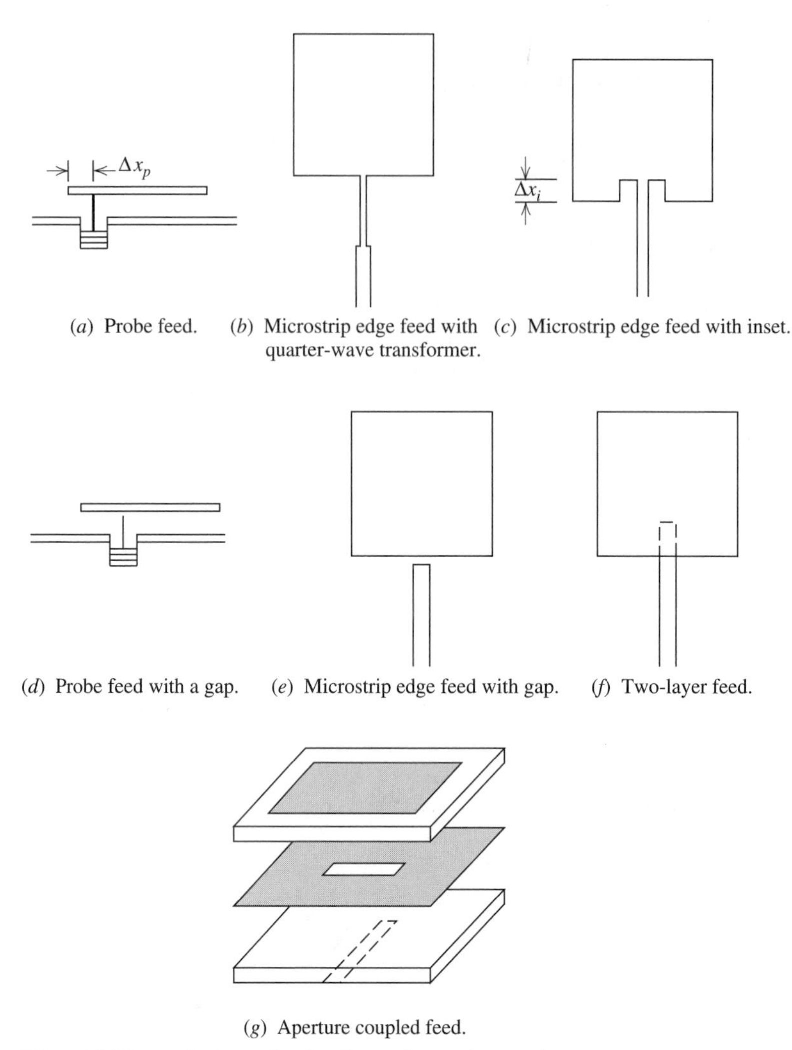

(*a*) Probe feed. (*b*) Microstrip edge feed with quarter-wave transformer. (*c*) Microstrip edge feed with inset.

(*d*) Probe feed with a gap. (*e*) Microstrip edge feed with gap. (*f*) Two-layer feed.

(*g*) Aperture coupled feed.

Figure 5-55 Techniques for feeding microstrip patch antennas.

that is a quarter-wavelength long based on the wavelength in the transmission line. This characteristic impedance of the matching section is given by

$$Z_o' = \sqrt{Z_A Z_o} \qquad \textit{quarter-wave transformer} \tag{5-76}$$

In general, the characteristic impedance of a microstrip line is decreased by increasing the strip width, much as loss resistance is inversely proportional to wire diameter; see (1-175). That is, the wider the strip, the lower the characteristic impedance.

Another type of microstrip feed is the inset feed shown in Fig. 5-55*c*, which offers the advantage of being planar and easily etched as well as providing adjustable input impedance through inset geometry changes. The input resistance of (5-75) is multiplied by the factor of $\cos^2 \dfrac{\pi\,\Delta x_i}{L}$ [33]. However, large input impedance changes that are required for high-permittivity substrates demand significant inset depths, which affects cross polarization and radiation pattern shape.

The direct feeds of Figs. 5-55*a* through 5-55*c* have a narrow bandwidth that can only be increased by increasing the substrate thickness, which has the drawback of increasing the power in the waves trapped along the surface, which are called *surface waves*. Electromagnetic coupled feeds (also called proximity, noncontacting, or gap feeds) do not contact the patch and have at least two design parameters. They also have the advantage of being less sensitive to etching errors. For each direct feed in Figs. 5-55*a* through 5-55*c*, there is gap feed counterpart shown in Figs. 5-55*d* through 5-55*f*. The probe feed with a gap in Fig. 5-55*d* has the advantages of coaxial feeds. Also, the gap capacitance partially cancels the probe inductance, permitting thicker substates. The microstrip feed with a gap in Fig. 5-55*e* is entirely planar and easy to etch. However, in high-permittivity designs, the gap distance may become small. The two-layer feed of Fig. 5-55*f* is a recent technique that is especially useful in microstrip arrays with a top layer for the patches and a second layer for the microstrip feed network.

The aperture-coupled feed of Fig. 5-55*g* is increasing in popularity. The upper substrate can be of low dielectric constant to promote radiation and a lower substrate containing the feed can be of high dielectric constant to enhance binding of the fields to the feed lines. This leads to increased bandwidth. Another advantage is that the central ground plane acts to isolate the feed system from the patches.

Materials such as PTFE composites and alumina are available for the dielectric substrate with ε_r ranging from 1 to about 25, with around 2.5 being most popular [33; 38, Chap. 3]. The selection of dielectric type is based on its loss, temperature and dimensional stability, uniformity of manufacturing (especially ε_r variations), and available sheet sizes and thicknesses.

The bandwidth and efficiency of a patch are increased by increasing substrate thickness t and by lowering ε_r. The associated penalty in array applications is an increase in side lobes and cross polarization as a result of surface waves across an array of patches. This is a fundamental design tradeoff. Bandwidth is often the ultimate limiting performance parameter and can be found from the following simple empirical formula for impedance bandwidth [36]:

$$B = 3.77 \frac{\varepsilon_r - 1}{\varepsilon_r^2} \frac{W}{L} \frac{t}{\lambda} \quad , \quad \frac{t}{\lambda} \ll 1 \tag{5-77}$$

where bandwidth is defined as fractional bandwidth relative to the center frequency for a VSWR less than 2:1. For example, at an operating frequency of 3.78 GHz, a square patch on a substrate of $\varepsilon_r = 2.2$ with $t = 1/16$ in. (1.59 mm) $= 0.02\lambda$, the bandwidth from (5-77) is 1.9%. The bandwidth dependence on thickness t is evident in (5-77).

Another rectangular patch antenna encountered in practice is the quarter-wave element that has $L \approx \lambda_d/4$ and is formed by placing shorting bins from the patch to the ground plane at $x = 0$ in Fig. 5-54*a* and eliminating the patch metalization for $x > 0$. The current peak (electric field null) is then in the same position as the half-wave patch.

EXAMPLE 5-2 *Half-Wave, Square Microstrip Patch Antenna*

A square, half-wave patch was designed to resonant at 3.03 GHz ($\lambda = 9.9$ cm) on a $t = 0.114$ cm (45 mils) thick substrate with $\varepsilon_r = 2.35$. From (5-72),

$$L = W = 0.49 \frac{\lambda}{\sqrt{\varepsilon_r}} = 0.49 \frac{9.9}{\sqrt{2.35}} = 3.16 \text{ cm} \tag{5-78}$$

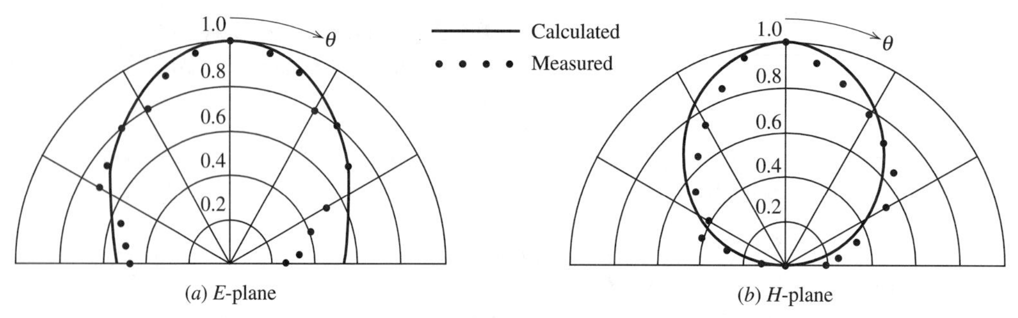

Figure 5-56 Radiation patterns for the square microstrip patch of Example 5-2 calculated (curves) using (5-74) and measured (points).

From (5-75), the input impedance is $Z_A = 368\ \Omega$, which compares to a measured value of $316\ \Omega$ at resonance. The measured resonant frequency of 3.01 GHz compares very well with the design frequency of 3.03 GHz. The radiation patterns in the principal planes are given by (5-74). These patterns are plotted in Fig. 5-56 together with data measured at Virginia Tech. The agreement is very good for the simple theory employed.

Other Patch Shapes. There are many patch shapes for special purposes [35]. Important among these are patches for creating circular polarization. Circular polarization can be achieved by feeding the corner of a square patch, by feeding adjacent orthogonal edges of a square patch 90° out-of-phase, or by using a pentagonal shaped patch. A number of software packages are available for analyzing the currents, impedance, and radiation from microstrip elements of most any shape as well as small arrays of patches with the feed network.

5.8.2 Microstrip Arrays

Arrays of microstrip antennas offer the advantage that the feed network as well as the radiating elements can be printed, often by fabrication on the same single layer printed circuit board. More sophisticated implementations are finding wide use in many system applications that fully integrate microstrip radiating elements and feed lines along with the transmitting and receiving circuitry. In fact, antenna technology is following an evolutionary path similar to that of electronics, moving from discrete devices that are individually connected to an antenna element toward full integration where chips are integrated with the feed lines and radiators.

Microstrip antennas are often used in one of many possible array configurations. Microstrip arrays are very popular for fixed-beam applications because the radiating elements and feed network can be fabricated on a single-layer printed circuit board using low-cost lithographic techniques. Interelement spacings for fixed-beam applications are usually chosen to be less than free-space wavelength (λ) to avoid grating lobes and greater than $\lambda/2$ to provide sufficient room for the feed lines, to achieve higher gain for a given number of elements, and to reduce mutual coupling. The active element patterns shown in Fig. 3-30 are very similar, indicating that mutual coupling effects are not significant for those microstrip patches that are spaced 0.57λ apart.

Largely as the result of pioneering work for sophisticated military radar, phase-scanned microstrip arrays can be produced with monolithic microwave integrated circuit (MMIC) techniques that fabricate amplifiers, phase shifters, and other devices on the same substrate. As with any array, the array geometry can be linear, planar, or conformal and the feed system can be parallel, series, or hybrid. Parallel-fed, planar microstrip arrays are examined here.

We illustrate microstrip array design using a broadside-beam planar array of $N = 4^n$ elements, where n is a positive integer. The goal is to achieve maximum gain, so uniform amplitude and phase to each element are required. This design problem illustrates impedance- and phase-matching techniques using a microstrip feed network. Figure 5-57*a* shows the fundamental 2×2 subarray unit that can be used to build up very large arrays. Four patches are fed at their edges by microstrip lines that are of equal length from the subarray center (point C) to preserve equal excitation phase. The operating frequency is 10 GHz ($\lambda = 3$ cm) and the substrate has $\varepsilon_r = 2.2$, so $\lambda_d = \lambda/\sqrt{\varepsilon_r} = 2.02$ cm. The element spacing in both principal directions is $d = 0.8\lambda = 2.4$ cm. The length of each half-wave patch from (5-72) is $L = 0.49\lambda_d = 0.99$ cm. The desired input impedance is 200 Ω, so (5-75) can be used to solve for the patch width:

$$W = \sqrt{\frac{90 \dfrac{\varepsilon_r^2}{\varepsilon_r - 1}}{Z_A}}\, L = \sqrt{\frac{363}{200}} \cdot 0.99 = 1.33 \text{ cm} \tag{5-79}$$

Impedance matching is accomplished by connecting a 200-Ω characteristic impedance microstrip line to a patch. The parallel feed network shown in Fig. 5-57*a* uses two divide-by-two operations to reach each element, yielding the desired 50-Ω input impedance at point C. For example, for the upper two elements, since $Z_A = 200$ Ω is matched to the lines with $Z_{o1} = 200$ Ω, the left and right lines present impedances of 200 Ω at point B. Their parallel combination yields 100 Ω, which is matched to the line $Z_{o2} = 100$ Ω. This line impedance at point C is still 100 Ω. Combining this in parallel with the lower two elements gives 50 Ω. A probe could be connected from the ground plane backing the subarray at point C. Or, as shown in Fig. 5-57*b*, another microstrip line can be used to connect to other similar subarrays to build up a large array. A quarter-wave transformer of $Z_{o3} = \sqrt{Z_C Z_{o4}} = \sqrt{50 \cdot 100} = 70.7$ Ω is used for impedance matching; this, of course, introduces bandwidth limitations.

Arrays similar to the one shown in Fig. 5-57 have been constructed and measured for $N = 16$, 64, 256, and 1024 elements [39]. The gain of the array with spacings of $d = 0.8\lambda$ is easily computed using

$$\begin{aligned} G &= \varepsilon_{ap} \frac{4\pi}{\lambda^2} A_p = \varepsilon_{ap} \frac{4\pi}{\lambda^2} L_x L_y = \varepsilon_{ap} \frac{4\pi}{\lambda^2} (\sqrt{N}\, d)(\sqrt{N}\, d) = \varepsilon_{ap} \frac{4\pi}{\lambda^2} N d^2 \\ &= \varepsilon_{ap} \frac{4\pi}{\lambda^2} N(0.8\lambda)^2 = 8.04 N \varepsilon_{ap} \end{aligned} \tag{5-80}$$

Since the array is uniformly excited, if there were no losses, ε_{ap} would be close to 100%. However, there are losses due to radiation from the transmission lines, surface waves, and dissipation in the lines that reduce the aperture efficiency. Measured gain was close to that computed using (5-80) with $\varepsilon_{ap} = 0.5$ for arrays up to 1024 elements [39]. This is competitive with a conventional aperture antenna such as a reflector of the same area. However, arrays with many more than 1024 elements experience significant dissipative loss in the feed network, resulting in efficiencies less than 50%.

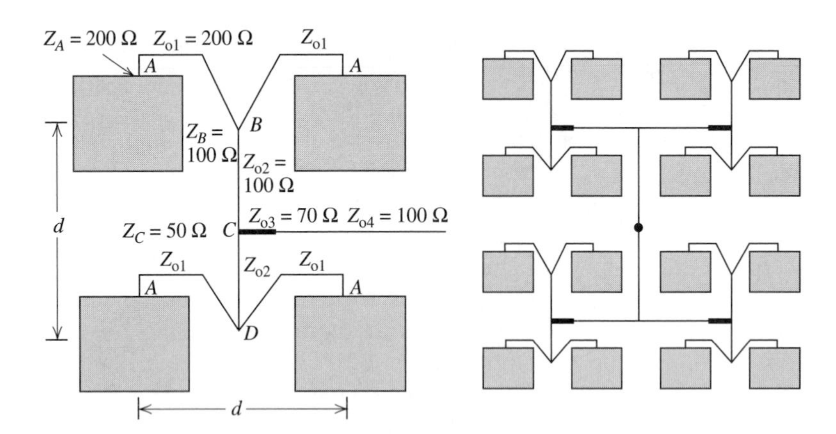

(*a*) The four-element subarray. (*b*) A 16-element array formed from subarrays.

Figure 5-57 A planar microstrip array with a feed network that produces equal amplitude and phase element excitations.

REFERENCES

1. R. W. P. King, "The Linear Antenna–Eighty Years of Progress," *Proc. IEEE*, Vol. 55, pp. 2–16, Jan. 1967.
2. R. W. P. King, *The Theory of Linear Antennas*, Harvard University Press, Cambridge, MA, 1956.
3. R. W. P. King, *Tables of Antenna Characteristics*, Plenum, New York, 1961.
4. J. D. Kraus, *Antennas*, 2nd ed., McGraw-Hill, New York, 1988.
5. R. W. P. King, H. R. Mimno, and H. H. Wing, *Transmission Lines, Antennas, and Waveguides*, Dover, New York, 1965, p. 86.
6. M. Abramowitz and I. A. Stegun, *Handbook of Mathematical Functions*, NBS Applied Math. Series 55, U.S. Government Printing Office, Washington, DC, 1964, Chap. 5.
7. E. Jordan and K. Balmain, *Electromagnetic Waves and Radiating Systems*, 2nd ed., Prentice-Hall, Englewood Cliffs, NJ, 1968, Chap. 14.
8. R. F. Schwartz, "Input Impedance of a Dipole or Monopole," *Microwave J.*, Vol. 15, p. 22, Dec. 1972.
9. "IEEE Standard Definitions of Terms for Antennas," *IEEE Trans. Antennas & Propagation*, Vol. AP-22, Jan. 1974.
10. Y. T. Lo and S. W. Lee, Eds., *Antenna Handbook*, Van Nostrand Reinhold, New York, pp. 27–21, 1988.
11. H. Jasik, Ed., *Antenna Engineering Handbook*, McGraw-Hill, New York, 1961, Sec. 3.3.
12. E. Wolff, *Antenna Analysis*, Wiley, New York, 1967, p. 67.
13. E. Jordan and K. Balmain, *Electromagnetic Waves and Radiating Systems*, 2nd ed., Prentice-Hall, Englewood Cliffs, NJ, 1968, p. 402.
14. R. C. Johnson, Ed., *Antenna Engineering Handbook*, McGraw Hill, New York, 1993, Chap. 43.
15. W. N. Caron, *Antenna Impedance Matching*, American Radio Relay League, Newington, CT, 1989.
16. G. A. Thiele, "Analysis of Yagi–Uda Type Antennas," *IEEE Trans. Antennas & Propagation*, Vol. AP-17, pp. 24–31, Jan. 1969.
17. H. Yagi, "Beam Transmission of Ultra-short Waves," *Proc. IRE*, Vol. 16, p. 715, 1928.
18. S. Uda and Y. Mushiake, *Yagi–Uda Antenna*, Saski Printing and Publishing Company, Ltd., Sendai, Japan, 1954.
19. H. W. Ehrenspeck and H. Poehler, "A New Method for Obtaining Maximum Gain for Yagi Antennas," *IRE Trans. Antennas & Propagation*, Vol. AP-7, pp. 379–386, Oct. 1959.
20. H. E. Green, "Design Data for Short and Medium Length Yagi–Uda Arrays," *Institution of Engineers (Australia), Elec. Eng. Trans.*, pp. 1–8, March 1966.
21. P. Viezbicke, "Yagi Antenna Design," NBS Technical Note 688, U.S. Government Printing Office, Washington, DC, Dec. 1976.
22. J. L. Lawson, *Yagi Antenna Design*, American Radio Relay League, Newington, CT, 1986.

23. J. H. Reisert, "How to Design Yagi Antennas," *Ham Radio Magazine*, pp. 22–31, Aug. 1977.
24. D. Proctor, "Graphs Simplify Corner Reflector Antenna Design," *Microwaves Magazine*, pp. 48–52, July 1975.
25. E. C. Jordan and K. G. Balmain, *Electromagnetic Waves and Radiating Systems*, 2nd ed., Prentice-Hall, Englewood Cliffs, NJ, 1968, Chap. 16.
26. E. K. Miller et al., "Analysis of Wire Antennas in the Presence of a Conducting Half-Space. Part I. The Vertical Antenna in Free Space," *Canadian J. Phys.*, Vol. 50, pp. 879–888, 1972.
27. E. K. Miller et al., "Analysis of Wire Antennas in the Presence of a Conducting Half-Space. Part II. The Horizontal Antenna in Free Space," *Canadian J. Phys.*, Vol. 50, pp. 2614–2627, 1972.
28. B. W. Griffith, *Radio-Electronic Transmission Fundamentals*, McGraw-Hill, New York, 1962, Chap. 43.
29. W. L. Weeks, *Antenna Engineering*, McGraw-Hill, New York, 1968, Sec. 2.6.
30. D. H. Werner, "An Exact Integration Procedure for Vector Potentials of Thin Circular Loop Antennas," *IEEE Trans. Antennas & Propagation*, Vol. 44, pp. 157–165, Feb. 1996.
31. J. D. Mahony, "Approximations to the Radiation Resistance and Directivity of Circular-Loop Antennas," *IEEE Antennas & Propagation Magazine*, Vol. 36, pp. 52–55, Aug. 1994.
32. R. A. Burberry, *VHF and UHF Antennas*, Peter Peregrinus Ltd., London, 1992, Chap. 4.
33. K. R. Carver and J. W. Mink, "Microstrip Antenna Technology," *IEEE Trans. Antennas & Propagation*, Vol. AP-29, pp. 2–24, Jan. 1981.
34. R. E. Munson, "Conformal Microstrip Antennas and Microstrip Phased Arrays," *IEEE Trans. Antennas & Propagation*, Vol. AP-22, pp. 74–78, Jan. 1974.
35. D. H. Schaubert, "Microstrip Antennas," *Electromagnetics*, Vol. 12, pp. 381–401, July–December 1992.
36. D. R. Jackson and N. G. Alexopoulos, "Simple Approximate Formulas for Input Resistance, Bandwidth, and Efficiency of a Resonant Rectangular Patch," *IEEE Trans. Antennas & Propagation*, Vol. 3, pp. 407–410, March 1991.
37. D. R. Jackson, S. A. Long, J. T. Williams, and V. B. Davis, "Computer-Aided Design of Rectangular Microstrip Antennas," Ch. 5 in *Advances in Microstrip and Printed Antennas*, edited by K.-F. Lee, Wiley, New York, 1997.
38. P. Bhartia, K. V. S. Rao, and R. S. Tomar, *Millimeter-Wave Microstrip and Printed Circuit Antennas*, Artech House, Norwood, MA, 1991.
39. E. Levine, G. Malanurd, S. Shtrikman, and D. Treves, "A Study of Microstrip Array Antennas with the Feed Network," *IEEE Trans. Antennas & Propagation*, Vol. 37, pp. 426–433, April 1989.

PROBLEMS

5.1-1 Use the integral from (F-11) in (5-3) to prove (5-4).

5.1-2 Starting with (5-6), show that for $L \ll \lambda$, the radiation pattern of a dipole reduces to that of a short dipole, sin θ.

5.1-3 a. The outputs from two collinear, closely spaced, half-wave dipoles are added together as indicated by a summing device in the figure below. The transmission lines from the antennas to the summer are of equal length. Write the pattern $F_a(\theta)$ of this antenna system using array techniques.

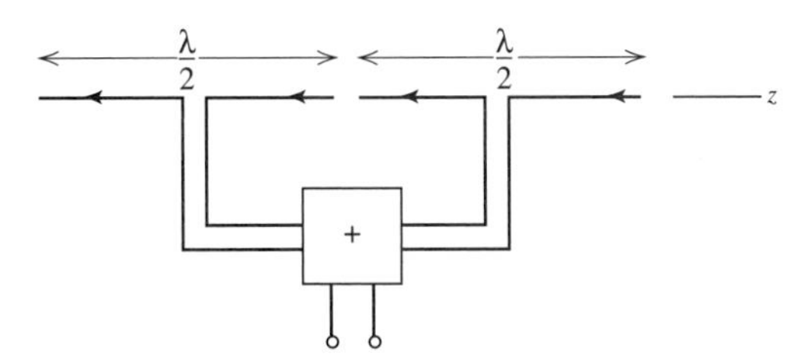

b. Now consider a center-fed, full-wave dipole that is along the z-axis. Write its pattern expression $F_b(\theta)$.

c. Now draw the current distributions $I_a(z)$ and $I_b(z)$ for both antennas. From these current distributions, can you make a statement about the patterns from the two antennas? Return to the pattern expressions and prove your statement mathematically.

5.1-4 The center-fed, full-wave dipole is rarely used because it has a current minimum at the feed point. If it is instead fed as shown below, sketch the current distribution. Also, rough sketch how you think the pattern should look, and explain how you obtained it.

$\leftarrow \frac{\lambda}{4} \rightarrow \leftarrow \frac{3\lambda}{4} \rightarrow$ — z

5.1-5 a. The array of Prob. 5.1-3(a) is parallel fed, in-phase array. Show how the parallel wire transmission lines are connected to perform the summing function. Also put current arrows on each wire.

b. Consider an array similar to that of part (a) except now the array elements (half-wave dipoles) are fed 180° out-of-phase. Show how the transmission lines are arranged to accomplish this subtraction function. Again, show the currents on each wire.

5.1-6 Use the results of the cosine-tapered current distribution in Sec. 4.2 to derive the pattern of a half-wave dipole in (5-7).

5.1-7 Verify that the normalization constant in (5-9) is 0.7148 for the pattern of a $3\lambda/2$ dipole. What are the angles θ_o in degrees for maximum radiation?

5.1-8 A resonant half-wave dipole is to be made for receiving TV Channel 7 of frequency 177 MHz. If $\frac{1}{2}$-in.-diameter tubular aluminum is used, how long (in centimeters) should the antenna be?

5.1-9 A four-element collinear array of half-wavelength spaced, half-wave dipoles is placed along the z-axis. All elements are fed with equal amplitude and phase.

a. Determine the complete radiation pattern $F(\theta)$ for the array.

b. Plot a sketch of the pattern in linear, polar form using array concepts.

5.1-10 Repeat Prob. 5.1-9 except now the half-wave dipole elements are parallel to each other and the x-axis, and are phased for ordinary endfire. The centers of the four elements are located on the z-axis and spaced a half-wavelength apart.

5.1-11 Use (5-1) and (5-4) to calculate and plot the current distribution and far-field pattern for dipoles of length 2.0 and 2.5λ. Compare with Figs. 5-3 and 5-4.

5.1-12 a. Show that the power radiated by a center-fed dipole of arbitrary length L with a sinusoidal current is

$$P = \frac{\eta I_m^2}{4\pi}\left\{0.5772 + \ln(\beta L) - \text{Ci}(\beta L) + \frac{1}{2}\sin(\beta L)[\text{Si}(2\beta L) - 2\,\text{Si}(\beta L)]\right.$$
$$\left. + \frac{1}{2}\cos(\beta L)\left[0.5772 + \ln\left(\frac{\beta L}{2}\right) + \text{Ci}(2\beta L) - 2\,\text{Ci}(\beta L)\right]\right\}$$

b. Derive an expression for the directivity and then plot directivity as a function of dipole length for L from 0 to 3λ.

5.1-13 Use the length reduction procedure for half-wave resonance in Table 5-2 to calculate the resonant frequencies of the two dipoles in Fig. 5-7.

5.1-14 Design an optimum directivity vee dipole to have a directivity of 6 dB.

5.1-15 To show that the vee dipole results of (5-23) and (5-24) give roughly the correct results for a full-wave straight wire dipole, use $D = 2.41$ and determine γ.

5.2-1 a. It is desired to have a simple formula for the length of a thin-wire half-wave folded dipole antenna. Show that it is $L(\text{cm}) = 14{,}250/f(\text{MHz})$.

b. Determine the length in centimeters of half-wave folded dipoles for practical application as receiving antennas for each VHF TV channel and the FM broadcast band (100 MHz). Tabulate results.

5.2-2 Calculate the input impedance of a folded dipole of length $L = 0.4\lambda$, wire size $2a = 0.001\lambda$, and wire spacing $d = 12.5a$ using the transmission line model. Compare your results to values from Fig. 5-16.

5.3-1 A *receiving* antenna with a real impedance R_L attached to its terminals has the equivalent circuit shown. Prove that maximum power transfer to the load for a fixed real antenna impedance R_A occurs for $R_L = R_A$.

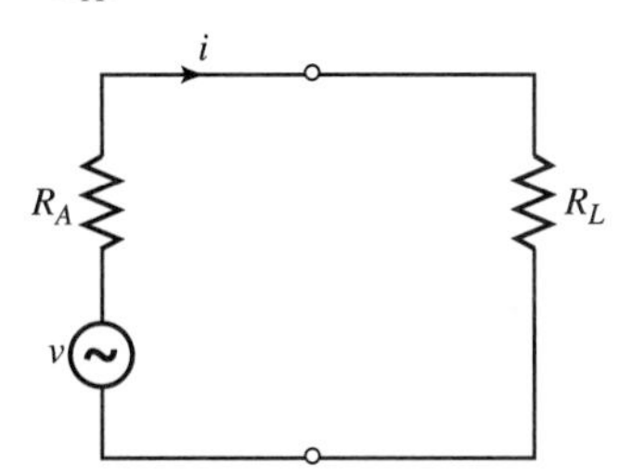

5.3-2 A transmitter with a real impedance of R_t is connected to a lossless transmission line of real characteristic impedance R_o and then to an antenna of real input impedance R_o.

a. Derive an expression for the transmit efficiency, that is, power delivered to antenna/total power dissipated. Neglect any mismatch effects.

b. Find the percent efficiency for $R_t = R_o$, $R_t = 0.5R_o$, and $R_t = 0.1R_o$.

5.3-3 The antenna shown is operated over a perfect ground plane. Its purpose is to enhance radiation in the xy-plane over that of a single quarter-wave monopole.

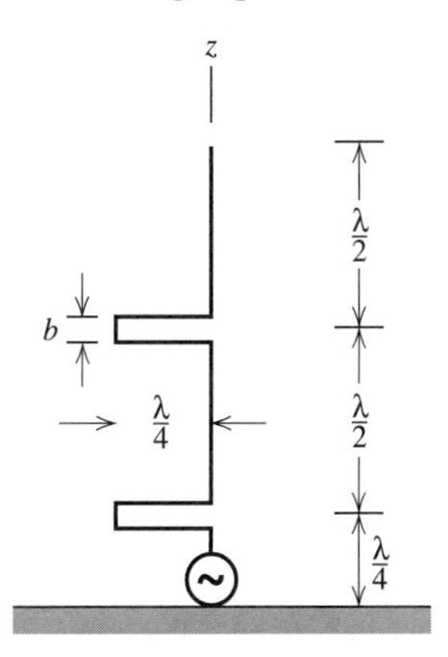

a. Determine and sketch the current distribution. Assume $b \ll \lambda$.

b. What is the purpose of the quarter-wavelength "stubs"?

c. Use array concepts to obtain a linear polar plot of the radiation pattern in a plane containing the z-axis.

d. Give a rough estimate of the input impedance for matching purposes.

5.4-1 Use array theory to analyze the array of Fig. 5-30*a* where the pattern of each element is that of a half-wave dipole.

a. Plot the H-plane pattern and compare to Fig. 5-30*b*.

b. Plot the E-plane pattern.

5.4-2 Numerical methods reveal that the currents on the elements of the two-element parasitic array of Fig. 5-30*a* are nearly sinusoidal and the current amplitudes and phases at each element center are $1.0\angle -88°$ for the driver and $0.994\angle 81.1°$ for the parasite. Use simple array theory to obtain and plot the H-plane pattern in linear, polar form.

5.4-3 Phasor diagrams are often helpful in obtaining a rough idea about how arrays perform. To illustrate, use phasor diagrams to obtain the relative far-zone field values in the endfire directions of the two-element parasitic array of Prob. 5.4-2 (i.e., find the front-to-back ratio). To do this, find the total phasor at each element location including the spatial phase delay due to the element separation. Assume the amplitudes of each element are unity and the phases are $-88°$ for the driver and $81.1°$ for the parasite.

5.4-4 Design a 10.2-dBd gain Yagi for operation at 50.1 MHz. The parasitic elements are insulated from the metal boom. The diameter of the elements is 0.0021λ.

5.4-5 Design a 14.2-dBd gain Yagi for operation at 432 MHz. The parasitic elements are mounted through a metal boom making electrical contact with it. The diameter of the elements is 0.00343λ. The boom diameter is 0.0275λ.

5.4-6 *Construction project—a 10¢ Yagi.* This project is designed to demonstrate how a high gain antenna can be built for under 10¢! Locate a channel on your (or a cooperating friend's) TV receiver that has marginal reception, such as a snowy picture when a modest antenna (like rabbit ears) is used. If it happens to be Channel 13, you can use the design values from Prob. 5.4-5. If not, repeat the calculations for the channel you have chosen. The construction phase proceeds as follows. Find a large rigid piece of corrugated cardboard and trim it so that it is several centimeters longer than the total array length and about 5 cm narrower than the director length. Now locate several thick coat hangers. Straighten them as much as possible and cut them to the lengths required for the reflector and directors. The feed element is a folded dipole constructed from a piece of twin-lead transmission line. Cut it to a length that is a little longer than the driver dimension. Strip the ends and solder the two wires at each end together such that the overall length is equal to the driver dimension. Next cut one wire of the driver at the center of the folded dipole and solder the ends to a long piece of twin-lead that serves as a transmission line for the antenna. Lay out all element positions on the cardboard by marking appropriately. Tape the folded dipole onto the cardboard at the driver location. The coat hanger parasitic elements are positioned by merely inserting them into the corrugations in the cardboard. Now connect the transmission line to the receiver. Rotate the antenna and note the effect on the reception. Large performance differences should be observed. Note that it may be necessary to elevate the antenna by placing it in the attic, for example. With this construction, it is very easy to change the element spacings by placing the coat hanger elements into difference corrugations. Very little difference will be observed for small distance changes. Normally, the best performance is achieved for horizontal polarization, that is, elements parallel to the ground.

5.4-7 *Construction project—a slightly more expensive Yagi.* A fairly rugged Yagi antenna can be constructed using the following technique. Select a TV channel with marginal reception and design a Yagi for that frequency. The materials required for this project are a 1×2 in. board of length slightly greater than the overall length of the array and a few meters of aluminum wire (usually No. 8 AWG). Trim and straighten wires for the reflector and directors. Drill holes in the wooden mast at the appropriate positions for the reflector and the directors. The holes should be just slightly greater than the wire diameter. Be sure all holes are along a straight line. The driver is a folded dipole oriented such that the plane of the dipole is perpendicular to the line of the array. Drill one hole in the mast about 2 cm above the array line. At the same distance below the array line, drill in from each side of the mast about 0.5 cm. Cut a piece of wire more than twice the length of the driver. Push it through the top hole and center it. Bend the wire at the required length at each end and fold it back to the mast. Now carefully trim away any excess wire such that the wire ends can just be forced into the shallow holes and still form a symmetric folded dipole. Now wrap the bared ends of a twin-lead transmission line to the ends of the folded dipole close to the mast (at the feed point). Be sure to get a good mechanical contact. Also leave a tab of polyethylene where you stripped the twin lead. Small wire brads can be wedged between the wire ends at the feed point and the mast, and at the same time pinch the twin-lead connection between the antenna wire and the brad. Solder the feed point connections. Tack the polyethylene tab to the bottom of the mast to provide strain relief. Insert the remaining elements into their holes, center them, and nail brads into the hole alongside the wires to secure their positions. The construction is now complete and you can connect the transmission line to the receiver and test the reception. Try several antenna locations and orientations.

5.4-8 A two-element Yagi has a current on the driven element of $1\angle 164°$ and a current on the parasitic element of $0.5\angle 238°$. The spacing between the elements is 0.2λ. Does the parasitic element act like a director or reflector element? Use a phasor diagram to show why.

5.4-9 The input impedance of a driven element in the presence of a short-circuited parasitic element is $40\angle 45°$ and is $60\angle 30°$ when the terminals of the parasitic element are open-circuited. What is the approximate self-impedance of the driven element? Why is your answer only approximate?

5.4-10 A two-element Yagi with 0.2λ element spacing has a current on the driven element of $1\angle 254°$ and a current on the parasitic element of $0.6\angle -32°$. Does the parasitic element act like a director or reflector element? Why?

5.5-1 a. Calculate and plot the magnitude of the array factor in (5-48) and verify that the maximum value of $|AF(\theta = 90°, \phi = 0°)|$ is obtained when $s = 0.5\lambda$.

b. Reason that this must be the case, using Fig. 5-43.

5.6-1 A resonant, half-wave, thin, vertical dipole is operated a half-wavelength above a perfect ground plane. Calculate the input impedance. Use the results in Sec. 3.6.

5.6-2 Derive an expression for the directivity of an ideal (infinitesimal) dipole as a function of its height h above a perfect ground plane. The dipole is oriented perpendicular to the ground plane. Make use of the results in Secs. 2.3 and 3.4.

5.6-3 A short dipole is a quarter-wavelength above a perfect ground plane. Use simple array theory for the dipole and its image to obtain polar plot sketches of the E- and H-plane patterns when the dipole is oriented (a) vertically and (b) horizontally.

5.6-4 Repeat Prob. 5.6-3 for a short dipole a half-wavelength above a perfect ground plane.

5.6-5 A horizontal short dipole is a quarter-wavelength above a planar real earth and is operating at 1 MHz. The conductivity of the earth is $\sigma = 12 \times 10^{-3}$ S/m and the relative dielectric constant is $\varepsilon_r = 15$. For this frequency, σ, and ε_r, we can approximate $|\Gamma_H|$ by 0.9 and the phase of Γ_H by $-190°$ for all θ.

a. Calculate and plot the H-plane elevation pattern in polar form in the upper half-space.

b. Compare the pattern with that of the short dipole over a perfectly conducting ground plane (i.e., the results of Prob. 5.6-3*b*).

5.6-6 A quarter-wave resonant monopole is to be used as a transmitting antenna at 1 MHz. A radial system of 120 radials is to be used. If 97% efficiency is to be achieved, how long must the radial wires be? Neglect any tower ohmic resistance.

5.7-1 *The uniform circular loop antenna.* A circular loop in the xy-plane with its center at the origin and a radius b carries a uniform amplitude, uniform phase current given by

$$\mathbf{I} = I_o \hat{\boldsymbol{\phi}}'$$

a. Due to symmetry, the pattern will not be a function of ϕ and $\mathbf{A}$ will have only a ϕ-component. Using these facts, show that

$$\mathbf{A} = \hat{\boldsymbol{\phi}}\, \mu \frac{e^{-j\beta r}}{4\pi r} I_o b \int_0^{2\pi} \cos\phi' e^{j\beta b \sin\theta \cos\phi'}\, d\phi'$$

in the far field. Use symmetry to reason that $\phi = 0$ can be assumed and only a ϕ-component exists.

b. Find an expression for E_ϕ. *Hint:* Use (F-7).

c. Show that this result reduces to that for a small loop antenna in (2-53). *Hint:* $J_1(x) \approx x/2$ for $x \gg 1$.

5.7-2 Show that (5-62) yields (5-63). To perform the integrations, decompose the functions $\cos(\beta x')$ and $\sin(\beta y')$ into sums of exponential functions using (E-6) and (E-7).

5.7-3 Compute the input reactance of a square loop antenna with a 0.2λ perimeter using small loop analysis and compare to the value from Fig. 5-53*b*.

5.7-4 This problem compares circular and square large loop antennas.

a. For the geometry of Fig. 5-51, use a moment method code (see Chap. 10) to evaluate directivity and input impedance for $L = 1, 1.5$, and 2λ.

b. Run the code to determine the resonant value of L that is close to 1λ. Give the directivity and input impedance, and plot the principal plane patterns.

c. Repeat (a) and (b) for a circular loop with the same perimeter-length values and oriented in a similar fashion with respect to the coordinate axes. Use the same wire radius.

d. Tabulate performance values for all L values for both antennas. Comment on similarities.

5.7-5 In a communication link, a half-wavelength folded dipole parallel to the z-axis is used to transmit and a square, one-wavelength loop is used to receive. Sketch a perspective view of the antennas in a common coordinate system for maximum power transfer; include the orientations of the antennas and their feed points.

5.7-6 *UHF TV antenna.* Use a moment method code (see Chap. 10 and Appendix G) to evaluate a popular UHF TV antenna that is a circular loop with an 18-cm diameter and is made of 2-mm-diameter aluminum wire. Give the following performance measures in the center of TV channel 37: input impedance, gain, and radiation patterns.

5.8-1 Show how the normalized principal plane patterns of (5-74) follow from (5-73). Discuss the electric field polarization in each plane.

5.8-2 A square microstrip patch with $L = W = 4.02$ cm is printed on an 0.159-cm-thick substrate with $\varepsilon_r = 2.55$. Find the resonant frequency, input impedance at resonance for an edge feed, and bandwidth.

5.8-3 A probe-fed square microstrip patch antenna on a 0.3-cm-thick substrate of $\varepsilon_r = 4.53$ is to be designed to operate at 3.72 GHz. Find the patch length at resonance. Find the probe location for a 50-Ω input impedance.

Chapter 6

Broadband Antennas

In many applications, an antenna must operate effectively over a wide range of frequencies. An antenna with wide bandwidth is referred to as a **broadband antenna**. The term "broadband" is a relative measure of bandwidth and varies with the circumstances. We shall be specific in our definition of broadband. Bandwidth is computed in one of two ways. Let f_U and f_L be the upper and lower frequencies of operation for which satisfactory performance is obtained. The center (or sometimes the design frequency) is denoted as f_C. Then bandwidth as a percent of the center frequency B_p is

$$B_p = \frac{f_U - f_L}{f_C} \times 100\% \tag{6-1}$$

Bandwidth is also defined as a ratio B_r by

$$B_r = \frac{f_U}{f_L} \tag{6-2}$$

The bandwidth of narrow band antennas is usually expressed as a percent using (6-1), whereas wideband antennas are quoted as a ratio using (6-2).

In the previous chapter, we saw that resonant antennas have small bandwidths. For example, the half-wave dipoles in Fig. 5-7 have bandwidths of 8 and 16% (f_U and f_L were determined by the VSWR = 2.0 points). On the other hand, antennas that have traveling waves on them (Sec. 6.1) rather than standing waves (as in resonant antennas) operate over wider frequency ranges. The definition of a broadband antenna is somewhat arbitrary and depends on the particular antenna, but we shall adopt a working definition. *If the impedance and the pattern of an antenna do not change significantly over about an octave* ($f_U/f_L = 2$) *or more, we will classify it as a broadband antenna.*

As we will see in this chapter, broadband antennas usually require structures that do not emphasize abrupt changes in the physical dimensions involved, but instead utilize materials with smooth boundaries. Smooth physical structures tend to produce patterns and input impedances that also change smoothly with frequency. This simple concept is very prominent in broadband antennas.

6.1 TRAVELING-WAVE WIRE ANTENNAS

The wire antennas we have discussed thus far have been resonant structures. The wave traveling outward from the feed point to the end of the wire is reflected, setting up a standing-wave-type current distribution. This can be seen by examining the expression for the current in (5-1) for the top half of the dipole that can be written as

$$I_m \sin\left[\beta\left(\frac{L}{2} - z\right)\right] = \frac{I_m}{2j} e^{j(\beta L/2)}(e^{-j\beta z} - e^{j\beta z}) \tag{6-3}$$

The first term in brackets can be taken to represent an outward traveling wave and the second term a reflected wave. The minus sign is the current reflection coefficient at an open circuit.

If the reflected wave is not strongly present on an antenna, it is referred to as a **traveling-wave antenna**. A traveling-wave antenna acts as a guiding structure for traveling waves, whereas a resonant antenna supports standing waves. Traveling waves can be created by using matched loads at the ends to prevent reflections. Also, very long antennas may dissipate most of the power, leading to small reflected waves by virtue of the fact that very little power is incident on the ends. In this section, several wire forms of traveling-wave antennas will be discussed. Some of the antennas in this section are essentially the traveling-wave counterparts of resonant wire antennas presented in Chap. 5. They tend to be broadband with bandwidths of as much as 2:1.

The simplest traveling-wave wire antenna is a straight wire carrying a pure traveling wave, referred to as the **traveling-wave long wire antenna**. A long wire is one that is greater than one-half wavelength long. The traveling-wave long wire is shown in Fig. 6-1 with a matched load R_L to prevent reflections from the wire end. Exact analysis of this structure, as well as others to be presented in this section, is formidable. We shall make several simplifying assumptions that permit pattern calculations that do not differ greatly from real performance. First, the ground plane effects will be ignored and we will assume that the antenna operates in free space. A traveling-wave long wire operated in the presence of an imperfect ground plane is called a **Beverage antenna**, or **wave antenna**. The ground plane may be accounted for in certain cases by using the techniques of the previous chapter. Second, the details of the feed are assumed to be unimportant. In Fig. 6-1, the long wire is shown being fed from a coaxial transmission line as one practical method. The vertical section of length d is assumed not to radiate, which is approximately true for $d \ll L$. Finally, we assume that the radiative and ohmic losses along the wire are small. When attenuation is neglected, the current amplitude is constant and the phase velocity is that of free space [1]. We can then write

$$I_t(z) = I_m e^{-j\beta z} \tag{6-4}$$

which represents an unattenuated traveling wave propagating in the $+z$-direction with the phase constant β of free space.

The current of (6-4) is that of a uniform line source with a linear phase constant

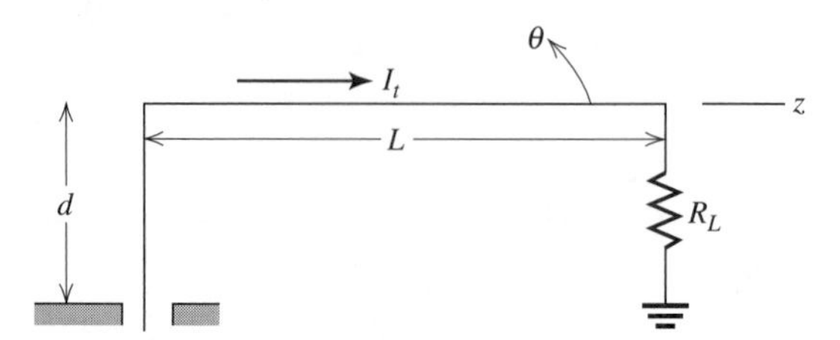

Figure 6-1 Traveling-wave long wire antenna.

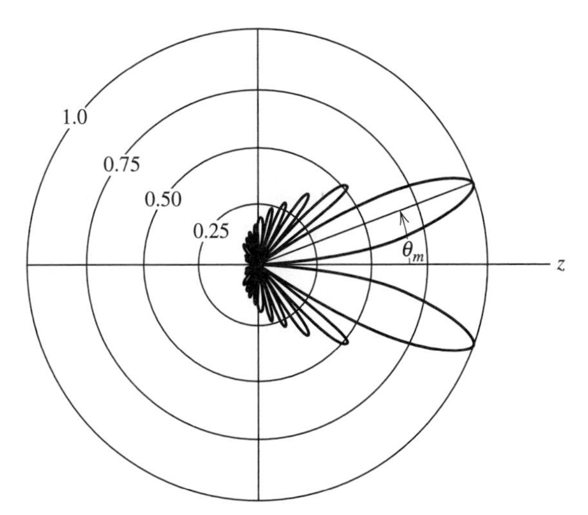

Figure 6-2 Pattern of a traveling-wave long wire antenna. $L = 6\lambda$ and $\theta_m = 20°$.

of $\beta_o = -\beta$. From (4-6), $\beta_o = -\beta \cos \theta_o$, so the pattern factor maximum radiation angle (not including element factor effects) is $\theta_o = 0°$, which implies an endfire pattern. The complete radiation pattern from (4-8) is

$$F(\theta) = K \sin \theta \frac{\sin[(\beta L/2)(1 - \cos \theta)]}{(\beta L/2)(1 - \cos \theta)} \tag{6-5}$$

where K is a normalization constant that depends on the length L. The polar pattern for $L = 6\lambda$ is shown in Fig. 6-2. The length $L = n\lambda$ results in n forward lobes in the angular range $0 < \theta_m < 90°$. In this example, $n = 6$. The element factor $\sin \theta$ forces a null in the endfire direction. Hence, instead of having a single endfire lobe (which the pattern factor produces), the "main beam" is a rotationally symmetric cone about the z-axis. The maximum radiation angle in this case is $\theta_m(L = 6\lambda) = 20.1°$. In general, it is a function of L. Solving (6-5) for θ_m as a function L produces the plot of Fig. 6-3. An approximate expression for the angle of maximum radiation is [2]

$$\theta_m = \cos^{-1}\left(1 - \frac{0.371}{L/\lambda}\right) \tag{6-6}$$

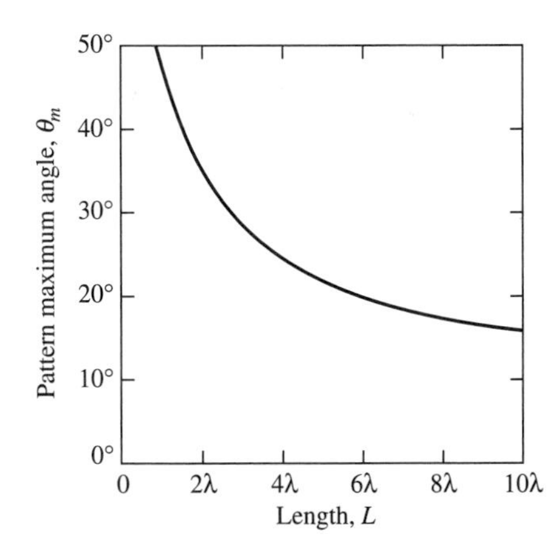

Figure 6-3 Pattern maximum angle for a traveling-wave long wire antenna of length L operating in free space. See (6-5).

The beam direction values from Fig. 6-3 or (6-6) for a traveling-wave long wire of length L may be used to calculate an approximate beam direction for a standing-wave straight wire antenna (i.e., dipole). For example, θ_m for $L = 3\lambda/2$ from Fig. 6-3 is 40° and θ_m for the dipole of Fig. 5-4*d* is 42.6°. As L increases, the traveling-wave and standing-wave antenna main beam maximum angles approach each other [3]. The standing-wave wire antenna is distinguished from its traveling-wave counterpart by the presence of a second major lobe in the reverse direction; see Fig. 5-4*d*. This can be seen by noting that the traveling-wave current of (6-4) corresponds to the first term of the standing-wave current of (6-3). The second term of (6-3), which is the reflected wave, produces a pattern similar in shape but oppositely directed. Thus, a traveling-wave antenna has a beam with a maximum in the $\theta = \theta_m$ direction and a standing-wave antenna of the same length has an additional beam in the $\theta = 180° - \theta_m$ direction.

The input impedance of a traveling-wave antenna is always predominantly real. This can be understood by recalling that the impedance of a pure traveling wave on a low-loss transmission line is equal to the (real) characteristic impedance of the transmission line. Antennas that support traveling waves operate in a similar manner. The radiation resistance of a traveling-wave long wire antenna is 200 to 300 Ω (see Prob. 6.1-5). The termination resistance should equal the value of the radiation resistance.

The resonant vee antenna discussed in Sec. 5.1.2 can be made into a traveling-wave antenna by terminating the wire ends with matched loads. The **traveling-wave vee antenna** is shown in Fig. 6-4. The pattern due to each arm separately is expressed by (6-3), an example of which is shown in Fig. 6-2. From Fig. 6-4, it is seen that when $\alpha \approx \theta_m$, the beam maxima from each arm of the vee will line up in the forward direction. A more accurate analysis of vee (see Prob. 6.1-8) includes the spatial separation effects of the arms. Pattern calculations as a function of α reveal that a good vee pattern is obtained when

$$\alpha \approx 0.8\theta_m \tag{6-7}$$

where θ_m is found from Fig. 6-3 or (6-4). For $L = 6\lambda$, $\theta_m = 20°$ from Fig. 6-3 and (6-5) yields $\alpha \approx 16°$; the pattern for a vee with this geometry is shown in Fig. 6-4. The large side lobes arise from portions of the beams from each half of the vee that

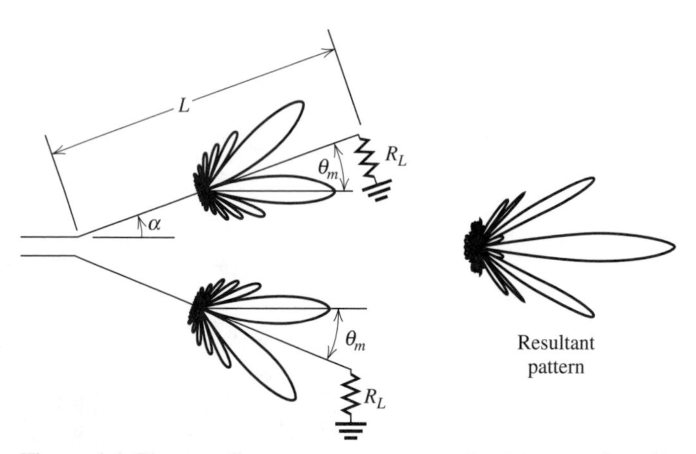

Figure 6-4 The traveling-wave vee antenna. In this case, $L = 6\lambda$ and $\alpha = 0.8\theta_m = 16°$.

do not line up along axis. The pattern of the vee out of the plane of the vee is rather complicated due to the merging of the conical beams for each half of the vee. The traveling-wave vee provides relatively high gain for a bent wire structure.

By extending the ideas of a traveling-wave vee antenna, we obtain a **rhombic antenna** as shown in Fig. 6-5. The operation of this antenna is visualized most easily by viewing it as a transmission line that has been spread apart and consequently the characteristic impedance is increased. The load resistor R_L is of such a value as to match the transmission line. The antenna carries outward traveling waves that are absorbed in the matched load. Since the separation between the lines is large relative to a wavelength, the structure will radiate. If designed properly, a directive pattern with a single beam in the z-direction can be obtained.

A rhombic antenna operating in free space can be modeled as two traveling-wave vee antennas put together. If we choose $\alpha = 0.8\theta_m$ as for the vee, the beams of the rhombic in Fig. 6-5 numbered 2, 3, 5, and 8 will be aligned. Again, θ_m follows from Fig. 6-3. Due to the spatial separation of the two vees, the rhombic pattern will not be the same as that of a single vee [3]. (See Prob. 6.1-9.)

The effects of a rhombic operating above a real earth ground can be included by the techniques of the previous section. For a rhombic that is oriented horizontally, the reflection coefficient Γ_H is approximately -1 and the real earth may be modeled as a perfect conductor; Fig. 5-47 illustrates that this assumption has a minor effect for horizontal antennas. The array factor of a rhombic a distance h above a perfect ground plane produces a null along the ground plane. There are several designs for rhombics above a ground plane in the literature [2–5]. One such design is for the alignment of the major lobe at a specific elevation angle. Then the rhombus angle α and the elevation angle of the main beam are equal, and the height above ground is given by

$$h = \frac{\lambda}{4 \sin \alpha} \tag{6-8}$$

and the length of each leg is

$$L = \frac{0.371\lambda}{\sin^2 \alpha} \tag{6-9}$$

For example, if $\alpha = 14.4°$, then $L = 6\lambda$ and $h = 1\lambda$. Rhombic impedances are typically on the order of 600 to 800 Ω.

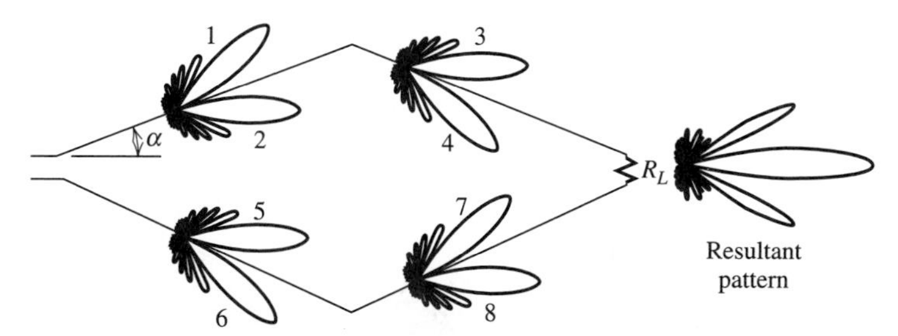

Figure 6-5 The rhombic antenna. Each side is of length L. Component beams 2 and 3, and 5 and 8 line up to form the main beam of the resultant pattern. In this case, $L = 6\lambda$ and $\alpha = 16°$.

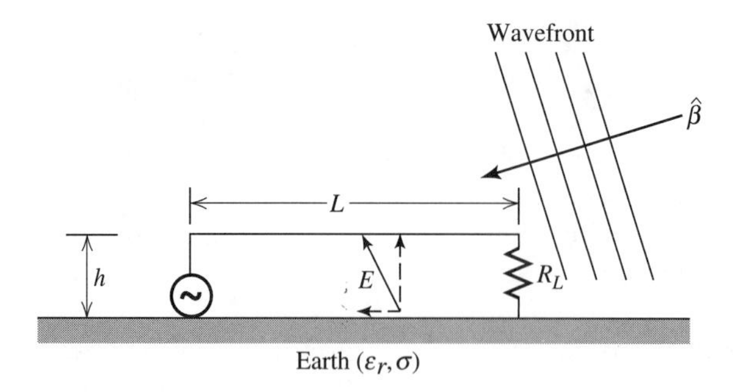

Figure 6-6 Beverage receiving antenna on an imperfect ground plane.

The efficiency of the rhombic antenna is decreased significantly because of the matched termination. The power that is not radiated is absorbed in the load R_L. However, this loss of power is essentially that which would have appeared in a large back lobe as a result of reflected current if the matched load were not present. The traveling wave feature not only improves the pattern but also produces wider impedance bandwidth. Well-designed traveling-wave antennas have input impedances that have little reactance since there is little or no reflected power.

The traveling-wave antennas discussed above have been examined in free space. One traveling-wave antenna that requires the imperfectly conducting properties of real earth is the Beverage antenna shown in Fig. 6-6. The height h is a small fraction of a wavelength, and the length L is usually between 2 and 10 wavelengths. The incoming vertically polarized plane wave in Fig. 6-6 produces a horizontal component of the electric field that is not totally shorted out by the imperfectly conducting earth. It is this horizontal component that is responsible for inducing a current on the antenna conductor of length L. Alternatively, the Beverage and its image in the lossy earth may be viewed as an unbalanced transmission line. As was noted in the folded dipole discussion in Sec. 5.2, unbalanced transmission lines can radiate.

Figure 6-7 shows the current on a Beverage antenna of length 2.18λ. The curve was calculated by the rigorous Sommerfeld theory for antennas in proximity to the imperfectly conducting earth [6, 7]. Because there are large dissipative losses in the earth as well as radiative losses, the current shows a definite decay from the feed

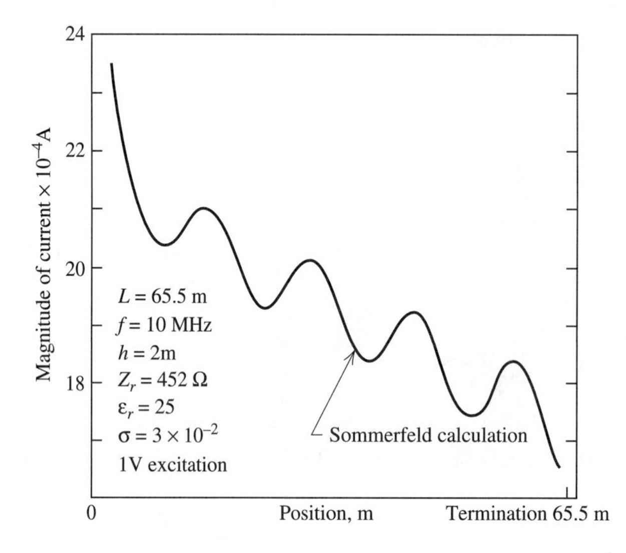

Figure 6-7 Current distribution on a Beverage antenna, $L = 2.18\lambda$, $h = 0.067\lambda$.

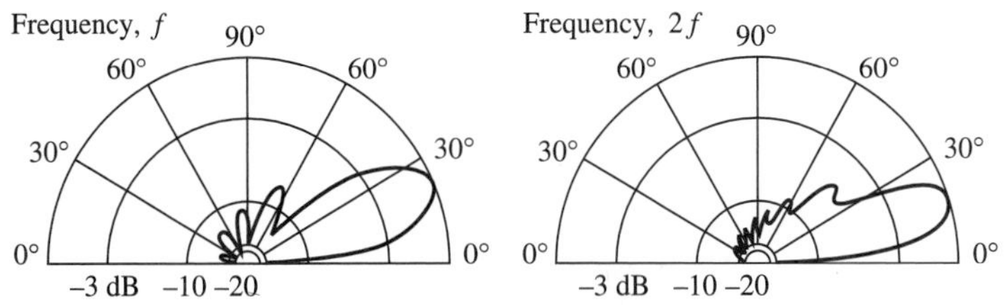

Figure 6-8 Generic elevation patterns for a Beverage antenna.

end to the load end. This portion of the current distribution can be accounted for by a modification of (6-4):

$$I_t(z) = I_m e^{-\alpha z} e^{-j\beta z} \tag{6-10}$$

The current also shows a standing wave superimposed on the major portion of the current distribution. This is evidence of a small reflected wave from the load end. The reflected wave also appears to diminish in strength as it moves toward the feed end. The reflected wave can be represented by a relationship similar to (6-10). The approximate theory presented in Sec. 5.6 cannot be used to obtain the current distribution in Fig. 6-7 because the Beverage interacts too strongly with the earth due to its close proximity to the ground and its long horizontal length. Instead, the rigorous but complicated Sommerfeld theory [6] must be used.

Figure 6-8 shows some generic elevation plane patterns for a Beverage antenna used by the military as a tactical field antenna. The antenna is particularly well-suited for this purpose because it does not need to be elevated much above the ground. Note several things about the patterns. The higher-frequency pattern has a narrower beamwidth in both the elevation and azimuthal planes, and has a lower angle of radiation as expected. Further, the back lobe radiation is low, particularly in the higher-frequency case where the antenna is electrically twice as long and there is more opportunity for dissipative loss.

Often, the feed end of the Beverage is higher than the load end to achieve as low a radiation angle relative to the ground (take-off angle) as possible. The Beverage antenna is usually used in the LF and lower HF portions of the frequency spectrum. It is believed to have been first used on Long Island in the early days of radiotelephone for trans-Atlantic communication between the United States and London at 50 and 60 kHz. The Beverage was the first antenna to use the traveling-wave principle.

6.2 HELICAL ANTENNAS

If a conductor is wound into a helical shape and fed properly, it is referred to as a **helical antenna**, or simply as a **helix**. The typical geometry for a helix is shown in Fig. 6-9. If one turn of the helix is uncoiled, the relationships among the various

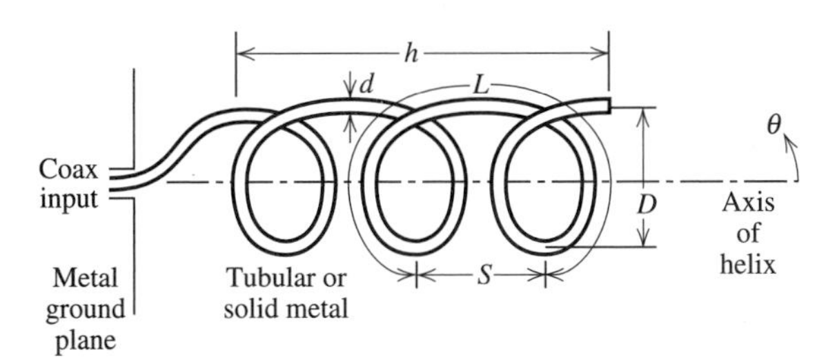

Figure 6-9 Geometry and dimensions of a helical antenna. This is a left-hand wound helix.

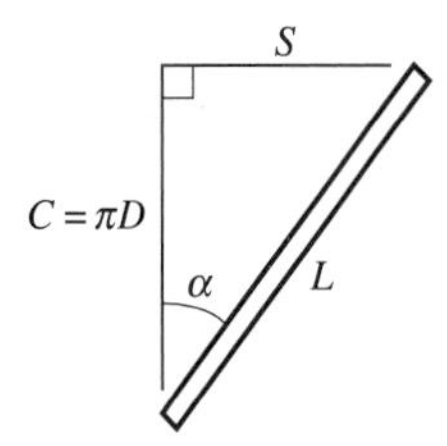

Figure 6-10 One uncoiled turn of a helix.

helix parameters are revealed, as shown in Fig. 6-10. The symbols used to describe the helix are defined as follows:

$$D = \text{diameter of helix} \quad \text{(between centers of coil material)}$$

$$C = \text{circumference of helix} = \pi D$$

$$S = \text{spacing between turns} = C \tan \alpha$$

$$\alpha = \text{pitch angle} = \tan^{-1} \frac{S}{C}$$

$$L = \text{length of one turn} = \sqrt{C^2 + S^2}$$

$$N = \text{number of turns}$$

$$L_w = \text{length of helix coil} = NL$$

$$h = \text{height} = \text{axial length} = NS$$

$$d = \text{diameter of helix conductor}$$

Note that when $S = 0$ ($\alpha = 0°$), the helix reduces to a loop antenna, and when $D = 0$ ($\alpha = 90°$), it reduces to a linear antenna.

The helix can be operated in two modes: the normal mode and the axial mode. The normal mode yields radiation that is most intense normal to the axis of the helix. This occurs when the helix diameter is small compared to a wavelength. The axial mode provides a radiation maximum along the axis of the helix. When the helix circumference is on the order of a wavelength, the axial mode will result. We introduce both the normal mode and axial mode helices in this section. See [8] and [9] for more details.

6.2.1 Normal Mode Helix Antenna

In the normal mode of operation, the radiated field is maximum in a direction normal to the helix axis and for certain geometries, in theory, will emit circularly polarized waves. For normal mode operation, the dimensions of the helix must be small compared to a wavelength, that is, $D \ll \lambda$ and usually $L \ll \lambda$ as well. The normal mode helix is electrically small and thus its efficiency is low.

Since the helix is small, the current is assumed to be constant in magnitude and phase over its length. The far-field pattern is independent of the number of turns and may be obtained by examining one turn. One turn can be approximated as a small loop and ideal dipole as shown in Fig. 6-11. The far-zone electric field of the ideal dipole from (1-72a) is

$$\mathbf{E}_D = j\omega\mu IS \frac{e^{-j\beta r}}{4\pi r} \sin\theta \, \hat{\boldsymbol{\theta}} \tag{6-11}$$

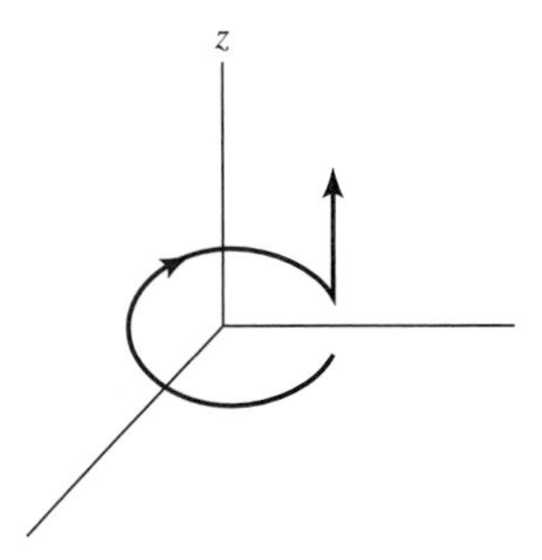

Figure 6-11 One turn of a normal mode helix approximated as a small loop and an ideal dipole.

where S, the spacing between helical turns, is the length of the ideal dipole in Fig. 6-11. The far-zone electric field of the small loop from (2-53) is

$$\mathbf{E}_L = \eta\beta^2 \frac{\pi}{4} D^2 I \frac{e^{-j\beta r}}{4\pi r} \sin\theta \, \hat{\boldsymbol{\phi}} \tag{6-12}$$

where $\pi D^2/4$ is the area of the loop. The total radiation field for one turn, as modeled in Fig. 6-11, is the vector sum of the fields in (6-11) and (6-12). Note that both components have a $\sin\theta$ pattern (see Fig. 6-12) and they are 90° out-of-phase. The ratio of the electric field components is

$$\frac{|E_\theta|}{|E_\phi|} = \frac{4\omega\mu S}{\sqrt{(\mu/\varepsilon)}\omega\sqrt{\mu\varepsilon}(2\pi/\lambda)\pi D^2} = \frac{2S\lambda}{\pi^2 D^2} = \frac{2\dfrac{S}{\lambda}}{\left(\dfrac{C}{\lambda}\right)^2} \tag{6-13}$$

This equals the axial ratio of the polarization ellipse when greater than unity and the inverse of axial ratio when less than unity; see Sec. 1.10. Limiting cases are 0 (with $S = 0$) corresponding to a small loop with horizontal polarization and ∞ (with $D = 0$) corresponding to a short dipole with vertical polarization.

Since the (perpendicular) linear components are 90° out-of-phase, circular polarization is obtained if the axial ratio is unity. This occurs for

$$C = \pi D = \sqrt{2S\lambda} \tag{6-14}$$

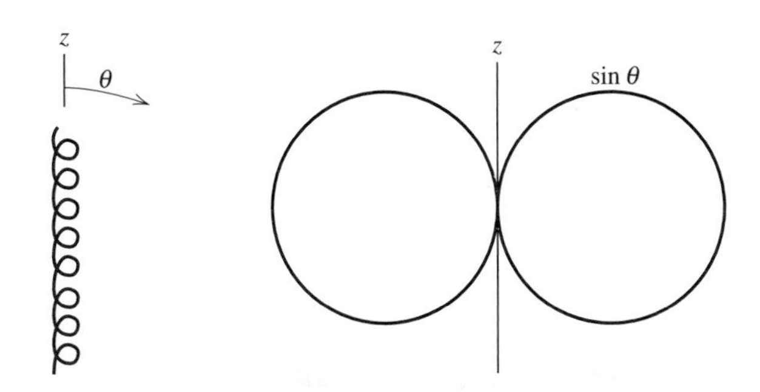

(*a*) Geometry. (*b*) Radiation pattern of both $|E_\theta|$ and $|E_\phi|$.

Figure 6-12 The normal mode helix and its radiation pattern.

which was found by setting (6-13) equal to unity. This circular polarization is obtained in all directions, except of course where the pattern is zero (along the axis of the helix). From Fig. 6-10, it is seen that

$$L \sin \alpha = S \qquad \text{or} \qquad \alpha = \sin^{-1} \frac{S}{L} \tag{6-15}$$

and

$$C^2 + S^2 = L^2 \tag{6-16}$$

For circular polarization in the normal mode, the circumference of the helix given by (6-14) used in (6-16) gives

$$S_{CP}^2 + 2S_{CP}\lambda - L^2 = 0 \tag{6-17}$$

This is a quadratic equation that may be solved for S as

$$S_{CP} = \frac{-2\lambda \pm \sqrt{4\lambda^2 + 4L^2}}{2} = \lambda\left[-1 \pm \sqrt{1 + \left(\frac{L}{\lambda}\right)^2}\right] \tag{6-18}$$

Choosing the plus sign to keep the physical length S positive and substituting into (6-15) yields the pitch angle required for circular polarization:

$$\alpha_{CP} = \sin^{-1}\left[\frac{-1 + \sqrt{1 + (L/\lambda)^2}}{L/\lambda}\right] \tag{6-19}$$

Usually, the normal mode helix is oriented vertically and operated such that the ratio in (6-13) is greater than unity, leading to predominantly vertically polarized radiation. This antenna is very popular in small transceivers such as handheld personal radios. For these applications, the wire length L_w is about a quarter-wavelength and the antenna is operated as a monopole fed against a ground plane. In this context, it is often referred to as a **normal mode helix antenna (NMHA)**, or **resonant (quarter-wave) stub helix antenna**. The pattern is, of course, nearly omnidirectional. The advantages of the stub helix over a conventional straight-wire monopole of the same height is that the helix acts as an inductor, tending to cancel the capacitance inherent in electrically short antennas. The current along the wire of the helix is approximately sinusoidal. The radiation resistance of a resonant stub helix above a perfect ground plane for heights under $\lambda/8$ is [9]

$$R_r \approx 640\left(\frac{h}{\lambda}\right)^2 \ \Omega \qquad \textit{resonant, stub helix} \tag{6-20}$$

The counterpart short monopole from (2-19) has $R_r = 395(h/\lambda)^2\ \Omega$. Since electrically short antennas suffer from low radiation resistance, the higher radiation resistance of the stub helix is another improvement over the conventional short (straight wire) monopole. Antennas on handheld radios often have a center section that telescopes up, operating as a 1/2 or 5/8 wavelength antenna. The stub helix is activated when the telescoping section is retracted. An impedance-matching network is usually included.

EXAMPLE 6-1 *A Stub Helix Antenna for Handheld Cellular Radios*

Consider a stub helix operating in the cellular telephone band at 883 MHz ($\lambda = 34$ cm). The four-turn helix is 2.25 in. (5.7 cm) high, or $h = 0.168\lambda$, and 0.2 in. (0.5 cm) in diameter, giving a circumference of $C = 0.046\lambda$. The turn spacing is $S = h/N = 5.7\text{ cm}/4 = 1.43\text{ cm} = 0.042\lambda$.

The helix length is $L_w = [(NC)^2 + h^2]^{1/2} = 0.25\lambda$, which is of resonant length.
The axial ratio from (6-13) is

$$|\text{AR}| = \frac{2\left(\dfrac{S}{\lambda}\right)}{\left(\dfrac{C}{\lambda}\right)^2} = \frac{2(0.042)}{(0.046)^2} = 38$$

This confirms the nearly vertical linear nature of the polarization. The radiation resistance from (6-20) is $R_r = 640(0.168)^2 = 18\ \Omega$. A straight monopole of the same height would have $R_r = 395\ (0.168)^2 = 11\ \Omega$.

Another popular resonant compact helix antenna, that is smaller than an axial mode helix, uses four half-wavelength long windings and produces a single, broad endfire beam with circular polarization. This **resonant quadrafilar helix antenna**, or **volute antenna**, is fed at the open end such that the two orthogonal pairs of helices are 90° out-of-phase. The windings are shorted to the ground plane at the other end. This antenna does not qualify as broadband since it has only a few percent bandwidth [9].

6.2.2 Axial Mode Helix Antenna

Axial mode helices are used when a moderate gain of up to about 15 dB and circular polarization are required. The relatively small cross section of the helix makes it popular at UHF frequencies, where it is widely used for satellite communications. In this section, we consider the axial mode, monofilar (single winding) helix antenna in detail with emphasis on design considerations. In the axial mode, the helix radiates as an endfire antenna with a single main beam along the axis of the helix (the $+z$-direction in Fig. 6-9). The radiation is close to circularly polarized near the axis. Further, the main beam narrows as turns are added to the helix. The axial mode occurs when the helix circumference C is on the order of one wavelength. Helices with a few turns perform well over the frequency range corresponding to

$$\frac{3}{4}\lambda \leq C \leq \frac{4}{3}\lambda \tag{6-21}$$

This gives a bandwidth ratio of

$$B_r = \frac{f_U}{f_L} = \frac{c/\lambda_U}{c/\lambda_L} = \frac{4/3}{3/4} = \frac{16}{9} = 1.78 \tag{6-22}$$

This is close to the conventional definition of a 2:1 bandwidth ratio for a wideband antenna. For long helices, the upper operating frequency is lower than $4\lambda/3$, reducing the bandwidth below 1.78.

Kraus [8] performed the pioneering work on the axial mode helix and provided a simple explanation of its operation as well as empirical formulas for pattern, gain, polarization, and impedance. Subsequent experiments [10, 11] produced more accurate models for helix antenna performance. The development that follows is based on these works.

An approximate model for the axial mode helix that offers a simple explanation for its operation assumes that the helix carries a pure traveling wave that travels outward from the feed. The electric field associated with this traveling wave rotates in a circle, producing radiation that is nearly circularly polarized off the end of the helix.

In contrast to the normal mode helix, which has a current that is nearly uniform in phase over the helix winding, the phase of the axial mode helix current shifts continuously along the helix, which is characteristic of a traveling wave. Since the circumference is close to one wavelength, the current at opposite points on a turn are about 180° out-of-phase. This cancels the current direction reversal introduced by the half-turn. Thus, the radiation from opposite points on the helix is nearly in-phase, leading to reinforcement along the axis in the far field. This radiation mechanism closely parallels that of the one-wavelength loop discussed in Sec. 5.7.

The radiation pattern of the axial mode helix can be modeled using array theory, with each turn being an element of the array. The element pattern is cos θ, approximating the pattern of a one-wavelength loop; see Fig. 5-52*b*. The array factor is that of N equally spaced elements with spacing S and progressive phase α_h. The current magnitudes on the turns are taken as uniform in this simple model. Then the array factor is given by (3-33) and the total pattern is

$$F(\theta) = K \cos\theta \frac{\sin(N\psi/2)}{N \sin(\psi/2)} \tag{6-23}$$

$$\psi = \beta S \cos\theta + \alpha_h \tag{6-24}$$

where K is a normalization constant. (See Fig. 9-18 for a measured pattern.)

The traveling wave along the helix produces an endfire beam along the helix axis (z-axis). Suppose initially that the helix can be modeled as an ordinary endfire array. Then a main beam maximum occurs in the $\theta = 0$ direction for $\psi = 0$, which yields $\alpha_h = -\beta S$ from (6-24); also see (3-36). The $-\beta S$ phase is phase delay due to axial propagation corresponding to the distance S along the axis for one turn. However, the current wave follows the helix. This introduces another -2π of phase shift since the circumference is about a wavelength. Thus for ordinary endfire, $\alpha_h = -\beta S - 2\pi$. Interestingly, it turns out that the traveling-wave mode on the axial mode helix corresponds to nearly a naturally occurring Hansen–Woodyard increased directivity-type endfire array. This effect is accounted for with an additional $-\pi/N$ phase delay over the ordinary endfire case; see (3-49). Thus, the element-to-element phase shift is

$$\alpha_h = -\left(\beta S + 2\pi + \frac{\pi}{N}\right) \tag{6-25}$$

This phase shift leads to a value for the phase velocity of the traveling wave. To see this, we write the phase shift of the wave in one transit around a turn of length L as

$$\alpha_h = -\beta_h L \tag{6-26}$$

where β_h is the phase constant associated with wave propagation along the helical conductor. Equating this to (6-25) gives

$$\beta_h = \frac{1}{L}\left(\beta S + 2\pi + \frac{\pi}{N}\right) \tag{6-27}$$

The velocity factor (phase velocity relative to the free-space velocity of light) is

$$p = \frac{v}{c} = \frac{\omega/c}{\omega/v} = \frac{\beta}{\beta_h} \tag{6-28}$$

where v is the phase velocity of the traveling wave along the helical conductor. Using (6-27) in (6-28) yields

$$p = \frac{L/\lambda}{S/\lambda + (2N + 1)/2N} \tag{6-29}$$

A typical configuration is $C = \lambda$, $\alpha = 12°$, and $N = 12$. Then $S = C \tan \alpha = 0.213\lambda$, $L = \sqrt{C^2 + S^2} = 1.022\lambda$, and $p = 0.815$. Therefore, the traveling wave has a phase velocity less than that if it were a plane wave in free space. Such a wave is referred to as a *slow wave*. Another remarkable feature of the helix is that as the helix parameters vary over rather large ranges ($5° < \alpha < 20°$ and $\frac{3}{4}\lambda < C < \frac{4}{3}\lambda$), the phase velocity adjusts automatically to maintain increased directivity.

Returning to the pattern calculation, we see that the main beam maximum occurs for $\theta = 0$ and from (6-24) and (6-25), $\psi = -2\pi - \pi/N$. Then (6-23) is

$$F(\theta = 0) = K \frac{\sin(-N\pi - \pi/2)}{N \sin(-\pi - \pi/2N)} = \frac{K(-1)^{N+1}}{N \sin(\pi/2N)} \tag{6-30}$$

Normalizing such that the maximum is unity yields $K = (-1)^{N+1} N \sin(\pi/2N)$, and the final pattern function is

$$F(\theta) = (-1)^{N+1} \sin \frac{\pi}{2N} \cos \theta \frac{\sin(N\psi/2)}{\sin(\psi/2)} \tag{6-31}$$

where

$$\psi = \beta S(\cos \theta - 1) - 2\pi - \frac{\pi}{N} \tag{6-32}$$

This pattern expression applies to both E_θ and E_ϕ.

The electrical performance of the axial mode helix is, of course, influenced by its geometric parameter values. The helix operates well over a wide range of pitch angles, but is optimum for $12° < \alpha < 14°$ and helices are usually constructed with $\alpha = 13°$. As more turns are added, gain increases and the polarization axial ratio decreases. A series of empirical formulas based on extensive measured data were initially proposed by Kraus [8] and later modified by King and Wong [11] for the prediction of electrical performance from the helix geometry parameters. The approximate formulas presented here are based on these results and are useful in the design of axial mode helix antennas with $3/4 < C/\lambda < 4/3$, $12° < \alpha < 15°$, and $N > 3$.

The half-power beamwidth in degrees is given by

$$\text{HP} = \frac{65°}{\dfrac{C}{\lambda}\sqrt{N \dfrac{S}{\lambda}}} \tag{6-33}$$

This applies to all planes since the main beam is nearly symmetric. The gain formula then follows from (7-95) as

$$G = \frac{26{,}000}{\text{HP}^2} = 6.2\left(\frac{C}{\lambda}\right)^2 N \frac{S}{\lambda} \tag{6-34}$$

This formula indicates that gain is proportional to f^3. Experiments show this to be most accurate for $N \approx 10$ [11]. Gain based on (6-34) increases linearly with the number of turns; that is, doubling N will increase gain by 3 dB. This result is slightly optimistic. Although (6-34) indicates that gain is a strong function of the circumference, it must be remembered that the helix is constructed to maintain $C \sim \lambda$. Experiments show that gain is peak for $C \approx 1.1\lambda$ [11].

The classical formula for axial ratio is [8]

$$|\mathrm{AR}| = \frac{2N+1}{2N} \tag{6-35}$$

indicating that the quality of circular polarization increases with the number of turns. This formula is approximate at best. Measured data give $|\mathrm{AR}| < 1.2$ for $0.8 < C/\lambda < 1.2$, corresponding to $N > 2.5$ from (6-35) [11]. The axial ratio can be improved by tapering the last few turns at the end of the helix [12]. The sense of polarization is determined by the sense of the helix windings as shown in Fig. 6-13. A left- (right-) hand wound helix is left- (right-) hand sensed polarized.

The classical formula for input resistance is

$$R_A = 140 \frac{C}{\lambda}\ \Omega \tag{6-36}$$

The input impedance is real-valued due to the nearly pure traveling-wave behavior of a properly designed helix. This simple formula must be regarded as only an approximation since the input impedance of actual helix antennas is affected by details of the feed. However, the input impedance does remain nearly resistive over a wide bandwidth.

An axial mode helix performs well and is represented approximately by the foregoing empirical relations when the increased directivity condition is satisfied. This occurs over the bandwidth of (6-22) for helices of a few turns. However, long helices have a reduced upper operating frequency; for example, f_U corresponds to about $C \approx \lambda$ for $N = 50$ [10].

If only a pure traveling wave exists on the helix, the ground plane would have little effect. However, other modes are present, including a wave reflected from the end of the helix that returns to the feed region. This makes the ground plane geometry important. An approximate guideline is that the ground plane should be at least $3\lambda/4$ in diameter. Ground structures such as cups or cones are often used in place of a larger ground plane [13]. The inner conductor of the coaxial connector is attached to the helix and the outer conductor is connected to the ground structure as indicated in Fig. 6-13. The conductor diameter is frequently selected to provide a rigid, self-supporting structure. Helix winding conductor diameter d usually is between 0.005λ and 0.05λ. A 50-Ω input impedance can be achieved by including an impedance transformer in the feed or by adjusting the location of the wire from the coax to the helix winding.

There are many variations of the helix antenna. For example, tapering the helix by gradually reducing the diameter near the end of the helix improves impedance,

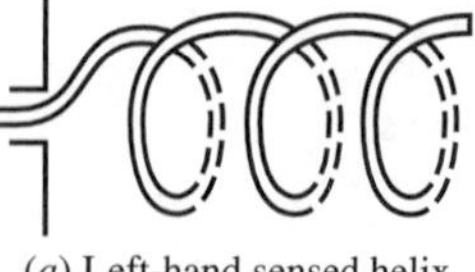

(*a*) Left-hand sensed helix

(*b*) Right-hand sensed helix

Figure 6-13 Left- and right-hand wound helices. For the axial mode helix, the sense of the windings determines the sense of polarization of the antenna.

pattern, and polarization [9, 14]. A compact volume can be achieved at the sacrifice of bandwidth with a helix that conforms to a spherical surface rather than a cylinder [15].

EXAMPLE 6-2 *A 10-Turn Axial Mode Helix Antenna*

The helix antenna is rather easy to construct and will perform approximately as predicted by the simple theory presented in this section, as will be demonstrated in this example. Calculations are compared to experimental results for a 10-turn helix constructed for a center frequency of 8 GHz ($\lambda = 3.75$ cm), where the helix was designed to have a circumference of $C = 0.92\lambda = 3.45$ cm. The helix was built with a pitch angle of $\alpha = 13°$. The spacing between turns is then $S = C \tan \alpha = 3.45 \tan 13° = 0.796$ cm. The measured radiation patterns for the two principal planes are shown in Figs. 6-14*a* and 6-14*b* [10]. These patterns are nearly alike and compare well to the pattern computed from simple theory with (6-31), which is plotted in Fig. 6-14*c*. The beamwidth of the measured patterns is about 44°. For comparison, the computed pattern of Fig. 6-14*c* has HP = 39°, and the approximate empirical formula of (6-33) gives a beamwidth of

$$\text{HP} = \frac{65°}{\dfrac{C}{\lambda}\sqrt{N\dfrac{S}{\lambda}}} = \frac{65°}{0.92\sqrt{10(0.212)}} = 48° \tag{6-37}$$

The gain predicted by (6-34) is

$$G = 6.2\left(\frac{C}{\lambda}\right)^2 N\frac{S}{\lambda} = 6.2(0.92)^2 10(0.212) = 11.1 = 10.5 \text{ dB} \tag{6-38}$$

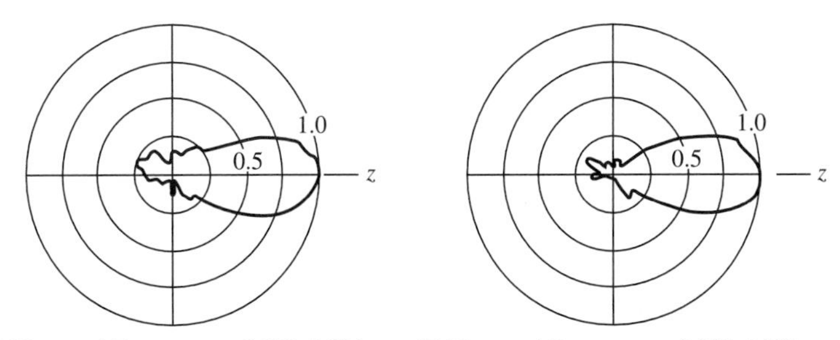

(*a*) Measured E_θ pattern at 8 GHz [10]. (*b*) Measured E_ϕ pattern at 8 GHz [10].

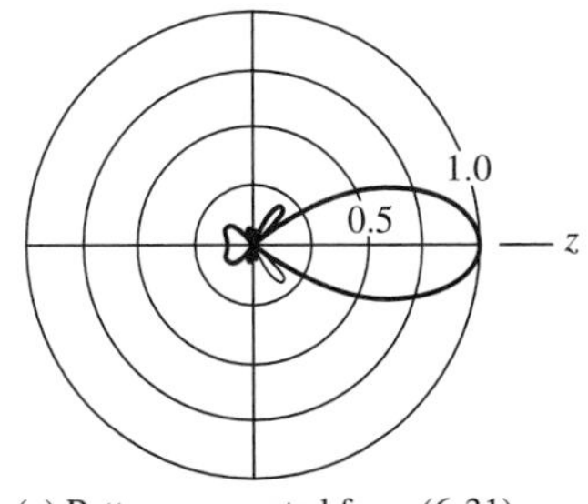

(*c*) Pattern computed from (6-31).

Figure 6-14 Radiation patterns of a 10-turn axial mode helix with $C = \lambda$ and $\alpha = 13°$ (Example 6-2).

6.3 BICONICAL ANTENNAS

The bandwidth of a simple dipole antenna can be increased by using thicker wire as indicated in Fig. 5-7. This concept can be extended to further increase bandwidth if the conductors are flared to form a biconical structure. Then the fixed wire diameter is replaced by a smoothly varying diameter and a fixed angle (of the conical surfaces). In this section, the idealized biconical antenna is considered first, followed by two practical forms—the finite biconical antenna and the discone.

6.3.1 Infinite Biconical Antenna

If the conducting halves of an antenna are two infinite conical conducting surfaces end-to-end, but with a finite gap at the feed point, the **infinite biconical antenna** of Fig. 6-15 results. Since the structure is infinite, it can be analyzed as a transmission line. With a time-varying voltage applied across the gap, currents will flow radially out from the gap along the surface of the conductors. These currents, in turn, create an encirculating magnetic field H_ϕ. If we assume a TEM transmission line mode (all fields transverse to direction of propagation), the electric field will be perpendicular to the magnetic field and be θ-directed. When the potential on the top cone is positive and the bottom cone is negative, the electric field lines extend from the top to the bottom cone as indicated in Fig. 6-15.

In the region between the cones, $\mathbf{J} = 0$, $\mathbf{H} = H_\phi \hat{\boldsymbol{\phi}}$, and $\mathbf{E} = E_\theta \hat{\boldsymbol{\theta}}$. Then Ampere's law, $\nabla \times \mathbf{H} = j\omega\varepsilon \mathbf{E} + \mathbf{J}$, reduces to

$$\frac{1}{r \sin \theta} \frac{\partial}{\partial \theta} (\sin \theta \, H_\phi) = j\omega\varepsilon E_r = 0 \tag{6-39}$$

for the r-component and

$$-\frac{1}{r} \frac{\partial}{\partial r} (r H_\phi) = j\omega\varepsilon E_\theta \tag{6-40}$$

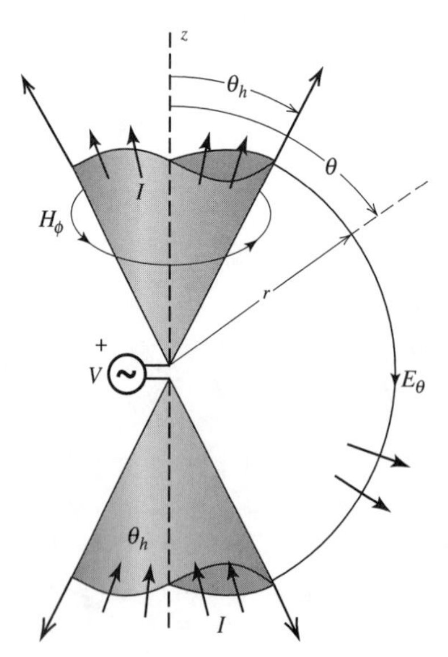

Figure 6-15 Infinite biconical antenna. The field components and current are shown.

for the θ-component. From (6-39), we see that $\partial/\partial\theta(\sin\theta H_\phi) = 0$ so

$$H_\phi \propto \frac{1}{\sin\theta} \tag{6-41}$$

Since the structure acts as a guide for spherical waves, we can write (6-41) as

$$H_\phi = H_o \frac{e^{-j\beta r}}{4\pi r} \frac{1}{\sin\theta} \tag{6-42}$$

Then, substituting this into (6-40), we obtain

$$\begin{aligned} E_\theta &= \frac{-1}{j\omega\varepsilon} \frac{1}{r} \frac{H_o}{4\pi \sin\theta} \frac{\partial}{\partial r}(e^{-j\beta r}) = \frac{\beta H_o}{\omega\varepsilon} \frac{1}{r} \frac{e^{-j\beta r}}{4\pi} \frac{1}{\sin\theta} \\ &= \eta H_o \frac{e^{-j\beta r}}{4\pi r} \frac{1}{\sin\theta} \end{aligned} \tag{6-43}$$

This equation is simply $E_\theta = \eta H_\phi$, which confirms our statement that the wave is TEM. The field components vary as $1/\sin\theta$, so the radiation pattern is

$$F(\theta) = \frac{\sin\theta_h}{\sin\theta}, \qquad \theta_h < \theta < \pi - \theta_h \tag{6-44}$$

which is normalized to unity at its maxima on the conductor surfaces. This pattern is plotted in Fig. 6-16.

In order to determine the input impedance, we first find the terminal voltage and current. Referring to Fig. 6-15, we see the voltage is found by integrating along a constant radius r and it is

$$V(r) = \int_{\theta_h}^{\pi-\theta_h} E_\theta r\, d\theta \tag{6-45}$$

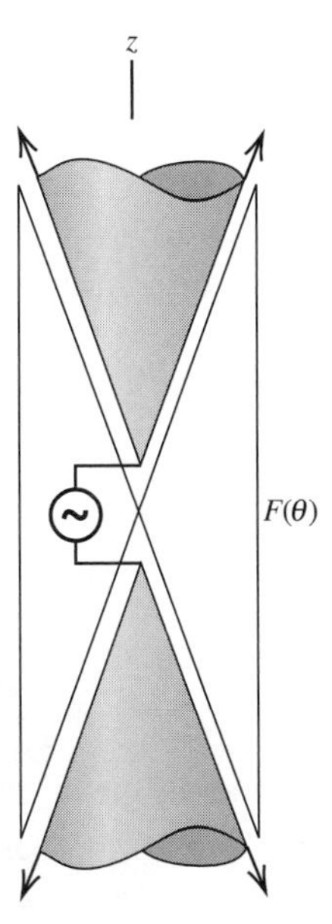

Figure 6-16 Radiation pattern of an infinite biconical antenna.

This can be performed for any r since the cones are equipotential surfaces. Substituting (6-43) into the above equation yields

$$V(r) = \frac{\eta H_o}{4\pi} e^{-j\beta r} \int_{\theta_h}^{\pi-\theta_h} \frac{d\theta}{\sin\theta} = \frac{\eta H_o}{4\pi} e^{-j\beta r} \left[\ln \left| \tan \frac{\theta}{2} \right| \right]_{\theta_h}^{\pi-\theta_h}$$

$$= \frac{\eta H_o}{2\pi} e^{-j\beta r} \ln\left(\cot \frac{\theta_h}{2} \right) \tag{6-46}$$

The boundary condition on H_ϕ at the conductor surface is $J_s = H_\phi$. The total current on one cone is found by integrating the current density J_s around the cone as shown in Fig. 6-15, so

$$I(r) = \int_0^{2\pi} H_\phi r \sin\theta \, d\phi = 2\pi r H_\phi \sin\theta \tag{6-47}$$

Substituting (6-42) in the above gives

$$I(r) = \frac{H_o}{2} e^{-j\beta r} \tag{6-48}$$

The characteristic impedance at any point r, from (6-46) and (6-48), is

$$Z_o = \frac{V(r)}{I(r)} = \frac{\eta}{\pi} \ln\left(\cot \frac{\theta_h}{2} \right) \tag{6-49}$$

Since this is not a function of r, it must be also the impedance at the input ($r = 0$). Thus, using $\eta \approx 120\pi$ in (6-49) gives the input impedance

$$Z_A = Z_o = 120 \ln\left(\cot \frac{\theta_h}{2} \right) \ \Omega \tag{6-50}$$

For θ_h less than 20°,

$$Z_A = Z_o \approx 120 \ln\left(\frac{2}{\theta_h} \right) \ \Omega \tag{6-51}$$

where θ_h is in radians. The input impedance is real because there is only a pure traveling wave. Since the structure is infinite, there are no discontinuities present to cause reflections setting up standing waves, which would show up as a reactive component in the impedance (except at a few resonance points). If $\theta_h = 1°$, $Z_A = 568 + j0\ \Omega$. If $\theta_h = 50°$, $Z_A = 91 + j0\ \Omega$.

If one cone is flared all the way out to form a perfect ground plane, a single infinite cone above a ground plane results. This monopole version of the infinite bicone then has an input impedance which is half that of the infinite bicone.

6.3.2 Finite Biconical Antenna

A practical biconical antenna is made by ending the two cones of the infinite bicone. This **finite biconical antenna** is shown in Fig. 6-17. Inside an imaginary sphere of radius h just enclosing the antenna, TEM waves exist together with higher-order modes created at the ends of the cones. These higher-order modes are the major

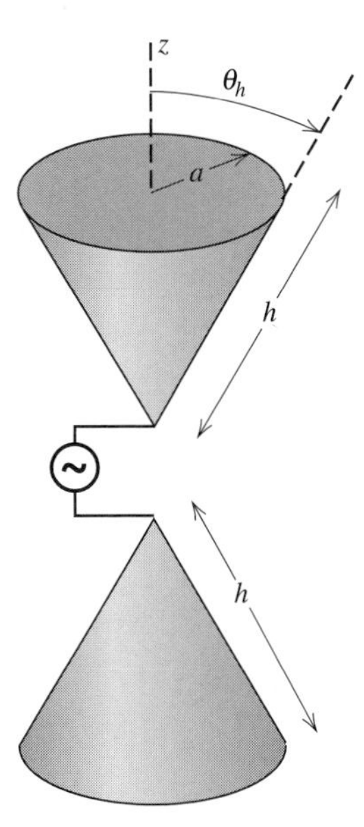

Figure 6-17 Finite biconical antenna.

contributors to the antenna reactance. The ends of the cones cause reflections that set up standing waves that lead to a complex input impedance.

The reactive part of the input impedance can be held to a minimum over a progressively wider bandwidth by increasing the angle θ_h in Fig. 6-17. At the same time, the real part of the input impedance becomes less sensitive to changing frequency (or changing h in Fig. 6-17). This is illustrated by measured data in Fig. 6-18 [3] for a conical monopole where the antenna impedance is plotted versus the height of the monopole L_h. These data clearly show that one can achieve the 2:1 impedance bandwidth necessary for one part of our definition of a broadband antenna. This is our first example of an antenna that can be more dependent on an angle in its geometry description than on its length. Frequency-independent antennas, considered later in this chapter, exploit this property. Another property that we will observe in many broadband and frequency-independent antennas is that some important dimension must be at least $\lambda/4$. Careful examination of Fig. 6-18 reveals that the impedance bandwidth starts when the height of the conical monopole is about $\lambda/4$ and extends upward beyond $\lambda/2$. The pattern of a conical monopole or finite biconical for small cone angles is very similar to that of an ordinary monopole or dipole of the same length.

A much simpler alternative to the finite biconical antenna is the common "**bow-tie**" antenna (shown later in Fig. 6-32). It offers less weight and less cost to build, but will have a somewhat more sensitive input impedance to changing frequency than the finite biconical.

6.3.3 Discone Antenna

If one cone of the finite biconical antenna is replaced with a disk-shaped ground plane, the structure becomes a disk-cone, or **discone**, antenna (see Fig. 6-19). The discone antenna was developed by Kandoian [16] in 1945, followed several years later by experimental design studies [17, 18]. It is used (like a vertical dipole) for vertical polarization and nearly uniform azimuth coverage (i.e., an omnidirectional

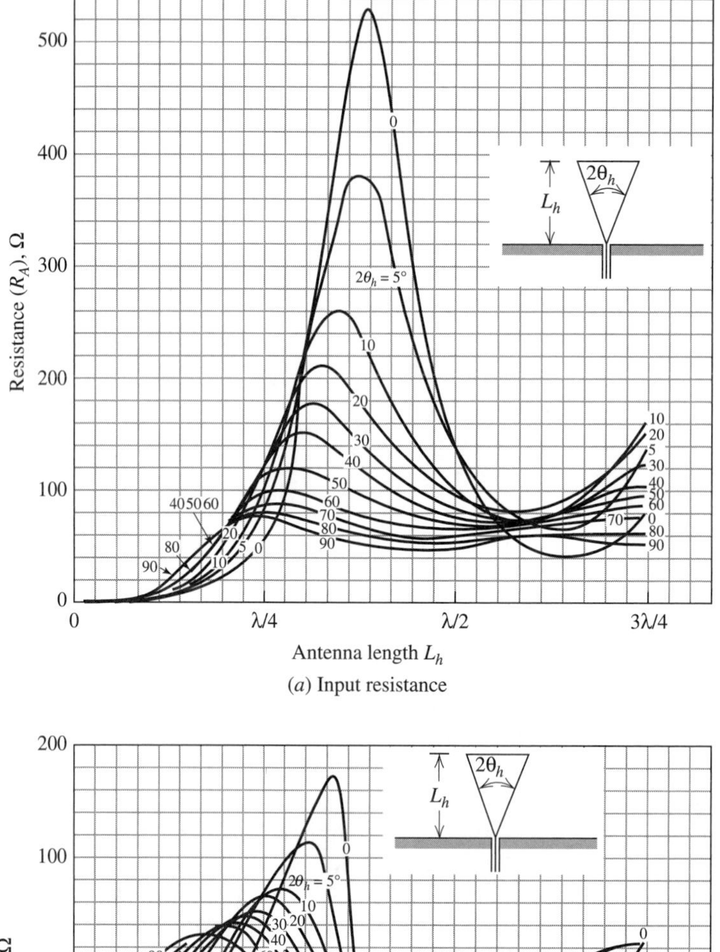

(a) Input resistance

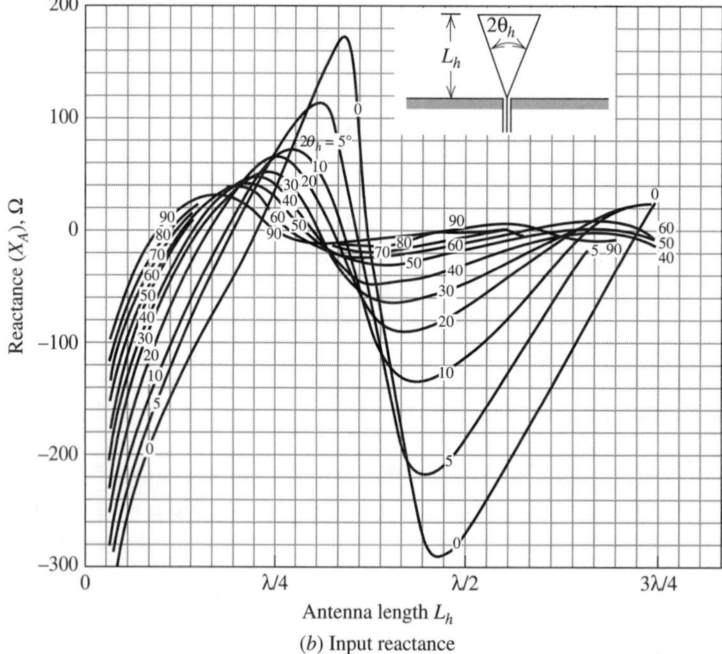

(b) Input reactance

Figure 6-18 Measured input impedance of a conical monopole with flare angle $2\theta_h \leq 90°$ versus monopole height L_h. (From Jasik [3]. © 1961 McGraw Hill. Used with permission of McGraw Hill Book Co.)

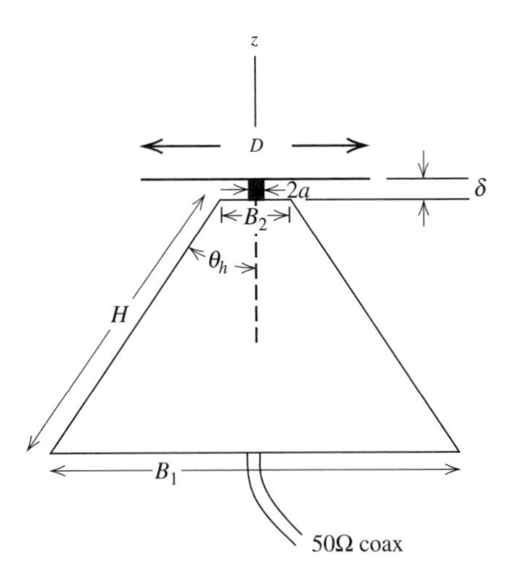

Figure 6-19 Discone antenna. Typical dimensions are $H \sim 0.7\lambda$, $B_1 \sim 0.6\lambda$, $D \sim 0.4\lambda$, and $\delta \ll D$.

pattern). The discone offers satisfactory operation over a wide frequency range (several octaves) while maintaining acceptable pattern and impedance properties.

The discone is constructed as shown in Fig. 6-19. The outer conductor of the coaxial transmission line is connected to the cone and the inner conductor is attached to the disk ground plane. The cone and disk can be either solid metal or radial wires. Ideally, the pattern between the ground plane and the cone is that of the infinite bicone. This omnidirectional pattern is well suited to broadcast applications.

The discone antenna can be designed for broadband impedance performance (typically 50 Ω), while maintaining acceptable pattern behavior with frequency [18]. Typical center frequency dimensions are $H = 0.7\lambda$, $B = 0.6\lambda$, $D = 0.4\lambda$, $\theta_h = 25°$, and $\delta \ll D$. For example, the discone with the patterns of Fig. 6-20 has a center frequency of 1 GHz ($\lambda = 30$ cm). So at 1 GHz, $H = 21.3$ cm $= 0.71\lambda$, $B_1 = 19.3$ cm $= 0.64\lambda$, and $\theta_h \approx \sin^{-1}[(B/2)/H] = 27°$. Nail [17] has given optimum design formulas of $D = 0.7B_1$ and $\delta = 0.3B_2$, independent of H and θ_h.

The pattern performance over a 3:1 bandwidth is revealed in Fig. 6-20. At low frequencies, the structure is small relative to a wavelength, and the pattern is not too different from that of a short dipole (see Fig. 6-20*a*). As frequency increases, the electrical size of the ground plane increases and the pattern is confined more to the lower half-space (see Fig. 6-20*b*). For further increases in frequency, the antenna behavior approaches that of an infinite structure. For example, at 1500 MHz, the pattern (of Fig. 6-20*c*) is very close to that of the monopole version of the infinite biconical antenna in Fig. 6-15. Measurements with several disk parameters D and spacings between the cone δ showed that the patterns are insensitive to these parameters [18].

Nail's optimum design formulas [17] are for $B_2 \approx \lambda_U/75$ at the highest operating frequency of the antenna and $\delta \ll D$. For larger values of B_2 and $\delta \approx 0.5B_2$, it has been found that Nail's equations need to be altered when an N-connector is used between the skirt and the disk [19]. In this case, $\delta = 0.5B_2$, $2a = 0.33B_2$, $D = 0.75B_1$, $L = 1.15\lambda_L$ based on experimental measurements. It has been reported in [19] that a VSWR below 1.5:1 over an octave bandwidth is easily achievable and $45° < 2\theta_h < 75°$ yields the best results.

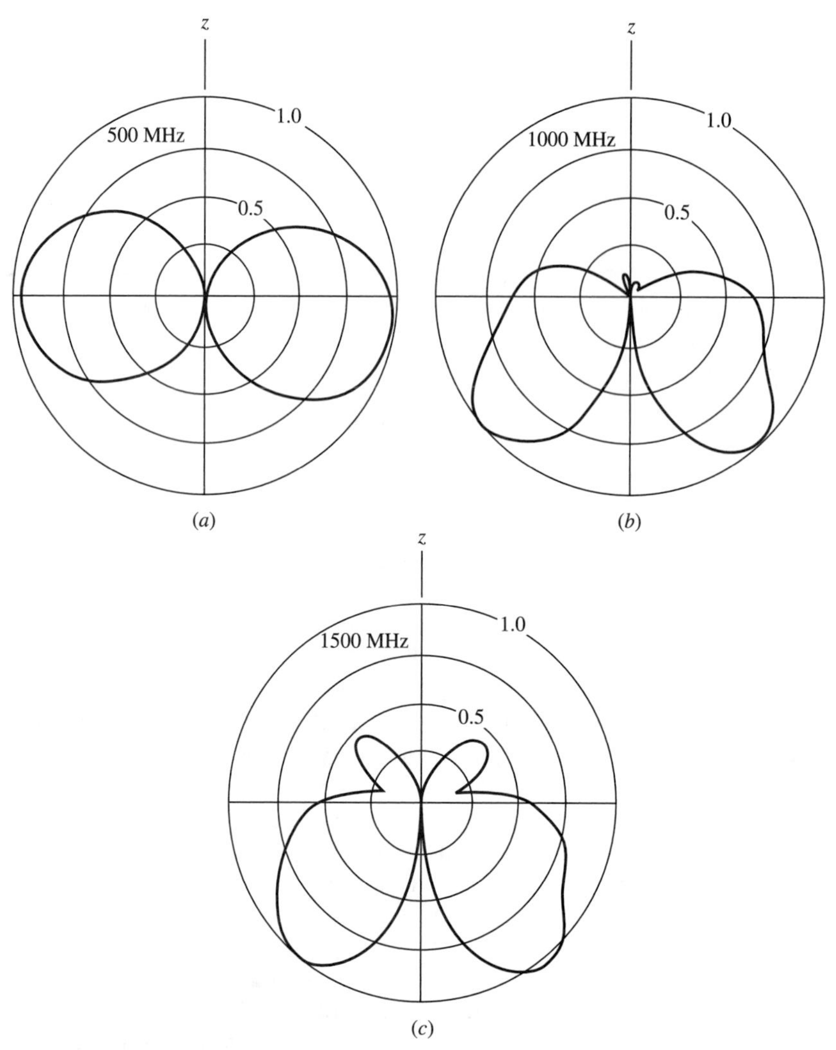

Figure 6-20 Measured patterns of a discone antenna for $H = 21.3$ cm, $B_1 = 19.3$ cm, and $\theta_h = 25°$.

6.4 SLEEVE ANTENNAS

In Sec. 5.1, we saw that the dipole antenna is very frequency-sensitive and its bandwidth is much less than the octave bandwidth provided by the antennas studied previously in this chapter. However, the addition of a sleeve to a dipole or monopole can increase the bandwidth to more than an octave. In this section, we will briefly examine a few forms of the **sleeve antenna**, which incorporates a tubular conductor sleeve around an internal radiating element. Emphasis will be placed on practical configurations.

6.4.1 Sleeve Monopoles

A **sleeve monopole** configuration is shown in Fig. 6-21*a* fed from a coaxial transmission line. The sleeve exterior acts as a radiating element and the interior of the

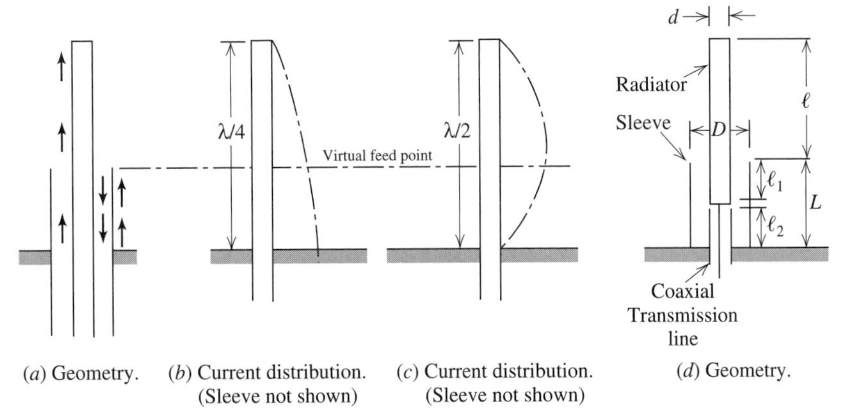

Figure 6-21 Sleeve monopole configurations: Arrows in (*a*) indicate polarity when $\ell + L \leq \lambda/2$. Different current distributions on the center conductor are shown in (*b*) and (*c*).

sleeve acts as the outer conductor of the feed coaxial transmission line. In principle, the length of the sleeve may be any portion of the total length of the monopole from zero (no sleeve) to where the sleeve constitutes the entire radiating portion of the antenna. However, in practice, the sleeve is usually about $\frac{1}{3}$ to $\frac{1}{2}$ the height of the monopole. The reason for this is apparent from Figs. 6-21*b* and 6-21*c*, which suggest that the current at the virtual feed point changes only slightly as the overall monopole height varies from $\lambda/4$ to $\lambda/2$. Thus, the impedance remains somewhat constant over at least an octave. As for an ordinary monopole with no sleeve, the antenna dimensions affect the impedance more than the pattern.

Consider Fig. 6-21*d*. The first sleeve monopole resonance occurs at a frequency where the monopole length $\ell + L$ is approximately $\lambda/4$. Design proceeds by locating this first resonance near the lower end of the frequency band, thereby fixing the total physical length $\ell + L$. The remaining design variable is ℓ/L. It has been found experimentally that a value of $\ell/L = 2.25$ yields optimum (nearly constant with frequency) radiation patterns over a 4:1 band [20]. The value of ℓ/L has little effect for $\ell + L \leq \lambda/2$ since the current on the outside of the sleeve will have approximately the same phase as that on the top portion of the monopole itself, as suggested by the arrows of Fig. 6-21*a*. However, for longer electrical lengths the ratio ℓ/L becomes very important and has a marked effect on the radiation pattern, since the current on the outside of the sleeve will not necessarily be in-phase with that on the top portion of the monopole. Some typical specifications for optimum performance are given in Table 6-1. In some applications, the VSWR may be too high, requiring a matching network.

Table 6-1 Specifications for Optimum Pattern Design of a Sleeve Monopole

Pattern bandwidth	4:1
$\ell + L$	$\lambda/4$ at low end of band
ℓ/L	2.25
D/d	3.0
VSWR	less than 8:1

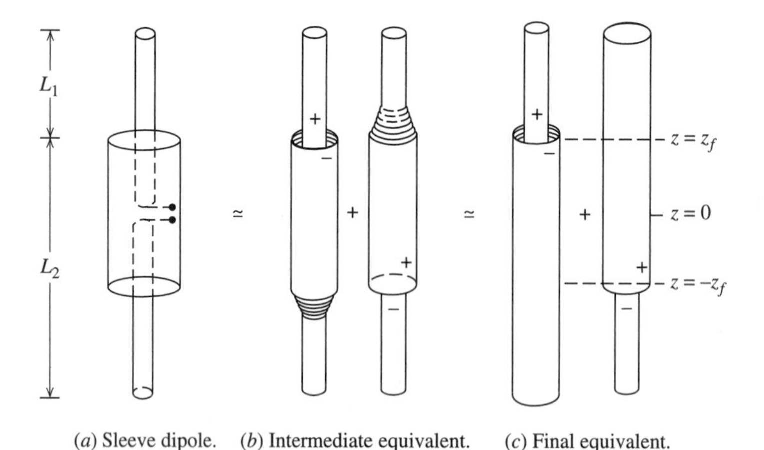

(*a*) Sleeve dipole. (*b*) Intermediate equivalent. (*c*) Final equivalent.

Figure 6-22 Sleeve dipole configuration and approximate equivalents [21].

6.4.2 Sleeve Dipoles

The sleeve monopole has a counterpart in the **sleeve dipole** antenna that is shown in Fig. 6-22*a*. An approximate impedance analysis of the sleeve dipole can be carried out according to Figs. 6-22*b* and 6-22*c* where the doubly driven structure of Fig. 6-22*a* is replaced by the pair of asymmetrically driven structures in Fig. 6-22*b*. The change in diameters on the longer arms is neglected, resulting in Fig. 6-22*c*. The current at the input to the sleeve (virtual feed) $I_A(z_f)$ is then approximately the sum of the currents at the point $z = z_f$ from the two configurations in Fig. 6-22*c*. For the left half of Fig. 6-22*c*, the current at the virtual feed in that asymmetrical structure is $I_{as}(z_f)$. The current at the same point due to the excitation in the lower half of the sleeve dipole (i.e., the right half of Fig. 6-22*c*) is identical to the current at the point $-z_f$ when the excitation is at the point z_f since the two structures are physically equivalent. Thus,

$$I_A(z_f) \approx I_{as}(z_f) + I_{as}(-z_f) \tag{6-52}$$

The input admittance to the sleeve is then

$$Y_A \approx \frac{I_{as}(z_f) + I_{as}(-z_f)}{V_A} \approx Y_{as}\left[1 + \frac{I_{as}(z_f)}{I_{as}(-z_f)}\right] \tag{6-53}$$

where

$$Y_{as} \approx \frac{2}{(Z_1 + Z_2)} \tag{6-54}$$

and where Z_1 is the impedance of a symmetrical antenna of half-length L_1 and Z_2 is the impedance of a symmetrical antenna with half-length L_2 [21]. Equation (6-54) is useful for estimating the impedance of asymmetrical dipoles such as that in Prob. 5.1-4.

The sleeve dipole of Fig. 6-22*a* can be approximated with an **open-sleeve dipole** in which the tubular sleeve is replaced by two conductors close to either side of the driven element as shown in Fig. 6-23. The length of the parasites (simulated sleeve)

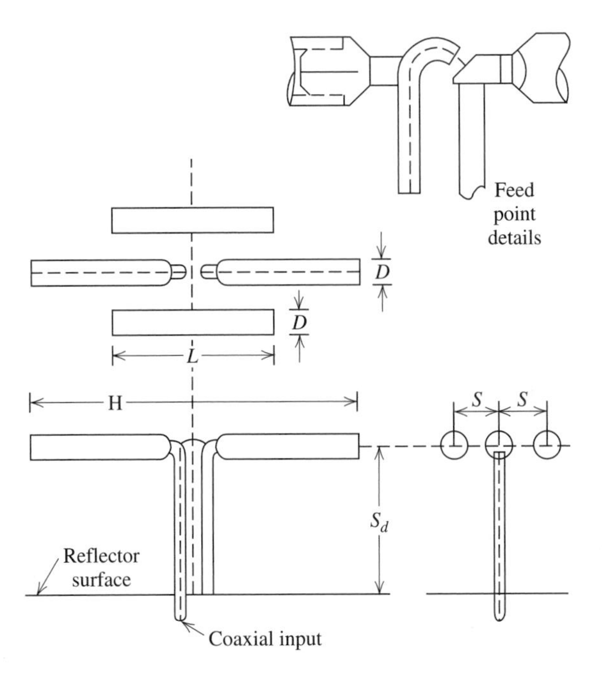

Figure 6-23 The open-sleeve dipole antenna with a flat reflector shown in front, top, and side views.

is approximately one-half that of the center-fed dipole. The open-sleeve dipole, which we will describe from an experimental viewpoint, is operated in front of a flat reflector, or ground plane [22]. The results are also applicable to sleeve dipoles without a flat reflector present.

The antenna was designed for the 225- to 400-MHz frequency band. The dipole to reflector spacing S_d was chosen to be 0.29λ at 400 MHz to avoid the deterioration of the radiation pattern that occurs for larger spacings. All the dimensions required for the design of the open-sleeve dipole are given in Table 6-2. These design values yield low VSWR over a wide bandwidth. This is illustrated in Fig. 6-24 by a comparison of the VSWR characteristics of a conventional (unsleeved) dipole and an open-sleeve dipole with a diameter D of 2.9 cm. Although these results do not represent exhaustive design data for the open-sleeve dipole, they do serve as a starting point in the design of open-sleeve dipoles with or without a reflector present.

Table 6-2 Electrical Dimensions of an Open-Sleeve Dipole with a Reflector for Lowest VSWR

Parameter (see Fig. 6-23)	Electrical Dimension at Lowest Frequency (225 MHz)	Electrical Dimension at Highest Frequency (400 MHz)
D	0.026λ	0.047λ
H	0.385λ	0.684λ
L	0.216λ	0.385λ
S	0.0381λ	0.0677λ
S_d	0.163λ	0.29λ

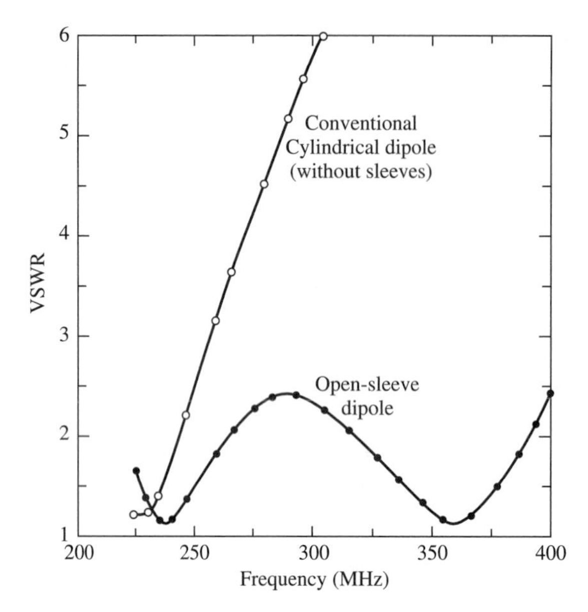

Figure 6-24 Comparison between the VSWR response of a conventional (unsleeved) cylindrical dipole and an open-sleeve dipole both with a diameter D of 2.9 cm [22].

6.5 PRINCIPLES OF FREQUENCY-INDEPENDENT ANTENNAS

Aperture antennas, to be discussed in the next chapter, are capable of bandwidths of 2:1 or more, but the main beam narrows as frequency is increased. Often, it is desirable to have the pattern of an antenna remain constant over a very wide range of frequencies. An antenna with a bandwidth of about 10:1 or more is referred to as a **frequency-independent antenna**. The purest form of a frequency-independent antenna has constant pattern, impedance, polarization, and phase center with frequency. Few antennas meet these criteria. The axial mode helix has constant impedance and phase center location over a bandwidth of about 2:1, but the main beam narrows with increasing frequency. The principles of frequency independence are discussed in this section and antennas capable of 10:1 bandwidth are introduced in the next two sections.

The biconical antenna represents the emergence of frequency-independent behavior. In Sec. 6.3, we found that the input impedance and pattern of the infinite biconical antenna were independent of frequency. This is precisely the behavior we desire. The feature of the biconical structure that is responsible for frequency independence is the emphasis on angles and the complete avoidance of finite lengths. This is verified by the observation that when the bicone is truncated to form the finite biconical antenna of Fig. 6-17, its bandwidth is limited. Of course, in general, if no finite lengths are present on an antenna, the structure would have to be infinite. Rumsey [23] noted that, in practice, frequency-independent antennas are designed to minimize finite lengths and maximize angular dependence. The concept of angle emphasis has been exploited to produce a family of frequency-independent antennas.

There is another property in addition to angle emphasis that leads to frequency-independent behavior, that of self-complementarity. Consider a metal antenna with input impedance Z_{metal}. A dual structure can be formed by replacing the metal with air and replacing air with metal. The resulting *complementary* antenna has input impedance Z_{air}. Complementary antennas are similar to a positive and negative in photography. An example is a ribbon dipole and its complement, the slot antenna, shown in Fig. 6-25.

Babinet's principle can be used to find the impedance of complementary antennas. Babinet's principle for optics states that a source of light behind complementary thin conducting sheets produces lit regions on the source-free side that when superposed give a completely lit region, just as would exist without the sheets present. Extending this to electromagnetics leads to the following important relationship for the input impedances of complementary antennas [24, p. 16]:

$$Z_{metal} Z_{air} = \frac{\eta^2}{4} = \frac{(376.7)^2}{4} = 35{,}475.7\ \Omega \qquad (6\text{-}55)$$

This assumes that no dielectric or magnetic materials are present; if so, the proper η must be used in place of the free-space value. If the dipole of Fig. 6-25*a* is resonated by reducing its length slightly below a half-wavelength, its impedance is $Z_{metal} = Z_{dipole} = 70\ \Omega$. Then from (6-55), the impedance for the slot antenna of Fig. 6-25*b*, Z_{air}, is

$$Z_{slot} = \frac{\eta^2}{4Z_{dipole}} = \frac{35{,}475.7}{4(70)} = 506.8\ \Omega \qquad \textit{half-wave slot antenna} \qquad (6\text{-}56)$$

Problem 6.5-2 addresses the slot that is complementary to the ideal dipole.

The product of the impedances of two complementary antennas is the constant $\eta^2/4$. If the antenna is its own complement, frequency-independent impedance behavior is achieved. This is the *self-complementary* property, in which the antenna and its complement are identical. A self-complementary structure can be made to exactly overlay its complement through translation and/or rotation. The value of impedance follows directly from (6-56), as noted by Mushiake [24]:

$$Z_{metal} = Z_{air} = \frac{\eta}{2} = 188.5\ \Omega \qquad \textit{self-complementary antenna} \qquad (6\text{-}57)$$

The frequency-independent impedance of (6-57) is the second design principle for frequency-independent antennas; that is, self-complementary antennas tend to be frequency-independent. It turns out, however, that many antennas that are not self-complementary still have small impedance variations with frequency.

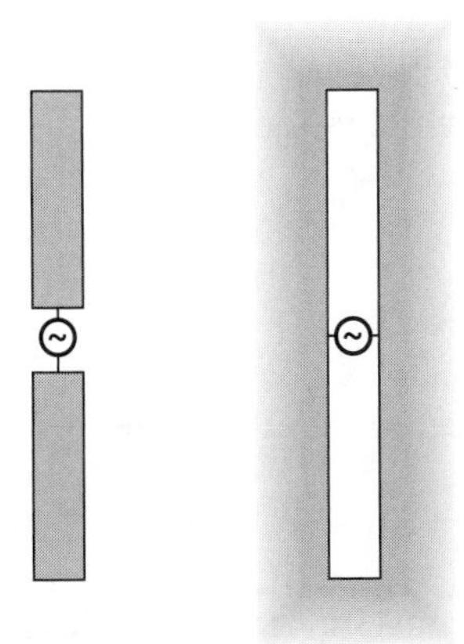

(*a*) Ribbon dipole (*b*) Slot antenna **Figure 6-25** Complementary dipole and slot antennas.

Our study of the antennas in this chapter has led to a number of characteristics that are likely to produce broadband behavior. Before moving on to antenna types that yield extremely wide bandwidth in the following sections, we summarize these properties. The characteristics that yield broadband behavior are:

1. ***Emphasis on angles rather than lengths.*** Examples are the helix in Sec. 6.2 and the spiral in Sec. 6.6, which both avoid fixed physical length elements and produce wide bandwidth.
2. ***Self-complementary structures.*** The equiangular spiral of Fig. 6-27 is an example.
3. ***Thick metal–"fatter is better."*** Increasing the wire diameter of even resonant antennas such as a dipole widens its bandwidth; see Fig. 5-7. The biconical antenna is the ultimate fat dipole and has wide bandwidth; the biconical antenna also emphasizes angles. The bow-tie antenna of Fig. 6-32 is another example.

Ideally, frequency-independent antennas should display all three of these properties. It is found in practice that successful wide bandwidth designs emphasize these properties, but in many cases strict adherence is not required. For example, we will see in Sec. 6.7 that some log-periodic antennas deviate from the self-complementary principle and still have wide bandwidth. The usual penalty for doing this is that the impedance will vary with frequency and not be constant as predicted with (6-57). This may not be a serious problem in many applications.

A distinguishing feature of frequency-independent antennas is their *self-scaling* behavior. Most radiation takes place from that portion of the frequency-independent antenna where its width is a half-wavelength or the circumference is one wavelength—the so-called *active region*. Radiation is maximum perpendicular to the plane of the structure and can be explained in a fashion similar to the one-wavelength loop discussed Sec. 5.7. As frequency decreases, the active region moves to a larger portion of the antenna, where the width is a half-wavelength. The characteristics of angle emphasis and using thick metal yield structures that provide regions where the current can adjust as the frequency changes.

Frequency-independent antennas can be divided into two types: spiral antennas and log-periodic antennas. Spirals are discussed in the next section and log-periodics are treated in the following section.

6.6 SPIRAL ANTENNAS

Spiral antennas and their variations are usually constructed to be either exactly or nearly self-complementary. This yields extremely wide bandwidth, up to 40:1. Historically, the equiangular spiral was invented first, so we begin our discussions with it [25, 26].

6.6.1 Equiangular Spiral Antenna

The equiangular spiral curve shown in Fig. 6-26 is represented by the generating equation

$$r = r_0 e^{a\phi} \tag{6-58}$$

where r_0 is the radius for $\phi = 0$ and a is a constant controlling the flare rate of the spiral. The spiral of Fig. 6-26 is right-handed. Left-handed spirals can be generated

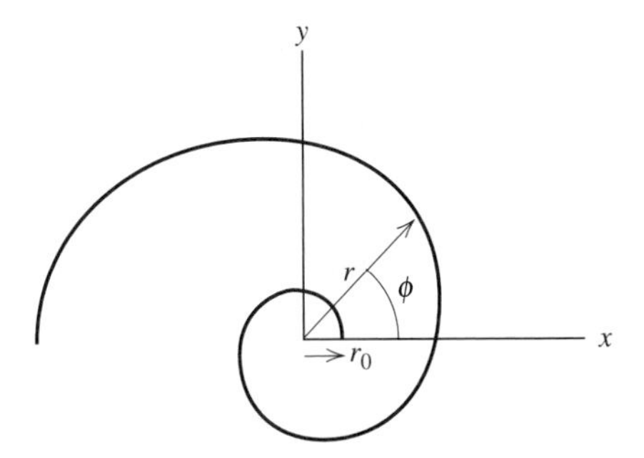

Figure 6-26 Equiangular spiral curve with $r = r_0 e^{a\phi}$ and $r_0 = 0.311$ cm and $a = 0.221$.

using negative values of a, or by simply turning over the spiral of Fig. 6-26. The equiangular spiral curve is used to create the antenna of Fig. 6-27, which is referred to as the **planar equiangular spiral antenna**. The four edges of the metallic region each have an equation for their curves of the form in (6-58). In particular, edge no. 1 is exactly that of Fig. 6-26, so $r_1 = r_0 e^{a\phi}$. Edge no. 2 has the same spiral curve but rotated through the angle δ, so $r_2 = r_0 e^{a(\phi-\delta)}$. The other half of the antenna has edges that make the structure symmetric; that is, rotating one spiral arm by one-half turn brings it into congruence with the other arm. So, $r_3 = r_0 e^{a(\phi-\pi)}$ and $r_4 = r_0 e^{a(\phi-\pi-\delta)}$. The structure of Fig. 6-27 is self-complementary, so $\delta = \pi/2$. It does not have to be constructed this way, but pattern symmetry is best for the self-complementary case.

The impedance, pattern, and polarization of the planar equiangular spiral antenna remain nearly constant over a wide range of frequencies. The feed point at the center, the overall radius, and the flare rate affect the performance. The flare rate a is more conveniently represented through *expansion ratio* ε, which is the increase factor of the radius for one turn of the spiral:

$$\varepsilon = \frac{r(\phi + 2\pi)}{r(\phi)} = \frac{r_0 e^{a(\phi+2\pi)}}{r_0 e^{a\phi}} = e^{a2\pi} \tag{6-59}$$

A typical value for ε is 4, and then from (6-59), $a = 0.221$. The frequency at the upper end of the operating band f_U is determined by the feed structure. The min-

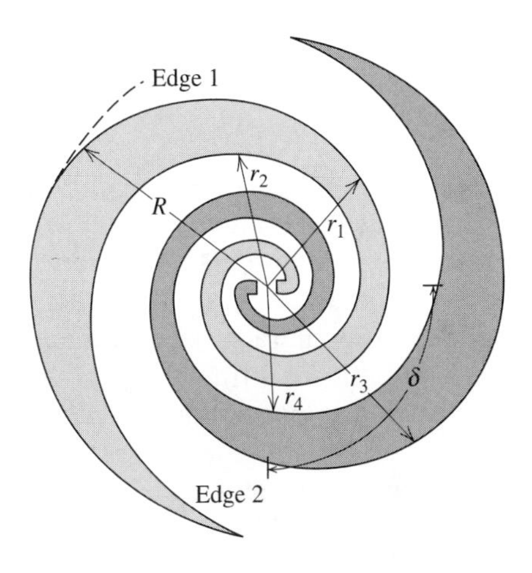

Figure 6-27 Planar equiangular spiral antenna for the self-complementary case with $\delta = 90°$.

imum radius r_0 is about a quarter-wavelength at f_U for an expansion ratio ε of 4 [23]. A nearly equivalent criterion is a circumference in the feed region of $2\pi r_0 = \lambda_U = c/f_U$. Of course, the spiral terminates at this point and is connected to the feed transmission line. The low-frequency limit is set by overall radius R, which is roughly a quarter-wavelength at f_L. Alternatively, the circumference of a circle just enclosing the spiral can be used to set the low-frequency limit through $C = 2\pi R = \lambda_L$.

Spirals with one-half to three turns have been found experimentally to be relatively insensitive to the parameters a and δ [25]. One and one-half turns is about optimum. For example, again consider a one and one-half turn spiral with $a = 0.221$ as shown in Fig. 6-27. Each edge curve is of the form in (6-59), so the maximum radius is $R = r(\phi = 3\pi) = r_0 e^{0.221(3\pi)} = 8.03 r_0$. This equals $\lambda_L/4$, where λ_L is the wavelength at the lower band edge frequency. At the feed point, $r = r(\phi = 0) = r_0 e^0 = r_0$, and this equals $\lambda_U/4$ where λ_U is the wavelength at the upper band edge. The bandwidth is then $f_U/f_L = \lambda_L/\lambda_U = \lambda_L/4/\lambda_U/4 = 8.03$. This 8:1 bandwidth is typical; however, bandwidths of 40:1 can be obtained.

Based on (6-57), the self-complementary equiangular spiral should have an input impedance value of $188.5 + j0\ \Omega$. In practice, the measured impedance values tend to be lower than this (about $120 + j0\ \Omega$), due to the finite thickness of the metallization and the presence of the coaxial feed line that is wound along one arm toward the feed at the center [25]. A feed of this type is referred to as an *infinite balun*. The balance function arises because any currents that are excited on the outside of the coax travel out from the feed point at the center, acting essentially like the currents on the arm and radiating upon reaching the active region. To maintain symmetry, a dummy coax is often attached to the second arm.

The radiation pattern of the self-complementary planar equiangular spiral antenna is bidirectional with two wide beams broadside to the plane of the spiral. The field pattern is approximately $\cos\theta$ when the z-axis is normal to the plane of the spiral. The half-power beamwidth is, thus, approximately 90°. The polarization of the radiation is close to circular over wide angles, out to as far as 70° from broadside. The sense of the polarization is determined by the sense of the flare of the spiral. For example, the spiral of Fig. 6-27 radiates in the right-hand sense for directions out of the page and left-hand sense for directions into the page.

6.6.2 Archimedean Spiral Antenna

Another form of the planar spiral is the **Archimedean spiral antenna** shown in Fig. 6-28. This antenna, as are many spiral antennas, is easily constructed using printed circuit techniques. The equation of the two spirals in Fig. 6-28 are

$$r = r_0\phi \qquad \text{and} \qquad r_0(\phi - \pi) \tag{6-60}$$

The Archimedean spiral is linearly proportional to the polar angle rather than an exponential for the equiangular spiral, and thus flares much more slowly.

The simple geometry of the Archimedean spiral antenna affords an opportunity to explain an important operating principle in frequency-independent antennas. This is the "band" description of radiation that is characterized by an active region responsible for radiation. Between the feed point of a frequency-independent antenna and the active region, currents exist in a transmission line mode and fields arising from them cancel in the far field. The active region occurs on that portion of the antenna that is one wavelength in circumference for curved structures or has half-wavelength-long elements in antennas with straight wires or edges. Beyond the

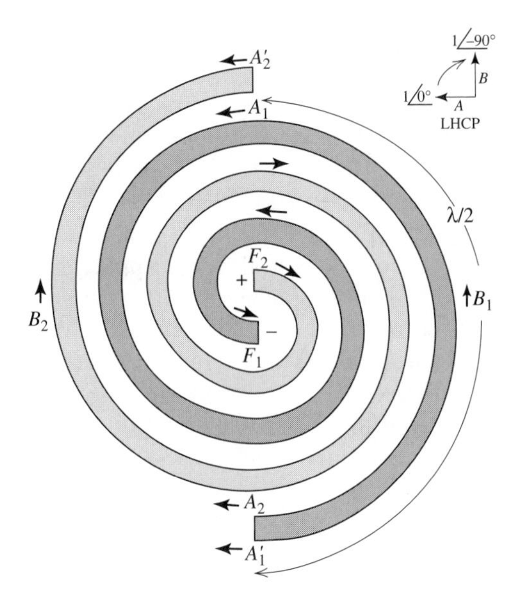

Figure 6-28 The Archimedean spiral antenna. The outside circumference in this case is one-wavelength and, thus, is the active region. The inset is a vector diagram for the radiated electric fields, showing that the outward radiation is left-hand circularly polarized.

active region, currents are small, having lost power to radiation in the active region. The antenna effectively behaves as if it is infinite in extent. Of course, the active region moves around the antenna with frequency. Since the geometry of a spiral is smooth, as frequency is reduced and the active region shifts to locations farther out on the spiral, the electrical performance remains unchanged. Hence, self-scaling occurs and frequency-independent behavior results.

We now give a physical explanation of how spiral antennas operate using Fig. 6-28. The arms are fed 180° out-of-phase at points F_1 and F_2. This is represented with oppositely directed current arrows. The current is inward for arm no. 1 $(-)$ and outward for arm no. 2 $(+)$. The lengths of the arms out to A, F_1A_1 and F_2A_2, are equal so the phase shifts from the feed to A are identical, preserving the current directions as shown in Fig. 6-28. The active region where the circumference is one wavelength contains points labeled with an A or B. It can be assumed that the current magnitudes over this region are nearly the same. The phase, however, shifts as the traveling waves progress along the arms. Since the circumference is electrically large in the active region, phase must be accounted for. The phase shifts 180° between A_1 and A_1' and between A_2 and A_2', because of the $\lambda/2$ differential path length. Adjacent points on different arms (A_1, A_2', and A_2, A_1') are now in-phase because the 180° phase shift counters the direction reversal introduced by the half-turn. In addition, the points opposite these pairs are in-phase; that is, A_1, A_2' are in-phase with A_2, A_1'. This in-phase condition leads to reinforcement of electric fields in the broadside direction, giving a radiation maximum. Interior to the active region, the electrical distance along different arms to adjacent points is not electrically large, preserving the antiphase condition due to the excitation. This is a transmission line mode and radiation is low. Often, resistive loads are added to the ends of the spiral to prevent reflection of the remaining traveling waves.

The final aspect that requires explanation is the circular polarization property. In the active region, points that are one-quarter turn around the spiral are 90° out-of-phase. For example, the phase at point B_1 lags that at point A_1 by 90°. In addition, the currents are orthogonal in space. The current magnitudes are also nearly equal.

Thus, all conditions are satisfied for circular polarized radiation: The radiated fields (created by the currents) are orthogonal, equal in magnitude, and 90° out-of-phase. As indicated by the vector diagram insert in Fig. 6-28, the wave is left-hand circular polarized. The left-hand sense results from the left-hand winding of the spiral. This is for radiation out of the page. Viewed from the other side of the page, the spiral is right-hand wound and thus produces RHCP.

Based on the above discussion, it is apparent that the spiral produces a broad main beam perpendicular to the plane of the spiral. Most applications require a unidirectional beam. This is created by backing the spiral with a ground plane. The most common construction approach is to use a metallic cavity behind the spiral, forming a **cavity-backed Archimedean spiral antenna**. This introduces a fixed physical length (the distance to the ground plane), thereby altering the true frequency-independent behavior. This is corrected in most commercial units by loading the cavity with absorbing material to reduce resonance effects; this, however, introduces loss. Typical performance parameter values for the cavity-backed Archimedean spiral are $HP = 75°$, $|AR| = 1$ dB, $G = 5$ dB over a 10:1 bandwidth or more. The input impedance is about 120 Ω, and is nearly real. The performance of the equiangular spiral is similar to that for the Archimedean spiral.

Very wideband antennas such as spirals are balanced structures. They are normally connected to a coaxial cable, which is an unbalanced structure. Therefore, a balun must be included with the spiral feed; see Sec. 5.3 for discussions of balun

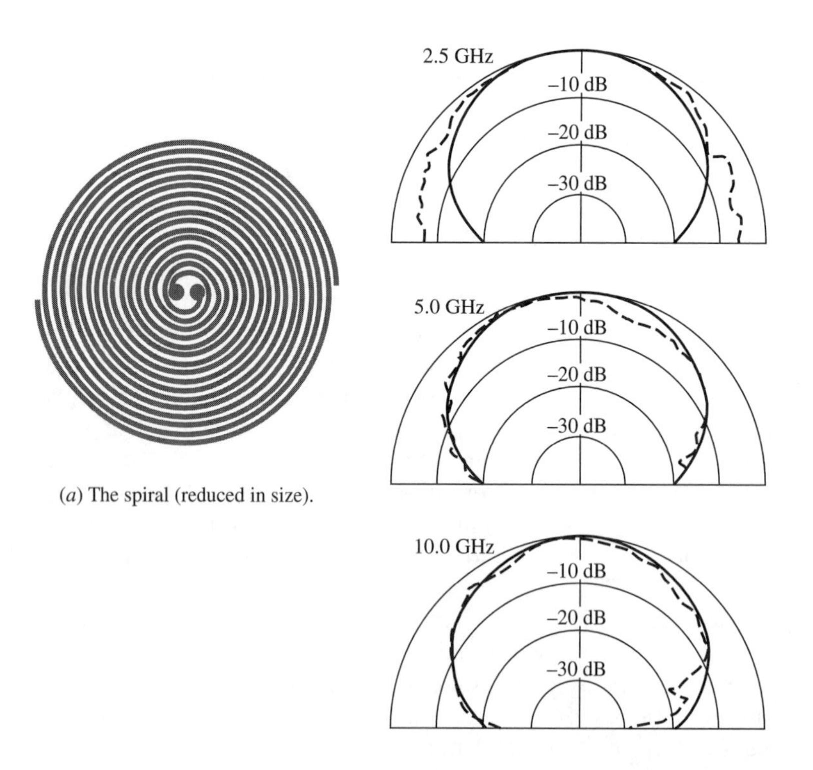

(*a*) The spiral (reduced in size).

(*b*) Radiation patterns: measured (dashed curve) and computed (solid curve) using $\cos^{5.8}(0.53\,\theta)$.

Figure 6-29 A 4:1 bandwidth cavity-backed Archimedean spiral antenna.

principles. One such wideband balun is the *tapered-coax wideband balun* that is formed by gradually cutting away the outer conductor of the coaxial cable, leaving a parallel wire line to attach to the spiral arms [27] as shown in Fig. 5-27*b*.

The pattern of the cavity-backed Archimedean spiral can be modeled by the following empirically derived function [28]:

$$F(\theta) = \cos^{5.8}(0.53\ \theta) \tag{6-61}$$

This pattern model has HP = 74°. Figure 6-29*a* shows an experimental model that has a diameter of 5.4 cm. This corresponds to a low-frequency cutoff of $f_L = c/\lambda = c/\pi D = 30/\pi 5.4 = 1.77$ GHz. The measured patterns are shown in Fig. 6-29*b* for three frequencies: 2.5 (slightly above cutoff), 5, and 10 GHz. Also plotted is the pattern model of (6-61). Note that the pattern remains nearly constant over a 4:1 bandwidth, characteristic of frequency-independent antennas. The test antenna patterns in Fig. 9-15 are for an Archimedian spiral.

6.6.3 Conical Equiangular Spiral Antenna

Nonplanar forms of spiral antennas are used to produce a single main beam, thereby avoiding a backing cavity. For example, the planar equiangular spiral antenna conformed to a conical surface forms the **conical equiangular spiral antenna** shown in Fig. 6-30. The equation for a conical equiangular spiral curve is

$$r = e^{(a \sin \theta_h)\phi} \tag{6-62}$$

The planar spiral is a special case of this with $\theta_h = 90°$. The equations for the edges of one spiral of metal are that of (6-62) for r_1, and $r_2 = e^{(a \sin \theta_h)(\phi - \delta)}$, and $\delta = \pi/2$ for the self-complementary case. The other spiral arm is produced by a 180° rotation. The edges of the arms maintain a constant angle α with a radial line for any cone half-angle θ_h [23]:

$$a = \cot \alpha \tag{6-63}$$

The conical equiangular spiral antenna has a single main beam that is directed off the cone tip in the $-z$-direction. A self-complementary shape yields the best

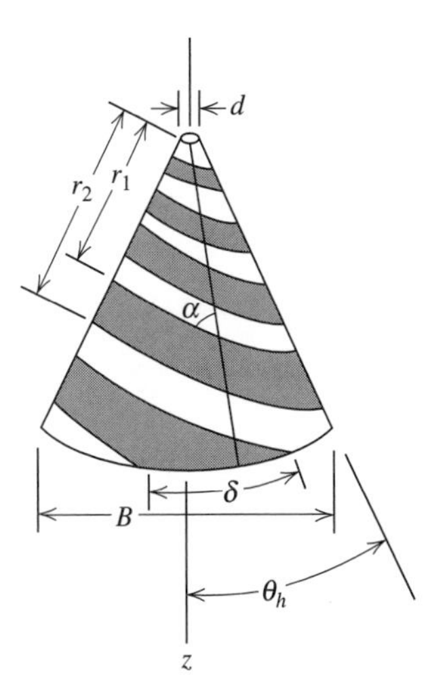

Figure 6-30 The conical equiangular spiral antenna.

radiation patterns. Typical patterns for $\theta_h \leq 15°$ and α about 70° have a broad main beam with a maximum in the $\theta = 180°$ direction and a half-power beamwidth of about 80°. Since the structure is rotationally symmetric, the pattern is also nearly rotationally symmetric. The polarization of the radiated field is very close to circular in all directions, with the sense determined by the sense of the spiral winding. However, the polarization ellipticity increases with the off-axis angle.

The impedance [23] can be approximated by the relation

$$Z_0 \approx 300 - 1.5\delta(\text{degrees})\ \Omega \tag{6-64}$$

where δ(degrees) is the angle δ of Fig. 6-30 in degrees. For the self-complementary case, δ is 90° and from (6-61) $Z_o \approx 165\ \Omega$, which is close to the 188.5-Ω theoretical value of (6-57). The impedance is not affected significantly by θ_h or α.

The design of the conical equiangular spiral antenna is rather simple and proceeds as follows [29]. The upper frequency f_U of the operating band occurs when the truncated apex diameter is a quarter-wavelength; that is, $\lambda_U/4$. The lower band edge frequency f_L is determined by the base diameter B and occurs for $B = 3\lambda_L/8$. θ_h is usually less than 15° and α about 70°. For $\theta_h = 10°$ and $\alpha = 73°$, the front-to-back ratio is 15 dB and the polarization axial ratio on the axis is below 3 dB.

6.6.4 Related Configurations

Spiral antennas can be operated in other than the fundamental odd mode described above, where the feed point terminals are excited in antiphase. Higher-order modes are possible and have an active region where the circumference is an odd multiple (3, 5, . . .) of a wavelength. Even modes can be created by feeding the arms in-phase, leading to a null broadside to the spiral. This pattern is useful in direction finding (DF) [30].

Square versions of the planar Archimedean spiral offer size reduction of 22%. This is based on the increase in perimeter from πD to $4D$. See [31] for discussions of square spirals as well as other variations of the spiral antenna.

A broadband antenna that is related to the spiral is the **sinuous antenna** [32]. As can be seen from Fig. 6-31, the sinuous antenna is more complicated than the spiral antenna. However, it offers more flexible polarization uses. Two opposite arm pairs produce orthogonal linear polarizations. These pairs can be used separately for polarization diversity or for transmit/receive operation. Or, the two-arm-pair out-

Figure 6-31 The sinuous antenna.

puts can be combined to produce simultaneous LHCP and RHCP. The operating principles are very similar to those for the planar spirals.

6.7 LOG-PERIODIC ANTENNAS

The spiral antennas of the previous section illustrate the principle that emphasis on angles will lead to a broadband antenna. Although spiral antennas are not complex structures, construction would be simplified if simple geometries, involving circular or straight edges, could be utilized. Antennas of this type are discussed in this section. To see how the ideas develop, first consider the **bow-tie antenna** (also called the **bifin antenna**) of Fig. 6-32. It is the planar version of the finite biconical antenna (see Fig. 6-17). It has a bidirectional pattern with broad main beams perpendicular to the plane of the antenna. It is also linearly polarized. The bow-tie antenna is used as a receiving antenna for UHF TV channels, frequently with a wire grid ground plane behind it to reduce the back lobe. Since currents are abruptly terminated at the ends of the fins, the antenna has limited bandwidth. As we shall see shortly, by modifying the simple bow-tie antenna as shown in Fig. 6-33, the currents will then die off more rapidly with distance from the feed point. The introduction of periodically positioned teeth distinguishes this antenna as one of a broad class of log-periodic antennas. A **log-periodic antenna** is an antenna having a structural geometry such that its impedance and radiation characteristics repeat periodically as the logarithm of frequency. In practice, the variations over the frequency band of operation are minor, and log-periodic antennas are usually considered to be frequency-independent antennas.

Most of the work on frequency-independent antennas took place at the University of Illinois in the late 1950s and the 1960s [33]. A series of antennas were developed through many experiments. (For an excellent historical discussion of this evolution, see [34].) Several geometries were examined, and those that produced broadband behavior led to the determination of the properties necessary for wide bandwidth.

Frequency-independent spiral antennas were discussed in the previous section. In this section, we outline the development of the log-periodic antenna family. The metamorphosis of the log-periodic produced the log-periodic dipole antenna, which is made up of only straight wire segments.

One of the first log-periodic antennas was the **log-periodic toothed planar antenna** shown in Fig. 6-33. It is similar to the bow-tie antenna except for the teeth. The teeth act to disturb the currents that would flow if the antenna were of bow-tie-type construction. Currents flow out along the teeth and, except at the frequency limits,

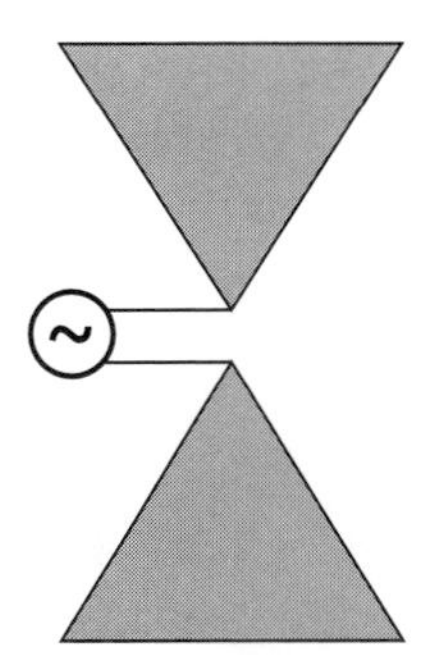

Figure 6-32 The bow-tie antenna.

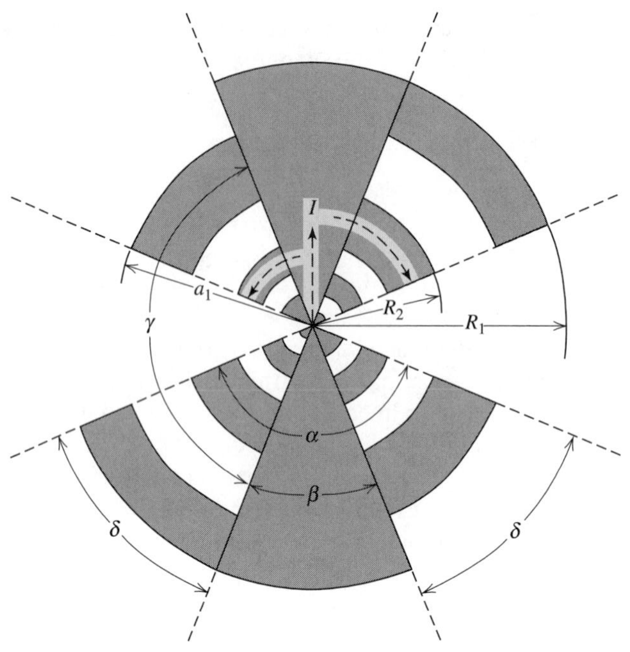

Figure 6-33 Log-periodic toothed planar antenna (self-complementary). Midband currents shown on top half only.

are not significant at the ends of the antenna. The rather unusual shape of this antenna is explained by examining the planar equiangular spiral antenna. Along a radial line from the center of the spiral, the positions of the far (or near) edges of a conductor from (6-58) are

$$r_n = r(\phi + n2\pi) = r_0 e^{a(\phi + n2\pi)} \tag{6-65}$$

The ratio of the $n + 1$th position to the nth position is

$$\frac{r_{n+1}}{r_n} = \frac{r_0 e^{a(\phi + (n+1)2\pi)}}{r_0 e^{a(\phi + n2\pi)}} = e^{a2\pi} = \varepsilon \tag{6-66}$$

which is the expansion ratio of (6-59). This is a constant, and thus the distances (or period) of the edges are of constant ratio for the planar spiral. For the structure of Fig. 6-33, the ratio of edge distances is also a constant and is given by the following scale factor:

$$\tau = \frac{R_{n+1}}{R_n} < 1 \tag{6-67}$$

The slot width is expressed by

$$\sigma = \frac{a_n}{R_n} < 1 \tag{6-68}$$

These relations are true for any n. The parameter τ gives the period of the structure. We would thus expect periodic pattern and impedance behavior with the same period. In other words, if frequencies f_{n+1} and f_n from adjacent periods lead to identical performance, then

$$\frac{f_n}{f_{n+1}} = \tau, \qquad f_n < f_{n+1} \tag{6-69}$$

Forming $f_{n+1} = f_n/\tau$ from this equation and taking the logarithm of both sides, we have

$$\log f_{n+1} = \log f_n + \log(1/\tau) \tag{6-70}$$

Thus, the performance is periodic in a logarithmic fashion and, hence, the name log-periodic antenna. All log-periodic antennas have this property.

If the teeth sizes of the log-periodic toothed planar antenna are adjusted properly, the structure can be made self-complementary. From Fig. 6-33, we see that in general (whether self-complementary or not)

$$\gamma + \beta = 180° \qquad \text{and} \qquad \beta + 2\delta = \alpha \tag{6-71}$$

If the structure is self-complementary (as shown),

$$\alpha = \gamma \qquad \text{and} \qquad \beta = \delta \tag{6-72}$$

Substituting (6-72) into (6-71) yields $\alpha + \beta = 180°$ and $\beta + 2\beta = \alpha$. Solving these two equations gives

$$\alpha = 135° \qquad \text{and} \qquad \beta = 45° \tag{6-73}$$

for a self-complementary log-periodic toothed planar antenna. As we saw in the previous section, an antenna that is self-complementary tends to be broadband and has an input impedance of 188.5 Ω.

If the widths of the teeth and gaps are equalized, $\sigma = a_n/R_n = R_{n+1}/a_n$. Using (6-67) and solving for σ give

$$\sigma = \sqrt{\tau} \tag{6-74}$$

This relationship and the self-complementary feature are popular in practice.

The properties of the log-periodic toothed planar antenna depend on τ. It has been found experimentally that the half-power beamwidth increases with increasing values of τ [23], increasing from about 30° at $\tau = 0.2$ to about 75° at $\tau = 0.9$. The pattern has two lobes with maxima in each normal direction to the plane of the antenna. The radiation is linearly polarized parallel to the teeth edges. This is perpendicular to what it would be if there were no teeth ($\delta = 0$), in which case the antenna would be a bow tie. The fact that transverse current flow dominates over radial current flow is significant. Most of the current appears on teeth that are about a quarter-wavelength long (the active region). This, we have seen, is key to achieving wide bandwidths. The frequency limits of operation are set by the frequencies where the largest and smallest teeth are a quarter-wavelength long.

The log-periodic toothed planar antenna should have a performance (impedance and pattern) that repeats periodically with frequency with period τ given by (6-67). The self-complementary version of the antenna, although not producing frequency-independent operation, does lead to performance that does not vary greatly for frequencies between periods, that is, for $f_n < f < f_{n+1}$. In fact, measurements have produced nearly identical patterns over a 10:1 bandwidth [23].

The **log-periodic toothed wedge antenna** of Fig. 6-34 is a unidirectional pattern form of its planar version in Fig. 6-33, in which the included angle ψ is 180°. A single broad main beam exists in the $+z$-direction. The patterns are nearly frequency-independent for $30° < \psi < 60°$. The polarization is linear and y-directed for an on-axis radiation, as indicated in Fig. 6-34. There is a small cross-polarized component (x-directed) arising from the radial current mode, as found in a biconical antenna. Typically, this cross-polarized component is 18 dB down from the copolarized (y-directed) component on-axis, indicating a strong excitation of the transverse current mode associated with frequency-independent behavior. The bandwidth of the wedge version is similar to the sheet version, but the input impedance is reduced for decreasing ψ. For the planar case ($\psi = 180°$), the self-complementary antenna, which should have an impedance of 188.5 Ω, has an impedance of about 165 Ω, whereas the wedge form with $\psi = 30°$ has a 70-Ω impedance. As ψ is decreased, the impedance variation over a period of the structure (frequency ratio of τ) increases. For example, a 3:1 variation occurs for $\psi = 60°$ relative to the geometric mean [23].

From a construction standpoint, it would be desirable if the toothed antennas could be made with straight edges. This simplification of the structure turns out to be of little consequence in the performance of the antenna. This is another major step in the development of the log-periodic antenna. As an example, if the tooth edges of the log-periodic toothed planar antenna in Fig. 6-33 are replaced by straight edges, the **log-periodic toothed trapezoid antenna** of Fig. 6-35 results. The performance of this antenna is similar to its curved edge version in Fig. 6-33. A **log-periodic toothed trapezoid wedge antenna** can be formed by bending the planar version into a wedge, creating an antenna similar to that of Fig. 6-34. In fact, the patterns of the two wedge forms (curved edge and trapezoid) are similar, but the trapezoid version has better impedance performance with only about a 1.6:1 variation over a period for $\psi = 60°$ [23].

The solid metal (or sheet) antennas we have described are practical for short wavelengths, but for low frequencies the required structures can become rather impractical. It turns out that the sheet antennas can be replaced by a wire version in which thin wires are shaped to follow the edges of the sheet antenna. An example of this major structural simplification is that of Fig. 6-36*a*, which is the wire version of Fig. 6-35. This **log-periodic trapezoid wire antenna** can also be bent at the apex to form a wedge that produces a unidirectional pattern. The **log-periodic trapezoid**

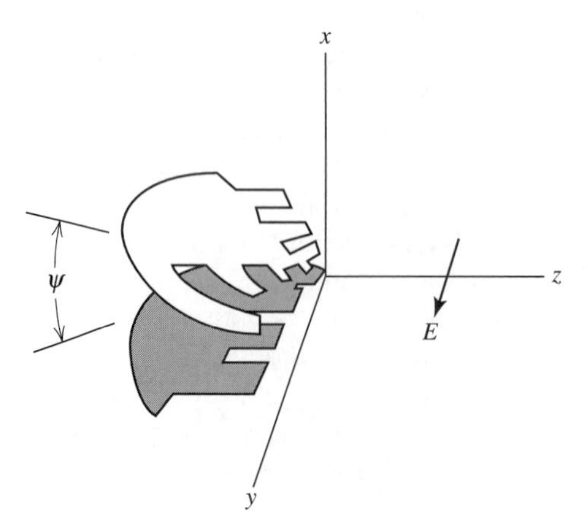

Figure 6-34 Log-periodic toothed wedge antenna.

Figure 6-35 Log-periodic toothed trapezoid antenna.

wedge wire antenna has a performance similar to its sheet version. Measurements for a wedge angle $\psi = 45°$ have yielded E- and H-plane half-power beamwidths of 66°, a gain of 9.2 dB, and a front-to-back ratio of 12.3 dB. The average input impedance has been measured as 110 Ω with a VSWR of 1.45 over a 10:1 band [35]. As with other wedge log-periodics, the main beam maximum is straight off the apex and the radiation is linearly polarized.

Other even simpler log-periodic wire antennas exist in both planar and wedge shapes. The **log-periodic zig-zag wire antenna** of Fig. 6-36*b* is an example.

The final phase in this metamorphosis of log-periodic antennas is the use of only parallel wire segments. This is the **log-periodic dipole array** of Fig. 6-37 [36, 37]. The log-periodic dipole array (LPDA) is a series-fed array of parallel wire dipoles of successively increasing lengths outward from the feed point at the apex. Note that the interconnecting feed lines cross over between adjacent elements. This can be explained by noting that the LPDA of Fig. 6-37 resembles the toothed trapezoid of Fig. 6-35 when folded on itself, making a wedge with zero included angle. The two center fins of metal then form a parallel transmission line with the teeth coming out from them on alternate sides of the fins. This alternate arm geometry occurs for all wedge log-periodic antennas.

A particularly successful method of constructing an LPDA is shown in Fig. 6-38. A coaxial transmission line is run through the inside of one of the feed conductors.

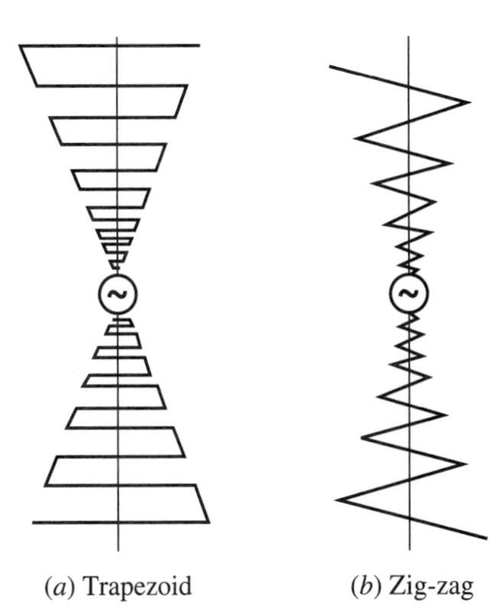

Figure 6-36 Log-periodic wire antenna configurations.

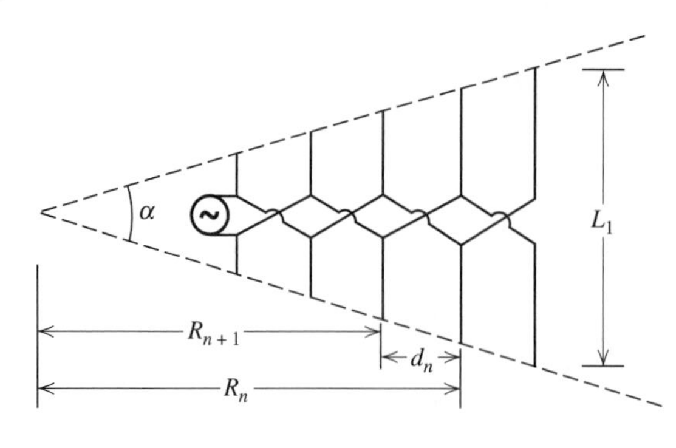

Figure 6-37 Log-periodic dipole array geometry.

The outer conductor of the coax is attached to that conductor and the inner conductor of the coax is connected to the other conductor of the LPDA transmission line.

As shown in Fig. 6-37, a wedge of enclosed angle α bounds the dipole lengths. The scale factor τ for the LPDA is

$$\tau = \frac{R_{n+1}}{R_n} < 1 \tag{6-75}$$

Right triangles of enclosed angle $\alpha/2$ reveal that

$$\tan\frac{\alpha}{2} = \frac{L_n/2}{R_n} = \frac{L_{n+1}/2}{R_{n+1}} \tag{6-76}$$

Thus,

$$\frac{L_1}{R_1} = \cdots = \frac{L_n}{R_n} = \frac{L_{n+1}}{R_{n+1}} = \cdots = \frac{L_N}{R_N} \tag{6-77}$$

Using this result in (6-75) gives

$$\tau = \frac{R_{n+1}}{R_n} = \frac{L_{n+1}}{L_n} \tag{6-78}$$

Thus, the ratio of successive element positions equals the ratio of successive dipole lengths.

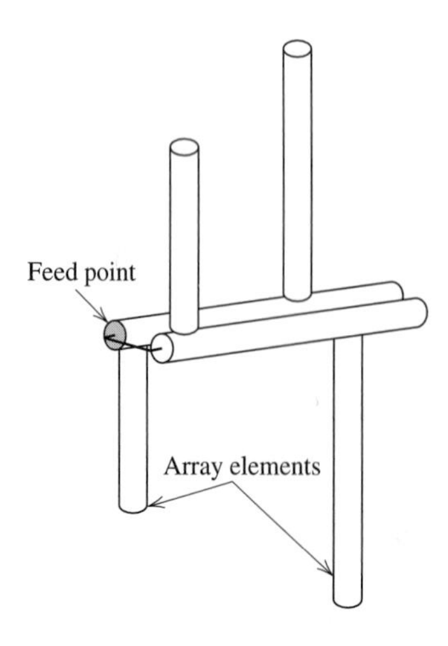

Figure 6-38 Construction details of the log-periodic dipole array.

The spacing factor for the LPDA is defined as

$$\sigma = \frac{d_n}{2L_n} \tag{6-79}$$

where the element spacings as shown in Fig. 6-37 are given by

$$d_n = R_n - R_{n+1} \tag{6-80}$$

But $R_{n+1} = \tau R_n$, so

$$d_n = R_n - \tau R_n = (1 - \tau)R_n \tag{6-81}$$

From (6-76), $R_n = L_n/2 \tan(\alpha/2)$. Using this in (6-81) yields

$$d_n = (1 - \tau)\frac{L_n}{2\tan(\alpha/2)} \tag{6-82}$$

Substituting this in (6-79) gives

$$\sigma = \frac{d_n}{2L_n} = \frac{1 - \tau}{4\tan(\alpha/2)} \tag{6-83}$$

or,

$$\alpha = 2\tan^{-1}\left(\frac{1 - \tau}{4\sigma}\right) \tag{6-84}$$

Combining (6-83) with (6-78), we note that all dimensions are scaled by

$$\tau = \frac{R_{n+1}}{R_n} = \frac{L_{n+1}}{L_n} = \frac{d_{n+1}}{d_n} \tag{6-85}$$

As we have seen with other log-periodic antennas, there is also an active region for the LPDA, where the few dipoles near the one that is a half-wavelength long support much more current than do the other radiating elements. It is convenient to view the LPDA operation as being similar to that of a Yagi–Uda antenna. The longer dipole behind the most active dipole (with largest current) behaves as a reflector and the adjacent shorter dipole in front acts as a director. The radiation is then off of the apex. The wedge enclosing the antenna forms an arrow pointing in the direction of the main beam maximum.

As the operating frequency changes, the active region shifts to a different portion of the antenna. The frequency limits of the operational band are roughly determined by the frequencies at which the longest and shortest dipoles are half-wave resonant, that is,

$$L_1 \approx \frac{\lambda_L}{2} \quad \text{and} \quad L_N \approx \frac{\lambda_U}{2} \tag{6-86}$$

where λ_L and λ_U are the wavelengths corresponding to the lower and upper frequency limits. Since the active region is not confined completely to one dipole, often dipoles are added to each end of the array to ensure adequate performance over the band. The number of additional dipoles required is a function of τ and σ [38, 39]. But for noncritical applications, (6-86) is sufficient.

The pattern, gain, and impedance of an LPDA depend on the design parameters τ and σ. Since the LPDA is a very popular broadband antenna of simple construction, low cost, and light weight, we will give the design details and illustrate them by examples. Gain contours are plotted in Fig. 6-39 as a function of τ and σ [37].

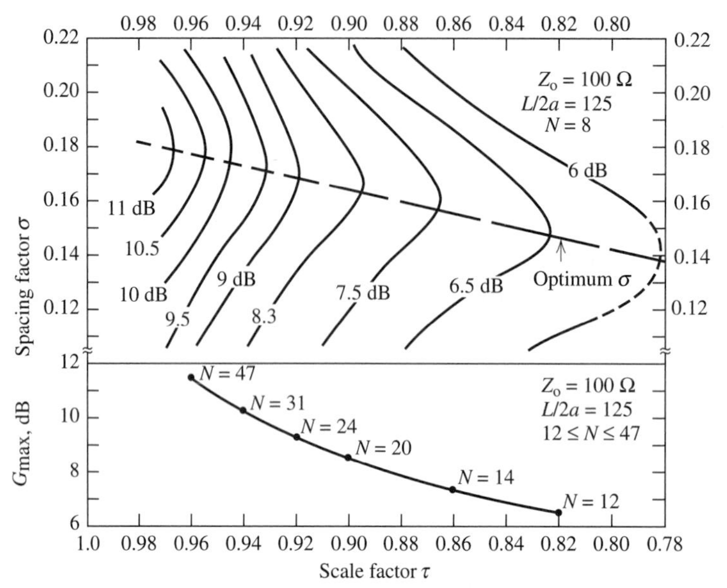

Figure 6-39 Gain of a log-periodic dipole array. {Contours (top) adapted from Carrel [37]. Maximum gain curve (bottom) derived from data in [38]}.

Note that high gain requires a large value of τ, which means a very slow expansion, that is, a LPDA of large overall length. Gain is only slightly affected by the dipole thickness. It increases about 0.2 dB for a doubling of the thickness [23]. Gain is also affected by the feeder impedance [38] and tends to decrease as the feeder impedance is increased above 100 Ω.

Figure 6-39 (top portion) shows the variation in gain of an LPDA with σ and τ. The curves are a modification of those originally presented by Carrel [37] that have been found to have a gain that is erroneously high [38, 40, 41]. In [42], Carrel's curves are reduced uniformly by 1 dB and in [43] uniformly by 1.5 dB. Based on data in [38, 41, 44], it appears that Carrel's original curves were more in error (for gain) for lower values of τ than for higher values. Thus, the 11-dB and 6-dB contours in Fig. 6-39 are 1 dB and 2 dB lower, respectively, than those in Carrel.

The bottom portion of Fig. 6-39 shows a gain curve that is derived from data in [38] where N, the number of dipoles, varies from 12 to 47 (unlike Carrel's modified contours above for which $N = 8$). Notice that the value of G_{max} is greater than the value of the gain contour at the optimum σ line in the top portion of Fig. 6-39. The G_{max} vs. τ curve probably represents an upper bound on the LPDA gain that can be achieved in practice for feeder impedances of 100 Ω or greater.

Further details on the design and calculations for the LPDA are available in the literature [38–41, 43]. Also, the LPDA can be constructed in a size-reduced form or by using printed circuit techniques [46–49].

EXAMPLE 6-2 Optimum Design of a 54- to 216-MHz Log-Periodic Dipole Antenna

It is desired to have an antenna that operates over the entire VHF-TV and FM broadcast bands, which span the 54- to 216-MHz frequency range for a 4:1 bandwidth. Suppose the design gain is chosen to be 6.5 dB. The corresponding values of τ and σ for optimum design from Fig. 6-39 are

$$\tau = 0.822 \qquad \text{and} \qquad \sigma = 0.149 \tag{6-87}$$

Then from (6-84), we have

$$\alpha = 2 \tan^{-1}\left[\frac{1 - 0.822}{4(0.149)}\right] = 33.3° \tag{6-88}$$

The length of the longest dipole is determined first. At the lowest frequency of operation (54 MHz), the dipole length from (6-86) should be near a half-wavelength, so

$$L_1 = 0.5\lambda_L = 0.5(5.55) = 2.78 \text{ m} \tag{6-89}$$

The shortest dipole length should be on the order of $L_U = 0.5\lambda_U = 0.694$ m at 216 MHz. The LPDA element lengths are computed until a length on the order of 0.694 m is reached. To be specific, element lengths are found from L_1 using $L_{n+1} = \tau L_n$. For example,

$$L_2 = \tau L_1 = (0.822)(2.78) = 2.29 \text{ m}$$

and

$$L_3 = \tau L_2 = (0.822)(2.29) = 1.88 \text{ m}$$

Completing this process leads to

$$\begin{aligned} &L_1 = 2.78 \text{ m}, \quad L_2 = 2.29 \text{ m}, \quad L_3 = 1.88 \text{ m}, \quad L_4 = 1.54 \text{ m} \\ &L_5 = 1.27 \text{ m}, \quad L_6 = 1.04 \text{ m}, \quad L_7 = 0.858 \text{ m}, \quad L_8 = 0.705 \text{ m} \\ &L_9 = 0.579 \text{ m} \end{aligned} \tag{6-90}$$

The array was terminated with nine elements since $L_9 = 0.579$ m is less than the 0.694 m length for the highest operating frequency. Elements could be added to either end to improve performance at the band edges.

The element spacings for this example are found from (6-83) as

$$d_n = 2\sigma L_n = 2(0.149)L_n = 0.298L_n \tag{6-91}$$

Using the element lengths of (6-90) gives

$$\begin{aligned} &d_1 = 0.828 \text{ m}, \quad d_2 = 0.682 \text{ m}, \quad d_3 = 0.560 \text{ m}, \quad d_4 = 0.459 \text{ m} \\ &d_5 = 0.378 \text{ m}, \quad d_6 = 0.310 \text{ m}, \quad d_7 = 0.256 \text{ m}, \quad d_8 = 0.210 \text{ m} \end{aligned} \tag{6-92}$$

These dipole lengths and spacings completely specify the LPDA, as shown in Fig. 6-37. The total length of the array is the sum of the spacings in (6-92), which gives a 3.683 m. The outline of the antenna fits into an angular sector of angle $\alpha = 33.3°$.

EXAMPLE 6-3 *Characteristics of a 200- to 600-MHz LPDA*

In this example, we examine the gain, pattern, impedance, and current distribution of a LPDA as a function of frequency. Suppose it is to be constructed for operation over the 200- to 600-MHz band. For optimum performance and a design goal of 9 dB gain, we see from Fig. 6-39 that $\tau = 0.917$ and $\sigma = 0.169$. The lowest frequency of operation (200 MHz) has a wavelength of $\lambda_L = 1.5$ m, so the first element has a length of $L_1 = \lambda_L/2 = 0.75$ m. The length of the shortest element should be on the order of a half-wavelength at 600 MHz, and $\lambda_U/2 = 0.500 \text{ m}/2 = 0.250$ m. Using the design techniques illustrated in the previous example and four extra elements at the narrow end gives the 18-element LPDA shown in Fig. 6-40. (The antenna geometry details are left as a problem.)

The LPDA of Fig. 6-40 was modeled using the computer techniques of Sec. 10.10.2. The

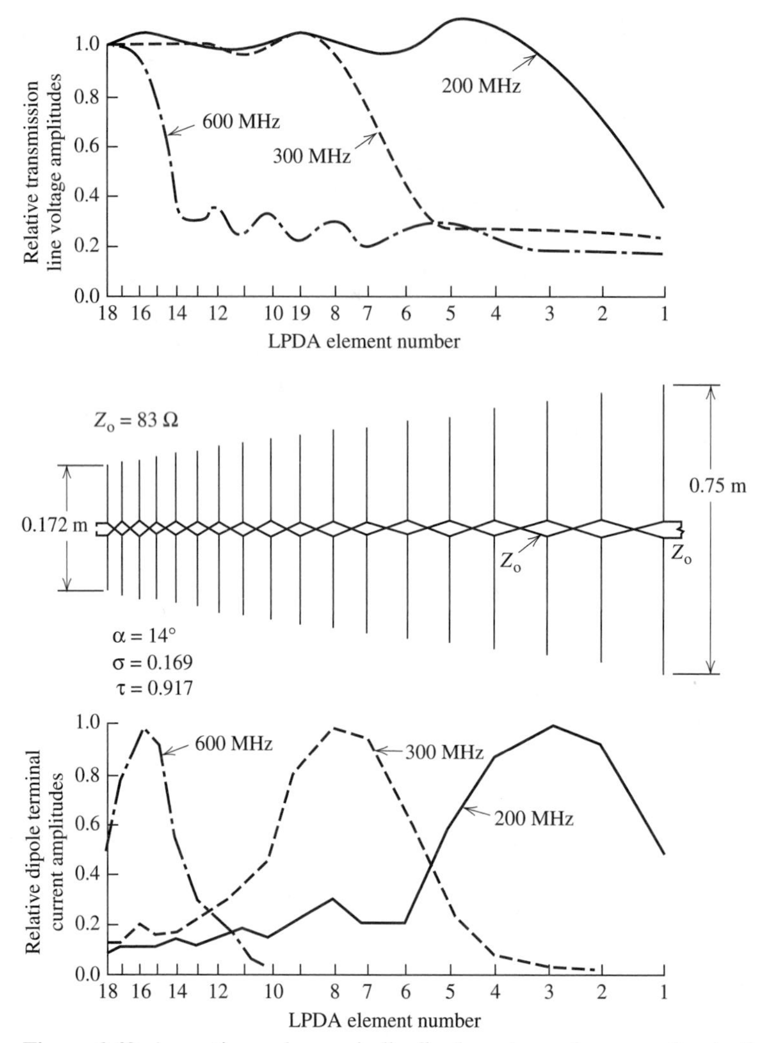

Figure 6-40 An optimum log-periodic dipole antenna for operation in the 200- to 600-MHz band (Example 6-3). (Top) Voltage distribution on the 83-Ω transmission line. (Middle) The geometry. (Bottom) Relative dipole terminal current amplitudes.

resulting dipole terminal currents at the band edges and one midway frequency are shown in Fig. 6-40. Also shown are the voltage amplitude distributions on the transmission line. These currents and voltages illustrate the active region behavior we have mentioned several times. For example, at 200 MHz there are three dipoles with strong currents on them and a total of five with significant currents. This is also true for other frequencies in the operating band, with the active region shifted to some other portion of the antenna as seen in Fig. 6-40. At the high-frequency limit, element 14 is about a half-wavelength long and the extra four elements provide support for the active region at 600 MHz.

The gain, pattern, and impedance behavior as a function of frequency are shown in Fig. 6-41. At 150 MHz, the gain is considerably less than the 9-dB design value due to the large back lobe. Also, the input impedance has a substantial imaginary part. This inferior performance is, of course, caused by insufficient antenna length required for proper support of the active region at that frequency. At the lower band edge of 200 MHz, however, the pattern has little back radiation, the gain is approaching the design goal, and the input impedance has a small imaginary part. Similarly, at 650 MHz the performance is only slightly inferior to that at the upper band edge of 600 MHz because of the added elements. At intermediate frequencies between the band edges, the gain, pattern, and impedance remain reasonably constant, indicating frequency-independent behavior. Figures 6-41*c* and 6-41*d* are typical of intermediate frequencies. The fact that the calculated gain exceeds the 9 dB value indicated by the contour is due to a combination of at least three factors: 1) a different feeder impedance (below 100 ohms) is used; 2) the ratio of $L/2a$ being different than 125; 3) the estimated gain reduction applied here to Carrel's original contours being, perhaps, slightly excessive. Each of these three factors could conceivably account for 0.1–0.2 dB.

The termination used in this example is purely resistive and equal to the characteristic impedance of the transmission line, but reactive terminations can also be used. The use of a reactive termination can lead to unwanted resonances on the LPDA caused by energy being

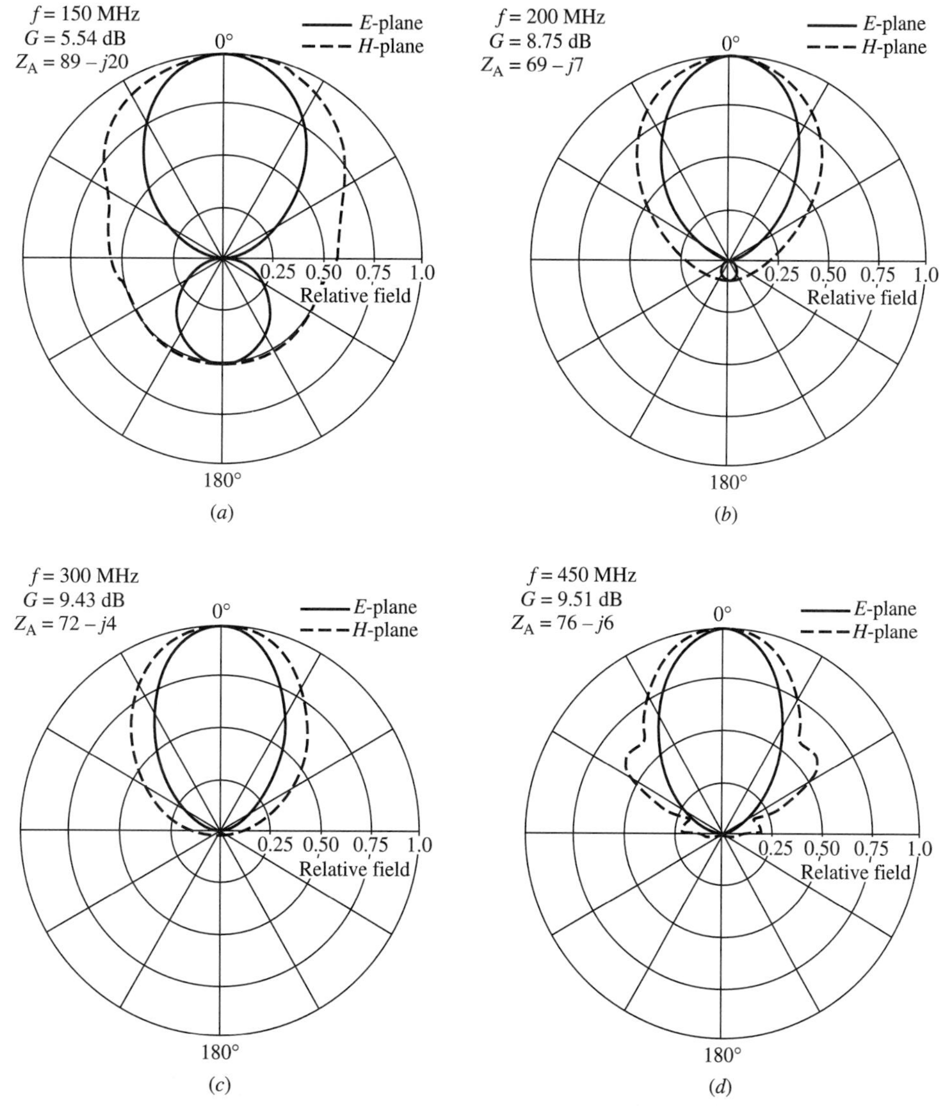

Figure 6-41 Radiation patterns at several frequencies for the log-periodic dipole antenna of Example 6-3. The gain and impedance values are also given. (*a*) 150 MHz. (*b*) 200 MHz. (*c*) 300 MHz. (*d*) 450 MHz. (*e*) 600 MHz. (*f*) 650 MHz.

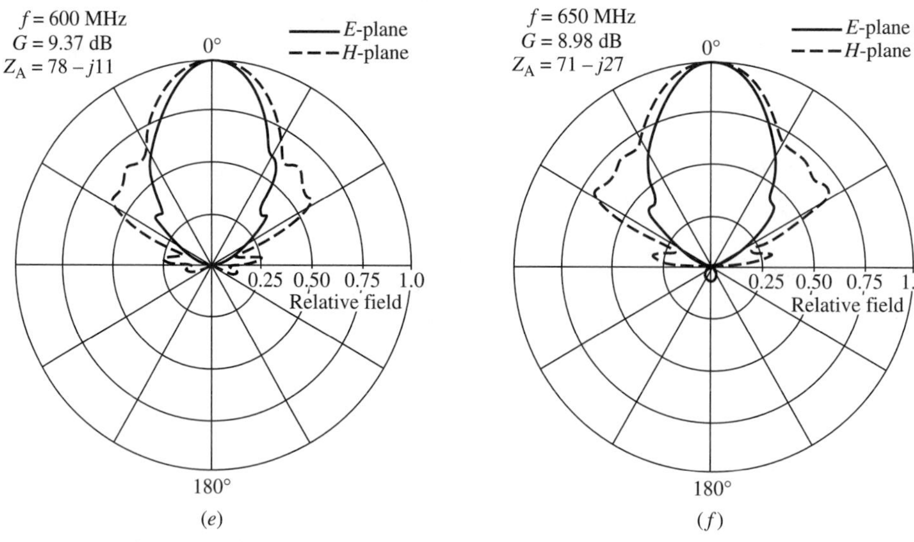

Figure 6-41 (continued)

trapped between the termination and the stop region on the termination side of the active region. These high Q resonances can be eliminated by using a termination that is at least slightly resistive or by using a relatively high value for the LPDA transmission line impedance (e.g., $Z_o \sim 150\ \Omega$) since this will cause the dipole elements to more heavily load the line. This makes the active region more efficient [49], with the result that there is relatively little energy left to propagate past the active region and cause a strong resonance effect on the radiation pattern.

REFERENCES

1. C. H. Walter, *Traveling Wave Antennas*, McGraw-Hill, New York, 1965; Dover, New York, 1970; Peninsula Publishing, Los Altos, CA, 1990, Sec. 8.2.
2. E. A. Wolff, *Antenna Analysis*, Wiley, New York, 1966, Chap. 8.
3. H. Jasik, Ed., *Antenna Engineering Handbook*, McGraw-Hill, New York, 1961, Chap. 6.
4. S. A. Schelkunoff and H. T. Friis, *Antenna Theory and Practice*, Wiley, New York, 1952, Chap. 14.
5. J. D. Kraus, *Antennas*, McGraw-Hill, New York, 1950, Chap. 14.
6. E. K. Miller, et al., "On Increasing the Radiation Efficiency of Beverage-Type Antennas," Lawrence Livermore National Lab, UCRL-52300, July 1977.
7. G. J. Burke, et al., "Computer Modeling of Antennas near the Ground," Lawrence Livermore National Lab, UCID-18626, May 1980.
8. J. D. Kraus, *Antennas*, 2nd ed., McGraw-Hill, New York, 1988, Chap. 7.
9. R. C. Johnson, Ed., *Antenna Engineering Handbook*, 3rd ed., McGraw-Hill, New York, 1993, Chap. 13.
10. T. S. Maclean and R. G. Kouyoumjian, "The Bandwidth of Helical Antennas," *IRE Trans. Antennas & Propagation*, Vol. AP-7, pp. S379–S386, Dec. 1959.
11. H. E. King and J. L. Wong, "Characteristics of 1 to 8 Wavelength Uniform Helical Antennas," *IEEE Trans. Antennas & Propagation*, Vol AP-28, pp. 291–296, March 1980.
12. H. Nakano, *Helical and Spiral Antennas*, Research Studies Press, Wiley, New York, 1987.
13. K. R. Carver, "The Helicone: A Circularly Polarized Antenna with Low Side Lobe Level," *Proc. IEEE*, Vol. 55, p. 559, April 1967.
14. R. G. Vaughan and J. Bach Anderson, "Polarization Properties of the Axial Mode Helix Antenna," *IEEE Trans. Antennas & Propagation*, Vol. AP-33, pp. 10–20, Jan. 1985.
15. A Safaai-Jazi and J. C. Cardoso, "Radiation Characteristics of a Spherical Helical Antenna," *IEEE Proc.—Microwave, Antennas, & Propagation*, Vol. 143, pp. 7–12, Feb. 1996.

16. A. G. Kandoian, "Three New Antenna Types and Their Applications," *Proc. IRE*, Vol. 34, pp. 70W–75W, Feb. 1946.
17. J. J. Nail, "Designing Discone Antennas," *Electronics*, Vol. 26, pp. 167–169, Aug. 1953.
18. T. H. Crowley and W. Marsh, "Discone Type Antennas," Ohio State Research Foundation Tech. Rept. 510-13, June 30, 1954.
19. T. Rappaport, "Discone Design Using Simple *N*-Connector Feed," *IEEE Antennas & Propagation* Newsl., Feb. 1988.
20. A. J. Poggio and P. E. Mayes, "Pattern Bandwidth Optimization of the Sleeve Monopole Antenna," *IEEE Trans. Antennas & Propagation*, Vol. AP-14, pp. 643–645, Sept. 1966.
21. W. L. Weeks, *Antenna Engineering*, McGraw-Hill, New York, 1968, Sec. 4.4.4.
22. H. E. King and J. L. Wong, "An Experimental Study of a Balun-Fed, Open-Sleeve Dipole in Front of a Metallic Reflector," Aerospace Corp. Rept. No. TR-0172 (2158)-2, Feb. 1972.
23. V. Rumsey, *Frequency Independent Antennas*, Academic Press, New York, 1966.
24. Y. Mushiake, *Self-Complementary Antennas*, Springer-Verlag, Berlin, 1996.
25. J. D. Dyson, "The Equiangular Spiral Antenna," *IRE Trans. Antennas & Propagation*, Vol. AP-7, pp. 181–187, April 1959.
26. R. C. Johnson, Ed., *Antenna Engineering Handbook*, 3rd. ed. McGraw-Hill, New York, 1993, Chap. 14.
27. J. W. Duncan and V. P. Minerava, "100-1 Bandwidth Balun Transformer," *Proc. IRE*, Vol. 44, pp. 31–35, Jan. 1960.
28. D. G. Shively and W.L. Stutzman, "Wideband Arrays with Variable Element Sizes," *IEE Proc.*, Part H, Vol. 137, pp. 238–240, Aug. 1990.
29. J. D. Dyson, "The Unidirectional Equiangular Spiral Antenna," *IRE Trans. Antennas & Propagation*, Vol. AP-7, pp. 329–334, Oct. 1959.
30. S. E. Lipsky, *Microwave Passive Direction Finding*, Wiley, New York, 1987, Sec. 3.2.
31. H. Nakano, *Helical and Spiral Antennas—A Numerical Approach*, Research Studies Press, Wiley, New York, 1987.
32. T.-T. Chu and H. G. Oltman, "The Sinuous Antenna," *Microwave Syst. News & Communication Technology*, Vol. 18, pp. 40–48, June 1988.
33. R. H. DuHammel and D. E. Isbell, "Broadband Logarithmically Periodic Antenna Structures," *IRE Intern. Conv. Record*, pp. 119–128, 1957.
34. W. L. Weeks, *Antenna Engineering*, McGraw-Hill, New York, 1968, Sec. 7.2.
35. R. S. Elliott, "A View of Frequency Independent Antennas," *Microwave J.*, Vol. 5, pp. 61–68, Dec. 1962.
36. D. E. Isbell, "Log Periodic Dipole Arrays," *IRE Trans. Antennas & Propagation*, Vol. AP-8, pp. 260–267, May 1960.
37. R. Carrel, "The Design of Log-Periodic Dipole Antennas," *IRE Interna. Convention Rec., Part 1, pp. 61–75, 1961.*
38. G. DeVito and G. B. Stracca, "Comments on the Design of Log-Periodic Dipole Antennas," *IEEE Trans. Antennas & Propagation*, Vol. AP-21, pp. 303–308, May 1973.
39. Carl E. Smith, *Log Periodic Antenna Design Handbook*, Smith Electronics, Inc., Cleveland, OH, 1966.
40. G. DeVito and G. B. Stracca, "Further Comments on the Design of Log-Periodic Dipole Antennas," *IEEE Trans. Antennas & Propagation*, Vol. AP-22, pp. 714–718, Sept. 1974.
41. P. C. Butson and G. T. Thompson, "A Note on the Calculation of the Gain of Log-Periodic Dipole Antennas," *IEEE Trans. Antennas & Propagation*, Vol. AP-14, pp. 105–106, Jan. 1976.
42. C. A. Balanis, *Antenna Theory and Analysis*, 2nd ed., Wiley, New York, 1997, p. 561.
43. J. K. Breakall and R. A. R. Solis, "A New Design Method for Low Side Lobe Level Log-Periodic Dipole Antennas," *Appl. Computational Electromagnetics Soc. J.*, Vol. 11, pp. 9–15, Nov. 1996.
44. W. L. Stutzman and G. A. Thiele, *Antenna Theory and Design*, 1st ed., Wiley, New York, 1981, pp. 299–302.
45. M. T. Ma, *Theory and Application of Antenna Arrays*, Wiley, New York, 1974, Chap. 5.
46. D. F. DiFonzo, "Reduced Size Log Periodic Antennas," *Microwave J.*, Vol. 7, pp. 37–42, Dec. 1964.
47. S. C. Kuo, "Size-Reduced Log-Periodic Dipole Array Antenna," *Microwave J.,* Vol. 15, pp. 27–33, Dec. 1972.
48. C. K. Campbell, et al., "Design of a Stripline Log-Periodic Dipole Antenna," *IEEE Trans. Antennas & Propagation*, Vol. AP-25, pp. 718–721, Sept. 1977.
49. C. C. Bantin and K. G. Balmain, "Study of Compressed Log-Periodic Dipole Antennas," *IEEE Trans. Antennas & Propagation*, Vol. AP-18, pp. 195–203, March 1970.

PROBLEMS

6.1-1 Verify that the maximum of the radiation from a traveling-wave long wire antenna that is 6λ long occurs at an angle of 20.1° from the wire.

6.1-2 Compare the approximate beam maximum angle formula of (6-6) for a traveling-wave long wire with the values of Fig. 6.3 for $L/\lambda = 1, 3, 6, 10$.

6.1-3 Show that the power radiated from a traveling-wave long wire antenna is

$$P = 30I_m^2\left[2.108 + \ln\left(\frac{L}{\lambda}\right) - \text{Ci}(2\beta L) + \frac{\sin(2\beta L)}{2\beta L}\right]$$

Use (4-8), (6-5), and (1-130).

6.1-4 Use the radiated power expression from Prob. 6.1-3 for a traveling-wave long wire to:

a. Derive the directivity expression

$$D = \frac{1.69\cot^2\left[\dfrac{1}{2}\cos^{-1}\left(1 - \dfrac{0.371}{L/\lambda}\right)\right]}{2.108 + \ln\left(\dfrac{L}{\lambda}\right) - \text{Ci}(2\beta L) + \dfrac{\sin(2\beta L)}{2\beta L}}$$

b. Evaluate the directivity for $L/\lambda = 2, 5, 10, 20$. $\text{Ci}(2\beta L)$ is approximately zero for these values of L.

6.1-5 Use the radiated power expression from Prob. 6.1-3 for a traveling-wave long wire to:

a. Find an expression for the radiation resistance.

b. Evaluate the radiation resistance for $L/\lambda = 2, 5, 10, 20$. $\text{Ci}(2\beta L)$ is approximately zero for these values of L.

6.1-6 Plot the linear, polar plot of a traveling-wave long wire antenna that is eight wavelengths long.

6.1-7 To be completely general, the traveling-wave long wire antenna has a current distribution given by

$$I_t(z) = I_m e^{-az} e^{-j\beta_o z}$$

where a is the attenuation coefficient representing radiation and ohmic losses. β_o is the phase constant and is related to the velocity factor $p = v/c$ as $\beta_o = \beta/p$.

a. Derive the pattern function

$$F(\theta) = K\sin\theta\,\frac{\sinh\left[\dfrac{aL}{2} + j\dfrac{\beta L}{2}\left(\dfrac{1}{p} - \cos\theta\right)\right]}{\dfrac{aL}{2} + j\dfrac{\beta L}{2}\left(\dfrac{1}{p} - \cos\theta\right)}$$

b. Show that this reduces to (6-5) for $a = 0$ and $p = 1$.

c. Plot the polar pattern for $a = 0$ and $L = 6\lambda$, for $p = 1.0, 0.75, 0.5$.

6.1-8 *Travel-wave vee antenna.*

a. Place the zero-phase reference point at the vertex of the vee antenna of Fig. 6-4, and derive the radiation pattern as

$$F_V(\theta) = K_V[F_1(\theta) - F_2(\theta)]$$

where

$$F_1(\theta) = e^{j(\beta L/2)[-1+\cos(\theta-\alpha)]}\sin(\theta - \alpha)\,\frac{\sin[(\beta L/2)(1 - \cos(\theta - \alpha))]}{(\beta L/2)(1 - \cos(\theta - \alpha))}$$

and $F_2(\theta)$ is the same as $F_1(\theta)$ except $-\alpha$ is replaced by α. This pattern expression is valid only in the plane of the vee.

b. Plot the polar pattern in Fig. 6-4 for $L = 6\lambda$ and $\alpha = 16°$.

6.1-9 *Rhombic antenna.*

a. Show that the pattern of the rhombic in Fig. 6-5 is

$$F_R(\theta) = K_R\{F_1(\theta) - F_2(\theta) + e^{-j\beta L}[F_3 - F_4]\}$$

where $F_3 = e^{j\beta L \cos(\theta-\alpha)}F_2$ and $F_4 = e^{j\beta L \cos(\theta+\alpha)}F_1$. F_1 and F_2 are given in Prob. 6.1-8. This expression is valid only in the plane of the rhombic.

b. Plot the polar pattern in Fig. 6-5 for $L = 6\lambda$ and $\alpha = 16°$.

6.1-10 A rhombic antenna above ground is to be designed for a main beam maximum at an elevation angle of 20°. Determine the rhombic configuration required.

6.1-11 a. Write an expression for the total current on the Beverage antenna in Fig. 6-7.

b. Determine α in (6-10) for the Beverage in Fig. 6-7.

6.2-1 Compare the radiation resistances of the resonant stub helix to a short monopole for the height values of 0.01, 0.05, 0.08, and 0.1λ.

6.2-2 Find the radiation resistance of a six-turn resonant stub helix that is 2 cm high and operates at 850 MHz.

6.2-3 An unfurlable helix was built with an overall length of 78.7 cm, a diameter of 4.84 cm, and a pitch angle of 11.7°. The center frequency of operation is 1.7 GHz. Calculate the number of turns, the gain in decibels, the half-power beamwidth in degrees, and the axial ratio for the helix.

6.2-4 It is desired to achieve a right-hand circularly polarized wave at 475 MHz having a half-power beamwidth of 39°. One of the easiest ways to do this is with a helix antenna. It is to be built with a pitch angle of 12.5°, and the circumference of one turn is to be one wavelength at the center frequency of operation.

a. Calculate the number of turns needed.

b. What is the directivity in decibels?

c. What is the axial ratio of the on-axis fields?

d. Over what range of frequencies will these parameters remain relatively constant?

e. Find the input impedance at the design frequency and at the ends of the band.

f. Evaluate HP at the band ends.

6.2-5 A commercially available axial mode helix antenna has six turns made of 0.95-cm aluminum tubing supported by fiberglass insulators attached to a 3.8-cm aluminum shaft. The band of operation is 300 to 520 MHz. The mechanical characteristics are as follows: length of helix, 118 cm; diameter of helix (center to center), 23.2 cm; and ground screen diameter, 89 cm.

a. Determine the pitch angle α.

b. Compute the gain in decibels at edges of the frequency band of operation.

6.2-6 A 12-turn axial mode helix has a circumference of 0.197 m, a pitch angle of 8.53°, and operates at 1525 MHz. Calculate and plot the radiation pattern in linear-polar form.

6.2-7 A helix antenna has five turns and a pitch angle of 12°. It is operated such that its circumference is one wavelength. (a) Use simple array theory techniques to derive and accurately sketch the radiation pattern. (b) Calculate the half-power beamwidth, not based on results from (a).

6.2-8 One turn of an axial mode helix radiates similarly to a one-wavelength loop antenna. Explain why, then, that the helix antenna radiates circular polarization and the loop radiates linear polarization.

6.3-1 Calculate the input impedance for infinite biconical antennas of the following cone half-angles: 0.1°, 1°, 10°, 20°, 50°.

6.3-2 Show that the radiated power of the infinite biconical antenna is

$$P = 4\pi\sqrt{\mu/\varepsilon}H_o^2 \ln[\cot(\theta_h/2)]$$

and that the directivity is

$$D = \frac{1}{\sin^2\theta \ln[\cot(\theta_h/2)]}$$

6.4-1 *Construction project.* Select a frequency for which you have laboratory equipment to measure impedance (probably in the VHF or UHF range). Construct both an optimum open-sleeve dipole and its ordinary dipole version. (Alternatively, monopoles may be constructed.) Measure the input impedance of both antennas over a 2:1 frequency range about the center frequency. (Alternatively, measure the VSWR.)

6.4-2 Use (6-54) and Figs. 5-5 and 5-6 to estimate the input impedance of the asymmetrically fed dipole in Prob. 5.1-4.

6.5-1 Find the input impedance of a slot that is complementary to a half-wavelength long ribbon dipole with impedance $73 + j42.5\ \Omega$.

6.5-2 The far-field components for a short slot with uniform electric field and of length $\Delta z \ll \lambda$ along the z-axis are

$$E_\phi = -\frac{V_o\,\Delta z}{2\pi}\, j\beta\, \frac{e^{-j\beta r}}{r} \sin\theta$$

$$H_\theta = \frac{V_o\,\Delta z}{2\pi\eta}\, j\beta\, \frac{e^{-j\beta r}}{r} \sin\theta$$

where V_0 is the excitation voltage across the center of the slot.

a. Find the input radiation resistance.

b. Verify that (6-55) is satisfied using the appropriate complementary antenna.

6.5-3 Frequency-independent antennas have constant HP with frequency. Explain this in terms of the formula $\text{HP} = K\lambda/L$.

6.6-1 Design an equiangular spiral antenna for operation over the band 450 to 900 MHz.

6.6-2 *Construction project.* Construct the equiangular spiral antenna of the previous problem using aluminum foil glued to cardboard. Test its performance with a receiver (perhaps a television).

6.7-1 Design a self-complementary log-periodic toothed planar antenna for operation from 400 MHz to 2 GHz with a half-power beamwidth of 70°.

6.7-2 A log-periodic dipole array is to be designed to cover the frequency range 84 to 200 MHz and have 7.5-dB gain. Give the required element lengths and spacings for optimal design.

6.7-3 Evaluate the dipole lengths and spacings for the LPDA of Example 6-3.

6.7-4 Design an optimum LPDA to operate from 470 MHz to 890 MHz with 9-dB gain. Add one extra element to each end over that required by (6-86).

6.7-5 What is the physical length of the LPDA in Example 6-3?

Chapter 7

Aperture Antennas

Three of the four classes of antennas (electrically small, resonant, and broadband), which are summarized in Fig. 1-6, have been discussed. In this chapter, we treat the fourth and final antenna type, the **aperture antenna**. Part of the structure of an aperture antenna is an aperture, or opening, through which electromagnetic waves flow. An aperture antenna operating as a receiver "collects" waves via the aperture. Analogies in acoustics are the megaphone and the parabolic microphone, which uses a parabolic reflector to focus sound waves on a microphone at the focal point. Also, the pupil of the human eye is an aperture for optical frequency electromagnetic waves. At radio frequencies, horns and reflectors are examples of aperture antennas; see Fig. 1-6. Aperture antennas are in common use at UHF frequencies and above. They are the antenna of choice in applications requiring very high gain. A distinguishing feature of large aperture antennas is the increase in gain with the operating frequency. The gain of an aperture antenna increases with the square of frequency if aperture efficiency is constant with frequency; see (2-89). Another feature is the nearly real-valued input impedance.

Since all receiving antennas act as collectors of waves, an effective aperture can be defined for every antenna; see Sec. 2.5. However, this chapter deals with antennas that have an obvious physical aperture. In the first section, general principles are developed for calculating the radiation patterns from any aperture antenna. Subsequent discussions focus on rectangular and circular aperture shapes. The properties of specific antennas such as horns and circular parabolic reflectors then follow naturally. As in preceding chapters, theoretical derivations lead to an accurate description of the antenna parameters, as well as design techniques. Both rigorous and approximate methods of gain calculation are also presented in this chapter.

7.1 RADIATION FROM APERTURES AND HUYGENS' PRINCIPLE

Although aperture antennas were not widely used until the World War II period, the basic concepts were available in 1690 when Huygens explained, in a simple way, the bending (or diffraction) of light waves around an object. This was accomplished by viewing each point of a wave front as a secondary source of spherical waves. The next wave front is the envelope of these secondary waves in the forward direction. Some 150 years after Huygens' contribution, Fresnel recognized that the phase shift between wave fronts is computed from the distance between wave fronts ΔL by the familiar relation $\beta\,\Delta L$. Figure 7-1 shows how a plane wave and a spherical wave

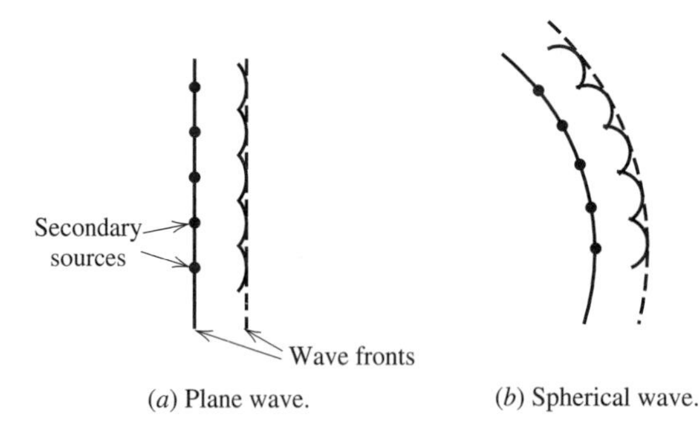

Figure 7-1 Secondary waves used to construct successive wavefronts.

can be constructed from secondary waves; also see Figs. 12-1 and 12-2. The envelope of secondary waves forms the new wave front. Geometrical optics (ray tracing) predicts that light shining through a slit in a screen will have a lit region and a completely dark shadow region with a sharp boundary between them. Geometrical optics works well only for apertures that are very large relative to a wavelength. The secondary source concept shows that the secondary waves will spread out away from the aperture and there will be a smooth blending of the lit and shadow regions. This diffraction effect is illustrated in Fig. 7-2 for a slit in an opaque screen with a plane wave incident on it.

Huygens' principle evolved into a mathematical form referred to as the *equivalence principle* (or, *field equivalence principle*). The field equivalence principle replaces an aperture antenna with equivalent currents that produce radiation fields equivalent to those from the antenna. The equivalence principle is derived by observing that *a* solution to Maxwell's equations and the boundary conditions, which all electromagnetic problems must satisfy, is *the* solution. This follows from the *uniqueness theorem* in mathematics which states that a solution that satisfies a differential equation (e.g., Maxwell's equations) and the boundary conditions is unique. We now use this concept to set up equivalent current relations for use in analyzing aperture antennas.

In the original problem of Fig. 7-3*a*, the fields that satisfy Maxwell's equations in the region exterior to volume V and that satisfy boundary conditions along S are unique.[1] As long as the sources exterior to V and the boundary conditions along S

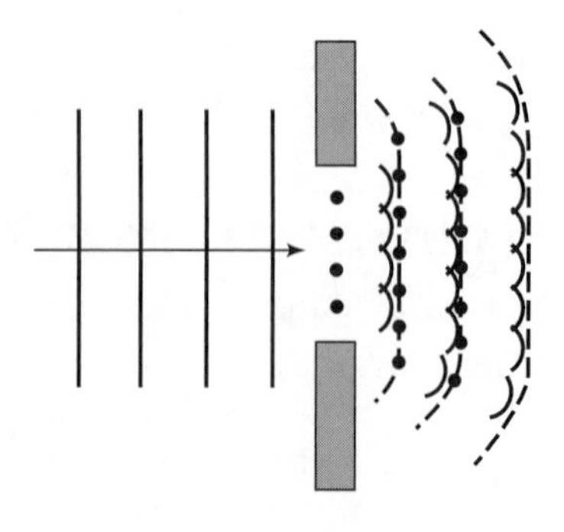

Figure 7-2 Plane wave incident on a slit in a screen. The edge diffraction leads to spreading of the radiation from the slit.

[1]In this chapter, the uppercase symbols V and S will be used to denote volume and surface.

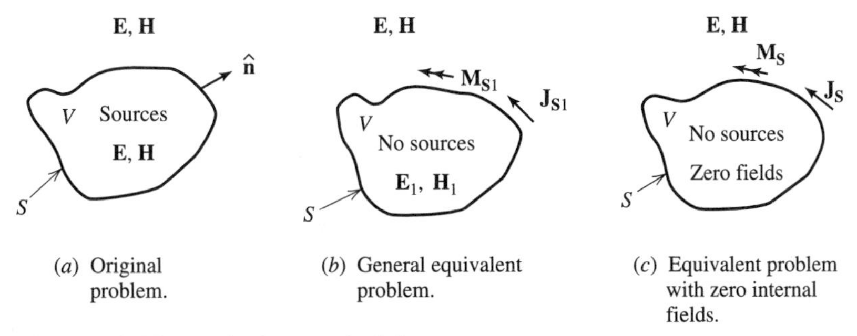

(*a*) Original problem.

(*b*) General equivalent problem.

(*c*) Equivalent problem with zero internal fields.

Figure 7-3 The equivalence principle.

are not changed, the solution $(\mathbf{E}, \mathbf{H})$ will not change. In the equivalent problem, the sources exterior to V are not changed, since there are none. Also, the boundary conditions are not changed, as will now be explained. In the original problem, the fields along the boundary are $\mathbf{E}(S)$ and $\mathbf{H}(S)$. In the equivalent problem of Fig. 7-3*b*, the original sources (e.g., the antenna structure) have been removed, altering the fields internal to S, denoted as $\mathbf{E}_1$ and $\mathbf{H}_1$. In order for the fields external to S to remain the same, equivalent currents must be introduced to satisfy the discontinuity of the fields across S. These equivalent currents are found from the boundary conditions of (1-22) and (1-23) as

$$\mathbf{J}_{S1} = \hat{\mathbf{n}} \times [\mathbf{H} - \mathbf{H}_1] \quad \text{on } S \tag{7-1a}$$

$$\mathbf{M}_{S1} = [\mathbf{E} - \mathbf{E}_1] \times \hat{\mathbf{n}} \quad \text{on } S \tag{7-1b}$$

where $(\mathbf{E}_1, \mathbf{H}_1)$ and $(\mathbf{E}, \mathbf{H})$ are the fields internal and external to S; see Fig. 7-3*b*. These equivalent currents, which are obtained from only a knowledge of the tangential fields over S, can be used to find the fields external to S. However, the fields on S required to determine the equivalent currents are unknown. Also, we do not know how to find the external fields from the equivalent currents. We now address these difficulties, starting with the second one.

Since the internal fields $(\mathbf{E}_1, \mathbf{H}_1)$ are arbitrary, we choose them to be zero for simplicity; see Fig. 7-3*c*. Then (7-1) becomes

$$\mathbf{J}_S = \hat{\mathbf{n}} \times \mathbf{H}(S) \tag{7-2a}$$

$$\mathbf{M}_S = \mathbf{E}(S) \times \hat{\mathbf{n}} \tag{7-2b}$$

where $\mathbf{E}(S)$ and $\mathbf{H}(S)$ are the fields over the surface S. This zero internal field formulation is referred to as *Love's equivalence principle.*

Since the fields inside S are zero in the equivalent problem of Fig. 7-3*c*, we are free to introduce materials inside S. If a perfect electric conductor is placed along S, $\mathbf{J}_S$ will vanish. The explanation is often given that the electric current is "shorted out" by the conductor. This leaves a magnetic current density $\mathbf{M}_S$ radiating in the presence of the electric conductor producing the external fields. Similarly, a perfect magnetic conductor can be introduced along S to eliminate $\mathbf{M}_S$, leaving only $\mathbf{J}_S$. Thus, we have two more equivalent formulations. They are $\mathbf{M}_S$ in the presence of a perfect electric conductor over S and $\mathbf{J}_S$ in the presence of a perfect magnetic conductor over S. Both yield the correct fields external to S. However, these problems are difficult to solve as long as S is a general surface. Note that if the real antenna contains conducting portions, then the $\mathbf{J_S}$ equals the actual current density over that portion and the aperture portion contains both $\mathbf{J_S}$ and $\mathbf{M_S}$ (before any fictitious conductors are introduced).

If the surface S is large in terms of a wavelength and the curvature of S is small, image theory can be applied locally across the surface to solve for the currents operating in the presence of the introduced conductors; this is exploited in Chap. 12. However, since the selection of S is for convenience and we are interested in radiation problems, we can extend S to infinity. Since S must be a closed surface, S includes the infinite plane along $z = 0$ and closes at infinity to enclose sources in the region $z < 0$ of Fig. 7-4. In this case, we can apply image theory with no approximation and easily solve many practical problems with planar apertures.

We can apply image theory to the planar surface S to simplify the solution procedure. To do this, we find the fields due to currents $\mathbf{J_S}$ and $\mathbf{M_S}$ on the $z = 0$ plane as shown in Fig. 7-4*a*; this will be followed by the application of image theory to reduce the solution formulation by one-half. Although not all antenna problems have a planar aperture that can be placed in the *xy*-plane, an equivalent planar aperture surface S can be set up. This will be fruitful if the tangential fields over S can be obtained; more will be said about this later. First, we need to solve for fields in $z > 0$ due to the equivalent currents of Fig. 7-4*a*.

S
$\mathbf{J_S}$
$\mathbf{M_S}$
E, H
z

(*a*) Both equivalent surface current densities acting in free space.

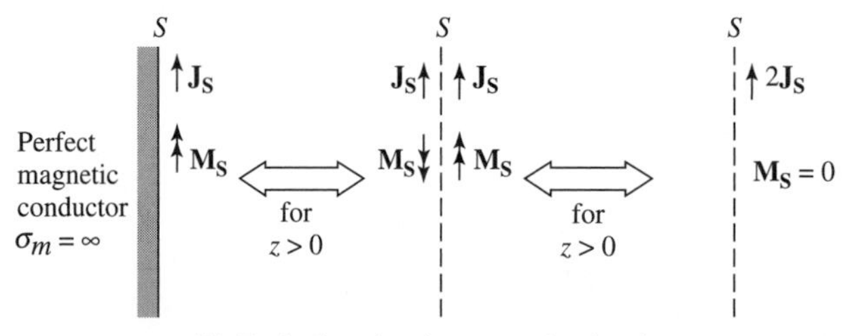

(*b*) Equivalent electric current density alone.

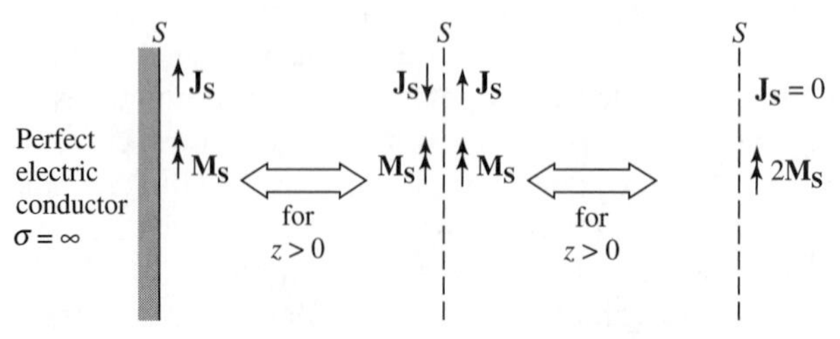

(*c*) Equivalent magnetic current density alone.

Figure 7-4 Equivalent current configurations for a planar aperture surface. The antenna located in $z < 0$ has been removed and three different equivalent current sets introduced as shown.

The fields (**E**, **H**) in the region $z > 0$, in general, are found by first evaluating **A** using (1-58) and finding **E** and **H** from (1-47) and (1-37). In this case, the equivalent currents $\mathbf{J_S}$ and $\mathbf{M_S}$ will yield the exact fields everywhere in $z > 0$. However, we restrict our solution to the far-field region appropriate to antenna problems. Then we can use the much simpler procedure of Sec. 1.7.4 that we have used many times to solve radiation problems. Now we slightly recast the formulation for the case of planar surface current densities in the xy-plane. First, the magnetic vector potential is found from the form of (1-101) appropriate to the geometry of Fig. 7-4*a*:

$$\mathbf{A} = \mu \frac{e^{-j\beta r}}{4\pi r} \iint_S \mathbf{J_S}(\mathbf{r}') e^{j\beta \hat{\mathbf{r}} \cdot \mathbf{r}'} \, dS' \tag{7-3}$$

The far-zone electric field from (1-105) is

$$\mathbf{E_A} = -j\omega(A_\theta \hat{\boldsymbol{\theta}} + A_\phi \hat{\boldsymbol{\phi}}) \tag{7-4}$$

The subscript A indicates that this field arises from the magnetic vector potential **A**.

The electric vector potential **F** associated with the magnetic current density is found using the duality principle introduced in Sec. 2.4.1[2]:

$$\mathbf{F} = \varepsilon \frac{e^{-j\beta r}}{4\pi r} \iint_S \mathbf{M_S}(\mathbf{r}') e^{j\beta \hat{\mathbf{r}} \cdot \mathbf{r}'} \, dS' \tag{7-5}$$

The far-zone magnetic field arising from **F** is the dual of (7-4):

$$\mathbf{H_F} = -j\omega(F_\theta \hat{\boldsymbol{\theta}} + F_\phi \hat{\boldsymbol{\phi}}) \tag{7-6}$$

Since the solution is in the far field, the electric field associated with $\mathbf{H_F}$ can be found from the TEM relationship of (1-107) as

$$\mathbf{E_F} = \eta \mathbf{H_F} \times \hat{\mathbf{r}} = -j\omega\eta(F_\phi \hat{\boldsymbol{\theta}} - F_\theta \hat{\boldsymbol{\phi}}) \tag{7-7}$$

The total electric field is then found by summing the contributions from each current:

$$\mathbf{E} = \mathbf{E_A} + \mathbf{E_F} = -j\omega[(A_\theta + \eta F_\phi)\hat{\boldsymbol{\theta}} + (A_\phi - \eta F_\theta)\hat{\boldsymbol{\phi}}] \tag{7-8}$$

The equivalent system of Fig. 7-4*a* involves both the electric and magnetic current densities. Computations can be reduced considerably if image theory is used so that we only have to deal with one of the currents. First, we introduce a perfect magnetic planar conductor along surface S. The image currents shown in Fig. 7-4*b* are obtained by the duality of images in a perfect magnetic ground plane; that is, a magnetic current parallel to the plane has an oppositely-directed image and a parallel electric current has a similarly directed image. The fields for $z > 0$ are unchanged after removing the conducting plane and introducing the images, as shown in Fig. 7-4*b*. Since the currents and their images are adjacent to the plane S, we can add them vectorially to obtain the final equivalent system, which has a doubled electric surface current density and no magnetic surface current density. The radiation electric field for $z > 0$ is $2\mathbf{E_A}$. In a similar fashion, a perfect electric ground plane can be introduced along S as shown in the leftmost part of Fig. 7-4*c*. Image theory renders the images shown; see Fig. 2-9. These images acting together yield a zero

---

[2]The symbol **F** for magnetic vector potential should not be confused with the normalized radiation pattern function $F(\theta, \phi)$.

total electric surface current density and a magnetic surface current density of $2\mathbf{M_S}$. Then the radiation electric field for $z > 0$ is $2\mathbf{E_F}$.

We can now summarize the equivalence theorem in terms most suitable to radiation pattern calculations. First, an aperture plane is selected for the antenna, this is usually the physical aperture of the antenna but need not be. Coordinates are set up such that the aperture plane is the xy-plane and the $+z$-axis is the forward radiation direction. Then the radiation fields for $z > 0$ are found by one of the three equivalent systems of Fig. 7-4 as follows:

a. $\mathbf{J_S}$ and $\mathbf{M_S}$ on S (xy-plane)

$$\mathbf{E} = \mathbf{E_A} + \mathbf{E_F} \tag{7-9a}$$

with (7-3) in (7-4) and (7-5) in (7-6) and (7-7)

b. $2\mathbf{J_S}$ on S

$$\mathbf{E} = 2\mathbf{E_A} \tag{7-9b}$$

c. $2\mathbf{M_S}$ on S

$$\mathbf{E} = 2\mathbf{E_F} \tag{7-9c}$$

The procedures for finding radiation from equivalent aperture plane currents are now clear. It remains then to focus on determining those currents, which are established using (7-2).

So far, no approximations have been introduced other than the usual far-field approximations. Indeed, if the exact fields $\mathbf{E}(S)$ and/or $\mathbf{H}(S)$ are used in any of the above three procedures, exact far-field results will be obtained in the half-space $z > 0$. However, such exact knowledge of the fields over the entire plane S is rarely available. Usually, at best it is possible to obtain only an approximate knowledge of the fields over a finite portion of the infinite aperture plane. One such approach is the popular *physical optics approximation*, in which it is assumed that the aperture fields $\mathbf{E}_a$ and $\mathbf{H}_a$ are those of the incident wave. It is usually assumed that these fields exist over only some finite portion S_a of the infinite plane S and the fields elsewhere over S are zero. In most cases, the aperture surface S_a coincides with the physical aperture of the antenna. These approximations improve as the dimensions of the aperture relative to a wavelength increase.

The three solution procedures will now be simplified. Suppose that aperture fields $\mathbf{E}_a$ and $\mathbf{H}_a$, which exist over and are tangent to some portion of S_a of the infinite plane S, are known (perhaps by employing the physical optics approximation). The equivalent surface current densities follow from (7-2) as

$$\mathbf{J_S} = \hat{\mathbf{n}} \times \mathbf{H}_a \tag{7-10}$$

$$\mathbf{M_S} = \mathbf{E}_a \times \hat{\mathbf{n}} \tag{7-11}$$

on S_a and zero elsewhere. Using these in (7-3) and (7-5) gives

$$\mathbf{A} = \mu \frac{e^{-j\beta r}}{4\pi r} \hat{\mathbf{n}} \times \iint_{S_a} \mathbf{H}_a e^{j\beta \hat{\mathbf{r}} \cdot \mathbf{r}'} \, dS' \tag{7-12}$$

$$\mathbf{F} = -\varepsilon \frac{e^{-j\beta r}}{4\pi r} \hat{\mathbf{n}} \times \iint_{S_a} \mathbf{E}_a e^{j\beta \hat{\mathbf{r}} \cdot \mathbf{r}'} \, dS' \tag{7-13}$$

The integral in the above two equations is a two-dimenional Fourier transform. The two-dimensional Fourier transform of an aperture field plays an important role in radiation calculations for aperture antennas, in a way similar to the Fourier transform of the current distribution for line sources (see Chap. 4). We therefore make the following definitions for the integrals:

$$\mathbf{P} = \iint_{S_a} \mathbf{E}_a e^{j\beta\hat{\mathbf{r}}\cdot\mathbf{r}'}\, dS' \tag{7-14}$$

$$\mathbf{Q} = \iint_{S_a} \mathbf{H}_a e^{j\beta\hat{\mathbf{r}}\cdot\mathbf{r}'}\, dS' \tag{7-15}$$

The far-zone electric field based on both aperture fields can be written in a single expression often encountered in the literature [1]. The total electric field in terms of the potentials, from (7-4) and (7-7), is

$$\mathbf{E} = -j\omega\mathbf{A} - j\omega\eta\mathbf{F} \times \hat{\mathbf{r}} \tag{7-16}$$

where the r-component of the first term is to be neglected. Substituting in (7-12) and (7-13) and performing some manipulations yield

$$\mathbf{E} = -j\beta \frac{e^{-j\beta r}}{4\pi r} \hat{\mathbf{r}} \times \iint_{S_a} [\hat{\mathbf{n}} \times \mathbf{E}_a - \eta\, \hat{\mathbf{r}} \times (\hat{\mathbf{n}} \times \mathbf{H}_a)] e^{j\beta\hat{\mathbf{r}}\cdot\mathbf{r}'}\, dS' \tag{7-17}$$

This gives the full vector form of the radiated electric field from the aperture fields and is often called a vector *diffraction integral.* The term "diffraction" is used because the field found using (7-17) represents the superposition of all elements of the source distribution; this is in contrast to geometrical optics that traces rays from points on the antenna directly to observation points (see Sec. 12.1). The subsequent developments here are cast in terms of the Fourier transforms **P** and **Q**; this provides a more procedural, as well as instructive, approach.

The aperture surface S_a is in the xy-plane, so $\mathbf{r}' = x'\hat{\mathbf{x}} + y'\hat{\mathbf{y}}$. This with $\hat{\mathbf{r}}$ in spherical coordinates from (C-4) in (7-14) and (7-15) yield

$$P_x = \iint_{S_a} E_{ax}(x', y') e^{j\beta(x' \sin\theta\cos\phi + y' \sin\theta\sin\phi)}\, dx'\, dy' \tag{7-18a}$$

$$P_y = \iint_{S_a} E_{ay}(x', y') e^{j\beta(x' \sin\theta\cos\phi + y' \sin\theta\sin\phi)}\, dx'\, dy' \tag{7-18b}$$

$$Q_x = \iint_{S_a} H_{ax}(x', y') e^{j\beta(x' \sin\theta\cos\phi + y' \sin\theta\sin\phi)}\, dx'\, dy' \tag{7-19a}$$

$$Q_y = \iint_{S_a} H_{ay}(x', y') e^{j\beta(x' \sin\theta\cos\phi + y' \sin\theta\sin\phi)}\, dx'\, dy' \tag{7-19b}$$

Now, (7-12) and (7-13) together with $\hat{\mathbf{n}} = \hat{\mathbf{z}}$ reduce to

$$\mathbf{A} = \mu \frac{e^{-j\beta r}}{4\pi r} (-Q_y\hat{\mathbf{x}} + Q_x\hat{\mathbf{y}}) \tag{7-20}$$

$$\mathbf{F} = -\varepsilon \frac{e^{-j\beta r}}{4\pi r} (-P_y\hat{\mathbf{x}} + P_x\hat{\mathbf{y}}) \tag{7-21}$$

Expressing $\hat{\mathbf{x}}$ and $\hat{\mathbf{y}}$ in spherical coordinates as in (C-1) and (C-2), and retaining only the θ- and ϕ-components give

$$\mathbf{A} = \mu \frac{e^{-j\beta r}}{4\pi r} [\hat{\boldsymbol{\theta}} \cos\theta (Q_x \sin\phi - Q_y \cos\phi) + \hat{\boldsymbol{\phi}}(Q_x \cos\phi + Q_y \sin\phi)] \quad \text{(7-22)}$$

$$\mathbf{F} = -\varepsilon \frac{e^{-j\beta r}}{4\pi r} [\hat{\boldsymbol{\theta}} \cos\theta (P_x \sin\phi - P_y \cos\phi) + \hat{\boldsymbol{\phi}}(P_x \cos\phi + P_y \sin\phi)] \quad \text{(7-23)}$$

Using these in (7-8) yields the final radiation field components

$$\text{(a)}\ E_\theta = j\beta \frac{e^{-j\beta r}}{4\pi r} [P_x \cos\phi + P_y \sin\phi + \eta \cos\theta (Q_y \cos\phi - Q_x \sin\phi)] \quad \text{(7-24a)}$$

$$E_\phi = j\beta \frac{e^{-j\beta r}}{4\pi r} [\cos\theta (P_y \cos\phi - P_x \sin\phi) - \eta (Q_y \sin\phi + Q_x \cos\phi)] \quad \text{(7-24b)}$$

In a similar fashion, the other two equivalent systems reduce to

$$\text{(b)}\ E_\theta = j\beta\eta \frac{e^{-j\beta r}}{2\pi r} \cos\theta (Q_y \cos\phi - Q_x \sin\phi) \quad \text{(7-25a)}$$

$$E_\phi = -j\beta\eta \frac{e^{-j\beta r}}{2\pi r} (Q_y \sin\phi + Q_x \cos\phi) \quad \text{(7-25b)}$$

$$\text{(c)}\ E_\theta = j\beta \frac{e^{-j\beta r}}{2\pi r} (P_x \cos\phi + P_y \sin\phi) \quad \text{(7-26a)}$$

$$E_\phi = j\beta \frac{e^{-j\beta r}}{2\pi r} \cos\theta (P_y \cos\phi - P_x \sin\phi) \quad \text{(7-26b)}$$

If the exact aperture fields over the entire aperture plane are used, the three formulations of (7-24) to (7-26) each yield the same result. Use of the exact aperture fields leads to equal contributions arising from the electric and magnetic currents [2]. Therefore, the equivalent system using both current types, as in (7-24), gives zero total field for $z < 0$ because $\cos\theta$ is negative for $\pi/2 \le \theta \le \pi$, and the contributions cancel as guaranteed by the equivalence theorem. However, the single current systems of (7-25) and (7-26) do not yield zero fields for $z < 0$. This is an expected result since image theory was involved in the development of these, and identical fields are obtained only in the region $z > 0$.

The trigonometric functions appearing in (7-24) to (7-26) actually describe the projections of the aperture equivalent surface current densities onto the plane containing the far-field components (i.e., perpendicular to $\hat{\mathbf{r}}$). For aperture field expressions, the trigonometric functions that multiply the radiation integrals are often referred to as *obliquity factors*. The element factor $\sin\theta$ for line sources along the z-axis is an obliquity factor. For apertures that are several wavelengths in extent, the obliquity factors do not reduce the main beam and first few side lobes by a significant amount. Then the Fourier transform adequately describes the pattern, and the aperture antenna problem reduces to first finding the (scalar) far-field pattern from the Fourier transform of the aperture electric field magnitude. Polarization is determined from the component(s) of the aperture electric field tangent to a far-field sphere by projecting $\mathbf{E_A}$ on to the far-field sphere.

In practice, only approximate information about the aperture fields is available, such as obtained from the physical optics approximation. Then the three formulations give different results. The accuracy of the three results depends on the accuracy of the aperture fields, but the differences are usually not significant. For apertures mounted in a conducting ground plane, the aperture plane (except for the aperture itself) is well modeled as an infinite, perfectly conducting plane. Then the magnetic current (aperture electric field) formulation of (7-26) is preferred since the aperture electric field, and thus magnetic current, is zero outside the aperture because of the boundary condition of zero tangential electric field on the conductor. For apertures in free space, the dual current formulation of (7-24) is used. This is usually accompanied by the assumption that the aperture fields are related as a transverse electromagnetic (TEM) wave:

$$\mathbf{H}_a = \frac{1}{\eta}\hat{\mathbf{z}} \times \mathbf{E}_a \tag{7-27}$$

This implies that

$$\mathbf{Q} = \frac{1}{\eta}\hat{\mathbf{z}} \times \mathbf{P} \quad \text{or} \quad Q_x = -\frac{P_y}{\eta}, \quad Q_y = \frac{P_x}{\eta} \tag{7-28}$$

This assumption is valid for moderate- to high-gain antennas and is often applied with success even to apertures that are only a few wavelengths in extent. Using (7-28)in (7-24) leads to

$$E_\theta = j\beta \frac{e^{-j\beta r}}{2\pi r} \frac{1 + \cos\theta}{2} [P_x \cos\phi + P_y \sin\phi] \tag{7-29a}$$

$$E_\phi = j\beta \frac{e^{-j\beta r}}{2\pi r} \frac{1 + \cos\theta}{2} [P_y \cos\phi - P_x \sin\phi] \tag{7-29b}$$

which is a simpler form than (7-24). The factors in brackets are identical to those in (7-26). The obliquity factor of $(1 + \cos\theta)/2$ differs only slightly from the $\cos\theta$ obliquity factor in (7-26) for small values of θ, where radiation is significant for high-gain antennas. Unlike (7-26), (7-29) remains valid over all space (i.e., $0 < \theta < 180°$) because image theory was not employed and we can take surface S to enclose the antenna, since equivalent currents are zero except over the finite aperture. However, accuracy is likely to degrade for directions far out from the main beam. In summary, (7-26) should be used for apertures in ground planes and (7-29) should be used for aperture antennas in free space.

EXAMPLE 7-1 *Slit in an Infinite Conducting Plane*

The aperture antenna calculation procedures and the physical optics approximation can be illustrated rather simply for a plane wave normally incident on a slit in an infinite perfectly conducting plane as shown in Fig. 7-5. This is the same problem as in Fig. 4-6*a*, except for a coordinate system change. The physical optics approximation leads us to assume that the incident field $\mathbf{E}^i = \hat{\mathbf{y}}E_o e^{-j\beta z}$ associated with the plane wave propagating in the $+z$-direction renders the field over the physical aperture, so

$$\mathbf{E}_a = \begin{cases} \hat{\mathbf{y}}E_o & |y| \le \dfrac{L}{2}, \quad z = 0 \\ 0 & \text{elsewhere} \end{cases} \tag{7-30}$$

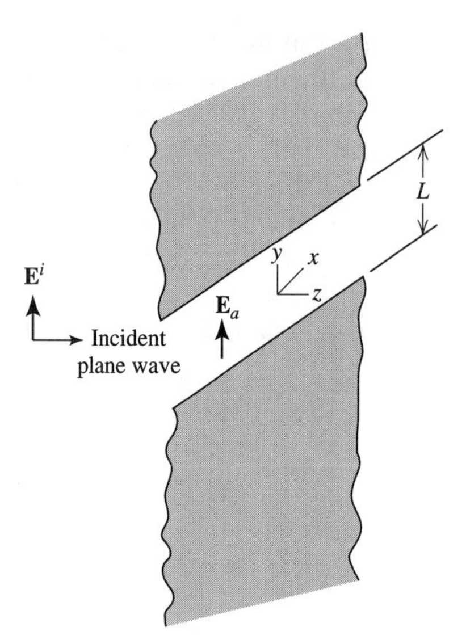

Figure 7-5 Plane wave incident on a slit in an infinite conducting plane. The slit is infinite in the x-direction and is L wide in the y-direction.

The magnetic current formulation is appropriate in this case because the aperture electric field is zero over the perfectly conducting portion of the aperture surface. This is essentially a one-dimensional problem because the aperture field is uniform in the x-direction; then the radiation fields will not change with position along the x-direction. We are thus concerned only with the yz-plane ($\phi = 90°$), and since the aperture field is only y-directed, (7-18) reduces to

$$\mathbf{P} = \hat{\mathbf{y}}P_y = \hat{\mathbf{y}} \int_{-L/2}^{L/2} E_o e^{j\beta y' \sin\theta}\, dy' = \hat{\mathbf{y}} E_o L \frac{\sin[(\beta L/2)\sin\theta]}{(\beta L/2)\sin\theta} \tag{7-31}$$

When normalized, the magnitude of this expression renders the following radiation pattern:

$$F(\theta) = \frac{\sin[(\beta L/2)\sin\theta]}{(\beta L/2)\sin\theta} \tag{7-32}$$

The polarization of the far field (electric field) is the tangent of the aperture electric field direction $\hat{\mathbf{y}}$ onto the far-field sphere (in the $\phi = 90°$ observation plane), which is the θ-component and there is no ϕ-component. Notice that (7-32) is nonzero at $\theta = 90°$; this is acceptable since E_θ can be normal to the conducting plane. A note of caution is in order for this example, which has an aperture that is infinite in one dimension. The problem is really two-dimensional rather than three-dimensional (equivalently, the aperture is one-dimensional rather than two-dimensional). Therefore, the complete electric field will *not* be given by (7-26). The spherical wave behavior of $e^{-j\beta r}/r$ (e.g., free-space Green's function) is replaced by the cylindrical wave behavior $e^{-j\beta r}/\sqrt{r}$. However, the one-dimensional Fourier transform as presented here yields the correct angular variation (pattern). The pattern based on this simple approach agrees well with that of more rigorous techniques [3].

7.2 RECTANGULAR APERTURES

There are several antenna types that have a physical aperture which is rectangular in shape. For example, many horn antennas have rectangular apertures. Another example is a rectangular slot in a metallic source structure such as a waveguide. In this section, we present some general principles about rectangular apertures that have uniform and tapered excitations. In Sec. 7.4, these principles are applied to rectangular aperture horn antennas.

7.2.1 The Uniform Rectangular Aperture

A general rectangular aperture is shown in Fig. 7-6. It is excited in an idealized fashion such that the aperture fields are confined to the L_x by L_y region. If the aperture fields are uniform in phase and amplitude across the physical aperture, it is referred to as a *uniform rectangular aperture*. Suppose the aperture electric field is y-polarized; then the uniform rectangular aperture electric field is

$$\mathbf{E}_a = E_o \hat{\mathbf{y}}, \qquad |x| \leq \frac{L_x}{2}, \qquad |y| \leq \frac{L_y}{2} \tag{7-33}$$

Then from (7-18b)

$$\begin{aligned} P_y &= E_o \int_{-L_x/2}^{L_x/2} e^{j\beta x' \sin\theta\cos\phi}\, dx' \int_{-L_y/2}^{L_y/2} e^{j\beta y' \sin\theta\sin\phi}\, dy' \\ &= E_o L_x L_y \frac{\sin[(\beta L_x/2)u]}{(\beta L_x/2)u} \frac{\sin[(\beta L_y/2)v]}{(\beta L_y/2)v} \end{aligned} \tag{7-34}$$

where we have introduced the pattern variables

$$u = \sin\theta\cos\phi, \qquad v = \sin\theta\sin\phi \tag{7-35}$$

The complete radiation fields are found from (7-26) as

$$E_\theta = j\beta \frac{e^{-j\beta r}}{2\pi r} E_o L_x L_y \sin\phi \frac{\sin[(\beta L_x/2)u]}{(\beta L_x/2)u} \frac{\sin[(\beta L_y/2)v]}{(\beta L_y/2)v} \tag{7-36a}$$

$$E_\phi = j\beta \frac{e^{-j\beta r}}{2\pi r} E_o L_x L_y \cos\theta\cos\phi \frac{\sin[(\beta L_x/2u]}{(\beta L_x/2)u} \frac{\sin[(\beta L_y/2)v]}{(\beta L_y/2)v} \tag{7-36b}$$

These fields are rather complicated functions of θ and ϕ, but fortunately they simplify in the principal planes. In the E-plane (yz-plane), $\phi = 90°$ (and 270°) and (7-36a) reduces to

$$E_\theta = j\beta \frac{e^{-j\beta r}}{2\pi r} E_o L_x L_y \frac{\sin[(\beta L_y/2)\sin\theta]}{(\beta L_y/2)\sin\theta} \qquad \textit{E-plane} \tag{7-37}$$

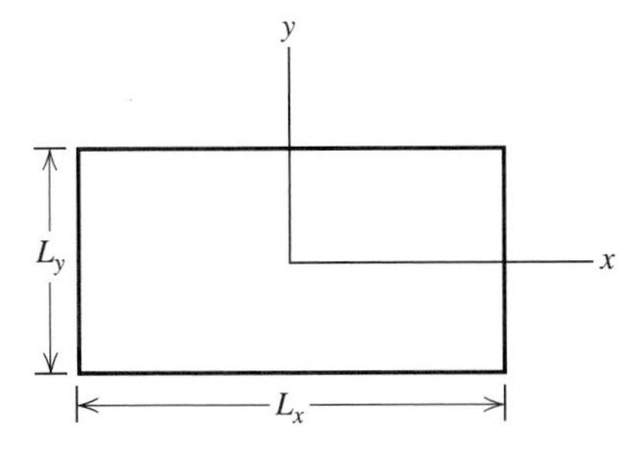

Figure 7-6 The rectangular aperture.

In the *H*-plane (*xz*-plane), $\phi = 0°$ (and 180°) and (7-36b) becomes

$$E_\phi = j\beta \frac{e^{-j\beta r}}{2\pi r} E_o L_x L_y \cos\theta \frac{\sin[\beta L_x/2)\sin\theta]}{(\beta L_x/2)\sin\theta} \qquad H\text{-plane} \qquad (7\text{-}38)$$

Note that E_ϕ goes to zero at $\theta = 90°$ where it is tangent to the perfect conductor introduced in the equivalent magnetic current formulation. The normalized forms of these principal plane patterns are

$$F_H(\theta) = \cos\theta \frac{\sin[(\beta L_x/2)\sin\theta]}{(\beta L_x/2)\sin\theta}, \qquad \phi = 0° \qquad (7\text{-}39)$$

$$F_E(\theta) = \frac{\sin[(\beta L_y/2)\sin\theta]}{(\beta L_y/2)\sin\theta}, \qquad \phi = 90° \qquad (7\text{-}40)$$

For large apertures (L_x, $L_y \gg \lambda$), the main beam is narrow, the $\cos\theta$ factor is negligible, and the principal plane patterns are both of the form $\sin(x)/x$ that we have encountered several times before, as, for example, with the uniform line source. By neglecting the obliquity factors in (7-36), the normalized pattern factor for the uniform rectangular aperture is

$$f(u, v) = \frac{\sin[(\beta L_x/2)u]}{(\beta L_x/2)u} \frac{\sin[(\beta L_y/2)v]}{(\beta L_y/2)v} \qquad (7\text{-}41)$$

which is the normalized version of P_y in (7-34). The half-power beamwidths in the principal planes follow from the line source result in (4-14). In the *xz*- and *yz*-planes, the beamwidth expressions are

$$\text{HP}_x = 0.886 \frac{\lambda}{L_x} \text{ rad} = 50.8 \frac{\lambda}{L_x} \text{ deg} \qquad (7\text{-}42\text{a})$$

$$\text{HP}_y = 0.886 \frac{\lambda}{L_y} \text{ rad} = 50.8 \frac{\lambda}{L_y} \text{ deg} \qquad (7\text{-}42\text{b})$$

Finally, we derive an expression for the directivity of a uniform rectangular aperture. Such calculations are greatly simplified by using the variables u and v. The transformation from θ and ϕ to u and v given by (7-35) is essentially a collapsing of the spherical surface of unit radius onto a planar surface through the equator, giving a circular disk of unit radius. The u, v disk is analogous to the azimuthal map projection used in cartography to show, for example, the northern hemisphere on a planar map; the globe is projected with the North Pole at the center and the azimuth (radial) lines give true compass directions. The visible region in u and v corresponding to $\theta \leq \pi/2$ is

$$u^2 + v^2 = \sin^2\theta \leq 1 \qquad (7\text{-}43)$$

which follows from (7-35).

The beam solid angle is found using

$$\Omega_A = \int_0^{2\pi} \int_0^{\pi/2} |F(\theta, \phi)|^2 \, d\Omega \qquad (7\text{-}44)$$

where only radiation for $\theta \leq \pi/2$ is considered. The beam solid angle can be evaluated by integrating over the entire visible region in terms of u and v. The projection of $d\Omega$ onto the u, v plane is given by $du\,dv = \cos\theta\, d\Omega$. From (7-43), it is seen that $\cos\theta = \sqrt{1 - u^2 - v^2}$. Therefore, $d\Omega = du\,dv/\sqrt{1 - u^2 - v^2}$ and (7-44) becomes

$$\Omega_A = \iint_{u^2+v^2\le 1} |F(u, v)|^2 \frac{du\,dv}{\sqrt{1 - u^2 - v^2}} \tag{7-45}$$

This is a general expression. For a large uniform phase aperture (L_x and $L_y \gg \lambda$), the radiation is concentrated in a narrow region about $u = v = 0$ ($\theta = 0$). Then the square root in (7-45) is approximately 1. Also, since the side lobes are very low, we can extend the limits to infinity without appreciably affecting the value of the integral.

Using these results and (7-41) for the uniform rectangular aperture in (7-45) yields

$$\Omega_A = \int_{-\infty}^{\infty} \frac{\sin^2[(\beta L_x/2)u]}{[(\beta L_x/2)u]^2}\, du \int_{-\infty}^{\infty} \frac{\sin^2[(\beta L_y/2)v]}{[(\beta L_y/2)v]^2}\, dv \tag{7-46}$$

The following change of variables:

$$u' = \frac{\beta L_x}{2} u = \frac{\beta L_x}{2} \sin\theta\cos\phi \tag{7-47a}$$

$$v' = \frac{\beta L_y}{2} v = \frac{\beta L_y}{2} \sin\theta\sin\phi \tag{7-47b}$$

leads to

$$\Omega_A = \frac{2}{\beta L_x}\frac{2}{\beta L_y} \int_{-\infty}^{\infty} \frac{\sin^2 u'}{(u')^2}\, du' \int_{-\infty}^{\infty} \frac{\sin^2 v'}{(v')^2}\, dv' \tag{7-48}$$

From (F-12) each integral above equals π, so

$$\Omega_A = \frac{4}{(2\pi/\lambda)^2 L_x L_y}\pi^2 = \frac{\lambda^2}{L_x L_y} \tag{7-49}$$

The directivity of the rectangular aperture with uniform amplitude and phase is then

$$D_u = \frac{4\pi}{\Omega_A} = \frac{4\pi}{\lambda^2} L_x L_y \tag{7-50}$$

From this expression, the physical area of the aperture can be identified as $A_p = L_x L_y$. Comparing this to $D = 4\pi A_{em}/\lambda^2$ from (2-84), we see that the maximum effective aperture A_{em} equals the physical aperture A_p for the uniform rectangular aperture. This is true for any shape aperture with uniform excitation. Also note that for ideal apertures, there are no ohmic losses (radiation efficiency $e_r = 1$), so gain equals directivity and $A_e = A_{em}$.

EXAMPLE 7-2 ***A 20λ × 10λ Uniform Rectangular Aperture***

The complete pattern for a uniform rectangular aperture that has $L_x = 20\lambda$ and $L_y = 10\lambda$ is from (7-41)

$$f(u, v) = \frac{\sin(20\pi u)}{20\pi u}\frac{\sin(10\pi v)}{10\pi v} \tag{7-51}$$

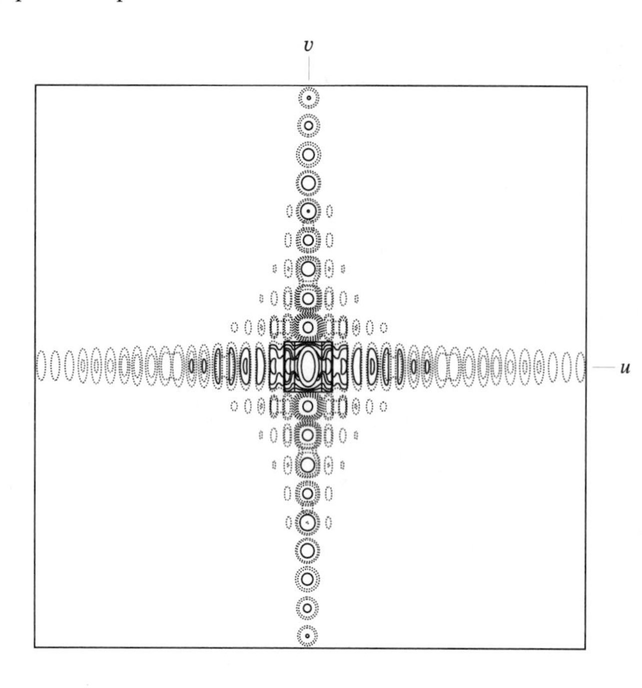

Figure 7-7 Contour plot of the pattern from a uniform amplitude, uniform phase rectangular aperture ($L_x = 20\lambda$, $L_y = 10\lambda$). The solid contour levels are 0, −5, −10, −15, −20, −25, −30 dB. The dashed contour levels are −35 and −40 dB. Principal plane profiles are shown in Fig. 7-8.

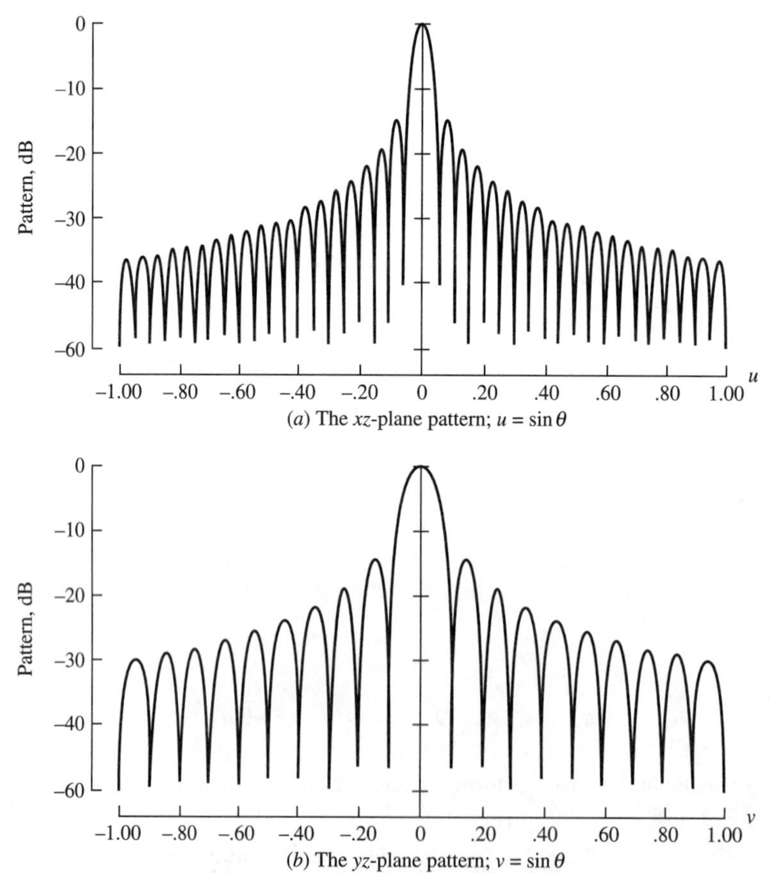

Figure 7-8 Principal plane patterns for a uniform amplitude, uniform phase rectangular aperture ($L_x = 20\lambda$, $L_y = 10\lambda$). The complete pattern is shown in Fig. 7-7.

The contour plot of this pattern is shown in Fig. 7-7. The principal plane patterns, which are profiles along the u and v axes of Fig. 7-7, are shown in Fig. 7-8. The aperture of Fig. 7-6 has a ratio $L_x/L_y = 2$ as in this example. Notice that the wide aperture dimension L_x leads to a narrow beamwidth in that direction (along the u-axis). The half-power beamwidth from (7-42) is $\text{HP}_x = 0.0443$ rad $= 2.54°$ in the xz-plane, and $\text{HP}_y = 0.0886$ rad $= 5.08°$ in the yz-plane. The directivity from (7-50) is $D = 4\pi(20\lambda)(10\lambda)/\lambda^2 = 2513 = 34$ dB.

7.2.2 Tapered Rectangular Apertures

In the previous section, we saw that the uniform rectangular aperture has an effective aperture equal to its physical aperture. In other words, uniform illumination leads to the most efficient use of the aperture area. It will be shown in Sec. 7.3 that uniform excitation amplitude for an aperture gives the highest directivity possible for all constant phase excitations of that aperture. In the antenna design problem, high directivity is not the only parameter to be considered. Frequently, low side lobes are important. As we saw in Chap. 4, the side lobes can be reduced by tapering the excitation amplitude toward the edges of a line source. This is also true for two-dimensional apertures. In fact, many of the line source results can be directly applied to aperture problems.

To simplify our general discussion of rectangular aperture distributions, we omit the polarization of the aperture electric field, so that E_a can represent either the x- or y-component of the aperture field. Then (7-18) becomes

$$P = \iint_{S_a} E_a(x', y')e^{j\beta ux'}e^{j\beta vy'}\, dx'\, dy' \tag{7-52}$$

Most practical aperture distributions are *separable* and can be expressed as a product of functions of each aperture variable alone:

$$E_a(x', y') = E_{a1}(x')E_{a2}(y') \tag{7-53}$$

Then (7-52) reduces to

$$P = \int_{-L_x/2}^{L_x/2} E_{a1}(x')e^{j\beta ux'}\, dx' \int_{-L_y/2}^{L_y/2} E_{a2}(y')e^{j\beta vy'}\, dy' \tag{7-54}$$

Each of these integrals is recognized as the pattern factor of a line source along the respective aperture directions. The normalized pattern factor for the rectangular aperture is then

$$f(u', v') = f_1(u')f_2(v') \tag{7-55}$$

where $f_1(u')$ and $f_2(v')$ arise from the first and second integrals in (7-54), which are essentially pattern factors of line source distributions along the x- and y-directions. Again, here we have neglected any obliquity factors. The uniform rectangular aperture result corresponding to (7-55) is (7-41). It is obtained directly from $\sin(u)/u$ of (4-9) by using u' of (7-47a) in place of u for $f_1(u')$ and v' of (7-47b) in place of u for $f_2(v')$. Note the different definition of u in Chap. 4 and this chapter.

Thus, the pattern expression for a rectangular aperture distribution that is separable, as in (7-53), is obtained by finding the patterns f_1 and f_2 corresponding to the distributions E_{a1} and E_{a2}, and then employing (7-55).

EXAMPLE 7-3 *The Open-Ended Rectangular Waveguide*

One of the smallest aperture antennas is the open-ended waveguide shown in Fig. 7-9. It requires no construction, since the antenna is the open end of a waveguide. It is often used as a probe, such as with a near-field measurement range (see Sec. 9.5), because of its compact size. When operated in the dominant TE_{10} mode, the aperture electric field is cosine-tapered in the x-direction with length $L_x = a$, similar to (4-23), and is uniform in the y-direction with length $L_y = b$. The radiation pattern then can be found from the corresponding line source results using $f(u', v') = f_1(u')f_2(v')$, where $f_1(u')$ is obtained from (4-27) and $f_2(v')$ is obtained from (4-9):

$$f(u', v') = \frac{\cos u'}{1 - [(2/\pi)u']^2} \frac{\sin v'}{v'} = \frac{\cos[(\beta L_x/2)u]}{1 - [(2/\pi)(\beta L_x/2)u]^2} \frac{\sin[(\beta L_y/2)v]}{(\beta L_y/2)v} \tag{7-56}$$

The vertical linear polarization of this antenna is evident from Fig. 7-9, which shows the components of the electric field in the far field. The complete far-field component expressions are easily obtained from (7-56) using the equivalent current formulations of (7-24) or (7-26). If the open-ended waveguide is surrounded by a large ground plane, (7-56) in (7-26) yields the following principal plane pattern results by a process identical to the development presented for (7-39) and (7-40):

$$F_H(\theta) = \cos\theta \frac{\cos\left[\frac{\beta a}{2}\sin\theta\right]}{1 - \left[\frac{2}{\pi}\frac{\beta a}{2}\sin\theta\right]^2} \qquad \phi = 0° \tag{7-57a}$$

open-ended waveguide on a ground plane, $0 < \theta < 90°$

$$F_E(\theta) = \frac{\sin\left[\frac{\beta b}{2}\sin\theta\right]}{\frac{\beta b}{2}\sin\theta} \qquad \phi = 90° \tag{7-57b}$$

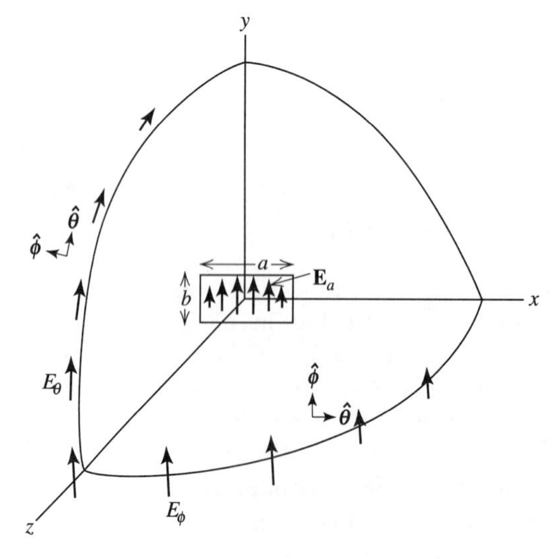

Figure 7-9 Geometry for an open-ended rectangular waveguide operating in the dominant TE_{10} mode as in Example 7-3. The aperture electric field $\mathbf{E}_a$ and radiated field components E_θ and E_ϕ are shown.

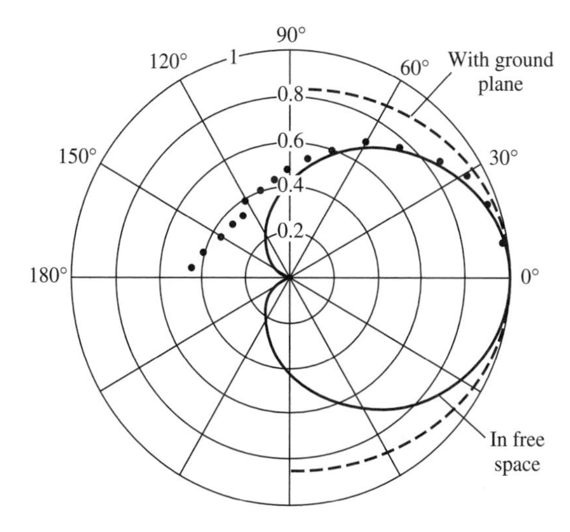

Figure 7-10 E-plane radiation patterns of the open-ended WR-90 waveguide of Example 7-3 operating at 9.32 GHz with the geometry of Fig. 7-9. Patterns are calculated using (7-58b) for free space (solid curve) and (7-57b) for a ground plane (dashed curve). Also shown is the measured pattern (dotted curve) for the open-ended waveguide in free space [4].

The magnetic current formulation is chosen because the ground plane is well represented by the perfect conductor used in the image theory model of Fig. 7-4*c*. If the waveguide radiates into free space, the complete expression of (7-24) is more appropriate and yields

$$F_H(\theta) = \frac{1 + \cos\theta}{2} \frac{\cos\left[\dfrac{\beta a}{2}\sin\theta\right]}{1 - \left[\dfrac{2}{\pi}\dfrac{\beta a}{2}\sin\theta\right]^2} \qquad \phi = 0° \qquad \text{(7-58a)}$$

open-ended waveguide in free space, $0 < \theta < 180°$

$$F_E(\theta) = \frac{1 + \cos\theta}{2} \frac{\sin\left[\dfrac{\beta b}{2}\sin\theta\right]}{\dfrac{\beta b}{2}\sin\theta} \qquad \phi = 90° \qquad \text{(7-58b)}$$

Note the difference in obliquity factors between (7-57) and (7-58). For (7-57), the boundary conditions on the ground plane at $\theta = 90°$ are satisfied as explained in association with (7-38). For operation in free space, (7-58) is valid all the way to $\theta = 180°$ where the $(1 + \cos\theta)/2$ obliquity factor takes the pattern to zero. However, an open-ended waveguide is an electrically small aperture and the theory we have used assumes the aperture to be large. Thus, both (7-57) and (7-58) are approximate. Figure 7-10 shows the E-plane patterns measured [1, p. 345] and calculated using (7-57b) and (7-58b) for a WR90 waveguide at 9.32 GHz. Agreement of the data measured in free space is better for the pattern based on (7-58b), as it should. Better results can be obtained by including the phase constant (β_g) of the waveguide, the reflection coefficient introduced by the discontinuity of the abrupt termination, and, most of all, the fringe currents on the waveguide walls [4]; see Prob. 7.2-4. The directivity of the open-ended waveguide is discussed in Example 7-4 and the half-power beamwidth in Prob. 7.2-3.

7.3 TECHNIQUES FOR EVALUATING GAIN

Aperture antennas are often selected for use in applications requiring high gain. It is, therefore, important to be able to evaluate gain as accurately as possible. In this section, techniques are presented for evaluating gain based on pattern information and on aperture field information. In addition, simple formulas are presented that provide approximate gain values. These techniques apply to a wide variety of antenna types.

7.3.1 Directivity

Two useful forms for directivity from (1-46) are

$$D = \frac{4\pi}{\Omega_A} \tag{7-59}$$

$$D = \frac{4\pi U_m}{P} \tag{7-60}$$

where U_m is the maximum of the radiation intensity defined through

$$U(\theta, \phi) = \frac{1}{2\eta}\,[|E_\theta|^2 + |E_\phi|^2]r^2 = U_m|F(\theta, \phi)|^2 \tag{7-61}$$

Ω_A is the beam solid angle:

$$\Omega_A = \int_0^{2\pi}\int_0^{\pi} |F(\theta, \phi)|^2 \sin\theta \, d\theta \, d\phi \tag{7-62}$$

The total radiated power P is evaluated by integrating (7-61) over all radiation space or simply $P = U_m\Omega_A$, as in (1-144). Accurate evaluation of directivity using (7-59) and (7-62) requires both a knowledge of the pattern over all angles θ, ϕ as well as integration of the pattern. If the pattern function is known, the integral in (7-62) can sometimes be evaluated analytically, but is usually found by numerical integration.

Pattern integration can be avoided when evaluating the directivity of aperture antennas. This approach is based on determining the radiated power in the aperture plane where it is easier to integrate. A knowledge of the aperture fields is required, of course. The formulation is simplified by assuming that the tangential aperture electric and magnetic fields are related as a TEM wave; see (7-27). This is justified by the good match to free space (e.g., low VSWR) that most aperture antennas exhibit, indicating real power flow as with a TEM wave. Using (7-27) in the general radiation field expression of (7-24) with (7-61) gives

$$U(\theta, \phi) = \frac{\beta^2}{32\pi^2\eta}\,(1 + \cos\theta)^2[|P_x|^2 + |P_y|^2] \tag{7-63}$$

The maximum value of this function, which corresponds to the main beam peak from (7-14), is

$$U_m = \frac{\beta^2}{8\pi^2\eta}\left|\iint_{S_a} \mathbf{E}_a \, dS'\right|^2 \tag{7-64}$$

since $\hat{\mathbf{r}} \cdot \mathbf{r}' = 0$ in the broadside case ($\theta = 0$) because $\hat{\mathbf{r}} = \hat{\mathbf{z}}$ and $\mathbf{r}'$ is in the xy-plane.

Integration of (7-63) to obtain P is, in general, rather difficult. This can be avoided by observing that the total power reaching the far field must have passed through the aperture. Within the validity of (7-27), the power density in the aperture is $|\mathbf{E}_a|^2/2\eta$ and we can determine the radiated power from

$$P = \frac{1}{2\eta} \iint_{S_a} |\mathbf{E}_a|^2 \, dS' \tag{7-65}$$

Substituting (7-64) and (7-65) in (7-60) gives a simplified, but powerful, directivity relationship:

$$\boxed{D = \frac{4\pi}{\lambda^2} \frac{\left| \iint_{S_a} \mathbf{E}_a \, dS' \right|^2}{\iint_{S_a} |\mathbf{E}_a|^2 \, dS'}} \tag{7-66}$$

This formula assumes the following: The pattern peak is directed broadside to the aperture, the aperture is large relative to a wavelength, and the aperture fields nearly form a plane wave. It turns out that the latter two conditions need not be strictly satisfied for good results to be obtained. Note the similarity of (7-66) to (3-93) for a half-wavelength spaced linear array.

If the aperture distribution is of uniform amplitude ($E_a = E_o$), then (7-66) reduces to

$$D_u = \frac{4\pi}{\lambda^2} A_p \tag{7-67}$$

where A_p is the physical aperture area. This was shown to be true for the rectangular aperture by direct evaluation; see (7-50). Further, (7-67) is a general result and implies that *the directivity of a uniform amplitude aperture is the highest obtainable from a uniform phase aperture.* This is true because the maximum of (7-66) occurs for a uniform illumination, which yields (7-67); see Prob. 7.3-2. The IEEE term for D_u is *standard directivity*.

EXAMPLE 7-4 Directivity of an Open-Ended Rectangular Waveguide

To illustrate the aperture field integration method of determining directivity, we return to the open-ended waveguide operating in the TE_{10} mode as described in Example 7-3 and illustrated in Fig. 7-9. The aperture field distribution is

$$\mathbf{E}_a = \hat{\mathbf{y}} E_o \cos \frac{\pi x'}{a}, \qquad -\frac{a}{2} \le x' \le \frac{a}{2}, \qquad -\frac{b}{2} \le y' \le \frac{b}{2} \tag{7-68}$$

where the waveguide (and, thus, the aperture) has wide and narrow dimensions of a and b. Then

$$\left| \iint_{S_a} \mathbf{E}_a \, dS' \right|^2 = \left(E_o \int_{-a/2}^{a/2} \cos \frac{\pi x'}{a} \, dx' \int_{-b/2}^{b/2} dy' \right)^2 = E_o^2 \left(\frac{2a}{\pi} \right)^2 b^2 \tag{7-69}$$

and

$$\iint_{S_a} |\mathbf{E}_a|^2 \, dS' = E_o^2 \int_{-a/2}^{a/2} \cos^2 \frac{\pi x'}{a} \, dx' \int_{-b/2}^{b/2} dy' = E_o^2 \frac{a}{2} b \tag{7-70}$$

Substituting these into (7-66) gives

$$D = \frac{4\pi}{\lambda^2}\left(\frac{8}{\pi^2}ab\right) = \frac{4\pi}{\lambda^2}(0.81)ab \tag{7-71}$$

This directivity is reduced by a factor of 0.81 (the aperture taper efficiency, ε_t) from that of the same aperture when uniformly illuminated as in (7-67). This formula provides only a rough approximation for a small aperture such as this. Accurate gain techniques for open-ended waveguides are available [4].

Most rectangular aperture distributions are separable, that is,

$$E_a(x, y) = E_{a1}(x)E_{a2}(y) \tag{7-72}$$

It can then be shown (see Prob. 7.3-15) that the directivity is also separable:

$$D = \pi D_x D_y \cos\theta_o \tag{7-73}$$

where

D_x, D_y = directivity of a line source with a relative current distribution of $E_{a1}(x)$, $E_{a2}(y)$

θ_o = main beam pointing direction relative to broadside

The $\cos\theta_o$ factor represents the projection of the aperture physical area onto the plane normal to the main beam maximum direction θ_o. This approximation is valid if the beam is not scanned within several beamwidths of endfire. The directivity of a uniform rectangular aperture for broadside ($\theta_o = 0$) can be expressed as follows using (4-21):

$$D_u = \pi\frac{2L_x}{\lambda}\frac{2L_y}{\lambda} = \frac{4\pi}{\lambda^2}L_xL_y \tag{7-74}$$

which is (7-50).

7.3.2 Gain and Efficiencies

Gain equals directivity reduced by the amount of power lost on the antenna structure; see Secs. 1.8 and 2.5 for previous discussions of gain. This is expressed using radiation efficiency from (1-159) as

$$G = e_r D \tag{7-75}$$

Another form follows from (7-60) with input power in place of radiated power since $P = e_r P_{\text{in}}$; also see (1-155):

$$G = \frac{4\pi U_m}{P_{\text{in}}} \tag{7-76}$$

This form is often used when evaluating antennas by numerical computation.

Since the directivity of an aperture antenna is directly proportional to its physical aperture area A_p, gain will be also:

$$\boxed{G = \frac{4\pi}{\lambda^2}A_e = \frac{4\pi}{\lambda^2}\varepsilon_{ap}A_p = \varepsilon_{ap}D_u} \tag{7-77}$$

where (2-89), (2-90), and (7-67) were used. A_e is the effective aperture and can be calculated through this equation for any antenna, including arrays. From this, we see that

$$\boxed{A_e = \varepsilon_{ap} A_p} \qquad 0 \le \varepsilon_{ap} \le 1 \tag{7-78}$$

Aperture efficiency ε_{ap} is a measure of how efficiently the antenna physical area is utilized. If ε_{ap} is known, it is a simple matter to calculate the gain of an aperture antenna of aperture area A_p using (7-77).

There are several contributions to the overall aperture efficiency. The following form shows the factors separately and is appropriate to general use:

$$\boxed{\varepsilon_{ap} = e_r \varepsilon_t \varepsilon_s \varepsilon_a} \tag{7-79}$$

All these factors have values from zero to unity. We discussed radiation efficiency e_r in Sec. 1.8; it represents all forms of dissipation on the antenna structure such as conductor losses. In most aperture antennas, these losses are very low, so $e_r \approx 1$ and

$$G \approx D \qquad \textit{most aperture antennas} \tag{7-80}$$

This may not hold if one of the following situations applies: The antenna size is less than a wavelength, a lossy transmission line or device is considered part of the antenna, or lossy materials are an integral part of the antenna such as a dielectric lens.

Aperture taper efficiency ε_t represents gain loss strictly due to the aperture amplitude distribution. It is also called the *utilization factor*. Often, the amplitude is tapered from the center to the edges of an aperture intentionally to reduce sidelobes. ε_t is the ratio of directivity computed with only the amplitude taper present D_t to the directivity of the same aperture uniformly illuminated D_u:

$$\varepsilon_t = \frac{D_t}{D_u} \qquad \text{or} \qquad D_t = \varepsilon_t D_u \tag{7-81}$$

Examples for line sources are given in Table 4-2. Also, in Example 7-4 we found $\varepsilon_t = 0.81$ for an open-ended waveguide.

Antennas that have a secondary radiating aperture illuminated by a primary (feed) antenna, such as a parabolic reflector, experience spillover loss due to power from the feed missing the radiating aperture. This *spillover efficiency* ε_s and aperture taper efficiency are the main sources of gain loss in most aperture antennas. The product $\varepsilon_t \varepsilon_s$ is called the *illumination efficiency* ε_i.

The remaining factor in (7-79) ε_a is *achievement efficiency* and can include many subefficiencies. More subefficiencies will be treated with reflector antennas in Sec. 7.5, but the following two are usually dominant:

$$\varepsilon_a = \varepsilon_{cr} \varepsilon_{ph} \ldots \tag{7-82}$$

Cross-polarization efficiency ε_{cr} represents loss due to power being radiated in a polarization state orthogonal to the intended polarization. *Phase-error efficiency* ε_{ph} represents loss due to nonuniform phase across the aperture.

Any of the efficiency factors can be expressed as a gain factor in decibels as

$$\varepsilon_n(\text{dB}) = 10 \log \varepsilon_n \tag{7-83}$$

Gain "loss" is negative of this. For example, the aperture taper efficiency for Example 7-4 is $\varepsilon_t = 0.81$, so $\varepsilon_t(\text{dB}) = -0.91$ dB and the gain loss is +0.91 dB. This is the only source of loss in this case. In general, (7-77) and (7-79) can be written in dB form as

$$G(\text{dB}) = 10 \log\left(\frac{4\pi}{\lambda^2} A_p\right) + e_r(\text{dB}) + \varepsilon_t(\text{dB}) + \varepsilon_s(\text{dB}) + \varepsilon_a(\text{dB}) \quad (7\text{-}84)$$

Recall that polarization mismatch factor p and impedance mismatch factor q are not included in aperture efficiency nor gain, but they play a role similar to the efficiency factors (as discussed in Sec. 2.5).

Another measure of antenna performance is *beam efficiency* ε_M. Instead of a gain loss, beam efficiency quantifies the solid angle extent of the main beam Ω_M relative to that of the entire pattern:

$$\varepsilon_M = \frac{\Omega_M}{\Omega_A} = \frac{\iint\limits_{\substack{\text{main}\\\text{beam}}} |F(\theta, \phi)|^2 \, d\Omega}{\iint\limits_{4\pi} |F(\theta, \phi)|^2 \, d\Omega} \quad (7\text{-}85)$$

7.3.3 Simple Directivity Formulas

It is often necessary to estimate the gain of an antenna, especially in system calculations. If the gain cannot be measured, simple gain equations can be used. The most direct and simplest approach is to use (7-77). The operating wavelength and physical aperture area are easily obtained. Aperture efficiency can sometimes be determined by using a theoretical model, as will be discussed for horns and reflectors later in this chapter. In many cases, it can be estimated. In general ε_{ap} ranges from 30 to 80% with 50% being a good overall value. Optimum gain pyramidal horns have an aperture efficiency near 50%. Parabolic reflector antennas have an efficiency of 55% or greater. Gain can be found by estimating the aperture efficiency. For example, a 30-dB gain antenna with an actual efficiency of 55% will have a gain error of 0.38 dB when an estimated efficiency of 60% is used.

It is very useful to have an approximate directivity expression that depends only on the half-power beamwidths of the principal plane patterns. This is expected to yield good results since we know that directivity varies inversely with the beam solid angle ($D = 4\pi/\Omega_A$) and the beam solid angle is primarily controlled by the main beam. Thus, we expect to find that $D \propto (\text{HP}_E \text{HP}_H)^{-1}$, where the product of the principal plane beamwidths approximates the beam solid angle. We now derive such a relation.

The directivity of a rectangular aperture with a separable distribution given by (7-73) for broadside operation ($\theta_o = 0$) is

$$D = \pi D_x D_y \quad (7\text{-}86)$$

where D_x and D_y are the directivities of a line source (or linear array) associated with the x and y aperture distribution variations. But we know from studying several linear current distributions, that these directivities are related to the aperture extents as

$$D_x = c_x \frac{2L_x}{\lambda}, \qquad D_y = c_y \frac{2L_y}{\lambda} \tag{7-87}$$

where directivity factors c_x and c_y are constants that vary slightly with the distributions $E_{a1}(x)$ and $E_{a2}(y)$. For uniform line sources, $c_x = c_y = 1$; see (4-21). Using (7-87) in (7-86) and rearranging give

$$D = \pi c_x \frac{2L_x}{\lambda} c_y \frac{2L_y}{\lambda} = \frac{4\pi c_x c_y k_x k_y}{\left(k_x \dfrac{\lambda}{L_x}\right)\left(k_y \dfrac{\lambda}{L_y}\right)} = \frac{4\pi c_x c_y k_x k_y}{\text{HP}_x \text{HP}_y} \tag{7-88}$$

The beamwidth factors k_x and k_y are constants associated with the following beamwidth formulas that we have used frequently (see Table 4-2):

$$\text{HP}_x = k_x \frac{\lambda}{L_x}, \qquad \text{HP}_y = k_y \frac{\lambda}{L_y} \tag{7-89}$$

For uniform line sources, $k_x = k_y = 0.886$. The numerator in (7-88) is the *directivity-beamwidth product*:

$$\text{DB} = 4\pi c_x c_y k_x k_y \cdot \left(\frac{180}{\pi}\right)^2 \quad (\text{deg}^2) \tag{7-90}$$

It is similar to the gain-bandwidth product that is commonly used to characterize circuit devices. It remains relatively constant under a variety of operating circumstances because as the amplitude is tapered, constants c_x, c_y decrease due to aperture taper efficiency reduction, but the constants k_x, k_y increase due to beam broadening and nearly cancel the decrease in c_x, c_y.

If we could determine the value of DB, (7-88) would be our desired simple expression for directivity. For uniform line sources,

$$\text{DB}_u = 4\pi(1)(1)(0.886)(0.886) = 9.86 \text{ rad}^2 = 32{,}383 \text{ deg}^2 \tag{7-91}$$

Then (7-88) becomes

$$D_{u_{\text{rect}}} = \frac{32{,}383}{\text{HP}_{E^\circ}\,\text{HP}_{H^\circ}} \tag{7-92}$$

where HP_{E° and HP_{H° are the principal plane beamwidths in degrees. Although this is based on a uniform rectangular aperture, it produces accurate results for any pattern with a moderately narrow major lobe and with minor lobes present. This relation can be used for scanned beams if the beamwidths are those of the scanned beam [5]. For uniform circular apertures,

$$\text{DB}_{u_{\text{cir}}} = D \cdot \text{HP}^2 = \frac{4\pi}{\lambda^2}\pi a^2 \left(1.02 \frac{\lambda}{2a}\right)^2 \left(\frac{180}{\pi}\right)^2 = 33{,}709 \text{ deg}^2 \tag{7-93}$$

where (7-171) and (7-172) were used. This is very close to the directivity-beamwidth product of 32,383 for uniform rectangular apertures. As the amplitude taper from the center to the perimeter of a circular aperture is changed, DB varies from 33,709 to about 39,000 deg^2 [6]. For a rectangular aperture with a cosine amplitude taper in the H-plane and uniform phase, as found in the open-ended waveguide of Example 7-4, DB is 35,230 deg^2.

Some simple mathematical models for patterns have no minor lobes. Examples are Gaussian and $\cos^q \theta$ beams; see Probs. 7.3-18 and 7.3-19. Also see Sec. 9.6.2. Then all radiation is contained in the main lobe and there are no minor lobes and the beam solid angle is well approximated as the product of the principal plane beamdwidths; that is, $\Omega_A \approx \text{HP}_E\text{HP}_H$. (See Fig. 1-18). And the directivity is

$$D = \frac{4\pi}{\Omega_A} \approx \frac{4\pi}{\text{HP}_E\text{HP}_H} = \frac{41{,}253}{\text{HP}_{E^\circ}\text{HP}_{H^\circ}} \tag{7-94}$$

This formula can work well for low-directivity antennas. For example, a half-wave dipole has $\text{HP}_{E^\circ} = 78$, $\text{HP}_{H^\circ} = 360$ and from (7-94) $D \approx 1.47$, which is close to the correct value of 1.64. Other simple formulas have been proposed, but they are mostly for special cases [7–10].

In practice, antennas produce radiation patterns with significant power content in their minor lobes. In addition, the main beam rolls off more slowly than idealized patterns such as the $\sin u/u$ pattern for a uniform line source. Thus, the solid angle of the main beam is larger than that predicted by the product of the principal plane beamwidths. This together with the increased solid angle in the minor lobes compared to ideal models leads to directivity reduction. A variation of (7-92) that compares well to high-gain antennas used in practice is

$$G = \frac{26{,}000}{\text{HP}_{E^\circ}\text{HP}_{H^\circ}} \tag{7-95}$$

Gain is used here instead of directivity because it applies directly to antenna practice. Performance data quoted by manufacturers usually is the measured gain because it relates directly to how the antenna will be used. In most cases, ohmic losses are very small so that $G \approx D$.

The first point to recognize when using the foregoing simple formulas for finding directivity or gain based only on half-power beamwidths is that accurate results cannot be expected. There simply is not enough information contained in the half-power beamwidth values. Generally speaking though, we can summarize the application of the simple formulas as follows. For antennas with very low directivity and usually without side lobes, (7-94) is the appropriate choice. When working with theoretical models of antennas with appreciable electrical size, (7-92) yields good directivity values. Finally, (7-95) yields very usable gain values for practical high-gain antennas. The numerator of (7-95) is smaller than that of (7-94) because of several influences on achieved gain, such as broad regions of side lobes that do not decay to negligible values as in the theoretical aperture models. The following examples illustrate the use of (7-95).

EXAMPLE 7-5 *Pyramidal Horn Antenna*

A pyramidal horn antenna (see Fig. 7-18*a*), with a rectangular aperture of width A and height B, designed for optimum gain has an aperture efficiency of 51%; so from (7-77)

$$G = 0.51 \frac{4\pi}{\lambda^2} AB \tag{7-96}$$

As a specific example, a "standard gain horn" operating from 33 to 50 GHz has a measured gain of 24.7 dB (G = 295.1) at 40 GHz (λ = 0.75 cm). The aperture dimensions of this horn are A = 5.54 cm and B = 4.55 cm. Using these values of λ, A, and B in (7-96) gives G = 287.2 = 24.6 dB. The gain can also be estimated from the principal plane half-power beamwidths, measured at 40 GHz to be HP_{E° = 9° and HP_{H° = 10°. Then (7-95) yields G = 288.9 = 24.6 dB. The gain values from both of these methods agree very well with the measured gain of 24.7 dB.

EXAMPLE 7-6 *Circular Parabolic Reflector Antenna*

The aperture efficiency of a typical circular parabolic reflector antenna with diameter D is 55%, so (7-77) becomes

$$G = \varepsilon_{ap}\frac{4\pi}{\lambda^2}A_p = 0.55\frac{4\pi}{\lambda^2}\left(\pi\frac{D^2}{4}\right) = 5.43\frac{D^2}{\lambda^2} \tag{7-97}$$

For a specific example, a 3.66-m (12-ft) circular reflector operating at 11.7 GHz (λ = 2.564 cm) has a measured value of G = 50.4 dB and $HP_{E^\circ} = HP_{H^\circ}$ = 0.5°. Again, we will check our estimation formulas. First, (7-97) gives $G = 5.43(366/2.564)^2$ = 110,644 = 50.4 dB. Next, (7-95) yields $G = 26{,}000/(0.5)^2$ = 104,000 = 50.2 dB. Both of these estimates are in good agreement with the measured gain.

7.4 RECTANGULAR HORN ANTENNAS

Horn antennas are extremely popular antennas in the microwave region above about 1 GHz. Horns provide high gain, low VSWR, relatively wide bandwidth, low weight, and they are rather easy to construct. As an additional benefit, the theoretical calculations for horn antennas are achieved very closely in practice.

The three basic types of horn antennas that utilize rectangular geometry are illustrated in Fig. 7-11. These horns are fed by a rectangular waveguide that is ori-

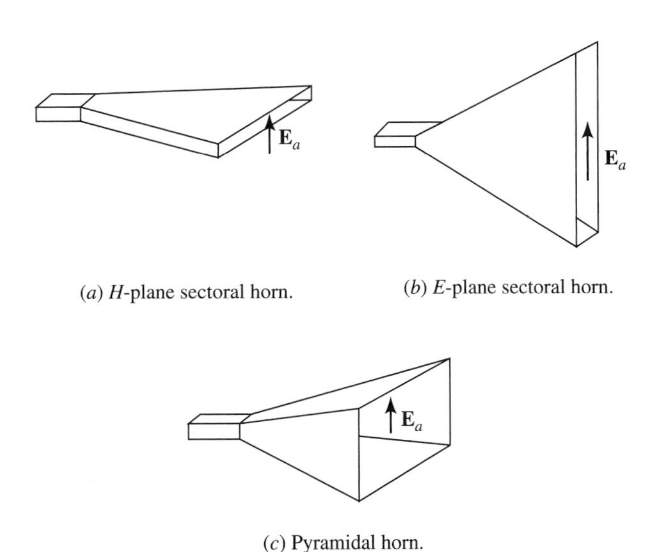

(*a*) *H*-plane sectoral horn. (*b*) *E*-plane sectoral horn.

(*c*) Pyramidal horn.

Figure 7-11 Rectangular horn antennas.

ented with its broad wall horizontal. For dominant waveguide mode excitation, the E-plane is then vertical and the H-plane horizontal. If the horn serves to flare the broad wall dimension and leave the narrow wall of the waveguide unchanged, it is called an ***H*-plane sectoral horn antenna** as shown in Fig. 7-11*a*. On the other hand, if the horn serves to flare only in the E-plane dimension, it is called an ***E*-plane sectoral horn antenna** and is shown in Fig. 7-11*b*. When both waveguide dimensions are flared, it is referred to as a **pyramidal horn antenna**, which is shown in Fig. 7-11*c*.

The operation of a horn antenna can be viewed as analogous to a megaphone, which is an acoustic horn radiator providing directivity for sound waves. The electromagnetic horn acts as a transition from the waveguide mode to the free-space mode. This transition reduces reflected waves and emphasizes the traveling waves. This traveling wave behavior, as we have seen with other antennas, leads to low VSWR and wide bandwidth.

Aperture antennas are among the oldest antennas. Heinrich Hertz experimented with microwave parabolic cylinder antennas in 1888. The Indian Physicist J. Chunder Bose operated a pyramidal horn, which he called a "collecting funnel," at 60 GHz in 1897. The horn antenna has been in widespread use since the 1940s; see [11] for a collection of many papers on horn antennas.

A characteristic of the horn antenna that we have not encountered until now is that the longer path length from the connecting waveguide to the edge of the horn aperture compared to the aperture center in the flare plane introduces a phase delay across the aperture. This aperture "phase error" is not present in antennas such as an open-ended waveguide and complicates the analysis. Phase errors occur in several areas of antennas and they are treated in this section on rectangular horn antennas. In addition to the rectangular horns, conical horn antennas are common. There are also special-purpose horns including those with a dielectric or metallic plate lens in the aperture to correct for the phase error and those with metallic ridges inside the horn to increase bandwidth [12]. One of the most important applications for horn antennas is as a feed for a reflector antenna. Popular feed horns have corrugations on the inside walls; these are discussed in Sec. 7.7.

7.4.1 The *H*-Plane Sectoral Horn Antenna

The H-plane sectoral horn of Fig. 7-12*a* is fed from a rectangular waveguide of interior dimensions a and b, with a the broadwall dimension. The aperture is of width A in the H-plane and height b in the E-plane. The H-plane cross section

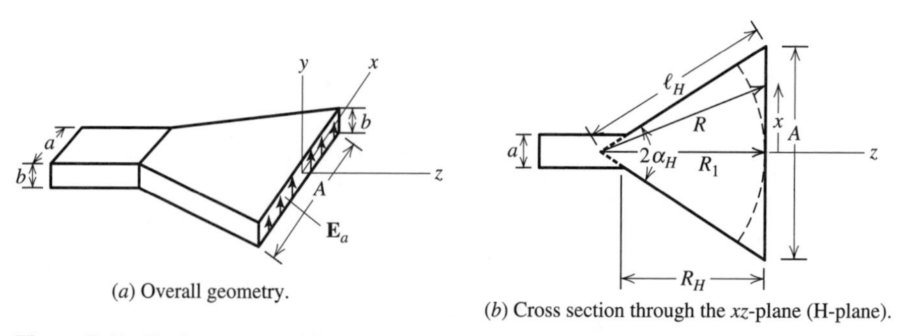

Figure 7-12 H-plane sectoral horn antenna.

of Fig. 7-12*b* reveals the geometrical parameters. The following relationships for the geometry will be of use in subsequent analysis:

$$\ell_H^2 = R_1^2 + \left(\frac{A}{2}\right)^2 \tag{7-98}$$

$$\alpha_H = \tan^{-1}\left(\frac{A}{2R_1}\right) \tag{7-99}$$

$$R_H = (A - a)\sqrt{\left(\frac{\ell_H}{A}\right)^2 - \frac{1}{4}} \tag{7-100}$$

The above relations follow directly from Fig. 7-12*b* and it is an exercise to prove (7-100). The dimensions A and R_H (or ℓ_H or R_1) must be determined to allow construction of the horn. We first investigate the principles of operation and then present design procedures for determining the horn dimensions.

The key to solving aperture antenna problems is to find the tangential fields over the aperture. The aperture plane for the H-plane sectoral horn shown in Fig. 7-12*a* is in the xy-plane. The aperture fields, of course, arise from the attached waveguide. As is usually the case in practice, we will assume that the waveguide carries the dominant TE_{10} rectangular waveguide mode. The transverse fields in the waveguide are then given by

$$E_y = E_{og} \cos\frac{\pi x}{a}\, e^{-j\beta_g z} \tag{7-101a}$$

$$H_x = -\frac{E_y}{Z_g} \tag{7-101b}$$

where $Z_g = \eta[1 - (\lambda/2a)^2]^{-1/2}$ is the waveguide characteristic impedance. The fields arriving at the aperture are essentially an expanded version of these waveguide fields. However, the waves arriving at different points in the aperture are not in-phase because of the different path lengths. We will now determine this phase distribution.

The path length R from the (virtual) horn apex in the waveguide to the horn aperture increases toward the horn mouth edges. Thus, waves arriving at aperture positions displaced from the aperture center lag in phase relative to those arriving at the center. The phase constant changes from that in the waveguide β_g to the free-space constant β as waves progress down the horn. But for relatively large horns, the phase constant for waves in the vicinity of the aperture is approximately that of free space. The aperture phase variation in the x-direction is then given by

$$e^{-j\beta(R-R_1)} \tag{7-102}$$

The aperture phase is uniform in the y-direction. An approximate form for R using Fig. 7-12*b* is

$$R = \sqrt{R_1^2 + x^2} = R_1\left[1 + \left(\frac{x}{R_1}\right)^2\right]^{1/2} \tag{7-103a}$$

$$\approx R_1\left[1 + \frac{1}{2}\left(\frac{x}{R_1}\right)^2\right] \tag{7-103b}$$

for $x \ll R_1$ that holds if $A/2 \ll R_1$. Then

$$R - R_1 \approx \frac{1}{2}\frac{x^2}{R_1} \tag{7-104}$$

The amplitude distribution is an expanded version of that in the waveguide, so it is a cosine taper in the x-direction. Using this fact and (7-104) in (7-102) leads to the aperture electric field distribution

$$E_{ay} = E_o \cos\frac{\pi x}{A}\, e^{-j(\beta/2R_1)x^2} \tag{7-105}$$

inside the aperture and zero elsewhere. Note that $E_{ay}(x = \pm A/2) = 0$ as required by boundary conditions. The phase distribution is often referred to as a quadratic phase error, since the deviation from a uniform phase condition varies as the square of the distance from the aperture center. This result can be derived more rigorously by representing the horn as a radial waveguide [13].

The quadratic phase error complicates the radiation integral; however, the result is worth the effort. Substituting (7-105) into (7-18b) yields

$$P_y = E_o \int_{-A/2}^{A/2} \cos\frac{\pi x'}{A}\, e^{-j(\beta/2R_1)x'^2} e^{j\beta u x'}\, dx' \int_{-b/2}^{b/2} e^{j\beta v y'}\, dy' \tag{7-106}$$

After considerable work, this reduces to

$$P_y = E_o\left[\frac{1}{2}\sqrt{\frac{\pi R_1}{\beta}}\, I(\theta, \phi)\right]\left\{b\,\frac{\sin[(\beta b/2)\sin\theta\sin\phi]}{(\beta b/2)\sin\theta\sin\phi}\right\} \tag{7-107}$$

where the factors in brackets correspond to each of the integrals in (7-106). The second factor is that for a uniform line source. The first involves the function

$$\begin{aligned} I(\theta, \phi) = {} & e^{j(R_1/2\beta)(\beta\sin\theta\cos\phi + \pi/A)^2}[C(s_2') - jS(s_2') - C(s_1') + jS(s_1')] \\ & + e^{j(R_1/2\beta)(\beta\sin\theta\cos\phi - \pi/A)^2}[C(t_2') - jS(t_2') - C(t_1') + jS(t_1')] \end{aligned} \tag{7-108}$$

where

$$\begin{aligned} s_1' &= \sqrt{\frac{1}{\pi\beta R_1}}\left(-\frac{\beta A}{2} - R_1\beta u - \frac{\pi R_1}{A}\right) \\ s_2' &= \sqrt{\frac{1}{\pi\beta R_1}}\left(\frac{\beta A}{2} - R_1\beta u - \frac{\pi R_1}{A}\right) \\ t_1' &= \sqrt{\frac{1}{\pi\beta R_1}}\left(-\frac{\beta A}{2} - R_1\beta u + \frac{\pi R_1}{A}\right) \\ t_2' &= \sqrt{\frac{1}{\pi\beta R_1}}\left(\frac{\beta A}{2} - R_1\beta u + \frac{\pi R_1}{A}\right) \end{aligned} \tag{7-109}$$

and the functions $C(x)$ and $S(x)$ are Fresnel integrals defined in (F-17) and tabulated in [14].

The total radiation fields can now be obtained. Using (7-29) gives the far-zone electric field components

$$E_\theta = j\beta\,\frac{e^{-j\beta r}}{4\pi r}(1 + \cos\theta)\sin\phi\, P_y \tag{7-110a}$$

$$E_\phi = j\beta\,\frac{e^{-j\beta r}}{4\pi r}(1 + \cos\theta)\cos\phi\, P_y \tag{7-110b}$$

These together with (7-107) give the complete radiated electric field

$$\mathbf{E} = j\beta E_o b \sqrt{\frac{\pi R_1}{\beta}} \frac{e^{-j\beta r}}{4\pi r} \left(\frac{1 + \cos\theta}{2}\right)(\hat{\boldsymbol{\theta}} \sin\phi + \hat{\boldsymbol{\phi}} \cos\phi) \cdot \frac{\sin[(\beta b/2) \sin\theta \sin\phi]}{(\beta b/2) \sin\theta \sin\phi} I(\theta, \phi) \tag{7-111}$$

where $I(\theta, \phi)$ is still given by (7-108).

The complete radiation expression is rather cumbersome, so we will examine the principal plane patterns. In the E-plane, $\phi = 90°$ and the normalized form of (7-111) is

$$F_E(\theta) = \frac{1 + \cos\theta}{2} \frac{\sin[(\beta b/2) \sin\theta]}{(\beta b/2) \sin\theta} \tag{7-112}$$

The second factor is the pattern of a uniform line source of length b along the y-axis, as one would expect from the aperture distribution.

In the H-plane, $\phi = 0°$ and the normalized H-plane pattern is

$$F_H(\theta) = \frac{1 + \cos\theta}{2} f_H(\theta) = \frac{1 + \cos\theta}{2} \frac{I(\theta, \phi = 0°)}{I(\theta = 0°, \phi = 0°)} \tag{7-113}$$

The H-plane pattern can be displayed rather simply using universal radiation pattern plots that are based on the maximum phase error across the aperture. The aperture distribution phase error as a function of position x from (7-105) is

$$\delta = \frac{\beta}{2R_1} x^2 \tag{7-114}$$

Since the maximum value of x is $A/2$, the maximum phase error is

$$\delta_{\max} = \frac{\beta}{2R_1}\left(\frac{A}{2}\right)^2 = 2\pi \frac{A^2}{8\lambda R_1} = 2\pi t \tag{7-115}$$

where t is defined to be

$$t = \frac{A^2}{8\lambda R_1} = \frac{1}{8}\left(\frac{A}{\lambda}\right)^2 \frac{1}{R_1/\lambda} \tag{7-116}$$

The function $I(\theta, \phi = 0°)$ in (7-108) can be expressed in terms of t as

$$\begin{aligned} I(\theta, \phi = 0°) = {} & e^{j(\pi/8t)[(A/\lambda)\sin\theta + 1/2]^2}[C(s_2) - jS(s_2) - C(s_1) + jS(s_1)] \\ & + e^{j(\pi/8t)[(A/\lambda)\sin\theta - 1/2]^2}[C(t_2) - jS(t_2) - C(t_1) + jS(t_1)] \end{aligned} \tag{7-117}$$

where

$$\begin{aligned} s_1 &= 2\sqrt{t}\left[-1 - \frac{1}{4t}\left(\frac{A}{\lambda}\sin\theta\right) - \frac{1}{8t}\right] \\ s_2 &= 2\sqrt{t}\left[1 - \frac{1}{4t}\left(\frac{A}{\lambda}\sin\theta\right) - \frac{1}{8t}\right] \\ t_1 &= 2\sqrt{t}\left[-1 - \frac{1}{4t}\left(\frac{A}{\lambda}\sin\theta\right) + \frac{1}{8t}\right] \\ t_2 &= 2\sqrt{t}\left[1 - \frac{1}{4t}\left(\frac{A}{\lambda}\sin\theta\right) + \frac{1}{8t}\right] \end{aligned} \tag{7-118}$$

This function is plotted in Fig. 7-13 for various values of t. It is normalized to the main beam peak for a zero phase error condition, which displays the directivity loss (reduction of the main beam peak) as the maximum phase error $2\pi t$ increases.

The curves in Fig. 7-13 are universal pattern plots from which antenna patterns can be derived for specific values of A, b, and λ. The H-plane plots (solid curves) are a function of $(A/\lambda)\sin\theta$. The E-plane plot (dashed curve) is the second factor of (7-112), and the abscissa for it is $(b/\lambda)\sin\theta$. The factor $(1+\cos\theta)/2$ that appears in both pattern functions (7-112) and (7-113) is not included in Fig. 7-13. For most situations, it has a small effect on the total pattern and may be neglected. Its effect, however, is easily included by adding $20\log[(1+\cos\theta)/2]$ to the corresponding pattern value from the universal pattern. Note that the E-plane plot of Fig. 7-13 has the -13.3-dB side lobe level of a uniform line source pattern, and the H-plane constant phase ($t=0$) plot has the -23-dB side lobe level of a cosine-tapered line source pattern. As the phase error increases, the H-plane pattern beamwidth and side lobes increase.

The pattern of $f_H(\theta)$ in (7-113) can be evaluated with excellent results using a mathematics application computer package to perform the numerical integration:

$$f_H(\theta) \propto \int_{-A/2}^{A/2} \cos\frac{\pi x'}{A}\, e^{-j\beta\sqrt{R_1^2+x'^2}} e^{j\beta\sin\theta\, x'}\, dx' \tag{7-119}$$

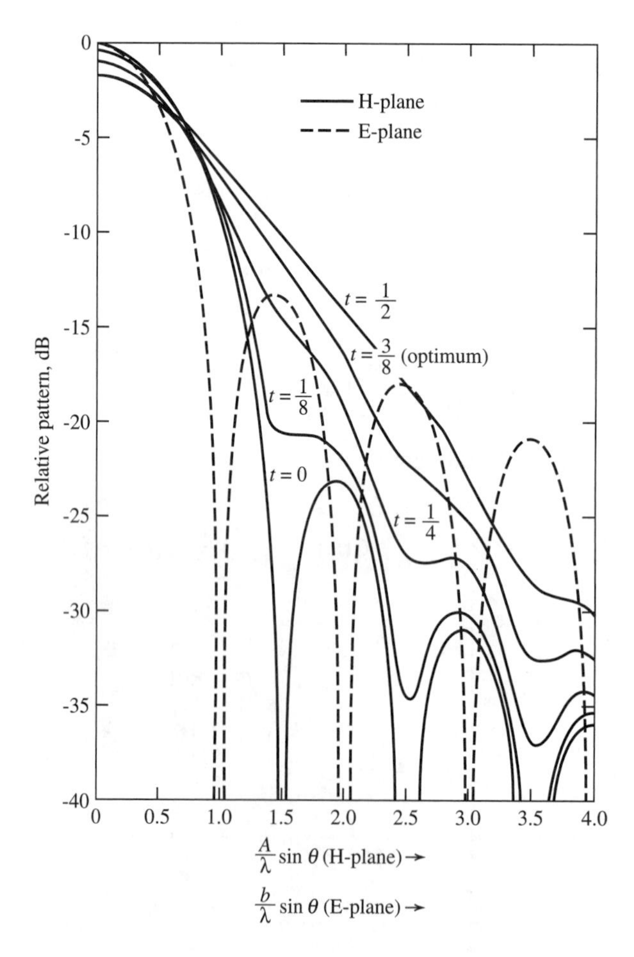

Figure 7-13 Universal radiation patterns for the principal planes of an H-plane sectoral horn as shown in Fig. 7-12. The factor $(1+\cos\theta)/2$ is not included.

where (7-106) was used with the exact phase error expression of (7-103a) rather than the quadratic approximation of (7-103b). This avoids the foregoing complicated expressions and permits inclusion of the exact phase.

The directivity for an H-plane sectoral horn is obtained from the aperture integration method of (7-66) as

$$D_H = \frac{b}{\lambda}\frac{32}{\pi}\left(\frac{A}{\lambda}\right)\varepsilon_{ph}^H = \frac{4\pi}{\lambda^2}\,\varepsilon_t\,\varepsilon_{ph}^H\,Ab \tag{7-120a}$$

where

$$\varepsilon_t = \frac{8}{\pi^2} \tag{7-120b}$$

$$\varepsilon_{ph}^H = \frac{\pi^2}{64t}\{[C(p_1) - C(p_2)]^2 + [S(p_1) - S(p_2)]^2\} \tag{7-120c}$$

$$p_1 = 2\sqrt{t}\left[1 + \frac{1}{8t}\right], \qquad p_2 = 2\sqrt{t}\left[-1 + \frac{1}{8t}\right] \tag{7-120d}$$

Note that $p_1 = -s_1' = t_2'$ and $p_2 = s_2' = -t_1'$ from (7-108) for $u = 0$. This expression explicitly shows the two efficiency factors associated with aperture taper and phase, ε_t and ε_{ph}^H.

A family of universal directivity curves is given in Fig. 7-14, where $\lambda D_H/b$ is plotted versus A/λ for various values of R_1/λ. Notice that for a given axial length R_1, there is an optimum aperture width A corresponding to the peak of the appro-

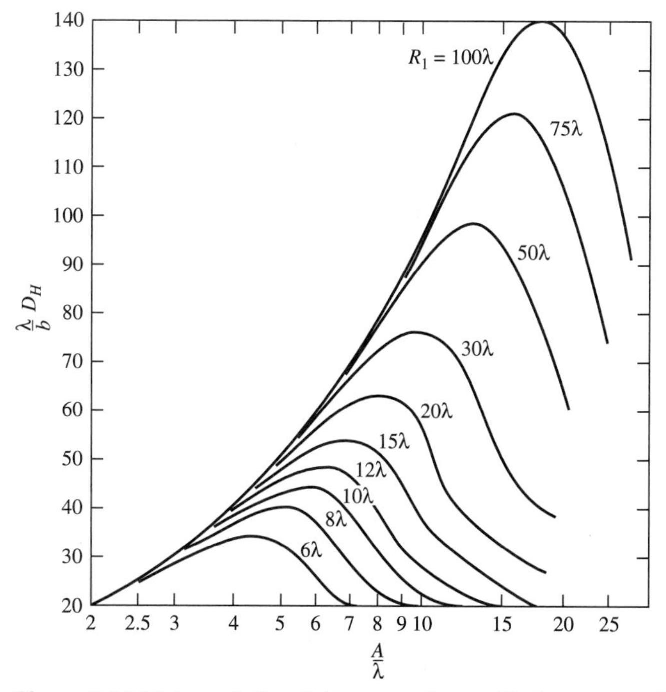

Figure 7-14 Universal directivity curves for an H-plane sectoral horn.

priate curve. The values of A/λ corresponding to optimum operation plotted versus R_1/λ produce a smooth curve with the equation $A/\lambda = \sqrt{3R_1/\lambda}$, giving

$$A = \sqrt{3\lambda R_1} \qquad optimum \tag{7-121}$$

For example, the value of A/λ for the peak of the $R_1/\lambda = 30$ curve of Fig. 7-14 is 9.5, and from (7-121), $A/\lambda = \sqrt{3R_1/\lambda} = \sqrt{3(30)} = 9.49$.

The optimum phase error parameter value corresponding to optimum directivity is found from (7-116) with (7-121) as

$$t_{op} = \frac{A^2}{8\lambda R_1} = \frac{3}{8} \qquad optimum \tag{7-122}$$

The optimum behavior of the directivity curves can be explained rather simply. For a fixed axial length, as the aperture width A is increased from a small value, the directivity increases by virtue of the increased aperture area. Optimum performance is reached when $t = t_{op} = 3/8$, which corresponds to a phase lag at the aperture edges ($x = \pm A/2$) of $\delta_{max} = 2\pi t_{op} = 3\pi/4 = 135°$. As A is increased beyond the optimum point, the phase deviations across the aperture lead to cancellations in the far field and decreased directivity, as can be seen from the pattern plots of Fig. 7-13.

The foregoing analysis can be performed without approximation by using numerical techniques together with the exact phase error (7-103a), as with the radiation integral in (7-119). However, it is easier to use a corrected phase error parameter that includes the effects of the exact phase error. The exact pattern is obtained if the value of t in (7-118) is replaced by the following [15]:

$$t_e = \left(\frac{A}{\lambda}\right)^2 \frac{1}{8t}\left\{\left[1 + \left(\frac{\lambda}{A}\right)^2 16\, t^2\right]^{1/2} - 1\right\} \qquad exact \tag{7-123}$$

If the phase error is not large and the aperture is more than a few wavelengths, then $t_e \approx t$. For example, the optimum case with a 3λ aperture ($A/\lambda = 3$) has an exact phase error parameter of $t_e = 0.354$, which is close to the approximate value of 0.375. Similarly, exact phase error conditions for directivity are obtained by replacing t with t_e in (7-120).

The half-power beamwidth for optimum performance can be determined from the pattern plot of Fig. 7-13 for $t = 3/8$. The 3-dB down point on the main beam occurs for $(A/\lambda)\sin\theta_H = 0.68$, so the H-plane beamwidth for an optimum H-plane sectoral horn is $2\theta_H = \sin^{-1}(0.68\lambda/A)$; and for $A \gg \lambda$,

$$\mathrm{HP}_H \approx 1.36\frac{\lambda}{A} = 78°\frac{\lambda}{A} \qquad optimum \tag{7-124}$$

7.4.2 The *E*-Plane Sectoral Horn Antenna

A rectangular horn antenna can also be formed by flaring the feed waveguide in the E-plane. The resulting horn is referred to as an ***E*-plane sectoral horn antenna** as shown in Fig. 7-15. The geometrical relationships for this horn are

$$\ell_E^2 = R_2^2 + \left(\frac{B}{2}\right)^2 \tag{7-125}$$

$$\alpha_E = \tan^{-1}\left(\frac{B}{2R_2}\right) \tag{7-126}$$

$$R_E = (B - b)\sqrt{\left(\frac{\ell_E}{B}\right)^2 - \frac{1}{4}} \tag{7-127}$$

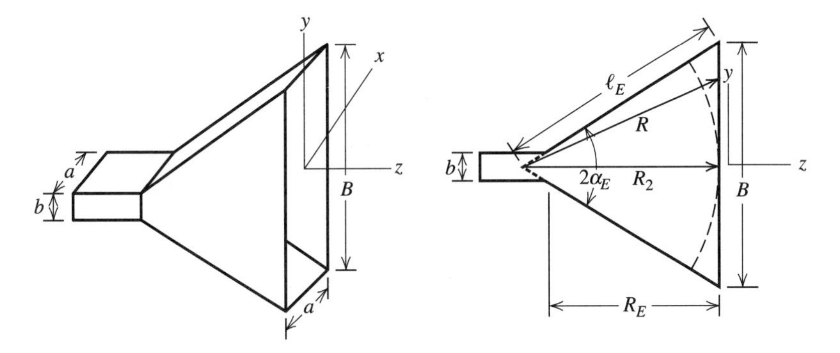

(*a*) Overall geometry. (*b*) Cross section through the *yz*-plane (*E*-plane).

Figure 7-15 *E*-plane sectoral horn antenna.

A similar line of reasoning as employed for the *H*-plane horn leads to the following aperture electric field distribution for the *E*-plane horn:

$$E_{ay} = E_o \cos \frac{\pi x}{a} e^{-j(\beta/2R_2)y^2} \tag{7-128}$$

The same steps as used with the *H*-plane sectoral horn yield the radiation field:

$$\begin{aligned}\mathbf{E} = j\beta E_o \sqrt{\frac{\pi R_2}{\beta}} \frac{4a}{\pi} \frac{e^{-j\beta r}}{4\pi r} e^{j(\beta R_2/2)v^2}(\hat{\boldsymbol{\theta}} \sin \phi + \hat{\boldsymbol{\phi}} \cos \phi) \\ \cdot \frac{1 + \cos \theta}{2} \frac{\cos[(\beta a/2)u]}{1 - [(\beta a/\pi)u]^2} [C(r_2) - jS(r_2) - C(r_1) + jS(r_1)]\end{aligned} \tag{7-129a}$$

where

$$r_1 = \sqrt{\frac{\beta}{\pi R_2}} \left(-\frac{B}{2} - R_2 v \right), \qquad r_2 = \sqrt{\frac{\beta}{\pi R_2}} \left(\frac{B}{2} - R_2 v \right) \tag{7-129b}$$

The normalized *H*-plane pattern follows from this with $\phi = 0°$ as

$$F_H(\theta) = \frac{1 + \cos \theta}{2} \frac{\cos[(\beta a/2) \sin \theta]}{1 - [(\beta a/\pi) \sin \theta]^2} \tag{7-130}$$

The second factor in this expression is the pattern of a uniform phase, cosine amplitude tapered line source of length *a*.

The aperture phase error in the *E*-plane is approximated with the quadratic phase error in (7-128) as $\delta = (\beta/2R_2)y^2$. The maximum phase error occurs for $y = \pm B/2$, giving $\delta_{max} = (\beta/2R_2)(B/2)^2 = 2\pi(B^2/8\lambda R_2) = 2\pi s$, where we define the phase error parameter *s* as

$$s = \frac{B^2}{8\lambda R_2} = \frac{1}{8} \left(\frac{B}{\lambda} \right)^2 \frac{1}{R_2/\lambda} \tag{7-131}$$

The *E*-plane pattern magnitude from (7-129) with $\phi = 90°$ can be expressed in terms of *s* as

$$\begin{aligned}|F_E(\theta)| &= \frac{1 + \cos \theta}{2} |f_E(\theta)| \\ &= \frac{1 + \cos \theta}{2} \left\{ \frac{[C(r_4) - C(r_3)]^2 + [S(r_4) - S(r_3)]^2}{4[C^2(2\sqrt{s}) + S^2(2\sqrt{s})]} \right\}^{1/2}\end{aligned} \tag{7-132a}$$

where

$$r_3 = 2\sqrt{s}\left[-1 - \frac{1}{4s}\left(\frac{B}{\lambda}\sin\theta\right)\right], \qquad r_4 = 2\sqrt{s}\left[1 - \frac{1}{4s}\left(\frac{B}{\lambda}\sin\theta\right)\right] \tag{7-132b}$$

Similar to (7-119) for the H-plane sectoral horn, the pattern of an E-plane sectoral horn can be evaluated by direct numerical integration; $f_E(\theta)$ in (7-132a) is found from

$$f_E(\theta) \propto \int_{-B/2}^{B/2} e^{-j\beta\sqrt{R_2^2+y'^2}} e^{j\beta \sin\theta\, y'}\, dy' \tag{7-133}$$

The universal patterns for the E-plane sectoral horn are plotted in Fig. 7-16. The E-plane patterns (solid curves) for various values of s are not normalized to 0 dB at the maximum point, but rather are given relative to the no-phase error case, which is $s = 0$ corresponding to a uniform line source. The H-plane pattern (dashed curve) is that of a cosine-tapered line source, which is the second factor of (7-130). The factor $(1 + \cos\theta)/2$ is not included in these plots.

The directivity of the E-plane sectoral horn found from (7-66) is

$$D_E = \frac{a}{\lambda}\frac{32}{\pi}\frac{B}{\lambda}\varepsilon_{ph}^E = \frac{4\pi}{\lambda^2}\varepsilon_t\, \varepsilon_{ph}^E\, aB \tag{7-134a}$$

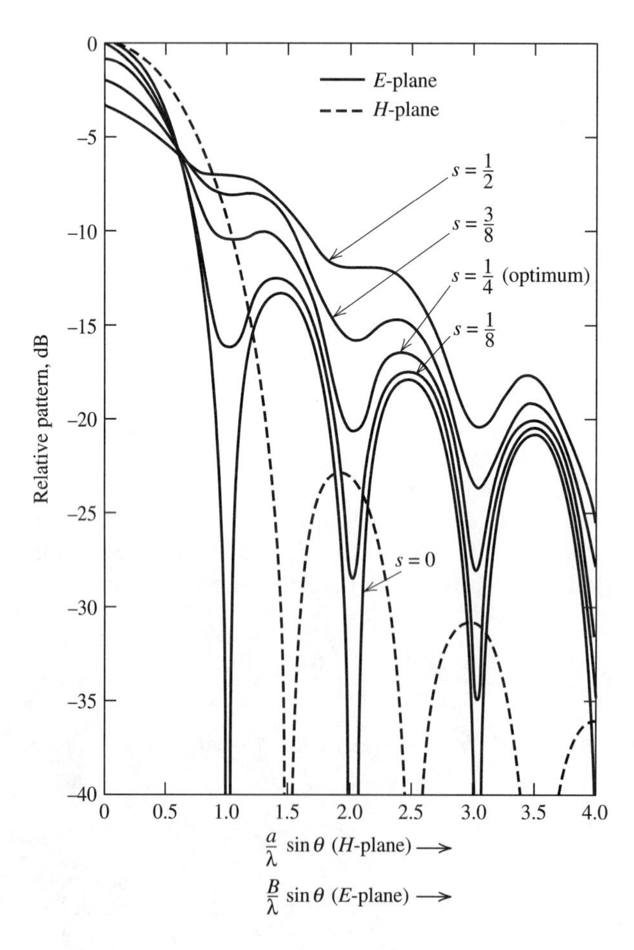

Figure 7-16 Universal radiation patterns for the principal planes of an E-plane sectoral horn antenna as shown in Fig. 7-15. The factor $(1 + \cos\theta)/2$ is not included.

where

$$\varepsilon_t = \frac{8}{\pi^2} \tag{7-134b}$$

$$\varepsilon_{ph}^E = \frac{C^2(q) + S^2(q)}{q^2} \tag{7-134c}$$

$$q = \frac{B}{\sqrt{2\lambda R_2}} = 2\sqrt{s} \tag{7-134d}$$

A family of universal directivity curves $\lambda D_E/a$ for various values of R_2/λ is given in Fig. 7-17 as a function of B/λ. The peak of each curve corresponds to optimum directivity for the value of R_2. A curve fit to pairs of values of B/λ and R_2/λ for optimum conditions yields

$$B = \sqrt{2\lambda R_2} \qquad \textit{optimum} \tag{7-135}$$

The corresponding value of s is

$$s_{\text{op}} = \frac{B^2}{8\lambda R_2} = \frac{1}{4} \qquad \textit{optimum} \tag{7-136}$$

Exact phase error conditions corresponding to spherical wave fronts in the aperture plane can be included easily, replacing s by s_e [15]:

$$s_e = \left(\frac{B}{\lambda}\right)^2 \frac{1}{8s}\left\{\left[1 + \left(\frac{\lambda}{B}\right)^2 16\,s^2\right]^{1/2} - 1\right\} \qquad \textit{exact} \tag{7-137}$$

That is, the pattern and directivity expressions of (7-132) and (7-134) are made exact by using (7-137). However, in practice, accuracy cannot be expected when the ap-

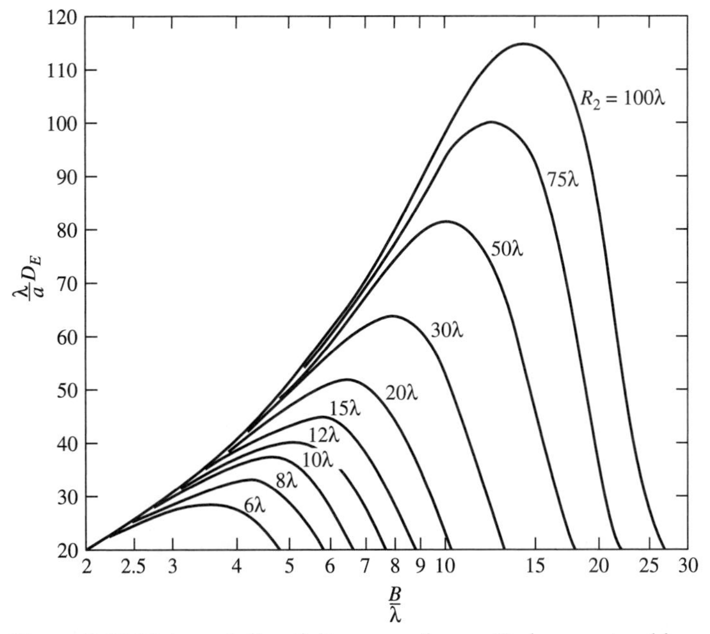

Figure 7-17 Universal directivity curves for an E-plane sectoral horn.

erture is small because the aperture fields are not well approximated by free-space conditions.

The half-power beamwidth relationship for the optimum horn follows from the $s = 1/4$ plot in Fig. 7-16 and is

$$\mathrm{HP}_E = 2 \sin^{-1} \frac{0.47}{B/\lambda} \approx 0.94 \frac{\lambda}{B} = 54° \frac{\lambda}{B} \qquad \textit{optimum} \tag{7-138}$$

Gain for horn antennas nearly equals directivity, that is, $G_E \approx D_E$ and $G_H \approx D_H$. The gain of an E-plane sectoral horn has been shown to be more accurately given by [16]

$$G_E = \frac{16aB}{\lambda^2(1 + \lambda_g/\lambda)} \frac{C^2(q_1) + S^2(q_1)}{q_1^2} e^{\pi(a/\lambda)(1-\lambda/\lambda_g)} \tag{7-139}$$

where $\lambda_g = \lambda/\sqrt{1 - (\lambda/2a)^2}$ is the wavelength of the dominant mode in the waveguide feeding the antenna and $q_1 = B[\sqrt{2\lambda_g \ell_E} \cos(\alpha_E/2)]^{-1}$. This expression yields values that agree quite well with experimental results. The values from (7-134) are less than those of (7-139) by 20% or more.

7.4.3 The Pyramidal Horn Antenna

Probably the most popular form of the rectangular horn antenna is the **pyramidal horn antenna**. As shown in Fig. 7-18, it is flared in both the E- and H-planes. This configuration will lead to narrow beamwidths in both principal planes, forming a pencil beam. The aperture electric field is obtained by combining the results for H- and E-plane sectoral horns from (7-105) and (7-128) giving

$$E_{ay} = E_o \cos\left(\frac{\pi x}{A}\right) e^{-j(\beta/2)(x^2/R_1 + y^2/R_2)} \tag{7-140}$$

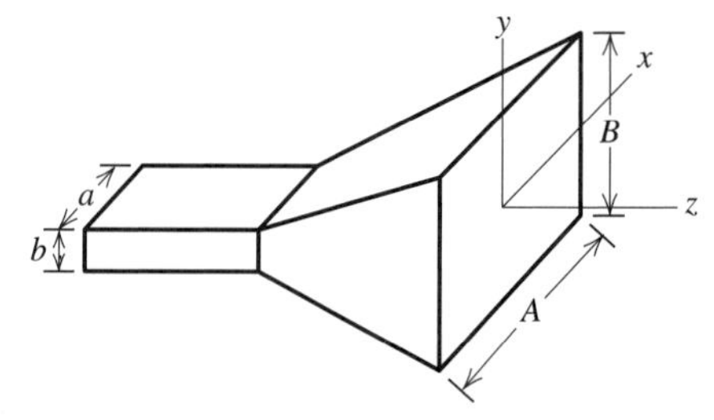

(a) Overall geometry.

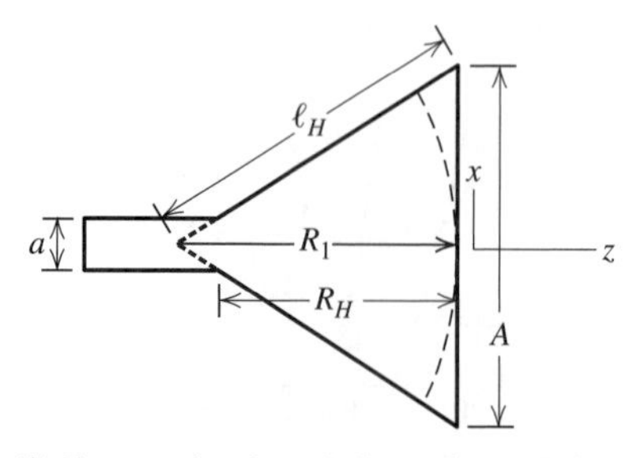

(b) Cross section through the xz-plane (H-plane). (c) Cross section through the yz-plane (E-plane).

Figure 7-18 Pyramidal horn antenna.

Following a procedure similar to that used for the sectoral horns will yield a general radiation field expression. The principal plane patterns are the same as those obtained from the sectoral horn calculations because the aperture distribution is separable as in (7-72). To be precise, the E- and H-plane patterns of the pyramidal horn equal the E-plane pattern of the E-plane sectoral horn and the H-plane pattern of the H-plane sectoral horn. Therefore, the E-plane pattern of the pyramidal horn can be found from the universal pattern plots (solid curves) of Fig. 7-16, and the H-plane pattern can be found from the solid curves of Fig. 7-13.

Since pyramidal horns are used as gain standards at microwave frequencies, accurate gain evaluation is important. The directivity of the pyramidal horn is found rather simply from

$$D_p = \frac{\pi}{32}\left(\frac{\lambda}{a} D_E\right)\left(\frac{\lambda}{b} D_H\right) \tag{7-141}$$

See Prob. 7.4-15. The terms in parentheses are obtained directly from the directivity curves for sectoral horns of Figs. 7-14 and 7-17, respectively. Gain values computed with (7-141) agree very well with experiment for sufficiently large horns. It includes the geometrical optics fields and singly diffracted fields from the horn edges. The inclusion of multiple diffraction and diffraction at the edges arising from reflections from the horn interior leads to small oscillations in the gain about that predicted by (7-141) as a function of frequency, and in agreement with experimental results [17].

It is instructive to examine the aperture efficiency contributions for horns. The radiation efficiency e_r is close to unity, so we can take gain to be equal to directivity; see (7-80). The two efficiencies that must be considered are the aperture taper efficiency ε_t and phase efficiency ε_{ph}:

$$\varepsilon_{ap} = \varepsilon_t \varepsilon_{ph} = \varepsilon_t \varepsilon_{ph}^E \varepsilon_{ph}^H \tag{7-142}$$

where we decomposed total phase efficiency into factors due to phase errors in the E- and H-planes. Gain is then expressed from (7-77) as

$$G = \frac{4\pi}{\lambda^2}\varepsilon_{ap}AB = \frac{4\pi}{\lambda^2}\varepsilon_t AB \varepsilon_{ph}^E \varepsilon_{ph}^H = G_o \varepsilon_{ph}^E \varepsilon_{ph}^H \tag{7-143}$$

where G_o is the gain without a phase error effect and includes aperture taper efficiency, which was found in Example 7-4 to be $\varepsilon_t = 0.81$. The phase error efficiencies can be found by evaluating directivity for the sectoral horns and removing the known taper efficiency. The results of this process are plotted in Fig. 7-19 as a function of phase error parameters s and t. The aperture efficiencies for optimum sectoral horns with $s = 0.25$ and $t = 0.375$ are

$$\varepsilon_{ap}^E = 0.649, \qquad \varepsilon_{ap}^H = 0.643 \qquad \textit{optimum} \tag{7-144}$$

Both include $\varepsilon_t = 0.81$. So,

$$\varepsilon_{ph}^E = \frac{\varepsilon_{ap}^E}{\varepsilon_t} = 0.80, \qquad \varepsilon_{ph}^H = \frac{\varepsilon_{ap}^H}{\varepsilon_t} = 0.79 \tag{7-145}$$

The aperture efficiency of an optimum pyramidal horn from (7-142) is

$$\varepsilon_{ap}^P = \varepsilon_t \varepsilon_{ph}^E \varepsilon_{ph}^H = 0.81(0.80)(0.79) = 0.51 \tag{7-146}$$

It is common to use an aperture efficiency value of 50% for optimum gain pyramidal horns.

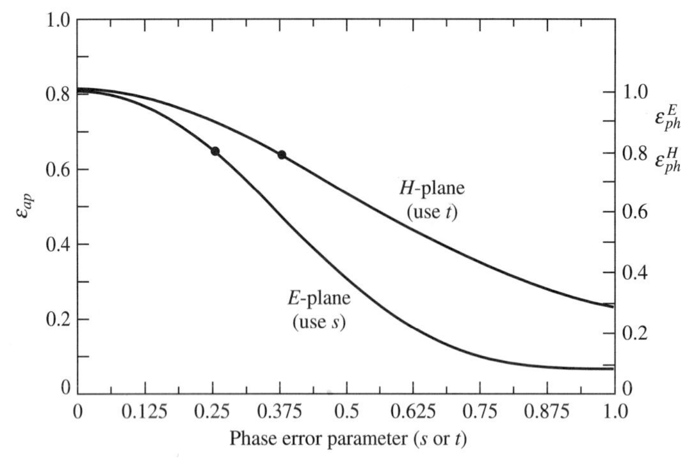

Figure 7-19 Aperture efficiencies for E- and H-plane sectoral horns (left ordinate) and phase efficiencies associated with E- and H-plane flares (right ordinate).

The gain of an optimum gain pyramidal horn from (7-146) in (7-143) is

$$G = 0.51 \frac{4\pi}{\lambda^2} AB \qquad \textit{optimum pyramidal horn} \tag{7-147}$$

It is popular to express horn gain in dB form by taking 10 log of (7-143):

$$G_{\text{dB}} = G_{o,\text{dB}} + \varepsilon_{ph,\text{dB}}^E(s) + \varepsilon_{ph,\text{dB}}^H(t) \tag{7-148}$$

The last two terms are gain reduction factors associated with the phase errors of (7-120c) and (7-134c). These phase efficiencies, before taking 10 log, can be approximated with simple formulas [18]:

$$\begin{aligned} \varepsilon_{ph}^E(s) &= \frac{1}{4s}\left[C^2(2\sqrt{s}) + S^2(2\sqrt{s})\right] \\ &\approx 1.00329 - 0.11911s - 2.75224s^2 \end{aligned} \tag{7-149}$$

$$\begin{aligned} \varepsilon_{ph}^H(t) &= \frac{\pi^2}{64t}\{[C(p_1) - C(p_2)]^2 + [S(p_1) - S(p_2)]^2\} \\ &\approx 1.00323 - 0.08784t - 1.27048t^2 \end{aligned} \tag{7-150}$$

The approximate formulas are valid from zero up to at least $s = 0.262$ and $t = 0.397$. For example, $s = 0.25$ and $t = 0.375$ in the approximate formulas give the values in (7-144), which are the points shown in Fig. 7-19. Increased accuracy is obtained if the exact phase error parameters in (7-123) and (7-137) are used.

Many applications for horns require a specified gain to be realized at a known operating frequency. Usually, the optimum gain design approach is used because it renders the shortest axial length for the specified gain. We now derive the single design equation that permits determination of the optimum horn geometry for the specified gain. The procedure includes the connecting waveguide internal dimensions a and b as well as the horn dimensions. There are three conditions that must be satisfied. The first two are that the phase error in the E- and H-planes be those associated with optimum performance. The third is that the structure of the pyramidal horn be physically realizable and properly mate to the connecting waveguide. This can be seen from Fig. 7-18 to be

$$R_E = R_H = R_p \tag{7-151}$$

From similar triangles in Fig. 7-18,

$$\frac{R_1}{R_H} = \frac{A/2}{A/2 - a/2} = \frac{A}{A - a} \tag{7-152}$$

$$\frac{R_2}{R_E} = \frac{B/2}{B/2 - b/2} = \frac{B}{B - b} \tag{7-153}$$

Imposing the optimum performance in the E-plane through (7-135) and substituting (7-153), we obtain

$$B = \sqrt{\frac{2\lambda R_E B}{B - b}} \quad \text{or} \quad B^2 - bB - 2\lambda R_E = 0 \tag{7-154}$$

which is a quadratic equation with one solution as follows:

$$B = \frac{1}{2}\left(b + \sqrt{b^2 + 8\lambda R_E}\right) \tag{7-155}$$

The second solution yields the impossible case of negative B and is ignored. Similarly, the optimum performance condition for the H-plane of (7-122) together with (7-152) yields

$$R_H = \frac{A - a}{A} R_1 = \frac{A - a}{A}\left(\frac{A^2}{3\lambda}\right) = \frac{A - a}{3\lambda} A \tag{7-156}$$

Imposing the physical realization condition of (7-151) with (7-156) in (7-155) gives

$$B = \frac{1}{2}\left(b + \sqrt{b^2 + \frac{8A(A - a)}{3}}\right) \tag{7-157}$$

Linking this to the specified gain G gives

$$G = \frac{4\pi}{\lambda^2}\varepsilon_{ap} AB = \frac{4\pi}{\lambda^2}\varepsilon_{ap} A \frac{1}{2}\left(b + \sqrt{b^2 + \frac{8A(A - a)}{3}}\right) \tag{7-158}$$

Expanding to form a fourth-order equation in A gives the desired design equation [19]:

$$\boxed{A^4 - aA^3 + \frac{3bG\lambda^2}{8\pi\varepsilon_{ap}} A = \frac{3G^2\lambda^4}{32\pi^2\varepsilon_{ap}^2}} \quad \textit{optimum pyramidal horn design equation} \tag{7-159}$$

It is possible to solve this quartic equation for its roots, but it is rather involved and the solution is easily obtained using a numerical equation solver routine. Alternatively, it can be solved by trial and error using a first guess approximation of

$$A_1 = 0.45\lambda\sqrt{G} \tag{7-160}$$

We now summarize the steps in the optimum horn design procedure:

Step 1: Specify the desired gain G at the operating wavelength λ and specify the connecting waveguide dimensions a and b.

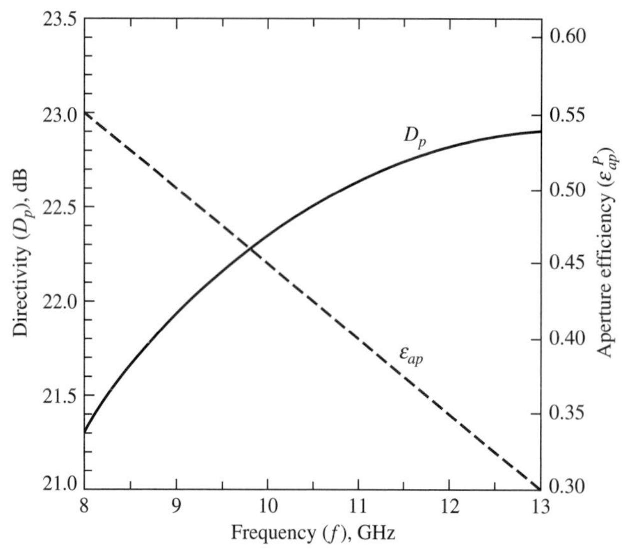

Figure 7-20 Directivity and aperture efficiency of the standard gain rectangular horn of Example 7-7.

Step 2: Solve (7-159) for A using $\varepsilon_{ap} = 0.51$.

Step 3: Find the remaining horn dimensions as follows: B from (7-147); R_1 from (7-121); R_H from (7-152); ℓ_H from (7-98); R_2 from (7-135); R_E from (7-153); and ℓ_E from (7-125).

Step 4: The correct solution can be verified by checking to see if R_E equals R_H and by evaluating (7-131) and (7-116) to see if $s = 0.25$ and $t = 0.375$.

Horn antennas operate well over a bandwidth of about 50%. However, performance is optimum only at the design frequency. Figure 7-20 is a gain curve for the "standard gain horn" in the 8.2 to 12.4 GHz band that is considered in Example 7-7. Note that gain increases with frequency, which is characteristic of aperture antennas. The curve is not a straight line as might seem to be the case from the explicit frequency-squared dependence in (7-77). This is because aperture efficiency decreases with frequency due to increasing phase errors, as shown in Fig. 7-20. Thus, an optimum gain horn is only "optimum" at its design frequency.

Before closing this section with an example of optimum horn design, we comment on the assumption that phase error arises from a spherical phase front in the aperture and the wavelength there equals that of free space. A solution technique for rectangular horns is available that uses a gradual change in phase velocity from the waveguide to the aperture by treating each point as a section of an infinitely long waveguide of that width [20]. However, for all but short horns with small apertures, gain does not differ noticeably from the foregoing design approach.

EXAMPLE 7-7 *Design of an Optimum Gain Pyramidal Horn Antenna*

Commercial "standard gain" pyramidal horn antennas are available to cover the frequency band from 8.2 to 12.4 GHz (X-band). They are fed from a WR90 waveguide with $a = 0.9$ in. $= 2.286$ cm and $b = 0.4$ in. $= 1.016$ cm. As the gain curve in Fig. 7-20 indicates, the aperture efficiency decreases rapidly with frequency. Therefore, the optimum design point is chosen near the low end of the band to provide more uniform gain over the whole band. Gain reduction due to increased phase errors at the high end of the band is significant, but the aperture is much larger electrically so gain actually increases.

For this design example, we choose the optimum design point to be at 8.75 GHz, where aperture efficiency is 51%. The desired gain is $G = 21.75$ dB or $10^{2.175} = 149.6$ at 8.75 GHz ($\lambda = 3.43$ cm). The design equation of (7-159) is solved by trial and error beginning with $A_1 = 18.9$ cm using (7-160). Step 3 of the design procedure gives all remaining horn dimensions:

$$\begin{aligned} A &= 18.61 \text{ cm}, & B &= 14.75 \text{ cm} \\ R_1 &= 33.67 \text{ cm}, & R_2 &= 31.72 \text{ cm} \\ \ell_H &= 34.93 \text{ cm}, & \ell_E &= 32.56 \text{ cm} \\ R_H &= 29.53 \text{ cm}, & R_E &= 29.53 \text{ cm} \end{aligned}$$

These values are verified by noting that $R_p = R_E = R_H$ and by evaluating (7-131) and (7-116) to obtain the optimum values of $s = 0.25$ and $t = 0.375$. The gain value is verified using the universal directivity curves $R_2/\lambda = 9.3$ and $B/\lambda = 4.3$ with Fig. 7-17, giving $\lambda D_E/a = 36$, and $R_1/\lambda = 10.1$ and $A/\lambda = 5.4$ with Fig. 7-14, giving $\lambda D_H/b = 43$. Then from (7-141),

$$D_p = \frac{\pi}{32}\left(\frac{\lambda}{a}D_E\right)\left(\frac{\lambda}{b}D_H\right) = \frac{\pi}{32}(36)(43) = 152 = 21.8 \text{ dB}$$

which is very close to the design goal of 21.75 dB. Accurate evaluation of directivity using (7-120c) and (7-134c) in (7-148) with $s_{\text{op}} = 0.25$ and $t_{\text{op}} = 0.375$ gives a value of 21.79 dB. The exact phase errors for this geometry are $s_e = 0.247$ and $t_e = 0.368$ from (7-137) and (7-123); they lead to a directivity of 21.85 dB. The directivity as a function of frequency is plotted in Fig. 7-20; see Prob. 7.4-17.

The complete radiation patterns at 8.75 GHz are plotted in Fig. 7-21 including the $(1 + \cos\theta)/2$ factor. The half-power beamwidths are

$$\text{HP}_E = 12.4°, \qquad \text{HP}_H = 14.2°$$

These agree exactly with the predicted values based on (7-138) and (7-124). Thus, the simple half-power beamwidth formulas of (7-138) and (7-124) give good results for optimum horns. The first side lobe of the E-plane and H-plane patterns in Fig. 7-21 are located at 16° and 44° with values of −9.4 and −32.5 dB, respectively. The E- and H-plane first side lobe values without the $(1 + \cos\theta)/2$ factor included are −9.2 and −31.2 dB, respectively, and can be found in Figs. 7-16 and 7-13.

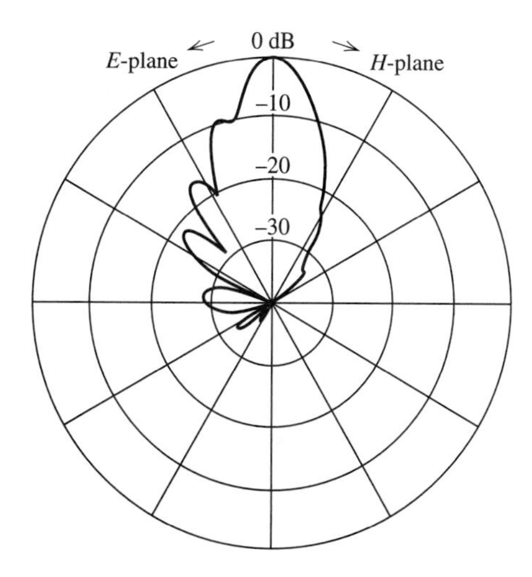

Figure 7-21 Principal plane patterns for the optimum pyramidal horn antenna of Example 7-7 at 8.75 GHz. The patterns include the $(1 + \cos\theta)/2$ factor. $\text{HP}_E = 12.4°$ and $\text{HP}_H = 14.2°$.

The gain (directivity) can also be estimated directly from the beamwidths using (7-95):

$$G \approx \frac{26{,}000}{\text{HP}_{E^\circ}\text{HP}_{H^\circ}} = \frac{26{,}000}{(12.4^\circ)(14.2^\circ)} = 21.7 \text{ dB}$$

which is close to the design value of 21.75 dB.

7.5 CIRCULAR APERTURES

An antenna that has a physical aperture opening with a circular shape is said to have a *circular aperture.* Various forms of circular aperture antennas are encountered in practice. In this section, we discuss ideal circular aperture distributions with uniform and tapered amplitudes. This is followed in the next section by a study of parabolic reflector antennas that are the most popular circular aperture antennas.

7.5.1 The Uniform Circular Aperture

A general circular aperture is shown in Fig. 7-22. If the aperture distribution amplitude is constant, it is referred to as a *uniform circular aperture.* This is approximated by a circular hole in a conducting sheet with a uniform plane wave incident from behind. Suppose the aperture electric field is x-directed, or

$$\mathbf{E}_a = \hat{\mathbf{x}} E_o \qquad \rho' \leq a \tag{7-161}$$

Then (7-14) gives

$$\mathbf{P} = \hat{\mathbf{x}} E_o \iint_{S_a} e^{j\beta \hat{\mathbf{r}} \cdot \mathbf{r}'} \, dS' \tag{7-162}$$

From Fig. 7-22, it is seen that

$$\mathbf{r}' = \rho' \cos \phi' \hat{\mathbf{x}} + \rho' \sin \phi' \hat{\mathbf{y}} \tag{7-163}$$

This with (C-4) yields

$$\begin{aligned} \hat{\mathbf{r}} \bullet \mathbf{r}' &= \rho' \sin \theta (\cos \phi \cos \phi' + \sin \phi \sin \phi') \\ &= \rho' \sin \theta \cos(\phi - \phi') \end{aligned} \tag{7-164}$$

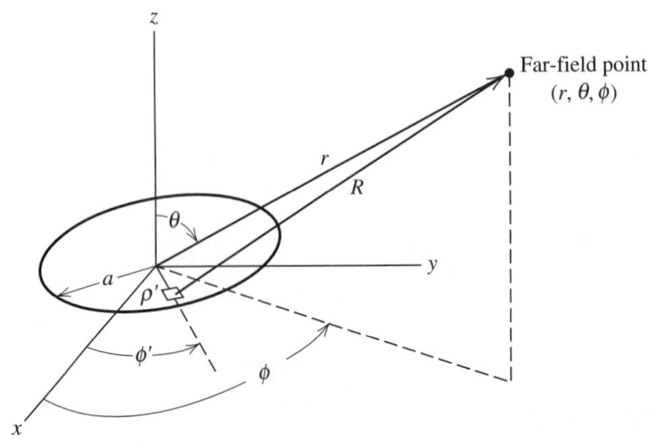

Figure 7-22 The circular aperture.

Hence, (7-162) becomes

$$\mathbf{P} = \hat{\mathbf{x}}E_o \int_0^a \left[\int_0^{2\pi} e^{j\beta\rho' \sin\theta \cos(\phi-\phi')}\, d\phi' \right] \rho'\, d\rho'$$
$$= \hat{\mathbf{x}}E_o 2\pi \int_0^a \rho' J_0(\beta\rho' \sin\theta)\, d\rho' \tag{7-165}$$

where (F-6) was used for the ϕ' integration. $J_0(x)$ is a Bessel function of the first kind and zero order, which is unity at $x = 0$ and is a decaying oscillatory function for increasing x. The ρ' integration can be performed using

$$\int xJ_0(x)\, dx = xJ_1(x) \tag{7-166}$$

which follows from (F-9). $J_1(x)$ is a Bessel function of the first kind and first order, which is zero for $x = 0$ and is a decaying oscillatory function for increasing x. Transforming variables as $x = \beta\rho' \sin\theta$ and using (7-166) in (7-165) yield

$$\mathbf{P} = \hat{\mathbf{x}}E_o 2\pi \frac{a}{\beta \sin\theta} J_1(\beta a \sin\theta) = \hat{\mathbf{x}}P_x \tag{7-167}$$

The equivalent magnetic current formulation of (7-26) gives

$$\mathbf{E} = (\hat{\boldsymbol{\theta}} \cos\phi - \hat{\boldsymbol{\phi}} \sin\phi \cos\theta) j\beta \frac{e^{j\beta r}}{2\pi r} P_x$$
$$= \hat{\mathbf{p}} E_o \pi a^2 j\beta \frac{e^{-j\beta r}}{2\pi r} f(\theta) \tag{7-168}$$

where the polarization vector is

$$\hat{\mathbf{p}} = \hat{\boldsymbol{\theta}} \cos\phi - \hat{\boldsymbol{\phi}} \sin\phi \cos\theta \tag{7-169}$$

and the relative variation of the radiation integral P_x normalized to unity maximum at $\theta = 0°$ is

$$f(\theta) = \frac{2J_1(\beta a \sin\theta)}{\beta a \sin\theta} \tag{7-170}$$

$f(\theta)$ is independent of ϕ due to the circular symmetry of the aperture distribution. In the E-plane, $\phi = 0°$ and (7-169) becomes $\hat{\mathbf{p}} = \hat{\boldsymbol{\theta}}$, and $f(\theta)$ represents the E_θ component. In the H-plane, $\phi = 90°$ and (7-169) reduces to $\hat{\mathbf{p}} = -\hat{\boldsymbol{\phi}} \cos\theta$, and $f(\theta)$ multiplied by $\cos\theta$ represents the E_ϕ component. This $\cos\theta$ factor ensures that the electric field goes to zero at $\theta = 90°$ as required by the boundary condition on the tangential electric field on the ground plane.

For large apertures, $f(\theta)$ gives a narrow main beam in the $\theta = 0°$ direction for the uniform phase aperture we are considering here. Thus, near the main beam, $\cos\theta \approx 1$ since θ is small and (7-169) gives $\hat{\mathbf{p}} \approx \hat{\boldsymbol{\theta}} \cos\phi - \hat{\boldsymbol{\phi}} \sin\phi$, which is the projection of the aperture electric field vector $\hat{\mathbf{x}}$ tangent to the far field sphere; see (C-1). In this case, all θ dependence is contained in $f(\theta)$. An example of $f(\theta)$ is plotted in Fig. 7-23 in the uv-plane for $a = 5\lambda$ out to the limit of the visible region ($\theta = 90°$). A plot of the radiation pattern in any plane passing through the

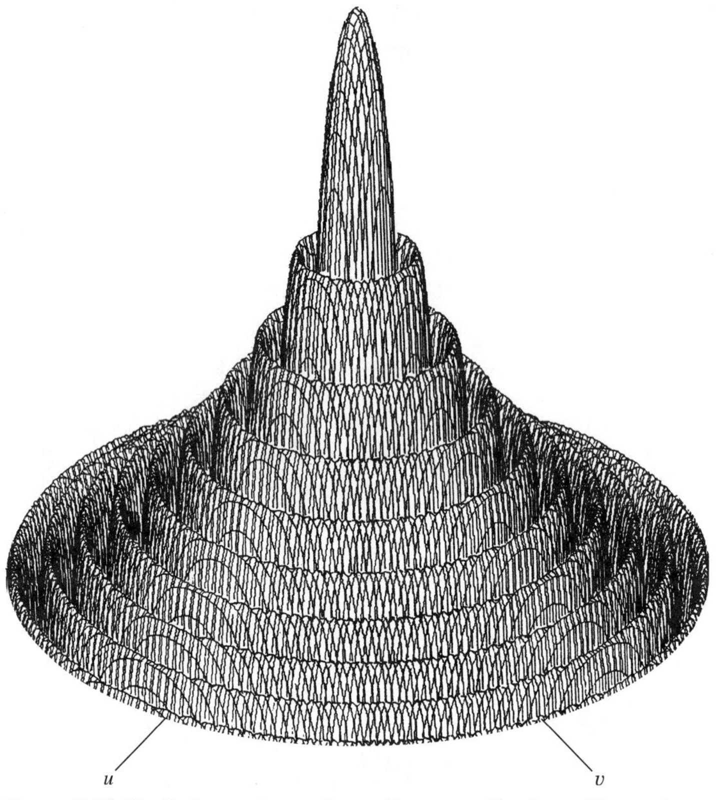

Figure 7-23 Radiation pattern of a uniform amplitude, uniform phase, 10-wavelength diameter circular aperture.

center of Fig. 7-23 is shown in Fig. 7-24. Note the similarity of this pattern, $2J_1(x)/x$, to the uniform line source pattern, $\sin x/x$.

The half-power point of (7-170) occurs at $\beta a \sin\theta = 1.6$, so the half-power beamwidth for $a \gg \lambda$ is

$$\mathrm{HP} = 2\theta_{\mathrm{HP}} = 2\sin^{-1}\frac{1.6}{\beta a} \approx 2\,\frac{1.6}{\pi}\,\frac{\lambda}{2a}$$

or

$$\mathrm{HP} = 1.02\,\frac{\lambda}{2a}\ \mathrm{rad} = 58.4\,\frac{\lambda}{2a}\ \mathrm{deg} \tag{7-171}$$

For the 10λ diameter example, HP $= 0.102$ rad $= 5.84°$. The side lobe level of any uniform circular aperture pattern is -17.6 dB. This can be seen in Fig. 7-24. Since the uniform circular aperture has uniform excitation amplitude, it has unity aperture taper efficiency and the directivity, from (7-66), is

$$D_u = \frac{4\pi}{\lambda^2}A_p = \frac{4\pi}{\lambda^2}\pi a^2 \tag{7-172}$$

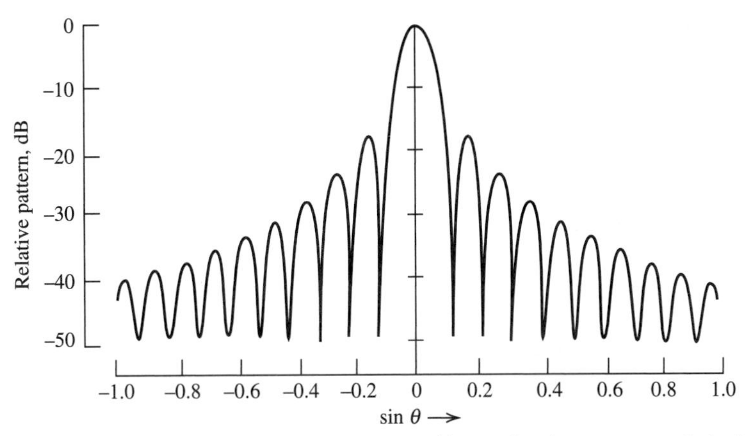

Figure 7-24 Pattern of a 10λ diameter uniform circular aperture. It is the pattern in any plane passing through the center of Fig. 7-23.

7.5.2 Tapered Circular Apertures

Many circular aperture antennas can be approximated as a radially symmetric circular aperture with an aperture field amplitude distribution that is tapered from the center of the aperture toward the edge. In practice, many circular aperture distributions are close to radially symmetric and do not vary with ϕ' (see Fig. 7-22). We shall assume this is the case, and again we will confine our attention to a broadside circular aperture that is large in terms of a wavelength. Then the pattern is well approximated by the unnormalized radiation integral

$$f_{\text{un}}(\theta) = \int_0^{2\pi} \int_0^a E_a(\rho') e^{j\beta\rho' \sin\theta \cos(\phi - \phi')} \rho' \, d\rho' \, d\phi' \tag{7-173}$$

Performing the integration over ϕ' with the aid of (F-6) leads to

$$f_{\text{un}}(\theta) = 2\pi \int_0^a E_a(\rho') \rho' J_0(\beta\rho' \sin\theta) \, d\rho' \tag{7-174}$$

This integral can be performed for various aperture tapers and normalized to obtain $f(\theta)$.

The properties of several common circular aperture tapers are given in Table 7-1. Similar data are available in the literature [6, 21, 22] including elliptical apertures [23]. Table 7-1 is analogous to Table 4-2 for line source distributions. The parabolic distribution ($n = 1$) of Table 7-1*a* provides a smooth taper from the aperture center to edge where the aperture field is zero. When $n = 0$, the distribution reduces to the uniform case where, of course, aperture taper efficiency is unity; see (7-81) and (7-172). The parabolic taper ($n = 1$) yields lower side lobes at the expense of wider beamwidth and reduced directivity compared to the uniform distribution. This effect is more pronounced for the parabolic squared ($n = 2$) distribution. The side-lobe level-beamwidth tradeoff can be customized by using the parabolic-on-a-pedestal aperture distribution in Table 7-1*b*. The pedestal height C is the edge (field) illumination relative to that at the center. This taper can be used to model illuminations commonly encountered with a circular reflector antenna, where the pedestal represents the fact that the feed antenna pattern is intercepted by the reflector only out to the reflector rim. Again, we observe that as the taper becomes more severe

Table 7-1 Characteristics of Tapered Circular Aperture Distributions

a. Parabolic taper

$$E_a(\rho') = \left[1 - \left(\frac{\rho'}{a}\right)^2\right]^n$$

$$f(\theta, n) = \frac{2^{n+1}(n+1)!\,J_{n+1}(\beta a \sin\theta)}{(\beta a \sin\theta)^{n+1}}$$

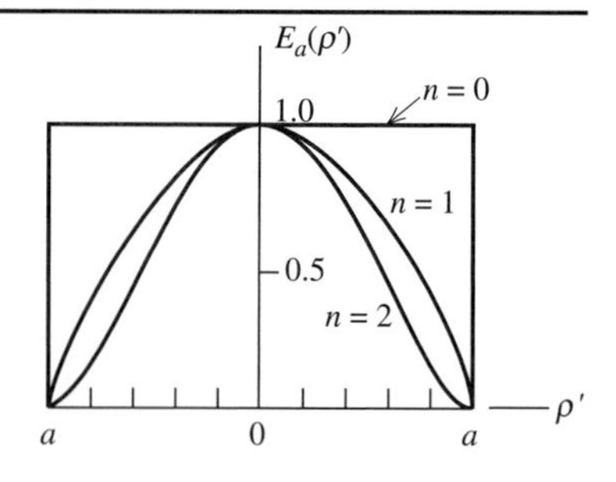

n	HP (rad)	Side Lobe Level (dB)	ε_t	Normalized Pattern $f(\theta, n)$	Distribution
0	$1.02\,\frac{\lambda}{2a}$	−17.6	1.00	$\frac{2J_1(\beta a \sin\theta)}{\beta a \sin\theta}$	Uniform
1	$1.27\,\frac{\lambda}{2a}$	−24.6	0.75	$\frac{8J_2(\beta a \sin\theta)}{(\beta a \sin\theta)^2}$	Parabolic
2	$1.47\,\frac{\lambda}{2a}$	−30.6	0.55	$\frac{48J_3(\beta a \sin\theta)}{(\beta a \sin\theta)^3}$	Parabolic squared

b. Parabolic taper on a pedestal

$$E_a(\rho') = C + (1 - C)\left[1 - \left(\frac{\rho'}{a}\right)^2\right]^n$$

$$f(\theta, n, C) = \frac{Cf(\theta, n = 0) + \frac{1 - C}{n + 1} f(\theta, n)}{C + \frac{1 - C}{n + 1}}$$

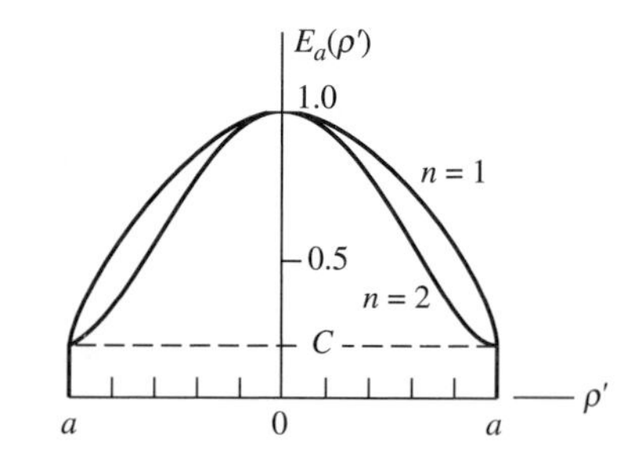

Edge Illumination		$n = 1$			$n = 2$		
C_{dB}	C	HP (rad)	Side Lobe Level (dB)	ε_t	HP (rad)	Side Lobe Level (dB)	ε_t
−8	0.398	$1.12\,\frac{\lambda}{2a}$	−21.5	0.942	$1.14\,\frac{\lambda}{2a}$	−24.7	0.918
−10	0.316	$1.14\,\frac{\lambda}{2a}$	−22.3	0.917	$1.17\,\frac{\lambda}{2a}$	−27.0	0.877
−12	0.251	$1.16\,\frac{\lambda}{2a}$	−22.9	0.893	$1.20\,\frac{\lambda}{2a}$	−29.5	0.834
−14	0.200	$1.17\,\frac{\lambda}{2a}$	−23.4	0.871	$1.23\,\frac{\lambda}{2a}$	−31.7	0.792
−16	0.158	$1.19\,\frac{\lambda}{2a}$	−23.8	0.850	$1.26\,\frac{\lambda}{2a}$	−33.5	0.754
−18	0.126	$1.20\,\frac{\lambda}{2a}$	−24.1	0.833	$1.29\,\frac{\lambda}{2a}$	−34.5	0.719
−20	0.100	$1.21\,\frac{\lambda}{2a}$	−24.3	0.817	$1.32\,\frac{\lambda}{2a}$	−34.7	0.690

Table 7-1 (continued)

b. Parabolic taper on a Pedestal (continued)

Interpolation equations for finding HP and ε_t when C_{dB} is between -8 and -20 dB:

Quantity	$n = 1$	$n = 2$
$\text{HP} = k\dfrac{\lambda}{2a}$	$k = -0.008C_{\text{dB}} + 1.06$	$k = -0.015C_{\text{dB}} + 1.02$
ε_t	$\varepsilon_t = 0.01C_{\text{dB}} + 1.02$	$\varepsilon_t = 0.019C_{\text{dB}} + 1.06$

(n increases or C decreases), the side-lobe level decreases while the beamwidth increases and directivity decreases. The data in Table 7-1 provide canonical forms for use in modeling parabolic reflector antennas discussed in the next section.

The directivity-bandwidth product DB_{cir} is found by a form of (7-90) appropriate to tapered circular apertures using $c_x c_y \rightarrow c^2 = \pi\varepsilon_t$ and $k_x k_y \rightarrow k^2$:

$$\text{DB}_{\text{cir}} = \pi^2 \varepsilon_t k^2 \left(\frac{180}{\pi}\right)^2 = 32{,}400\varepsilon_t k^2 \quad \text{deg}^2 \tag{7-175}$$

For the range of values given in Table 7-1b, this product remains nearly constant, leading to

$$D \approx \frac{\text{DB}_{\text{cir}}}{\text{HP}^2} = \frac{39{,}000}{\text{HP}^2} \tag{7-176}$$

where HP is the half-power beamwidth in all planes in degrees.

This section is closed by deriving the fundamental relation used to develop Table 7-1. The parabolic distribution used in (7-174) gives

$$f_{\text{un}}(\theta) = 2\pi \int_0^a \left[1 - \left(\frac{\rho'}{a}\right)^2\right]^n \rho' J_0(\beta\rho' \sin\theta)\, d\rho' \tag{7-177}$$

The integral can be evaluated using

$$\int_0^1 (1 - x^2)^n x J_0(bx)\, dx = \frac{2^n n!}{b^{n+1}} J_{n+1}(b) \tag{7-178}$$

by letting $x = \rho'/a$ and $b = \beta a \sin\theta$. Then (7-177) reduces to

$$f_{\text{un}}(\theta) = \frac{\pi a^2}{n+1} f(\theta, n) \tag{7-179}$$

where

$$f(\theta, n) = \frac{2^{n+1}(n+1)!\, J_{n+1}(\beta a \sin\theta)}{(\beta a \sin\theta)^{n+1}} \tag{7-180}$$

is the normalized pattern function. The patterns given in Table 7-1a follow from (7-180). The aperture taper efficiency is (see Prob. 7.5-4)

$$\varepsilon_t = \frac{\left[C + \dfrac{1-C}{n+1}\right]^2}{C^2 + \dfrac{2C(1-C)}{n+1} + \dfrac{(1-C)^2}{2n+1}} \tag{7-181}$$

7.6 REFLECTOR ANTENNAS

High-gain antennas are required for long-distance radio communication and high-resolution radar applications. Reflector systems are perhaps the most widely used high-gain antennas and routinely achieve gains far in excess of 30 dB in the microwave region. Such gains would be difficult to obtain with any other single antenna we have discussed thus far. In this section, we consider the more important reflector antenna configurations, with emphasis on those that have circular apertures.

7.6.1 Parabolic Reflector Antenna Principles

The simplest reflector antenna consists of two components: A reflecting surface that is large relative to a wavelength and a much smaller feed antenna. The most popular form is the **parabolic reflector antenna** shown in Fig. 7-25*a*. The reflector (or "dish") is a paraboloid of revolution. The intersection of the reflector with any plane containing the reflector axis (z-axis) forms a curve of the parabolic type shown in Fig. 7-25*b*. The equation describing the parabolic reflector surface shape in the rectangular form using (ρ', z_f) is

$$(\rho')^2 = 4F(F - z_f), \qquad \rho' \leq a \tag{7-182}$$

The apex of the dish corresponds to $\rho' = 0$ and $z_f = F$, and the edge of the dish to $\rho' = a$ and $z_f = F - a^2/4F$. For a given displacement ρ' from the axis of the reflector, the point R on the reflector surface is a distance r_f away from the *focal point O*. The parabolic curve can also be expressed in polar coordinates (r_f, θ_f) as

$$r_f = \frac{2F}{1 + \cos\theta_f} = F\sec^2\frac{\theta_f}{2} \tag{7-183}$$

Then the projection of this distance r_f onto the aperture plane is

$$\rho' = r_f \sin\theta_f = \frac{2F\sin\theta_f}{1 + \cos\theta_f} = 2F\tan\frac{\theta_f}{2} \tag{7-184}$$

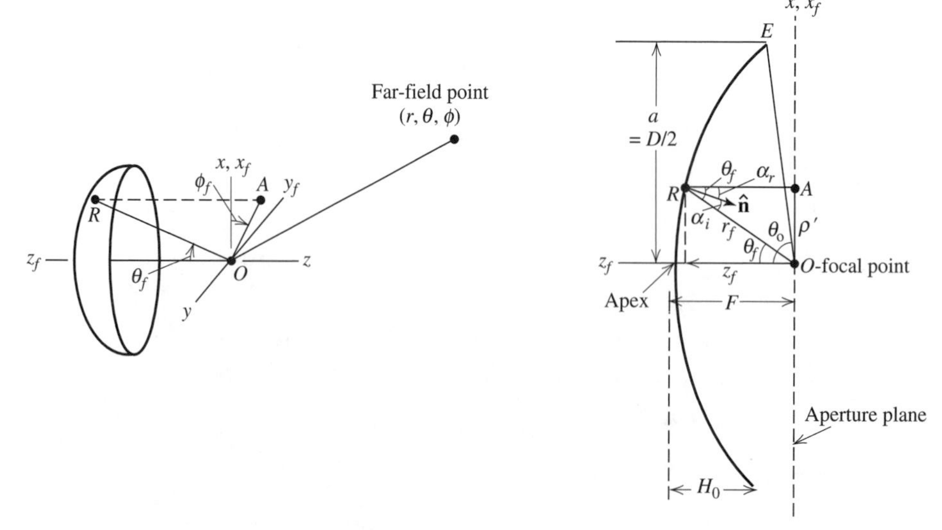

(*a*) Parabolic reflector and coordinate system (*b*) Cross section of the reflector in the *xz*-plane.

Figure 7-25 The axisymmetric parabolic reflector antenna.

At the apex ($\theta_f = 0°$), $r_f = F$, and $\rho' = 0$. At the reflector edge ($\theta_f = \theta_o$), $r_f = 2F/(1 + \cos\theta_o)$ and $\rho' = a$.

The axisymmetric parabolic reflector is completely specified with two parameters, the diameter D and focal length F. Equivalently, the reflector is often stated in terms of D and F/D, which give the size and shape (curvature rate), respectively. The "focal-length-to-diameter" ratio F/D represents the curvature rate of the dish. In the limit as F/D approaches infinity, the reflector becomes planar. A flat reflector "focuses" at infinity, and a normally incident plane wave is reflected back as a plane wave (i.e., it is focused at infinity). Shapes associated with commonly used reflectors are shown in Fig. 7-26. When F/D are 0.25, the focal point lies in the plane passing through the rim. As indicated in Fig. 7-25*b*, the angle from the feed axis (the z_f-axis) to the reflector rim is related to the F/D using (7-184) at point E ($\rho' = a$, $\theta_f = \theta_o$) as

$$\frac{F}{D} = \frac{1}{4\tan\dfrac{\theta_o}{2}} \tag{7-185a}$$

$$\theta_o = 2\tan^{-1}\left(\frac{1}{4\left(\dfrac{F}{D}\right)}\right) \tag{7-185b}$$

The reflector design problem consists primarily of matching the feed antenna pattern to the reflector. The usual goal is to have the feed pattern about 10 dB down in the direction of the rim, for example, $F_f(\theta_f = \theta_o) = -10$ dB. Feed antennas with this property can be constructed for the commonly used F/D values of 0.3 to 1.0 and are discussed in Sec. 7.7. The F/D choice also impacts on cross-polarization performance, as we shall see.

The focal distance of a reflector is easily calculated using diameter D and height H_0. This practical relation is found by solving (7-182) at the rim, where $\rho' = D/2$ and $z_f = F - H_0$, giving

$$F = \frac{D^2}{16H_0} \tag{7-186}$$

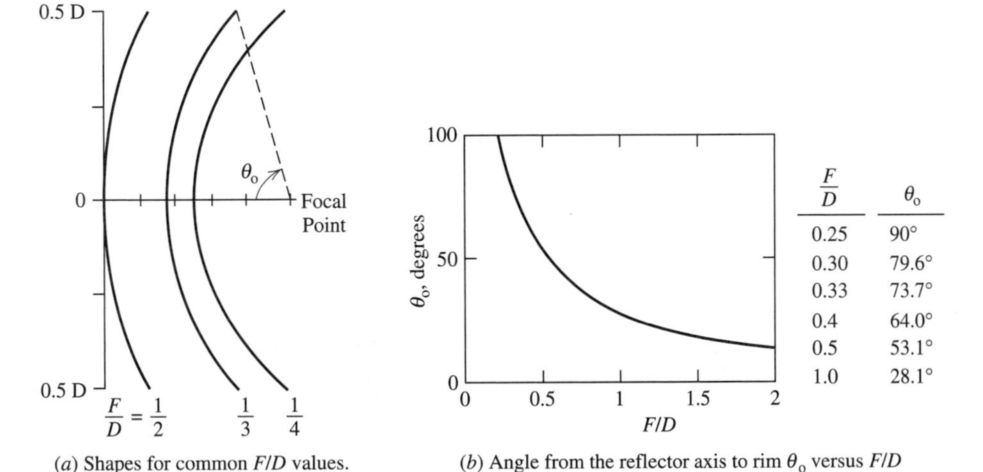

$\frac{F}{D}$	θ_o
0.25	90°
0.30	79.6°
0.33	73.7°
0.4	64.0°
0.5	53.1°
1.0	28.1°

(*a*) Shapes for common F/D values.

(*b*) Angle from the reflector axis to rim θ_o versus F/D

Figure 7-26 Curvatures of parabolic reflectors.

For example, when $F/D = 1/4$, this gives $H_0 = D/4$; thus, $H_0 = F$, which is evident from Fig. 7-26.

The following two very important properties make the parabolic reflector useful as an antenna:

1. All rays leaving the focal point O are collimated after reflection from the reflector and the reflected rays are parallel to the reflector axis (z-axis).
2. All path lengths from the focal point to the reflector and on to the aperture plane are the same and are equal to $2F$.

The terminology used here is that of geometrical optics (GO) which treats wave propagation as rays that are normal to the equiphase surface. For a point source at the focus, the wave fronts are spherical and all rays are along r_f shown in Fig. 7-25. GO principles will now be used to verify the above two properties.

The first property follows directly from the enforcement of the law of reflection on the reflector surface; that is, $\alpha_r = \alpha_i$ in Fig. 7-25*b*. To show this, we first determine the surface normal $\hat{\mathbf{n}}$ by evaluating the gradient of the parabolic curve equation, $C_p = F - r_f \cos^2(\theta_f/2) = 0$, based on (7-183) in feed coordinates:

$$\begin{aligned}\mathbf{N} = \nabla C_p &= \nabla\left(F - r_f \cos^2 \frac{\theta_f}{2}\right) \\ &= \left[\hat{\mathbf{r}}_\mathbf{f} \frac{\partial}{\partial r_f} + \hat{\boldsymbol{\theta}}_\mathbf{f} \frac{1}{r_f}\frac{\partial}{\partial \theta_f} + \hat{\boldsymbol{\phi}}_\mathbf{f} \frac{1}{r_f \sin\theta_f}\frac{\partial}{\partial \phi_f}\right] C_p \\ &= -\hat{\mathbf{r}}_\mathbf{f} \cos^2\frac{\theta_f}{2} + \hat{\boldsymbol{\theta}}_\mathbf{f} \cos\frac{\theta_f}{2}\sin\frac{\theta_f}{2} \end{aligned} \tag{7-187}$$

Normalizing using $N^2 = \mathbf{N} \bullet \mathbf{N} = \cos^2 \frac{\theta_f}{2}$ gives

$$\hat{\mathbf{n}} = \frac{\mathbf{N}}{N} = -\hat{\mathbf{r}}_\mathbf{f} \cos\frac{\theta_f}{2} + \hat{\boldsymbol{\theta}}_\mathbf{f} \sin\frac{\theta_f}{2} \tag{7-188}$$

The angles between the surface normal and the incident and reflected rays are then easily found from

$$\cos\alpha_i = -\hat{\mathbf{r}}_\mathbf{f} \bullet \hat{\mathbf{n}} = \cos\frac{\theta_f}{2} \tag{7-189a}$$

$$\begin{aligned}\cos\alpha_r &= \hat{\mathbf{z}} \bullet \hat{\mathbf{n}} = (-\hat{\mathbf{r}}_\mathbf{f}\cos\theta_f + \hat{\boldsymbol{\theta}}_\mathbf{f}\sin\theta_f) \bullet \hat{\mathbf{n}} \\ &= \cos\theta_f \cos\frac{\theta_f}{2} + \sin\theta_f \sin\frac{\theta_f}{2} \equiv \cos\frac{\theta_f}{2}\end{aligned} \tag{7-189b}$$

Comparing these two equations, we see that

$$\alpha_i = \alpha_r = \frac{\theta_f}{2} \tag{7-190}$$

proving that the law of reflection is satisfied. Also, note from Fig. 7-25*b* that $\alpha_i + \alpha_r = \theta_f$, which is consistent with (7-190).

The equal path length property follows from (7-183) as

$$\begin{aligned}\overline{OR} + \overline{RA} &= \text{total path length from focal point to aperture} \\ &= r_f + r_f\cos\theta_f = r_f(1 + \cos\theta_f) = 2F\end{aligned} \tag{7-191}$$

Since the total path length is constant ($2F$), the phase of waves arriving in the aperture plane from a point source at the focus will also be constant. Thus, the *parabolic reflector with a feed that has a point phase center at the focal point will produce uniform phase across the aperture plane.* The aperture amplitude distribution, however, will not be uniform.

Reflector antennas are analyzed by tracing rays to the aperture and setting up an aperture distribution that can be integrated to obtain the far-field pattern. Alternatively, an equivalent surface current over the reflector can be integrated. In either case, GO principles are used to determine the current distribution. Application of GO requires the following to be true:

a. The radius of curvature of the main reflector is large compared to a wavelength and the local region around each reflection point can be treated as planar.
b. The radius of curvature of the incoming wave from the feed is large and can be treated locally at the reflection point as a plane wave.

In addition, with metallic objects we make the following assumption:

c. The reflector acts as a perfect conductor so that the incident and reflected wave amplitudes are equal; in fact, $\Gamma = -1$.

The law of reflection applied to a reflector (e.g., see property 1 for parabolic reflectors) relies on these assumptions.

The parabolic reflector is inherently a very wideband antenna. The bandwidth of a reflector is determined at the low-frequency end by the size of the reflector; it should be at least several wavelengths in extent for GO principles to hold. At the high-frequency end, performance is limited by the smoothness of the reflector surface. Surface distortions must be much less than a wavelength to avoid phase errors in the aperture; see (7-235). In practice, the bandwidth of a reflector antenna system is usually limited by the bandwidth of the feed antenna rather than the reflector.

We now discuss techniques for analyzing reflector antennas. The techniques are not limited to reflectors of parabolic shape, but we consider only parabolic reflector cases.

GO/Aperture Distribution Method. The most basic method of analyzing reflector antennas is to use GO to determine the aperture field distribution and then find the far-field radiation pattern using the aperture theory developed in Sec. 7.5. This is done by tracing rays from the feed antenna to the aperture. First, we assume the feed is an isotropic radiator; the influence of the radiation pattern of a real feed antenna will be included later. Since all rays from the feed travel the same physical distance to the aperture, the aperture distribution of a parabolic reflector will be of uniform phase (this is true for all frequencies). However, there is a nonuniform amplitude distribution introduced. This is due to the fact that the power density of the rays leaving the isotropic feed falls off as $1/r_f^2$ since the wave is spherical. After reflection, there is no longer any spreading loss since the rays are parallel (i.e., focused at infinity), forming a section of plane wave. Hence, the aperture field intensity varies as $1/r_f$. This is proved more formally below.

Geometrical optics (see Section 12.1) assumes that power density in free space follows straight-line paths. Applied to this case, the power in a conical wedge of solid angle $d\Omega$ with cross-sectional angle $d\theta_f$ as shown in Fig. 7-27 will remain confined to that conical wedge as it progresses out from the feed. After reflection, the power associated with the increment $d\theta_f$ arrives at the aperture plane in a thin ring of thickness $d\rho'$ and area dA. The power leaving the feed, assumed to be

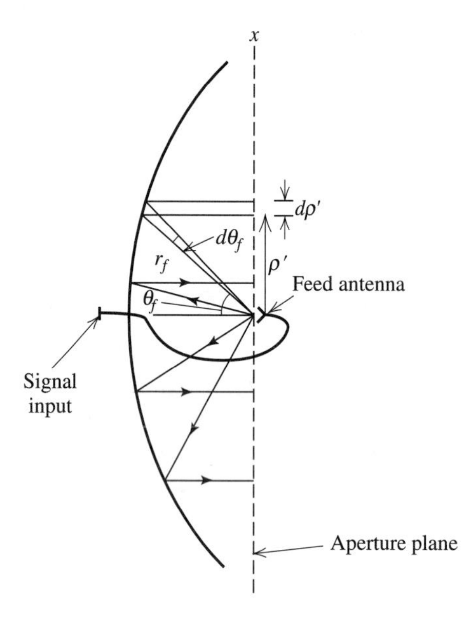

Figure 7-27 Axisymmetric, focus-fed parabolic reflector antenna in cross section.

isotropic, and arriving at the aperture is proportional to $P_t\, d\Omega$, where P_t is the transmit power. This power is distributed over area dA in the aperture plane. Thus, the power density in the aperture plane varies as

$$S_a(\rho') \propto \frac{P_t\, d\Omega}{dA} \propto \frac{d\Omega}{dA} \tag{7-192}$$

since P_t is a constant. After integration over ϕ_f, $d\Omega = 2\pi \sin \theta_f\, d\theta_f$ and $dA = 2\pi\rho'\, d\rho'$. So,

$$S_a(\rho') \propto \frac{2\pi \sin \theta_f\, d\theta_f}{2\pi\rho'\, d\rho'} = \frac{\sin \theta_f}{\rho'} \frac{d\theta_f}{d\rho'} \tag{7-193}$$

From (7-184),

$$\frac{d\rho'}{d\theta_f} = \frac{d}{d\theta_f}\left(2F \tan \frac{\theta_f}{2}\right) = F \sec^2 \frac{\theta_f}{2} = r_f \tag{7-194}$$

where (7-183) was used for the last equality. Then

$$\frac{d\theta_f}{d\rho'} = \frac{1}{r_f} \tag{7-195}$$

Hence, (7-192) with (7-184) and (7-195) becomes

$$S_a \propto \frac{\sin \theta_f}{r_f \sin \theta_f} \frac{1}{r_f} = \frac{1}{r_f^2} \tag{7-196}$$

This proves the spherical wave nature of the feed radiation and is referred to as *spherical spreading loss*. And since $E_a \propto \sqrt{S_a}$,

$$E_a(\theta_f) \propto \frac{1}{r_f} \tag{7-197}$$

Thus, there is a natural amplitude taper in the aperture caused by the curvature of the reflector.

If the primary (or feed) antenna is not isotropic, the effect of its normalized radiation pattern $F_f(\theta_f, \phi')$ can be included, using the coordinate system of Fig. 7-25 as

$$\mathbf{E}_a(\theta_f, \phi') = V_o e^{-j\beta 2F} \frac{F_f(\theta_f, \phi')}{r_f} \hat{\mathbf{u}}_\mathbf{r} \tag{7-198}$$

where $\hat{\mathbf{u}}_\mathbf{r}$ is the unit vector of the aperture electric field. The phase shift associated with the $2F$ path length from the focal point to the aperture plane is also included. The coordinates ρ' and ϕ' are appropriate for describing the aperture electric field. Feed coordinates r_f and θ_f are expressed in terms of the aperture coordinate ρ' using

$$r_f = \frac{4F^2 + \rho'^2}{4F} \tag{7-199a}$$

$$\theta_f = 2 \tan^{-1} \frac{\rho'}{2F} \tag{7-199b}$$

which follow from (7-183) and (7-184). These transformations can be used with (7-198) to obtain the aperture distribution at points (ρ', ϕ') from the feed antenna radiation pattern F_f. It remains only to find the polarization of the aperture electric field vector by determining $\hat{\mathbf{u}}_\mathbf{r}$ in (7-198). This follows by using the approximations that at the point of reflection, the reflector behaves as if planar [assumption (a)] and is perfectly conducting [assumption (c)]. Then the tangential component of the total electric field formed by the sum of the incident and reflected wave electric fields, $\mathbf{E_i} + \mathbf{E_r}$, is zero at the reflector. The law of reflection requires that $\hat{\mathbf{n}}$ bisect the incident and reflected rays; then, $\mathbf{E_i} + \mathbf{E_r} = 2(\hat{\mathbf{n}} \cdot \mathbf{E_i})\hat{\mathbf{n}}$, or

$$\mathbf{E_r} = 2(\hat{\mathbf{n}} \cdot \mathbf{E_i})\hat{\mathbf{n}} - \mathbf{E_i} \tag{7-200}$$

Since the reflector is assumed to be perfect [assumption (c) above], $|\mathbf{E_r}| = |\mathbf{E_i}|$; using this to normalize the above equation gives

$$\hat{\mathbf{u}}_\mathbf{r} = 2(\hat{\mathbf{n}} \cdot \hat{\mathbf{u}}_\mathbf{i})\hat{\mathbf{n}} - \hat{\mathbf{u}}_\mathbf{i} \tag{7-201}$$

where $\hat{\mathbf{u}}_\mathbf{r} = \mathbf{E_r}/|\mathbf{E_r}|$ and $\hat{\mathbf{u}}_\mathbf{i} = \mathbf{E_i}/|\mathbf{E_i}|$.

We can now write the radiation pattern function for the entire reflector system. The Fourier transform of the aperture distribution follows from (7-164) and (7-198) in (7-14):

$$\mathbf{P} = V_o \int_0^{2\pi} \int_0^a \frac{F_f(\theta_f, \phi')}{r_f} \hat{\mathbf{u}}_\mathbf{r} e^{j\beta\rho' \sin\theta \cos(\phi - \phi')} \rho' \, d\rho' \, d\phi' \tag{7-202}$$

for a circular projected aperture of radius $a = D/2$. The complete radiation pattern then follows from (7-29). A single reflector fed from a feed antenna at the focal point is often called a *focus-fed* or *prime-focus* reflector antenna. The feed is the *primary antenna* and the reflector forms the *secondary antenna.* The feed pattern is then the *primary pattern* and the pattern from the antenna system as a whole is called the *secondary pattern.*

The uniform aperture phase associated with the GO formulation and the use of a real-valued feed pattern function $F_f(\theta_f, \phi')$ leads to a symmetric pattern function since the Fourier transform of a real function is symmetric. Thus, the GO formulation always renders a symmetric far-field pattern. However, for general situations, such as offset reflectors to be discussed in Sec. 7.6.3, reflector antennas have asymmetric patterns. A more accurate analysis technique is now introduced for this purpose.

PO/Surface Current Method. The theory developed in Sec. 1.7 indicates that we can integrate over a current distribution to obtain the far field. For reflector antennas, we use the current on the metallic reflector generated by the incident fields from the feed antenna. Using the general expression for the magnetic vector potential of (1-101) appropriate to a surface current in the general far-field electric field expression of (1-105) gives

$$\mathbf{E} = -j\omega\mu \frac{e^{-j\beta r}}{4\pi r} \iint_{S_r} [\mathbf{J_s} - (\mathbf{J_s} \cdot \hat{\mathbf{r}})\hat{\mathbf{r}}] e^{j\beta \hat{\mathbf{r}} \cdot \mathbf{r}'} \, dS' \tag{7-203}$$

where S_r is the surface of the reflector. This approach, of course, is viable only if the surface current $\mathbf{J_s}$ is known. The current is found using the *physical-optics* (PO) approximation that makes use of assumptions (a) to (c) used in GO analysis to relate the surface current to the incident field from the feed. That is, the wave arriving from the feed behaves locally as a plane wave and reflects from a locally-plane reflector that behaves as a perfect reflector. Then the incident magnetic field from the feed $\mathbf{H_i}$ and the magnetic field associated with the reflected wave $\mathbf{H_r}$ are related to the surface current from (1-26) as $J_s = H_{\text{tan}}$, where H_{tan} is the tangential component of the total magnetic field that is given by $\hat{\mathbf{n}} \times (\mathbf{H_i} + \mathbf{H_r})$. But for a perfect conductor, the reflected magnetic field equals the incident magnetic field, so we are led to

$$\mathbf{J_S} = \begin{cases} 2\hat{\mathbf{n}} \times \mathbf{H_i} & \text{over the front of the reflector} \\ 0 & \text{on the shadowed side of the reflector} \end{cases} \tag{7-204}$$

Thus, the PO approximation interprets the reflector surface S_r as having a nonzero current only over portions illuminated by the feed using ray tracing. Also, the discontinuity at the rim of the reflector separating the illuminated and shadowed regions is neglected. This effect as well as direct radiation and blockage/scattering effects due to the feed assembly can be treated separately.

The integral in (7-203) can be evaluated analytically for arbitrary symmetrical reflectors. Known as *Rusch's method*, it ushered in the era of modern reflector antenna analysis and remains the most popular approach for reflector analysis [24, 25]. However, the integration is usually performed numerically [26]. In addition, a Jacobian transformation is usually employed to evaluate the integral using aperture plane coordinates, avoiding direct integration over the curved reflector surface [27]. Series expansions are also employed for efficient integral evaluation [28]. A number of codes are available to evaluate reflector antenna performance. The powerful code GRASP [29] includes PO evaluation. PRAC (Parabolic Reflector Antenna Code) is a user-friendly program written for the personal computer and is described in Appendix G.

For axisymmetric reflectors, the GO/aperture integration (GO/AI) and PO/surface integration (PO/SI) methods yield identical results [30]. However, the two methods, as conventionally applied to offset reflectors, yield different results, with the GO/AI method being slightly inferior. In addition, the pattern accuracy of both formulations degrades beyond the main beam and first few side lobes; the pattern effects in this region are dominated by diffraction from the reflector rim.

Diffraction Effects. The GO/AI and PO/SI techniques just discussed produce accurate results for the main beam and first few side lobes. The pattern in the far-out side-lobe region is found by including diffraction (i.e., scattering) from the rim of the reflector (and subreflector in dual systems) and any other sharp edges. This is done by augmenting GO with diffraction effects through the use of the geometrical theory of diffraction (GTD) or by augmenting PO with a fringe current on the rim using the physical theory of diffraction (PTD). GTD and PTD are discussed in Chap. 12 and an example of diffraction from a reflector is treated in Sec. 12.16.

7.6.2 The Axisymmetric Parabolic Reflector Antenna

In this and the remaining subsections, we discuss commonly used reflector geometries and their properties. The most popular reflector antenna is the focus-fed, axisymmetric parabolic reflector illustrated in Fig. 7-27. The feed is located at the focal point and its main beam peak is directed toward the reflector center. Usually, the feed is some type of horn antenna as discussed in Sec. 7.7. A simple dipole feed can be used at UHF frequencies and below.

Consider a feed antenna that is linearly polarized along the x-axis (coincident with the x_f-axis) and pointed toward the reflector apex with E- and H-plane patterns $C_E(\theta_f)$ and $C_H(\theta_f)$, respectively. The aperture field produced by a feed that is represented by (7-239a) is found from (7-198) and (7-201) with (7-188) as

$$\begin{aligned}\mathbf{E}_a = V_o \frac{e^{-j\beta 2F}}{r_f} &\{-\hat{\mathbf{x}}[C_E(\theta_f)\cos^2\phi_f + C_H(\theta_f)\sin^2\phi_f] \\ &+ \hat{\mathbf{y}}[C_E(\theta_f) - C_H(\theta_f)]\sin\phi_f\cos\phi_f\}\end{aligned} \tag{7-205}$$

Note the use of pattern coordinate unit vectors. If the feed is balanced with a rotationally symmetric normalized pattern $F_f(\theta_f)$, (7-205) simplifies to $\mathbf{E}_a \sim -\hat{\mathbf{x}}F_f(\theta_f)$. So, cross-polarization (i.e., y-polarized field content) arises from the imbalance between the E- and H-plane copolarized patterns, C_E and C_H. Equation (7-205) also shows that cross-polarization in the aperture is maximum in the $\phi_f =$ 45°, 135° plane. This aperture cross-polarization causes far-field cross-polarization. The aperture distribution of (7-205) can be integrated to find the radiation. It is instructive to examine (7-205) for an x_f-polarized short dipole, which has $C_E = \cos\theta_f$ and $C_H = 1$:

$$\begin{aligned}\mathbf{E}_a = V_o \frac{e^{-j\beta 2F}}{r_f} &\{-\hat{\mathbf{x}}(\cos\theta_f\cos^2\phi_f + \sin^2\phi_f) \\ &- \hat{\mathbf{y}}(1 - \cos\theta_f)\sin\phi_f\cos\phi_f\}\end{aligned} \tag{7-206}$$

The field components are shown in Fig. 7-28. The bracketed expression reduces to $-\hat{\mathbf{x}}\cos\theta_f$ in the E-plane ($\phi_f = 0°$) and $-\hat{\mathbf{x}}$ in the H-plane ($\phi_f = 90°$). Thus, the

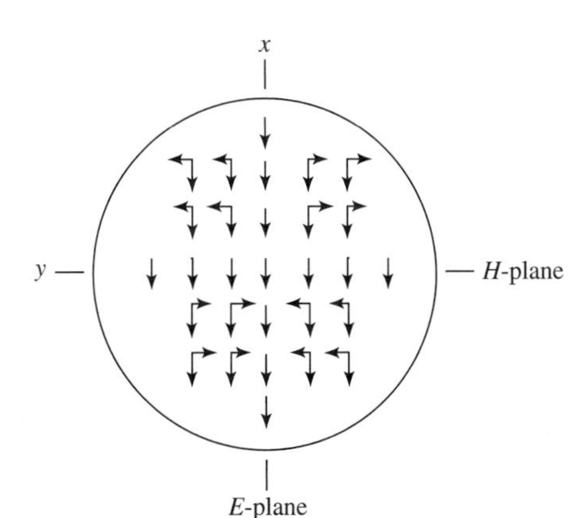

Figure 7-28 Electric field distribution in the aperture of a parabolic reflector for an x-polarized short dipole feed antenna. The electric field is decomposed into its x- and y-components. See (7-206).

aperture electric field is polarized parallel to the short dipole feed in the principal planes. Note that the aperture fields are inverted relative to the incident fields due to the reflection process. For nonprincipal planes, field components orthogonal to that of the feed (e.g., y-components) are present, giving cross-polarization. These cross-polarization properties are also true for the reflector radiation. The cross-polarization content in the aperture fields cancels in principal planes. This follows from the opposite phase of the cross-polarized components on opposite halves of the aperture. However, in nonprincipal planes, complete cancellation in the far field

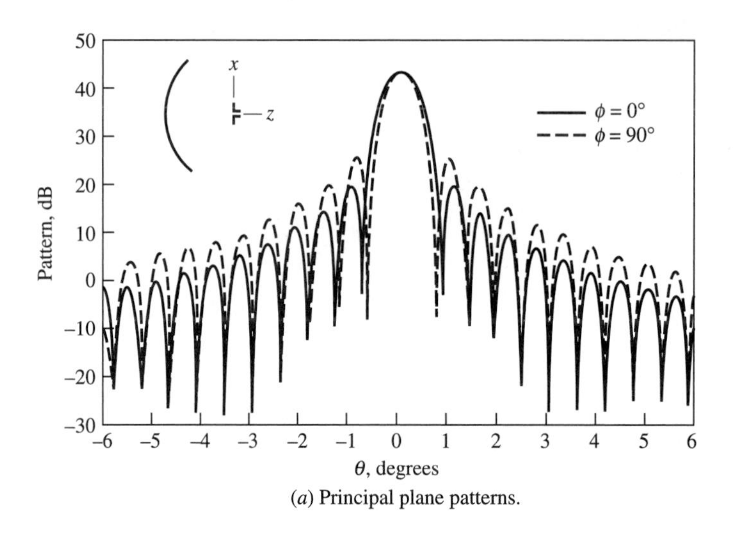

(a) Principal plane patterns.

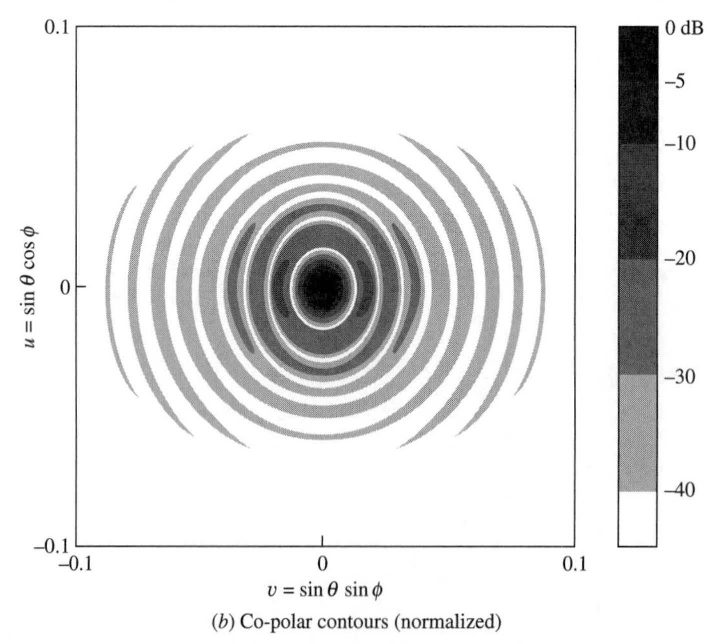

(b) Co-polar contours (normalized)

Figure 7-29 An axisymmetric parabolic reflector with diameter $D = 100\lambda$ and $F/D = 0.5$ fed by a half-wave dipole located at the focus. All data were computed using GRASP [29]. From [43] © 1993. Reprinted by permission of Artech House, Inc., Boston, MA.

does not occur. The largest cross-polarized components introduced by the reflector occur in the 45° planes. We also note a very important conclusion: *As F/D increases, cross-polarization decreases.* This follows by first noting from (7-185b) that as F/D increases, the maximum feed angle $\theta_f = \theta_o$ decreases and thus the second term in the (7-206) decreases, leading to reduced cross-polarization. In the limit of large F/D, the reflector becomes flat and does not introduce cross-polarization.

Figure 7-29 presents pattern data computed with the GRASP commercial reflector antenna code [29] using physical optics, surface current integration. A half-wave

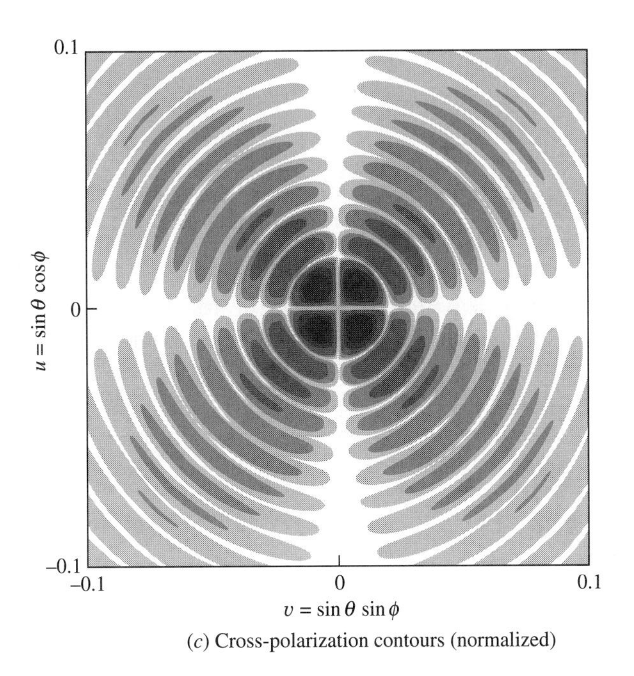

(*c*) Cross-polarization contours (normalized)

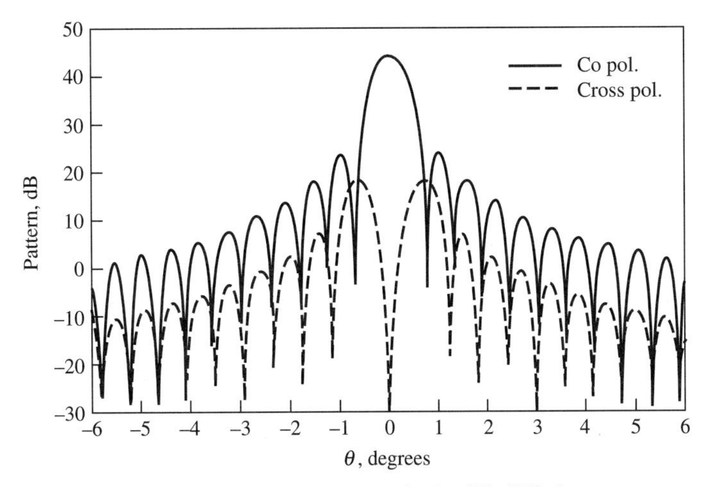

(d) Co- and cross-polarization patterns in $\phi = 45°$, 135° plane. The peak cross polarization is 26.3 dB below the co-polar peak.

Figure 7-29 (continued)

dipole, as is frequently encountered in practice, is used as the feed and behaves similar to the short dipole discussed above. Cross-polarization for reflectors is defined as ratio of the cross-polarization relative to the co-polarized pattern peak value. The peak cross-polarization is denoted as XPOL. Note the lack of cross-polarization in the principal planes in Fig. 7-29 and a cross-polar peak of XPOL $= -26.3$ dB in the $\phi = 45°$ plane. Ignoring the polarization vector and substituting (7-199) into (7-198) give an expression for the normalized *aperture illumination* in aperture coordinates:

$$E_{an}(\rho') = F_f\left(\theta_f = 2\tan^{-1}\frac{\rho'}{2F}\right)\frac{1}{1+\left(\frac{\rho'}{2F}\right)^2} \quad (7\text{-}207a)$$

$$= 20\log|F_f| - 20\log\left[1+\left(\frac{\rho'}{2F}\right)^2\right] \quad (\text{dB}) \quad (7\text{-}207b)$$

This permits direct evaluation of the electric field variation as a function of aperture radius ρ'. The second factor in (7-207a) is called the spherical spreading factor and represents the fact that the distance r_f from the focal point to the reflector increases with ρ'. As noted earlier, rays leaving the feed at the focal point spread out in all directions, leading to r_f^{-1} field variation. After reflection from the main reflector, the rays are collimated and no longer experience amplitude decay. The exact form of the aperture distribution has less influence on the pattern and directivity of a reflector than the *edge illumination EI*, or *edge taper ET*, that are found from (7-207b) for $\rho' = a$ as

$$EI = 20\log[E_{an}(\rho' = a)] = -FT - L_{\text{sph}} = -ET \quad (\text{dB}) \quad (7\text{-}208)$$

where

EI = edge illumination (dB) $= 20\log C$
ET = edge taper (dB) $= -EI$
FT = feed taper (at aperture edge) (dB) $= -20\log[F_f(\theta_o)]$
L_{sph} = spherical spreading loss at the aperture edge (dB)

$$= 20\log\left[1+\frac{1}{16\left(\frac{F}{D}\right)^2}\right] = -20\log\left[\frac{1+\cos\theta_o}{2}\right]$$

The above expression for L_{sph} shows that F/D influences the amount of spherical spreading loss. It varies from 0.5 to 6.0 dB for F/D, ranging from 1.0 down to 0.25; see Fig. 7-26.

Reflector antenna performance can be estimated by a simple process. First, the aperture distribution is obtained using (7-207). Next, a canonical distribution, such as presented in Table 7-1, is selected so that it approximates the aperture distribution. Then the performance parameters of Table 7-1 such as HP, SLL, and ε_t are evaluated. Interpolation can be used for intermediate values. This canonical distribution method is illustrated in Example 7-8. The approach is described in detail in [6], which also contains useful data for canonical distributions; however, [6] failed to include spherical spreading loss.

EXAMPLE 7-8 *A 28-GHz Parabolic Reflector Antenna*

An axisymmetric parabolic reflector antenna 1.22 m (4 ft) in diameter was used at Virginia Tech to receive signals at 28.56 GHz ($\lambda = 1.05$ cm) from a geostationary satellite. It has an $F/D = 0.5$. It is constructed of epoxy fiberglass with a reflective metal layer and has an rms surface accuracy of 0.2 mm. The feed antenna is a circular corrugated horn positioned at the focal point and supported with four thin-profile spars. The feed pattern is slightly asymmetric, but analysis using a canonical distribution yields good results. We assume the feed to be rotationally symmetric and equal to the measured E-plane feed pattern with the following beamwidths:

$$\mathrm{HP} = 56°, \qquad \mathrm{BW}_{-10\mathrm{dB}} = 104° \tag{7-209}$$

The angle from the center to the edge of the reflector from (7-185b) is $\theta_o = 53.1°$. The 10-dB down point will fall inside the reflector since $\mathrm{BW}_{-10\mathrm{dB}}/2 = 52°$. We assume the feed pattern has fallen to 11 dB down at the rim, that is, $FT = 11$ dB. The spherical spreading loss from (7-208b) is $L_{\mathrm{sph}} = 1.9$ dB. The edge taper from (7-208a) is $ET = 11 + 1.9 = 12.9$ dB. The corresponding edge illumination $EI = -12.9$ dB ($C = 0.2265$) falls between the values in Table 7-1. Linearly interpolating for the parabolic-squared taper yields the following results:

$$\mathrm{HP} = 1.214 \frac{\lambda}{D} = 0.599° \quad (0.605° \text{ measured}) \tag{7-210a}$$

$$\mathrm{SLL} = -30.5 \text{ dB} \qquad (-28.5 \text{ dB measured}) \tag{7-210b}$$

Note the very good agreement to measured values. The complete pattern using $f(\theta, n, C)$ from Table 7-1b is plotted in Fig. 7-30. Also shown for comparison is the measured pattern; it is identical to the ground computed pattern over the main beam. The gain is evaluated in Prob. 7.6-6.

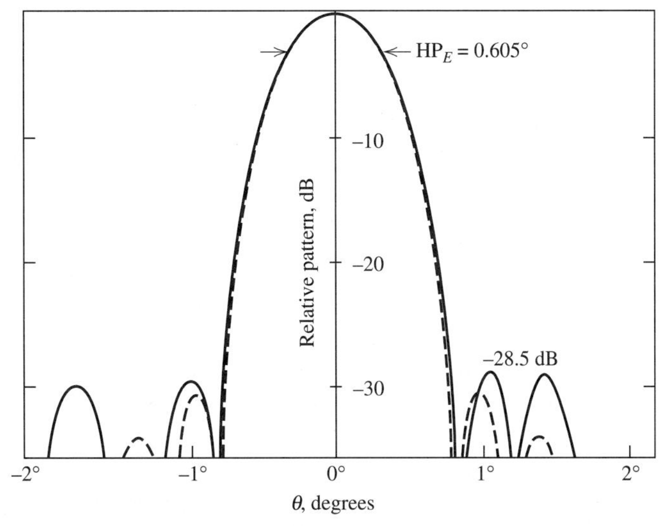

Figure 7-30 Measured (solid) E-plane pattern for the 1.22-m-diameter axisymmetric parabolic reflector at 28.56 GHz in Example 7-8. The computed (dashed) pattern is for a parabolic-squared circular aperture distribution on a -12.9-dB pedestal.

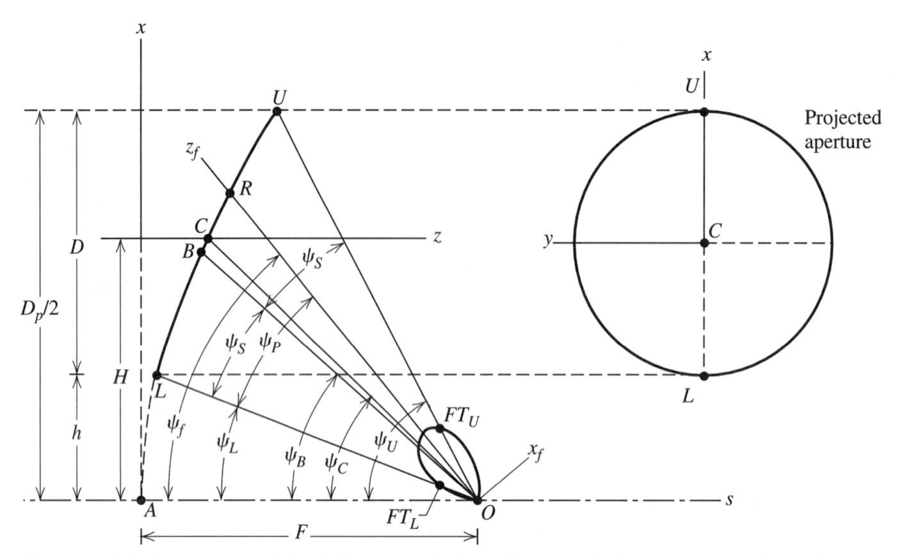

Figure 7-31 Geometry of the offset parabolic reflector of diameter D and focal length F. The axis of symmetry s bisects the parent parabolic curve of diameter D_p. Note that the axisymmetric case occurs when $H = 0$.

7.6.3 Offset Parabolic Reflectors

The blockage of the main reflector aperture by the feed assembly and associated support structure can be reduced or eliminated by using the offset reflector shown in Fig. 7-31. The properties of the offset reflector are similar to the axisymmetric counterpart formed by using the diameter of the *parent reflector* D_p. That is, the appropriate focal-length-to-diameter ratio to use for electrical performance is F/D_p. This degrades cross-polarization performance because $F/D_p < F/D$ for the offset reflector and cross-polarization levels rise with decreasing F/D_p for an unbalanced feed pointed toward the apex, that is, $\theta_f = 0°$. More will be said about cross-polarization in Sec. 7.6.5.

The analysis techniques explained in the previous section for axisymmetric geometries are general and are also used for offset reflectors. The GO/AI and PO/SI methods yield identical results if the integration surface is chosen to cap the reflector [30]. This is the natural choice for axisymmetric reflectors where the integration plane is selected to coincide with the physical aperture plane that contains the rim of the reflector, thus capping it. However, if the projected aperture of an offset reflector (as shown in Fig. 7-31) is used for the integration surface, GO/AI and PO/SI yield different results [31]. PO/SI is thought to produce more accurate patterns, especially for cross-polarization. In addition, the pattern accuracy of both formulations degrades beyond the main beam and first few side lobes. Pattern effects in this region are dominated by diffraction from the reflector rim; see Sec. 12.16.

Pencil beams are required in communication applications for high gain and in remote sensing for scene resolution. Offset reflectors are used not only to produce pencil beam patterns, as discussed so far, but also for *contoured beams*. Offset reflectors produce contour beams by using an array of feed horns or by shaping the main reflector (e.g., using a nonparabolic shape) [27]. Offset reflectors avoid blockage caused by hardware in the feed region created by a cluster of feed horns. A popular application for contour beams is on geostationary satellites that have antennas which produce a *footprint* conforming to a desired earth region such as a country or continent. The multiple feed antennas in the focal region, each creating a scanned beam according to the displacement from the focal point, are combined with amplitude and phase weighting to produce a custom-shaped main beam.

7.6.4 Dual Reflector Antennas

A subreflector can be introduced between the feed and main reflector of a single reflector antenna to form a *dual reflector*. The most popular dual reflector is the axisymmetric *Cassegrain reflector antenna* shown in Fig. 7-32. The main reflector is parabolic and the subreflector is hyperbolic. This geometry again produces a focused system. That is, rays associated with an incoming plane wave parallel to the axis of symmetry reflect from both reflectors and intersect at a point, the focal point F'. The *virtual focal point F* shown in Fig. 7-32 is the point from which transmitted rays appear to emanate with a spherical wave front after reflection from the subreflector. That is, the feed is mirrored in the subreflector.

A second form of the dual reflector antenna that offers perfect focusing is the *Gregorian reflector*. It has a concave rather than a convex subreflector that is located beyond the virtual focal point as shown in Fig. 7-32 and has an elliptical cross section. Both Cassegrain and Gregorian systems have their origins in optical telescopes and are named for their inventors. The subreflector for the Gregorian system being more distant from the main reflector requires more support structure. Both types of dual reflectors offer the major advantage of having the feed conveniently located near the apex of the main reflector. This provides convenient access to the feed region, reduces the support problem for feed hardware, and eliminates the long transmission line run, and associated losses, often used to reach the focal region of a prime focus reflector. Another advantage of dual reflectors over single reflectors is in low-noise earth terminal applications. The feed radiation not intercepted by the subreflector of a dual reflector (e.g., spillover) is directed toward the low-noise sky region rather than the more noisy ground as seen by the spillover of a single reflector.

The subreflector shapes used in the classical dual reflector configurations are described by conic sections. Figure 7-33 gives the geometry of the subreflectors in subreflector coordinates x_s and z_s; the complete subreflector surface is obtained by rotating the curve about the z_s-axis. The subreflector is determined by its di-

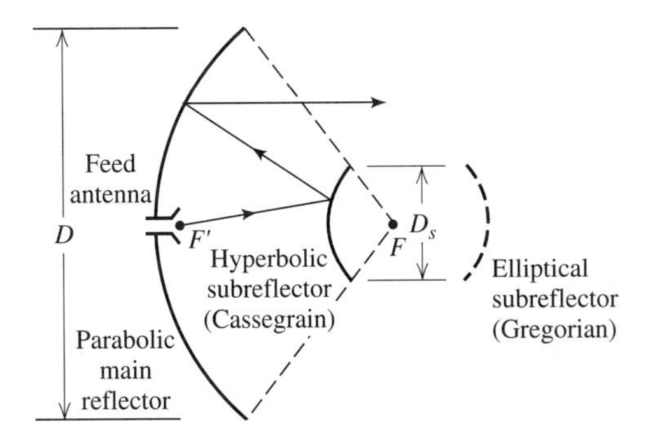

Figure 7-32 Classical axisymmetrical dual reflectors. The main reflector is parabolic and the subreflector is hyperbolic (elliptical) for the Cassegrain (Gregorian) reflector system.

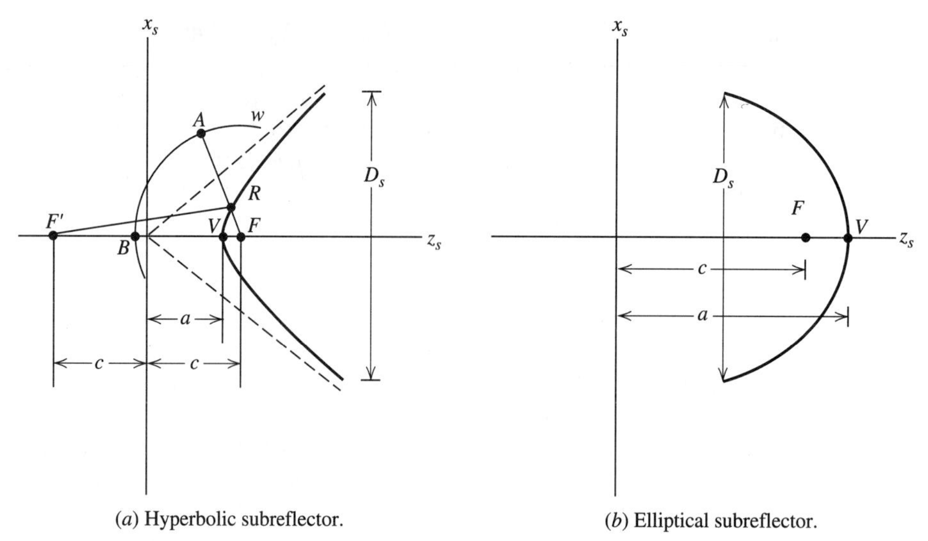

(*a*) Hyperbolic subreflector.

(*b*) Elliptical subreflector.

Figure 7-33 Geometry of classical subreflectors.

ameter D_s and *eccentricity* e. The shape is controlled by the eccentricity, which is defined as

$$e = \frac{c}{a} \begin{cases} >1 & \text{hyperbola (Cassegrain)} \\ <1 & \text{ellipse (Gregorian)} \end{cases} \tag{7-211}$$

Example shapes are $e = \infty$, planar; $e = 0$, circle (sphere); and $e = 1$, parabola. The equations of the subreflector surfaces are given by

$$\begin{aligned} \frac{z_s^2}{a^2} - \frac{x_s^2}{b^2} &= 1 \qquad b^2 = c^2 - a^2 \qquad \text{hyperbola} \\ \frac{z_s^2}{a^2} + \frac{x_s^2}{b^2} &= 1 \qquad b^2 = a^2 - c^2 \qquad \text{ellipse} \end{aligned} \tag{7-212}$$

The distances c and a are shown in Fig. 7-33. The required hyperbolic shape will be proved for the Cassegrain dual reflector. This derivation also illustrates how the subreflector operates.

The function of the hyperbolic subreflector is to convert the incoming wave from a feed antenna located at the focal point F' to a spherical wave front w that appears to originate from the virtual focal point F. For this to be true, the optical path (total distance) from F' to wavefront w must be constant. Enforcement of this condition determines the subreflector shape. As seen from Fig. 7-33*a*, the total distance from F' to A including reflection from the subreflector is

$$\overline{F'R} + \overline{RA} = \overline{F'V} + \overline{VB} = c + a + \overline{VB} \tag{7-213}$$

But

$$\overline{RA} = \overline{FA} - \overline{FR} = \overline{FB} - \overline{FR} \tag{7-214}$$

where $\overline{FB} = \overline{FA}$ was based on the exiting wave being spherical. This in (7-213) leads to

$$\overline{F'R} - \overline{FR} = c + a - (\overline{FB} - \overline{VB}) = c + a - (c - a) = 2a \tag{7-215}$$

This result coincides with the following definition of a hyperbola: A hyperbola is the locus of a point that moves such that the difference of its distances, $\overline{F'R} - \overline{FR}$, from two fixed points, F' and F, is equal to a constant, $2a$.

Dual reflectors can be modeled with a single *equivalent parabolic reflector* as shown in Fig. 7-34 for the axisymmetric Cassegrain reflector [32]. The equivalent parabola has the same diameter ($D_e = D$), but a focal length (F_e) longer than that of the main reflector (F):

$$F_e = \frac{e + 1}{e - 1} F = MF \tag{7-216}$$

where M is called the *magnification*. From (7-211), $e > 1$ for the hyperbolic subreflector of a Cassegrain reflector, so $M > 1$ and $F_e > F$. This increased effective focal length provides several advantages, which are noted by examining the equivalent (single) parabolic reflector with diameter D and focal length F_e, and thus larger focal-length-to-diameter ratio than the actual dual reflector. First, as noted in Sec. 7.6.2, cross-polarization in the far-field pattern improves with larger focal length-to-diameter ratio. Second, there is less spherical spreading loss at the rim of the main reflector; see (7-207). Finally, main beam scan performance is improved. This follows because the larger the focal-length-to-diameter ratio is, the less the radiation pattern deteriorates as the feed antenna is laterally displaced (in the $x_s y_s$-plane). This is explained by examining the limiting case of an infinite focal length-to-diameter ratio case (e.g., a flat main reflector) in which no deterioration occurs for reflection off normal incidence. The equivalent parabola concept applies to dual offset reflectors also [27, 33].

In a single reflector antenna system, the phase front from the feed antenna is converted to the desired exiting phase front. This is usually a spherical to planar conversion as accomplished with a parabolic reflector. Limited aperture amplitude control is accomplished through feed taper and F/D; see (7-208). This is true for traditional single and dual reflectors with a parabolic main reflector. However, if both reflector shapes in dual reflectors are allowed to be "shaped," both the aperture amplitude and phase can be controlled. In the usual synthesis case, the subreflector has a highly tapered illumination to reduce spillover and is shaped to spread the reflected rays out for uniform amplitude. This requires shaping of the main reflector. The topic of dual reflector synthesis is rather advanced, but can be

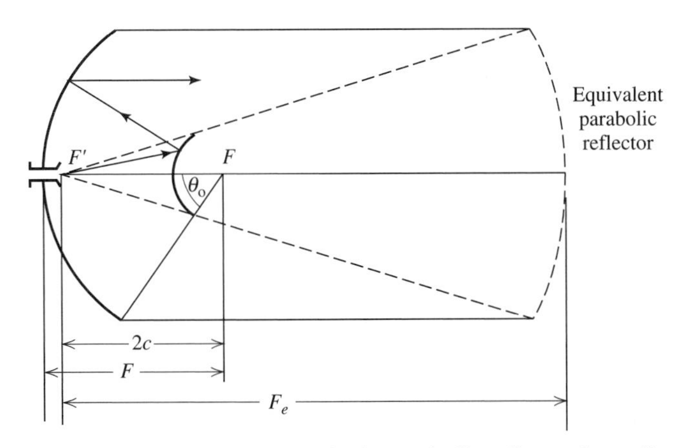

Figure 7-34 The equivalent single parabolic reflector for a Cassegrain dual reflector.

understood with the following simplified explanation [24, 34–38]. In principle, the shapes of both the main reflector and the subreflector can be determined exactly for axisymmetric systems [36] and for offset dual reflectors [38]. As in the above-mentioned case, the shapes of dual reflectors can be determined to yield uniform aperture amplitude and phase, giving maximum aperture utilization. The design problem is to convert the rather broad feed antenna radiation pattern to a nearly uniform amplitude and phase aperture distribution while keeping spillover acceptable. The concept is to underilluminate the subreflector in order to reduce its spillover and then increase its curvature over classical shape to direct reflected rays to edges of the main reflector. Within the limit of geometrical optics, main reflector spillover is avoided with proper reflector sizing. Shaping of the subreflector changes the total path length from the feed to the aperture. But that can be nearly compensated for by shaping the main reflector to correct for phase errors introduced by subreflector shaping. The amount of main reflector shape change is on the same order as subreflector shape change because both introduce about the same phase error. However, subreflector shaping almost completely controls the aperture amplitude distribution. This sequential shaping does not yield an exact solution but avoids the difficult mathematics associated with the exact approach, which requires simultaneous solution for the reflector shapes. Dual-shaped reflectors are in wide use for high-gain axisymmetric systems such as satellite communications earth terminals with main reflectors larger than a few meters in diameter. Shaped offset dual reflectors for smaller systems are also being used. One of the highest reported aperture efficiencies (85%) was achieved with a 1.5-m main reflector in a shaped offset dual reflector configuration operated at 31 GHz [39]. Dual reflectors can also be designed for low side lobes [40].

For the synthesis of dual-shaped reflectors, it is common to use geometrical optics (GO) during the design process. The previous discussion of shaped reflectors for high efficiency is a good example. After the reflector shapes are determined using GO-based synthesis, a computer code is then employed to accurately determine performance parameters such as gain, side-lobe level, and cross-polarization. There are two popular approaches for the computer analysis of dual reflectors: GTD/PO that uses the geometrical theory of diffraction (GTD is treated in Chap. 12) on the subreflector and PO on the main reflector, and PO/PO that uses PO on both the subreflector and main reflector. For small subreflectors less than about 10λ, PO/PO is thought to give more accurate results [27]. GTD/PO is usually used for electrically large reflector systems because PO/PO computations are very time-consuming.

7.6.5 Cross-Polarization and Scanning Properties of Reflector Antennas

Reflectors are used in many situations, including dual polarized operation that permits two communication channels on the same frequency. This requires the two polarizations to be nearly orthogonal. A second popular use for reflectors is to scan a single main beam or to form multiple narrow beams with the same large main reflector by displacing the feed(s) from the focal point. Both cross-polarization and beam scanning are discussed in this section.

Analysis of the axisymmetric parabolic reflector in Sec. 7.6.2 showed that for a purely polarized but unbalanced feed, such as a dipole, reflector-induced cross-polarization is zero in the principal planes and is maximum in the 45° planes; see Fig. 7-29. A balanced feed with a rotationally symmetric pattern (see Sec. 7.7) positioned with its perfect phase center at the focus produces no far-field cross-polarization based on GO analysis [42]. However, PO analysis of an axisymmetric reflector with a balanced feed does yield a small cross-polarized pattern. In practice, the feed is usually responsible for the dominant contribution to cross-polarization from axisymmetric reflectors.

As noted in Sec. 7.6.2, the cross-polarization of axisymmetric reflectors decreases with increasing F/D. That is, as the dish curvature reduces, less cross-polarization is introduced. This is easily remembered because as F/D becomes larger, the reflector becomes flatter and a flat reflector does not depolarize. If the feed antenna pattern is rotationally symmetric (e.g., balanced) and purely polarized, cross-polarization is reduced significantly, but a residual level remains due to axial (z-directed) currents on the reflector.

Cross-polarization behavior for offset reflectors is more complex. The general geometry is shown in Fig. 7-31. As explained in Sec. 7.6.3, the parent reflector diameter must be used for electrical performance. Since F/D of the actual offset reflector is greater in F/D_p for the parent reflector, cross-polarization will be worse in an offset reflector. This assumes that the feed is directed toward the apex; that is, $\psi_f = 0°$. This would lead to considerable spillover. Instead, the feed is usually pointed so that its axis bisects the angle subtended by the reflector ($\psi_f = \psi_B$) or the ray along the feed axis arrives in the center of the projected aperture ($\psi_f = \psi_C$). The influence of feed pointing angle is illustrated in Fig. 7-35 for an offset reflector with a diameter of 85.5λ, offset height $h = D/8$, and $F/D = 0.3$. The feed is linearly polarized with a circularly symmetric pattern (e.g., balanced feed) and a 10-dB beamwidth of 70°. This geometry yields results that are representative of offset reflectors used in practice. In this example, the bisector angle is $\psi_B = 45.1°$ and the angle of the central ray is $\psi_C = 49.7°$. Note in Fig. 7-35 that cross-polarization performance degrades as the feed pointing angle increases. However, gain shows a broad peak centered on $\psi_f = 47°$, which is between ψ_B and ψ_C. Side-lobe level is fairly constant with pointing angle as long as $\psi_f > 30°$. It turns out that the feed pointing angle $\psi_f = \psi_E$ that leads to equal edge illuminations ($EI_L = EI_U$) produces very low side lobes with only small penalties in gain and cross-polarization [42]. In

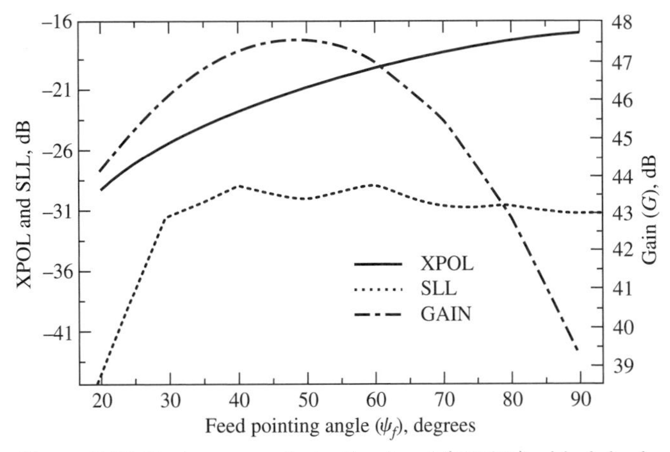

Figure 7-35 Peak cross-polarization level (XPOL), side-lobe level (SLL), and gain (G) as a function of feed pointing angle (ψ_f) for an offset reflector (see Fig. 7-31) with $D = 85.5\lambda$, $F/D_p = 0.3$, and $h = D/8$. The bisector angle is $\psi_B = 45.1°$. The central ray angle is $\psi_C = 49.7°$. The feed is balanced and linearly polarized. From [42] © 1993. Reprinted by permission of IEEE.

this example, $\psi_E = 49.6°$, where SLL $= -30.1$ dB. An important fundamental limitation on offset reflectors is apparent from Fig. 7-35. For maximum gain (i.e., aperture efficiency), cross-polarization is on the order of -23 dB, which is unacceptably high in many applications.

Cross-polarization performance of single offset reflectors is summarized in Fig. 7-36 [43, 44]. The cross-polarization for a balanced feed linearly polarized will be maximum in the plane of asymmetry (yz-plane) and zero in the plane of symmetry (xz-plane) [43–45]. The polarization in the (secondary) pattern of a reflector antenna is influenced by both the cross-polarization of the (primary) feed pattern, XPOL_F, as well as that introduced by the reflector, XPOL_R. So far, we have neglected any feed cross-polarization, that is, $\text{XPOL}_F = 0$. Since cross-polarization requires a code such as a physical optics code for exact evaluation, an approximate technique is very useful. A worst-case estimate of cross-polarization of the reflector system is found by adding the contributions [46]:

$$\text{XPOL}_S = \text{XPOL}_F + \text{XPOL}_R \tag{7-217}$$

For example, a feed with -30-dB cross-polarization ($\text{XPOL}_F = 0.0316$) used in a reflector with -23-dB cross-polarization ($\text{XPOL}_R = 0.0708$) gives $\text{XPOL}_S = 0.1024 = -19.8$ dB.

The situation is much different for a circularly polarized (CP), balanced feed as indicated in Fig. 7-36. There is no cross-polarization; however, the main beam "squints" off axis in the yz-plane [43–47]. The main beam rotates (or squints) to opposite sides of the reflector axis for opposite senses of CP. Note that with CP feeds, the sense of CP radiated from the reflector will be opposite that of the feed due to the sense reversal encountered during reflection.

The advantages of the offset reflector of Sec. 7.6.3 and the dual reflector of Sec. 7.6.4 can be combined in the form of a *dual offset reflector*. Aperture blockage is eliminated and the subreflector introduces a second design variable for reducing cross-polarization far below that of a single offset reflector. The performance of a dual offset reflector can be evaluated in a manner similar to that for the axisym-

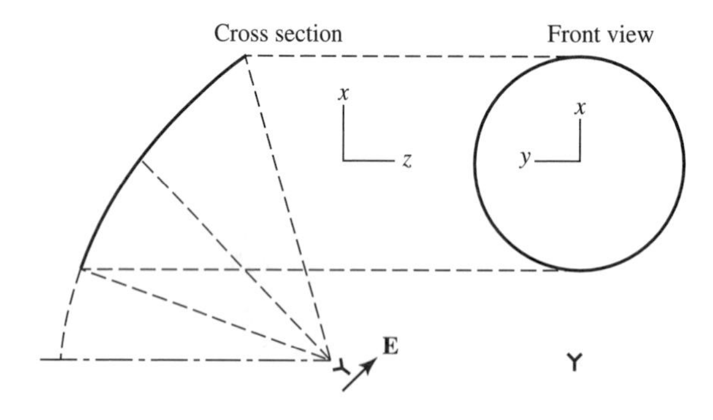

Cross-polarization properties

Feed	xz-plane (plane of symmetry, plane of offset)	yz-plane (plane of asymmetry)
Ideal *LP*	No cross-polarization	Cross-polarization
Ideal *CP*	No cross-polarization	No cross-polarization Beam squint

Figure 7-36 Summary of cross-polarization level properties of an offset parabolic reflector with a balanced feed. From [43] © 1993. Reprinted by permission of Artech House, Inc., Boston, MA.

metric dual reflector by using an equivalent single paraboloid with the same diameter as the main reflector and a focal length given by $F_e = MF$ from (7-216). If the dual reflector geometry is chosen such that the feed axis of the original system is coincident with the equivalent paraboloid axis, cross-polarization will be minimum.

Reflector antennas designed for high gain are focused systems. That is, an incoming wave parallel to the axis of the main reflector will be focused to a small region near the focal point. This leads to a simple antenna system, but limits beam scanning possibilities if rotating the entire reflector system is to be avoided. Some beam scanning is possible by displacing the feed off the focal point. This can be understood by considering a planar reflector with a small feed antenna that is displaced (and tilted back to aim at the reflector) from the axis perpendicular to the reflector. The reflected wave, or main beam from the reflector, will exit at an angle equal to the displacement angle. A similar effect applies to parabolic reflectors as shown in Fig. 7-37, where the feed antenna is laterally displaced a distance δ in the focal plane. If the reflector is flat ($F/D = \infty$), the angle of the beam scan angle θ_B equals the feed tilt angle θ_F. For curved reflectors ($F < \infty$), the beam scan angle will be less than the feed tilt angle. Scanning is quantified with *beam deviation factor* (BDF):

$$\text{BDF} = \frac{\theta_B}{\theta_F} \tag{7-218}$$

BDF is maximum at unity for a flat reflector and decreases with decreasing F/D or F/D_p for axisymmetric and offset reflectors, respectively. The following approximate expression can be used for small displacements δ [27, 48]:

$$\text{BDF} = \frac{1 + 0.36\left[4\dfrac{F}{D}\right]^{-2}}{1 + \left[4\dfrac{F}{D}\right]^{-2}} \tag{7-219}$$

Lateral feed displacement introduces a planar phase front tilted with respect to the aperture plane that is responsible for scanning the beam in a direction opposite to the displacement, as indicated in Fig. 7-37. However, nonlinear phase as a function of position in the aperture plane is also introduced, leading to pattern distortion, including beam broadening, and gain loss [49]. These effects worsen with increasing displacement and lower F/D values. One characteristic pattern distortion is a high first side lobe called the *coma lobe* on the reflector axis side of the main beam.

Multiple reflectors can also be scanned by displacement of the feed off the focal point. Dual reflectors offer the advantage of a longer effective focal length through the equivalent parabolic reflector; see (7-216). This leads to better scan perfor-

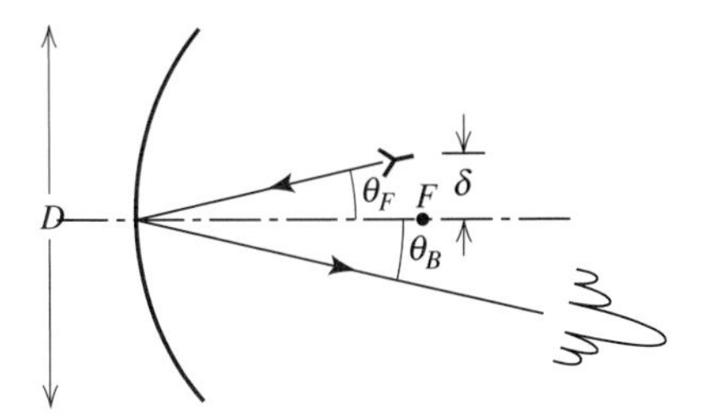

Figure 7-37 Beam scanning of a reflector antenna by feed displacement.

mance. Application of an equivalent single paraboloid provides approximate results for small scan angles with dual offset reflectors [50]. Advanced techniques can be used with tri-reflectors to minimize motion during scan [51, 52].

7.6.6 Gain Calculations for Reflector Antennas

Aperture antennas usually have an obvious physical aperture of area A_p through which energy passes on its way to the far field. The maximum achievable gain for an aperture antenna from (7-67) and (7-75) is

$$G_{\max} = D_u = \frac{4\pi}{\lambda^2} A_p \tag{7-220}$$

This gain is possible only under the ideal circumstances of a uniform amplitude, uniform phase antenna with no spillover or ohmic losses present. In practice, these conditions are not satisfied and gain is decreased from ideal, as represented through the following:

$$G = \varepsilon_{ap} D_u = \varepsilon_{ap} \frac{4\pi}{\lambda^2} A_p \tag{7-221}$$

where ε_{ap} is aperture efficiency and $0 \le \varepsilon_{ap} \le 1$; see (7-77) and (7-78). Since wavelength and physical aperture area are easily found, the study of gain reduces to one of aperture efficiency, which can be expressed as a product of subefficiencies:

$$\varepsilon_{ap} = e_r \varepsilon_t \varepsilon_s \varepsilon_a \tag{7-222}$$

where

e_r = radiation efficiency
ε_t = aperture taper efficiency
ε_s = spillover efficiency
ε_a = achievement efficiency

We now explain each of these efficiencies.

Aperture taper efficiency is obtained by working with that portion of the power that reaches the aperture. That is, if we ignore achievement and spillover losses, (7-66) for a circular reflector aperture of radius a leads to

$$\varepsilon_t = \frac{1}{\pi a^2} \frac{\left| \int_0^{2\pi} \int_0^a E_a(\rho', \phi') \rho' \, d\rho' \, d\phi' \right|^2}{\int_0^{2\pi} \int_0^a |E_a(\rho', \phi')|^2 \rho' \, d\rho' \, d\phi'} \tag{7-223}$$

This expression can be written directly in terms of the known feed antenna pattern by transforming to feed angles:

$$\varepsilon_t = \frac{4F^2}{\pi a^2} \frac{\left| \int_0^{2\pi} \int_0^{\theta_o} F_f(\theta_f, \phi') \tan \dfrac{\theta_f}{2} \, d\theta_f \, d\phi' \right|^2}{\int_0^{2\pi} \int_0^{\theta_o} |F_f(\theta_f, \phi')|^2 \sin \theta_f \, d\theta_f \, d\phi'} \tag{7-224}$$

The aperture taper efficiency can be evaluated from the feed pattern using this expression.

The feed antenna radiation pattern has the greatest influence on reflector antenna gain by its control over the aperture distribution and aperture taper efficiency, as discussed in Sec. 7.6.2. Since the feed pattern will extend beyond the rim of the reflector, the associated power will not be redirected by the reflector into the main beam and consequently gain is reduced. This is referred to as *spillover* and the associated efficiency factor is called *spillover efficiency* ε_s, which is defined as the fraction of power radiated by the feed that is intercepted by the main reflector of a single reflector or the subreflector of a dual reflector.

Spillover efficiency measures that portion of the feed pattern that is intercepted by the main reflector (and redirected through the aperture into the main beam) relative to the total feed power:

$$\varepsilon_s = \frac{\displaystyle\int_0^{2\pi}\int_0^{\theta_o} |F_f(\theta_f, \phi')|^2 \sin\theta_f \, d\theta_f \, d\phi'}{\displaystyle\int_0^{2\pi}\int_0^{\pi} |F_f(\theta_f, \phi')|^2 \sin\theta_f \, d\theta_f \, d\phi'} \tag{7-225}$$

Notice that the numerator involves an integral over the feed pattern only out to the angular extent of the reflector, whereas the denominator integral extends over the entire feed pattern.

The reflector design problem reduces to a tradeoff between aperture taper and spillover through feed antenna choice. A broad feed pattern introduces little amplitude taper across the aperture, but there will be a significant spillover as illustrated in Fig. 7-38*a*. The spillover problem is solved by using a feed with a narrow pattern as illustrated in Fig. 7-38*b*. However, now the feed pattern taper is large, leading to low aperture taper efficiency.

Taper and spillover efficiencies can be combined to form *illumination efficiency* ε_i to completely account for feed pattern and main reflector effects. That is, ε_i yields total aperture efficiency under ideal circumstances of no ohmic losses ($e_r = 1$) and no achievement losses ($\varepsilon_a = 1$). Multiplying (7-224) and (7-225) and using $a = 2F\tan(\theta_o/2)$ from (7-185a) lead to

$$\varepsilon_i = \varepsilon_t \varepsilon_s = \frac{G_f}{4\pi^2}\cot^2\frac{\theta_o}{2}\left|\int_0^{2\pi}\int_0^{\theta_o} F_f(\theta_f, \phi')\tan\frac{\theta_f}{2}\, d\theta_f\, d\phi'\right|^2 \tag{7-226}$$

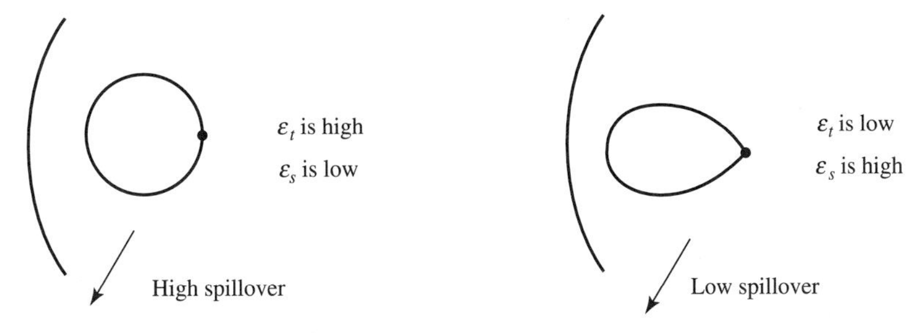

(*a*) Broad feed pattern giving high aperture taper efficiency but low spillover efficiency.

(*b*) Narrow feed pattern giving high spillover efficiency but low aperture taper efficiency.

Figure 7-38 Illustration of the influence of the feed antenna pattern on reflector aperture taper and spillover.

Here, we have made use of the following expression for the gain of the feed:

$$G_f = \frac{4\pi}{\displaystyle\int_0^{2\pi}\int_0^{\pi} |F_f(\theta_f, \phi')|^2 \sin\theta_f \, d\theta_f \, d\phi'} \tag{7-227}$$

This is actually feed directivity, but feed ohmic losses are included in e_r for the entire reflector system.

An ideal, and unrealizable, feed antenna pattern would compensate for spherical spreading loss by increasing with angle off axis and then abruptly falling to zero in the direction of the rim to avoid spillover. This pattern follows from (7-196) and (7-194) as

$$F_f(\theta_f, \phi_f) = \begin{cases} \cos^2\dfrac{\theta_o}{2}\sec^2\dfrac{\theta_f}{2} & \theta_f \le \theta_o \\ 0 & \theta_f > \theta_o \end{cases} \qquad \textit{ideal feed} \tag{7-228}$$

which is normalized to a peak of unity at $\theta_f = \theta_o$. Using this in (7-226) yields an efficiency of $\varepsilon_i = 1$; see Prob. 7.6-12. Thus, the ideal feed pattern of (7-228) will lead to 100% aperture efficiency if no ohmic or achievement losses are present. The ideal feed must, however, be infinitely large in order to produce the discontinuous pattern.

Usually, (7-226) cannot be evaluated analytically and must either be estimated based on canonical distributions or evaluated numerically. However, there is one feed pattern function that is used to model the patterns of real feeds such as conical corrugated horns and that can be handled analytically. This pattern, which is discussed in detail in Sec. 7.7.4, is the rotationally symmetric pattern of

$$F_f(\theta_f) = \begin{cases} \cos^q\theta_f & \theta_f \le \dfrac{\pi}{2} \\ 0 & \theta_f > \dfrac{\pi}{2} \end{cases} \tag{7-229}$$

The evaluation of (7-225) and (7-227) using this feed model is straightforward and yields

$$\varepsilon_s = 1 - \cos^{2q+1}\theta_o \tag{7-230}$$

$$G_f = 2(2q+1) \tag{7-231}$$

The evaluation of ε_t in (7-226) is more difficult. Expressions for ε_i follow for a few q values [1, p. 425]:

$$\varepsilon_i = \cot^2\frac{\theta_o}{2}\cdot\begin{cases} 24\left[\sin^2\dfrac{\theta_o}{2} + \ln\left(\cos\dfrac{\theta_o}{2}\right)\right]^2 & q = 1 \\ 40\left[\sin^4\dfrac{\theta_o}{2} + \ln\left(\cos\dfrac{\theta_o}{2}\right)\right]^2 & q = 2 \\ 14\left[\dfrac{1}{2}\sin^2\theta_o + \dfrac{1}{3}(1-\cos\theta_o)^3 + 2\ln\left(\cos\dfrac{\theta_o}{2}\right)\right]^2 & q = 3 \end{cases} \tag{7-232}$$

The edge illumination EI for this distribution from (7-208a) is

$$EI = \frac{1 + \cos\theta_o}{2}\cos^q\theta_o \tag{7-233}$$

The q-value of 2 is representative of situations encountered in practice. The taper and spillover efficiencies and their product found from (7-230) and (7-232) are plotted in Fig. 7-39. The tradeoff between taper and spillover is evident. The peak value is about $\varepsilon_i = 82\%$ and occurs for an edge illumination of about $EI_{\text{dB}} = -11$ dB. Thus, we arrive at a general rule: *Peak aperture efficiency of a parabolic reflector occurs for an edge illumination of about −11 dB, or* $E_{an}(\rho' = a) = 0.28$. It turns out that the peak illumination efficiency for q-values of 1 to 4 is near 82%; see Prob. 7.6-16. In practice, the highest achievable aperture efficiency for a single reflector using a nearly rotationally symmetric feed pattern is about 75%. If simple feeds such as an open-ended waveguide are used, the aperture efficiency is about 60%. We now examine the remaining efficiencies responsible for gain reduction.

The several factors that reduce gain for practical implementation reasons are lumped together into achievement efficiency, which is expressed using subefficiencies as

$$\varepsilon_a = \varepsilon_{rs}\varepsilon_{cr}\varepsilon_{blk}\varepsilon_{\phi r}\varepsilon_{\phi f} \tag{7-234}$$

where

ε_{rs} = random surface error efficiency
ε_{cr} = cross-polarization efficiency
ε_{blk} = aperture blockage efficiency
$\varepsilon_{\phi r}$ = reflector phase error efficiency
$\varepsilon_{\phi f}$ = feed phase error efficiency

All these efficiencies can range from 0 to 1, but for properly designed systems they are just slightly less than unity. We now discuss them.

Random surface deviations from the ideal shape of a reflector cause gain reduction and side-lobe increase. This is due to the distortions in the aperture phase because of the consequent departure from equal ray path lengths of a focused reflector system. *Random surface error efficiency* ε_{rs} is the efficiency factor associated

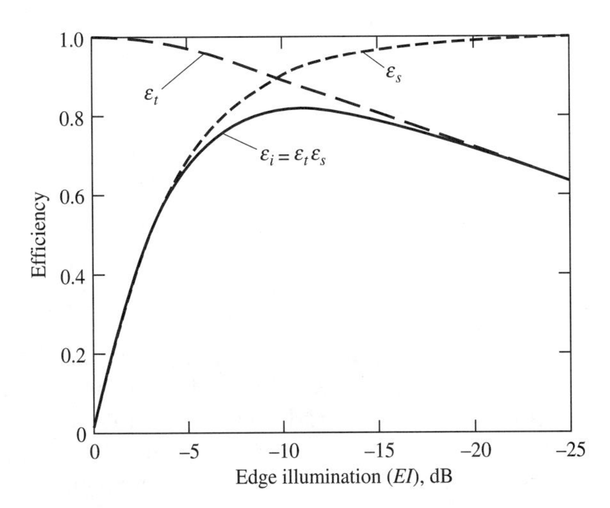

Figure 7-39 Aperture taper ε_t, spillover ε_s, and illumination ε_i efficiencies for a $\cos^2\theta_f$ feed pattern ($q = 2$) as a function of edge illumination EI.

with gain loss from random reflector surface errors. This efficiency can be expressed in terms of the rms surface deviation δ, which is approximately one-third of the peak-to-peak error. For surface errors that are not large and have a correlation length that is small compared to the reflector size,

$$\varepsilon_{rs} = e^{-(4\pi\delta/\lambda)^2} = 685.8(\delta/\lambda)^2 \text{ dB} \tag{7-235}$$

$2\pi/\lambda$ converts the surface errors to phase and the additional factor of 2 accounts for the two-way path of the reflected rays. This result was derived for a flat reflector with Gaussian distributed errors, but it works well in practice [53]. It can be seen from (7-235) that for $\delta \ll \lambda$, the efficiency is nearly 100%. For a fixed random error δ, as frequency increases such that δ varies from 0.01λ and 0.1λ, the efficiency decreases from 0.98 to 0.21. The corresponding gain loss from (7-235) is 0.07 to 6.9 dB. We conclude that random error loss is in transition for δ near 0.01λ. Smaller errors are negligible, whereas larger errors can be a significant problem. The manufacturing techniques for a reflector determine its surface accuracy. Machined metal reflectors are the most accurate with δ near 0.04 mm (0.001 in.). Mass production of reflectors that are a few meters in diameter or less using presses and molds yields slightly larger errors and this accuracy degrades for larger reflectors.

Cross-polarization efficiency ε_{cr} has contributions due to the reflector(s) and the feed antenna. The former is usually small (except for offset reflectors) and is neglected. Feed antennas have a component that is orthogonal to the desired polarization. The associated power ends up in the far field and is wasted—hence, a gain loss. Typical feeds yield ε_{cr} values from 96 to 99% [54], corresponding to gain losses of from 0.2 to 0.04 dB.

Structures placed in front of a reflector such as the feed, subreflector, and support hardware will block rays exiting the aperture and scatter power into the side-lobe region. A simple approximate formula is available for *aperture blockage efficiency* [54]:

$$\varepsilon_{bk} = \left[1 - \frac{1}{\varepsilon_t}\frac{A_b}{A_p}\right]^2 \tag{7-236}$$

where A_b is the blockage area projected onto the physical aperture of area A_p. The square is present because of gain loss due to a decrease of on-axis power by blockage and due to the increase in off-axis power by redirection of the same power into the side lobes. The aperture taper efficiency ε_t is included to weight the central area more heavily where blockage is usually present. For optimum operation, ε_t is about 0.89; see Prob. 7.6-16. Then for blockages of $A_b/A_p = 1$, 5, and 10%, $\varepsilon_{blk} = 0.98$, 0.89, and 0.79, respectively.

Under ideal circumstances, reflector antennas have uniform aperture phase. As with horn antennas, phase errors in the aperture plane lead to gain loss and pattern deterioration [54]. Phase errors arise for the following reasons:

a. ***Displacement of the phase center of the feed antenna off the focal point.*** The reflector is said to be defocused. Lateral displacement causes beam scanning as discussed in Sec. 7.6.5. Often, these errors can be corrected by repositioning the phase center of the feed antenna to the focal point.

b. ***Deterministic deviations of the reflector(s) from design shapes.*** For example, a single reflector that deviates from a paraboloidal shape with a "potato chip" distortion over the entire reflector will produce a smooth phase error over the aperture. Forces such as wind, temperature gradients, and differential gravity effects in addition to manufacturing defects cause deterministic errors whose efficiency is represented with $\varepsilon_{\phi r}$. Only sophisticated techniques such as array feeds are capable of correcting for deterministic errors [55].

c. ***An imperfect feed antenna phase center.*** The loss is represented by $\varepsilon_{\phi f}$ and can be partially compensated by feed repositioning [54].

d. ***Random surface error effects.*** These effects cannot be corrected. The associated efficiency factor ε_{rs} is given in (7-235).

The first three listed effects can be combined into phase-error efficiency ε_{ph}. Since random phase errors are usually the dominant effect, ε_{rs} is shown explicitly in (7-234).

The diffraction effects mentioned in Sec. 7.6.1 also cause gain loss, but are usually small compared to spillover loss [54]. This and other sources of gain loss not specifically mentioned are included in ε_a.

It is important to remember that a reflector antenna usually includes some processing components such as an orthomode transducer (OMT) to separate orthogonal polarizations at the feed. These components are lossy and reduce the gain. Their losses, along with other losses such as radome loss, are all included in e_r. Systems using offset reflectors usually place the upconverter/downconverter hardware immediately behind the feed horn because aperture blockage is not a problem. This greatly reduces the RF transmission line loss compared to an axisymmetric reflector with a transmission line running from the feed to the rear of the reflector.

Although highly approximate, it is helpful to have a "typical" aperture efficiency value for a reflector antenna. For many applications, it can be approximated by

$$\varepsilon_{ap} \approx 0.65 \tag{7-237}$$

7.6.7 Other Reflector Antennas

The principles of reflecting surfaces for focusing have been employed in optical telescopes for several centuries. The reflector antenna, however, did not appear until 1888 when Hertz used a cylindrical parabolic zinc mirror, fed with a dipole along the focal line and connected to a spark-gap generator as shown in Fig. 1-1*a*. Several other scientists investigated reflectors shortly after Hertz's work. But the use of reflector antennas did not fully emerge until shortly before World War II when in 1937 Grote Reber constructed a 9.1-m-diameter prime-focus, reflector antenna for radio astronomy. A more detailed history of reflector antennas is found in [56].

Single and dual parabolic reflectors, as described in this chapter, were developed roughly from World War II through 1960. Since that period, modifications to the basic reflector types have been introduced for the purpose of increasing aperture efficiency or for special antenna pattern-shaping applications to produce a pencil beam, a fan beam, a shaped main beam, low side lobes, or multiple main beams. In this section, we introduce a few of the many types of reflector antennas that are in common use.

A parabolic reflector with a circular perimeter and a simple feed at the focal point as in Fig. 7-27 is used to produce a pencil-beam pattern that is rotationally symmetric. As we have seen, the configuration can be axisymmetric or offset, and subreflectors can be used to form a multiple reflector. There are many applications for a high-gain reflector antenna with different beamwidths in the principal planes. An example is shown in Fig. 7-40*a*, which is a single parabolic reflector with wider horizontal than vertical aperture extent. This produces a narrower main beam in the horizontal plane as needed for VSAT (Very Small Aperture Terminal) satellite communications. The narrow beam is in the geostationary satellite arc to avoid interference between adjacent satellites. The feed antenna must have a broader pattern in the horizontal plane for proper dish illumination. A pattern with different principal plane beamwidths can also be produced with a *parabolic cylinder* as shown in Fig. 7-40*b*, which has a parabolic cross section in one plane and a line cross section parallel to the reflector axis. The narrow beamwidth is in the plane containing the reflector axis and requires a feed that extends along the focal line. The corner reflector antenna discussed in Sec. 5.5.3 is a simplified version of the parabolic cylinder that uses flat metallic sides. The *parabolic torus* of Fig. 7-40*c* is, in a sense, a curved version of the parabolic cylinder, having a parabolic and circular cross sections in the principal planes. A popular application for the parabolic torus employs multiple feeds located along the focal arc to produce separate beams for receiving different satellites with a single earth terminal antenna. Aperture efficiency is sacrificed, but there is a cost savings over using several antennas. The *spherical reflector* of Fig. 7-40*d*, with a circular cross section in all planes containing the reflector axis, produces a pencil beam but with lower aperture efficiency than a parabolic version due to nonuniform aperture phase; equivalently, there is a focal region rather than a focal point. However, the feed can be moved over the focal region to scan the beam with lower gain loss than experienced when displacing a feed from the focal point of a parabolic reflector. The *horn reflector* antenna of Fig. 7-40*e* is formed by joining a horn to an offset parabolic reflector. It is very popular for terrestrial microwave communication links because of its low side and back lobes.

Finally, we mention that *shaped reflectors* are used to produce *shaped beams* for either optimum power distribution in desired directions or to reduce power in directions of interference. Geometrical optics-based techniques are usually used for synthesizing shaped beams [57].

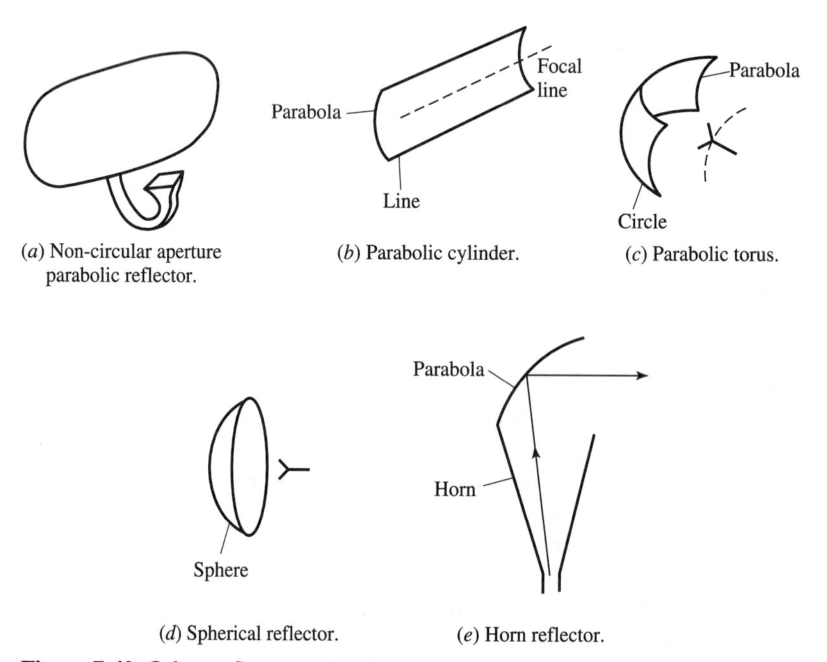

(*a*) Non-circular aperture parabolic reflector.

(*b*) Parabolic cylinder.

(*c*) Parabolic torus.

(*d*) Spherical reflector.

(*e*) Horn reflector.

Figure 7-40 Other reflector antenna types.

7.7 FEED ANTENNAS FOR REFLECTORS

A reflector antenna system feed must be fed properly in order to realize maximum performance, such as high aperture efficiency. The feed is referred to as the primary antenna and the reflector is called the secondary radiator. This section presents design principles and the types of commonly used feed antennas.

7.7.1 Field Representations

The electric field from a feed antenna can be expressed in general, following the geometry of Fig. 7-25, as

$$\mathbf{E}_f = V_o \frac{e^{-j\beta r_f}}{r_f} [U_f(\theta_f, \phi_f)\hat{\boldsymbol{\theta}}_\mathbf{f} + V_f(\theta_f, \phi_f)\hat{\boldsymbol{\phi}}_f] \tag{7-238}$$

Rarely are U_f and V_f known for all angles. Instead, usually only the principal plane patterns are available: $U_f(\theta_f, \phi_f = 0) = C_E(\theta_f)$ in the E-plane and $V_f(\theta_f, \phi_f = 90°) = C_H(\theta_f)$ in the H-plane. Then the field from the feed is found approximately for any angle ϕ_f by interpolation. If the feed is purely linearly polarized, it can be modeled in terms of its principal plane patterns as

$$\mathbf{E}_f = V_o \frac{e^{-j\beta r_f}}{r_f} [C_E(\theta_f) \cos \phi_f \, \hat{\boldsymbol{\theta}}_\mathbf{f} - C_H(\theta_f) \sin \phi_f \, \hat{\boldsymbol{\phi}}_\mathbf{f}] \qquad x_f\text{-polarized} \tag{7-239a}$$

or

$$\mathbf{E}_f = V_o \frac{e^{-j\beta r_f}}{r_f} [C_E(\theta_f) \sin \phi_f \, \hat{\boldsymbol{\theta}}_\mathbf{f} + C_H(\theta_f) \cos \phi_f \, \hat{\boldsymbol{\phi}}_\mathbf{f}] \qquad y_f\text{-polarized} \tag{7-239b}$$

As an example, the E-plane and H-plane of an x_f-polarized short dipole are

$$C_E(\theta_f) = \cos \theta_f, \qquad C_H(\theta_f) = 1 \qquad x_f\text{-polarized short dipole} \tag{7-240}$$

As illustrated in Fig. 7-28 and with (7-206), cross-polarization is present in the aperture of an axisymmetric reflector fed with a dipole antenna and the resulting far-field patterns contain cross-polarization except in the principal planes.

An axisymmetric reflector will have a rotationally symmetric secondary pattern and very low cross-polarization if it is fed with a rotationally symmetric feed pattern:

$$C_E(\theta_f) = C_H(\theta_f) \qquad \text{balanced feed} \tag{7-241}$$

A feed that creates such a pattern is referred to as a *balanced feed*. The field representations of (7-239) for a balanced feed reduce to

$$\begin{aligned}\mathbf{E}_f^v &= V_o \frac{e^{-j\beta r_f}}{r_f} F_f(\theta_f)[\cos \phi_f \, \hat{\boldsymbol{\theta}}_\mathbf{f} - \sin \phi_f \, \hat{\boldsymbol{\phi}}_\mathbf{f}] \\ &= V_o \frac{e^{-j\beta r_f}}{r_f} F_f(\theta_f)\hat{\mathbf{v}} = E_v \hat{\mathbf{v}} \qquad x_f\text{-polarized}\end{aligned} \tag{7-242a}$$

$$\begin{aligned}\mathbf{E}_f^h &= V_o \frac{e^{-j\beta r_f}}{r_f} F_f(\theta_f)[\sin \phi_f \, \hat{\boldsymbol{\theta}}_\mathbf{f} + \cos \phi_f \, \hat{\boldsymbol{\phi}}_\mathbf{f}] \\ &= V_o \frac{e^{-j\beta r_f}}{r_f} F_f(\theta_f)\hat{\mathbf{h}} = E_h \hat{\mathbf{h}} \qquad y_f\text{-polarized}\end{aligned} \tag{7-242b}$$

These correspond to vertical (v) and horizontal (h) feed polarizations with pure linear polarizations in the $x_f z_f$- and $y_f z_f$-planes, respectively. Note that they have a rotationally symmetric pattern $F_f(\theta_f)$. Also, there is no cross-polarization since, for example, with the vertically polarized feed $\mathbf{E}_f^v \cdot \hat{\mathbf{h}} = E_v \hat{\mathbf{v}} \cdot \hat{\mathbf{h}} = 0$.

The aperture electric field for a balanced x_f-polarized feed from (7-205) is

$$\mathbf{E}_a = -\hat{\mathbf{x}} V_o \frac{e^{-j\beta 2F}}{r_f} F_f(\theta_f) \tag{7-243}$$

This corresponds to the GO model of (7-198) and has no cross-polarization. However, there will be a small amount of off-axis cross-polarization in the secondary pattern that is not accounted for here and arising from the axial currents (z-directed) on the surface of the reflector.

7.7.2 Matching the Feed to the Reflector

There are two equivalent viewpoints that can be used to select a feed for proper illumination of a reflector for high aperture efficiency: Matching the feed pattern to the reflector or matching of the feed antenna aperture distribution to the focal field distribution. We discuss these approaches in this section.

As noted previously, the feed pattern is matched to the reflector when its pattern gives about a −11-dB edge illumination. The governing equation for axisymmetric reflectors, (7-208), can be solved using the popular $\cos^q \theta_f$ feed pattern model to determine the required half-power and −10-dB beamwidths. The result is plotted in Fig. 7-41. These curves are very useful in reflector design. Also shown in Fig. 7-41 is the reflector edge angle θ_o from Fig. 7-26*b*.

The focal field matching approach involves a plane wave incident on the reflector. In the limit of infinite frequency, the rays converge to the focal point. In practice, the received fields extend over a finite region near the focal point, resulting in a focal plane distribution (FPD). It turns out that the FPD is approximately the Fourier transform of the aperture plane distribution (APD), with increasing accuracy with larger *F*/*D* [58]. So, the uniform APD created by the incident plane wave leads to a sin(u)/u form FPD and 100% aperture efficiency. The purpose of the feed is to capture the FPD. In fact, if the aperture distribution of the feed antenna placed in the focal plane matches the FPD, the aperture efficiency will be 100%. However, a feed of infinite extent would be required to collect all the fields. This "ideal feed"

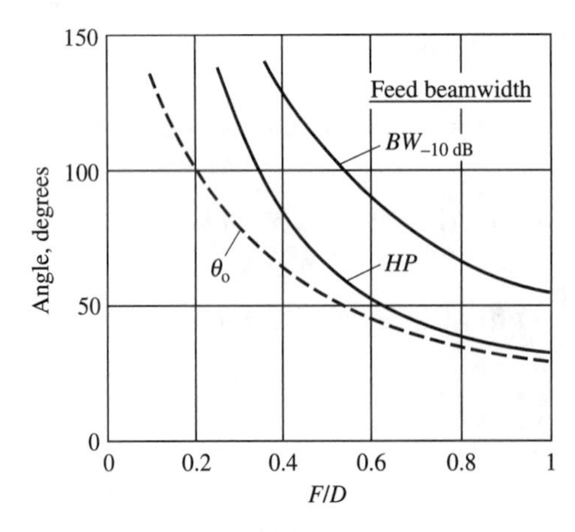

Figure 7-41 Angle from the axis of an axisymmetric parabolic reflector to the rim θ_o (dashed curve) and the required feed pattern beamwidths (solid curves) to produce an −11-dB edge illumination as a function of *F*/*D*.

for an axisymmetric parabolic reflector has a point phase center and pattern that is rotationally symmetric, extending over a cone only out to the reflector rim and compensating for spherical spreading loss. This is expressed functionally in normalized form in (7-228). The pattern discontinuity at the reflector rim ($\theta_f = \theta_o$) gives the required uniform APD and zero outside. This pattern is impossible to achieve. The Fourier transform gives a FPD with infinite extent, requiring an infinite-sized feed to realize.

A classical feed for producing the purely linearly polarized aperture distribution as in (7-243), yielding low cross-polarization in the secondary pattern, is the *Huygen's source.* Its rotationally symmetric pattern leads to high efficiency when feeding an axisymmetric parabolic reflector. The development of the Huygen's source begins by reexamining the aperture fields created by a short dipole feeding a parabolic reflector. The electric fields of an x_f-polarized short dipole in (7-206) have cross-polarized components as indicated in Fig. 7-28. This means that the total aperture electric field has outward curvature as shown in Fig. 7-42*a*. Opposite curvature fields as in Fig. 7-42*b* are created by a y_f-directed magnetic dipole (see Sec. 2.4.2) at the focal point. The combination of crossed electrically-small electric and magnetic dipoles produces the purely linearly polarized field of Fig. 7-42*c*. The Huygen's source aperture fields can be derived using (7-243) for a short dipole and their dual form for a magnetic dipole feed; see Prob. 7.7-4. The magnetic current required to produce an electric field from the magnetic dipole equaling that for the electric dipole is found by equating the components of (1-72b) and (2-44), yielding $I^m = \eta I$. Practical Huygen's sources are discussed in Sec. 7.7.4.

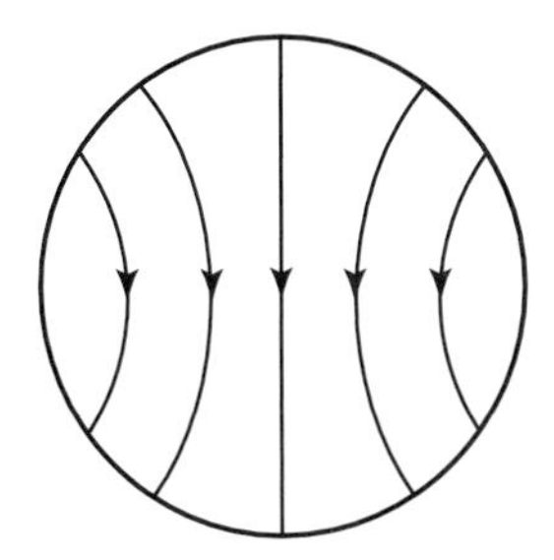

(*a*) Aperture electric field when fed with short dipole.

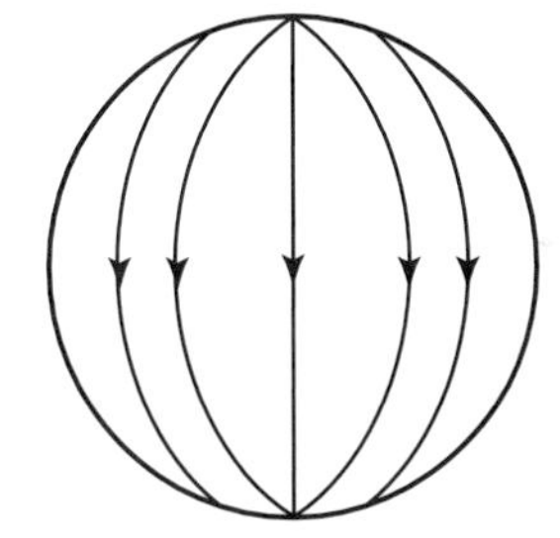

(*b*) Aperture electric field when fed with an electrically small magnetic dipole.

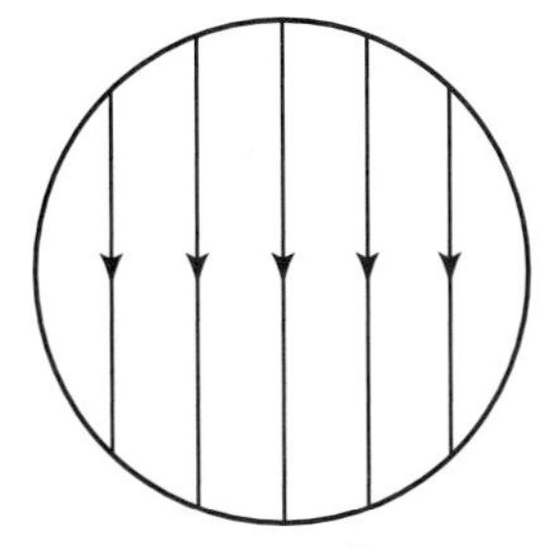

(*c*) Aperture electric field when fed with crossed electric and magnetic dipoles.

Figure 7-42 Aperture electric fields of an axisymmetric parabolic antenna with various feeds.

7.7.3 A General Feed Model

A popular representation for feed patterns is the $\cos^q \theta_f$ pattern given by

$$C_E(\theta_f) = \cos^{q_E} \theta_f, \qquad C_H(\theta_f) = \cos^{q_H} \theta_f, \qquad \theta_f < \pi/2 \tag{7-244}$$

which for a balanced feed reduces to

$$C_E(\theta_f) = C_H(\theta_f) = F_f(\theta_f) = \cos^q \theta_f, \qquad \theta_f < \pi/2 \tag{7-245}$$

The value of q (or q_E and q_H) is chosen to match the pattern of a real feed antenna at one point in addition to the unity beam peak:

$$q = \frac{\log[F_f(\theta_f')]}{\log(\cos \theta_f')} \tag{7-246}$$

where θ_f' is the match point, such as the −3- or −10-dB pattern point or θ_o. An advantage of using the simple pattern form of (7-245) is that it can be used to evaluate important parameters such as the feed antenna directivity:

$$G_f = \frac{2(2q_E + 1)(2q_H + 1)}{q_E + q_H + 1} \tag{7-247}$$

which reduces to $2(2q + 1)$ as in (7-231) for a balanced feed.

We now have all the tools to formulate a simple procedure for designing an axisymmetric reflector using the following steps:

1. ***Determine the reflector diameter.*** The diameter to achieve a required gain is found using (7-77) if an aperture efficiency value can be assumed. If the beamwidth is specified, the diameter is found by solving the following for D:

$$\text{HP} = 1.18 \frac{\lambda}{D} \text{ rad} \tag{7-248}$$

   which is a good approximation for reflectors with a −11-dB edge illumination.
2. ***Choose F/D.*** The normal range of F/D values is 0.3 to 1.0. Higher values lead to better cross-polarization performance, but require a narrower feed pattern and, hence, physically larger feed antenna.
3. ***Determine the required feed pattern.*** The edge illumination is specified for a desired performance and the q-value for a $\cos^q \theta_f$ feed model is found by solving (7-233) for q:

$$q = \frac{\log\left[EI\left(1 + \dfrac{1}{16(F/D)^2}\right)\right]}{\log\left[\cos\left(2 \tan^{-1} \dfrac{1}{4(F/D)}\right)\right]} \tag{7-249}$$

   $EI = 0.28(-11 \text{ dB})$ is used for optimum gain.

The final step in the complete design process is to select a feed antenna that approximates the $\cos^q \theta_f$ pattern with the q-value found from (7-249). The next two subsections address feed design. This subsection is closed with a comprehensive example.

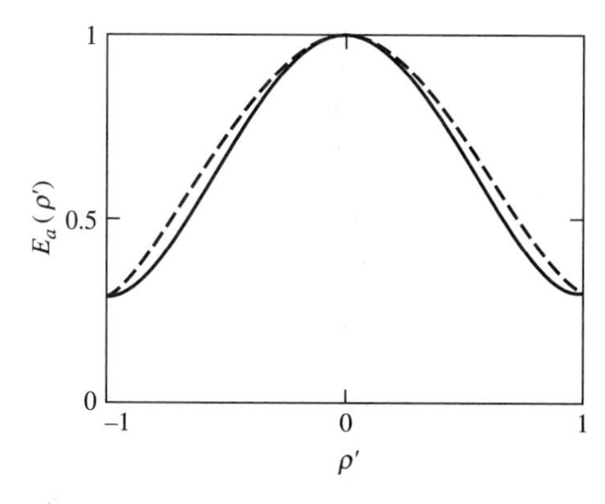

Figure 7-43 Aperture field distribution for axisymmetric parabolic reflector of Example 7-9 (dashed curve) along with the parabolic-squared-on-a-pedestal distribution with $C = 0.28$ (solid curve).

EXAMPLE 7-9 Design of an Axisymmetric Reflector Antenna

Suppose that a high-gain, narrow beam antenna is required at 10 GHz. The axisymmetric parabolic reflector antenna is a good choice. To achieve a 1° half-power beamwidth, the required diameter follows from (7-248) as

$$D = \frac{1.18\lambda}{\text{HP}\,\dfrac{\pi}{180^\circ}} = \frac{1.18(0.03\text{ m})}{1 \cdot \dfrac{\pi}{180^\circ}} = 2.0\text{ m}$$

The F/D is chosen to be 0.5 for low cross-polarization. Solving (7-249) for the optimum case of $EI = 0.28$ gives a value of q near 2. The edge illumination value is verified using (7-208) with $\theta_o = 53.1°$.

$$\begin{aligned} EI &= -FT - L_{\text{sph}} = 20\log(\cos^q \theta_o) + 20\log[(1 + \cos\theta_o)/2] \\ &= -8.86 - 1.93 = -10.79\text{ dB} \approx -11\text{ dB} \end{aligned}$$

The aperture distribution based on (7-208) is plotted in Fig. 7-43 along with a parabolic-squared taper on a pedestal with $C = 0.28$. The agreement suggests that the parabolic-squared tapered circular aperture model works well for reflectors. The illumination efficiency follows from (7-232) for $q = 2$ as $\varepsilon_i = 0.82$. The spillover efficiency from (7-230) is

$$\varepsilon_s = 1 - \cos^{2q+1}\theta_o = 1 - \cos^5 53.1° = 0.92$$

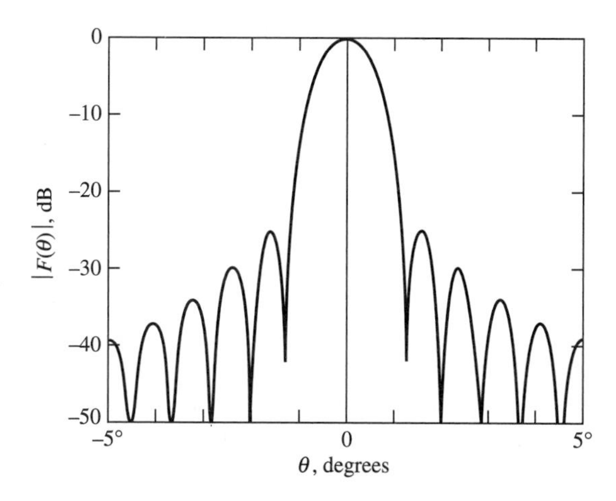

Figure 7-44 Pattern for the 2-m axisymmetric parabolic reflector antenna of Example 7-9 computed using the PRAC code.

So,

$$\varepsilon_t = \varepsilon_i / \varepsilon_s = 0.82/0.92 = 0.89$$

This is consistent with Table 7-1*b* for the parabolic-squared on a −11-dB pedestal distribution. The radiation pattern in the 45° plane computed using PRAC (see Appendix G) is shown in Fig. 7-44. See Prob. 7.7-3 for gain evaluation.

7.7.4 Feed Antennas Used in Practice

The ideal feed of (7-228) produces a uniform amplitude and phase distribution that compensates for spherical spreading loss and does not have spillover. However, it cannot be realized in practice. A practical feed is smaller than a few wavelengths in diameter and has a broad pattern, which can usually be modeled with a $\cos^q \theta_f$ pattern. If high aperture efficiency is desired, a feed is selected that has the following characteristics:

1. The feed pattern should be rotationally symmetric, or balanced, as in (7-245).
2. The feed pattern should be such that the reflector edge illumination is about −11 dB, as discussed in Sec. 7.7.2.
3. The feed should have a point phase center and the phase center should be positioned at the focal point of the reflector.
4. The feed should be small in order to reduce blockage; it is usually on the order of a wavelength in diameter.
5. The feed should have low cross-polarization, usually below −30 dB.
6. The above characteristics should hold over the desired operational frequency band.

Usually, the feed is responsible for limiting the performance of a reflector antenna system.

The simplest feed antenna is a dipole, which is often combined with some type of metallic backing to reduce direct feed radiation in the direction of the reflector main beam. This not only reduces aperture efficiency, but also leads to significant cross-polarization since it is the unbalance in the principal plane patterns that most strongly influences cross-polarization, as seen in (7-205). Dipoles are widely used as feeds for reflectors operating in the UHF range. Aperture efficiency is, however, low. For example, the illumination efficiency of the dipole-fed reflector in Fig. 7-29 is only 24%. At frequencies above a few GHz, waveguide and small horn antennas are used.

The open-ended rectangular waveguide and rectangular horn antennas operating in the dominant TE_{10} mode, discussed in Secs. 7.2 and 7.4, respectively, are used as feeds. Circular waveguides and conical horns operating in the dominant TE_{11} mode are also used as feeds and provide more symmetric principal plane patterns. Next, we discuss these two feed antennas, followed by a discussion of multimode feeds.

The *open-ended circular waveguide* has cross-polarization below −30 dB. It is small in size, with diameters from 0.8 to 1.15λ and −10-dB beamwidths of about 140 to 104°, respectively. The *E*- and *H*-plane beamwidths are not greatly different [32]. Equal principal plane beamwidths occur for a diameter of 0.96λ where $BW_{-10\,dB} = 118°$. This provides a good match to a reflector with $\theta_o \approx 59°$, or $F/D \approx 0.44$. An axisymmetric parabolic reflector with this feed has $\varepsilon_i = 0.74$ [32].

Conical horn antennas behave similarly to pyramidal horn antennas and display optimum gain with 52 to 56% aperture efficiency [59]. The half-power beamwidths under the condition of optimum gain are $HP_E = 1.05\lambda/d_f$ and $HP_H = 1.22\lambda/d_f$ [60], which can be used with the $\cos^q \theta_f$ feed model to perform reflector design.

The simple feeds discussed above operate in their dominant mode (TE_{10} for rectangular and TE_{11} for circular) and have unbalanced principal plane patterns. This is due to the markedly different amplitude distributions that are uniform in the *E*-plane and taper to zero in the *H*-plane. Since the aperture phase errors caused by the spherical phase fronts are strongly frequency-dependent, equal principal plane patterns can be obtained only over a narrow frequency range. Wider bandwidth-balanced feeds with low cross-polarization can be achieved by introducing higher-order modes, leading to a *multimode horn feed.*

There are several forms of multimode feed horns [59]. Here, we consider the most popular form, the *dual mode (conical) horn* or *Potter horn* [61]. The operating principle of the dual mode horn is similar to the Huygens' source of Fig. 7-42. In addition to the dominant TE_{11} mode of the conical horn, a TM_{11} mode is generated internal to the horn that has little effect on the *H*-plane pattern, but with proper amplitude and phase will alter the TE_{11} mode field distribution in the *E*-plane to be nearly like that in the *H*-plane. The electric fields of the separate modes as well as their combination are shown in Fig. 7-45*a*. Note that the modes reinforce in the central region of the feed aperture and cancel around the aperture perimeter, giving the desired circular symmetry and pure linear polarization. Conversion from the TE_{11} mode to the TM_{11} mode can be accomplished with an iris, dielectric ring, flare, or, as shown in Fig. 7-45*b*, with a step. Proper TM_{11} mode amplitude is controlled by the step size and phase is controlled by the distance *d*. The horn diameter must be greater than 1.3λ and has $HP \approx 1.26\lambda/d_f$; thus, $HP < 55°$ and the feed is usually used with large *F*/*D* reflectors [59]. Bandwidths of 10% are typical.

The limited bandwidth of the dual mode horn can be overcome while still achieving axial beam symmetry, low side lobes, and low cross-polarization by using a *hybrid mode feed.* Here, the mixture of TE_{11} and TM_{11} modes occurs in a natural way and propagates with a common phase velocity, forming what is known as a hybrid HE_{11} mode. This leads to bandwidths of 1.6:1 or more. The most popular hybrid mode feed is the *corrugated (conical) horn.* Some variation of the corrugated horn is used as the feed for most of today's microwave reflector antennas. There is no exact formulation for the corrugated horn, but considerable design data are available [59, 62]. The basic principle is to provide the same boundary conditions around the inside of the horn. This is accomplished using corrugations (or grooves

(*a*) Horn aperture electric field distribution.

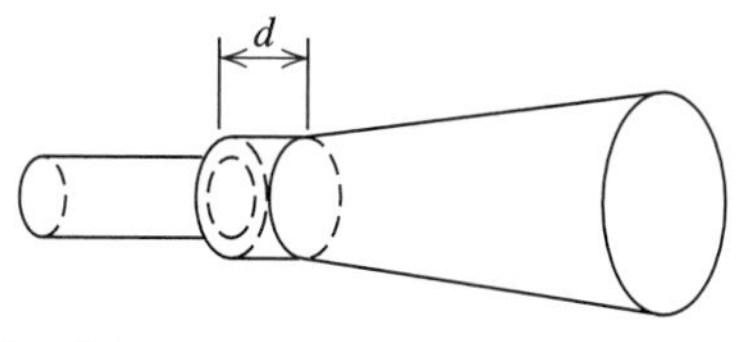

(*b*) A dual mode horn that uses a step to generate the TM_{11} mode.

Figure 7-45 The dual mode feed horn antenna.

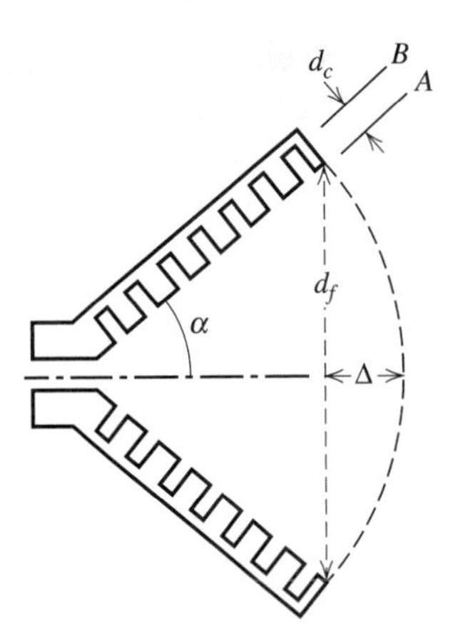

Figure 7-46 The corrugated feed horn antenna.

or teeth) as shown in Fig. 7-46. When the corrugation depths d_c are a quarter-wavelength, the short circuit at the bottom (B) is transformed to an open circuit at the surface (A), choking off current. If there are several corrugations per wavelength, the surface appears to be uniform. The axial current conditions are equivalent to no azimuthal magnetic field. Similarly, an azimuthal electric field is not possible due to the shorting effect of the teeth edges. Thus, all fields decay to zero at the walls, yielding symmetric horn aperture fields and, consequently, a far-field pattern from the horn that is symmetric down to as low as -25 dB. This symmetry along with low side lobes leads to low spillover when the horn is used as a feed.

The corrugated horn in Fig. 7-46 is often called a *scalar horn* because of its field direction independence. Horns with flare angles α from 0 to 90° are used in practice, but the term scalar horn is usually reserved for the large flare angle cases, which we consider further.

The corrugated horn has the desirable feature for feed antennas of a phase center that is stable with frequency. Of course, the phase center of the feed is positioned at the focal point of a reflector system for maximum gain. The phase center of the corrugated horn is at the horn aperture center for small Δ and moves along the axis toward the throat as Δ increases, becoming fixed at the horn apex for $\Delta > 0.7\lambda$ [59, Chap. 15]. It can be shown from the geometry in Fig. 7-46 that

$$\Delta = \frac{d_f}{2} \tan \frac{\alpha}{2} \tag{7-250}$$

A simple model is not available for the patterns of corrugated horns, but curves relating the beamwidth of the feed to its geometry (d_f and α) are available [27]. Also, the following equation can be used to design a corrugated horn feed:

$$\text{BW}_{-12\text{ dB}} \approx 0.8\ \alpha \tag{7-251}$$

REFERENCES

1. S. Silver, Ed., *Microwave Antenna Theory and Design*, MIT Radiation Laboratory Series, Vol. 12, McGraw-Hill, New York, 1949.
2. R. E. Collin and F. J. Zucker, Eds., *Antenna Theory*, Part 1, McGraw-Hill, New York, 1969, p. 73.
3. M. S. Leong, P. S. Kooi, and Chandra, "Radiation from a Flanged Parallel-Plate Waveguide: Solution by Moment Method with Inclusion of Edge Condition," *IEE Proc.*, Part H, Vol. 135, pp. 249–255, Aug. 1988.

4. A. D. Yaghjian, "Approximate Formulas for the Far Field and Gain of Open-Ended Rectangular Waveguide," *IEEE Trans. Ant. & Prop.*, Vol. AP-32, pp. 378–384, April 1984.
5. R. C. Hansen, Ed., *Microwave Scanning Antennas*, Vol. II, Academic Press, New York, 1964, p. 45.
6. A. F. Sciambi, "Instant Antenna Patterns," *Microwaves*, Vol. 5, pp. 48–60, June 1966.
7. R. J. Stegen, "The Gain-Beamwidth Product of an Antenna," *IEEE Trans. Ant. & Prop.*, Vol. AP-12, pp. 505–506, July 1964.
8. C. T. Tai and C. S. Pereira, "An Approximate Formula for Calculating the Directivity of an Antenna," *IEEE Trans. Ant. & Prop.*, Vol. AP-24, pp. 235–236, March 1976.
9. N. A. McDonald, "Approximate Relation between Directivity and Beamwidth for Broadside Collinear Arrays," *IEEE Trans. Ant. & Prop*, Vol. AP-26, pp. 340–341, March 1978.
10. D. Pozar, "Directivity of Omnidirectional Antennas," *IEEE Ant. & Prop. Magazine*, Vol. 35, pp. 50–51, Oct. 1993.
11. A. W. Love, Ed., *Electromagnetic Horn Antennas*, IEEE Press, New York, 1976.
12. C. J. Sletten, Ed., *Reflector and Lens Antennas*, Artech House, Norwood, MA, 1988.
13. R. E. Collin and F. J. Zucker, Eds., *Antenna Theory*, Part 1, McGraw-Hill, New York, 1969, p. 636.
14. M. Abramowitz and I. Stegun, Eds., *Handbook of Mathematical Functions*, NBS Applied Mathematics Series 55, U.S. Printing Office, 1964, Chap. 7.
15. M. J. Maybell and P. S. Simon, "Pyramidal Horn Gain Calculation with Improved Accuracy," *IEEE Trans. Ant. & Prop.*, Vol. 41, pp. 884–889, July 1993.
16. E. V. Jull and L. E. Allan, "Gain on an *E*-Plane Sectoral Horn—A Failure of the Kirchhoff Theory and a New Proposal," *IEEE Trans. Ant. & Prop.*, Vol. AP-22, pp. 221–226, March 1974.
17. E. V. Jull, "Errors in the Predicted Gain of Pyramidal Horns," *IEEE Trans. Ant. & Prop.*, Vol. AP-21, pp. 25–31, Jan. 1973.
18. J. Aurand, "Pyramidal Horns, Part 1: Analysis of Directivity as a Function of Aperture Phase Errors," *IEEE Ant. Prop. Society Newsletter*, Vol. 31, pp. 33–34, June 1989.
19. J. Aurand, "Pyramidal Horns, Part 2: Design of Horns for Any Desired Gain and Aperture Phase Error," *IEEE Ant. & Prop. Society Newsletter*, Vol. 31, pp. 25–27, Aug. 1989.
20. D. Hawkins and F. Thompson, "Modifications to the Theory of Waveguide Horns," *IEE Proc.-H*, Vol. 140, pp. 381–386, Oct. 1993.
21. J. F. Ramsey, "Lambda Functions Describe Antenna/Diffraction Patterns," *Microwaves*, Vol. 6, pp. 69–104, June 1967.
22. C. J. Sletten, Ed., *Reflector and Lens Antennas*, Artech House, Norwood, MA, 1988, Sec. 1.3.
23. D.-W. Duan and Y. Rahmat-Samii, "A Generalized Three-Parameter (3-P) Aperture Distribution for Antenna Applications," *IEEE Trans. Ant. & Prop.*, Vol. 40, pp. 697–713, June 1992.
24. W. V. T. Rusch and P. D. Potter, *Analysis of Reflector Antennas*, Academic Press, New York, 1970.
25. C. Scott, *Modern Methods of Reflector Antenna Analysis and Design*, Artech House: Boston, MA, 1990.
26. W. L. Stutzman, S. W. Gilmore, and S. H. Stewart, "Numerical Evaluation of Radiation Integrals for Reflector Antenna Analysis Including a New Measure of Accuracy," *IEEE Trans. Ant. & Prop.*, Vol. 36, pp. 1018–1023, July 1988.
27. Y. Rahmat-Samii, *Antenna Handbook: Theory, Applications and Design*, edited by Y. T. Lo and S. W. Lee, Van Nostrand Reinhold, New York, 1988, Chap. 15.
28. Y. Rahmat-Samii and V. Galindo-Israel, "Shaped Reflector Antenna Analysis Using the Jacobi–Bessel Series," *IEEE Trans. Ant. & Prop.*, Vol. AP-28, pp. 425–435, July 1980.
29. "GRASP—Single and Dual Reflector Antenna Program Package," TICRA Eng., Copenhagen, Denmark.
30. A. D. Yaghijan, "Equivalence of Surface Current and Aperture Field Integration for Reflector Antennas," *IEEE Trans. Ant. & Prop.*, Vol. AP-32, pp. 1355–1358, Dec. 1984.
31. Y. Rahmat-Samii, "A Comparison between GO/Aperture-Field Integration and Physical-Optics Methods for Offset Reflectors," *IEEE Trans. Ant. & Prop.*, Vol. AP-32, pp. 301–306, March 1984.
32. R. E. Collin, *Antennas and Radiowave Propagation*, McGraw-Hill, New York, 1985, Chap. 4.
33. W. V. T. Rusch, A. Prata, Y. Rahmat-Samii, and R. A. Shore, "Derivation and Application of the Equivalent Paraboloid for Classical Offset Cassegrain and Gregorian Antennas," *IEEE Trans. Ant. & Prop.*, Vol. 38, pp. 1141–1149, Aug. 1990.
34. V. Galindo, "Design of Dual-Reflector Antennas with Arbitrary Phase and Amplitude Distributions," *IEEE Trans. Ant. & Prop.*, Vol. AP-12, pp. 403–408, July 1964.
35. W. F. Williams, "High Efficiency Antenna Reflector," *Microwave J.*, Vol. 8, pp. 79–82, July 1965.
36. R. Mittra and V. Galindo-Israel, "Shaped Dual Reflector Synthesis," *IEEE Ant. & Prop. Soc. Newsletter*, Vol. 22, pp. 5–9, Aug. 1980.
37. P. J. Wood, *Reflector Antenna Analysis and Design*, Peter Peregrinus, London, 1980.

38. P.-S. Kildal, "Synthesis of Multireflector Antennas by Kinematic and Dynamic Ray Tracing," *IEEE Trans. Ant. & Prop.*, Vol. 38, pp. 1587–1599, Oct. 1990.
39. A. Cha, "An Offset Dual Shaped Reflector with 84.5 Percent Efficiency," *IEEE Trans. Ant. & Prop.*, Vol. AP-31, pp. 896–903, Nov. 1983.
40. S. V. Parekh and J. H. Cook, "Reshaped Subreflectors Reduce Antenna Sidelobes," *Microwaves*, Vol. 16, pp. 170–173, May 1977.
41. I. Koffman, "Feed Polarization for Parallel Currents in Reflectors Generated by Conic Sections," *IEEE Trans. Ant. & Prop.*, Vol. AP-14, pp. 37–40, Jan. 1966.
42. M. A. Terada and W. L. Stutzman, "Design of Offset Parabolic Reflector Antennas for Low Cross Polarization and Low Side Lobes," *IEEE Ant. & Prop. Magazine*, Vol. 35, pp. 46–49, Dec. 1993.
43. W. L. Stutzman, *Polarization in Electromagnetic Systems*, Artech House, 1993.
44. M. A. Terada and W. L. Stutzman, "Cross Polarization and Beam Squint in Single and Dual Offset Reflector Antennas," *Electromagnetics*, Vol. 16, pp. 633–650, Nov./Dec., 1996.
45. T.-S. Chu and R. H. Turin, "Depolarization Properties of Offset Reflector Antennas," *IEEE Trans. Ant. & Prop.*, Vol. AP-21, pp. 339–345, May 1973.
46. M. A. Terada and W. L. Stutzman, "Computer-Aided Design of Reflector Antennas," *Microwave J.*, Vol. 38, pp. 64–73, Aug. 1995.
47. D. W. Duan and Y. Rahmat-Samii, "Beam Squint Determination in Conic-Section Reflector Antennas with Circularly Polarized Feeds," *IEEE Trans. Ant. & Prop.*, Vol. 39, pp. 612–619, May 1973.
48. Y. T. Lo, "On the Beam Deviation Factor of a Parabolic Reflector," *IEEE Trans. Ant. & Prop.*, Vol. AP-8, pp. 347–349, May 1960.
49. J. Ruze, "Lateral-Feed Displacement in a Paraboloid," *IEEE Trans. Ant. & Prop.*, Vol. AP-13, pp. 660–665, Sept. 1965.
50. Y. Rahmat-Samii and V. Galindo-Israel, "Scan Performance of Dual Offset Reflector Antennas for Satellite Communications," *Radio Sci*, Vol. 16, pp. 1093–1099, Nov./Dec. 1981.
51. P. C. Werntz, W. L. Stutzman, and K. Takamizawa, "A High-Gain Trireflector Antenna for Beam Scanning," *IEEE Trans. Ant. & Prop.*, Vol. 42, pp. 1205–1214, Sept. 1994.
52. B. Shen and W. L. Stutzman, "A Scanning Spherical Tri-Reflector Antenna with a Moving Flat Mirror," *IEEE Trans. Ant. & Prop.*, Vol. 43, pp. 270–276, March 1995.
53. J. Ruze, "Antenna Tolerance Theory—A Review," *Proc. IEEE*, Vol. 54, pp. 633–640, April 1966.
54. A. W. Rudge et al., Eds., *The Handbook of Antenna Design*, Vol. I, Van Nostrand Reinhold, New York, 1988, Chap. 3.
55. W. T. Smith and W. L. Stutzman, "A Pattern Synthesis Technique for Array Feeds to Improve Radiated Performance of Large Distorted Reflector Antennas," *IEEE Trans. Ant. & Prop.*, Vol. 40, pp. 57–62, Jan. 1992.
56. R. C. Hansen, Ed., *Microwave Scanning Antennas, Vol. I—Apertures*, Academic Press, New York, 1964, Chap. 2.
57. B. S. Wescott, *Shaped Reflector Antenna Design*, Research Studies Press, Letchworth, England, 1983.
58. A. W. Rudge and M. J. Withers, "New Techniques for Beam Steering with Fixed Parabolic Reflectors," *Proc. IEE*, Vol. 118, pp. 857–863, July 1971.
59. R. C. Johnson, Ed., *Antenna Engineering Handbook*, 3rd ed., McGraw-Hill, New York, 1993.
60. A. P. King, "The Radiation Characteristics of Conical Horn Antennas," *Proc. IRE*, Vol. 38, pp. 249–251, March 1950.
61. P. D. Potter, "A New Horn Antenna with Suppressed Side Lobes and Equal Beamwidths," *Microwave J.*, Vol. 6, pp. 71–78, June 1963.
62. P. J. B. Clarricoats and A. D. Olver, *Corrugated Horns for Microwave Antennas*, Peter Peregrinus, London, 1984.

PROBLEMS

7.1-1 An ideal dipole with an infinitesimal current centered along the z-axis has only a θ-component of **E**. If this uniform current element is now rotated to line up with the x-axis, there will be both θ- and ϕ-components. Make the necessary changes to the far-zone **E** expression for the z-directed current case to obtain the far-zone **E** expression for the x-directed current case. Sketch the **E** and **H** field orientations (not the pattern) for the x-directed ideal dipole in the xz- and yz-planes.

7.1-2 Use the principle of duality to derive (7-5) and (7-6) from (7-3) and (7-4), respectively.

7.1-3 Show that (7-24) follows from (7-22) and (7-23).

7.1-4 Show how (7-26) follows from (7-6).
7.1-5 Show how (7-24) follows from (7-17).
7.1-6 If the incident field in Example 7-1 is x-polarized, write radiation field E_ϕ and the pattern $F(\theta)$. Your answer will be that of (4-17) with a coordinate change. Is the appropriate boundary condition for E_ϕ satisfied on the conducting plane?
7.1-7 Derive (5-73).
7.2-1 Derive (7-34).
7.2-2 Use geometric arguments to prove that $du\, dv = \cos\theta\, d\Omega$, where $d\Omega = \sin\theta\, d\theta\, d\phi$; that is, sketch a hemisphere and project the intersection of the differential $d\Omega$ with the hemisphere onto the uv-plane.
7.2-3 For the open-ended waveguide of Example 7-3:

a. Numerically evaluate the pattern expressions to verify the results in Fig. 7-10. Plot the two computed patterns.

b. Numerically evaluate and plot the H-plane patterns of (7-57a) and (7-58a). (Experimental data are available in [4].)

c. Find the HPs for the free-space case and compare to that computed using (4-28) and (7-42). Explain the differences.

7.2-4 The open-ended waveguide of Fig. 7-9 can be more accurately modeled by including the reflection coefficient at the aperture Γ and the waveguide phase constant $\beta_g = 2\pi/\lambda_g$, where $\lambda_g = \lambda/\sqrt{1 - (\lambda/2a)^2}$. The aperture fields are given by [4]

$$E_y = E_o(1 + \Gamma)\cos\frac{\pi x}{a}\, e^{-j\beta_g z} \qquad \text{and} \qquad H_x = -\frac{E_o}{Z_o}(1 - \Gamma)\cos\frac{\pi x}{a}\, e^{-j\beta_g z}$$

where $Z_o = \omega\mu/\beta_g$.

a. Derive complete expressions for the far-field electric field components. Use the magnetic and electric current equivalent current formulation.

b. Write the normalized E- and H-plane pattern expressions.

c. Write the normalized obliquity factor in the principal planes for the case when the waveguide is matched ($\Gamma = 0$) and the guide phase velocity is that of free space.

7.3-1 Prove (7-63).
7.3-2 Prove that the uniform amplitude aperture excitation yields the highest directivity of all uniform phase excitations. *Hint:* Use (7-66) and the Schwarz inequality

$$\left|\iint_S fg\, dS'\right|^2 \le \iint_S f^2\, dS' \iint_S g^2\, dS'$$

for any functions f and g. Let $g = 1$ and f equal the aperture field.
7.3-3 Show that the aperture taper efficiency is $\frac{2}{3}$ for a rectangular aperture with a uniform amplitude distribution in one direction and a cosine-squared distribution in the other.
7.3-4 A rectangular aperture (L_x by L_y) has a field distribution that is cosine-tapered in both the x- and y-directions. Derive the directivity expression. What is the aperture taper efficiency?
7.3-5 Compute the directivity in decibels for a rectangular aperture with $L_x = 10\lambda$ and $L_y = 20\lambda$ for (a) a completely uniform aperture illumination and (b) a cosine amplitude taper in one direction and a uniform taper in the other aperture direction.
7.3-6 Evaluate the aperture taper efficiency for a triangular tapered, rectangular aperture distribution where

$$\mathbf{E}_a(x', y') = \left[1 - \frac{2|x'|}{L_x}\right]\hat{\mathbf{x}}, \qquad |x'| \le \frac{L_x}{2}, \qquad |y'| \le \frac{L_y}{2}$$

7.3-7 Strictly speaking, is the uniform aperture distribution physically realizable? Why?
7.3-8 An antenna operating at 150 MHz has a physical aperture area of 100 m^2, a gain of 23 dB, and a directivity of 23.5 dB. Compute (a) effective aperture A_e, (b) maximum effective

aperture A_{em}, (c) aperture efficiency ε_{ap}, (d) radiation efficiency e_r, and (e) aperture taper efficiency ε_t.

7.3-9 Verify the last of (7-94).

7.3-10 The general antenna gain relation of (7-77) includes a frequency-squared dependence. However, the class of frequency-independent antennas displays nearly constant gain over large frequency variations. Explain this apparent paradox.

7.3-11 A horn antenna with a 185.5 × 137.4 cm rectangular aperture has the following measured parameter values at 0.44 GHz: $\text{HP}_{E^\circ} = 30°$, $\text{HP}_{H^\circ} = 27°$, and $G = 15.5$ dB.

a. Compute the aperture efficiency.

b. Estimate the gain from the measured half-power beamwidths.

7.3-12 Repeat Prob. 7.3-11 for a horn with a 28.85 × 21.39 cm aperture and $\text{HP}_{E^\circ} = 12°$, $\text{HP}_{H^\circ} = 13°$, and $G = 22.1$ dB at 6.3 GHz.

7.3-13 A 3.66-m (12-ft) diameter circular parabolic reflector operates at 460 MHz. The measured parameters of this antenna are $G = 22.2$ dB and $\text{HP}_{E^\circ} = \text{HP}_{H^\circ} = 12.5°$. Estimate the gain using both (7-97) and (7-95).

7.3-14 Estimate the gain of a circular parabolic reflector operating at 28.56 GHz in two ways:

a. Using only its size, which is 1.22 m (4 ft) in diameter.

b. Using only the measured half-power beamwidths, which are $\text{HP}_{E^\circ} = 0.605°$ and $\text{HP}_{H^\circ} = 0.556°$.

7.3-15 a. Prove (7-86) for a separable distribution using (7-72) and Prob. 4.2-11.

b. Using $\varepsilon_{ap} = \varepsilon_{apx}\varepsilon_{apy}$ and assuming $e_r = 1$, write expressions for ε_{apx} and ε_{apy}.

c. For a general aperture distribution, show that

$$\varepsilon_t = \frac{1}{A_p} \frac{\left[\iint |E_a|\, dS\right]^2}{\iint |E_a|^2\, dS}$$

7.3-16 Show that the directivity-beamwidth product for a uniform phase rectangular aperture with a cosine amplitude taper in the H-plane and uniform amplitude in the E-plane is 35,230 deg^2.

7.3-17 A geostationary satellite is 42,000 km from the center of the earth. If the −3-dB pattern points fall near the edge of the earth, find an approximate value for the spacecraft antenna gain. Note that the result is independent of frequency.

7.3-18 A Gaussian power pattern of half-power beamwidth HP in degrees is

$$P_n(\theta) = e^{-(4 \ln 2)(\theta/\text{HP})^2}$$

Derive the following approximate directivity expression for narrow beam Gaussian patterns:

$$D \approx \frac{36{,}407}{\text{HP}^2}$$

7.3-19 Horn antennas used as feeds for reflectors have patterns that are well approximated by

$$F(\theta) = \cos^q(\theta), \qquad 0 \leq \theta \leq \pi/2$$

a. Derive the directivity expression $D = 2(2q + 1)$.

b. Compare directivity values computed using (7-92) and (7-94) for $q = 0, 1, 5, 10$, and 50; tabulate the results.

7.4-1 Derive the expression for R_H in (7-100).

7.4-2 Derive the H-plane sectoral horn radiation field expression (7-107) and (7-108) by changing to complex exponentials and then completing the square in the exponents in the integrand.

7.4-3 In the H-plane pattern expression of (7-117) and (7-118) for an H-plane sectoral horn:

a. Show that s_1 follows from s_1' of (7-109).

b. Show that the phase term $(\pi/8t)[(A/\lambda)\sin\theta + 1/2]^2$ follows from the corresponding term in (7-108).

7.4-4 Derive the directivity formula of (7-120) for an H-plane sectoral horn from (7-66). The numerator in (7-66) can be evaluated using (7-106) through (7-108).

7.4-5 Use a computer program to evaluate Fresnel integrals. Compute $C(x)$ and $S(x)$ for $x = 0, 1, 2, 3, 4,$ and 5. Tabulate the values along with those from a math table, giving the deviation from the known values.

7.4-6 The H-plane pattern for an H-plane sectoral horn arises from the first integral in (7-106).

a. First evaluate this integral for a no phase error condition.

b. Show that the on-axis value of the H-plane pattern relative to the on-axis value of the zero phase error case is given by

$$\frac{\pi}{16\sqrt{t}}\, I(\theta = 0°, \phi = 0°)$$

c. Evaluate this for $t = \frac{1}{8}, \frac{1}{4}, \frac{3}{8}$, and $\frac{1}{2}$ and compare to the values from Fig. 7-13.

7.4-7 An H-plane sectoral horn antenna has an axial length of 5λ and a flare half-angle α_H of 12.6°.

a. Plot the polar plot of the H-plane radiation pattern in decibels.

b. Compute the directivity function $\lambda D_H/b$ using (7-120) and compare to that obtained from Fig. 7-14.

c. Since the aperture is not large relative to a wavelength, use the zero phase error directivity formula of (7-71) to compute $\lambda D_H/b$.

7.4-8 Design an optimum H-plane sectoral horn antenna with 12.15-dB gain at 10 GHz. It is fed with a WR90 waveguide.

a. Find the horn dimensions employing Fig. 7-14.

b. Draw the H-plane horn geometry to scale.

c. Use (7-120) to compute the directivity as a check.

7.4-9 Repeat Prob. 7.4-8(a), except use (7-120) and (7-121) instead of Fig. 7-14.

7.4-10 Derive the E-plane sectoral horn far-zone electric field expression of (7-129).

7.4-11 Show how the E-plane pattern magnitude expression for an E-plane sectoral horn of (7-132) follows from (7-129).

7.4-12 Use physical reasoning to explain why the phase error parameters of optimum E-plane and H-plane sectoral horns are different.

7.4-13 An E-plane sectoral horn antenna is attached to a WR90 waveguide. Determine the horn dimensions for a half-power beamwidth of 11° in the E-plane and an optimum gain of 14.9 dB at 10 GHz.

7.4-14 An E-plane sectoral horn has an E-plane aperture height of 24.0 cm and a half-flare angle of 16.5°. It is attached to a WR284 waveguide. Compute the gain at 3.75 GHz (a) using (7-134a) and (b) using (7-139).

7.4-15 Derive (7-141) by starting with (7-77) and using $\varepsilon_{ap} = \varepsilon_t \varepsilon_{ap}^E \varepsilon_{ap}^H$.

7.4-16 Start with $A = 18.61$ cm and verify all horn dimensions given in Example 7-7.

7.4-17 The aperture efficiency in Fig. 7-20 for the pyramidal horn of Example 7-7 is based on the aperture quadratic phase error approximation.

a. Find the aperture efficiency by the direct evaluation of (7-120c) and (7-134c) from 8 to 13 GHz to verify Fig. 7-20. Then repeat, using the exact phase errors (7-123) and (7-137). Compare these results.

b. Evaluate aperture efficiency at 8, 10, and 13 GHz using the approximate formulas in (7-149) and (7-150). Compare to results from (a); tabulate values.

7.4-18 Explain why an optimum horn is designed for about 50% aperture efficiency at a frequency near the low end of its operating band as in Example 7-7.

7.4-19 Design an optimum gain pyramidal horn antenna connected to a WR90 waveguide with 20-dB gain at 10 GHz. (a) Give all horn dimensions, and (b) evaluate the directivity at 10 GHz using the exact phase errors.

7.4-20 A manufactured standard gain horn antenna operates from 18 to 26.5 GHz and has a WR42 waveguide input. The gain is 24.7 dB at 24 GHz.

a. Use optimum gain design principles to determine the horn geometry values in centimeters.

b. Plot the E- and H-plane polar patterns in decibels including the $(1 + \cos\theta)/2$ factor.

c. Determine the half-power beamwidths from pattern calculations and compare to the simple formula values.

d. Evaluate the gain at the design frequency using the exact phase errors. Give the aperture efficiency value.

e. Compare the gain to that calculated using the approximation in (7-95) for both beamwidth values found in (c).

7.4-21 Repeat Prob. 7.4-20 for a pyramidal horn designed for optimum operation at 1 GHz and a gain of 15.45 dB. It is connected to a WR975 waveguide with $a = 9.75$ in. $= 24.765$ cm and $b = 4.875$ in. $= 12.3825$ cm.

7.4-22 For Example 7-7, compute the half-power beamwidths using line source models from Chap. 4 for the same amplitude tapers as in the horn aperture. Explain why there are deviations from the values in Example 7-7.

7.4-23 Derive the following relationship that must be satisfied for a physical realizable pyramidal horn antenna:

$$\frac{R_1}{\lambda} = \frac{1 - \dfrac{b/\lambda}{B/\lambda}}{1 - \dfrac{a/\lambda}{A/\lambda}} \frac{R_2}{\lambda}$$

7.4-24 *A square main beam horn antenna.* It is often desirable to have equal principal plane half-power beamwidths. This problem develops a design technique for a so-called square main beam pyramidal horn. If optimum design techniques under the condition of a square main beam are used, the resulting horn dimensions will render a horn that cannot be constructed. To avoid this problem, we can design for a square main beam and aim for *near* optimum conditions. To do this we first determine the aperture dimensions that give the desired beamwidths and optimum operation. Then the axial lengths are adjusted to provide a physically realizable structure. This will probably not move the operating point too far from optimum. Follow this procedure to design a square main beam horn at 8 GHz with 12° beamwidths and fed by a WR90 waveguide.

a. Determine A/λ and B/λ.

b. Use the results of Prob. 7.4-23 for adjusting the axial lengths. Do this to keep the fractional increase or decrease of both the same, that is, use

$$\frac{R_1}{\lambda} = \frac{R_{1\text{op}}}{\lambda} f \qquad \text{and} \qquad \frac{R_2}{\lambda} = \frac{R_{2\text{op}}}{\lambda} \frac{1}{f}$$

and solve for the constant f.

c. Evaluate the final phase error parameters t and s.

d. Give the horn dimensions in centimeters.

e. Evaluate the gain.

f. Compute the aperture efficiency.

7.5-1 Write the radiated electric field expression analogous to (7-168) using the equivalent current formulation that includes both electric and magnetic surface current densities.

7.5-2 Verify that the uniform circular aperture pattern of (7-170) is unity for $\theta = 0$.

7.5-3 Derive the pattern expression $f(\theta, n, C)$ in Table 7-1*b* for a parabolic taper on a pedestal.

7.5-4 For a parabolic-on-a-pedestal circular aperture distribution, (a) derive (7-181) using (7-66), and (b) evaluate ε_t for $n = 1$ and 2 for a -10-dB edge taper.

7.5-5 For a tapered circular aperture, (a) prove that $c = c_x c_y = \sqrt{\pi \varepsilon_t}/2$, and (b) show by examining the values in Table 7-1*b* that $\text{DB}_{\text{cir}} \approx 39{,}000$ deg^2.

7.6-1 Derive (7-199a).

7.6-2 By vector diagram sketches, show that the components of the incident and reflected electric fields tangent to a parabolic reflector cancel and that (7-200) holds.

7.6-3 Plot the edge illumination in decibels for a circular parabolic reflector due to spherical spreading loss only (i.e., the feed is isotropic) as a function of F/D from 0 and 1.

7.6-4 A commercially available parabolic reflector antenna operating at 2.1 GHz has an aperture diameter of 1.83 m (6 ft). Compute the gain in decibels.

7.6-5 A commercially available parabolic reflector antenna operating at 11.2 GHz has an aperture diameter of 3.66 m (12 ft). Compute the gain in decibels.

7.6-6 Analyze the reflector in Example 7-8 using a reflector computer code (see Appendix G). Model the feed using a $\cos^q \theta_f$ pattern with a 10-dB beamwidth of 104°. Tabulate the values for HP, SLL, G, and ε_{ap}. Include values for the canonical distribution approach. Plot the pattern in decibels.

7.6-7 A commercial axisymmetric reflector antenna used for Ku-band satellite reception (11.95 GHz midband) is 2.4 m in diameter and has $F/D = 0.37$. Assuming a $\cos^q \theta_f$ feed pattern, (a) use a canonical aperture distribution to determine reflector performance, and (b) use a reflector code (see Appendix G) to evaluate performance. Tabulate results from (a) and (b) including G, HP, and SLL.

7.6-8 Use a reflector code (see Appendix G) to determine the following performance parameters for an offset parabolic reflector with $D = 100\lambda$, $H = 70\lambda$, $F/D_p = 0.466$, $\psi_f = 34.72°$, and a $\cos^q \theta_f$ feed with $q = 13.0897$: (a) gain, (b) SLL, (c), XPOL peak location, (d) XPOL peak value in decibels relative to the main beam peak. (e) Plot the pattern in decibels out to 3° in the plane normal to the offset plane (i.e., $\phi = 90°$).

7.6-9 A popular commercial offset parabolic reflector antenna for receiving direct broadcast television (12.45-GHz midband) has the following geometric parameters: $D = 45.70$ cm (18 in.), $F = 26.23$ cm, $D_p = 94.00$ cm, and $H = 24.15$ cm. The beamwidth between -10dB points on the feed pattern is 80.8° and the peak of the feed pattern is aimed 49.5° from the reflector axis. Use a computer program to evaluate the radiation pattern in the principal planes. Tabulate the values of G, HPs, SLLs and XPOLs.

7.6-10 Derive (7-224) and (7-226).

7.6-11 Derive (7-230) and (7-231).

7.6-12 Prove that the ideal feed of (7-228) produces 100% aperture efficiency.

7.6-13 (a) Derive an expression for the aperture efficiency of an axisymmetric reflector fed with an isotropic feed antenna, and (b) evaluate for $F/D = 0.25$, 0.5, and 1.

7.6-14 Derive the illumination efficiency expression (7-232) for a $\cos^q \theta_f$ feed pattern for $q = 2$.

7.6-15 A geostationary satellite transmits at 4 GHz using a parabolic reflector antenna. The peak of the beam is directed toward the center of the earth disk and the -3dB pattern points fall on the edge of the earth. Find the gain in decibels and the diameter of the spacecraft antenna in meters. (Earth radius = 6,400 km; distance from the center of Earth to orbit = 42,000 km.)

7.6-16 This problem serves to verify the claim that the -11-dB edge illumination yields about $\varepsilon_i \approx 0.82$ under a variety of axisymmetric reflector system cases. For $\cos^2 \theta_f$ feed patterns as in (7-229) and values of $q = 1, 2$, and 3, find the F/D value of the optimum gain axisymmetric reflector. Tabulate the following for each q value: F/D, $2\theta_o$, feed $\text{BW}_{-10\,\text{dB}}$, ε_s, ε_i found using (7-232), and ε_t.

7.6-17 Compute the blockage efficiency for a reflector of optimum gain for $A_b/A_p = 0.1$, 1, 2, 5, and 10%.

7.6-18 A subreflector in a Cassegrain dual reflector has a diameter that is 10% of the main reflector diameter. Find the aperture blockage efficiency assuming optimum operation.

7.7-1 Find the half-power and -10-dB beamwidths of a $\cos^q \theta_f$ feed pattern required to produce an edge illumination of -11 dB in an axisymmetric reflector with $F/D = 0.4$. Give the value of q.

7.7-2 Plot the aperture electric field amplitude distribution for an axisymmetric reflector with

$F/D = 0.3$ and a $\cos^2 \theta_f$ feed pattern. Also, show on the same plot the parabolic-squared aperture distribution with the same edge illumination.

7.7-3 For the reflector of Example 7-9, (a) calculate the gain in decibels using aperture efficiency, and (b) use a reflector code to find the gain and compare to the value from (a).

7.7-4 Derive the aperture electric field expression for the Huygens' source of Fig. 7-42.

7.7-5 A commercial offset parabolic reflector antenna with a diameter of 1.8 m is used for Ku-band satellite communications. It is just fully offset (that is, $h = 0$) and $F/D_p = 0.305$. The feed has a -10-dB beamwidth of 76.8°. For the middle of the transmit band at 14.25 GHz, (a) determine the feed pointing angle that produces nearly equal edge illumination at the upper and lower reflector edges. (b) Use a reflector code to evaluate the reflector performance. (c) Find a canonical distribution that approximates the aperture distribution. Tabulate values from (b) and (c) for as many of the following parameters as possible: G, HP, SLL, XPOL, and ε_{ap}.

7.7-6 An optimum gain conical horn is used to feed an axisymmetric parabolic reflector with $F/D = 0.44$. Using HP $\approx 1.14 d_f/\lambda$ as an average beamwidth expression, find the d_f/λ value for maximum illumination efficiency.

7.7-7 Derive (7-247).

Chapter 8

Antenna Synthesis

Thus far in this book, attention has been focused on antenna analysis and design. The analysis problem is one of determining the radiation pattern and impedance of a given antenna structure. Antenna design is the determination of the hardware characteristics (lengths, angles, etc.) for a specific antenna to produce a desired pattern and/or impedance. Antenna synthesis is similar to antenna design and, in fact, the terms are occasionally used interchangeably. However, antenna synthesis, in its broadest sense, is one of first specifying the desired radiation pattern and then using a systematic method or combination of methods to arrive at an antenna configuration that produces a pattern which acceptably approximates the desired pattern, as well as satisfying other system constraints. Hence, antenna synthesis, in general, does not depend on an *a priori* selection of the antenna type. Unfortunately, there is no single synthesis method that yields the "optimum" antenna for the given system specifications. There are, however, several synthesis methods for different classes of antenna types. In this chapter, we discuss the more useful synthesis methods in current use. The discussion serves as an introduction to the topic of synthesis and should provide a foundation for studying more advanced treatments [1–3].

8.1 THE SYNTHESIS PROBLEM

8.1.1 Formulation of the Synthesis Problem

We will pose the antenna synthesis problem as one of determining the excitation of a given antenna type that leads to a radiation pattern which suitably approximates a desired pattern. The desired pattern can vary widely depending on the application and has the variables are listed in Table 8-1. To illustrate, consider a communication satellite in synchronous orbit that is required to generate separate antenna beams for the western United States and for Alaska. Two main beams are required, both shaped for nearly uniform illumination of each region. Also low side lobes may be specified to minimize interference over other regions of the earth, but higher side lobes could be permitted for directions not toward the earth. This type of pattern has multiple shaped main beams and a shaped side-lobe envelope.

The antenna itself can take many forms as listed in Table 8-1. Antenna type refers to the geometry of the antenna and consists of continuity, shape, and size. The performance of an antenna is used to define the antenna classes in Fig. 1-6 and includes the performance parameters listed in Table 1-1. Performance parameters other than pattern shape can be included in the synthesis problem specifications. In

Table 8-1 Antenna Synthesis Variables

Antenna Type Variables	Radiation Pattern Variables
Continuity	Main beam region
Continuous	Narrow main beam
Discrete—array	Single beam
Shape	Multiple beams
Linear	Shaped beam
Planar	Side-lobe region
Conformal	Nominal side lobes
Three-dimensional	Low side lobes
Size	Shaped side lobes

this chapter, we consider the problem of pattern synthesis. The remaining performance parameters are considered elsewhere in the book. A general synthesis procedure would yield the antenna type and its excitation that produces the best approximation to specified performance values including the desired pattern shape. No such general synthesis method exists. Instead, synthesis methods have been developed for each antenna type. The discussion of synthesis in this chapter is divided between continuous and discrete (array) antenna types. Before addressing these methods, we present further general remarks.

If the radiation electric field components E_θ and E_ϕ are specified in the synthesis problem, a secondary synthesis problem can be formulated in terms of antenna aperture field transform components. For example, the aperture magnetic equivalent surface current solution of (7-26) can be solved, giving

$$\begin{bmatrix} P_x \\ P_y \end{bmatrix} = \left(j\beta \frac{e^{-j\beta r}}{2\pi r} \right)^{-1} \begin{bmatrix} \cos\phi & \sin\phi \\ -\cos\theta\sin\phi & \cos\theta\cos\phi \end{bmatrix}^{-1} \begin{bmatrix} E_\theta \\ E_\phi \end{bmatrix} \tag{8-1}$$

This can be used to obtain P_x and P_y from specified functions E_θ and E_ϕ. The problem is then of synthesizing desired functions P_x and P_y, which are Fourier transforms of the aperture electric field components; see (7-18). The process is similar for each of P_x and P_y. Therefore, we let $f(\theta, \phi)$ be the normalized pattern factor for either and frame our discussions using $f(\theta, \phi)$. As another example, consider a line source along the z-axis. If $F_d(\theta)$ is the normalized desired radiation pattern, then the desired pattern factor is

$$f_d(\theta) = \frac{F_d(\theta)}{\sin\theta} \tag{8-2}$$

This chapter discusses synthesis of the pattern function $f(\theta, \phi)$ that provides an approximation to the desired pattern $f_d(\theta, \phi)$. The pattern synthesis techniques will be presented for one-dimensional formulations with a geometry yielding $f(\theta)$. That is, the continuous form (line sources) and the discrete form (linear arrays) will be treated. However, these results can be applied to two-dimensional antennas such as planar aperture and planar array antennas. Direct application of the methods is possible if the two-dimensional aperture distribution is separable (see Sec. 7.2.2). Then the synthesized pattern function f is used to represent each principal plane pattern. Synthesis methods can be separated by antenna or pattern type. Only a few methods exist that can be applied to a variety of antenna and pattern types [4]. Usually, synthesis methods for shaped beam patterns are completely different from those for low side-lobe, narrow beam patterns, so we will separate the methods by pattern type. Line source and linear array synthesis principles with applications to shaped beam patterns are detailed in Secs. 8.2 and 8.3. Low side-lobe, narrow main beam methods are presented in Sec. 8.4.

8.1.2 Synthesis Principles

The radiation electric field from a line source of current (actual or equivalent) along the z-axis and of length L is given by (4-1) for the geometry of Fig. 1-14*a*. For synthesis problems, we are only interested in the relative pattern variations. Furthermore, the element factor $g(\theta) = \sin\theta$ is accounted for separately; for narrow-beam, broadside line sources, it is negligible. The normalized pattern factor of a line source follows from (4-31) as[1]

$$f(\theta) = \frac{1}{\lambda}\int_{-L/2}^{L/2} i(z)e^{j\beta z\cos\theta}\,dz \tag{8-3}$$

where $i(z)$ is the normalized form of the current function $I(z)$, and it is usually normalized such that (8-3) produces a pattern $f(\theta)$ that is unity at its maximum. The linear phase shift that scans the main beam is contained in $i(z)$; for example, see (4-3). For convenience, we define

$$w = \cos\theta \qquad \text{and} \qquad s = \frac{z}{\lambda} \tag{8-4}$$

and w is related to u in Chap. 4 through $u = (\beta L/2)w$. Then (8-3) becomes

$$f(w) = \int_{-L/2\lambda}^{L/2\lambda} i(s)e^{j2\pi ws}\,ds \tag{8-5}$$

This equation forms the relationship between the relative current distribution $i(s)$ and the normalized pattern factor $f(w)$.

Since the current distribution $i(z)$ extends only over the length L, [that is, $i(s)$ is zero for $|s| > L/2\lambda$], the limits of the integral in (8-5) can be extended to infinity, giving

$$f(w) = \int_{-\infty}^{\infty} i(s)e^{j2\pi ws}\,ds \tag{8-6}$$

This is recognized as a Fourier transform. The corresponding inverse Fourier transform is

$$i(s) = \int_{-\infty}^{\infty} f(w)e^{-j2\pi sw}\,dw \tag{8-7}$$

(See Prob. 8.1-1).

It is important to understand the requirements on a current distribution in order to achieve a pattern shape. This is useful in synthesis and in finding explanations for pattern abnormalities. The (linear) Fourier transform relationship between the current and pattern developed in Sec. 4.3 can be used to infer the general properties shown in Table 8-2. These principles also apply to arrays.

The current distribution and the pattern functions can be described mathemati-

---

[1]Frequently, the z-axis is selected to be normal to the line source, in which case $\cos\theta$ in (8-3) becomes $\sin\theta$.

Table 8-2 Symmetry Properties of Current Distributions and Patterns

Definitions:

$i(s) = i_r(s) + ji_i(s)$	Real and imaginary
$i(s) = \lvert i(s)\rvert e^{j\phi(s)} = A(s)e^{j\phi(s)}$	Amplitude and phase
$i(s) = i_e(s) + i_o(s)$ $i_e(-s) = i_e(s), \quad i_o(-s) = -i_o(s)$	Even and odd
$\lvert i(-s)\rvert = \lvert i(s)\rvert$	Symmetric

Properties:

Pattern	Required Current Distribution
1. Real pattern $f(w) = f_r(w) + j0$	$i(-s) = i^*(s)$: $A(-s) = A(s)$ Sym. amp. $\phi(-s) = -\phi(s)$ Odd phase
2. Symmetric pattern $\lvert f(-w)\rvert = \lvert f(w)\rvert$	$A(s)$ symmetric and $\phi(s)$ even; that is, $i(s)$ even *or* $A(s)$ asymmetric and $\phi(s)$ constant
3. Asymmetric pattern $\lvert f(-w)\rvert \neq \lvert f(w)\rvert$	$A(s)$ symmetric and $\phi(s)$ odd, nonzero *or* $A(s)$ asymmetric and $\phi(s)$ nonconstant

cally in terms of real and imaginary, amplitude and phase, or even and odd parts as indicated in Table 8-2. A current or pattern function is said to be symmetric if its magnitude is mirror-imaged about the origin. Property 1 states that *a real-valued pattern results if and only if the current amplitude is symmetric and the phase is odd.* Real patterns are often used in synthesis for mathematical simplicity, but in general a pattern can be complex-valued. A symmetric pattern is obtained if either of the two conditions on the current shown in Property 2 of Table 8-2 is satisfied. Note the important special case that *a real current distribution produces a symmetric pattern.* This follows from the fact that a current with a symmetric amplitude and zero phase satisfies the first condition of Property 2. Allowing phase to float by synthesizing a power pattern $|f(w)|^2$ instead of a field pattern $f(w)$ introduces an extra degree of freedom, but changes the nature of the synthesis problem from linear to nonlinear; see [1] for a discussion of power pattern synthesis methods. It is questionable that this additional problem complexity is warranted; so we confine ourselves to real patterns.

Often, an asymmetric pattern is required. Property 3 in Table 8-2 shows that *an asymmetric pattern can be obtained only through the use of aperture phase control.* An important application is the steering of a symmetric narrow beam pattern off broadside. This is achieved by a linear phase taper, which is an odd phase function.

8.2 LINE SOURCE SHAPED BEAM SYNTHESIS METHODS

8.2.1 The Fourier Transform Method

The Fourier transform pair relationship for the pattern and current of (8-6) and (8-7) suggests a synthesis method. If $f_d(w)$ is the *desired pattern*, the corresponding current distribution $i_d(s)$ is found rather easily from (8-7) as

$$i_d(s) = \int_{-\infty}^{\infty} f_d(w)e^{-j2\pi sw}\, dw \tag{8-8}$$

This is very direct, but unfortunately, the resulting $i_d(s)$ will *not*, in general, be confined to $|s| \le L/2\lambda$ as required; it will usually be, in fact, of infinite extent. An approximate solution can be obtained by truncating $i_d(s)$, giving the synthesized current distribution as follows:

$$i(s) = \begin{cases} i_d(s) & |s| \le \dfrac{L}{2\lambda} \\ 0 & |s| > \dfrac{L}{2\lambda} \end{cases} \tag{8-9}$$

The current $i(s)$ produces an approximate pattern $f(w)$ from (8-6). The current $i_d(s)$ extending over all s produces the pattern $f_d(w)$ exactly.

The Fourier transform synthesized pattern yields the least mean-square error (MSE), or least mean-squared deviation from the desired pattern, over the entire w-axis. The mean-squared error

$$\text{MSE} = \int_{-\infty}^{\infty} |f(w) - f_d(w)|^2 \, dw \tag{8-10}$$

with $f(w)$ corresponding to $i(s)$ in (8-9), is the smallest of all patterns arising from line sources of length L. The Fourier transform synthesized pattern, however, does not provide minimum mean-squared deviation in the visible region.

EXAMPLE 8-1 *Fourier Transform Synthesis of a Sector Pattern*

A *sector pattern* is a shaped beam pattern that, ideally, has uniform radiation over the main beam (a sector of space) and zero side lobes. Such patterns are popular for search applications where vehicles are located by establishing communications or by a radar echo in the sector of space occupied by the antenna pattern main beam. As a specific example, let the desired pattern be

$$f_d(\theta)(= \begin{cases} 1 & \cos^{-1} c \le \theta \le \cos^{-1}(-c) \\ 0 & \text{elsewhere} \end{cases} \tag{8-11a}$$

or, equivalently,

$$f_d(w) = \begin{cases} 1 & |w| \le c \\ 0 & \text{elsewhere} \end{cases} \tag{8-11b}$$

$f_d(w)$ is shown in Fig. 8-1*a* by the dashed curve. Using (8-11b) in (8-8) and (8-9) gives

$$i(s) = 2c \, \frac{\sin(2\pi cs)}{2\pi cs}, \qquad |s| \le \frac{L}{2\lambda} \tag{8-12}$$

If this sin $(x)/x$ function were not truncated, its Fourier transform (its pattern) would be exactly the sector pattern of (8-11). The actual pattern from (8-6) using (8-12) is

$$f(w) = \frac{1}{\pi}\left\{ \text{Si}\left[\frac{L}{\lambda}\,\pi(w + c)\right] - \text{Si}\left[\frac{L}{\lambda}\,\pi(w - c)\right]\right\} \tag{8-13}$$

where Si is the sine integral of (F-13). Alternate means of evaluating $f(w)$ include direct numerical integration or numerical Fourier transform. This synthesized sector pattern is plotted in Fig. 8-1*a* for $c = 0.5$ and $L = 10\lambda$. The pattern is plotted in linear form, rather than in decibels, to emphasis the details of the main beam. Note the oscillations about the desired pattern on the main beam, called ripple, and the nonzero side lobes. This appearance of main

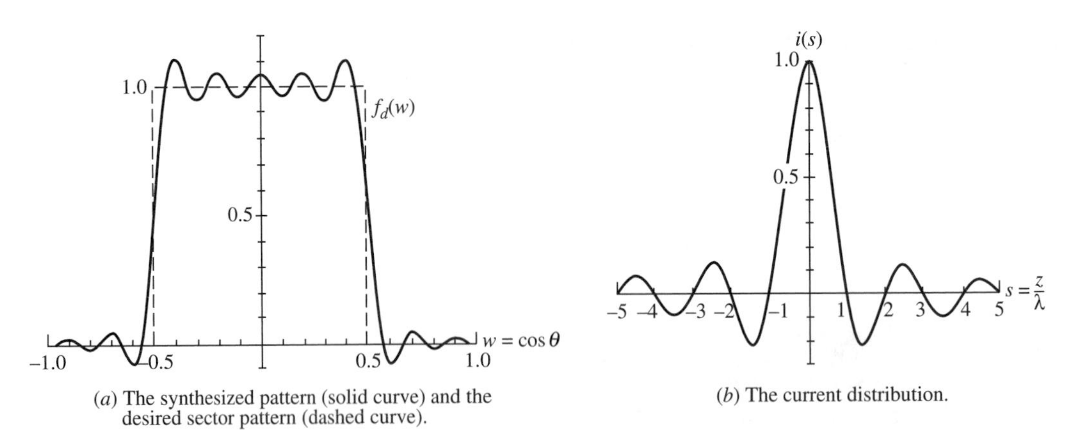

(*a*) The synthesized pattern (solid curve) and the desired sector pattern (dashed curve).

(*b*) The current distribution.

Figure 8-1 Fourier transform synthesis of a sector pattern using a 10λ line source (Example 8-1).

beam ripple and side lobes is typical of any synthesized pattern. The current distribution of (8-12) is plotted in Fig. 8-1*b*.

8.2.2 The Woodward–Lawson Sampling Method

A particularly convenient way to synthesize a radiation pattern is to specify values of the pattern at various points, that is, to sample the pattern. The Woodward–Lawson method is the most popular of the sampling methods [5, 6]. It is based on decomposition of the source current distribution into a sum of uniform amplitude, linear phase sources:

$$i_n(s) = \frac{a_n}{L/\lambda}\, e^{-j2\pi w_n s}, \qquad |s| \leq \frac{L}{2\lambda} \tag{8-14}$$

The pattern corresponding to this component current, from (8-6), is

$$f_n(w) = a_n \operatorname{Sa}\left[\pi \frac{L}{\lambda}(w - w_n)\right] \tag{8-15}$$

where the sampling function Sa(x) is defined as $\operatorname{Sa}(x) = \sin(x)/x$. This component pattern has a maximum of a_n centered at $w = w_n$. The current component phase coefficient w_n in (8-14) controls the location of the component pattern maximum, and the current component amplitude coefficient a_n controls the component pattern amplitude.

In the Woodward–Lawson method, the total current excitation is composed of a sum of $2M + 1$ component currents as

$$i(s) = \sum_{n=-M}^{M} i_n(x) = \frac{1}{L/\lambda} \sum_{n=-M}^{M} a_n e^{-j2\pi w_n s} \tag{8-16a}$$

where

$$w_n = \frac{n}{L/\lambda}, \qquad |n| \leq M, \qquad |w_n| \leq 1.0 \tag{8-16b}$$

The pattern corresponding to this current is

$$f(w) = \sum_{n=-M}^{M} f_n(w) = \sum_{n=-M}^{M} a_n \text{ Sa}\left[\pi \frac{L}{\lambda}(w - w_n)\right]$$
$$= \sum_{n=-M}^{M} a_n \text{ Sa}\left[\pi\left(\frac{L}{\lambda} w - n\right)\right] \tag{8-17}$$

At pattern points $w = w_n = n\lambda/L$, we have $f(w = w_n) = a_n$. Thus, the pattern can be made to have specified values a_n, called pattern *sample values*, at the pattern locations w_n of (8-16b), called *sample points*. The pattern sample values are chosen to equal the values of the desired pattern at the sample points:

$$a_n = f_d(w = w_n) \tag{8-18}$$

The Woodward–Lawson synthesis procedure is very easy to visualize. The current distribution required to produce a pattern with values a_n at locations w_n is that of (8-16).

The Woodward–Lawson sampling method can be made more flexible by noting that as long as adjacent samples are separated by the sampling interval $\Delta w = \lambda/L$, the pattern values at the sample joints are still uncorrelated, that is, (8-18) holds. The total number of samples is chosen such that the visible region is just covered; samples located outside the visible region could lead to superdirective results. Since the visible region is of extent 2 and $\Delta w = \lambda/L$, the number of samples $2M + 1$ is on the order of $2/(\lambda/L)$, or M is on the order of L/λ.

EXAMPLE 8-2 *Woodward–Lawson Line Source Synthesis of a Sector Pattern*

The sector pattern of Example 8-1 is now to be synthesized with a 10-wavelength-long line source using the Woodward–Lawson method. Sampling this pattern according to $a_n = f_d(w = w_n)$ with sample locations $w_n = n\lambda/L = 0.1n$ gives the values in Table 8-3. The sample value at the discontinuity ($w = 0.5$) could be selected as 1, 0.5, or 0 according to the specific application. Using $a_{\pm 5} = 1$ gives the widest main beam, whereas $a_{\pm 5} = 0$ gives the narrowest.

Table 8-3 Sample Locations and Sample Values for a 10λ Woodward–Lawson Sector Pattern (Example 8-2)

n	Sample Location w_n	Pattern Sample Value a_n
0	0	1
±1	±0.1	1
±2	±0.2	1
±3	±0.3	1
±4	±0.4	1
±5	±0.5	0.5
±6	±0.6	0
±7	±0.7	0
±8	±0.8	0
±9	±0.9	0
±10	±1.0	0

In this case, we choose $a_{\pm 5} = 0.5$ as a compromise. The synthesized pattern is computed using the sample values and locations of Table 8-3 in (8-17) and is plotted in Figure 8-2*a*. The sample points are indicated by dots.

To illustrate the sampling nature of the Woodward–Lawson method, two sampling functions from the sum in (8-17) are shown in Fig. 8-2*b* for sample locations $w_{-1} = -0.1$ and $w_0 = 0$. Note that when one sampling function is maximum, the other is zero, thus making the samples independent. Further, each sampling function is zero at all sample locations $w_n = n\lambda/L$, except at its maximum. When all samples are included, the value of the total synthesized pattern at locations w_n is completely determined by the Sa function centered at that location. This is the beauty of the Woodward–Lawson sampling method.

Note that the Woodward–Lawson pattern of Fig. 8-2*a* is a better approximation to the desired pattern (in the visible region) than that of the Fourier transform method in Fig. 8-1*a*, both generated from a 10-wavelength line source. Detailed comparisons of all the sector pattern examples are presented in Sec. 8.3.3.

The current distribution corresponding to the sector pattern of this example is plotted in Fig. 8-2*c*. It was obtained from (8-16). Note the similarity to the current distribution in Fig. 8-1*b* for the Fourier transform method. This occurs because the Fourier transform of any pattern is the antenna current distribution. Since the patterns in Examples 8-1 and 8-2 are

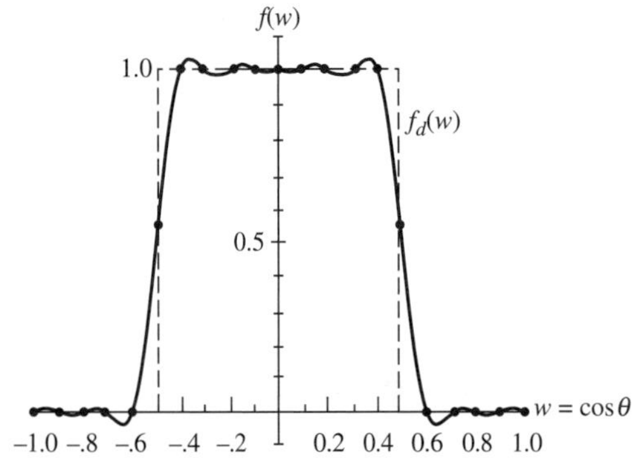

(*a*) The synthesized pattern (solid curve) and the desired pattern (dashed curve). The dots indicate the sample values and locations.

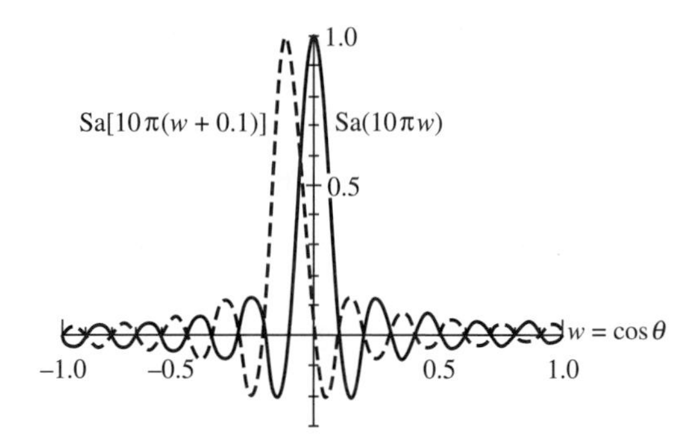

(*b*) Two component patterns at sample locations $w_{-1} = -0.1$ and $w_0 = 0$.

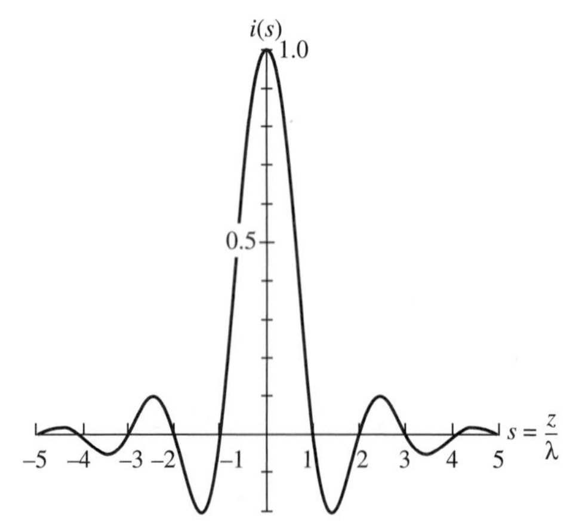

(c) The current distribution corresponding to the synthesized pattern.

Figure 8-2 Woodward–Lawson synthesis of a sector pattern using a 10λ line source (Example 8-2).

both close to a sector pattern, their Fourier transforms (currents) must be close to that of an ideal sector pattern, which is $\sin(\pi s)/\pi s$ in these examples.

8.3 LINEAR ARRAY SHAPED BEAM SYNTHESIS METHODS

In this section, the Fourier series and Woodward–Lawson methods for equally spaced linear arrays are discussed. These two important pattern synthesis methods are the array counterparts of the Fourier transform and Woodward–Lawson methods of the previous section. Before presenting these methods, we model the array configuration for use with any synthesis method.

Consider an equally spaced linear array along the z-axis with interelement spacings d. For simplicity, the physical center of the array is located at the origin. The total number of elements in the array P can be either even (then let $P = 2N$) or odd (then let $P = 2N + 1$). For an odd element number, the element locations are given by

$$z_m = md, \qquad |m| \leq N \tag{8-19}$$

and $P = 2N + 1$. The corresponding array factor is

$$f(w) = \sum_{m=-N}^{N} i_m e^{j2\pi m(d/\lambda)w} \tag{8-20}$$

where i_m are the element currents and again $w = \cos\theta$. This expression is similar to (3-54).

For an even number of elements, the element positions are

$$\begin{aligned} z_m &= \frac{2m-1}{2}d, \qquad 1 \leq m \leq N \\ z_{-m} &= -\frac{2m-1}{2}d, \qquad -N \leq -m \leq -1 \end{aligned} \tag{8-21}$$

and $P = 2N$. The corresponding array factor is

$$f(w) = \sum_{m=1}^{N} \left(i_{-m} e^{-j\pi(2m-1)(d/\lambda)w} + i_m e^{j\pi(2m-1)(d/\lambda)w}\right) \tag{8-22}$$

for P even.

For comparison to a line source, the total array length is defined as

$$L = Pd \tag{8-23}$$

This definition applies to both the even and odd element cases, and it includes a distance $d/2$ beyond each end element.

8.3.1 The Fourier Series Method

The array factor resulting from an array of identical discrete radiators (elements) is, of course, the sum over the currents for each element weighted by the spatial phase delay from each element to the far-field point. This array factor summation can be made to be of a form that is very similar to a Fourier series, just as the radiation integral for a continuous source resembles a Fourier transform (see Sec. 8.2.1). To see how this correspondence comes about, we first observe that a function

$f_d(w)$, the desired pattern function, can be expanded into a Fourier series in the interval $-\lambda/2d < w < \lambda/2d$ as

$$f_d(w) = \sum_{m=-\infty}^{\infty} b_m e^{j2\pi m(d/\lambda)w} \tag{8-24}$$

where

$$b_m = \frac{d}{\lambda} \int_{-\lambda/2d}^{\lambda/2d} f_d(w) e^{-j2\pi m(d/\lambda)w}\, dw \tag{8-25}$$

If we identify d as the spacing between elements of an equally spaced linear array and $w = \cos\theta$ where θ is the angle from the line of the array, the sum in (8-24) is recognized as the array factor of an array with an infinite number of elements with currents b_m.

An infinite array is, of course, not practical, but truncating the series (8-24) to a finite number of terms produces the following approximation to $f_d(w)$:

$$f(w) = \sum_{m=-N}^{N} b_m e^{j2\pi m(d/\lambda)w} \tag{8-26}$$

If we let the currents of each element in the array equal the Fourier series coefficients, that is,

$$i_m = b_m, \qquad |m| \leq N \tag{8-27}$$

then (8-26) is identical to (8-20), the array factor for an array with an odd number of elements.

The Fourier series synthesis procedure is, then, to use element excitations i_m equal to the Fourier series coefficients b_m calculated from the desired pattern f_d, as in (8-25). The array factor f arising from these element currents is an approximation to the desired pattern. This Fourier series synthesized pattern provides the least mean-squared error [see (8-10)] over the region $-\lambda/2d < w < \lambda/2d$. If the elements are half-wavelength spaced ($d = \lambda/2$), this region is exactly the visible region ($-1 < w < 1$, or $0 < \theta < \pi$).

A similar line of reasoning leads to the results for an even number of elements. In this case, the Fourier series coefficient currents are

$$\begin{aligned} i_m = b_m &= \frac{d}{\lambda} \int_{-\lambda/2d}^{\lambda/2d} f_d(w) e^{-j\pi(2m-1)(d/\lambda)w}\, dw, \qquad m \geq 1 \\ i_{-m} = b_{-m} &= \frac{d}{\lambda} \int_{-\lambda/2d}^{\lambda/2d} f_d(w) e^{j\pi(2m-1)(d/\lambda)w}\, dw, \qquad -m \leq -1 \end{aligned} \tag{8-28}$$

for P even. The synthesized pattern is given by (8-22). Note that if N is infinite, (8-22) together with (8-28) is the Fourier series expansion of f_d; that is, $f(w) = f_d(w)$.

EXAMPLE 8-3 Fourier Series Synthesis of a Sector Pattern

For an equally spaced linear array with an even number of elements and $c < \lambda/2d$, the sector pattern of (8-11) in (8-28) yields excitation currents

$$i_m = i_{-m} = 2\frac{d}{\lambda} c \,\mathrm{Sa}\left[\pi(2m-1)\frac{d}{\lambda}c\right], \qquad 1 \leq m \leq N \tag{8-29}$$

Table 8-4 Array Positions and Currents for a Fourier Series Synthesized Linear Array of 20 Half-Wavelength Spaced Elements for a Sector Pattern (Example 8-3)

Element Number m	Element Position z_m	Excitation Current i_m
± 1	$\pm 0.25\lambda$	0.4502
± 2	$\pm 0.75\lambda$	0.1501
± 3	$\pm 1.25\lambda$	−0.0900
± 4	$\pm 1.75\lambda$	−0.0643
± 5	$\pm 2.25\lambda$	0.0500
± 6	$\pm 2.75\lambda$	0.0409
± 7	$\pm 3.25\lambda$	−0.0346
± 8	$\pm 3.75\lambda$	−0.0300
± 9	$\pm 4.25\lambda$	0.0265
± 10	$\pm 4.75\lambda$	0.0237

Since these currents are symmetric, the array factor of (8-22) reduces to

$$f(w) = 2 \sum_{m=1}^{N} i_m \cos\left[\pi(2m - 1)\frac{d}{\lambda} w\right] \tag{8-30}$$

which is a real function. Note this is a special case of symmetry Property 1 in Table 8-2.

The specific case of $c = 0.5$, $d/\lambda = 0.5$, and 20 elements ($N = 10$) has an array length $L = Pd = 10\lambda$ and excitation currents from (8-29) given by

$$i_m = i_{-m} = \frac{1}{2} \operatorname{Sa}\left[\frac{\pi}{4}(2m - 1)\right], \qquad 1 \le m \le 10 \tag{8-31}$$

These excitation values are listed in Table 8-4, together with the element positions from (8-21). When these are used in the pattern expression (8-30), the pattern shown in Fig. 8-3 is produced.

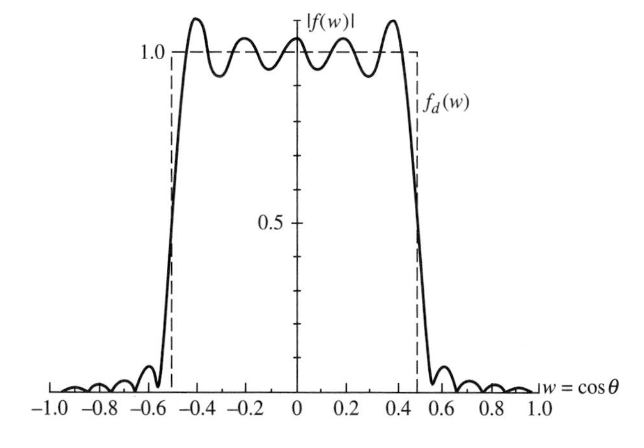

Figure 8-3 Fourier series synthesized array factor for a 20-element, $\lambda/2$ spaced linear array (Example 8-3). The desired pattern (dashed curve) is a sector pattern.

8.3.2 The Woodward–Lawson Sampling Method

The Woodward–Lawson sampling method for linear arrays is analogous to the Woodward–Lawson sampling method for line sources (see Sec. 8.2.2). In the array case, the synthesized array factor is the superposition of array factors from uniform amplitude, linear phase arrays:

$$f(w) = \sum_{n=-M}^{M} a_n \frac{\sin[(P/2)(w - w_n)(2\pi/\lambda)\, d]}{P \sin[\frac{1}{2}(w - w_n)(2\pi/\lambda)\, d]} \tag{8-32}$$

where the sample values are

$$a_n = f_d(w = w_n) \tag{8-33}$$

and the sample points are

$$w_n = n\frac{\lambda}{Pd} = \frac{n}{L/\lambda}, \qquad |n| \leq M, \qquad |w_n| \leq 1.0 \tag{8-34}$$

The element currents required to give this pattern are found from

$$i_m = \frac{1}{P} \sum_{n=-M}^{M} a_n e^{-j2\pi(z_m/\lambda)w_n} \tag{8-35}$$

These results hold for arrays with either an even or odd number of elements.

EXAMPLE 8-4 ***Woodward–Lawson Array Synthesis of a Sector Pattern***

Again, the sector pattern of (8-11) with $c = 0.5$ is to be synthesized, this time with a 20-element, half-wavelength spaced linear array using the Woodward–Lawson method. The sample locations from (8-34) are $w_n = 0.1n$. Thus, the sample locations and values are the same as for Example 8-2 and are given in Table 8-3. Using these and element positions z_m from (8-21) in (8-35) yields the array currents of Table 8-5. The pattern can be generated from either the Woodward–Lawson pattern expression of (8-32), or by direct array computation using (8-30), which is the version of (8-22) for the symmetric case, and the array parameters of Table 8-5. The pattern is plotted in Fig. 8-4.

Table 8-5 Array Element Currents and Positions Synthesized from the Woodward–Lawson Method for a Sector Pattern (Example 8-4)

Element Number m	Element Position z_m	Excitation Current i_m
±1	±0.25λ	0.44923
±2	±0.75λ	0.14727
±3	±1.25λ	−0.08536
±4	±1.75λ	−0.05770
±5	±2.25λ	0.04140
±6	±2.75λ	0.03020
±7	±3.25λ	−0.02167
±8	±3.75λ	−0.01464
±9	±4.25λ	0.00849
±10	±4.75λ	0.00278

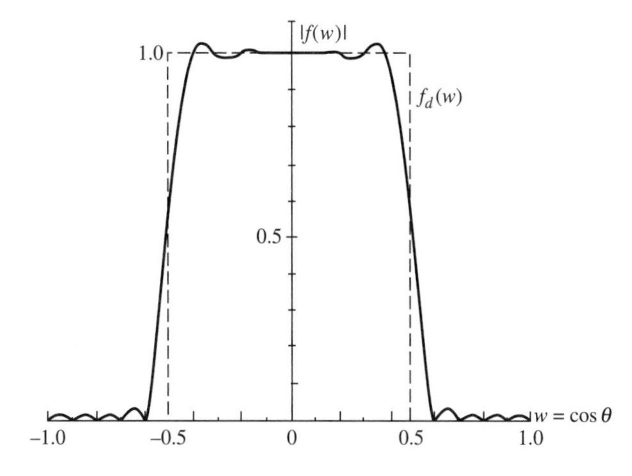

Figure 8-4 Woodward–Lawson synthesized array factor for a 20-element, $\lambda/2$ spaced linear array (Example 8-4). The desired pattern (dashed curve) is a sector pattern.

8.3.3 Comparison of Shaped Beam Synthesis Methods

Most shaped beam antenna patterns have three distinct types of pattern regions: side lobe, main beam, and transition. The side-lobe region is easily recognized, and the *side-lobe level*, SLL, is defined from

$$\text{SLL} = 20\log\left|\frac{\text{value of the highest side-lobe peak}}{\text{maximum of desired pattern}}\right| \tag{8-36}$$

over the side-lobe region. The quality of fit to the desired pattern $f_d(w)$ by the synthesized pattern $f(w)$ over the main beam is measured by the *ripple* R, which is defined as

$$R = 20\left\{\text{maximum}\left|\log\left|\frac{f(w)}{f_d(w)}\right|\right|\right\} \quad \text{dB} \tag{8-37}$$

over the main beam. Also of interest is the region between the main beam and side-lobe region, referred to as the transition region. In many applications, such as direction finding, it is desirable to have the main beam fall off very sharply into the side-lobe region. To quantify this, *transition width* T is introduced and defined as

$$T = |w_{f=0.9} - w_{f=0.1}| \tag{8-38}$$

where $w_{f=0.9}$ and $w_{f=0.1}$ are the values of w where the synthesized pattern f equals 90 and 10% of the local discontinuity in the desired pattern. For unsymmetrical, single beam patterns, there are two transition regions with different transition widths. Transition width is analogous to rise time in time-signal analysis.

The shaped beam synthesis methods we have discussed in this and the previous section can be compared rather easily using SLL, R, and T. The sector pattern results of Examples 8-1 to 8-4 are presented in Table 8-6. A few general trends can be extracted from the table. The Woodward–Lawson methods (for both line sources and arrays) tend to produce low side lobes and low main beam ripple at some sacrifice in transition width. On the other hand, Fourier methods yield somewhat inferior side-lobe levels and ripples. The Fourier series synthesized pattern gives very sharp rolloff from the main beam to the side-lobe region; that is, small transition width.

Table 8-6 Comparison of Synthesized Sector Patterns ($c = 0.5$, $L = 10\lambda$)

Method	Type	Example Number	Figure Number	Side-Lobe Level, SLL (dB)	Ripple, R (dB)	Transition Width, T
Fourier transform	10λ line source	8-1	8-1*a*	−21.9	0.83	0.0893
Woodward–Lawson	10λ line source	8-2	8-2*a*	−29.8	0.27	0.1303
Fourier series	20-element $\lambda/2$ spaced array	8-3	8-3	−22.6	0.87	0.0941
Woodward–Lawson	20-element $\lambda/2$ spaced array	8-4	8-4	−29.6	0.27	0.1343

8.4 LOW SIDE-LOBE, NARROW MAIN BEAM METHODS

The synthesis methods presented in the previous two sections are most useful for shaping the main beam of an antenna pattern. Another major class of pattern synthesis methods is that for achieving a narrow main beam accompanied by low side lobes. Patterns of this type have many applications, such as in point-to-point communications and direction finding. In this section, we discuss the two most important narrow main beam, low side-lobe methods: the Dolph–Chebyshev method for linear arrays and the Taylor line source method. These two methods are closely related and the Dolph–Chebyshev method is presented first to simplify the development.

8.4.1 The Dolph–Chebyshev Linear Array Method

In Sec. 3.5, several excitations of equally spaced, linear rays were examined. It was found that as the current amplitude taper from the center to the edges of the array increased, the side-lobe level decreased, but with an accompanying increase in the width of the main beam. In most applications, it is desirable to have both a narrow main beam as well as low side lobes. It would, therefore, be useful to have a pattern with an optimum compromise between beamwidth and side-lobe level. In other words, for a specified beamwidth the side-lobe level would be as low as possible; or vice versa, for a specified side-lobe level the beamwidth would be as narrow as possible. In this section, a method for achieving this is presented for broadside, linear arrays with equal spacings that are equal to or greater than a half-wavelength.

As might be expected, optimum beamwidth-side-lobe level performance occurs when there are as many side lobes in the visible region as possible and they have the same level. Dolph [7] recognized that Chebyshev polynomials possess this property, and he applied them to the synthesis problem. It is important to be familiar with Chebyshev polynomials, so we shall give a brief treatment of them before proceeding to synthesis.

The Chebyshev (often spelled "Tchebyscheff") polynomials are defined by

$$T_n(x) = \begin{cases} (-1)^n \cosh(n \cosh^{-1}|x|), & x < -1 \\ \cos(n \cos^{-1} x), & -1 < x < 1 \\ \cosh(n \cosh^{-1} x), & x > 1 \end{cases} \tag{8-39}$$

A few of the lower-order polynomials are

$$
\begin{aligned}
T_0(x) &= 1 \\
T_1(x) &= x \\
T_2(x) &= 2x^2 - 1 \\
T_3(x) &= 4x^3 - 3x \\
T_4(x) &= 8x^4 - 8x^2 + 1
\end{aligned}
\tag{8-40}
$$

Higher-order polynomials can be generated from the recursive formula

$$T_{n+1}(x) = 2xT_n(x) - T_{n-1}(x) \tag{8-41}$$

or by letting $\delta = \cos^{-1} x$ and expanding $\cos m\delta$ in powers of $\cos \delta$. For example, $T_3(x) = \cos(3 \cos^{-1} x) = \cos 3\delta = 4 \cos^3 \delta - 3 \cos \delta$ from (D-13). Hence, $T_3(x) = 4x^3 - 3x$. A few polynomials are plotted in Fig. 8-5.

Some important general properties of Chebyshev polynomials follow from (8-39) or Fig. 8-5. The even-ordered polynomials are even, that is, $T_n(-x) = T_n(x)$ for n even, and the odd-ordered ones are odd, that is, $T_n(-x) = -T_n(x)$ for n odd. All polynomials pass through the point (1, 1). In the range $-1 \le x \le 1$, the polynomial values lie between -1 and 1, and the maximum magnitude is always unity there. All zeros (roots) of the polynomials also lie in $-1 \le x \le 1$.

The equal amplitude oscillations of Chebyshev polynomials in the region $|x| \le 1$ is the desired property for equal side lobes. Also, the polynomial nature of the functions makes them suitable for array factors since an array factor can be written as a polynomial. The connection between arrays and Chebyshev polynomials is established by considering a symmetrically excited, broadside array for which (see Table 8-2)

$$i_{-m} = i_m \tag{8-42}$$

Symmetrical excitation leads to a real-valued array factor that, from (8-20) and (8-22), is given by

$$
f(\psi) = \begin{cases} i_0 + 2 \sum\limits_{m=1}^{N} i_m \cos m\psi & P \text{ odd} \\ 2 \sum\limits_{m=1}^{N} i_m \cos\left[(2m-1)\dfrac{\psi}{2}\right] & P \text{ even} \end{cases} \tag{8-43}
$$

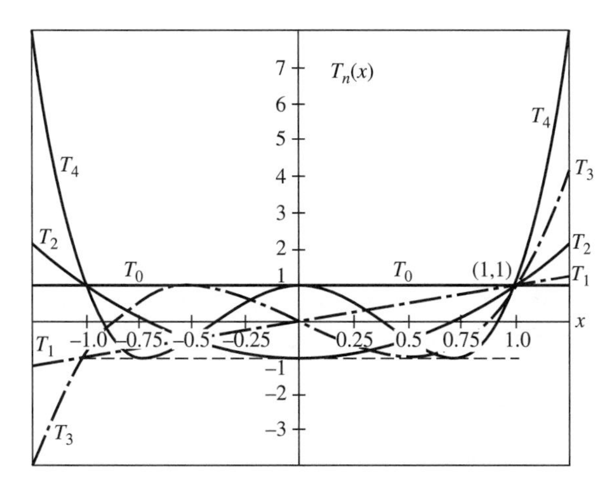

Figure 8-5 Chebyshev polynomials $T_0(x)$, $T_1(x)$, $T_2(x)$, $T_3(x)$, and $T_4(x)$.

where $\psi = 2\pi(d/\lambda)w$. This array factor (for odd or even P) is a sum of $\cos(m\psi/2)$ terms for m up to $P - 1$. But each term containing $\cos(m\psi/2)$ can be written as a sum of terms with powers of $\cos(\psi/2)$ up to m, through the use of trigonometric identities. Therefore, the array factor is expressible as a sum of terms with powers of $\cos(\psi/2)$ up to $P - 1$.

By choosing an appropriate transformation between x and ψ, the array factor and Chebyshev polynomial will be identical. The transformation

$$x = x_o \cos\frac{\psi}{2} \tag{8-44}$$

and the correspondence

$$f(\psi) = T_{P-1}\left(x_o \cos\frac{\psi}{2}\right) \tag{8-45}$$

will yield a polynomial in powers of $\cos(\psi/2)$ matching that of the array factor. The main beam maximum value of R occurs for $\theta = 90°$, or $\psi = 0$, for a broadside array.[2] Then (8-44) indicates that $x = x_o$ at the main beam maximum. The visible region extends from $\theta = 0°$ to $180°$, or $\psi = 2\pi(d/\lambda)$ to $-2\pi(d/\lambda)$. These limits correspond to $x = x_o \cos(\pi d/\lambda)$; for half-wavelength spacing, the limits are $x = 0$. Thus, for $d = \lambda/2$, the visible region begins at $x = 0$, or $\theta = 0°$, and x increases as θ does until x_o (the main beam maximum point) is reached and retraces back to $x = 0$, or $\theta = 180°$ (see Fig. 8-6).

The main beam-to-side lobe ratio R is the value of the array factor at the main beam maximum, since the side-lobe level magnitude is unity (see Fig. 8-6). The side-lobe level is thus $1/R$, or

$$\text{SLL} = -20 \log R \qquad \text{dB} \tag{8-46}$$

Evaluating (8-45) at the main beam maximum gives

$$R = T_{P-1}(x_o) = \cosh[(P - 1)\cosh^{-1} x_o] \tag{8-47}$$

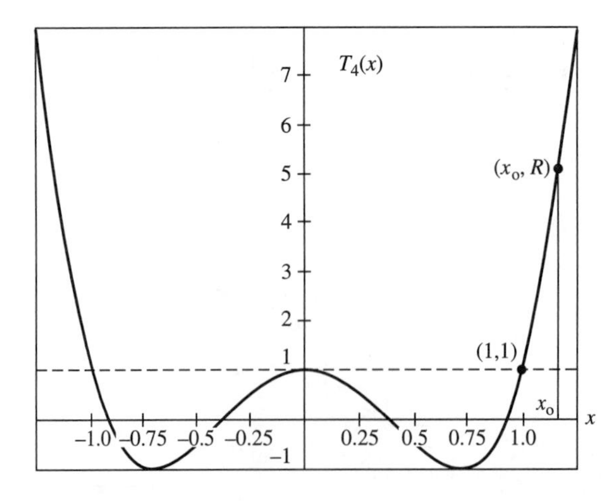

Figure 8-6 Chebyshev polynomial $T_4(x)$.

[2]The symbol f is usually reserved for a pattern normalized to a maximum value of unity, but for the Dolph–Chebyshev array, it is more convenient to normalize the array factor f to a maximum value of R.

from (8-39). Or, solving for x_o, we get

$$x_o = \cosh\left(\frac{1}{P-1}\cosh^{-1} R\right) \tag{8-48}$$

The design procedure can now be summarized. For a given side-lobe ratio, R can be determined from (8-46), leading to x_o from (8-48). The array factor is then given by (8-45), or it can be computed from (8-43) directly from the current values. The excitation currents are found by comparison between the array factor of (8-43) and the Chebyshev polynomial of (8-45). This synthesis procedure will be illustrated in Example 8-5. The currents are difficult to find by this method for arrays larger than a few elements. However, direct solution techniques and tabulated values are available [8].

The Dolph–Chebyshev array design procedure provides the lowest side-lobe pattern for a specified beamwidth. However, these solutions only depend on the number of elements, not their spacing (and thus total array length). Practical design seeks the narrowest beamwidth solution within the Dolph–Chebyshev family. This result is obtained by including as many side lobes as possible in the visible region without letting the grating lobe emerge to a level above the design side-lobe level. A general expression for the optimum spacing of the Dolph–Chebyshev array, with isotropic elements, that gives the narrowest beamwidth possible for a specified side-lobe level and given number elements is [8]

$$d_{\text{opt}} = \lambda\left[1 - \frac{\cos^{-1}\frac{1}{\gamma}}{\pi}\right] \qquad \textit{broadside} \tag{8-49a}$$

$$d_{\text{opt}} = \frac{\lambda}{2}\left[1 - \frac{\cos^{-1}\frac{1}{\gamma}}{\pi}\right] \qquad \textit{endfire} \tag{8-49b}$$

where

$$\gamma = \cosh\left[\frac{1}{P-1}\ln\left(R + \sqrt{R^2 - 1}\right)\right] \tag{8-49c}$$

An interesting result is that the directivity is identical for broadside and endfire operation for both half-wavelength and optimum spacings [9]. Although the beamwidth is broader at endfire than at broadside, the main beam is a fan beam at broadside, leading to the same solid angle (and, thus, the same directivity) as for the pencil-shaped endfire beam.

The half-power beamwidth of a Dolph–Chebyshev array, in general, is given by

$$\text{HP} = \pi - 2\cos^{-1}\frac{\psi_h}{\beta d} \qquad \textit{broadside} \tag{8-50a}$$

$$\text{HP} = \cos^{-1}\left(1 - \frac{\psi_h}{\beta d}\right) \qquad \textit{endfire} \tag{8-50b}$$

where

$$\psi_h = 2\cos^{-1}\left\{\frac{\cosh\left[\frac{1}{P-1}\cosh^{-1}\frac{R}{\sqrt{2}}\right]}{\cosh\left[\frac{1}{P-1}\cosh^{-1} R\right]}\right\} \tag{8-50c}$$

and an approximate form for the broadside case is

$$\text{HP} \approx \sqrt{\frac{1}{\pi}\ln(2R)}\,\frac{\lambda}{L} \qquad \textit{broadside} \tag{8-50d}$$

This can be rewritten with a factor that shows the beam broadening relative to the uniform case, that is, HP $\approx 0.886\lambda/L$; the beam broadening factor is

$$b_{\text{HP}} = \sqrt{\frac{1}{\pi}\ln(2R)}\ /\ 0.886 = 0.637\sqrt{\ln(2R)} \tag{8-50e}$$

The following approximate formula gives directivity using the half-power beamwidth at broadside found from (8-50a) or (8-50d) [10]:

$$D \approx \frac{2R^2}{1 + R^2\,\text{HP}} \tag{8-51}$$

This directivity result can be used for general situations. In fact, for optimum spacings directivity is exactly the same at endfire and broadside. Thus, the half-power beamwidth of (8-50a) or (8-50d) for broadside should be used in (8-51) for all scan cases.

Before closing this discussion of low side-lobe arrays with two examples, we point out that arbitrarily high directivity (narrow beamwidth) can be obtained from an array of fixed length. However, this requires currents with very high amplitudes and alternating signs [11]. This is clearly impractical and leads to narrow bandwidth and a high sensitivity to the accuracy of excitation.

EXAMPLE 8-5 *A Five-Element, Broadside, −20-dB Side-Lobe, Half-Wavelength Spaced Dolph–Chebyshev Array*

For a five-element array ($P = 5$, $N = 2$), the array factor from (8-43) is

$$f(\psi) = i_0 + 2i_1 \cos\psi + 2i_2 \cos 2\psi \tag{8-52}$$

where $\psi = 2\pi(d/\lambda)\cos\theta = \pi\cos\theta$ for $d = \lambda/2$. Using $\cos(2\psi/2) = 2\cos^2(\psi/2) - 1$ from (D-12) and $\cos(4\psi/2) = 8\cos^4(\psi/2) - 8\cos^2(\psi/2) + 1$ from (D-14), the array factor can be written as

$$f(\psi) = (i_0 - 2i_1 + 2i_2) + (4i_1 - 16i_2)\cos^2\frac{\psi}{2} + 16i_2\cos^4\frac{\psi}{2} \tag{8-53}$$

And from (8-40),

$$T_4(x) = 1 - 8x^2 + 8x^4 = 1 - 8x_o^2\cos^2\frac{\psi}{2} + 8x_o^4\cos^4\frac{\psi}{2} \tag{8-54}$$

where (8-44) was used in the second step. Now, the currents are found by successively equating the coefficients of like terms of (8-53) and (8-54). From the $\cos^4(\psi/2)$ term,

$$i_2 = \tfrac{1}{2}x_o^4 \tag{8-55}$$

The $\cos^2(\psi/2)$ term yields

$$i_1 = 4i_2 - 2x_o^2 = 2x_o^4 - 2x_o^2 \tag{8-56}$$

using (8-55). The final term gives

$$i_0 = -2i_2 + 2i_1 + 1 = 3x_o^4 - 4x_o^2 + 1 \tag{8-57}$$

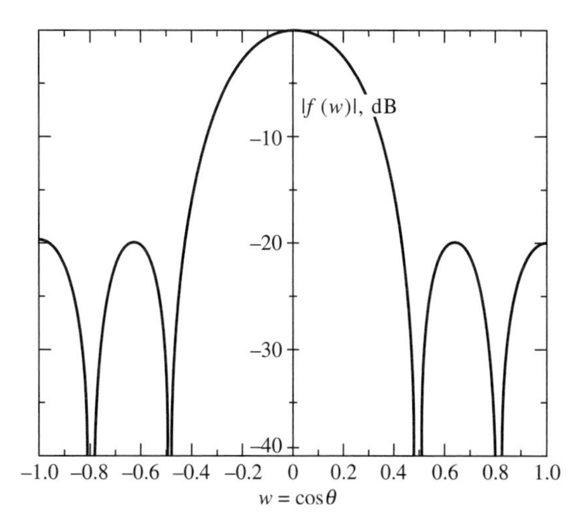

Figure 8-7 Dolph–Chebyshev synthesized array factor for a five-element, λ/2 spaced, broadside array with −20-dB side lobes (Example 8-5).

using (8-55) and (8-56). The current values will be completely determined when x_o is evaluated. This is accomplished by first finding the main beam-to-side lobe ratio from (8-46) using the specified −20-dB side-lobe level;

$$R = 10^{-\text{SLL}/20} = 10 \tag{8-58}$$

Then from (8-48) with $P = 5$ and $R = 10$,

$$x_o = 1.293 \tag{8-59}$$

The element currents from (8-55) to (8-57) with (8-59) are

$$i_2 = i_{-2} = 1.3975, \qquad i_1 = i_{-1} = 2.2465, \qquad i_0 = 2.6978 \tag{8-60}$$

These currents yield a main beam maximum of $R = 10$ and unity side lobes. Normalizing these to unity edge currents gives a 1:1.61:1.93:1.61:1 current distribution. The currents of (8-60) in (8-52) lead to the pattern in Fig. 8-7, which was normalized to 0 dB on the main beam maximum. The half-power beamwidth from (8-50a) is 23.7°, which was also found by direct evaluation of the polar pattern in Fig. 3-23*d*. The directivity for equi-phased, half-wavelength spaced arrays can be obtained from (3-93). For this example, the directivity is

$$D = \frac{\left|\sum_{m=-2}^{2} i_m\right|^2}{\sum_{m=-2}^{2} i_m^2} = 4.69 \tag{8-61}$$

The directivity found from (8-51) is 4.72.

EXAMPLE 8-6 *Optimum 10-Element, −30-dB Side-Lobe Dolph–Chebyshev Endfire Array*

Using $P = 10$ and $R = 10^{-(-30)/20} = 31.62$ in (8-49a) yields

$$d_{\text{opt}} = 0.4292\lambda \tag{8-62}$$

The element current amplitudes are [8]

$$1:1.67:2.60:3.41:3.88:3.88:3.41:2.60:1.67:1 \tag{8-63}$$

These with interelement phase shift $\alpha = -\beta d \cos(0°) = -154.5°$ yield the pattern in Fig. 8-8. The half-power beamwidth from (8-50a) for broadside operation in 7.58° and the approxi-

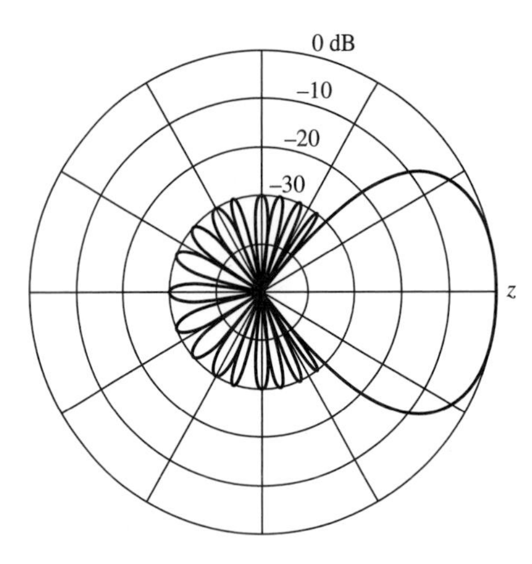

Figure 8-8 Polar pattern for the optimum endfire Dolph–Chebyshev, 10 element −30-dB side-lobe array of Example 8-6.

mation of (8-50d) gives 7.67°. The directivity cannot be computed as in (8-61), which holds only for half-wavelength spaced elements; instead, (8-51) is used with $\text{HP}_{\text{broad}} = 7.58°$ (or 7.67°), giving $D = 15.00$ (or 14.83). The actual half-power beamwidth from (8-50b) is $\text{HP}_{\text{end}} = 29.8°$.

8.4.2 The Taylor Line Source Method

Although the Dolph–Chebyshev array does yield the highest directivity and narrowest beamwidth, the constant side-lobe envelope leads to a high reactive energy condition, especially for large arrays. This means high-Q, or low bandwidth, operation [3]. This situation can be avoided by first designing a line source with nearly constant side lobes and using the current values at element locations in an array configuration that produces a very similar pattern; see Prob. 8.4-10.

The optimum narrow beam pattern from a line source antenna occurs when all side lobes are of equal level, just as in the array case. The required functional form, as we have seen, is that of the Chebyshev polynomial. The Chebyshev polynomial $T_N(x)$ has $N - 1$ equal level "side lobes" in the region $-1 < x < 1$, and for $|x| > 1$, its magnitude increases monotonically. A change of variables will transform the Chebyshev polynomial into the desired pattern form; that is, with a zero slope main beam maximum at $x = 0$ and equal level side lobes. The new function resulting from the variable change is

$$P_{2N}(x) = T_N(x_o - a^2x^2) \tag{8-64}$$

where a is a constant and

$$x = \frac{L}{\lambda}\cos\theta = \frac{L}{\lambda}w \tag{8-65}$$

At the pattern maximum,

$$P_{2N}(w = 0) = T_N(x_o) = R \tag{8-66}$$

which is the main beam-to-side lobe ratio. A plot of (8-64) for $N = 4$ is shown in Fig. 8-9; it is the transformed version of Fig. 8-6.

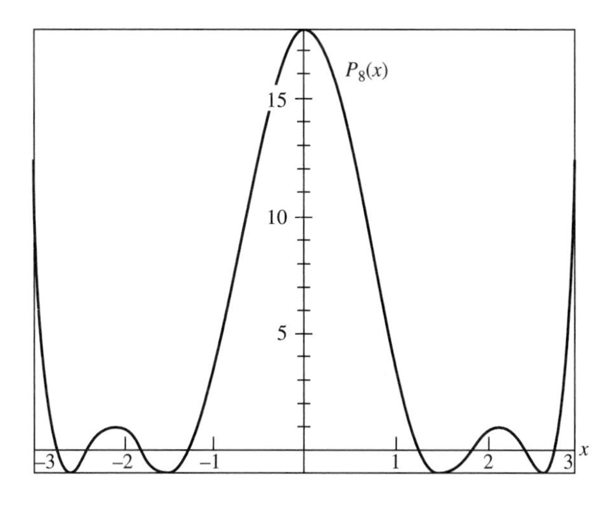

Figure 8-9 Transformed Chebyshev polynomial $P_8(x) = T_4(x_o - a^2x^2)$. Values of $a = 0.55536$ and $x_o = 1.42553$ corresponding to Example 8-6 were used.

From (8-39), we have in the side-lobe region

$$P_{2N}(x) = \cos[N \cos^{-1}(x_o - a^2x^2)], \qquad |x_o - a^2x^2| < 1 \tag{8-67}$$

The zeros of this function occur when the cosine argument equals $(2n - 1)\pi/2$, or when the values of x are as follows:

$$x_n = \pm\frac{1}{a}\sqrt{x_o - \cos\frac{(2n-1)\pi}{2N}}, \qquad |n| \geq 1 \tag{8-68}$$

where the plus sign is used for zero locations on the positive x-axis and $x_{-n} = -x_n$. In the main beam region, from (8-39),

$$P_{2N}(x) = \cosh[N \cosh^{-1}(x_o - a^2x^2)], \qquad |x_o - a^2x^2| > 1 \tag{8-69}$$

The main beam maximum value of P_{2N} is R and occurs for $x = 0$; see (8-65) and (8-66). Solving (8-69) for x_o at the main beam maximum yields

$$x_o = \cosh\left(\frac{1}{N}\cosh^{-1} R\right) \tag{8-70}$$

It is convenient to introduce A such that

$$A = \frac{1}{\pi}\cosh^{-1} R \tag{8-71}$$

so then

$$x_o = \cosh\frac{\pi A}{N} \tag{8-72}$$

In order to have all side-lobe levels equal, we let N approach infinity, but simultaneously the argument of P_{2N} is changed to keep the first nulls stationary, thus leaving the beamwidth unchanged. For large N, $x_o = \cosh(\pi A/N) \approx 1 + \frac{1}{2}(\pi A/N)^2$ and

$$\cos\frac{(2n-1)\pi}{2N} \approx 1 - \frac{1}{2}\left[\frac{(2n-1)\pi}{2N}\right)^2$$

and using these in (8-68) gives

$$x_n = \pm\frac{1}{a}\frac{\pi}{\sqrt{2N}}\sqrt{A^2 + (n - \tfrac{1}{2})^2}, \qquad N \to \infty \tag{8-73}$$

By letting

$$a = \frac{\pi}{\sqrt{2N}} \tag{8-74}$$

the first zero location remains fixed as N increases. Then

$$x_n = \pm\sqrt{A^2 + (n - \tfrac{1}{2})^2} \tag{8-75}$$

The pattern factor is a polynomial in x with an infinite number of roots x_n and can be expressed as a product of factors $(x - x_n)$ for n from $-\infty$ to $+\infty$. And since $x_{-n} = -x_n$, the pattern is

$$\prod_{n=1}^{\infty}(x^2 - x_n^2) = \prod_{n=1}^{\infty}[x^2 - A^2 - (n - \tfrac{1}{2})^2] \tag{8-76}$$

Normalizing this to unity at $x = 0$ gives

$$f(x) = \frac{\displaystyle\prod_{n=1}^{\infty}\left[1 - \frac{x^2 - A^2}{(n - \frac{1}{2})^2}\right]}{\displaystyle\prod_{n=1}^{\infty}\left[1 + \frac{A^2}{(n - \frac{1}{2})^2}\right]} = \frac{\cos(\pi\sqrt{x^2 - A^2})}{\cosh \pi A} \tag{8-77}$$

The last step above utilizes the closed-form expression for the infinite products. Using (8-65) and (8-71) in (8-77) gives the pattern in w as

$$f(w) = \frac{\cos\{\pi\sqrt{[(L/\lambda)w]^2 - A^2}\}}{R} \tag{8-78}$$

Note that this is normalized to unity at the maximum ($w = 0$) and oscillates between $-1/R$ and $1/R$ in the side-lobe region. For large w, the argument of the cosine function in (8-78) is approximately $\pi wL/\lambda$, so the zero locations of the pattern are $w_n \approx \pm\lambda(n - \frac{1}{2})/L$ or $x_n \approx \pm(n - \frac{1}{2})$, and thus they are regularly spaced. Also note that for $w < \lambda A/L$, the cosine argument of (8-78) is imaginary and since $\cos(j\theta) = \cosh\theta$, (8-78) is more conveniently expressed as

$$f(w) = \frac{\cosh\{\pi\sqrt{A^2 - [(L/\lambda)w]^2}\}}{\cosh \pi A} \tag{8-79}$$

This pattern is that of the *ideal Taylor line source* [12]. It is a function of A that is found from the side-lobe level; see (8-46) and (8-71). The line source is "ideal" in the sense that equi-level side lobes extend to infinity in pattern space, thus leading to infinite power. The required source excitation, in turn, must possess infinite power and, in fact, will have singularities at each end of the line source.

An approximate realization of the ideal Taylor line source, referred to as the *Taylor line source*, nearly equals the first few side lobes but has decreasing far-out side lobes [12]. The decaying side-lobe envelope removes the infinite power difficulty encountered with the ideal Taylor line source. The Taylor line source pattern is again a polynomial in x, but with zero locations given by

$$x_n = \begin{cases} \pm\sigma\sqrt{A^2 + (n - \frac{1}{2})^2} & 1 \leq n < \bar{n} \\ \pm n & \bar{n} \leq n < \infty \end{cases} \tag{8-80}$$

The zeros for $n < \bar{n}$ are those of the ideal line source in (8-75) scaled by the factor σ. The far-out side lobes for $n \geq \bar{n}$ are located at the integer x positions. The zero arrangement for a $\sin(\pi x)/\pi x$ pattern is $x = \pm n$ for $n \geq 1$, so the Taylor pattern far-out side lobes are those of the $\sin(\pi x)/\pi x$ pattern. The scaling parameter σ is determined by making the zero location expressions in (8-80) identical for $n = \bar{n}$, which yields

$$\sigma = \frac{\bar{n}}{\sqrt{A^2 + (\bar{n} - \frac{1}{2})^2}} \tag{8-81}$$

From the zero locations of (8-80), we write the approximate Taylor line source pattern as

$$f(x, A, \bar{n}) = \frac{\sin \pi x}{\pi x} \prod_{n=1}^{\bar{n}-1} \frac{1 - (x/x_n)^2}{1 - (x/n)^2} \tag{8-82}$$

The side lobes are nearly constant at the value $1/R$ out to $x = \bar{n}$ and decay as $1/x$ beyond $x = \bar{n}$. The pattern in terms of $w = \cos\theta$ is

$$f(w, A, \bar{n}) = \frac{\sin(\pi L w/\lambda)}{\pi L w/\lambda} \prod_{n=1}^{\bar{n}-1} \frac{1 - (w/w_n)^2}{1 - (Lw/\lambda n)^2} \tag{8-83}$$

where the pattern zero locations on the w-axis are

$$w_n = \begin{cases} \pm\dfrac{\lambda}{L}\sigma\sqrt{A^2 + (n - \frac{1}{2})^2} & 1 \leq n < \bar{n} \\ \pm\dfrac{\lambda}{L} n & \bar{n} \leq n < \infty \end{cases} \tag{8-84}$$

with σ given by (8-81).

The Taylor line source is actually a pattern of the Woodward–Lawson family. We show how this comes about and also determine the sample values and locations. First, assume that the required source excitation can be expanded in a Fourier series as

$$i(s) = \frac{\lambda}{L} \sum_{n=-\infty}^{\infty} a_n e^{-j2\pi(\lambda/L)ns}, \qquad |s| \leq \frac{L}{2\lambda} \tag{8-85}$$

The corresponding pattern from (8-17) is

$$f(w) = \sum_{n=-\infty}^{\infty} a_n \,\mathrm{Sa}\left[\left(w - \frac{\lambda}{L} n\right) \frac{L}{\lambda} \pi\right] \tag{8-86}$$

where the sample locations are identified as

$$w_n^s = \frac{\lambda}{L} n \tag{8-87}$$

The infinite expansion of (8-86) gives the exact pattern if the sample values are (see Prob. 8.4-7)

$$a_n = f(w = w_n^s) = f(n, A, \bar{n}) \tag{8-88}$$

But the pattern zeros correspond to the sample locations of (8-87) for $|n| \geq \bar{n}$ since $x_n = n$, or $w_n = (\lambda/L)n$ for $|n| \geq \bar{n}$ from (8-80). Thus,

$$a_n = 0 \qquad \text{for} \qquad |n| \geq \bar{n} \tag{8-89}$$

Using (8-88) and (8-89) in (8-86) gives the pattern expression

$$f(w) = \sum_{n=-\bar{n}+1}^{\bar{n}-1} f(n, A, \bar{n}) \, \text{Sa}\left[(w - w_n^s) \frac{L}{\lambda} \pi\right] \tag{8-90}$$

The required current distribution from (8-85) is

$$i(s) = \frac{\lambda}{L}\left[1 + 2 \sum_{n=1}^{\bar{n}-1} f(n, A, \bar{n}) \cos\left(2\pi \frac{\lambda}{L} ns\right)\right] \tag{8-91}$$

The coefficients $f(n, A, \bar{n})$ are the samples of Taylor line source pattern for $x = n$ and $n < \bar{n}$. They are found from

$$f(n, A, \bar{n}) = \begin{cases} \dfrac{[(\bar{n} - 1)!]^2}{(\bar{n} - 1 + n)!(\bar{n} - 1 - n)!} \displaystyle\prod_{m=1}^{\bar{n}-1} \left(1 - \frac{n^2}{x_m^2}\right) & |n| < \bar{n} \\ 0 & |n| \geq \bar{n} \end{cases} \tag{8-92}$$

and $f(-n, A, \bar{n}) = f(n, A, \bar{n})$. Tables of the coefficient values are also available [13, Appendix I]. These coefficients together with (8-90) and (8-91) determine the Taylor line source pattern and current.

The half-power beamwidth expression is obtained rather easily for the ideal pattern. Evaluating (8-79) at the half-power points yields

$$\frac{1}{\sqrt{2}} = \frac{1}{R} \cosh\left(\pi\sqrt{A^2 - \left(\frac{L}{\lambda} w_{\text{HP}}\right)^2}\right) \tag{8-93}$$

Solving this gives the two solutions

$$w_{\text{HP}} = \pm\frac{\lambda}{L\pi}\left[(\cosh^{-1} R)^2 - \left(\cosh^{-1} \frac{R}{\sqrt{2}}\right)^2\right]^{1/2} \tag{8-94}$$

The half-power beamwidth in w is then

$$\text{HP}_{w_i} = 2|w_{\text{HP}}| = \frac{\lambda 2}{L\pi}\left[(\cosh^{-1} R)^2 - \left(\cosh^{-1} \frac{R}{\sqrt{2}}\right)^2\right]^{1/2} \tag{8-95}$$

The angle from broadside is $\gamma = \theta - 90°$, so $w = \cos\theta = \cos(\gamma + 90°) = -\sin\gamma$ and $\gamma = -\sin^{-1} w$. The half-power beamwidth based on the ideal Taylor line source is

$$\begin{aligned} \text{HP}_i &= |\theta_{\text{HP}_{i\,\text{left}}} - \theta_{\text{HP}_{i\,\text{right}}}| = |\gamma_{\text{HP}_{i\,\text{left}}} - \gamma_{\text{HP}_{i\,\text{right}}}| \\ &= |\sin^{-1} w_{\text{HP}^+} - \sin^{-1} w_{\text{HP}^-}| = 2|\sin^{-1} w_{\text{HP}}| \\ &= 2\sin^{-1}\left\{\frac{\lambda}{L\pi}\left[\left(\cosh^{-1} R\right)^2 - \left(\cosh^{-1} \frac{R}{\sqrt{2}}\right)^2\right]^{1/2}\right\} \end{aligned} \tag{8-96}$$

where w_{HP^+} and w_{HP^-} are the two solutions of (8-94). The beamwidth for the approximate Taylor line source is given approximately by [13]

$$\mathrm{HP}_w \approx \sigma \mathrm{HP}_{w_i} \tag{8-97}$$

and in θ by

$$\mathrm{HP} \approx 2 \sin^{-1}\left\{\frac{\lambda\sigma}{L\pi}\left[(\cosh^{-1} R)^2 - \left(\cosh^{-1}\frac{R}{\sqrt{2}}\right)^2\right]^{1/2}\right\} \tag{8-98}$$

EXAMPLE 8-7 ***A 10-Wavelength Taylor Line Source with −25-dB Side Lobes and $\bar{n} = 5$***

The side-lobe ratio is

$$R = 10^{-\mathrm{SLL}/20} = 10^{1.25} = 17.7828 \tag{8-99}$$

From (8-71),

$$A = \frac{1}{\pi}\cosh^{-1} R = 1.13655 \tag{8-100}$$

Then from (8-81),

$$\sigma = \frac{\bar{n}}{\sqrt{A^2 + (\bar{n} - \frac{1}{2})^2}} = 1.07728 \tag{8-101}$$

If we use values of A and σ, the zero locations x_n can be calculated from (8-80), and then the sample coefficients follow from (8-92) as given in Table 8-7. The sample locations from (8-87) are also tabulated. The pattern and current distribution can now be computed from (8-90) and (8-91) with the sample values and locations of Table 8-7. The resulting pattern and current distribution are plotted in Fig. 8-10. The side-lobe decay envelope for the far-out side lobes of the pattern is shown in Fig. 8-10*a*. The half-power beamwidths from (8-95) to (8-98) are

$$\mathrm{HP}_{wi} = 0.0978, \qquad \mathrm{HP}_i = 5.606° \tag{8-102}$$

and

$$\mathrm{HP}_w \approx 0.1054, \qquad \mathrm{HP} \approx 6.039° \tag{8-103}$$

In this case, the ideal Taylor line source beamwidth is very close to that of the approximate Taylor line source. The half-power beamwidth HP_w is indicated in Fig. 8-10*a*.

Table 8-7 Sample Values and Locations for the Taylor Line Source of Example 8-7 ($L = 10\lambda, \bar{n} = 5$)

n	$a_n = f(n, A, \bar{n})$ $= f(n, 1.13655, 5)$	w_n^s
0	1.000000	0
±1	0.221477	±0.1
±2	−0.005370	±0.2
±3	−0.006621	±0.3
±4	0.004917	±0.4

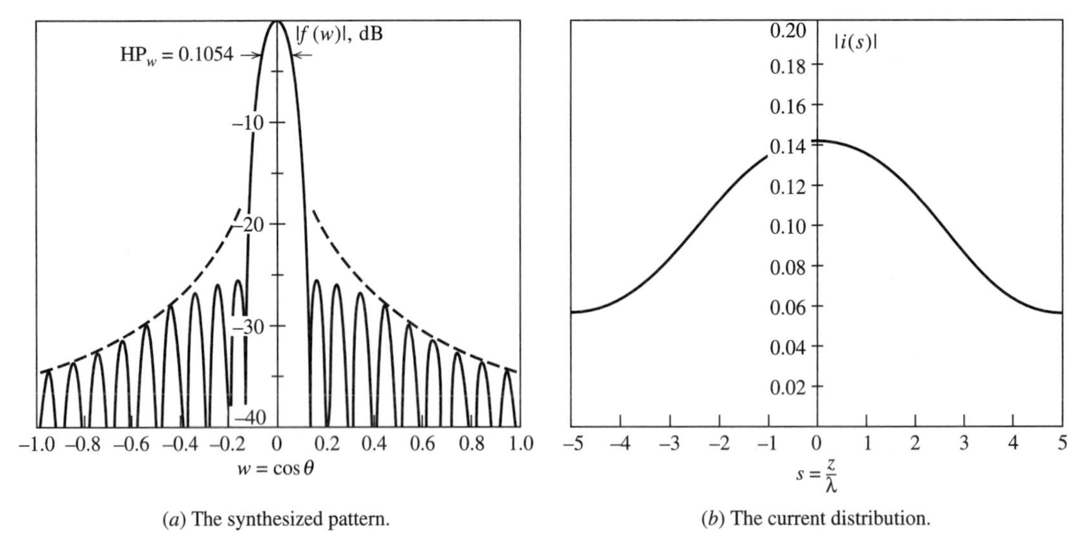

(*a*) The synthesized pattern. (*b*) The current distribution.

Figure 8-10 A 10λ Taylor line source with -25-dB side lobes and $\bar{n} = 5$ (Example 8-7).

8.5 PERSPECTIVE

This chapter introduced several techniques for synthesizing both shaped main beam and low side-lobe radiation patterns. The methods were chosen for their simplicity and because they are in widespread use. Many other techniques have been developed and, with foundations laid in this chapter, the student can pursue more advanced approaches. We mention a few here.

The array factor for an equally spaced array can be written as a polynomial in the variable $Z = e^{j\psi}$. Properties of the array polynomial can be related to its roots [14, 15]. Modern techniques exist to derive the array polynomial, and associated array excitation, for synthesizing custom-shaped patterns [16]. There are also techniques for synthesizing narrow main beam, low side-lobe patterns for planar arrays and apertures [2, 3].

We close by mentioning one technique that can be applied to all antenna types and patterns. The *iterative sampling method* [4] is an extension of the Woodward–Lawson method that can be applied to apertures or arrays, planar or linear. It can be used for shaped main beams and/or low side lobes as well as other custom pattern types. It is based on the simple concept of adding correction beams similar to the Woodward–Lawson component patterns, positioned and weighted to reduce the deviation of the actual pattern from the desired pattern. This is done successively until the desired effect is achieved, such as reducing ripple on the main beam or lowering side lobes.

REFERENCES

1. R. E. Collin and F. J. Zucker, Eds., *Antenna Theory*, Part 1, McGraw-Hill, New York, 1969, Chap. 7.
2. R. C. Hansen, *Handbook of Antenna Design*, Vol. 2, edited by A. W. Rudge et al., IEE/Peregrinus, London, 1983, Chap 9 and 10.
3. R. C. Hansen, "Array Pattern Control and Synthesis," *Proc. IEEE*, Vol. 80, pp. 141–151, Jan. 1992.
4. W. L. Stutzman and E. L. Coffey, "Radiation Pattern Synthesis of Planar Antennas Using the Iterative Sampling Method," *IEEE Trans. Ant. & Prop.*, Vol. AP-23, pp. 764–769.

5. P. M. Woodward, "A Method for Calculating the Field over a Plane Aperture Required to Produce a Given Polar Diagram," *J. IEE*, Vol. 93, Part IIIA, pp. 1554–1558, 1946.
6. P. M. Woodward and J. D. Lawson, "The Theoretical Precision with Which an Arbitrary Radiation Pattern May be Obtained from a Source of Finite Extent," *J. IEE*, Vol. 95, Part II, pp. 363–370, Sept. 1948.
7. C. L. Dolph, "A Current Distribution for Broadside Arrays Which Optimizes the Relationship between Beam Width and Side-lobe Level," *Proc. IRE*, Vol. 34, pp. 335–348, June 1946.
8. A Safaai-Jazi, "A New Formulation of the Design of Chebyshev Arrays," *IEEE Trans. Ant. & Prop.*, Vol. 42, pp. 439–443, March 1994.
9. A Safaai-Jazi, "Directivity of Chebyshev Arrays with Arbitrary Element spacing," *Electronics Lett.*, Vol. 31, pp. 772–774, 11 May 1995.
10. R. J. Mailloux, *Phased Array Antenna Handbook*, Artech House, Boston, MA, 1994, Chap. 3.
11. N. Yaru, "A Note on Super-Gain Antenna Arrays," *Proc. IRE*,Vol. 39, pp. 1081–1085, Sept. 1951.
12. T. T. Taylor, "Design of Line-Source Antennas for Narrow Beamwidth and Low Sidelobes," *IRE Trans. Ant. & Prop.*, Vol. AP-3, pp. 16–28, Jan. 1955.
13. R. C. Hansen, Ed., *Microwave Scanning Antennas*, Vol. I, Academic Press, New York, 1964.
14. S. A. Schelkunoff, "A Mathematical Theory of Linear Arrays," *Bell System Tech. J.*, Vol. 22, pp. 80–107, Jan. 1943.
15. T. T. Taylor and J. R. Whinnery, "Application of Potential Theory to the Design of Linear Arrays," *J. Appl. Phy.*, Vol. 22, pp. 19–29, Jan. 1951.
16. H. J. Orchard, R. S. Elliott, and G. J. Stern, "Optimizing the Synthesis of Shaped Antenna Patterns," *IEE Proc.*, Vol. 132, Part H, pp. 63–68, Feb. 1985.

PROBLEMS

8.1-1 If $g(t)$ and $G(\omega)$ are a Fourier transform pair, then

$$G(\omega) = \int_{-\infty}^{\infty} g(t)e^{-j\omega t}\, dt \qquad \text{and} \qquad g(t) = \frac{1}{2\pi}\int_{-\infty}^{\infty} G(\omega)e^{j\omega t}\, d\omega$$

If g, t, G, and ω are replaced by f, u, i, and $2\pi s$, respectively, show that (8-6) and (8-7) follow from the above equations.

8.1-2 Decomposing a linear current distribution $i(s)$ into real/imaginary and even/odd parts and applying the Fourier transform, give the following pattern expression:

$$\begin{aligned} f(w) &= 2\int_0^{\infty} [i_{re}(s)\cos 2\pi ws - i_{io}(s)\sin 2\pi ws]\, ds \\ &\quad + j2\int_0^{\infty} [i_{ie}(s)\cos 2\pi ws + i_{ro}(s)\sin 2\pi ws]\, ds \end{aligned}$$

Use this to prove the following: (a) Property 1 in Table 8-2, and (b) Property 2 in Table 8-2 by first forming the magnitude of $f(w)$.

8.2-1 A narrow pencil beam pattern represented by a delta function is scanned to the location $\theta = 53.1°$. Find the required current distribution using Fourier line source synthesis.

8.2-2 Use (8-6) to find the pattern from a uniform amplitude, zero phase line source of length L centered on the z-axis.

8.2-3 Derive the current distribution $i_d(s)$ required to exactly produce the sector pattern of (8-11b).

8.2-4 a. Derive the Fourier transform synthesis pattern of (8-13) for a sector pattern.

b. Plot this pattern, thus verifying Fig. 8-1*a*. Numerical integration of the Fourier transform via computer may be easier than using (8-13).

8.2-5 Derive (8-15).

8.2-6 Repeat the Woodward–Lawson synthesis of the sector pattern of Example 8-2, but this time for a five-wavelength line source.

a. Plot the pattern in linear, rectangular form as a function of w.
b. Plot the current distribution.

8.2-7 A cosecant pattern (see Prob. 1.8-7 for a discussion of the cosecant pattern) is given by

$$f_d(w) = \begin{cases} 1 & 0 \le w \le 0.1 \\ \dfrac{0.1}{w} & 0.1 \le w \le 0.5 \\ 0 & \text{elsewhere} \end{cases}$$

Use the Woodward–Lawson method to synthesize an approximation to this pattern for a 10λ line source.

a. Plot the pattern in linear, rectangular form together with the desired pattern as a function of w.

b. Plot the required current amplitude and phase.

8.3-1 Discuss the conditions on $f_d(w)$ such that it can be represented by the Fourier series in (8-24).

8.3-2 Derive the Fourier series coefficient expression in (8-25).

8.3-3 Derive the element current expression (8-29) for the Fourier series synthesis of a sector pattern.

8.3-4 Synthesize a sector pattern with $c = 0.5$ using the Fourier series method as in Example 8-3 for an array of 20 elements that are spaced 0.6λ apart.

a. Determine the element locations and current values.

b. Plot the radiation pattern in linear, rectangular form as a function of w.

8.3-5 Repeat Prob. 8.3-4 for an array of 10 elements and half-wavelength spacings.

8.3-6 Use the Fourier series synthesis method to synthesize a sector pattern with $c = 0.5$ for an array of 21 half-wavelength spaced elements. Derive the general element current expression and evaluate for each element. Plot the pattern. Compare pattern parameters to the 20-element array result of Example 8-3.

8.3-7 Repeat the cosecant pattern synthesis as in Prob. 8.2-7a, using the Fourier series method for a 20-element, half-wavelength spaced array. Tabulate the element current values.

8.3-8 Show that the Woodward–Lawson sampling method pattern of (8-32) arises from the array factor with the currents of (8-35) for:

a. An odd number of elements. *Hint:* Use (8-19) and (8-20).

b. An even number of elements. *Hint:* Use (8-21) and (8-22).

8.3-9 Verify the array element positions and currents of Table 8-5 for the Woodward–Lawson synthesized sector pattern of Example 8-4.

8.3-10 Repeat the Woodward–Lawson synthesis as in Example 8-4 for a 10-element half-wavelength spaced array.

8.3-11 A collinear array of 18 half-wave dipole antennas is to be used to synthesize a sector pattern with a main beam sector over the region $70° \le \theta \le 110°$, that is, $F_d(\theta) = 1$ over this region and zero elsewhere.

a. For 0.65λ spacings, determine the input currents required for Woodward–Lawson synthesis of the complete pattern. Account for the element pattern.

b. Plot the total array pattern in linear, polar form as a function of θ.

8.3-12 Repeat Prob. 8.3-11 for a cosecant desired pattern, where $F_d(\theta)$ is 1 for $80° \le \theta \le 90°$, $\cos 80°/\cos\theta$ for $0° \le \theta \le 80°$, and zero elsewhere. Use 18 pattern samples.

8.4-1 For the five-element, broadside, −20-dB side-lobe, half-wavelength spaced Dolph–Chebyshev array of Example 8-5:

a. Obtain the pattern plot in logarithmic, rectangular form as a function of w.

b. Verify the side-lobe level and beamwidth from your pattern calculations.

8.4-2 Design a Dolph-Chebyshev broadside array of five, half-wavelength spaced elements for −30-dB side lobes.

a. Verify the current distribution as given in Fig. 3-23*e*.

b. Compute the directivity.

8.4-3 Design a broadside Dolph–Chebyshev array with six, 0.6λ spaced elements for -25-dB side lobes.

a. Obtain the element currents.
b. Plot the pattern in logarithmic, rectangular form.

8.4-4 Design a low side-lobe, broadside collinear array of half-wave dipoles. Use isotropic elements to design a Dolph–Chebyshev array with eight elements for the narrowest beamwidth and -20-dB side lobes. Evaluate and plot the patterns in polar-dB form with and without the element pattern.

8.4-5 Derive the ideal Taylor line source pattern results of (8-76) and (8-77).

8.4-6 Show how the approximate Taylor line source pattern of (8-82) follows from the zero locations.

8.4-7 The sampling theorem from time-signal analysis states that a signal $g(t)$ is exactly reconstructed from the time samples $g(m/2B)$ as

$$g(t) = \sum_{m=-\infty}^{\infty} g\left(\frac{m}{2B}\right) \mathrm{Sa}\left[2\pi B\left(t - \frac{m}{2B}\right)\right]$$

where B is the highest frequency component of the signal. Draw the appropriate analogies to antenna theory to obtain the sampled data pattern expression of (8-86).

8.4-8 Verify (8-94).

8.4-9 Compute the sample values a_n of Table 8-7 for the Taylor line source of Example 8-7.

8.4-10 Compute the half-power beamwidth values for the Taylor line source of Example 8-7. Compare your answers to those of (8-102) and (8-103).

8.4-11 An array antenna can be designed by choosing the element current excitations at the corresponding points of the continuous current from a line source synthesized for the desired pattern. This is illustrated in this problem with a narrow main beam, low side-lobe pattern. The Taylor line source of Example 8-7 has current values appropriate for a 20-element array given in the table.

Array Excitations for Problem 8.4-11

m	z_m/λ	i_m
± 1	± 0.25	0.14234
± 2	± 0.75	0.13833
± 3	± 1.25	0.13127
± 4	± 1.75	0.12175
± 5	± 2.25	0.10935
± 6	± 2.75	0.09429
± 7	± 3.25	0.07891
± 8	± 3.75	0.06676
± 9	± 4.25	0.05980
± 10	± 4.75	0.05720

a. Use these current values to obtain the array factor of the corresponding linear array.
b. Compare and comment on the half-power beamwidths and side-lobe levels of the array and line source patterns.

8.4-12 Design an eight-wavelength Taylor line source ($\bar{n} = 7$) with -30-dB side lobes.

a. Obtain and tabulate the sample values and locations.
b. Plot the pattern in rectangular-logarithmic form as a function of w.
c. Plot the current distribution.

8.4-13 Evaluate σ for several values of $\bar{n}$ for the case of a −25-dB side-lobe level. Using $\text{HP}_w \approx \sigma \text{HP}_{wi}$, explain the half-power beamwidth behavior as a function of $\bar{n}$.

8.4-14 Design an optimum broadside Dolph–Chebyshev array with 10 elements and −20-dB side lobes. With the same array geometry, find the element currents by sampling a −20-dB Taylor line source ($\bar{n} = 8$) current distribution with the same length. Plot the polar-dB patterns for both arrays and compare for (a) isotropic elements and (b) collinear half-wave dipoles along the line of the array.

8.4-15 *Effect of mutual coupling on array synthesis.* Use a moment method code to evaluate the array of Example 8-5; see Chap. 10 and Appendix G. Use resonated half-wavelength dipoles parallel to the x-axis with centers along the z-axis with voltage sources proportional to the desired currents. Compute the yz-plane pattern and compare to that of the example. Tabulate the side-lobe levels and currents for the two approaches.

Chapter 9

Antennas in Systems and Antenna Measurements

Antennas are used in communication, radar, and radiometer systems. These system topics as well as antenna measurements are discussed in this chapter. A knowledge of how antenna measurements are made is valuable for interpreting measured data. Also, the study of antenna measurements increases understanding of quantities such as pattern, gain, and polarization.

9.1 RECEIVING PROPERTIES OF ANTENNAS

Antennas are used to receive signals from distant sources by converting the arriving power density to a current on a transmission line that connects to a receiver. It is important to carefully account for all losses since the signals are often very weak. Essential to this is proper modeling of the receiving antenna. This section addresses various receiving antenna models and discusses losses associated with the system configuration, including polarization and impedance mismatch losses. Several basic definitions that will be useful are found in Secs. 1.8 and 1.9.

The signal power received by an antenna is proportional to its gain in the direction of the signal, as will be developed in (9-54):

$$G(\theta, \phi) = G|F(\theta, \phi)|^2 \tag{9-1}$$

For receiving antennas, gain is better expressed in terms of effective aperture by generalizing (2-89):

$$G(\theta, \phi) = \frac{4\pi}{\lambda^2} A_e(\theta, \phi) \tag{9-2}$$

Effective aperture $A_e(\theta, \phi)$ is a very important antenna parameter that can be thought of as the "collecting area" of the antenna. It is a measure of the ability of an antenna to collect power from space around the antenna and deliver it to a terminating device. It depends on the direction of arrival of the incident wave, so for a signal arriving at the antenna from direction (θ, ϕ), the effective aperture is $A_e(\theta, \phi)$ and $A_e|F(\theta, \phi)|^2$. For a wave arriving from the direction of the radiation pattern maximum, $A_e(\theta_{max}, \phi_{max}) = A_e$ since the maximum of $|F(\theta, \phi)|^2$ is 1. Although effective aperture is more intuitive for receiving antennas, it also applies to transmitting antennas.

The power received by an antenna with effective aperture A_{er} from (2-93) is $P_r = SA_{er}$ when the pattern maximum is aimed toward the incoming signal. Ohmic losses on the antenna are included in A_{er}; however, the effects of polarization and impedance mismatch are not. The definitions of gain and effective aperture could be modified to include mismatch effects. However, the gain value would then only be useful for those particular operating conditions. The power delivered to load attached to the receiving antenna from (2-96) is

$$P_D = pqP_r \tag{9-3}$$

The polarization efficiency p is nearly 1 in most system applications since adjustments can be made to achieve an approximate polarization match between the incoming wave and the receiving antenna. Impedance match may be more difficult to achieve. Thus, losses due to impedance mismatch are usually present in operational systems, so the impedance mismatch factor q is usually less than 1. Also, when performing gain measurements, care must be taken to impedance match the test antenna. We now discuss impedance and polarization mismatch factors in more detail.

Impedance Mismatch. The power delivered to the load resistor R_L from Fig. 2-18b is

$$P_D = \frac{1}{2}|I_A|^2 R_L = \frac{1}{2}\frac{|V|^2}{(R_A + R_L)^2 + (X_A + X_L)^2} R_L \tag{9-4}$$

Maximum power will be transferred to R_L when a conjugate impedance match exists:

$$R_L = R_A, \qquad X_L = -X_A \qquad \textit{conjugate match} \tag{9-5}$$

Then

$$P_{D_{\max}} = \frac{1}{8}\frac{|V|^2}{R_A} \tag{9-6}$$

The fraction of power delivered, or *impedance mismatch factor*, is, in general, given by the ratio of (9-4) to (9-6):

$$q = \frac{P_D}{P_{D_{\max}}} = \frac{4R_A R_L}{(R_A + R_L)^2 + (X_A + X_L)^2} \tag{9-7}$$

When matched as in (9-5), this reduces to $q = 1$. In the usual situation of a transmission line with characteristic impedance Z_o connected to the antenna,

$$q = \frac{4R_A Z_o}{(R_A + Z_o)^2 + X_A^2}, \qquad R_L = Z_o, X_L = 0 \tag{9-8}$$

In many cases, the antenna impedance is not known, but instead the voltage standing wave ratio (VSWR) has been measured. Since the magnitude of the reflection coefficient can be found from VSWR, the fraction of power traveling down the line is

$$\begin{aligned} q &= 1 - |\Gamma|^2 \\ &= 1 - \left|\frac{Z_o - Z_A}{Z_o + Z_A}\right|^2 = 1 - \left[\frac{\text{VSWR} - 1}{\text{VSWR} + 1}\right]^2 \end{aligned} \tag{9-9}$$

When the antenna is matched to the transmission line, VSWR = 1 and $Z_A = Z_o$, giving $q = 1$ that indicates there is no mismatch loss. For a large mismatch (and large value of VSWR), q approaches zero. Antenna gain for a given polarization reduced by impedance mismatch (i.e., qG) is directly measurable and is called *realized gain.*

Polarization Mismatch. Polarization principles were introduced in Sec. 1.10 and we can use them to determine polarization mismatch. The *polarization efficiency* (or *polarization mismatch factor*) varies from 0 to 1 as the incoming wave and receiving antenna vary from completely mismatched in polarization to completely matched. A complete match ($p = 1$) exists when the wave and antenna polarization states are identical. A complete mismatch ($p = 0$) occurs when the wave and antenna are *cross-polarized.* Examples of cross-polarized states are orthogonal linear states such as horizontal and vertical linear polarizations, and right-hand and left-hand circular polarizations.

The interaction of the incident wave electric field $\mathbf{E^i}$ with the receiving antenna is facilitated by the concept of *vector effective length* of an antenna $\mathbf{h}$, which is defined through

$$V_A = \mathbf{E^i} \cdot \mathbf{h}^* \tag{9-10}$$

V_A is the open circuit voltage across the antenna terminals, which follows from Fig. 2-18 with Z_L removed. The receiving antenna relation of (9-10) applies to any antenna and is very intuitive. The dot product gives the projection of the incident field vector $\mathbf{E^i}$ in volts per meter onto the vector effective length $\mathbf{h}$ in meters, resulting in the output voltage V_A in volts. For example, if both the wave and antenna are linearly polarized and aligned, maximum output voltage will result. Vector effective length describes both the phase and polarization properties of the antenna. The complex conjugate is used because $\mathbf{h}$ is associated with the transmitting case and (9-10) is a receiving relationship. That is, the complex conjugate acts to reverse the reference direction; see Prob. 9.1-4.

EXAMPLE 9-1 *Vector Effective Length of an Ideal Dipole*

As an example, consider the radiation electric field of an ideal dipole, which from (1-72a) is

$$\mathbf{E} = \frac{j\omega\mu I}{4\pi}\frac{e^{-j\beta r}}{r}\Delta z \sin\theta\,\hat{\boldsymbol{\theta}} \tag{9-11}$$

Since $\mathbf{h}$ contains information on the size of the antenna and the angular dependence of the radiation pattern, we can write

$$\mathbf{E} = \frac{j\omega\mu I}{4\pi}\frac{e^{-j\beta r}}{r}\mathbf{h} \tag{9-12}$$

where

$$\mathbf{h} = \Delta z \sin\theta\,\hat{\boldsymbol{\theta}} \tag{9-13}$$

Note that the dimension of $\mathbf{h}$ is length and this equation has the obvious interpretation that the effective length of the ideal dipole is the projection of the physical length viewed from the angle θ. This, however, is not true in general. The vector effective length of a small loop antenna is treated in Prob. 9.1-5.

Polarization information for the wave and antenna is contained in $\mathbf{E^i}$ and $\mathbf{h}$, respectively, and polarization efficiency can be determined from them. Received power is proportional to the terminal voltage squared, which from (9-10) is $|\mathbf{E^i} \cdot \mathbf{h}^*|^2$. Normalizing yields the fraction of power received

$$p = \frac{|\mathbf{E^i} \cdot \mathbf{h}^*|^2}{|\mathbf{E^i}|^2 |\mathbf{h}|^2} = |\hat{\mathbf{e}}^{\mathbf{i}} \cdot \hat{\mathbf{h}}^*|^2 \tag{9-14}$$

where $\hat{\mathbf{e}}^{\mathbf{i}}$ and $\hat{\mathbf{h}}$ are the complex unit vectors for the incident wave and antenna vector length, respectively. $\hat{\mathbf{e}}^{\mathbf{i}}$ represents the polarization state of the incident wave and equals that of the distant transmitting antenna $\hat{\mathbf{h}}_{\mathbf{t}}$ if the intervening propagation medium does not depolarize the wave. Based on (9-14), a receiving antenna with vector length $\hat{\mathbf{h}}$ is said to be polarization-matched (i.e., *co-polarized*) or orthogonally polarized (i.e., *cross-polarized*) to the incoming wave when

$$|\hat{\mathbf{e}}^{\mathbf{i}} \cdot \hat{\mathbf{h}}^*| = 1 \qquad \textit{co-polarized} \tag{9-15a}$$

$$\hat{\mathbf{e}}^{\mathbf{i}} \cdot \hat{\mathbf{h}}^* = 0 \qquad \textit{cross-polarized} \tag{9-15b}$$

Polarization mismatch is easily evaluated using (9-14) because the wave and antenna polarizations are expressed in their own relative coordinates. That is, the wave polarization is expressed in xy-coordinates with the z-axis in the direction of wave propagation, and the antenna state uses xy-coordinates as shown in Fig. 9-1. The z-axis for antenna coordinates is directed away from the antenna since antenna polarization is always defined for transmission. Note that tilt angles are taken rela-

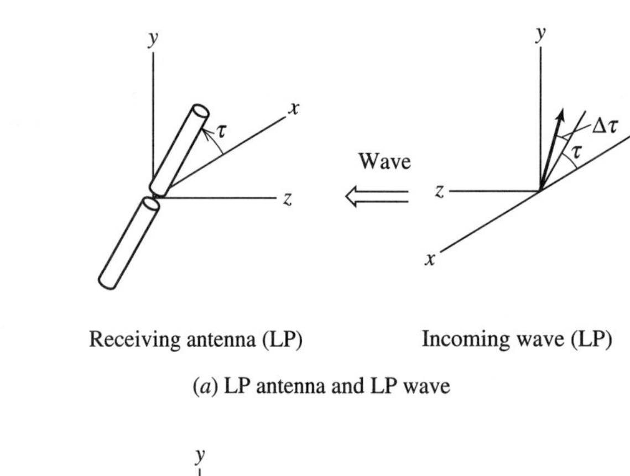

(*a*) LP antenna and LP wave

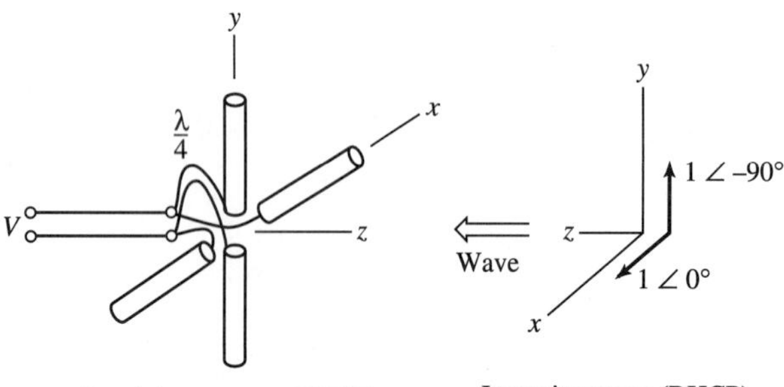

(*b*) RHCP antenna and RHCP wave

Figure 9-1 Illustration of reception of an incident wave with electric field $\mathbf{E^i}$ by a receiving antenna.

tive to the receiving antenna x-axis; see $\Delta\tau$ in Fig. 9-1a. Then the wave and antenna polarizations written in the form of (1-190) are

$$\hat{\mathbf{e}}^{\mathbf{i}} = \cos\gamma_i\,\hat{\mathbf{x}} + \sin\gamma_i\,e^{j\delta_i}\,\hat{\mathbf{y}} \tag{9-16a}$$

$$\hat{\mathbf{h}} = \cos\gamma\,\hat{\mathbf{x}} + \sin\gamma\,e^{j\delta}\,\hat{\mathbf{y}} \tag{9-16b}$$

where (γ_i, δ_i) and (γ, δ) are the polarization parameters associated with the incoming wave and the antenna in the arrival direction; see Fig. 1-24. The process of evaluating polarization efficiency is illustrated with the following examples.

EXAMPLE 9-2 *Reception of an LP Wave with an LP Antenna*

A linearly polarized (LP) incident wave with a tilt angle of τ_i illuminates an LP antenna at tilt angle τ as shown in Fig. 9-1, where a dipole is used to illustrate a general LP antenna. The wave arrives normal to the plane of the dipole antenna (xy-plane), corresponding to the usual operating situation for a receiving antenna. The angles γ_i and γ equal τ_i and τ, respectively, in the LP case; these follow from (1-191) with $\varepsilon = 0$ and $\varepsilon_i = 0$. Substituting these values in (9-16) permits evaluation of (9-14) as follows:

$$\begin{aligned} p &= |\hat{\mathbf{e}}^{\mathbf{i}} \cdot \hat{\mathbf{h}}^*|^2 = |(\cos\tau_i\,\hat{\mathbf{x}} + \sin\tau_i\,\hat{\mathbf{y}}) \cdot (\cos\tau\,\hat{\mathbf{x}} + \sin\tau\,\hat{\mathbf{y}})|^2 \\ &= |\cos\tau_i\cos\tau + \sin\tau_i\sin\tau|^2 = \cos^2(\tau_i - \tau) \\ &= \cos^2(\Delta\tau) \end{aligned} \tag{9-17}$$

Thus, polarization efficiency is a function only of the relative tilt angle $\Delta\tau$ when both the wave and antenna are LP. When $\Delta\tau = 0°$, the wave and antenna are aligned (e.g., $\mathbf{E}^{\mathbf{i}}$ is parallel to the dipole). The wave and antenna are copolarized; that is, they are polarization-matched and $p = 1$. When the wave and antenna are orthogonal, $\Delta\tau = 90°$ and (9-17) yields $p = 0$. Then the receiving antenna produces no output and the wave and antenna are cross-polarized. This is an idealized result because in practice, most antennas have some response to a wave state cross-polarized to their nominal co-polarized state.

EXAMPLE 9-3 *Reception of CP Wave with a CP Antenna*

Examination of circular polarization (CP) reveals the power of the complex vector formation of polarization states and the role of the complex conjugate in (9-14). Consider a right-hand circular polarization (RHCP) receiving antenna illustrated in Fig. 9-1*b* as crossed dipoles with a quarter-wave delay line. When operated as a transmitting antenna, this antenna has a polarization state in the $+z$-direction given by

$$\hat{\mathbf{h}} = \frac{1}{\sqrt{2}}(\hat{\mathbf{x}} - j\hat{\mathbf{y}}) \qquad \textit{RHCP} \tag{9-18}$$

since the magnitudes of the x- and y-components are equal and the y-component lags the x-component by 90°. This result also follows from (9-16b) with $\gamma = 45°$, $\delta = -90°$; see Sec. 1.10. Similarly, for a RHCP incident wave

$$\hat{\mathbf{e}}^{\mathbf{i}} = \frac{1}{\sqrt{2}}(\hat{\mathbf{x}} - j\hat{\mathbf{y}}) \qquad \textit{RHCP} \tag{9-19}$$

The polarization efficiency from (9-14) is then

$$p = |\hat{\mathbf{e}}^{\mathbf{i}} \cdot \hat{\mathbf{h}}^*|^2 = \left|\frac{1}{\sqrt{2}}(\hat{\mathbf{x}} - j\hat{\mathbf{y}}) \cdot \frac{1}{\sqrt{2}}(\hat{\mathbf{x}} - j\hat{\mathbf{y}})^*\right|^2 = 1 \tag{9-20}$$

and the wave is perfectly matched to the antenna. Note that the wave and antenna polarizations are expressed in their own coordinate systems as shown in Fig. 9-1*b*, with the z-direction for the wave and antenna polarizations taken in the wave propagation direction and in the direction of radiation when the antenna transmits, respectively. An analogy for the use of relative coordinates is two people being "matched" when shaking hands if they use their right hands. In the same fashion, a RHCP wave is matched to a RHCP receiving antenna. The result in (9-20) can also be explained by examining how the antenna responds to the incoming wave. The x-dipole produces a voltage of $1\angle 180°$; the 180° is included because of the opposite reference direction of the x-axes of the wave and antenna. The y-dipole is excited by $1\angle 90°$ and its output is delayed by 90° due to the quarter-wavelength section, producing a net $1\angle 180°$ excitation at the connecting transmission line terminals. Combining the voltage from the two dipoles gives $2\angle 180°$, indicating complete reinforcement of the x- and y-components. Therefore, the antenna is matched to the wave. Note that if the wave is left-hand CP, then the phase of the y-component of the wave is $+90°$ rather than $-90°$ and there is complete cancellation at the transmission line. For LHCP, the sign of the $\hat{\mathbf{y}}$-term in (9-19) would be positive and $p = 0$, indicating a cross-polarized situation.

EXAMPLE 9-4 *Reception of an LP Wave by a CP Antenna*

Now suppose the LP wave of Fig. 9-1*a* is incident on the CP antenna of Fig. 9-1*b*. The polarization efficiency is evaluated using $\hat{\mathbf{e}}^{\mathbf{i}}$ from Example 9-2 and $\hat{\mathbf{h}}$ from Example 9-3:

$$p = |\hat{\mathbf{e}}^{\mathbf{i}} \cdot \hat{\mathbf{h}}^*|^2 = \left|(\cos \tau_i \,\hat{\mathbf{x}} + \sin \tau_i \,\hat{\mathbf{y}}) \cdot \frac{1}{\sqrt{2}} (\hat{\mathbf{x}} - j\hat{\mathbf{y}})^*\right|^2 = \frac{1}{2} |\cos \tau_i + j \sin \tau_i|^2 = \frac{1}{2} \tag{9-21}$$

Thus, one-half of the power available from an LP wave is lost when received by a CP antenna. The same is true for a CP wave and an LP antenna. In most system applications, this 3-dB loss is significant and an antenna matched to the wave must be used. On the other hand, there are operational links with one antenna linear and the one circular. For example, if a spacecraft has a linearly polarized antenna, the effects due to spacecraft motion or Faraday rotation in the ionosphere on the incoming linearly polarized wave orientation angle will not lead to power level fluctuations if a circularly polarized receive antenna is used. Even though a 3-dB signal loss is encountered, the received signal remains constant.

9.2 ANTENNA NOISE TEMPERATURE AND RADIOMETRY

Receiving systems are vulnerable to noise and a major contribution is the receiving antenna, which collects noise from its surrounding environment. Antenna noise and radiometry are introduced in this section. In most situations, a receiving antenna is surrounded by a complex environment as shown in Fig. 9-2*a*. Any object (except a perfect reflector) that is above absolute zero temperature will radiate electromagnetic waves. An antenna picks up this radiation through its antenna pattern and produces noise power at its output. The equivalent terminal behavior is modeled in Fig. 9-2*b* by considering the radiation resistance of the antenna to be a noisy resistor at a temperature T_A. The *antenna temperature* T_A is not the actual physical temperature of the antenna, but is an equivalent temperature that produces the same noise power P_{NA} as the antenna operating in its surroundings. The noise power available from the noise resistor R_r in bandwidth Δf at temperature T_A is

$$P_{NA} = kT_A \,\Delta f \tag{9-22}$$

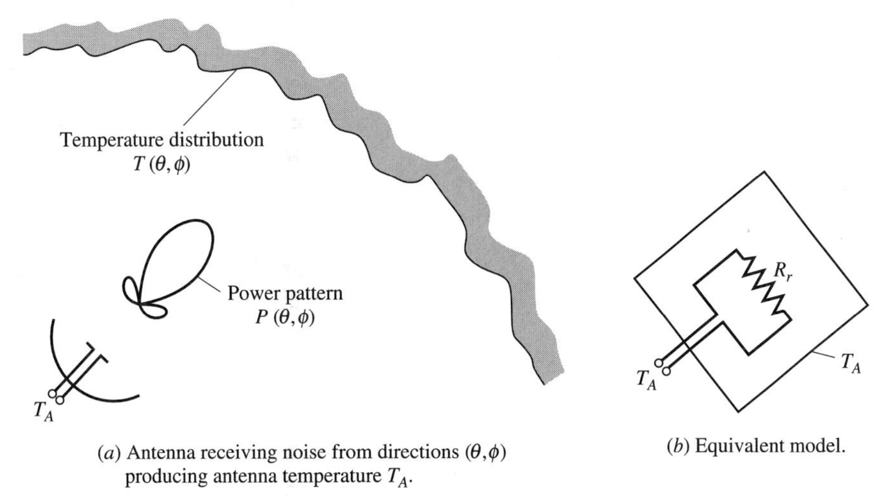

(*a*) Antenna receiving noise from directions (θ, ϕ) producing antenna temperature T_A.

(*b*) Equivalent model.

Figure 9-2 Antenna temperature.

where

P_{NA} = available power due to antenna noise (W)
k = Boltzmann's constant = 1.38×10^{-23} J K^{-1}
T_A = antenna temperature (K)
Δf = receiver bandwidth (Hz)

Such noise is often referred to as Nyquist or Johnson noise. The system noise power P_N is calculated using the total system noise temperature T_{sys} in place of T_A in (9-22) with $T_{sys} = T_A + T_r$, where T_r is the receiver noise temperature.

There are two motivations for studying the noise performance of antennas: for noise corruption to communications and active remote-sensing systems and for noise calculations of passive remote-sensing systems. In most communications and active remote sensing systems, the noise power level must be sufficiently below the signal level for proper operation. This is especially challenging in long distance communication systems with weak signals. In these cases, the system is evaluated through "carrier-to-noise ratio," which is determined from the signal power and the system noise power:

$$\text{CNR} = \frac{P_D}{P_N} \tag{9-23}$$

where $P_N = kT_{sys}\,\Delta f$.

A second reason for studying antenna temperature is for radiometry (passive remote sensing). A radiometer picks up noise from "hot" objects and can form an image by scanning through a noise scene with a narrow beam antenna. An example is a satellite microwave radiometer viewing precipitation on the earth's surface from space. Water particles are lossy and produce noise, just as does a noise resistor. In the radiometer case, noise from the desired passive object acts as the "signal" and is compared to undesired noise due to the background scene and receiver noise.

Noise power is found by first evaluating antenna temperature. As seen by Fig. 9-2*a*, T_A is found from the collection of noise through the scene temperature distribution $T(\theta, \phi)$ weighted by the response function of the antenna, the normalized power pattern $P(\theta, \phi)$. This is expressed mathematically by integrating over the temperature distribution:

$$T_A = \frac{1}{\Omega_A} \int_0^{\pi} \int_0^{2\pi} T(\theta, \phi) P(\theta, \phi) \, d\Omega \tag{9-24}$$

Figure 9-2*a* illustrates an earth terminal antenna looking at the sky, but (9-24) is completely general. The form of (9-24) can be rationalized by examining a few special cases. If the scene is of constant temperature T_o over all angles, T_o comes out of the integral and then

$$T_A = \frac{T_o}{\Omega_A} \int_0^{\pi} \int_0^{2\pi} P(\theta, \phi) \, d\Omega = \frac{T_o}{\Omega_A} \Omega_A = T_o \tag{9-25}$$

This is an expected result. The antenna is completely surrounded by noise of temperature T_o and its output antenna temperature equals T_o independent of the antenna pattern shape. For the case of a discrete source of small solid angular extent Ω_s and constant temperature T_s, if the antenna beam is directed toward its center and is broad compared to the source, $P(\theta, \phi) \approx 1$ over the source and then (9-24) reduces to

$$T_A = \frac{\Omega_s}{\Omega_A} T_s \qquad \textit{small discrete source} \tag{9-26}$$

In general, antenna noise power P_{NA} is found from (9-22), using T_A from (9-24) once the temperature distribution $T(\theta, \phi)$ is determined. Of course, this depends on the scene, but in general, $T(\theta, \phi)$ consists of two components: sky noise and ground noise. Ground noise temperature in most situations is well approximated for soils by the value of 290 K. Surfaces that are highly reflective have a ground temperature close to the temperature of the reflected sky noise. Smooth surfaces have high reflection for near-grazing incidence angles.

Unlike ground noise, sky noise is a strong function of frequency. Sky noise is made up of atmospheric, cosmic, and man-made noise. (See [1] for a review of natural radio noise and [2] for a discussion of antenna noise.) Atmospheric noise increases with decreasing frequency below 1 GHz and is primarily due to lightning, which propagates over large distances via ionospheric reflection below several MHz. Atmospheric noise increases with frequency above 10 GHz due to water vapor and hydrometer absorption; these depend on time, season, and location. It also increases with decreasing elevation angle. Atmospheric gases have strong, broad spectral lines, such as water vapor and oxygen lines at 22 and 60 GHz, respectively.

Cosmic noise originates from discrete sources such as the sun, moon, and "radio stars" as well as our galaxy, which has strong emissions for directions toward the galactic center. Galactic noise increases with decreasing frequency below 1 GHz. Man-made noise is produced by power lines, electric motors, etc., and usually can be ignored except in urban areas at low frequencies. Sky noise is very low for frequencies between 1 and 10 GHz, and can be as low as a few K for high elevation angles.

Of course, the antenna pattern strongly influences antenna temperature; see (9-24). The ground noise temperature contribution to antenna noise can be very low for high-gain antennas with low side lobes in the direction of the earth. Broad beam antennas, on the other hand, pick up a significant amount of ground noise as well as sky noise. A figure of merit used with satellite earth terminals is G/T_{sys}, which is the antenna gain divided by system noise temperature usually expressed in dB/K. It is desired to have high values of G to increase signal and to have low values of T_{sys} to decrease noise, giving high values of G/T_{sys}.

EXAMPLE 9-5 ***Direct Broadcast Satellite Reception***

Example 2-3 is revisited here for noise calculations. The receiver uses a 67-K noise temperature low-noise block downconverter. This is the dominant receiver contribution and, when combined with antenna temperature, leads to a system noise temperature of $T_{sys} = 125$ K. The noise power in the effective signal bandwidth $\Delta f = 20$ MHz is

$$\begin{aligned} P_N &= kT_{sys}\,\Delta f \\ &= 1.38 \times 10^{-23} \cdot 125 \cdot 20 \times 10^6 = 3.5 \times 10^{-14} \\ &= -134.6 \text{ dBW} \end{aligned} \tag{9-27}$$

Thus, the carrier-to-noise ratio from (9-23) and (9-27) is

$$\text{CNR (dB)} = P_D \text{ (dBW)} - P_N \text{ (dBW)} = -116.9 - (-134.6) = 17.7 \text{ dB} \tag{9-28}$$

where the received power value from (2-102) was used for P_D. This is a reasonable margin for proper operation.

9.3 RADAR

We now turn our attention to radar. Suppose an airplane is the target of a radar as shown in Fig. 9-3. We assume that the transmit and receive antennas are collocated, forming a *monostatic radar*, and are pointed such that the pattern maxima are directed toward the target. The power density incident on the target is then

$$S^i = \frac{P_t}{4\pi R^2} G_t = \frac{P_t A_{et}}{\lambda^2 R^2} \tag{9-29}$$

where (2-89) and (2-92) were used. The power intercepted by the target is proportional to the incident power density, so

$$P^i = \sigma S^i \tag{9-30}$$

where the proportionality constant σ is the *radar cross section* RCS (m^2) and is the equivalent area of the target based on the target reradiating the incident power isotropically. Although the incident power P^i is not really scattered isotropically, we are only concerned about the power scattered in the direction of the receiver

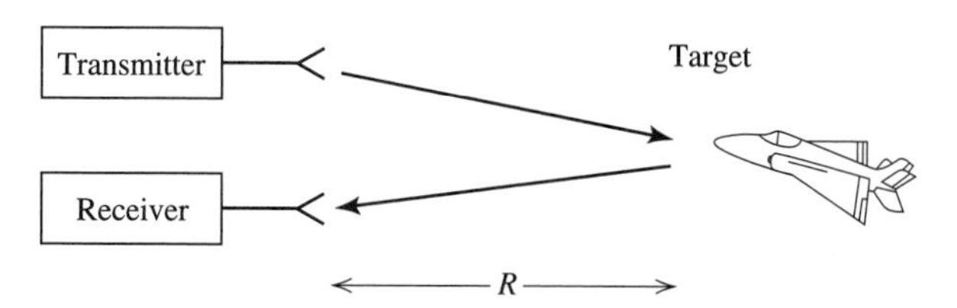

Figure 9-3 Radar example.

and can assume the target scatters isotropically. Because P^i appears to be scattered isotropically, the power density arriving at the receiver is

$$S^s = \frac{P^i}{4\pi R^2} \tag{9-31}$$

The power available at the receiver from (2-93) is

$$P_r = A_{er} S^s \tag{9-32}$$

Combining the above four equations gives

$$P_r = A_{er} \frac{\sigma S^i}{4\pi R^2} = P_t \frac{A_{er} A_{et}\, \sigma}{4\pi R^4 \lambda^2} \tag{9-33}$$

which is referred to as the *radar equation*. Using (2-89), we can rewrite this equation in a convenient form as

$$P_r = P_t \frac{\lambda^2 G_r G_t \sigma}{(4\pi)^3 R^4} \tag{9-34}$$

If the transmitting and receiving antennas are identical as is usually the case, $G_r G_t = G^2$. Polarization and impedance mismatch effects are included using (9-3).

Combining (9-30) and (9-31) forms a definition of radar cross section:

$$\sigma = \frac{4\pi R^2 S^s}{S^i} \tag{9-35}$$

which is the ratio of 4π times the radiation intensity $R^2 S^s$ in the receiver direction to the incident power density from the transmitter direction. Radar cross section for complex shaped scattering objects is a function of many variables, including incidence angle, frequency, polarization, and scattering angle.

EXAMPLE 9-6 ***Range of a Monostatic Radar***

The operational distance R for a radar is called the range. It depends on several parameters of the radar. In this example, we find the range for a radar with the following typical parameter values:

$$P_t = 100 \text{ kW}$$

$$G_t = G_r = 40 \text{ dB} = 10^4$$

$$f = 3 \text{ GHz}$$

$$\sigma = 1 \text{ m}^2$$

$$P_r = -100 \text{ dBm} = 10^{-13} \text{ W}$$

The maximum range for the radar is found by solving (9-34) for R:

$$R = \left[P_t \frac{\lambda^2 G^2 \sigma}{(4\pi)^3 P_r} \right]^{1/4} = \left[10^5 \frac{(0.1)^2 (10^4)(1)}{(4\pi)^3 (10^{-13})} \right]^{1/4} = 149.8 \text{ km}$$

9.4 RECIPROCITY AND ANTENNA MEASUREMENTS

The remainder of this chapter is devoted to antenna measurements. This study provides a deeper understanding of antennas, allows one to interpret measured data, and serves as an introduction to those who desire to make antenna measurements. The principles introduced here also apply to broader situations such as scattering measurements, but are directly primarily toward antenna measurements. The primary measured antenna characteristics are radiation pattern, gain, polarization, and impedance. The first three of these are discussed in the following sections. Impedance is usually measured with a network analyzer and was discussed in Secs. 1.9 and 5.3.

In this section, we show that the radiation pattern of an antenna is the same whether it is used as a transmitting antenna or receiving antenna. Reciprocity allows the calculation or measurement of an antenna pattern in either the transmit or receive case, whichever is more convenient. Practical considerations for the measurement of antenna patterns are also discussed in this section.

In order to show that transmit and receive patterns are identical, it is necessary to discuss reciprocity theorems. There are several forms reciprocity theorems take for electromagnetic field problems. We consider two forms of reciprocity for use in antenna problems. The Lorentz reciprocity theorem is discussed first. Let sources $\mathbf{J}_a$ and $\mathbf{M}_a$ produce fields $\mathbf{E}_a$ and $\mathbf{H}_a$ and sources $\mathbf{J}_b$ and $\mathbf{M}_b$ produce fields $\mathbf{E}_b$ and $\mathbf{H}_b$. See Fig. 9-4. The frequencies of all quantities are identical. The Lorentz reciprocity theorem that is derivable from Maxwell's equations (see Prob. 9.4-1) states that for isotropic media,

$$\iiint_{v_a} (\mathbf{E}_b \cdot \mathbf{J}_a - \mathbf{H}_b \cdot \mathbf{M}_a)\, dv' = \iiint_{v_b} (\mathbf{E}_a \cdot \mathbf{J}_b - \mathbf{H}_a \cdot \mathbf{M}_b)\, dv' \qquad (9\text{-}36)$$

The left-hand side is the reaction (a measure of the coupling) of the fields from sources b on sources a, and the right-hand side is the reaction of the fields from sources a on sources b. This is a very general expression, but it can be put into a more usable form. Let sources b consist of only an ideal electric dipole of vector length $\mathbf{p}$ located at point (x_p, y_p, z_p). Since the ideal dipole can be represented as an infinitesimal source and $\mathbf{M}_b$ is zero, (9-36) becomes[1]

$$\mathbf{E}_a(x_p, y_p, z_p) \cdot \mathbf{p} = \iiint_{v_a} (\mathbf{E}_b \cdot \mathbf{J}_a - \mathbf{H}_b \cdot \mathbf{M}_a)\, dv' \qquad (9\text{-}37)$$

This expression allows calculation of the electric field from sources a by evaluating the integral using known sources $\mathbf{J}_a$ and $\mathbf{M}_a$ and known ideal dipole fields $\mathbf{E}_b$ and $\mathbf{H}_b$ of (1-69) and (1-68), evaluated at the location of sources a. This can be performed for various orientations $\mathbf{p}$ of the ideal dipole, which is acting as a field probe.

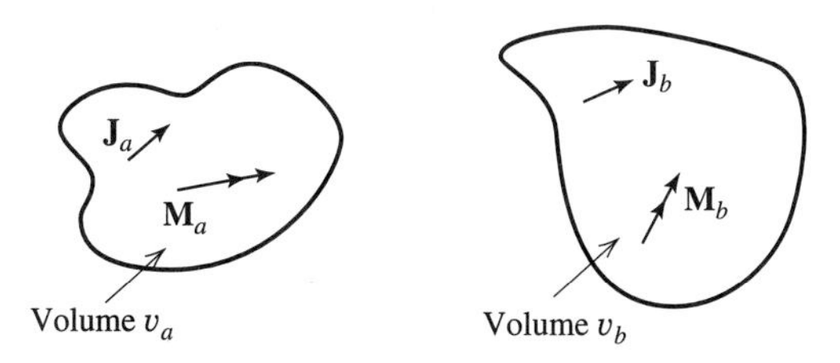

Figure 9-4 Source configuration for the Lorentz reciprocity theorem.

[1]The ideal dipole current could be written as $\mathbf{J}_b = \delta(x - x_p)\,\delta(y - y_p)\,\delta(z - z_p)\mathbf{p}$. This together with $\mathbf{M}_b = 0$ in (9-36) yields (9-37).

The Lorentz reciprocity theorem can also be used to derive a second reciprocity theorem using terminal voltages and currents. Suppose sources a and b are antennas excited with ideal (infinite impedance) current generators I_a and I_b. Since no magnetic sources are present, (9-36) reduces to

$$\iiint_{v_a} \mathbf{E}_b \cdot \mathbf{J}_a \, dv' = \iiint_{v_b} \mathbf{E}_a \cdot \mathbf{J}_b \, dv' \tag{9-38}$$

For perfectly conducting antennas, the electric fields will be zero over the antennas; however, voltages will be produced across the terminals. Taking the current to be constant in the terminal region and using the concept of $\int \mathbf{E} \cdot \mathbf{d\ell} = -V$, we see that (9-37) becomes

$$V_a^{oc} I_a = V_b^{oc} I_b \tag{9-39}$$

where V_a^{oc} is the open circuit voltage across the terminals of antenna a due to the field $\mathbf{E}_b$ generated by antenna b and, similarly, V_b^{oc} is the open circuit voltage at antenna b due to antenna a. Open circuit voltages have been used because of the infinite impedances of the generators. Rearranging (9-39) leads to a statement of reciprocity in circuit form

$$\frac{V_a^{oc}}{I_b} = \frac{V_b^{oc}}{I_a} \tag{9-40}$$

Several factors affect the voltage appearing at one antenna due to another antenna that is excited: the specific antennas used, the medium between the antennas with perhaps other objects present, and the relative orientation of the antennas. We can represent the general situation entirely in terms of circuit parameters using the following, which holds for any linear passive network:

$$V_a = Z_{aa} I_a + Z_{ab} I_b \tag{9-41a}$$

$$V_b = Z_{ba} I_a + Z_{bb} I_b \tag{9-41b}$$

where V_a, V_b, I_a, and I_b are the terminal voltages and currents of antennas a and b. If antenna a is excited with a generator of current I_a, the open circuit voltage appearing at the terminals of antenna b is $V_b|_{I_b=0}$. The transfer impedance Z_{ba} from (9-41b) with I_b zero is

$$Z_{ba} = \left. \frac{V_b}{I_a} \right|_{I_b=0} \tag{9-42}$$

If antenna b is excited with a generator of current I_b, the open circuit voltage appearing at the terminals of antenna a is $V_a|_{I_a=0}$. The transfer impedance Z_{ab} is, from (9-41a) with I_a zero,

$$Z_{ab} = \left. \frac{V_a}{I_b} \right|_{I_a=0} \tag{9-43}$$

Comparing (9-42) and (9-43) to (9-40), we see that

$$Z_{ab} = Z_{ba} = Z_m \tag{9-44}$$

where Z_m is the transfer (or mutual) impedance between the antennas. This can also be shown from the circuit formulation of (9-41) if the individual impedances are linear, passive, and bilateral. (See Probs. 9.4-3 and 9.4-4). This, in turn, is true if the medium and the antennas are linear, passive, and isotropic.

The significance of these results is now explained using the model of Fig. 9-5. If an ideal current source of current I excites antenna a, the open circuit voltage at the terminals of b from (9-42) is

$$V_b|_{I_b=0} = IZ_{ba} \tag{9-45}$$

If the same source is now applied to the terminals of antenna b, the open circuit voltage appearing at the terminals of antenna a from (9-43) is

$$V_a|_{I_a=0} = IZ_{ab} \tag{9-46}$$

But $Z_{ab} = Z_{ba}$, so the preceding two equations yield

$$V_a|_{I_a=0} = V_b|_{I_b=0} = V \tag{9-47}$$

Thus, the same excitation current will produce the same terminal voltage independent of which port is excited, as illustrated in Fig. 9-5. In other words, reciprocity states that the source and the measurer can be interchanged without changing the system response. The same is true of an ideal voltage source and short circuit terminal currents. These are familiar results from network theory.

The self-impedances of the antennas from (9-41) are

$$Z_{aa} = \left.\frac{V_a}{I_a}\right|_{I_b=0} \tag{9-48}$$

$$Z_{bb} = \left.\frac{V_b}{I_b}\right|_{I_a=0} \tag{9-49}$$

If antennas a and b are widely separated, which is the usual operating situation, Z_{aa} and Z_{bb} are much greater than $Z_{ab} = Z_{ba} = Z_m$. Thus, the input impedance to antenna a, for example, from (9-41a) is

$$Z_a = \frac{V_a}{I_a} = Z_{aa} + Z_{ab}\frac{I_b}{I_a} \approx Z_{aa} \tag{9-50}$$

Thus, if an antenna is isolated so that all objects including other antennas are far away and the antenna is lossless, the self-impedance equals its input impedance.

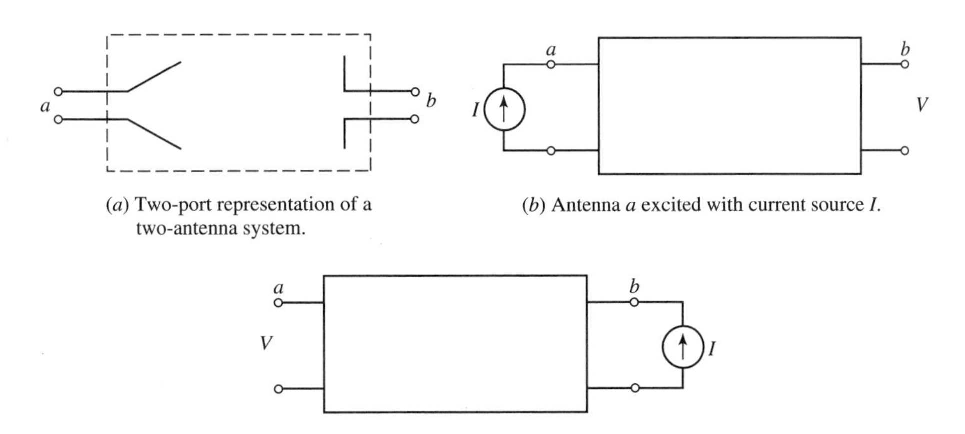

(a) Two-port representation of a two-antenna system.

(b) Antenna a excited with current source I.

(c) Antenna b excited with current source I.

Figure 9-5 Reciprocity for antennas. The output voltage V is the same in (b) and (c) for the same input current I.

Suppose antenna a is excited (i.e., acting as a transmitter) and the voltage produced at the terminals of antenna b is measured with an ideal voltmeter. If the antennas are separated so that they are in each other's far field, the transfer impedance Z_{ba} is actually the far-field (or radiation) pattern of antenna a if antenna b is moved around a on a constant radius as shown in Fig. 9-6a. As antenna b is moved, it is maintained with the same orientation and polarization relative to antenna a. The output voltage of b as a function of angle around antenna a gives the relative angular variation of the radiation from antenna a, that is, its radiation pattern. Examining (9-42), we see that this is really Z_{ba} (I_a is constant). Thus, Z_{ba} as a function of angle is the transmitting pattern of antenna a. If now antenna b is excited and antenna a acts as a receiver, the terminal voltage of antenna a is the receiving pattern of antenna a as antenna b is again moved around at a constant distance from antenna a; see Fig. 9-6b. Thus, Z_{ab} as a function of angle is the receiving pattern of antenna a. Since the transfer impedances are identical, we can conclude that *the transmit and receive patterns of an antenna are identical.* This is an important consequence of reciprocity.

The equality of the transmit and receive patterns of an antenna is not an unexpected result. This can be seen through the relation $G(\theta, \phi) = 4\pi A_e(\theta, \phi)/\lambda^2$ of (9-2), which relates the receiving characteristic of the antenna $A_e(\theta, \phi)$ for an incoming plane wave from angle (θ, ϕ) to the gain pattern value $G(\theta, \phi)$ in the direction (θ, ϕ) when the antenna transmits. The reciprocal property is of major practical importance. It permits the test antenna to be used in either a receive or transmit mode during pattern measurements. In practice, pattern measurements are usually made with the test antenna used in reception.

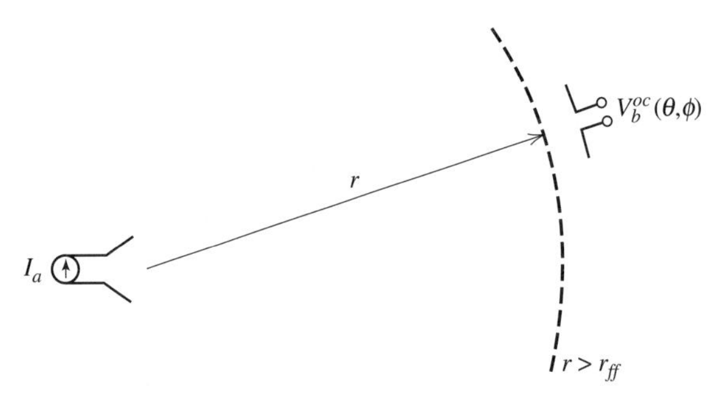

(a) The transmitting pattern of antenna a is $Z_{ba}(\theta,\phi) = V_b^{oc}(\theta,\phi)/I_a$.

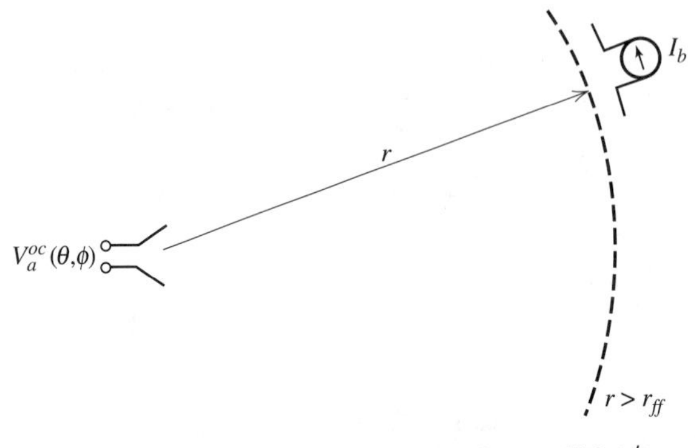

(b) The receiving pattern of antenna a is $Z_{ab}(\theta,\phi) = V_a^{oc}(\theta,\phi)/I_b$.

Figure 9-6 Antenna pattern reciprocity. The transmitting and receiving patterns of an antenna are identical because $Z_{ab}(\theta, \phi) = Z_{ba}(\theta, \phi) = Z_m(\theta, \phi)$.

It is important to note that reciprocity, as illustrated in Fig. 9-5 or through (9-44), is a general result. Also, in the case when the antennas are far removed from each other, $Z_m(\theta, \phi)$ is the far-field pattern. Of course, if an antenna contains any nonreciprocal components, reciprocity does not hold. An example is a ferrite isolator included in the antenna system.

9.5 PATTERN MEASUREMENT AND ANTENNA RANGES

An antenna pattern is a graphical representation of the field magnitude at a fixed distance from an antenna as a function of direction. With the antenna at the origin of a spherical coordinate system, radiation fields **E** and **H** are perpendicular to each other and both are transverse to the direction of propagation $\hat{\mathbf{r}}$. Also, the field intensities vary as r^{-1}. In antenna pattern discussions, electric field is used, but magnetic field behavior follows directly since its intensity is proportional to the electric field and its direction is perpendicular to **E** and $\hat{\mathbf{r}}$; see (1-107).

The radiated electric field is both a vector and a phasor. In general, it has two orthogonal components, E_θ and E_ϕ. These components are complex-valued and their relative magnitude and phase determine the polarization; see Sec. 1.10. For simple antennas, only one component is present. For example, the ideal dipole parallel to the z-axis has only an E_θ component as shown in Fig. 1-10. Measurement of the radiation pattern in this case is conceptualized by moving a receiving probe around the antenna as it transmits a constant signal a fixed distance away, r. The probe's orientation is maintained parallel to E_θ as shown in Fig. 9-7. The output of the probe varies in direct proportion to the intensity of the received field component arriving from direction (θ, ϕ). The pattern of the ideal dipole is $\sin\theta$; see Fig. 1-10. In general, antennas will have both E_θ and E_ϕ components and patterns are cut twice, once with the probe oriented parallel to E_θ and once with it parallel to E_ϕ.

Although we have conceptualized the measurement of a radiation pattern by moving a receiver over a sphere of constant radius, this is obviously an impractical way of making such measurements. The important feature is to maintain a constant large distance between the antennas and to vary the observation angle. This is accomplished by rotating the *test antenna*, or *antenna under test* (AUT), as illustrated in Fig. 9-8. By reciprocity, it makes no difference if the test antenna is operated as a receiver or transmitter, but usually the test antenna is used as a receiving antenna and we adopt this convention. The fields from the motionless source antenna provide a constant illumination of the test antenna whose output varies with its angular position. This leads to the rule that *it is the pattern of the rotated antenna that is being measured.*

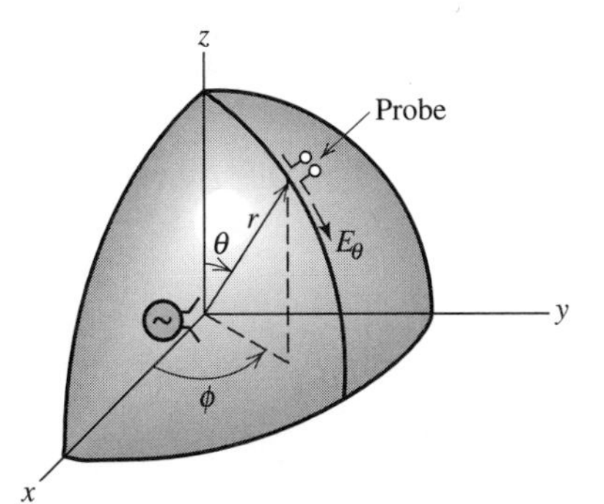

Figure 9-7 Pattern measurement conceptualized by movement of a probe antenna over the surface of a sphere in the far field of the antenna.

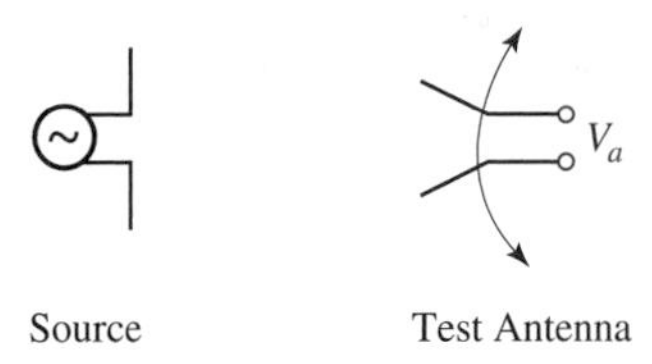

Figure 9-8 Radiation pattern measurement. The pattern of antenna a is proportional to the terminal voltage V_a, which is a function of the angular position of antenna a during rotation.

A complete representation of the radiation properties of an antenna would, of course, require measuring the radiation at all possible angles (θ, ϕ). This is rarely attempted and fortunately is not necessary. For most applications, the principal plane patterns are sufficient. See Fig. 1-10 for an illustration of the principal plane patterns using an ideal dipole.

There are many ways of displaying antenna patterns. For example, a principal plane pattern could be plotted in polar or rectangular form. In addition, the scale could be either linear or logarithmic (decibel). All combinations of plot type and scale type are used: polar-linear, polar-log, rectangular-linear, and rectangular-log. Figure 9-9 shows the same radiation pattern plotted in these four ways. Generally speaking, log plots are used for high-gain, low side-lobe patterns and linear plots are used when the main beam details are of primary interest. These antenna pattern representations can be recorded directly using commercially available measuring and recording equipment. When more detailed information is required, the results of several planar cuts can be put together to make a contour plot. It is important to appreciate that measured patterns are usually not perfectly symmetric even though the antenna structure appears to be symmetric and also nulls are often partially filled.

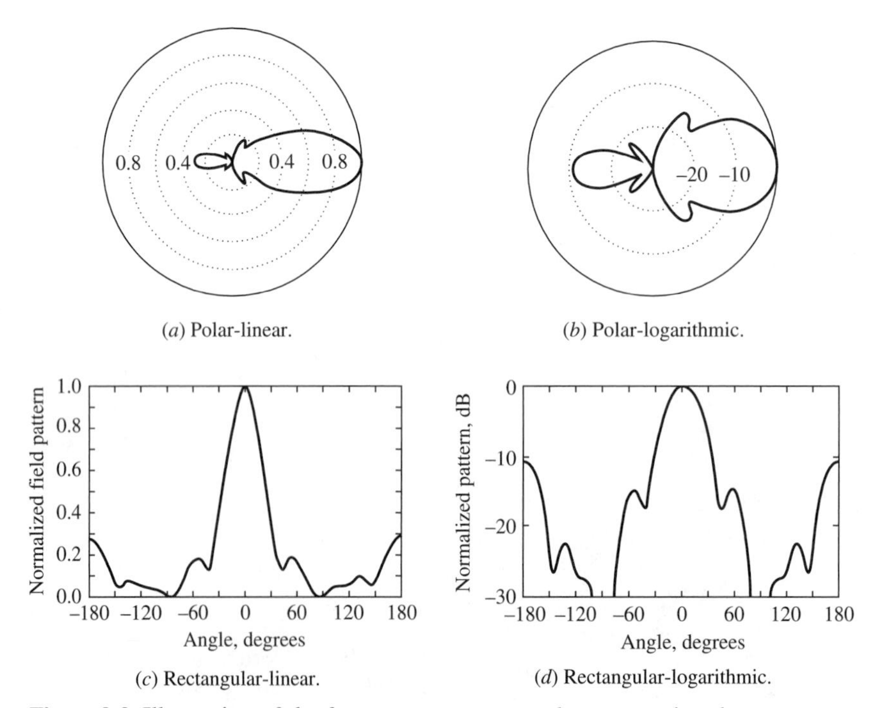

Figure 9-9 Illustration of the four antenna pattern plot types using the same pattern.

A facility used to measure antenna radiation characteristics is referred to as an *antenna range*. Often, the same range can also be used to measure scattering characteristics such as radar cross section. The entire measurement facility consists of the space (indoor or outdoor) for the source and test antennas, antenna positioners, a transmitter, a receiving system, and data display/recording equipment. In this section, we discuss the basic range layouts; see [3] for a complete discussion of antenna measurement techniques.

Table 9-1 lists the types of antenna ranges together with their characteristics and advantages and disadvantages. Most ranges are *free-space ranges* that are designed

Table 9-1 Characteristics of Antenna Ranges

Range Type	Description	Advantages	Disadvantages
FREE SPACE RANGES	Effects of all surroundings are suppressed to acceptable levels.		
Far-field Ranges			
Elevated range	Source and test antenna are placed on towers, buildings, hills, etc.	Low cost.	Requires real estate. May require towers. Outdoor weather.
Slant range	Either the source or test antenna is elevated.	Low cost.	Requires real estate. May require a tower. Outdoor weather.
Anechoic chamber	A room is lined with absorber material to suppress reflections.	Indoors.	Absorber and large room are costly.
Compact Range	The test antenna is illuminated by the collimated near field of a large reflector.	Small space.	A large reflector is required.
Near-field Range	The magnitude and phase of the near field of the test antenna are sampled and the far field is computed.	Very small space.	Accurate probe positioning is required. Accurate amplitude and phase are required. Time-consuming measurements. Computer-intensive.
GROUND REFLECTION RANGE	The ground between the source and test antennas is reflective, enhancing the indirect ray that interferes with the direct ray, giving a smooth test antenna illumination.	Test tower is short. Operates well at low frequencies (VHF).	Outdoor weather.

to have strong direct illumination of the test antenna with weak indirect illumination. First, we consider *far-field ranges* in which the source antenna is far from the test antenna. This can be accomplished by elevating either both or one of the source and test antennas giving an *elevated range* or *slant range*. For all antenna ranges, the site around the test antenna affects pattern measurement accuracy. The guiding principle is to have the line of sight (direct) path between the source and test antenna unblocked and as high above the ground (or floor) as practical. This yields large values for the angles α_t and α_r shown in Fig. 9-10. Then directive antennas will have indirect rays arising from specular reflection from the ground of reduced level because angles α_t and α_r usually correspond to side-lobe directions. In the elevated range of Fig. 9-10, the source and test antennas are approximately the same height, $h_t \approx h_r$. The slant range is similar to the elevated range except that only the source is elevated, leaving the test antenna conveniently located near the ground. When indoor rooms are used for a far-field range, the walls must be lined with absorbing material to reduce reflections. Frequently, the absorber is pyramidal-shaped to eliminate flat surfaces that reflect rays toward the test antenna.

In far-field ranges, the test antenna is located in the far field of the source antenna so that the incoming waves are nearly planar as indicated in Fig. 9-10. In fact, a common goal of all antenna ranges is to provide plane wave illumination of the test antenna. Deviations from uniform field illumination amplitude (i.e., magnitude) and phase across the test antenna aperture add to the inherent aperture taper of the test antenna, causing pattern measurement errors. In far-field ranges, the illumination field amplitude variation is determined by the radiation pattern of the source antenna. The effect of increased amplitude taper imposed on the test antenna aperture is to reduce the measured gain and change the side lobes close to the main beam. If the source antenna pattern peak is centered on the test antenna, as it should be in all cases, and the amplitude taper created by the source antenna pattern is -0.25 dB at the edges of the test antenna aperture, there will be a directivity (and thus, gain) reduction of 0.1 dB [4]. That is, the pattern point at angle $\alpha/2$ is 0.25 dB down from the peak; see Fig. 9-10. This is difficult to achieve for the wide variety of measurement situations on an antenna range, but in all cases the source antenna should be directed toward the test antenna and have a beamwidth that is as small

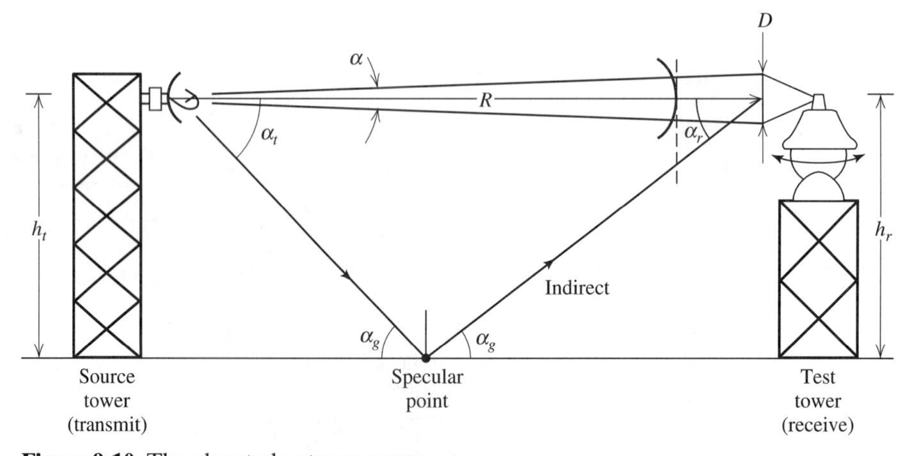

Figure 9-10 The elevated antenna range.

as possible to reduce illumination of the surroundings and to increase the received signal for adequate dynamic range. At the same time, the source antenna should not have a beamwidth that is narrow enough to impose an amplitude taper across the test antenna.

The instrumentation used with an antenna range varies from a simple signal source together with a relative power indicating subsystem to complete commercial systems with automatic data collection and display features. The signal source should be stable in power level and frequency. The receiving system should have a linear dynamic range of at least 40 dB. Both amplitude-only and amplitude-phase receiving systems are available. Also, a network analyzer can be used.

Phase errors are due to the fact that to achieve a planar phase front from a finite-sized source antenna, the source must be an infinite distance away from the test antenna. The spherical waves from the source antenna cause a phase error across the test antenna extent of D that behave exactly as the far-field distance phase error discussed in Sec. 1.7.3. There we found that spherical waves deviate from parallel rays with a 22.5° phase error ($\lambda/16$ distance error) at a distance of $2D^2/\lambda$. Thus, we can say that a phase error of 22.5° is created by the phase front curvature over a test antenna of extent D at a separation distance of

$$r_{ff} = \frac{2D^2}{\lambda} \tag{9-51}$$

The extent of the source antenna need not be included for pattern measurements, but the measurement distance should be more than doubled when the source and test antennas are of the same size to preserve gain accuracy [5].

The measurement distance of (9-51) is adequate for moderate-to-high-gain antennas if high accuracy is not required in the side-lobe levels. In general, the effects of reducing the measurement distance from infinity is to fill in the nulls between side lobes, increase the peak of the side lobes (mainly near the main beam), broaden the main beam, and reduce the main beam peak (implying a directivity reduction) [6]. For example, an antenna with the first side lobe 30 dB below the main beam peak (SLL $= -30$ dB) when measured at an infinite distance from the source has an error of 3 dB for a measurement distance of $2D^2/\lambda$; that is, the first side lobe is SLL $= -27$ dB [5, 7]. Gain is reduced about 0.1 dB at the measurement distance of $2D^2/\lambda$ for typical high-gain antennas. In the case of broad main beam antennas, the measurement distance should also be at least that of (9-51) to ensure the accurate measurement of pattern ripple [8].

Electrically large antennas require very large measurement distances. For example, the Deep Space Network 70-m reflector antenna at Goldstone, CA, operating at 2.3 GHz requires a measurement distance from (9-51) of 75 km! Conventional techniques cannot be used to measure such an antenna. However, a source flown in an airplane or available from a satellite can be used. Or, noise from a strong "radio star" can be used as a source together with a radiometer receiver.

The concept for the *compact range* is to place the test antenna close to a reflector antenna as shown in Fig. 9-11. This is possible because the near field of a reflector is collimated, giving a nearly flat phase front and an amplitude taper equal to that across the reflector aperture. Therefore, the phase error problem associated with far-field ranges is traded for an amplitude problem in a compact range. Improved

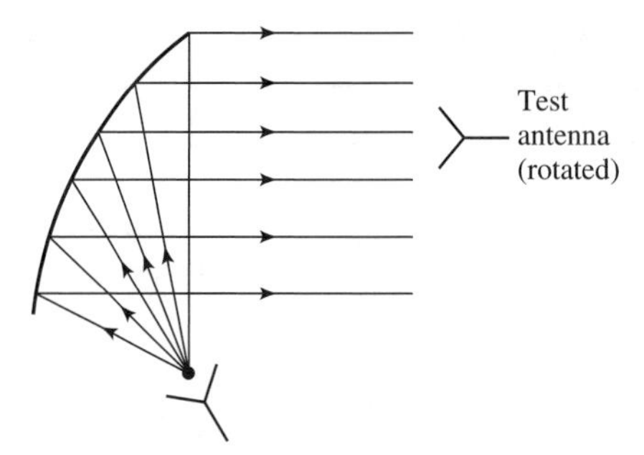

Figure 9-11 The compact range.

performance is possible with a dual reflector. A disadvantage of the compact range is that the reflector must be large—about three times larger than the test antenna.

The final antenna range type is the *near-field range* shown in Fig. 9-12.`Here, the test antenna acts as a transmitter and the amplitude and phase are sampled at regular intervals in the near field of the test antenna. The samples are weighted equally by the receiver, providing uniform amplitude and phase across the test antenna as required for accurate measurements. Radiation properties such as the pattern are then computed using a Fourier transform [9]. Accurate probe positioning is required for accurate pattern computation. The near-field range offers the benefit of having the aperture distribution data available for diagnostic use. For example, a dead element in an array antenna can be located.

The operating principle of the *ground reflection range* is completely different from that of free-space ranges. The source and test antenna heights are small and the ground between the towers is constructed to be flat and reflective, which causes the indirect ray to arrive with an amplitude close to that of the direct ray. The indirect ray path distance is not greatly different from that of the direct ray. This gives a slowly varying phase over the test region, which in turn gives a slowly varying interference pattern and a relatively constant field illumination over the test zone. A low test tower height is convenient for large test objects such as antennas on full-scale aircraft.

Rotation positioners are required in most of antenna ranges. Often, a simple *azimuth positioner* (or "turntable") is sufficient. An *elevation-over-azimuth positioner* as illustrated in Fig. 9-13*a* permits alignment with the source antenna placed

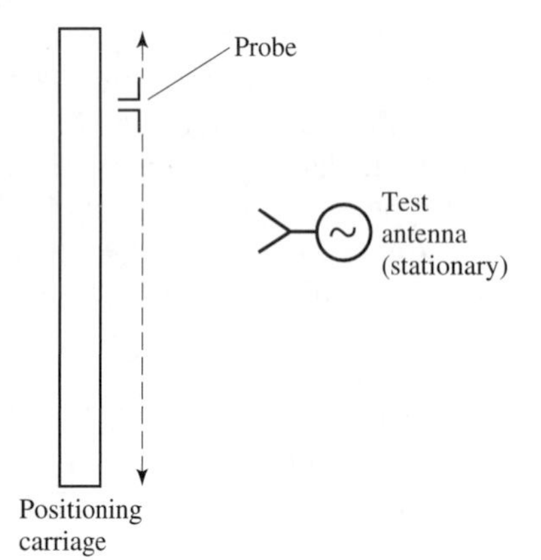

Figure 9-12 The near-field range.

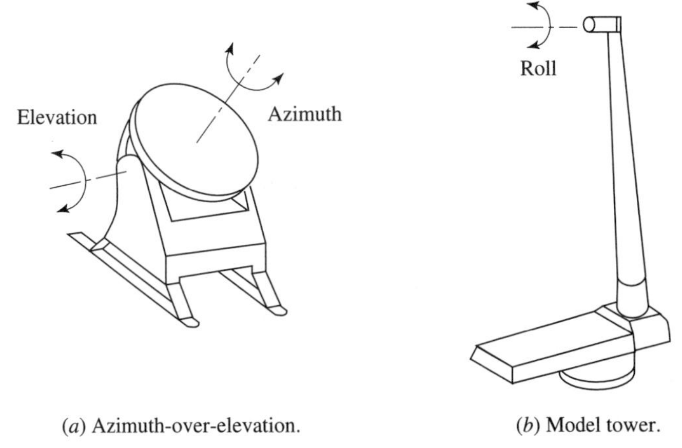

(*a*) Azimuth-over-elevation. (*b*) Model tower.

Figure 9-13 Antenna positioners for antenna testing.

at any height. Frequently, a *model tower* is used to position the test antenna over the axis of the rotation positioner; see Fig. 9-13*b*. Included with the model tower is a *roll positioner* that rotates the test antenna about its own axis for controlling the pattern cut. The source tower also often has a roll positioner for proper orientation of the source antenna polarization.

9.6 GAIN MEASUREMENT

Pattern measurement discussed in the previous section is a relative measurement that gives the angular variation of the test antenna's radiation. Gain is also needed to fully characterize the radiation properties of a test antenna. It is an absolute quantity and thus is more difficult to measure. Techniques exist to measure the gain of a test antenna with no *a priori* knowledge. However, most gain measurements are made using an antenna of known gain, called a *standard gain antenna* [4, Chap. 12]. The technique is called the *gain comparison* (or *gain transfer*) *method.* A transmitter of fixed input power P_t is connected to a suitable source antenna whose pattern peak is centered on the test antenna. Received power is measured for both the test antenna P_T and the standard gain antenna P_S, as illustrated in Fig. 9-14 by placing each antenna on the test positioner, pointing toward the source for peak output, and recording the received power levels. The gain of the test antenna is

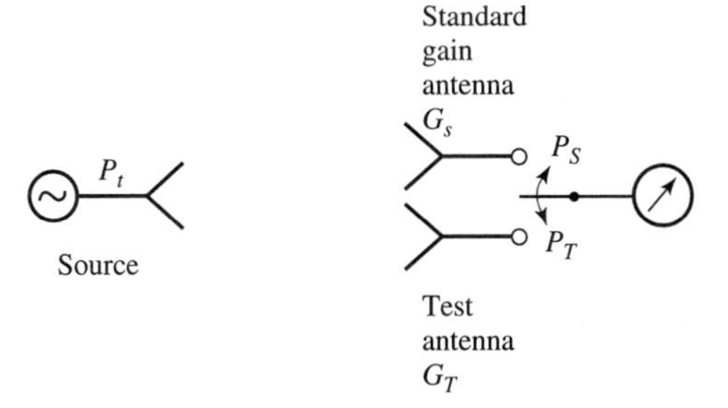

Figure 9-14 Measurement of the gain of a test antenna G_T using the gain comparison method based on the known gain of a standard gain antenna G_S and $G_T = (P_T/P_S)G_S$. From [10] © 1993. Reprinted by permission of Artech House, Inc., Boston, MA.

then easily computed from the gain of the standard gain antenna multiplied by the ratio of the received powers:

$$G_T = \frac{P_T}{P_S} G_S \tag{9-52}$$

This relation is more convenient when expressed in decibels:

$$G_T \text{ (dB)} = P_T \text{ (dBm)} - P_S \text{ (dBm)} + G_S \text{ (dB)} \tag{9-53}$$

This result is intuitive and simply says that the gain of the test antenna differs from that of the standard antenna by the difference in received powers from the test antenna and the standard antenna. A special case is when the received powers are equal ($P_T = P_S$); then the gain of the test antenna is identical to that of the standard gain antenna. This result is also easily derived by evaluating (2-99) for both the test and standard antenna cases and subtracting; the terms involving distance R, frequency f, and transmit power P_t are constants and drop out, leaving (9-53).

It is obvious from (9-53) that accurate gain measurement requires accurate power measurement. With modern receivers, this is often possible. An approach that does not rely on receiver linearity is the RF substitution method in which a precision attenuator is used to establish the power level change. That is, the attenuator is adjusted to bring the receiver to the same reading in both cases; then the difference in the corresponding attenuator settings equals P_T (dBm) $-$ P_S (dBm). Accuracy also depends directly on a knowledge of the gain of the standard gain antenna. Popular standard gain antennas are the half-wave dipole for UHF frequencies and below and the pyramidal horn for UHF frequencies and above. The gain of the dipole is 2.15 dB (see Fig. 2-6) and manufacturers of standard gain horns supply data on gain across the operating frequency range like Fig. 7-20.

Note that the term *gain* is synonymous with *absolute gain* or *peak gain*. Gain and pattern data can be merged into a *gain pattern* by multiplying gain by the normalized pattern:

$$G(\theta, \phi) = GP(\theta, \phi) = G|F(\theta, \phi)|^2 \tag{9-54}$$

Expressed in decibels (by taking 10 log), the unit of dBi is often used, indicating that the pattern has been referenced to an isotropic antenna.

EXAMPLE 9-7 ***Gain Measurement by Gain Comparison***

Suppose that a standard gain antenna has a gain of 63, or 18, dB. Following the measurement technique illustrated in Fig. 9-14, the measured powers are P_S = 3.16 mW or 5 dBm (5 dB above a milliwatt), and P_T = 31.6 mW, or 15 dBm. The gain of the test antenna is then G_T = (31.6/3.16)63 = 630, or in terms of decibels,

$$G_T \text{ (dB)} = P_T \text{ (dBm)} - P_S \text{ (dBm)} + G_S \text{ (dB)} = 15 - 5 + 18 = 28 \text{ dB} \tag{9-55}$$

9.6.1 Gain Measurement of CP Antennas

If good-quality circularly polarized (CP) source and standard gain antennas are available, the gain comparison method of Fig. 9-14 can be used. Frequently, though, the gain of elliptically polarized antennas is measured by using two orthogonal linearly polarized (LP) antennas, or customarily one LP antenna used in two orthogonal orientations. Suppose the gains are measured for vertical and horizontal LP cases. These partial gains, G_{Tv} and G_{Th}, are combined to give the total gain [10, 11]:

$$G_T\ (\text{dB}) = 10\log(G_{Tv} + G_{Th}) \quad [\text{dBic}] \tag{9-56}$$

This is referred to as the *partial gain method.* Any perpendicular orientations can be used because the power in an elliptically polarized wave is contained in the sum of any two orthogonal components. As a side note, we observe that a CP antenna performs this sum instantaneously. Therefore, the gain in (9-56) is relative to an ideal CP antenna. The unit dBic indicates gain relative to an isotropic, perfect CP antenna. Gain measurement accuracy depends on the purity of the source antenna. An LP standard gain antenna usually has an axial ratio of 40 dB or better and does not contribute significantly to gain error.

EXAMPLE 9-8 *Calculation of Gain Using the Partial Gain Method*

Figure 9-15 gives two patterns measured with an LP source antenna and a nominally CP test antenna, which is a cavity-backed spiral antenna operating at 1054 MHz. Also shown is the pattern of a standard gain horn, which has a gain at 1054 MHz of 14.15 dB based on the manufacturer's gain curve. The receiver gain setting and the source power were constant during these measurements. The peak gains for vertical and horizontal polarizations then are

$$G_{Tv}\ (\text{dB}) = 14.15 - 16.1 = -1.95\ \text{dB}, \qquad G_{Th}\ (\text{dB}) = 14.15 - 13.25 = 0.9\ \text{dB} \tag{9-57}$$

because the vertical and horizontal LP pattern peaks are 13.25 and 16.1 dB below the standard gain horn pattern peak, respectively. Then

$$G_{Tv} = 10^{-1.95/10} = 0.64, \qquad G_{Th} = 10^{0.9/10} = 1.23 \tag{9-58}$$

and (9-56) gives

$$G_T\ (\text{dB}) = 10\log(0.64 + 1.23) = 2.71\ \text{dBic} \tag{9-59}$$

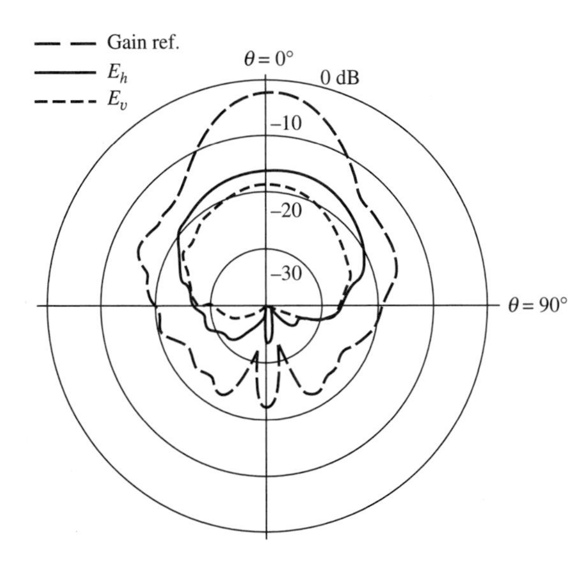

Figure 9-15 Illustration of measurement of the gain of a CP antenna using the partial gain method; see Example 9-7. The patterns are for an LP standard gain horn (long dashed curve) and the nominally CP antenna with the source vertically (solid curve) and horizontally (short dashed curve) polarized. From [10] © 1993. Reprinted by permission of Artech House, Inc., Boston, MA.

9.6.2 Gain Estimation

Frequently, gain can be estimated based on a knowledge of the pattern and the antenna operation. Since directivity is a pattern quantity, we can use pattern data to estimate directivity. That is, directivity from (1-146b) varies inversely with beam solid angle, $D = 4\pi/\Omega_A$. It is possible to record many pattern cuts and numerically integrate the pattern using (1-143) to find Ω_A. A simpler approach is to use the principal plane pattern half-power beamwidths to estimate beam solid angle. To do this, we first find *beam efficiency* from main beam solid angle Ω_M using (7-85):

$$\varepsilon_M = \frac{\Omega_M}{\Omega_A} \tag{9-60}$$

The main beam solid angle is well approximated as the product of the half-power beamwidths in the principal planes:

$$\Omega_M \approx \text{HP}_E \text{HP}_H \tag{9-61}$$

Then directivity can be estimated by combining these results:

$$D = \frac{4\pi}{\Omega_A} = \frac{4\pi\varepsilon_M}{\Omega_M} \approx \frac{4\pi\varepsilon_M}{\text{HP}_E\text{HP}_H} = \frac{41{,}253\varepsilon_M}{\text{HP}_{E^\circ}\text{HP}_{H^\circ}} \tag{9-62}$$

where HP_{E° and HP_{H° are the half-power beamwidths in the E- and H-planes expressed in degrees. Often, it is assumed that all power is in the main beam, giving $\varepsilon_M = 1$; see (7-94). Antennas, in practice, have a nonnegligible amount of power in the side lobes; a typical value for ε_M is 0.63. If no loss is present, $e_r = 1$ and the gain for antennas encountered in practice from (9-62) from (7-95) is

$$G = e_r D \approx D \approx \frac{26{,}000}{\text{HP}_{E^\circ}\text{HP}_{H^\circ}} \tag{9-63}$$

See Sec. 7.3 for more discussion on gain. It must be emphasized that this very approximate formula should be used for rough estimates when the only data available are the half-power beamwidths.

9.7 POLARIZATION MEASUREMENT

Quite often, the polarization of an antenna can be inferred from the geometry of the active portion of the antenna. For example, the ideal dipole in Fig. 1-10 is vertical linearly polarized since the radiating element is oriented vertically. For gain and co-polarized pattern measurements, the test antenna should be illuminated with a wave of the expected polarization of the antenna: in this case, a vertical linearly polarized wave. Real antennas always have a certain amount of power in the polarization orthogonal to the intended polarization. For the practical realization of the ideal dipole, there will be a small amount of horizontal linear polarization. Such cross-polarization arises from horizontal currents flowing on the antenna or nearby structures. Thus, a complete antenna measurement set includes characterization of the polarization properties of the test antenna. This is often accomplished by making pattern cuts in the E- and H-planes of the test antenna with it both co-polarized and cross-polarized to the source antenna. This is illustrated in Fig. 9-16 for the case of a nominally LP test antenna and an LP source antenna. Of course, the cross-polarized patterns will be much lower in level than the co-polarized patterns, and

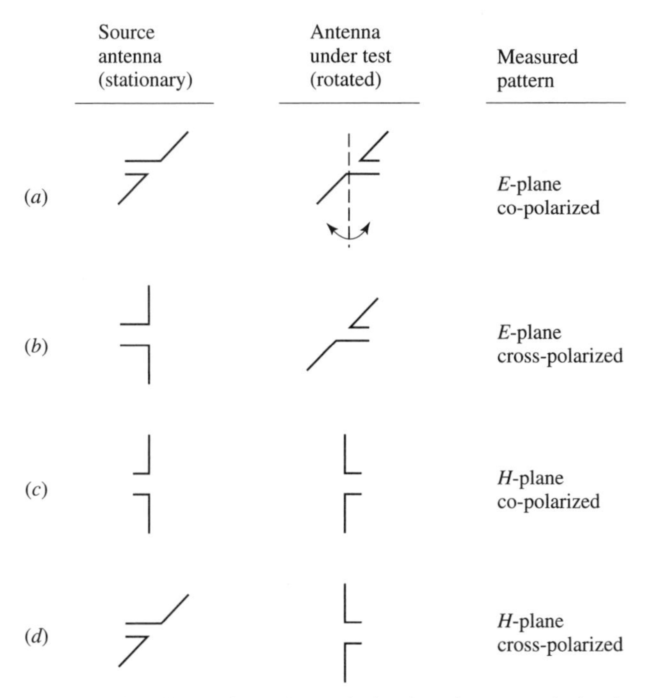

Figure 9-16 Illustration of copolarized and cross-polarized pattern measurement. The source antenna is LP and the test antenna operating in the receiving mode is nominally LP and is rotated about its axis. From [10] © 1993. Reprinted by permission of Artech House, Inc., Boston, MA.

will be zero for a perfect LP test antenna. Co- and cross-polarized patterns for reflector antennas are discussed in Sec. 7.6.5.

There are three measurement techniques used to characterize an antenna that is elliptically polarized but has an axial ratio that is not large (i.e., the polarization state is not close to pure LP). These methods are discussed in the remainder of this section [10].

9.7.1 Polarization Pattern Method

A *polarization pattern* is the amplitude response of an antenna as it is rotated about its roll axis. It can be measured at any fixed pattern rotation angle. The resulting pattern shown in Fig. 9-17 is a polar plot of the response of the test antenna as a function of the relative angle α between the illuminating LP wave orientation and a reference orientation of the antenna. Either the LP source antenna is rolled while the test antenna is stationary or vice versa. It is easier to explain the polarization pattern method with the test antenna operated as an elliptically polarized transmitting antenna and the receiving antenna as a linearly polarized probe. Reciprocity permits us to do this. The tip of the instantaneous electric field vector from the test antenna lies on the polarization ellipse and rotates at the frequency of the wave; that is, the electric vector completes f rotations around the ellipse per second. The output voltage of the LP probe is proportional to the peak projection of the electric field onto the LP orientation line at angle α. This is the distance OP in Fig. 9-17 projected from the tangent point T on the ellipse. The locus of points P as the LP

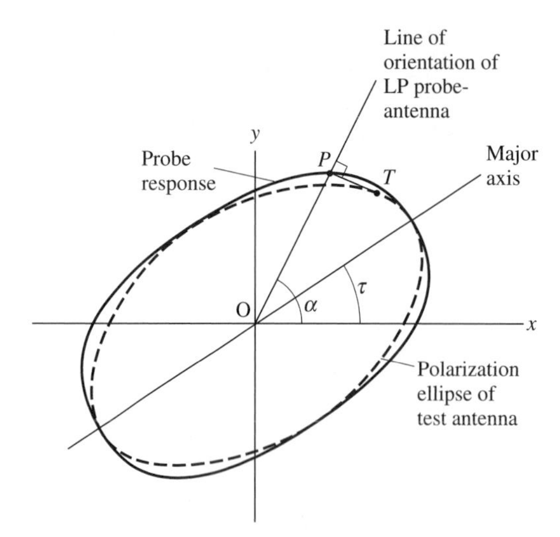

Figure 9-17 Polarization pattern (solid curve) of an elliptically polarized test antenna. It is the response of an LP receiving probe with orientation angle α to a transmitting test antenna with the polarization ellipse shown (dashed curve). From [10] © 1993. Reprinted by permission of Artech House, Inc., Boston, MA.

probe is rotated is fatter than the ellipse, which is also shown in Fig. 1-24. Of course, for a CP test antenna both curves in Fig. 9-17 are circular.

Note that the maximum and minimum of the polarization pattern are identical to the corresponding maximum and minimum of the polarization ellipse when scaled to the same size. Although the measured polarization pattern does not give the polarization ellipse, it does produce the axial ratio magnitude of the antenna polarization. It is also obvious from Fig. 9-17 that the tilt angle of the ellipse is determined as well. *The polarization pattern gives the axial ratio magnitude $|AR|$ and tilt angle τ of the polarization ellipse, but not the sense.* The sense can be determined by additional measurements. For example, two nominally CP antennas that are identical except for sense can be used as receiving antennas with the test antenna transmitting. The sense of the antenna with the greatest output is then the sense of the test antenna.

The polarization pattern method in many cases is a practical way to measure antenna polarization. If the test antenna is nearly circularly polarized, the axial ratio is near unity and measured results are insensitive to the purity of the LP probe. If the test antenna is exactly circular, tilt angle is irrelevant. In the case of a test antenna that is nearly linearly polarized, axial ratio measurement accuracy depends on the quality of the LP probe, which must have an axial ratio much greater than that of the test antenna.

9.7.2 Spinning Linear Method

The *spinning linear* (or *rotating source*) *method* provides a rapid measurement technique for determining the axial ratio magnitude as a function of pattern angle. The test antenna is rotated as in a conventional pattern measurement while an LP probe antenna (usually transmitting) is spun. The spin rate of the LP antenna should be such that the test antenna pattern does not change appreciably during one-half revolution of the LP antenna while the test antenna rotates slowly. An example pattern is shown in Fig. 9-18, which is a pattern of a helix antenna. Superimposed on the antenna pattern are rapid variations representing twice the rotation rate of the probe antenna. For logarithmic (dB) patterns as in Fig. 9-18, the axial ratio is

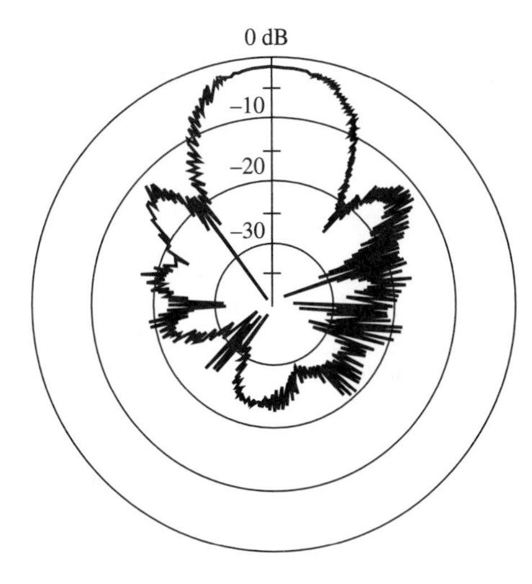

Figure 9-18 Axial ratio measurement as a function of pattern angle using the spinning linear method. The axial ratio is the difference in decibels between adjacent peaks and nulls. The test antenna is a helix antenna operating at X-band and the source is a rotating LP antenna. From [10] © 1993. Reprinted by permission of Artech House, Inc., Boston, MA.

the difference between adjacent maxima and minima at each angle. For example, at a pattern angle of 30° counterclockwise from the main beam axis, the maximum and minimum pattern envelopes are about −8 and −10 dB, corresponding to a 2-dB axial ratio.

Sense cannot be obtained using the spinning linear method. Tilt angle could, in theory, be obtained if probe orientation information were known accurately at the pattern points, but this is usually not done in practice.

9.7.3 Dual-Linear Pattern Method

A method related to the spinning linear method is the *dual-linear pattern method*. In this method, two patterns are measured for orthogonal orientations of the LP probe source antenna so that they align with the major and minor axes of the test antenna polarization ellipse. Figure 9-19 illustrates the resulting patterns for the same sample antenna as in Fig. 9-18 for the spinning linear method. For the same example pattern point at 30° counterclockwise from the beam peak, the two linear

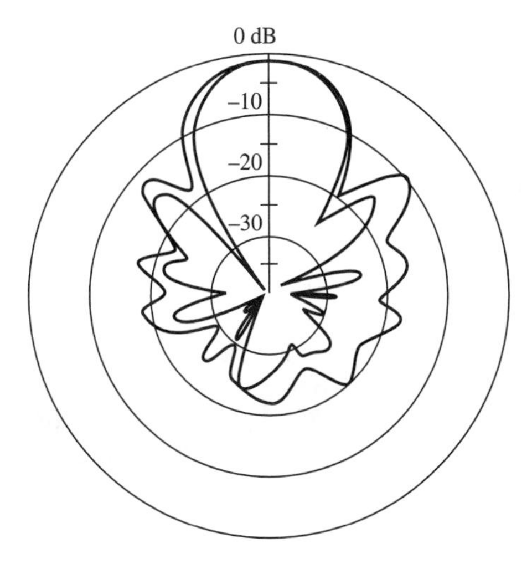

Figure 9-19 Axial ratio measurement as a function of pattern angle using the dual-linear pattern method. The axial ratio is the decibel difference between the two patterns that represent planes containing the major and minor axes of the test antenna polarization ellipse. The test antenna is identical to that in Fig. 9-18. From [10] © 1993. Reprinted by permission of Artech House, Inc., Boston, MA.

pattern values are −8 and −10 dB, again giving a 2-dB axial ratio value. Of course, the gains and other equipment settings must remain constant during the entire measurement period.

9.8 FIELD INTENSITY MEASUREMENTS

A very small receiving antenna can be used as a *field probe*. Probes are used when it is necessary to measure the spatial amplitude distribution of electromagnetic fields. The probe must be small relative to the structure whose fields are being measured in order to minimize the disturbances introduced by the probe itself. The electrically small dipole, in any of its practical forms discussed in Sec. 2.1, is used to probe electric fields. The small loop is used as a magnetic field probe.

Receiving antennas are also used to measure absolute field intensity. For example, it is often necessary to know the field intensity at a fixed distance from a transmit antenna. The antenna pattern can, of course, also be measured by moving a receiving probe around the transmitter at a fixed distance from it in the far field; this gives the relative field intensity variation. Such measurements are often required because the effects of terrain and the real earth surface are difficult to include in calculations. If the gain of the measuring antenna is known (it usually is) and the voltage developed across its terminals is measured, the field intensity incident upon the measuring antenna can be calculated. We now discuss this.

The same model as in Fig. 2-18*b* is used to derive field intensity. The power delivered to the terminating load is

$$P_D = \frac{1}{2}\frac{|V_A|^2}{R_L} = \frac{V_{A,\text{rms}}^2}{R_L} \tag{9-64}$$

where $V_{A,\text{rms}} = |V_A|/\sqrt{2}$ since V_A is a peak quantity. The field form of the delivered power expression from (9-3) and (2-88) is

$$P_D = pqSA_e = pq\frac{(E_{\text{rms}}^i)^2}{\eta}A_e \tag{9-65}$$

Equating these two relations yields

$$(E_{\text{rms}}^i)^2 = \eta\frac{V_{A,\text{rms}}^2}{pqR_L}\frac{1}{A_e} = \eta\frac{V_{A,\text{rms}}^2}{pqR_L}\frac{4\pi}{G\lambda^2} \tag{9-66}$$

where (2-89) was used for A_e. Converting wavelength to frequency using $\lambda = c/f$ and expressing the relation in decibels by taking 10 log of both sides gives

$$\begin{aligned} E_{\text{rms}}^i\ (\text{dB}\mu\text{V/m}) = {} & V_{A,\text{rms}}\ (\text{dB}\mu\text{V}) + 20\log f\ (\text{MHz}) - G\ (\text{dB}) \\ & - 10\log R_L - 10\log p - 10\log q - 12.8 \end{aligned} \tag{9-67}$$

This expression permits easy calculation of electric field intensity E_{rms}^i in decibels relative to 1 μV/m, using the voltage $V_{A,\text{rms}}$ in decibels relative to 1 μV, measured at the terminals of a probe antenna with gain G. Gain loss due to mispointing can also be included. For example, suppose the probe antenna has 6-dB gain and is pointed so that the incoming wave arrives from a direction on the receiving antenna pattern that is 2 dB below its maximum. Then 4-dB gain is used in (9-67) rather than the peak gain of 6 dB.

EXAMPLE 9-9 ***Sensitivity of an FM Receiver***

As an example, suppose the antenna and transmission line input impedances are both 300 Ω. Then (9-67) becomes [11]

$$E^i_{\text{rms}}\ (\text{dB}\mu\text{V/m}) = 20 \log f\ (\text{MHz}) - G\ (\text{dB}) + V_{A,\text{rms}}\ (\text{dB}\mu\text{V}) - 37.57 \qquad (9\text{-}68)$$

To be specific, consider a typical FM broadcast receiver with a sensitivity of 1 μV; that is, minimum satisfactory performance is produced when the value of $V_{A,\text{rms}}$ is 1 μV, or 0 dBμV. The most popular antenna for FM receivers is the half-wave folded dipole (see Sec. 5.2) that has a real impedance of about 300 Ω and a gain of 2.15 dB. At a frequency of 100 MHz, the incident field intensity required for minimum satisfactory performance from (9-68) is 0.28 dBμV/m, or 1.03 μV/m.

At frequencies below 1 GHz, antenna measurements are made by illuminating the test antenna with a known field intensity and measuring the terminal voltage. Antenna factor is used to quantify this measurement. *Antenna factor K* is defined as the ratio of the field intensity illuminating the antenna to the received voltage across the antenna terminals:

$$K = \frac{E^i}{V_A} \quad (\text{m}^{-1}) \qquad (9\text{-}69)$$

This is an electric field antenna factor; a corresponding one involving magnetic field intensity is also in use. Antenna factor is often used to determine receiver sensitivity. Then, (9-69) in decibel form using (9-67) becomes

$$E^i_{\text{rms}}\ (\text{dB}\mu\text{V/m}) = \text{receiver sensitivity} = V_{A,\text{rms}}\ (\text{dB}\mu\text{V}) + K\ (\text{dB/m}) \qquad (9\text{-}70\text{a})$$

where

$$\begin{aligned} K\ (\text{dB/m}) = 20 \log[f\ (\text{MHz})] - G\ (\text{dB}) - 10 \log R_L \\ - 10 \log p - 10 \log q - 12.8 \end{aligned} \qquad (9\text{-}70\text{b})$$

It is common to specialize this definition to $R_L = 50\ \Omega$, since that is the normal receiver input impedance. Antenna factor includes impedance mismatch effects and antenna gain. The polarizations of the wave and antenna are usually assumed to be matched (i.e., $q = 0$), which is the customary measurement situation.

EXAMPLE 9-10 ***Sensitivity of an FM Receiver***

We repeat Example 9-9 using antenna factor. Substituting $R_A = Z_o = 300\ \Omega$, $G = 1.64$ and $\lambda = 3$ m in (9-70b) gives

$$K = 20 \log(100) - 2.15 - 10 \log(300) - 0 - 0 - 12.8 = 0.28\ \text{dB/m} = 1.03\ \text{m}^{-1} \qquad (9\text{-}71)$$

Then for a 1-μV sensitivity, (9-70a) gives

$$E^i_{\text{rms}} = 0\ \text{dB}\mu\text{V} + 0.28\ \text{dB/m} = 0.28\ \text{dB}\mu\text{V/m} \qquad (9\text{-}72)$$

which is the result we obtained in Example 9-9.

REFERENCES

1. W. L. Flock and E. K. Smith, "Natural Radio Noise—A Mini-review," *IEEE Trans. Ant. & Prop.*, Vol. AP-32, pp. 762–767, July 1984.

2. R. S. Bokulic, "Use Basic Concepts to Determine Antenna Noise Temperature," *Microwaves & RF*, Vol. 30, pp. 107–115, March 1991.
3. G. E. Evans, *Antenna Measurement Techniques*, Artech House, Boston, MA, 1990.
4. *IEEE Test Procedures for Antennas*, IEEE Standard 149-1979, Piscataway, NJ, 1979, p. 19.
5. T. Uno and S. Adachi, "Range Distance Requirements for Large Antenna Measurements," *IEEE Trans. Ant. & Prop.*, Vol. 37, pp. 707–720, June 1989.
6. P. S. Hacker and H. E. Schrank, "Range Distance Requirements for Measuring Low and Ultra Low Sidelobe Antenna Patterns," *IEEE Trans. Ant. & Prop.*, Vol. AP-30, pp. 956–966, Sept. 1982.
7. R. C. Hansen, "Measurement Distance Effects on Low Sidelobe Patterns," *IEEE Trans. Ant. & Prop.*, Vol. AP-32, pp. 591–594, June 1994.
8. D. G. Hundt and W. L. Stutzman, "Pattern Measurement Distance for Broad Beam Antennas," *J. Electromagnetic Waves & App.*, Vol. 8, pp. 221–235, Feb. 1994.
9. D. Slater, *Near-Field Antenna Measurements*, Artech House, Boston, MA, 1991.
10. W. L. Stutzman, *Polarization in Electromagnetic Systems*, Artech House, Boston, MA, 1993. Chap. 4.
11. H. V. Carnagan, "Measure That Field Using Any Antenna," *Microwaves*, Vol. 14, pp. 45–47, July 1975.

PROBLEMS

9.1-1 Show that (9-8) follows from (9-9).

9.1-2 A transmitting antenna is not matched to the impedance of a connecting transmission line. The radiation intensity, or equivalently the power density at a specified distance, is reduced from the perfect impedance match case. Compute this reduction in decibels for mismatch situations that produce VSWR values on the transmission line of 1.01, 1.2, 2, and 10.

9.1-3 Find the complex unit vector $\hat{\mathbf{e}}$ for a right-hand elliptically polarized wave with an axial ratio of 2 dB and tilt angle $\tau = 45°$. Then compute the polarization efficiency for receiving antennas with the following polarizations: (a) horizontal linear, (b) vertical linear, (c) right-hand circular, (d) left-hand circular, (e) right-hand elliptical with AR (dB) $= 2$ and $\tau = 45°$, and (f) left-hand elliptical with AR (dB) $= 2$, and $\tau = 135°$.

9.1-4 Derive the vector effective length expression

$$\mathbf{h} = -j\beta\mu_{\text{eff}}NS \sin\theta\,\hat{\boldsymbol{\phi}}$$

for a small loop antenna oriented in the xy-plane with N turns, effective relative permeability μ_{eff}, and single turn area of S. Make use of (2-53).

9.3-1 A monostatic radar system (i.e., the transmitter and receiver are in the same location) illuminates a target which is a resonant half-wave dipole that has a radar cross section of approximately $0.85\lambda^2$. The radar operates at 10 GHz, the range is 1000 m, the gain of the transmit and receive antennas is 20 dB, and the transmit power is 1000 W. Compute the received power.

9.4-1 Let sources $\mathbf{J}_a$, $\mathbf{M}_a$, $\mathbf{J}_b$ and $\mathbf{M}_b$ all be of the same frequency in a linear medium. The following steps lead to the Lorentz reciprocity theorem:

a. Maxwell's equations for sources a are

$$\nabla \times \mathbf{E}_a = -j\omega\mu\mathbf{H}_a - \mathbf{M}_a$$

$$\nabla \times \mathbf{H}_a = j\omega\varepsilon\mathbf{E}_a + \mathbf{J}_a$$

Similar equations can be written for sources b. Manipulate these four equations and use the vector identity (C-19) to show that

$$\nabla \cdot (\mathbf{E}_a \times \mathbf{H}_b - \mathbf{E}_b \times \mathbf{H}_a) = \mathbf{E}_b \cdot \mathbf{J}_a + \mathbf{H}_a \cdot \mathbf{M}_b - \mathbf{H}_b \cdot \mathbf{M}_a - \mathbf{E}_a \cdot \mathbf{J}_b$$

b. Integrate the above equation over a volume v enclosing all sources, employ the divergence theorem (C-23) for the left-hand side, and let the volume extend to infinity. Then the fields arriving at the surface of the volume behave like spherical waves, and the TEM wave relationships can be employed to show that the left-hand side is zero, leading to a proof of (9-36).

9.4-2 Use the reciprocity theorem form of (9-37) to show that the distant field of *any* finite electric current distribution in free space can have no radial component.

9.4-3 Since any two-port network can be reduced to an equivalent T section, the general antenna system of Fig. 9-5a can be modeled as shown in the figure. First, excite terminals a with a current source I_a and find the open circuit output voltage $V_b|_{I_b=0}$. Then, excite terminals b with a current source I_b and find the open circuit output voltage $V_a|_{I_a=0}$. From these relationships, find Z_{ba} and Z_{ab}; they will, of course, be equal.

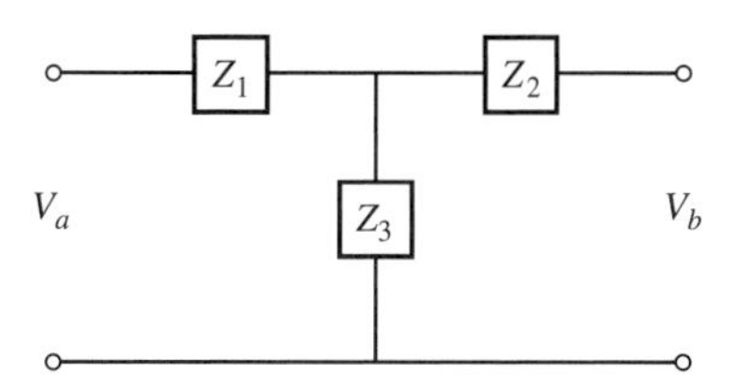

9.4-4 Write the voltage equations for the network representation of Prob. 9.4-3 and compare to (9-41) to show that the T network impedances are $Z_1 = Z_{aa} - Z_m$, $Z_2 = Z_{bb} - Z_m$, and $Z_3 = Z_m$.

9.4-5 If antennas a and b are identical, how is the network of Probs. 9.4-3 and 9.4-4 simplified?

9.4-6 Reciprocity can also be shown with voltage generators and short circuit currents:

a. Drive terminals a of the network in Prob. 9.4-3 with a voltage generator V_a and short circuit terminals b. Find the expression of $V_a/I_b|_{V_b=0}$ in terms of Z_1, Z_2, and Z_3.Then drive terminals b with voltage source V_b while short circuiting terminals a. Find $V_b/I_a|_{V_a=0}$. It should equal $V_a/I_b|_{V_b=0}$.

b. Find the same transfer impedance expressions in terms of Z_{aa}, Z_{bb}, Z_{ab}, and Z_{ba} from (9-41). Show that they are equal if $Z_{ab} = Z_{ba}$.

c. Using $Z_1 = Z_{aa} - Z_m$, $Z_2 = Z_{bb} - Z_m$, and $Z_3 = Z_m$ from Prob. 9.4-4, show that the transfer impedance expressions of (a) are the same as those of (b).

9.4-7 Using the model of Fig. 9-5a, excite antenna a with voltage V_a and prove that the power received in antenna b, which is terminated in load R_L, is proportional to $|Z_m|^2$ for antenna b in the far field of a.

9.5-1 *Anechoic chamber design.* An anechoic chamber with a separation distance between the source and test antenna of 7 m is to be used to measure the pattern of a 2-dB gain antenna. Measurements are to be made from 1 to 12 GHz and a receiver dynamic range of 45 dB is required for the pattern. The receiver has a sensitivity of −110 dBm at 1 GHz and −95 dBm at 12 GHz. The transmitter power is 10 dBm. Find the gain of the source antenna (constant over the band) in order to keep the received signal power above the receiver sensitivity by the dynamic range.

9.6-1 *Absolute gain measurement.* Gain can be measured without referencing to a standard gain antenna using one of the following two techniques. (a) The *three-antenna method* involves three antennas with unknown gains G_a, G_b, and G_c. Using (2-99), write the three equations representing the antenna range measurements. Discuss how they are solved. (b) Reduce the equations in (a) to one equation for the case of two identical antennas ($G_a = G_b = G_T$); this is the *two-antenna method.*

9.8-1 Derive (9-67) and (9-66).

9.8-2 A voltage of 200 μV (peak) is required at the input of an FM broadcast receiver for acceptable performance. The receiver input impedance is 300 Ω (real). The antenna is a linearly polarized folded dipole with an input impedance of 300 + $j0\Omega$ and has negligible loss. The antenna has a gain of 1.64 and is oriented for maximum received signal. The connecting transmission line is a 300-Ω twin lead. (a) What are the radiation and impedance efficiencies e_r and q? (b) If the radio station transmitting antenna is circularly polarized, find the minimum peak electric field strength incident on the receiver required for proper reception at 100 MHz.

9.8-3 (a) Derive an expression for the antenna factor using (9-66) in (9-69). (b) Use this relation to derive (9-70b). (c) Show that (a) reduces to the following popular formula for a matched, 50-Ω system:

$$K = \frac{9.73}{\lambda\sqrt{G}} \quad \text{m}^{-1}$$

9.8-4 Evaluate the antenna factor of a matched antenna operating at 30 MHz with a gain of 3 dB and terminated with a 50-Ω resistor.

Chapter 10

CEM for Antennas: The Method of Moments

10.1 INTRODUCTION TO CEM

In antenna analysis and design, two numerical methods in computational electromagnetics (CEM) stand out: the method of moments (MoM) and finite difference time-domain (FD-TD). Use of the former has been well established for several decades, whereas the potential of the latter in antenna work has only begun to be realized more recently. This chapter will present MoM and is followed by a discussion of FD-TD in the next chapter.

CEM is broadly defined as the discipline that intrinsically and routinely involves the use of a digital computer to obtain numerical results for electromagnetic problems [1]. It is a third tool available to electromagnetics engineers, the other two being mathematical analysis, which we have employed in the first seven chapters of this book, and experimental observation (Chap. 9). It is not uncommon to verify analysis results and CEM results with experimental results, nor is it uncommon to employ analysis and/or CEM to understand experimental results.

There are various ways to classify the assortment of techniques in CEM. Here, we choose to divide CEM into two major categories: numerical methods and high-frequency or asymptotic methods as shown in Fig. 10-1. For the most part, numerical techniques are used in the region where the size of the antenna or scatterer is on the order of the wavelength to a few tens of wavelengths as indicated in Fig. 10-2. On the other hand, high-frequency methods, which are considered in Chap. 12, are best suited to objects that are many wavelengths in extent.

In turn, there are various ways to classify numerical methods. Here, we choose to classify them as either differential-equation-based or integral-equation-based.

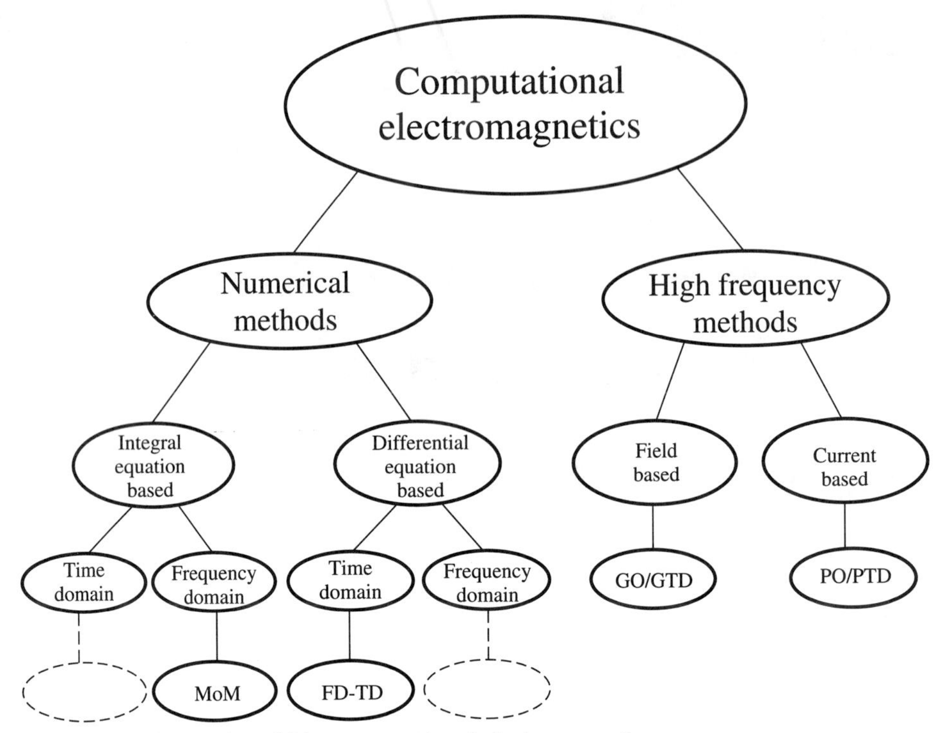

Figure 10-1 Categories within computational electromagnetics.

Both these categories can be subdivided into two parts: frequency domain and time domain. In this chapter, we investigate a technique that is integral-equation-based and in the frequency domain (i.e., the MoM). The next chapter investigates a technique that is differential-equation-based and in the time domain (i.e., FD-TD).

There are also techniques that are integral-equation-based and in the time domain (e.g., the space-time integral equation [2]) and techniques that are differential-equation-based and in the frequency domain (e.g., the finite-element methods [3]). Of these and other techniques, the finite-element methods (FEM) have seen the most use and this has been primarily in scattering problems, with microstrip patch antennas being a notable exception.

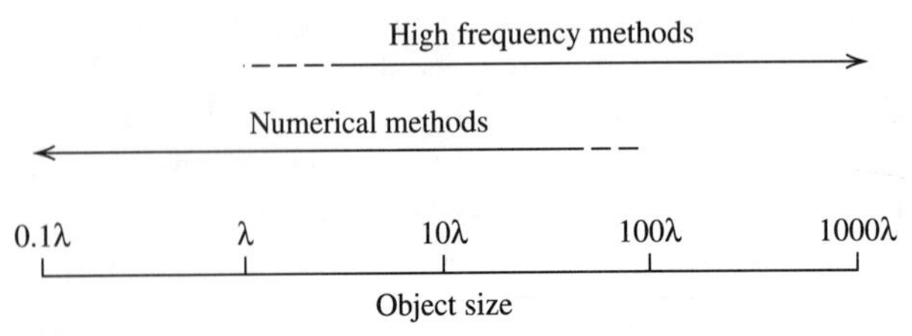

Figure 10-2 Regions of applicability for the major categories of Fig. 10-1.

10.2 INTRODUCTION TO THE METHOD OF MOMENTS

Thus far, we have studied a variety of antenna configurations, but for the most part we have assumed either that the current distribution was uniform (e.g., the ideal dipole) or sinusoidal. It was then a relatively straightforward procedure to obtain the near- and far-zone fields created by the current.

In this chapter, we eliminate the need for assuming the form of the current distribution. Naturally, this greatly expands the number of antenna configurations that can be investigated. Indeed, we are then able to study, for example, wire antennas of almost arbitrary configuration. The methods we use to do this are, therefore, very general methods capable of yielding answers whose accuracy is within the limit of experimental error. The potential price for using such powerful methods lies in the effort required to write the necessary software, the time required for computer execution, and the effort required for validation. Fortunately, cost-effective electromagnetic software is readily available and it is not necessary to write software from "scratch." However, a reasonable understanding of the principles on which the electromagnetic software is based is necessary in order to avoid its misuse and the misinterpretation of results.

Consider the wire antenna along the z-axis in Fig. 10-3. A generic form for an integral equation describing such an antenna is

$$-\int I(z')K(z, z')\, dz' = E^i(z) \tag{10-1}$$

The kernel $K(z, z')$ depends on the specific integral equation formulation used; the popular Pocklington form is presented in the next section.

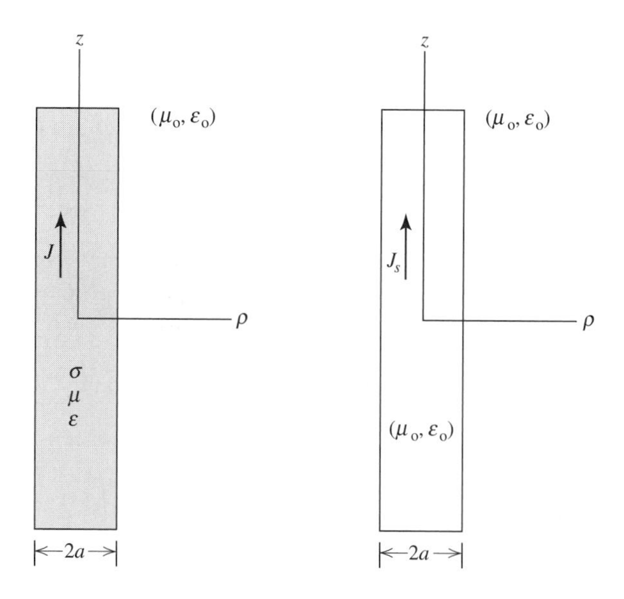

(*a*) Highly conducting wire with current density J.

(*b*) Surface equivalence model with equivalent surface current density J_s in free space.

Figure 10-3 Highly conducting thin wire along z-axis.

Electromagnetic radiation problems can always be expressed as an integral equation of the general form in (10-1) with an inhomogeneous source term on the right and the unknown within the integral. However, it was not until the availability of modern high-speed digital computers in the mid–1960s that it was feasible to solve most such equations. Since that time, many MoM procedures and codes have been developed [4–7].

MoM is a solution procedure for approximating an integral equation, such as that in (10-1), with a system of simultaneous linear algebraic equations in terms of the unknown current $I(z')$. Then, as we have seen in the previous chapters, once the current is known, it is a fairly straightforward procedure to determine the radiation pattern and impedance.

In this chapter, we set forth the basic principles involved in solving integral equations via MoM and demonstrate the procedure with several examples. The serious student is encouraged to use these basic principles to write a simple computer code, such as that suggested by Prob. 10.5-1.

10.3 POCKLINGTON'S INTEGRAL EQUATION

One of the common integral equations that arises in the treatment of wire antennas is that derived by Pocklington in 1897. It enabled him to show that the current distribution on thin wires is approximately sinusoidal and propagates with nearly the speed of light. To derive this equation, consider the wire of conductivity σ in Fig. 10-3a surrounded by free space (μ_o, ϵ_o). Assume the conductivity of the wire is high (e.g., copper) such that the current is largely confined to the surface of the wire. The equivalence model for the wire becomes that in Fig. 10-3b (see Prob. 10.3-1), where current on the material wire is replaced by an equivalent surface current in free space (i.e., the wire material is removed). This step is necessary so that the vector potential, which employs the free-space Green's function, can be used.

When the wire radius is much less than the wavelength, we may assume only z-directed currents are present. From the Lorentz gauge condition in (1-45),

$$\frac{\partial A_z}{\partial z} = -j\omega\varepsilon_o\mu_o\Phi \tag{10-2}$$

where Φ is the scalar potential and A_z is the z-component of the magnetic vector potential. If we use (1-40), the vector electric field arising from potentials is

$$\mathbf{E} = -j\omega\mathbf{A} - \nabla\Phi \tag{10-3}$$

which for the situation in Fig. 10-3 reduces to the scalar equation

$$E_z = -j\omega A_z - \frac{\partial\Phi}{\partial z} \tag{10-4}$$

Taking the derivative of (10-2) and substituting into (10-4) give

$$E_z = \frac{1}{j\omega\mu_o\varepsilon_o}\left(\frac{\partial^2 A_z}{\partial z^2} + \beta^2 A_z\right) \tag{10-5}$$

If we consider a z-directed volume current element $J\,dv'$,

$$dE_z = \frac{1}{j\omega\varepsilon_o}\left[\frac{\partial^2\psi(z, z')}{\partial z^2} + \beta^2\psi(z, z')\right]J\,dv' \tag{10-6}$$

where $\psi(z, z')$ is the free-space Green's function given in (1-56) as

$$\psi(z, z') = \frac{e^{-j\beta R}}{4\pi R} \tag{10-7}$$

and R is the distance between the observation point (x, y, z) and the source point (x', y', z') or

$$R = \sqrt{(x - x')^2 + (y - y')^2 + (z - z')^2} \tag{10-8}$$

The total contribution to the electric field is the integral over the wire volume:

$$E_z = -\frac{1}{j\omega\varepsilon_o} \iiint \left[\frac{\partial^2\psi(z, z')}{\partial z^2} + \beta^2\psi(z, z')\right] J \, dv' \tag{10-9}$$

We only need consider a volume distribution of current density if the wire is not of sufficiently high conductivity. If we assume the conductivity to be infinite, then the current is confined to the wire surface and (10-9) reduces to

$$E_z = -\frac{1}{j\omega\varepsilon_o} \oint_c \int_{-L/2}^{L/2} \left[\frac{\partial^2\psi(z, z')}{\partial z^2} + \beta^2\psi(z, z')\right] J_s \, dz' \, d\phi' \tag{10-10}$$

where c is the cross-sectional curve of the wire surface as shown in Fig. 10-4*a*. For wires of good conducting material, the assumption of a surface current is approximately true and leads to no complications. If one observes the surface current distribution from a point on the wire axis as in Fig. 10-4*b*, then

$$R = \sqrt{(z - z')^2 + a^2} \tag{10-11}$$

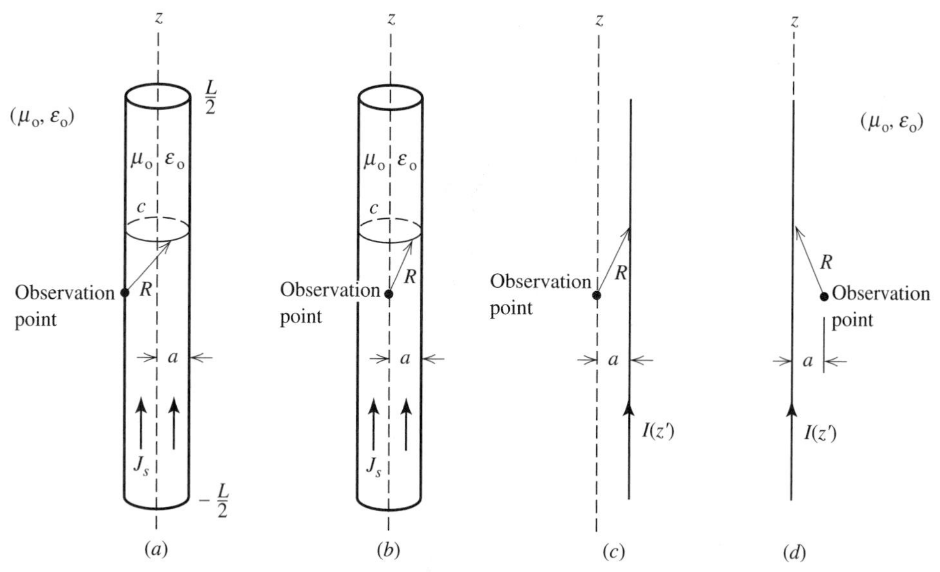

Figure 10-4 Theoretical models for a thin wire.
(*a*) Wire with equivalent surface current density J_s and observation point on the surface.
(*b*) Wire with surface current density J_s and observation point on the wire axis.
(*c*) Equivalent filamentary line source for the situation in (*b*).
(*d*) Alternate representation of (*c*).

For $a \ll \lambda$, the current distribution is nearly uniform with respect to ϕ', and (10-10) reduces to a line integral of (total) current. Thus,

$$E_z = \frac{1}{j\omega\varepsilon_o} \int_{-L/2}^{L/2} \left[\frac{\partial^2 \psi(z, z')}{\partial z^2} + \beta^2 \psi(z, z') \right] I(z')\, dz' \tag{10-12}$$

Note that the equivalent filamentary line source $I(z')$ is located a radial distance a from the observation point as in Figs. 10-4c and 10-4d and that we have not assumed the wire to be infinitely thin as was the case for dipoles studied in Chaps. 2 and 5.

In accordance with the surface equivalence principle of Sec. 7.1, we can denote the quantity E_z in (10-12) as the scattered field E_z^s. That is, E_z^s is the field radiated in free space by the equivalent current $I(z')$. The other field present is the incident or impressed field E_z^i. At the surface of the perfectly conducting wire and also interior to the wire, the sum of the tangential components of the scattered field and the incident field must be zero. Hence, $-E_z^s = +E_z^i$, and using (10-12) we write

$$\frac{-1}{j\omega\varepsilon_o} \int_{-L/2}^{L/2} I(z') \left[\frac{\partial^2 \psi(z, z')}{\partial z^2} + \beta^2 \psi(z, z') \right] dz' = E_z^i(z) \tag{10-13}$$

which is the type of integral equation derived by Pocklington and is of the general form used in (10-1).

Equation (10-13) is an integral equation of the first kind because the unknown $I(z')$ appears only inside the integral. It is known as an integral equation because a boundary condition is incorporated there. This is in contrast to (10-12) that is merely an expression for the so-called scattered field, which we can think of as that field radiated by a current independent of how the current was established (e.g., an impressed source on an antenna or incident plane wave).

Before we leave this section, it is worthwhile to summarize the important approximations that were used based on the assumption that $a \ll \lambda$.

1. Circumferential currents on the wire are negligible.
2. Enforcement of the boundary condition on the surface of the wire (Fig. 10-4a) was performed on the axis of the wire (Fig. 10-4b), and the surface current then "collapsed" into a filament (Fig. 10-4c). By using reciprocity, the current filament was placed on the axis of the wire and the observation point placed a distance "a" away from the filament (Fig. 10-4d).
3. The distance R given by (10-11) leads to the widely used thin wire kernel or reduced kernel. R can never be zero; hence, the kernel is never singular. However, it is nearly singular and care must sometimes be taken during integrations when $R \approx a$.

In the following section, we illustrate how an integral equation such as (10-13) is solved numerically and point out how the procedure is analogous to Kirchhoff's network equations as noted by Schelkunoff [8] many years ago.

10.4 INTEGRAL EQUATIONS AND KIRCHHOFF'S NETWORK EQUATIONS

One purpose of this section is to show the resemblance between integral equations of the type given in (10-13) and Kirchhoff's network equations:

$$\sum_{n=1}^{N} Z_{mn} I_n = V_m, \qquad m = 1, 2, 3, \ldots, N \tag{10-14}$$

Thus, we will solve the integral equation numerically by writing N equations in N unknowns just as we would do if we were solving an N mesh or N node circuit problem.

For convenience, let us write (10-13) in the form

$$-\int_{-L/2}^{L/2} I(z')K(z, z')\, dz' = E_z^i(z) \tag{10-15}$$

The first step in solving (10-15) is to approximate the unknown current by a series of known *expansion functions* F_n such that

$$I(z') = \sum_{n=1}^{N} I_n F_n(z') \tag{10-16}$$

where the I_n's are complex expansion coefficients and are unknown. To keep the discussion as simple as possible, we assume the expansion functions are a set of orthogonal pulse functions given by

$$F_n(z') = \begin{cases} 1 & \text{for } z' \text{ in } \Delta z_n' \\ 0 & \text{otherwise} \end{cases} \tag{10-17}$$

The expansion in terms of pulse functions is a "stairstep" approximation to the current distribution on the wire, where the wire is divided into N segments of length $\Delta z_n'$. See Fig. 10-5.

Substituting (10-16) into (10-15) gives

$$-\int_{-L/2}^{L/2} \sum_{n=1}^{N} I_n F_n(z')K(z_m, z')\, dz' \approx E_z^i(z_m) \tag{10-18}$$

where the subscript m on z_m indicates that the integral equation is being enforced at segment m. Note that the left side is only approximately equal to the right side because we have replaced the actual current distribution with an approximate distribution. Using (10-17) in (10-18) enables us to write

$$-\sum_{n=1}^{N} I_n \int_{\Delta z_n'} K(z_m, z')\, dz' \approx E_z^i(z_m) \tag{10-19}$$

For convenience, we let

$$f(z_m, z_n') = -\int_{\Delta z_n'} K(z_m, z')\, dz' \tag{10-20}$$

Then (10-16) and (10-17) in (10-15) yield

$$\begin{aligned}-\int_{-L/2}^{L/2} I(z')K(z_m, z')\, dz' &\approx I_1 f(z_m, z_1') + I_2 f(z_m, z_2') + \cdots + I_n f(z_m, z_n') \\ &\quad + \cdots + I_N f(z_m, z_N') \approx E_z^i(z_m)\end{aligned} \tag{10-21}$$

as illustrated in Fig. 10-5. A physical interpretation of this equation is as follows. The wire has been divided up into N segments, each of length $\Delta z_n' = \Delta z'$, with the current being an unknown constant over each segment. At the center of the mth segment, the sum of the scattered fields from all N segments is set equal to the incident field at the point z_m. The incident field is a known field arising from either a source located on the wire (transmitting case) or from a source located at a large distance (receiving case or radar scattering case). As we might surmize, if a more

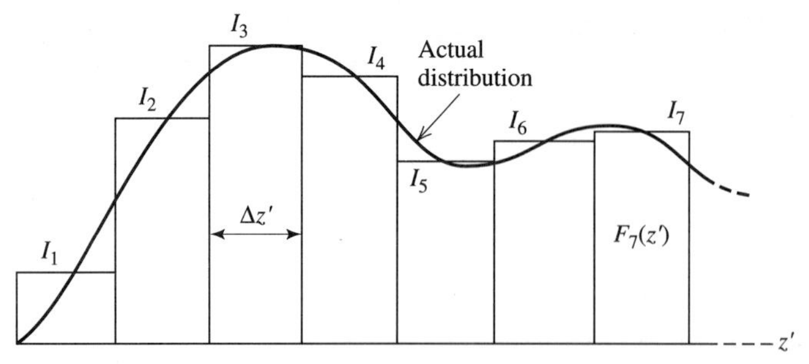

Figure 10-5 "Staircase" approximation to an actual current distribution.

accurate representation of $I(z')$ is required, then shorter segments (and a larger N) can be used. More will be said about this later.

Equation (10-21) leads to

$$\sum_{n=1}^{N} Z_{mn} I_n = V_m \tag{10-22}$$

where in this example situation

$$Z_{mn} = f(z_m, z_n') \tag{10-23}$$

and

$$V_m = E_z^i(z_m) \tag{10-24}$$

Note that we have achieved our goal of reducing the electromagnetic problem to (10-22), which is identical to the network formulation of (10-14). It should be mentioned, however, that in network problems Z_{mn} is known at the start, whereas in electromagnetic problems it is necessary to calculate Z_{mn} as we have shown in this elementary example.

So far, we have only generated one equation in N unknowns. We need $N - 1$ additional independent equations to solve for the N unknowns. To obtain these additional equations, we choose a different z_m for each equation. That is, we enforce the integral equation at N points on the axis of the wire. The process of doing this is called *point-matching*. It is a special case of the more general method of moments.

Point-matching at N points results in the following system of equations:

$$\begin{matrix} I_1 f(z_1, z_1') + I_2 f(z_1, z_2') + \cdots + I_N f(z_1, z_N') = E_z^i(z_1) \\ I_1 f(z_2, z_1') + I_2 f(z_2, z_2') + \cdots + I_N f(z_2, z_N') = E_z^i(z_2) \\ \vdots \qquad\qquad \vdots \qquad\qquad \vdots \qquad\qquad \vdots \\ I_1 f(z_N, z_1') + I_2 f(z_N, z_2') + \cdots + I_N f(z_N, z_N') = E_z^i(z_N) \end{matrix} \tag{10-25}$$

which can be written in matrix form as

$$\begin{bmatrix} f(z_1, z_1') & f(z_1, z_2') & \ldots & f(z_1, z_N') \\ f(z_2, z_1') & f(z_2, z_2') & \ldots & f(z_2, z_N') \\ \vdots & \vdots & & \vdots \\ f(z_N, z_1') & f(z_N, z_2') & \ldots & f(z_N, z_N') \end{bmatrix} \begin{bmatrix} I_1 \\ I_2 \\ \vdots \\ I_N \end{bmatrix} = \begin{bmatrix} E_z^i(z_1) \\ E_z^i(z_2) \\ \vdots \\ E_z^i(z_N) \end{bmatrix} \tag{10-26}$$

or in the compact notation as

$$[Z_{mn}][I_n] = [V_m] \tag{10-27}$$

where Z_{mn} and V_m are given by (10-23) and (10-24), respectively. We refer to the first index (m) as the match point index because it is associated with the observation point at which the mth equation is valid. The second index is the source point index since it is associated with the field from the nth segment or nth source. Because of the analogy to the network equations, the matrices $[Z_{mn}]$, $[I_n]$, and $[V_m]$ are referred to as *generalized* impedance, current, and voltage matrices, respectively. But this is only an analogy and thus the units of $[Z_{mn}]$, $[I_n]$, and $[V_m]$ need not necessarily be ohms, amperes, and volts, respectively. The analogy is not restricted to collinear segments as in the example treated here, but applies to arbitrary configurations of wires as well.

We can write the solution to (10-27) symbolically as

$$[I_n] = [Z_{mn}]^{-1}[V_m] \tag{10-28}$$

In practice, the explicit inverse $[Z_{mn}]^{-1}$ is not usually evaluated, but instead the system of equations is solved by one of several fairly standard matrix algorithms. Once $[I_n]$ is found, the approximate current distribution of (10-16) is known in discrete form and we can then proceed to determine impedance and radiation patterns or the radar cross section.

To summarize this section, we have obtained an elementary numerical solution to an integral equation of the form in (10-15). This was done by successively enforcing the integral equation at N different points, as illustrated in (10-25). For mathematical convenience and simplicity, the locations of the points were chosen to be at the center of the N equal-length segments into which the wire was divided. Strictly speaking, in order for the equations in (10-25) to be exact equalities, N must approach infinity. However, in practice we can obtain accurate solutions for the current distribution by allowing N to be sufficiently large as will be demonstrated in the next section.

10.5 SOURCE MODELING

Three source models are commonly used in the MoM. For transmitting antennas, the delta gap source and frill source produce the required incident field. For a receiving antenna or scatterer, the incident field is usually a plane wave. We examine all three in this section.

No doubt, the most used generator model in wire antenna theory is the *delta gap model*, shown in Fig. 10-6, which is often referred to as a *slice generator*. Although such sources do not exist in practice, they do permit surprisingly good calculations

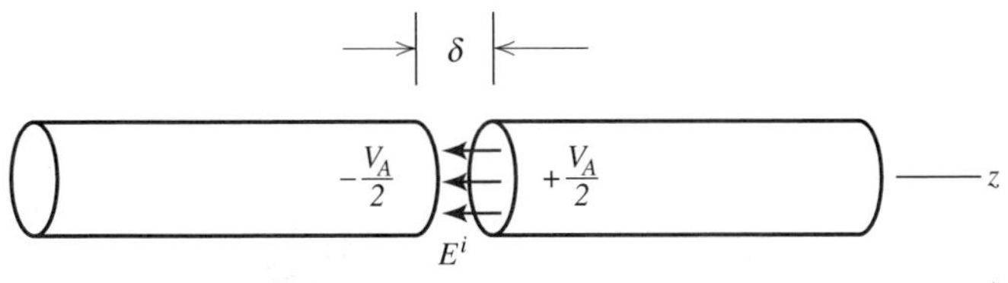

Figure 10-6 The delta gap source model with impressed field $E^i = V_A/\delta$.

to be made. The source arises from the assumption that a voltage is placed across the gap, giving rise to an impressed electric field $E^i = V/\delta$ confined entirely to the gap (i.e., no fringing). With reference to Fig. 10-6, the voltage across the gap is determined by the line integral of the electric field across the gap. The result is $V_A = +E^i\delta$. The voltage V_A applied across the gap is not to be confused with the elements V_m in the generalized voltage matrix $[V_m]$. For a delta gap source model, $V_m = E^i = V_A/\delta$ when the point-matching technique is used.

A second generator model, which has practical significance, is the so-called *frill generator*. Consider Fig. 10-7*a* that shows a coaxial line feeding a monopole on a ground plane. Assuming a purely dominant mode distribution (TEM) in the coaxial aperture and image theory, we can replace the ground plane and the coaxial aperture with a frill of magnetic current as shown in Fig. 10-7*b*. Since the assumed form of the electric field in the aperture is

$$E_{\rho'}(\rho') = \frac{1}{2\rho' \ln(b/a)} \tag{10-29}$$

the corresponding magnetic current distribution from $\mathbf{M} = 2\hat{\mathbf{n}} \times \mathbf{E}$ is

$$M_{\phi'} = 2E_{\rho'} = \frac{-1}{\rho' \ln(b/a)} \tag{10-30}$$

from which it can be shown that the electric field on the axis of the monopole is [7, 40]

$$E_z^i(0, z) = \frac{1}{2 \ln(b/a)} \left(\frac{e^{-j\beta R_1}}{R_1} - \frac{e^{-j\beta R_2}}{R_2} \right) \tag{10-31}$$

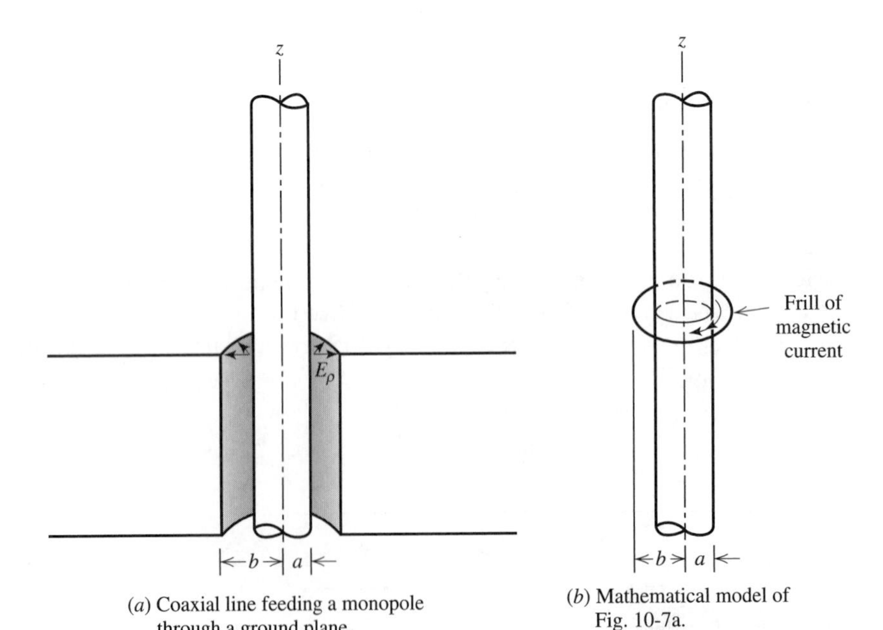

(*a*) Coaxial line feeding a monopole through a ground plane.

(*b*) Mathematical model of Fig. 10-7a.

Figure 10-7 Magnetic frill source.

where

$$R_1 = \sqrt{z^2 + a^2} \tag{10-32a}$$

$$R_2 = \sqrt{z^2 + b^2} \tag{10-32b}$$

if the frill center is at the coordinate origin.

The third source to consider is that of an incident plane wave. To obtain the elements in the generalized voltage matrix in this case, we need the tangential component of the incident field at the match points along the axis of the wire dipole. For our z-directed dipole of Fig. 10-4, this would be

$$E^i_{\tan} = \hat{\mathbf{z}} \cdot \mathbf{E} e^{j\beta z \cos\theta} \tag{10-33}$$

For example, for a unit amplitude plane wave normally incident on the z-directed dipole, the elements of the generalized voltage matrix are all $(1 + j0)$.

EXAMPLE 10-1 Point-Matching on a Short Dipole

The purpose of this example is to illustrate the application of (10-26). An objective is to use MoM to determine the input impedance Z_A of a short dipole with a length of 0.1λ and a radius of 0.005λ. For convenience of illustration, choose $N = 5$. With reference to Fig. 10-8, the elements of $[Z_{mn}]$ are calculated to be

$$[Z_{mn}] = 10^2 \begin{bmatrix} 679.5\angle{-89.99^\circ} & 292.6\angle 89.97^\circ & 33.03\angle 89.73^\circ & 9.75\angle 89.09^\circ & 4.24\angle 87.92^\circ \\ 292.6\angle 89.97^\circ & 679.5\angle{-89.99^\circ} & 292.6\angle 89.97^\circ & 33.03\angle 89.73^\circ & 9.75\angle 89.09^\circ \\ 33.03\angle 89.73^\circ & 292.6\angle 89.97^\circ & 679.5\angle{-89.99^\circ} & 292.6\angle 89.97^\circ & 33.03\angle 89.73^\circ \\ 9.75\angle 89.09^\circ & 33.03\angle 89.73^\circ & 292.6\angle 89.97^\circ & 679.5\angle{-89.99^\circ} & 292.6\angle 89.97^\circ \\ 4.24\angle 87.92^\circ & 9.75\angle 89.09^\circ & 33.03\angle 89.73^\circ & 292.6\angle 89.97^\circ & 679.5\angle{-89.99^\circ} \end{bmatrix}$$

For a 1-V excitation at the center of the short dipole (i.e., segment 3), the following voltage matrix $[V_m]$ is obtained using the frill source discussed in Sec. 10.5 with $b/a = 2.3$, and upon solving (10-28), the following current matrix $[I_n]$ is also obtained:

$$[V_m] = \begin{bmatrix} 0.484\angle{-0.31^\circ} \\ 3.128\angle{-0.04^\circ} \\ 67.938\angle{-0.002^\circ} \\ 3.128\angle{-0.04^\circ} \\ 0.484\angle{-0.31^\circ} \end{bmatrix}, \qquad [I_n] = 10^{-3} \begin{bmatrix} 0.78\angle 89.54^\circ \\ 1.48\angle 89.64^\circ \\ 2.35\angle 89.75^\circ \\ 1.48\angle 89.64^\circ \\ 0.78\angle 89.54^\circ \end{bmatrix}$$

On the other hand, if a 1-V delta gap excitation is used, $V_3 = 1/\Delta z = 1/0.02 = 50\angle 0^\circ$, and the resulting voltage and current matrices are

$$[V_m] = \begin{bmatrix} 0\ \angle 0^\circ \\ 0\ \angle 0^\circ \\ 50.0\angle 0^\circ \\ 0\ \angle 0^\circ \\ 0\ \angle 0^\circ \end{bmatrix}, \qquad [I_n] = 10^{-3} \begin{bmatrix} 0.52\angle 89.54^\circ \\ 0.98\angle 89.64^\circ \\ 1.63\angle 89.76^\circ \\ 0.98\angle 89.64^\circ \\ 0.52\angle 89.54^\circ \end{bmatrix}$$

Note that the current distribution decreases from the center toward the ends as expected. The input impedance for the frill may be found from $Z_A = V_A/I_3 = 1.0/(2.35 \times$

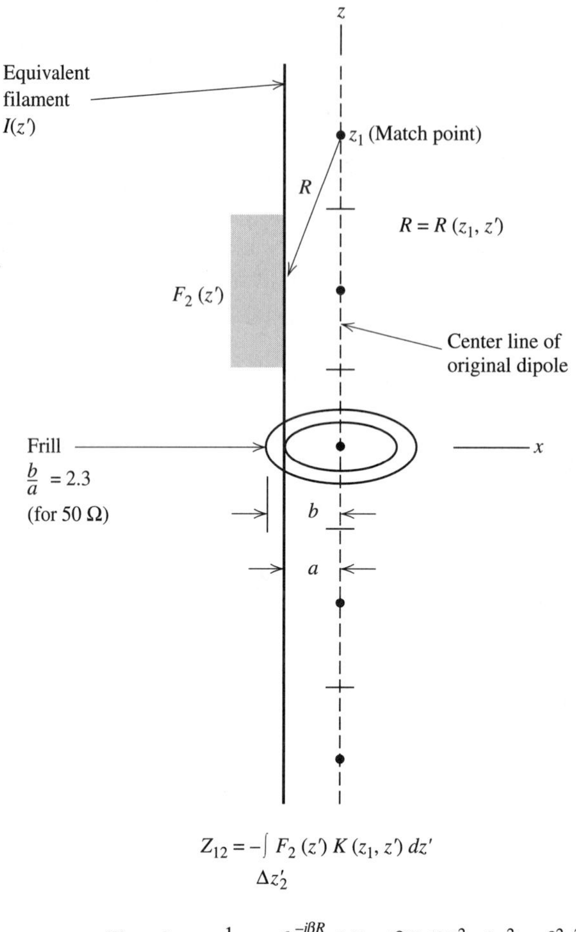

Figure 10-8 Calculation of Z_{12} for the short dipole in Example 10-1.

$10^{-3}\angle 89.75°) = 425.53 \angle -89.75° = 1.857 - j425.53\ \Omega$. Comparing with the thinner dipole in Figs. 5-5 and 5-6, we see that the input impedance of a 0.1λ long dipole also has a very small real part and a large negative reactive part. Further, the real part of $1.857\ \Omega$ compares fairly well with the approximate formula $20\pi^2(L/\lambda)^2 = 1.974\ \Omega$ even though only five segments were used here.

In the above example, a short dipole was represented by only five segments for the purposes of numerical illustration. To illustrate the behavior of a pulse point-matching solution to Pocklington's equation for a resonant size dipole as the number of segments is varied, consider Fig. 10-9. Figure 10-9 shows the input impedance of a dipole of length $L = 0.47\lambda$ as the number of segments varies from 10 to 120. Both the frill source and delta gap are used. For both sources, it is apparent that for N sufficiently large, the solution has converged to a final or reasonably stable result. In many instances, N cannot be made arbitrarily large without encountering a numerically unstable result. For example, the reactance of the delta gap source exhibits divergence for large N in Fig. 10-9*b*, but this should not be viewed as a general behavior of the delta gap since it does not necessarily occur for other MoM for-

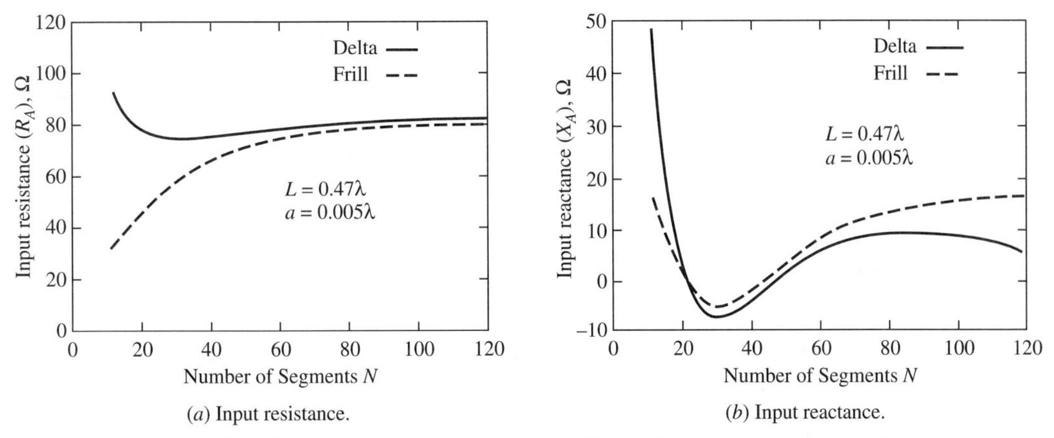

(*a*) Input resistance.

(*b*) Input reactance.

Figure 10-9 Curves showing convergence of input impedance as the number of pulse functions is increased for two different sources: the delta gap and the frill.

mulations (e.g., Fig. 10-13*b*) or for (10-13) if the order of differentiation and integration is interchanged.

A curve, such as those in Fig. 10-9, is well worth the effort since it clearly shows the convergence behavior of a solution. A comparison with experimental data is shown in Fig. 10-10.

To summarize this section, an elementary numerical solution to an integral equation of the form given in (10-15) was obtained by successively enforcing the integral

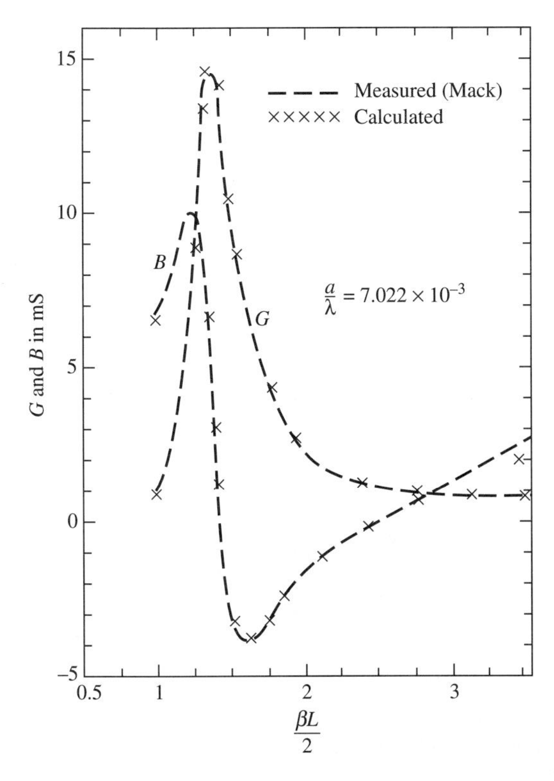

Figure 10-10 Comparison of measured dipole admittance with data calculated using pulse functions ($N = 100$) and a frill source.

equation at N different points as illustrated in (10-25). For mathematical convenience and simplicity, the segments were of equal length, and the match points were located at the center of each segment. Strictly speaking, in order for the equations (10-25) to be exact equalities, N must approach infinity. However, in practice, accurate solutions can be obtained by allowing N to be sufficiently large. In this regard, a convergence curve, such as that in Fig. 10-9, can be invaluable.

10.6 WEIGHTED RESIDUALS AND THE METHOD OF MOMENTS

Our objective in this section is to derive a moment method procedure more general than the point-matching method of the previous section. This is accomplished by using an approach known as the method of weighted residuals [10].

Consider the straight wire example of the previous section. Define the *residual R* to be the sum of the tangential components of the scattered and incident fields:

$$R = E^s_{\tan} + E^i_{\tan} \tag{10-34}$$

Clearly, we wish the residual to be zero and thereby satisfy the boundary condition. In our example of Sec. 10.4, with pulse expansion functions the residual is found from (10-19) to be

$$R(z) = -\sum_{n=1}^{N} I_n f(z, z'_n) + E^i_z(z) \tag{10-35}$$

Stated in terms of the electric field boundary condition, the residual is the sum of the tangential components of the scattered and incident fields at the wire surface. Equation (10-35) when evaluated for $z = z_m$ gives the residual at the mth match point where, of course, the residual must be zero since the solution for the I_n's was obtained subject to the electric field boundary condition at the N matching points. However, at points other than the match points, the total tangential electric field will not generally be zero as Fig. 10-11 indicates. Therefore, the residual for $z \neq z_m$,

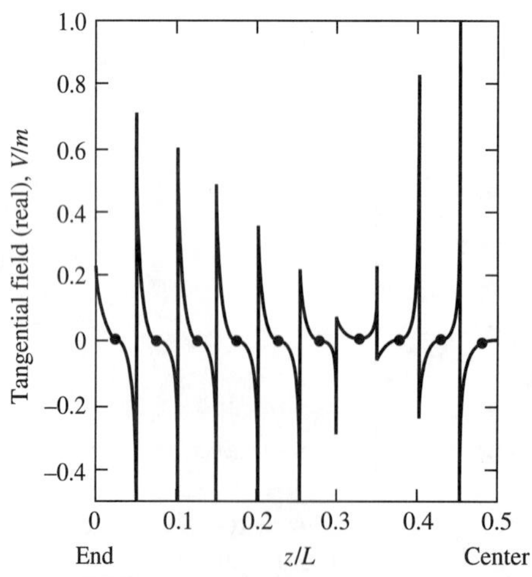

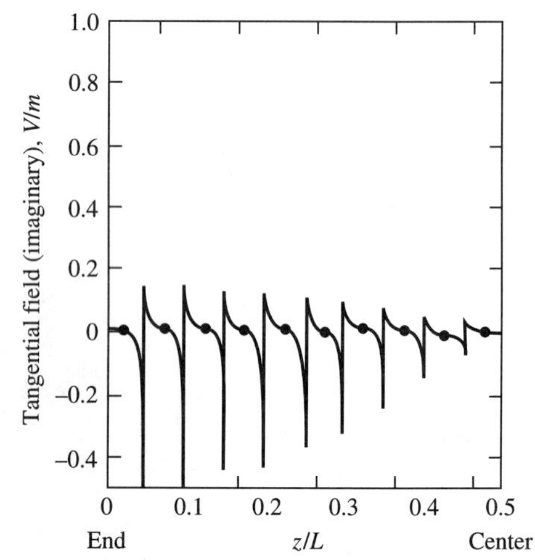

Figure 10-11 Normalized tangential electric field along one-half of a center-fed dipole with pulse expansion functions and delta weighting functions (courtesy of E. K. Miller). Dots indicate match point locations.

$m = 1, 2, 3, \ldots, N$, will not be zero either. Physically, we can view the point-matching procedure as a relaxation of the boundary condition such that it is only satisfied at specified points. In between those points, one can only hope that the boundary condition is not so badly violated that the solution is rendered useless. Thus, it is not surprising that as N is increased (within limits) the solution tends to improve as we saw in Fig. 10-9.

In the method of weighted residuals the I_n's are found such that the residual is forced to zero in a weighted average sense. So, in the wire problem of Fig. 10-3 the weighted integrals of the residual are set to zero as follows.

$$\int W_m(z) R(z)\, dz = 0, \qquad m = 1, 2, 3, \ldots, N \tag{10-36}$$

where $W_m(z)$ is called a *weighting* or *testing* function. Substituting (10-35) into (10-36) gives

$$-\int_{-L/2}^{L/2} W_m(z) \sum_{n=1}^{N} I_n f(z, z_n')\, dz + \int_{-L/2}^{L/2} W_m(z) E_z^i(z)\, dz = 0, \qquad m = 1, 2, 3, \ldots, N \tag{10-37}$$

If the weighting functions are Dirac delta functions

$$W_m(z) = \delta(z - z_m) \tag{10-38a}$$

then (10-37) reduces to (10-21). If the weighting functions are the pulse functions

$$W_m(z) = \begin{cases} 1 & \text{for } z \text{ in } \Delta z_m \\ 0 & \text{otherwise} \end{cases} \tag{10-38b}$$

then (10-37) becomes

$$-\sum_{n=1}^{N} I_n \int_{\Delta z_m} f(z, z_n')\, dz + \int_{\Delta z_m} E_z^i(z)\, dz = 0, \qquad m = 1, 2, 3, \ldots, N \tag{10-39}$$

It follows that

$$Z_{mn} = \int_{\Delta z_m} f(z, z_n')\, dz \tag{10-40a}$$

and

$$V_m = \int_{\Delta z_m} E_z^i(z)\, dz \tag{10-40b}$$

The current obtained from solving (10-40) will not necessarily be such that the sum of the scattered and incident fields (i.e., the residual) is zero everywhere along the surface of the wire, but the average over the wire will tend to be zero, presumably giving a more accurate current distribution for a given N than when the weight functions are delta functions. Actually, this may or may not be the case depending on the particular choice of expansion functions for the current and weighting (or testing) functions.

The question of how one chooses the expansion functions and weighting functions is certainly a valid one. It is, however, a question without a concise answer. But, as rules of thumb, it is desirable to choose expansion functions that closely resemble the anticipated form of the current on the wire and to use the same functions for the weighting functions as used for the expansion functions. There are exceptions to these rules including the pulse point-matching solution of Sec. 10.4. When the expansion function and the weight function are the same, the procedure is often referred to as *Galerkin's method*, which is closely related to variational methods [6, 7, 10].

EXAMPLE 10-2 *Galerkin's Method on a Short Dipole*

The purpose of this example is to repeat Example 10-1 using pulse functions for weight functions instead of delta functions. With reference to Fig. 10-12, the impedance matrix $[Z_{mn}]$ for this pulse-pulse Galerkin solution based on (10-40) is calculated to be

$$[Z_{mn}] = 10^2 \begin{bmatrix} 14.4\angle-89.99^\circ & 6.14\angle 89.97^\circ & 0.759\angle 89.76^\circ & 0.206\angle 89.14^\circ & 0.087\angle 87.98^\circ \\ 6.14\angle 89.97^\circ & 14.4\angle-89.99^\circ & 6.14\angle 89.97^\circ & 0.759\angle 89.76^\circ & 0.206\angle 89.14^\circ \\ 0.759\angle 89.76^\circ & 6.14\angle 89.97^\circ & 14.4\angle-89.99^\circ & 6.14\angle 89.97^\circ & 0.759\angle 89.76^\circ \\ 0.206\angle 89.14^\circ & 0.759\angle 89.76^\circ & 6.14\angle 89.97^\circ & 14.4\angle-89.99^\circ & 6.14\angle 89.97^\circ \\ 0.087\angle 87.98^\circ & 0.206\angle 89.14^\circ & 0.759\angle 89.76^\circ & 6.14\angle 89.97^\circ & 14.4\angle-89.99^\circ \end{bmatrix}$$

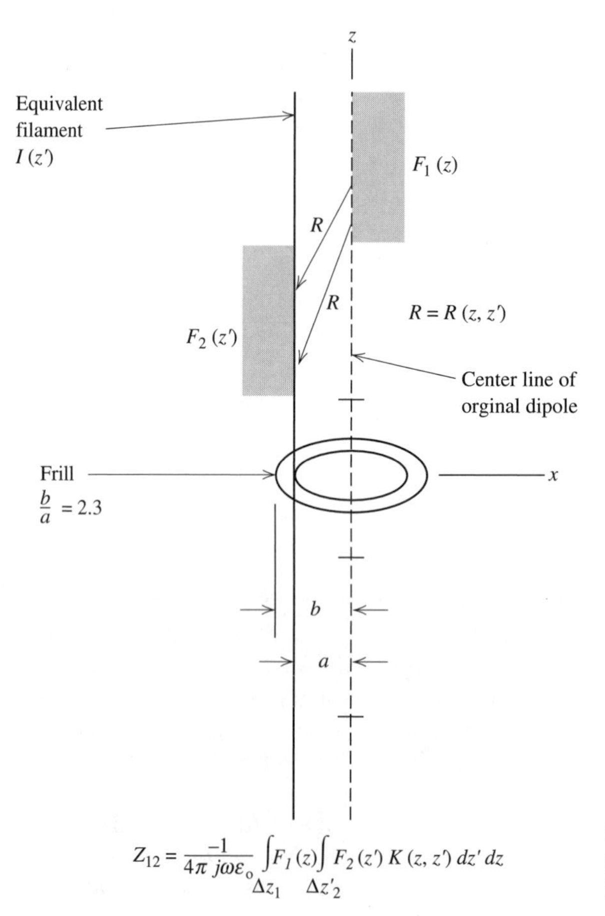

Figure 10-12 Calculation of Z_{12} for the short dipole in Example 10-2.

The voltage matrix $[V_m]$, using a 1-V frill source, and solution matrix $[I_n]$ are, respectively,

$$[V_m] = \begin{bmatrix} 0.011 \angle -0.280^\circ \\ 0.089 \angle -0.034^\circ \\ 0.791 \angle -0.003^\circ \\ 0.089 \angle -0.034^\circ \\ 0.011 \angle -0.280^\circ \end{bmatrix}, \qquad [I_n] = 10^{-3} \begin{bmatrix} 0.49 \angle 89.57^\circ \\ 0.91 \angle 89.66^\circ \\ 1.38 \angle 89.75^\circ \\ 0.91 \angle 89.66^\circ \\ 0.49 \angle 89.57^\circ \end{bmatrix}$$

On the other hand, if a 1-V delta gap excitation is used,

$$[V_m] = \begin{bmatrix} 0 \angle 0^\circ \\ 0 \angle 0^\circ \\ 1 \angle 0^\circ \\ 0 \angle 0^\circ \\ 0 \angle 0^\circ \end{bmatrix}, \qquad [I_n] = 10^{-3} \begin{bmatrix} 0.49 \angle 89.57^\circ \\ 0.91 \angle 89.67^\circ \\ 1.52 \angle 89.78^\circ \\ 0.91 \angle 89.67^\circ \\ 0.49 \angle 89.57^\circ \end{bmatrix}$$

Note that all these five matrices are different from those in Example 10-1. Of course, we would expect $[Z_{mn}]$ and $[V_m]$ to be different because they are computed by a different process. The reason $[I_n]$ is different is solely attributable to the fact that N is only 5. As in Example 10-1, a larger value of N is required in order to obtain a converged result. The input impedance based on the above current using the frill source is $Z_A = 3.162 - j724\ \Omega$, whereas for the delta gap source $Z_A = 2.526 - j658\ \Omega$. The impedance based on $N = 25$ is $Z_A = 2.35 - j556\ \Omega$.

Figure 10-13 shows the convergence of the input impedance for a dipole of length 0.47λ using pulse expansion functions and pulse weighting functions in Pocklington's equation. Comparing with Fig. 10-9, we see that the convergence is more rapid with pulse weights than delta weights, and the pulse-pulse formulation is less sensitive to the kind of source (i.e., frill or delta gap) than the pulse-delta formulation. In many formulations, as in the one here, the averaging process provided by nondelta weights tends to improve the rate of convergence and stability of the solution.

Next, we relate the quantities in the weighted residual integral to Kirchhoff's network equations, just as was done in Sec. 10.4. In doing so, let us generalize

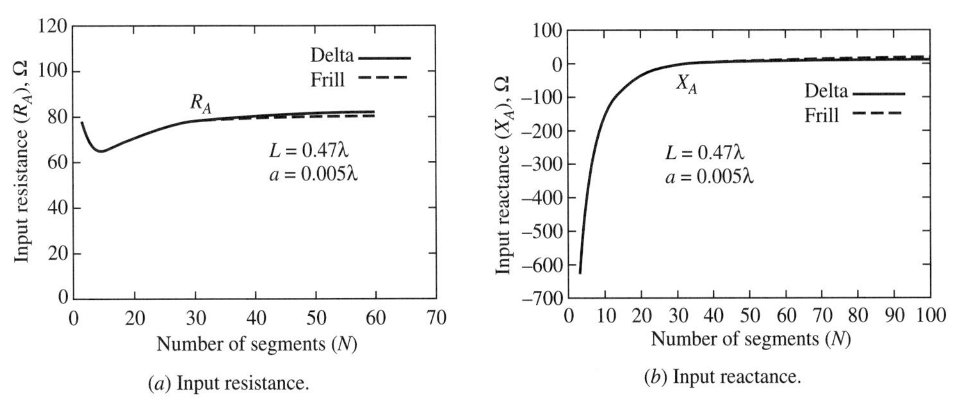

(*a*) Input resistance. (*b*) Input reactance.

Figure 10-13 Curves showing convergence of input impedance as the number of segments (pulse Galerkin) is increased for two different sources: the delta gap and the frill.

somewhat and consider a wire as shown in Fig. 10-14. In this case, the residual may be written as

$$\mathbf{R}(\ell) = \mathbf{E}^s_{\tan}(\ell) + \mathbf{E}^i_{\tan}(\ell) = \sum_{n=1}^{N} I_n \mathbf{E}^s_n(\ell) + \mathbf{E}^i_{\tan}(\ell) \tag{10-41}$$

and the weighted residual as

$$\int_{\text{along wire}} \mathbf{W}_m(\ell) \cdot \mathbf{R}(\ell)\, d\ell = 0 \tag{10-42}$$

so that

$$\sum_{n=1}^{N} I_n \int_{-\Delta\ell_m/2}^{\Delta\ell_m/2} \mathbf{W}_m(\ell) \cdot \mathbf{E}^s_n(\ell)\, d\ell + \int_{-\Delta\ell_m/2}^{\Delta\ell_m/2} \mathbf{W}_m(\ell) \cdot \mathbf{E}^i(\ell)\, d\ell = 0, \quad m = 1, 2, 3, \ldots, N \tag{10-43}$$

This equation can be viewed in the form of (10-14) and if the scattered field from the nth expansion function of the current is denoted as $\mathbf{E}^s_n(\ell)$, then the general mnth element in the generalized impedance matrix is

$$Z_{mn} = -\int_{-\ell_m/2}^{\ell_m/2} \mathbf{W}_m(\ell) \cdot \mathbf{E}^s_n(\ell)\, d\ell \tag{10-44}$$

and for the mth generalized voltage matrix element,

$$V_m = \int_{-\ell_m/2}^{\ell_m/2} \mathbf{W}_m(\ell) \cdot \mathbf{E}^i(\ell)\, d\ell \tag{10-45}$$

where $\mathbf{W}_m(\ell)$ is the mth testing function taken to be located interior to the wire as suggested in Fig. 10-12. Strictly speaking, the test function should be located at the wire surface (see Fig. 10-4*a*), in which case (10-44) and (10-45) would be double integrals over the surface. In placing the testing function on the axis, we are in a sense modifying the electric field boundary condition for the sake of mathematical simplification. In doing this, experience has shown that we are restricted to wires for which the radius is less than about 0.01λ. This is sufficient for most wire antenna or scattering problems. For thicker wires, a more exact formulation is available [11].

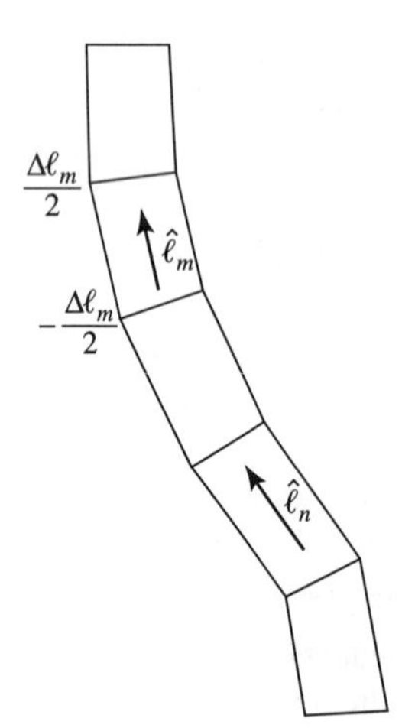

Figure 10-14 Segmented curved wire.

The process of expanding the unknown current $I(\ell')$ in a series of expansion functions and then generating N equations in N unknowns using the weighted residual integral of (10-42) is more commonly referred to in the electromagnetics literature as the method of moments [5–7, 9]. MoM is, as we have seen in this section, equivalent to the method of weighted residuals. If the testing or weighting functions are delta functions, then the specific MoM procedure is known as *point-matching*. This is also known as *collocation*. This was the procedure used to obtain the system of equations in (10-25). If both the test function and expansion function are the same, then the specific MoM procedure is known as *Galerkin's method*. A pulse-pulse Galerkin formulation was used in Example 10-2. There are functions other than the pulse function that have been shown to be useful. Some of these are discussed in Sec. 10.8. In the next section, we discuss two other approaches to the MoM: the reaction concept and the linear algebra formulation.

10.7 TWO ALTERNATIVE APPROACHES TO THE METHOD OF MOMENTS

In the previous sections of this chapter, MoM has been developed using an approach that makes takes advantage of concepts a student is likely to have previously experienced (e.g., Kirchhoff's network equations and the use of the "staircase approximation" to an integral in Sec. 10.4). Two other approaches to MoM are found in the literature. One has a physical interpretation (i.e., reaction) and the other is entirely mathematical (i.e., the linear algebra approach). This section will consider both of these approaches.

10.7.1 Reaction

In 1954, Rumsey introduced a physical observable (e.g., mass, length, charge, etc.) called *reaction* that permitted a general approach to boundary value problems in electromagnetic theory [12]. His approach resulted in the formulaton of the reaction integral equation. Equation (10-43) is really a special form of the reaction integral equation that applies to wire geometries. A rigorous derivation of the reaction integral equation can be derived using only principles of electromagnetic theory. The derivation is somewhat difficult to follow and so we will use inductive reasoning here, having derived (10-43) in the previous section by the relatively straightforward weighted residual approach.

Reaction is basically "a measure of the coupling" between one source and another. Thus, if we view the test function (weight function) as a *test source*, then the impedance matrix elements given by (10-44) may be taken as a calculation of the coupling between the mth test source and the scattered field from the nth expansion function or *actual source*. Similarly, the mth voltage matrix element in (10-45) can be interpreted as the coupling between the mth test source and the incident field. When referring to (10-45), for instance, we can say that we are "reacting" the mth test source with the incident field, or in the case of (10-44) that we are "reacting" the electric field from the nth actual source with the current on the mth test source.

We obtained (10-43) for a wire. The method of moments or the method of weighted residuals applies to geometries other than just wire geometries as indicated in Fig. 10-15a. Consider the equivalent situation in Fig. 10-15b. Let $(\mathbf{J}_m, \mathbf{M}_m)$ be the surface current densities of a test source and $(\mathbf{E}_m, \mathbf{H}_m)$ be the fields from the test source. The currents on the conducting body are both replaced by equivalent surface

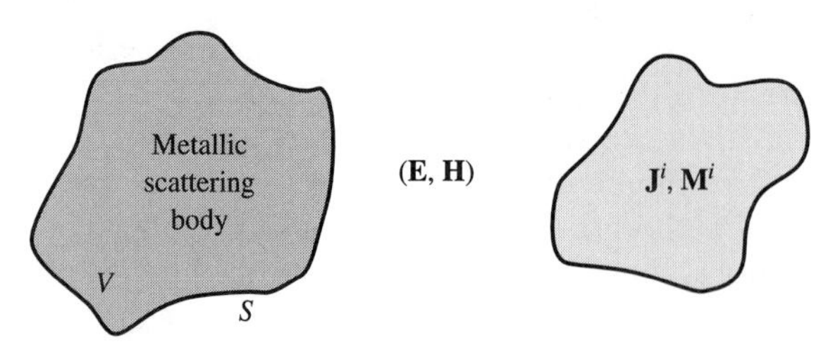

(a) Source current densities $\mathbf{J}^i$ and $\mathbf{M}^i$ acting in the presence of a metallic scattering body bounded by surface S create fields ($\mathbf{E}$, $\mathbf{H}$) exterior to S.

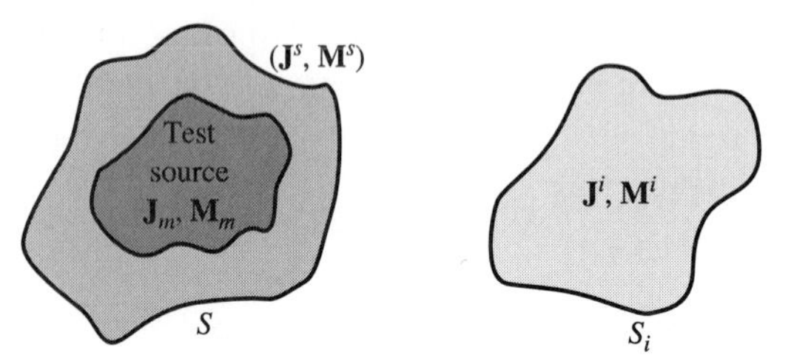

(b) Test source interior to surface S with equivalent currents ($\mathbf{J}^s$, $\mathbf{M}^s$) in free space.

Figure 10-15 Sources used in reaction concept.

currents ($\mathbf{J}^s$, $\mathbf{M}^s$) radiating the fields ($\mathbf{E}^s$, $\mathbf{H}^s$) in free space. The generalization of (10-43) then becomes

$$\iint_S (\mathbf{J}_m \cdot \mathbf{E}^s - \mathbf{M}_m \cdot \mathbf{H}^s)\, ds + \iint_S (\mathbf{J}_m \cdot \mathbf{E}^i - \mathbf{M}_m \cdot \mathbf{H}^i)\, ds = 0, \qquad m = 1, 2, 3, \ldots, N \tag{10-46}$$

The minus sign associated with $\mathbf{M}_m$ can be justified by referring to the reciprocity field theorems in Sec. 9.4. The physical interpretation of (10-46) is that we wish to have zero reaction (i.e., zero coupling) between the test source and the sum of the incident and scattered fields. Clearly, this is equivalent to the condition stated by (10-36). Nevertheless, the alternative physical interpretation offered by (10-46) and the reaction concept is a useful one and the student will find it used in the literature.

If we denote the fields from the nth expansion function of the actual current by ($\mathbf{E}_n^s$, $\mathbf{H}_n^s$), the sum of the N fields being ($\mathbf{E}^s$, $\mathbf{H}^s$), then we may write for the general mnth element in the generalized impedance matrix:

$$Z_{mn} = -\iint_S (\mathbf{J}_m \cdot \mathbf{E}_n^s - \mathbf{M}_m \cdot \mathbf{H}_n^s)\, ds \tag{10-47}$$

Similarly, we may write the general mth element in the voltage matrix:

$$V_m = \iint_S (\mathbf{J}_m \cdot \mathbf{E}^i - \mathbf{M}_m \cdot \mathbf{H}^i)\, ds \tag{10-48}$$

The incident field ($\mathbf{E}^i$, $\mathbf{H}^i$), which originates from the impressed currents $\mathbf{J}^i$ in Fig. 10-15a, may be the field from a source located on S (antenna transmitting situation) or be from a source located at a great distance from S (antenna receiving and radar scattering situation). The general relationships in (10-47) and (10-48) will be useful later for both wire and nonwire geometries.

10.7.2 Linear Algebra Formulation of MoM

Another way of approaching the general formulation of MoM is through the use of linear algebra. This an approach commonly found in the literature [6, 7]. Consider a general metallic body with a surface current density **J** on it. For simplicity, assume there are no magnetic currents. The extension of what follows to the case where both **J** and **M** are present can be deduced easily from the previous section on reaction.

The development of the linear algebra approach begins by requiring that the total tangential electric field be zero everywhere on the surface of the body, or that

$$0 = \mathbf{E}^s_{\tan} + \mathbf{E}^i_{\tan} \tag{10-49}$$

where $\mathbf{E}^s_{\tan}$ is the scattered electric field radiated by the current density **J** and $\mathbf{E}^i_{\tan}$ is the tangential component of the incident electric field due to a source located anywhere on or outside the body. We will drop the subscript *tan* since it will be understood that the tangential electric field boundary condition is being used.

Rewriting (10-49) in the form

$$-\mathbf{E}^s = \mathbf{E}^i \tag{10-50}$$

and defining the operator

$$L_{\text{op}}(\mathbf{J}) \equiv -\mathbf{E}^s \tag{10-51}$$

we can use the concept of linear vector spaces and operators to write the operator equation

$$L_{\text{op}}(\mathbf{J}) = \mathbf{E}^i \tag{10-52}$$

where L_{op} is an operator that must be determined for the problem of interest, $\mathbf{E}^i$ is a known excitation function or source, and **J** is the unknown response function to be determined. In the problems that are considered in this chapter, L_{op} is an integral operator operating on the current **J**.

For a given problem, we must determine the domain of definition of the operator or, in other words, the space of functions on which it operates and also the range of the operator or the functions resulting from the operation. In reality, the operator performs a mapping from some subset containing **J** to one containing $\mathbf{E}^i$. If the solution is to be unique, this mapping must be one to one.

Next, expand the response (solution) function **J** in a series of basis functions F_1, F_2, F_3, ... on a surface S and defined in the domain of L_{op}. That is,

$$\mathbf{J} = \sum_n I_n \mathbf{F}_n \tag{10-53}$$

where the coefficients I_n are, in general, complex. The I_n's are the unknown coefficients that are to be determined. Substituting (10-53) into (10-52) yields

$$L_{\text{op}}\left(\sum_n I_n \mathbf{F}_n\right) = \mathbf{E}^i \tag{10-54}$$

Using the linearity of L_{op}, we get

$$\sum_n I_n L_{op}(\mathbf{F}_n) = \mathbf{E}^i \tag{10-55}$$

The next step in the solution outlined above is to define a set of weighting functions $\mathbf{W}_1, \mathbf{W}_2, \ldots$ in the domain of L_{op} and then form the inner product:

$$\sum_n I_n \langle \mathbf{W}_m, L_{op}\mathbf{F}_n \rangle = \langle \mathbf{W}_m, \mathbf{E}^i \rangle \tag{10-56}$$

Note that if $\mathbf{W}_m$ is a delta function, (10-56) becomes the point-matching case, and if $\mathbf{W}_m = \mathbf{F}_m$ then (10-56) is a Galerkin formulation (e.g., Sec. 10.6). For the Galerkin formulation, write

$$\sum_n I_n \langle \mathbf{F}_m, L_{op}\mathbf{F}_n \rangle = \langle \mathbf{F}_m, \mathbf{E}^i \rangle \tag{10-57}$$

and the inner products appear as the reaction quantity mentioned earlier. Note that the basis functions $\mathbf{F}_n$ and the weight functions $\mathbf{F}_m$ represent currents. The inner product $\langle \mathbf{F}, \mathbf{E} \rangle$ is a scalar quantity obtained by integrating $\mathbf{F} \cdot \mathbf{E}$ over the surface under consideration. This particular inner product is called reaction. The inner product is defined such that the following conditions are satisfied:

$$\langle \mathbf{F}, \mathbf{E} \rangle = \langle \mathbf{E}, \mathbf{F} \rangle \tag{10-58a}$$

$$\langle \alpha\mathbf{F} + \beta\mathbf{F}, \mathbf{E} \rangle = \alpha\langle \mathbf{F}, \mathbf{E} \rangle + \beta\langle \mathbf{F}, \mathbf{E} \rangle \tag{10-58b}$$

if

$$\langle \mathbf{F}^*, \mathbf{F} \rangle > 0, \quad \text{then } \mathbf{F} \neq 0 \tag{10-58c}$$

if

$$\langle \mathbf{F}^*, \mathbf{F} \rangle = 0, \quad \text{then } \mathbf{F} = 0 \tag{10-58d}$$

where α and β are scalars and * denotes complex conjugation.

The third step is to calculate the various inner products given in (10-56) and thereby form the matrix equation

$$\begin{bmatrix} \langle \mathbf{F}_1, L_{op}\mathbf{F}_1 \rangle & \langle \mathbf{F}_1, L_{op}\mathbf{F}_2 \rangle & \cdots \\ \langle \mathbf{F}_2, L_{op}\mathbf{F}_1 \rangle & \cdot & \\ \cdot & & \cdot \\ \cdot & & \cdot \\ \cdot & & \cdot \end{bmatrix} \begin{bmatrix} I_1 \\ I_2 \\ \cdot \\ \cdot \\ \cdot \\ I_N \end{bmatrix} = \begin{bmatrix} \langle \mathbf{F}_1, \mathbf{E}^i \rangle \\ \langle \mathbf{F}_2, \mathbf{E}^i \rangle \\ \cdot \\ \cdot \\ \cdot \\ \langle \mathbf{F}_N, \mathbf{E}^i \rangle \end{bmatrix} \tag{10-59}$$

or, in more compact notation,

$$[Z_{mn}][I_n] = [V_m] \tag{10-60}$$

The procedure for obtaining a MoM solution in terms of linear algebra can be summarized in the following way:

1. Expand the unknown in a series of basis functions $\mathbf{F}_n$, spanning $\mathbf{J}$ in the domain of L_{op}.
2. Determine a suitable inner product and define a set of weighting functions.

3. Take the inner products and thereby form the matrix equation.
4. Solve the matrix equation for the unknown.

The first two steps are examined in more detail in Sec. 10.8.1.

EXAMPLE 10-3 ***Linear Algebra Interpretation of Sec. 10.6***

The purpose of this example is to interpret the formulation given in (10-40) in terms of linear algebra as discussed in this section. From (10-40a),

$$Z_{mn} = \int_{\Delta z_m} F_m(z) f(z, z_n') \, dz$$

where

$$f(z, z_n') = -\int_{\Delta z_n'} F_n(z') K(z, z') \, dz'$$

Thus,

$$Z_{mn} = \underbrace{\int_{\Delta z_m} F_m(z) \underbrace{\left[-\int_{\Delta z_n'} F_n(z') K(z, z') \, dz' \right]}_{L_{\text{op}}(\mathbf{F}_n)} dz}_{\langle \mathbf{F}_m, \, L_{\text{op}}(\mathbf{F}_n) \rangle}$$

where the integral operator is given by

$$L_{\text{op}} = -\int_{\Delta z_n'} K(z, z') \, dz'$$

From (10-40b), write

$$V_m = \underbrace{\int_{\Delta z_m} F_m(z) E_z^i(z) \, dz}_{\langle \mathbf{F}_m, \, \mathbf{E}_z^i \rangle}$$

Note that the inner products actually contain a dot product of two vectors as in (10-44) and (10-45), but the integral expressions above only contain scalars since the dot products have effectively already been performed.

10.8 FORMULATION AND COMPUTATIONAL CONSIDERATIONS

The development and use of a computer model of an electromagnetic problem can be divided into the following four steps:

1. Development of the *mathematical formulation* based on the physics of the problem, the object size in terms of λ, and mathematical principles (e.g., MoM, FD-TD, etc.)
2. *Coding* the mathematical formulation into a computer algorithm
3. *Validation* of the computer code
4. *Computation* to solve analysis and design problems

The following six subsections address issues in MoM affecting item 1 above. The last subsection addresses the important issue of validation.

10.8.1 Other Expansion and Weighting Functions

In Secs. 10.5 and 10.6, the pulse function was used as the expansion function and either the pulse function or delta function was used as the weighting function. The advantage of these functions lies in the simplicity they provide to the mathematical formulation and, hence, the coding. However, there are other functions commonly used. These include:

Triangle functions (piecewise linear): (10-61)

$$J(z) = \begin{cases} \dfrac{I_n(z_{n+1} - z) + I_{n+1}(z - z_n)}{\Delta z_n} & \text{for } z \text{ in } \Delta z_n \\ 0 \quad \text{otherwise} \end{cases} \tag{10-61}$$

Piecewise sinusoidal:

$$J(z) = \begin{cases} \dfrac{I_n \sin\beta(z_{n+1} - z) + I_{n+1}\sin\beta(z - z_n)}{\sin\beta\,\Delta z_n} & \text{for } z \text{ in } \Delta z_n \\ 0 \quad \text{otherwise} \end{cases} \tag{10-62}$$

Sinusoidal interpolation:

$$J(z) = \begin{cases} A_n + B_n \sin\beta(z - z_n) + C_n \cos\beta(z - z_n) & \text{for } z \text{ in } \Delta z_n \\ 0 \quad \text{otherwise} \end{cases} \tag{10-63}$$

where $\Delta z_n = z_{n+1} - z_n$.

The triangle functions were introduced in much of Harrington's early work and were used both as expansion and weighting functions (a Galerkin formulation). The triangle Galerkin formulation is also used in the *MININEC Professional* thin wire code [13].

Piecewise sinusoidal functions were first used by Richmond in a Galerkin formulation developed with the reaction integral equation [14]. These functions are computationally very efficient for wire geometries in free space, in part because the actual current distributions are nearly sinusoidal. A convergence curve for a dipole impedance is shown in Fig. 10-16. The rapid convergence is evident. The piecewise

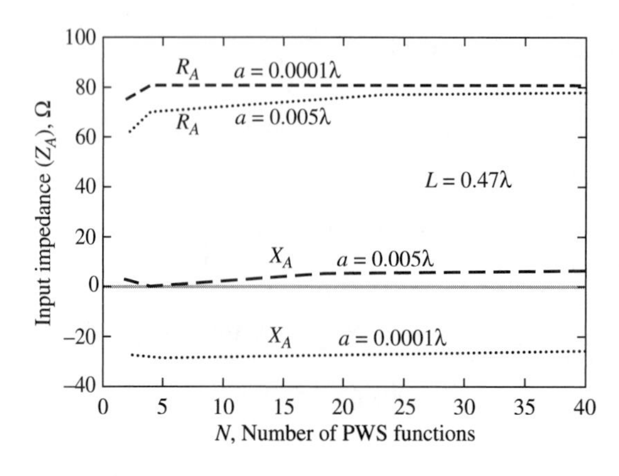

Figure 10-16 Input impedance convergence of a piecewise sinusoidal (PWS) Galerkin code [14] for two different wire radii.

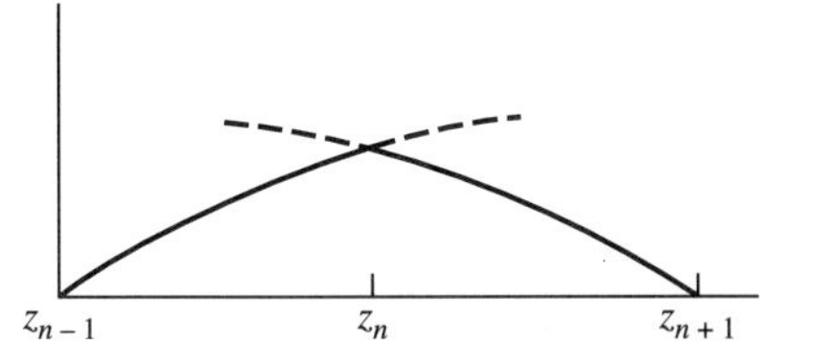

(*a*) Piecewise sinusoidal expansion function.

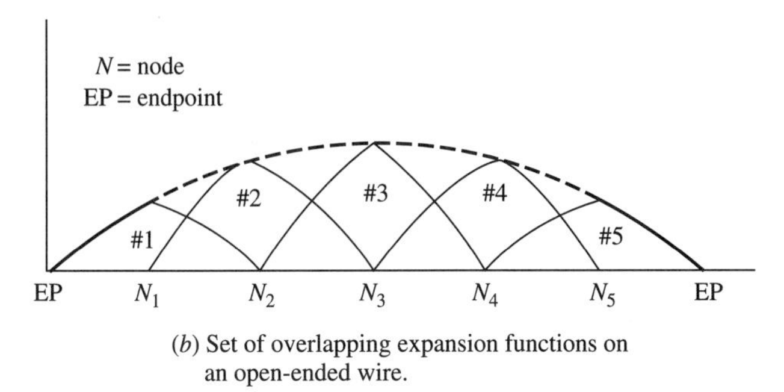

(*b*) Set of overlapping expansion functions on an open-ended wire.

Figure 10-17 Illustration of overlapping expansion functions using the piecewise sinusoid.

sinusoidal function is illustrated in Fig. 10-17*a*. Two segments are required to define the function. When several functions are used, as for the dipole suggested by Fig. 10-17*b*, each function overlaps with adjacent functions. The junction of two (or more) segments is called a node. In Fig. 10-17*b*, there are five nodes and there are five functions spanning six segments. Seven points (i.e., five nodes plus two endpoints) are required to define the six segments. On the other hand, if there are no endpoints as in a loop, the number of nodes, the number of segments, and the number of overlapping functions are all the same.

The sinusoidal interpolation function along with delta weighting functions is used in the *Numerical Electromagnetics Code* or *NEC* code as it is widely known [15].

The choice of functions has been the subject of research in past years. Some discussion of this may be found in [16] and [17]. The choice of functions is also influenced by a consideration of how to treat junctions of more than two wires. In the case of the (nonoverlapping) pulse function, no special consideration is required. In fact, Kirchhoff's current law will automatically be satisfied at a multiwire junction as a consequence of Maxwell's equations being satisfied. In the case of overlapping functions (e.g., triangle, piecewise sinusoid) at a junction of N wires, there are $N - 1$ independent currents (the Nth being determined by Kirchhoff's current law); therefore, only $N - 1$ functions are needed that go across the N-wire junction (see Prob. 10.13-3).

10.8.2 Other Electric Field Integral Equations for Wires

One form of an electric field integral equation (EFIE) is the Pocklington form in Sec. 10.3. Another form is the potential form used by Harrington [6] in his pioneering work. For z-directed wires, the potential form is

$$E^i = \int_{-L/2}^{L/2} \left[j\omega\mu_o I(z') - \frac{1}{j\omega\varepsilon_o} \frac{\partial I(z')}{\partial z'} \frac{\partial}{\partial z} \right] \frac{e^{-j\beta R}}{4\pi R} \, dz' \qquad (10\text{-}64)$$

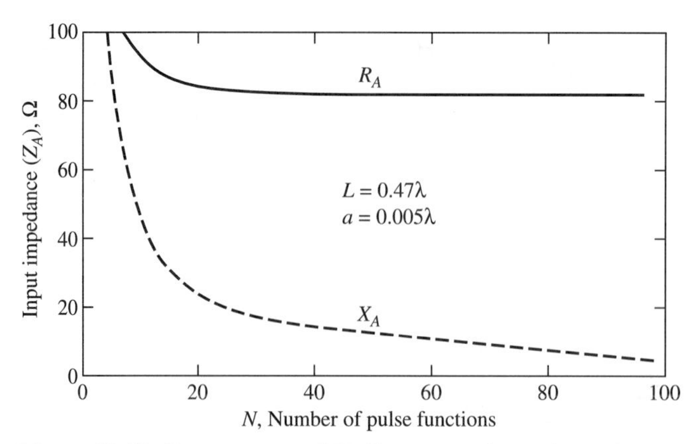

Figure 10-18 Convergence of Hallen's equation using pulse expansion functions and point-matching.

which is derived by using both the vector and scalar potentials (see Probs. 10.3-3 and 10.8-1). This form can provide more rapid convergence than the Pocklington form, which implies that convergence rates are dependent on the characteristics of the kernel as well as the choice of expansion and weighting functions.

A quite different appearing equation for wire antenna work is that due to Hallen (see Prob. 10.8-2). Hallen's integral equation for a z-directed wire antenna is

$$\int_{-L/2}^{L/2} I(z')\frac{e^{-j\beta R}}{4\pi R}\,dz' = -\frac{j}{\eta}\left(C_1 \cos \beta z + \frac{V_A}{2}\sin \beta|z|\right) \tag{10-65}$$

where V_A is the terminal voltage.

Hallen's equation has a simple kernel and is generally simpler to code than the either the Pocklington or potential forms. For N unknowns, there must be $N + 1$ equations because the constant C_1 is an unknown as well [17, 18]. Example 10-4 shows a sample impedance matrix, whereas Fig. 10-18 shows the rate of convergence for pulse expansion functions and delta weights.

EXAMPLE 10-4 *Subdomain Solution of Hallen's Equation*

The purpose of this example is to implement a pulse expansion function, point-matching solution to Hallen's equation for a z-directed wire dipole. The solution can be expressed as

$$\sum_{n=1}^{N} I_n Z_{mn} + C_1' \cos \beta z_m = V_m, \qquad m = 1, 2, \ldots, N+1$$

where

$$Z_{mn} = \int_{-\Delta z_n'/2}^{\Delta z_n'/2} \frac{e^{-j\beta R}}{R}\,dz', \qquad V_m = -j\frac{2\pi V_A}{\eta}\sin \beta|z|, \qquad C_1' = j\frac{4\pi}{\eta}C_1$$

and R is given by (10-11). Treating C_1' as an unknown rather than C_1 significantly improves the condition number of the matrix $[Z_{mn}']$ shown below. The Z_{mn} for $m \neq n$ are easily computed by numerical integration. For $m = n$, special care may be necessary for very small wire radii. The match points are chosen at the center of each pulse function, but an odd number of functions are required so that there is a match at the feed location. Further, it is necessary to have a match point at one end of the dipole (see Prob. 10.8-2). This is accomplished conceptually by employing a "phantom pulse" extending $\Delta z/2$ beyond one end of the dipole such that there is a match point at the end of the dipole.

For a dipole 0.1λ in length, radius 0.005λ, and three pulse functions, the following modified (because of the C_1' term) impedance matrix is written as

$$\begin{bmatrix} [Z_{mn}] & \vdots & Z'_{1,N+1} \\ & \vdots & \vdots \\ \hline Z'_{N+1,1} & \cdots & Z'_{N+1,N+1} \end{bmatrix}$$

where the block $[Z_{mn}]$ is toeplitz, the $N + 1$ column is given by

$$Z'_{i,N+1} = \cos(\beta z_i)$$

and the remaining elements $Z'_{N+1,i}$ are found from the numerical evaluation of the integral (above) in this example (see Fig. 10-19). Thus, for the short dipole numerical example being used here, the modified impedance matrix is

$$[Z'_{mn}] = \begin{bmatrix} 3.83\ \angle{-3.13^\circ} & 1.08\ \angle{-11.1^\circ} & 0.51\ \angle{-23.6^\circ} & 0.98\ \angle 0^\circ \\ 1.08\ \angle{-11.1^\circ} & 3.83\ \angle{-3.13^\circ} & 1.08\ \angle{-11.1^\circ} & 1.00\ \angle 0^\circ \\ 0.51\ \angle{-23.6^\circ} & 1.08\ \angle{-11.1^\circ} & 3.83\ \angle{-3.13^\circ} & 0.98\ \angle 0^\circ \\ 0.40\ \angle{-29.7^\circ} & 0.69\ \angle{-17.4^\circ} & 2.59\ \angle{-4.62^\circ} & 0.95\ \angle 0^\circ \end{bmatrix}$$

The voltage matrix is

$$[V_m] = 10^{-3} \begin{bmatrix} 3.46\ \angle{-90^\circ} \\ 0\ \angle{-90^\circ} \\ 3.46\ \angle{-90^\circ} \\ 5.15\ \angle{-90^\circ} \end{bmatrix}$$

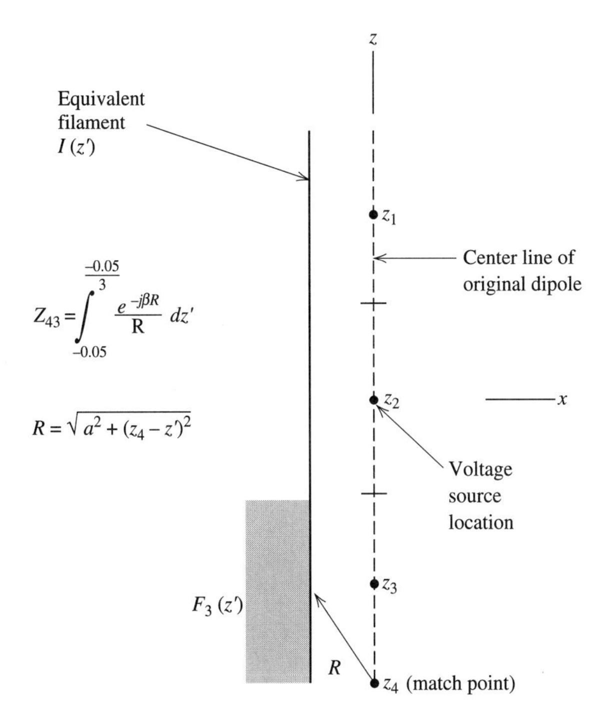

Figure 10-19 Calculation of Z_{43} for the short dipole in Example 10-4.

Solving for the current matrix gives

$$[I_n] = 10^{-3}\begin{bmatrix} 0.84 \angle 89.45^\circ \\ 1.98 \angle 89.69^\circ \\ 0.84 \angle 89.45^\circ \\ 9.41 \angle -95.02^\circ \end{bmatrix}$$

Note that I_4 is the value for the coefficient C_1' for which there is no further use. The input impedance is $1.0/(1.98 \times 10^{-3}\angle 89.69^\circ) = 2.75 - j504\ \Omega$. If five pulses were used instead of just three, the input impedance would be $2.1 - j489\ \Omega$, which more closely agrees with the approximate value of $1.974\ \Omega$ for the real part obtained from $20\pi^2(L/\lambda)^2$.

10.8.3 Computer Time Considerations

Historically, there have been two computer limitations to the use of MoM: (1) the amount of computer storage necessary for the N^2 impedance matrix elements, and (2) the amount of time to compute those N^2 elements and solve the resulting system of equations. Computer technology has significantly decreased the impact of the first limitation. Iterative methods have increased the speed of solution of a system of dense simultaneous equations, leaving the time required to compute the N^2 elements as a major limitation on the electrical size of an object that can be reasonably accommodated by MoM (see Fig. 10-2). Let us examine where the computation time is spent.

A square impedance matrix of N^2 elements is said to be of order N. Let N_i be the number of different source or incident fields (i.e., radar cross section is a function of incidence angle) associated with a given impedance matrix and let N_a be the number of observation points at which the field is to be computed from the current solution; then the time t for execution will be approximately given by [11]

$$t \simeq AN^2 + B_3N^3 + CN^2N_i + DNN_iN_a \tag{10-66}$$

where the algorithm and computer-dependent factors A, B, C, and D are

A = time required to compute a typical impedance matrix element

B_3N^3 = time required to solve $[Z_{mn}][I_n] = [V_m]$ for $[I_n]$ by matrix inversion $[I_n] = [Z_{mn}]^{-1}[V_m]$ for a system of order N

CN^2N_i = time required to perform the operation $[Z_{mn}]^{-1}[V_m]$ or its equivalent for each *new* $[V_m]$

DNN_iN_a = time for computing the far field from $[I_n]$

The second term in (10-66) dominates. However, it is unlikely that we would solve a large system of equations by finding the inverse. Instead, usually an algorithm such as Gauss–Jordan or Crout is used, in which case $B_3N^3 \to B_2N^2$ and we have

$$t \simeq AN^2 + B_2N^2 + CN^2N_i + DNN_iN_a \tag{10-67}$$

which is a significant reduction in the solution time required for large N. If iterative methods are used [17], more favorable reductions are possible.

If the impedance matrix is toeplitz (see Sec. 10.8.4), then $B_3N^3 \rightarrow B_1N^{5/3}$ and we have

$$t \simeq AN + B_1N^{5/3} + CN^2N_i + DNN_iN_a \tag{10-68}$$

for which there is a significant improvement in the first term as well as the second.

When the second term in (10-66) is on the order of N^2 or less, then usually the first term, which is associated with the time required to calculate the matrix elements, becomes the dominating factor. In the following subsections, we examine briefly some ways for reducing the total time required for the operations associated with the first two terms in (10-66).

10.8.4 Toeplitz Matrices

Certain types of problems produce impedance matrices where there is a systematic repetition in the matrix elements. Often, this repetition can be used to decrease the impact of both the first and second terms in (10-66). Consider the straight wire in Fig. 10-3. If the segments are of equal length, all the values of the N^2 matrix elements are contained in any one row of $[Z_{mn}]$, say, the first one. All other rows are merely a rearranged version of the first. The remaining elements can be obtained by the rearrangement algorithm:

$$Z_{mn} = Z_{1,|m-n|+1}, \qquad m \geq 2, \qquad n \geq 1 \tag{10-69}$$

Such a matrix is said to be a *toeplitz matrix*. Computer programs exist for solving toeplitz matrices that are considerably more efficient than those for solving a non-toeplitz matrix. For a toeplitz matrix, the first two terms in (10-66) become AN and $BN^{5/3}$, respectively, and the execution time in (10-66) is reduced as in (10-68).

Toeplitz matrices can arise in the treatment of certain wire geometries. These are the straight wire (see Examples 10-1 and 10-2), the circular loop, and the helix. A toeplitz matrix can also arise in the treatment of geometries other than the wire, but these are outside the scope of this chapter.

10.8.5 Block Toeplitz Matrices

Consider the linear array of parallel dipoles in Fig. 10-20. The impedance matrix that characterizes the array will be toeplitz by blocks or submatrices when the array elements are of the same length and equally spaced. Thus, if the impedance matrix for the array $[Z]_{\text{array}}$ is written in terms of submatrices $[S]$ as

$$[Z]_{\text{array}} = \begin{bmatrix} [S]_{11} & [S]_{12} & \cdots & [S]_{1J} \\ [S]_{21} & [S]_{22} & \cdots & [S]_{2J} \\ \vdots & & \ddots & \vdots \\ [S]_{J1} & \cdots & & [S]_{JJ} \end{bmatrix} = \begin{bmatrix} [S]_{11} & [S]_{12} & \cdots & [S]_{1J} \\ [S]_{12} & [S]_{11} & \cdots & [S]_{1(J-1)} \\ \vdots & & \ddots & \vdots \\ [S]_{1J} & [S]_{1(J-1)} & \cdots & [S]_{11} \end{bmatrix} \tag{10-70}$$

where $[S]_{ij} = [Z_{mn}]$, the entire impedance matrix is toeplitz by blocks. Thus, if one row of submatrices is known, the remaining submatrices may be filled by the algorithm

$$[S]_{ij} = [S]_{1,|i-j|+1}, \qquad i \geq 2, \qquad j \geq 1 \tag{10-71}$$

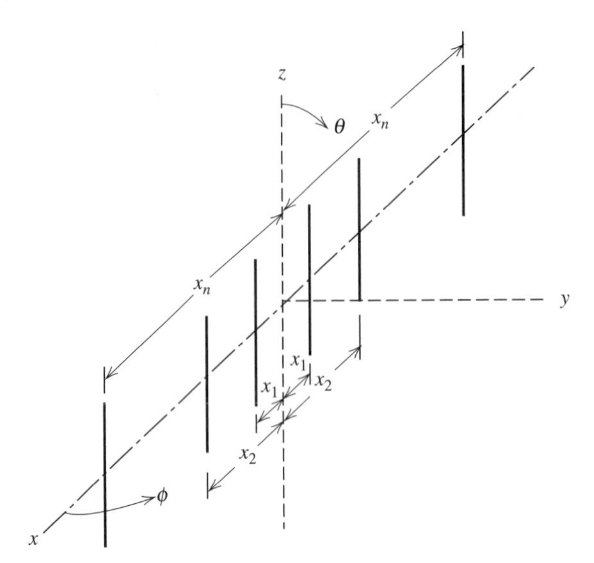

Figure 10-20 Linear array of parallel dipoles.

Consequently, the first term in (10-66) is of order N^2/J, where J is the number of independent submatrices. The second term in (10-66) will be of order $N^{9/5}$.

If the submatrices are themselves toeplitz, as they would be if all segments are of identical length and radius, then the matrix fill time is reduced even further. Computer programs exist for solving block toeplitz matrices. The potential savings in execution time for a problem that is block toeplitz over the same size nonblock toeplitz problem can be considerable.

10.8.6 Compressed Matrices

In certain problems, there will be a repetition in the values within $[I_n]$ due to the symmetry of the problem. If this can be recognized in advance, it can be used to advantage to compress the matrix from order N to order N/L, where L is the degree of symmetry.

Consider the following simple but very common example of symmetry suggested by Fig. 10-7*a*. Here, the monopole and its image will have a symmetrical current about the feed point. Suppose $I_n = I_{n+N/2}$; then we can write

$$\sum_{n=1}^{N/2} (Z_{mn} + Z_{m,n+N/2})I_n = V_m, \qquad m = 1, 2, 3, \ldots, N/2 \tag{10-72}$$

The solution of this compressed system of $N/2$ equations will yield the $N/2$ independent I_n's. From (10-67), we see that the solution time for the system is $B_2(N/2)^2$, or a reduction in time by a factor of 4 for this portion of the computing process. For higher degrees of symmetry, the savings in time would be even more considerable. For some large problems, it may be necessary to compress the matrix for another reason, namely storage requirements. It is possible that an impedance matrix may be so large that it cannot be stored in readily available core memory and that through symmetry it may be compressed to a reasonable size [7].

So, it is the execution time and computer storage that tend to limit the electrical size of problems that may be reasonably treated with MoM. In some of the following sections, we will approach certain situations in such a way that the impact of these two limitations is minimized.

10.8.7 Validation

A computer code must reliably produce accurate results if it is to be useful. Errors in the code are most likely to occur because of an encoding mistake, but can also occur due to an oversight in the mathematical formulation or cumulative numerical errors caused by large numbers of numerical operations and/or inadequate numerical precision. The investigation of the possible existence of errors such as these, as well as others, is the process of validation. We examine two kinds of validation: external validation and internal validation.

External validation usually means that the output of a code is compared against either experimental measurements or the output of an independent code (perhaps thereby validating both codes). Near-field quantities, such as antenna input impedance, are more critical gages than far-field quantities. Consequently, earlier figures in this chapter have used the input impedance as a check against both experimental data and results from other independent codes.

Internal validation usually implies that additional coding and/or computational effort is involved. Some examples of internal checks are:

1. A convergence check to establish that a limiting value is smoothly approached as N is increased
2. A power balance check where the power supplied by the incident field is equal to the radiated plus dissipated powers
3. A reciprocity check where source and observation points are interchanged
4. Boundary condition check where the applicable boundary condition (e.g., total tangential E is zero) is met

Of these four checks, only the last is both necessary and sufficient. The others are necessary but not sufficient checks.

In the case of a code that is written to handle a wide variety of geometries, frequencies, etc., a representative sample of such situations must be validated much like a new aircraft must be test flown in a variety of different configurations, speeds, weather conditions, etc., before it can be certified as safe for general use.

10.9 CALCULATION OF ANTENNA AND SCATTERER CHARACTERISTICS

Thus far, our discussion of MoM has been mainly concerned with acquiring a knowledge of some unknown current distribution. Now consider how we can obtain other information as well. But first, we should make one further remark about the currents derived from the solution of the matrix equation.

If pulse functions are used as the expansion functions in the point-matching technique, a knowledge of the current coefficients I_n means that the current distribution at the match points is known "precisely," if we assume, of course, that the solution has converged. In between the match points, we do not know the current but, since the distance between the match points is small in terms of the wavelength, one can simply fit a curve through the current values at the match points to obtain a good approximation of the current distribution along the wire.

In the case of overlapping functions, such as the piecewise sinusoidal or the triangle (see the previous section), a knowledge of the coefficients I_n again only means that the current is known at the junctions of the segments. Along the segments, we use the overlapping functions themselves to approximate the current distribution between segment junctions.

After we have determined the current distribution, the input or terminal current can be found by evaluating the current distribution at the antenna terminal location. In turn then, the input impedance may be calculated by dividing the terminal voltage by the terminal current. The calculation of accurate impedance data is a task that is somewhat sensitive to the model used for the feed point. Two such models were discussed in the Sec. 10.5.

Distributive loading, which arises when a wire is not perfectly conducting, may affect the current distribution in certain situations. For simplicity, consider a wire whose axis is parallel to the z-axis. When the wire has finite conductivity, we can relate the tangential electric field at the surface of the wire to the equivalent electric surface current density by the use of the surface impedance Z_s, which is defined [19] as the ratio of the tangential electric field strength at the surface of a conductor to the current density that flows as a result of that tangential electric field. Thus,

$$\mathbf{E} = Z_s \mathbf{J}_s \tag{10-73}$$

Using $\mathbf{M}_s = \mathbf{E} \times \hat{\mathbf{n}}$ and the relationship $\mathbf{J}_s = \hat{\mathbf{z}} I(z)/2\pi a$, we can write

$$\mathbf{M}_s = Z_s \mathbf{J}_s \times \hat{\boldsymbol{\rho}} = \frac{\hat{\boldsymbol{\phi}} Z_s I(z)}{2\pi a} \tag{10-74}$$

Writing the reaction integral equation from (10-46) and reciprocity as

$$-\iint_S (\mathbf{E}_m \cdot \mathbf{J}^s - \mathbf{H}_m \cdot \mathbf{M}^s)\, ds = V_m \tag{10-75}$$

and substituting (10-74) lead to

$$-\int I(z)[\hat{\mathbf{z}} \cdot \mathbf{E}_m - Z_s \hat{\boldsymbol{\phi}} \cdot \mathbf{H}_m]\, dz = V_m \tag{10-76}$$

Using (10-16) in (10-76), we can write the generalized impedance matrix element Z'_{mn}, modified for finite conducting wires, as

$$Z'_{mn} = \int_{\Delta z_n} F_n(z) \hat{\mathbf{z}} \cdot \mathbf{E}_m\, dz - Z_s \int_{\Delta z_n} F_n(z) \hat{\boldsymbol{\phi}} \cdot \mathbf{H}_m\, dz \tag{10-77}$$

From Ampere's law, a suitable approximation for $\mathbf{H}_m$ is

$$\hat{\boldsymbol{\phi}} \cdot \mathbf{H}_m = \frac{F_m(z)}{2\pi a} \tag{10-78}$$

and thus (10-75) can be written as

$$Z'_{mn} = Z_{mn} - \frac{Z_s}{2\pi a} \int_{(m,n)} F_n(z) F_m(z)\, dz \tag{10-79}$$

where region (m, n) is the wire surface shared by testing or weighting function m and expansion or basis function n. In the case of overlapping expansion functions, this region covers two intersecting segments if m and n are equal. When $m \neq n$, the shared region covers, at most, one wire segment. This means that distributive loading is accounted for by a modification of only the appropriate main diagonal elements, and those elements adjacent to the modified main diagonal elements if overlapping functions are being used. In the case of nonoverlapping functions, such as the pulse, only the main diagonal elements are modified.

The effect of either lumped loading (considered in Sec. 10.10) or distributive loading is to alter the current distribution on the wire antenna or scatterer. If we know the current distribution, the far field can be obtained by the classical methods used previously in this book (e.g., Sec. 1.7.4). To illustrate, consider again z-directed segments with pulse expansion functions of the current. Then from (4-2), we have

$$\mathbf{E} = \hat{\boldsymbol{\theta}} j\omega\mu \frac{e^{-j\beta r}}{4\pi r} \Delta z \sin\theta \sum_{n=0}^{N-1} I_n e^{j\beta z_n' \cos\theta} \tag{10-80}$$

where z_n' is the center of each short segment.

Once the far field is known, the gain may be determined from the general relationship

$$G(\theta, \phi) = \frac{[|E_\theta|^2 + |E_\phi|^2]r^2}{30|I_A|^2 R_A} \tag{10-81}$$

where R_A is the real part of the antenna input impedance. The directivity is obtained by replacing R_A with R_r, the radiation resistance.

The radar cross section may be found from (9-35) as

$$\sigma = \lim_{r\to\infty} 4\pi r^2 \frac{|E^s|^2}{|E^i|^2} \tag{10-82}$$

where E^s can be determined, for example, from (10-12). The radar cross section for a dipole scatterer is shown in Fig. 10-21.

The radiation efficiency is calculated using (1-174) as

$$e_r = \frac{R_r}{R_A} = \frac{R_r}{R_r + R_{\text{ohmic}}} \tag{10-83}$$

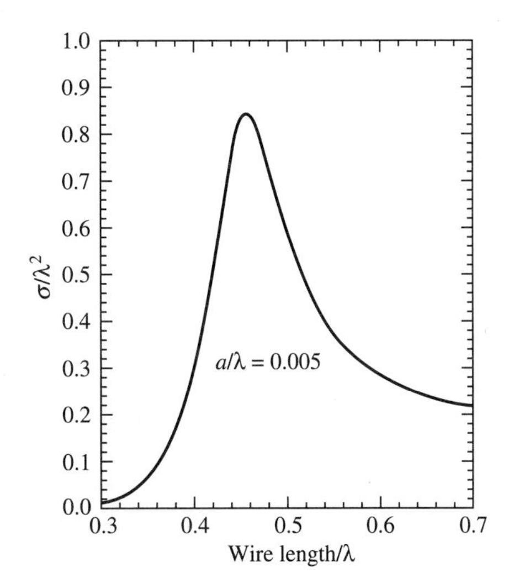

Figure 10-21 Monostatic radar cross section of a straight wire at normal incidence as a function of wire length.

R_{ohmic} is the loss resistance due to dissipative loading, either distributed or lumped (see Sec. 10.10.1). Alternatively, we could determine the radiated power by integrating the power density in the far field as we did in Chap. 1. However, the above method (10-83) is computationally more efficient.

From the discussion in this section, we can see, among other things, that refinements such as losses due to finite conductivity (distributive loading) or the effects of lumped loading can be included into a moment method solution in a fairly straightforward manner. In the next section, we investigate lumped loading further.

10.10 THE WIRE ANTENNA OR SCATTERER AS AN *N*-PORT NETWORK

In Sec. 10.4, we saw the resemblance between the simultaneous linear equation approximation of an integral equation and Kirchhoff's network equations. It follows that we may view the junction of two or more segments as a port in the usual circuit sense as indicated by Fig. 10-22*a*. At each port, we may place either series or parallel elements that are either passive or active. Series connections are treated on an impedance basis, whereas parallel connections are handled on an admittance basis. This section considers both types.

10.10.1 Series Connections

We already have considered a single generator placed at the junction of two wire segments (e.g., Sec. 10.5). The generator was in series with the implied port terminals located at the ends of the two adjacent segments. We could, of course, place as many generators on the wire as there are segment junctions. Thus, for an N-segment dipole, there would be $N - 1$ ports. If there is no generator or passive element across the port, the port is understood to be short-circuited.

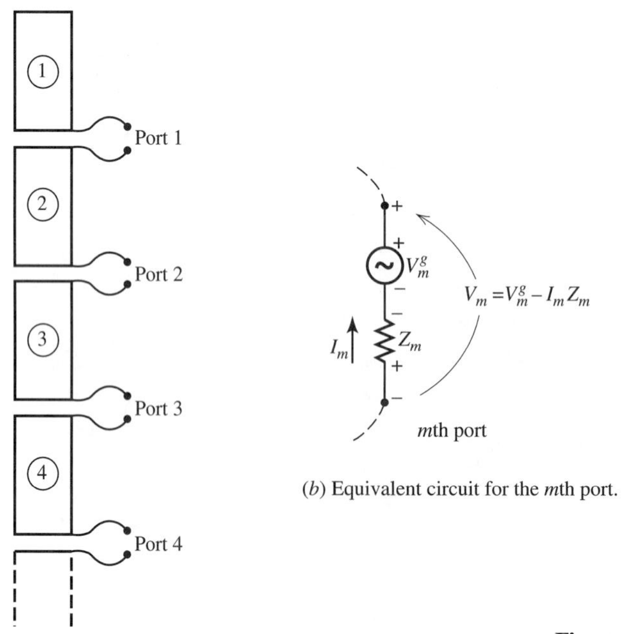

(*a*) $N - 1$ port terminal pairs.

(*b*) Equivalent circuit for the *m*th port.

Figure 10-22 N segment wire showing ports.

Previously in Sec. 10.9, we saw how distributed loading was accounted for in the moment method by modifying certain elements in the impedance matrix. Here, let us examine how lumped loading may be handled. If a load Z_m is inserted into a wire antenna at segment junction m having a current I_m, the total voltage acting at that point is

$$V_m = V_m^g - I_m Z_m \tag{10-84}$$

where V_m^g represents a voltage generator that may be located at point m in series with Z_m as indicated in Fig. 10-22*b*. In many cases, V_m^g will be zero. Considering the mth equation in a system of N linear equations, we can write

$$\sum_{n=1}^{N} Z_{mn} I_n = V_m^g - I_m Z_m \tag{10-85}$$

or

$$\sum_{n=1}^{N} Z'_{mn} I_n = V_m^g \tag{10-86}$$

where

$$Z'_{mm} = Z_{mm} + Z_m \tag{10-87}$$

Except for the diagonal elements, the new impedance matrix is the same as the original, or $Z'_{mn} = Z_{mn}$, $m \neq n$. Thus, the effect of lumped loading may be accounted for by simply adding the load impedances Z_m to the corresponding diagonal elements in the impedance matrix. The effects of lumped loading can be substantial. For example, resistive loading can be used to achieve increased bandwidth, but at the expense of lower efficiency. Or, lumped loading can be used for impedance matching as in the following illustration of a $\frac{3}{4}\lambda$ monopole.

Figure 10-23 shows a $\frac{3}{4}$ wavelength monopole antenna. A series inductance at the base improves the VSWR and the quarter wavelength stub between the (nominal) half-wave and quarter-wave sections provides the necessary phase shift for good

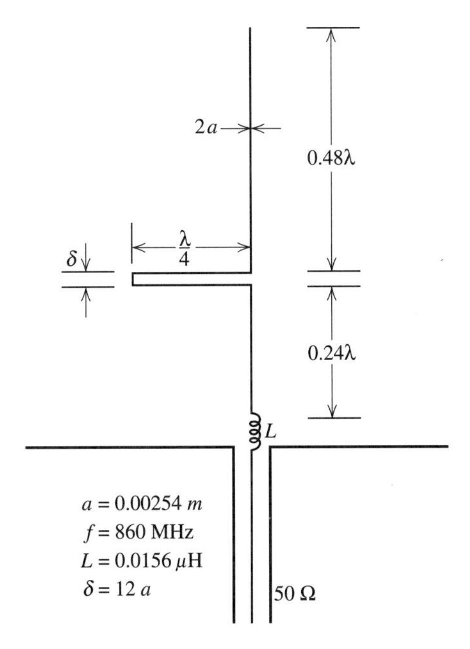

Figure 10-23 Three-quarter wavelength monopole with series loading.

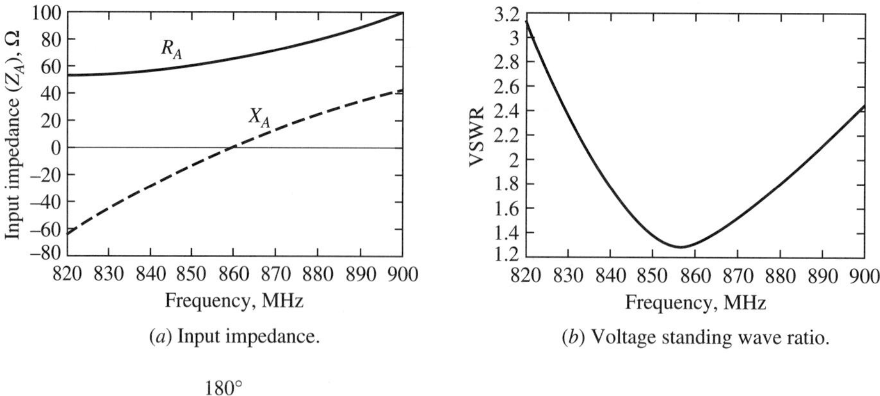

(*a*) Input impedance.

(*b*) Voltage standing wave ratio.

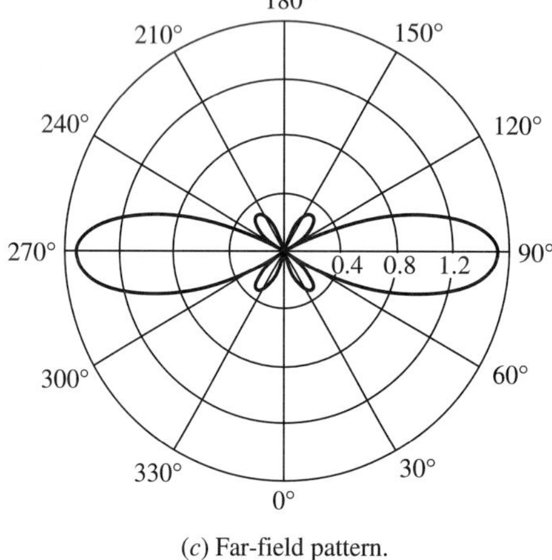

(*c*) Far-field pattern.

Figure 10-24 Performance of the three-quarter wavelength monopole in Fig. 10-23 [14].

pattern performance. Figure 10-24*a* shows the input impedance of the monopole. Note that the inductor at the base creates a zero reactance at 860 MHz. Figure 10-24*b* shows the VSWR referenced to 50 Ω. The VSWR curve could be made more symmetrical by slightly increasing the inductive reactance at the base. The field pattern is illustrated in Fig. 10-24*c*. Note the strong broadside radiation compared to the $3\lambda/2$ case in Fig. 5-4 without a phasing stub.

10.10.2 Parallel Connections

In the previous subsection, we saw how circuit elements, when connected in series at a given port, resulted in modification of certain entries in the open-circuit moment method impedance $[Z_{mn}]$. If, however, we connect one port in parallel with another as in a log-periodic antenna, then it is necessary to work with the short-circuit moment method admittance matrix $[Y_{mn}]$.

Consider Fig. 10-25 that shows a log-periodic dipole antenna (LPDA). The LPDA is viewed as the parallel connection of two N-port networks. One N-port represents the mutual coupling between N dipole antennas. The other represents the transmission line that interconnects the dipoles. Therefore, there is one *network* port for each of the dipoles in the system.

The approach is shown schematically in Fig. 10-26. The N-port labeled "antenna elements" includes the self- and mutual impedances between N unconnected dipole

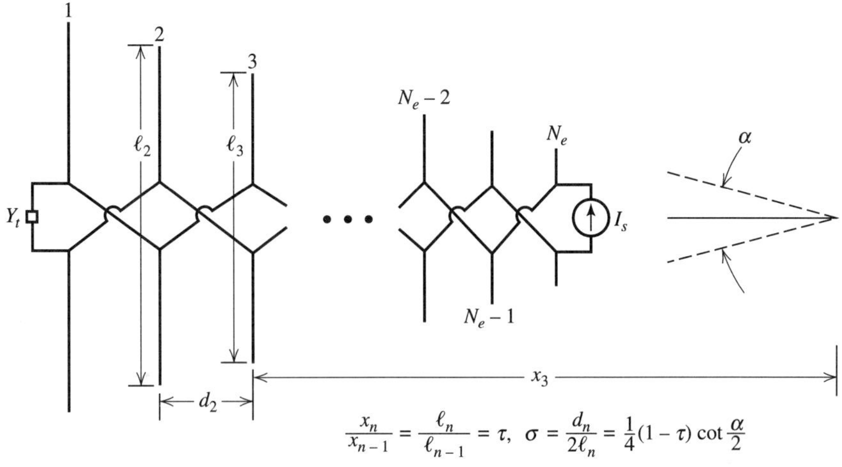

Figure 10-25 Log-periodic dipole antenna.

antennas located arbitrarily in space. The "transmission line" N-port represents the transmission line connecting the dipole antennas. Included in this network is the effect of reversing the polarity between successive dipoles. Note that there is a current source I_S on the LPDA. If there are N_e dipole elements on the antenna, then there are voltage sources applied on ports 1, 2, 3, . . . , N_e. Also, there is a terminating admittance on the LPDA antenna Y_t that exists at port 1. We do not know the numerical values of the applied voltage sources. Thus, they must be found before we can solve for the currents on the LPDA.

Let $[Y_A]$ and $[Z_A]$ be the short-circuit admittance matrix and open-circuit impedance matrix, respectively, for the "antenna elements" network, where $[Y_A] = [Z_A]^{-1}$. We note that $[Z_A]$ is *not* the moment method impedance matrix. An element of $[Z_A]$, say, $[Z_A]_{ij}$, represents the voltage induced on dipole i in the LPDA by a unit current on dipole j with all other dipoles open-circuited. Thus,

$$[Z_A]_{ij} = \frac{V_i}{I_j} \tag{10-88}$$

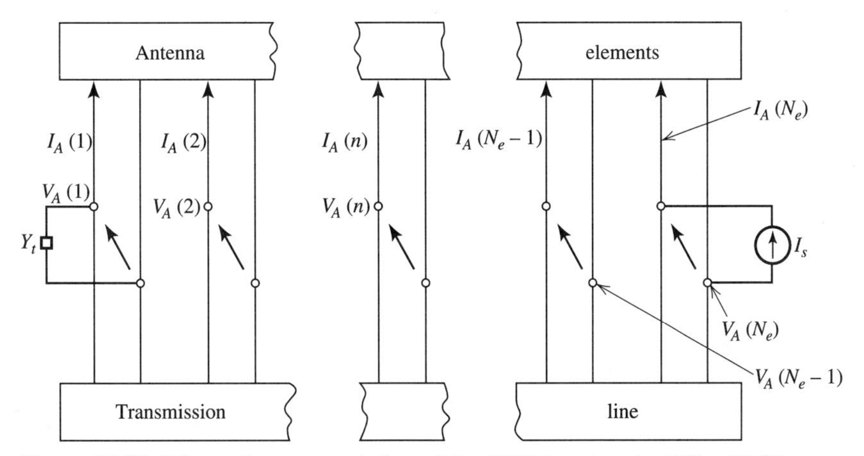

Figure 10-26 Schematic representation of the LPDA network of Fig. 10-25.

Let $[Y_T]$ be the short-circuit admittance matrix for the "transmission line" network. Let $[I_A]$ and $[V_A]$ be the column matrices representing the voltage and current at each port of the "antenna elements" network. Since the two networks are in parallel, the total current can be written as

$$[I_s] = [[Y_A] + [Y_T]][V_A] \tag{10-89}$$

where $[I_s]$ represents the applied current sources. The $[I_s]$ matrix contains all zero elements, except at the port where there is a current source I_s. The current source, of course, represents the excitation of the LPDA antenna. Note in (10-89) that we know the entry in $[I_s]$ but not the entries in $[V_A]$. These must be found so that the moment method column matrix $[V_m]$ can be constructed and the usual equation $[I_n] = [Z_{mn}]^{-1}[V_m] = [Y_{mn}][V_m]$ solved for the current distributions in the antenna dipole elements. But, before we can solve (10-89) for $[V_A]$ and construct $[V_m]$, we must know $[Y_A]$ and $[Y_T]$.

To obtain the elements of $[Y_A]$, we proceed as follows. Consider an LPDA with N_e dipoles and M expansion functions on each dipole. The moment method impedance matrix will be of order $N_e \times M$. To obtain the moment method admittance matrix $[Y_{mn}]$, we note that $[Y_{mn}] = [Z_{mn}]^{-1}$ and

$$[I] = [Z_{mn}]^{-1}[V] = [Y_{mn}][V] \tag{10-90}$$

or

$$I_m = \sum_{n=1}^{M\times N_e} Y_{mn} V_n, \qquad m = 1, 2, \ldots, M \times N_e \tag{10-91}$$

To obtain $[Y_A]$, we note that most of the V_n's will be zero since voltages are only applied by the transmission line on the center ports of each dipole in the LPDA. Suppose we rearrange the system of equations in (10-91) such that the first N_e entries in $[V]$ as well as $[I]$ correspond to the center ports of the dipoles in the LPDA. Then the currents at those ports containing a generator (i.e., antenna element ports) are related to the voltages at those ports by

$$I_j = \sum_{i=1}^{N_e} Y_{ji} V_i, \qquad j = 1, 2, \ldots, N_e \tag{10-92}$$

or

$$[I_A] = [Y_A][V_A] \tag{10-93}$$

where all the V_i's in $[V_A]$ will be nonzero. (See Prob. 10.10-1.) In finding $[Y_A]$ in this manner, we have done so without making approximations other than those appropriate to the moment method itself. Indeed, all mutual couplings are included and we are not limited to LPDAs of less than 2:1 bandwidth as in the treatments by Carrel [20] and Kyle [21].

To obtain the transmission line admittance matrix $[Y_T]$ in (10-89), we first recognize that $[Y_T]$ is the transmission line admittance matrix for a simple terminated transmission line with a port at the position where each dipole is connected. Since $[Y_T]$ is the short-circuit admittance matrix, a given element $(Y_T)_{ji}$ represents the current induced across port j (which is shorted) by a unit voltage at port i, with all other ports shorted. Thus, $(Y_T)_{ji}$ is nonzero only for $i - 1 \le j \le i + 1$.

It is possible to write the transmission line admittance matrix $[Y_T]$ in a straightforward fashion [21]. For a single LPDA, it is

$$[Y_T] = \begin{bmatrix} (Y_t - jY_o \cot \beta d_1) & -jY_o \csc \beta d_1 & 0 & \cdots & 0 \\ -jY_o \csc \beta d_1 & -jY_o(\cot \beta d_1 + \cot \beta d_2) & -jY_o \csc \beta d_2 & \cdots & 0 \\ 0 & -jY_o \csc \beta d_2 & -jY_o(\cot \beta d_2 + \cot \beta d_3) & \cdots & 0 \\ \cdots & \cdots & \cdots & \cdots & \cdots \\ 0 & 0 & 0 & -jY_o \csc \beta d_{N_e-1} & -jY_o \cot \beta d_{N_e-1} \end{bmatrix} \tag{10-94}$$

where Y_o is the transmission line characteristic admittance and β the propagation constant of the transmission line. (See Prob. 10.10-2.)

With the proper elements of both $[Y_A]$ and $[Y_T]$ in hand, the voltages $[V_A]$ acting at the driven port of each dipole are

$$[V_A] = [[Y_A] + [Y_T]]^{-1}[I_s] \tag{10-95}$$

where $[I_s]$ has one nonzero entry. With these voltages at each dipole, the moment method voltage matrix $[V_m]$ can be filled and the current distribution on each dipole in the LPDA obtained from

$$[I_n] = [Z_{mn}]^{-1}[V_m] \tag{10-96}$$

where the elements of $[V_A]$ are the nonzero elements of $[V_m]$ and the elements of $[I_n]$ are the complex coefficients associated with the expansion functions on the various dipole elements.

It is worthwhile to summarize the above procedure for analyzing the LPDA. First, the open-circuit impedance matrix $[Z_{mn}]$ was formed in the usual manner. By taking the inverse of $[Z_{mn}]$, the short-circuit admittance matrix was obtained. Next, the antenna elements admittance matrix $[Y_A]$ was formed from $[Y_{mn}]$ as in (10-91) and (10-92). Then $[Y_A]$ was added to the transmission line admittance matrix $[Y_T]$. Then the current generator shown in Figs. 10-25 and 10-26 was used in (10-95) to obtain the voltage $[V_A]$ acting at each dipole port. These voltages were then used to obtain the moments method voltage matrix $[V_m]$. Solution for the currents $[I_n]$ on each dipole in the LPDA followed according to (10-96). Patterns obtained using this procedure are given in Fig. 6-41 and agree with those in [22].

10.11 ANTENNA ARRAYS

The use of moment methods in the analysis and design of arrays of wire antennas (or scatterers) has significant advantages over the more classical methods used in treating arrays in that mutual coupling between array elements is taken completely into account (e.g., see the LPDA treatment in Sec. 10.10.2). Furthermore, no unrealistic assumptions need be made regarding the current distributions on the wires, and the array elements can be excited at any point(s) or be loaded at any point(s) along their lengths. Thus, the type of wire element array problem that can be considered is rather general. In this section, we examine several array configurations of parallel dipoles and illustrate some typical mutual coupling effects.

10.11.1 The Linear Array

Consider the linear array of parallel wire elements shown in Fig. 10-20. The elements need not be of the same length and radius or be equally spaced in order to be treated by MoM. Clearly, they could be quite arbitrarily configured and, in fact, need not even be parallel. However, in this subsection we wish to illustrate the effects of mutual coupling in a typical linear dipole array by comparing MoM results (using a voltage generator with an internal impedance of 72 Ω) with results suggested by the methods of Chap. 3 (i.e., current generator excitation). For this purpose, without loss of generality, we consider a linear array of 12 equally spaced ($d = \lambda/2$), parallel, center-fed, half-wave dipoles phased for a beam maximum 45° off broadside. Each dipole is divided into six segments and a 1-V generator is placed in series with a 72-Ω resistance at the center port of each dipole, the piecewise sinusoidal current amplitudes obtained using (10-62) and the methods of Sec. 10.7.1 are given in Table 10-1. We note that neither the feed point currents nor the input impedances [see (3-103)] are identical across the array. This is due to mutual coupling. Since the main beam is at $\phi_o = 45°$, there is no symmetry in the currents about the array center as there would be if the array were phased for broadside radiation.

The normalized patterns are shown in Figs. 10-27*a* and 10-27*b* along with the normalized pattern for uniform current excitation. In spite of the differences noted in Table 10-1, there is little difference seen in the three normalized patterns shown in Figs. 10-27*a* and 10-27*b*. There is, of course, some small difference in the directivity in the two cases. It is possible to synthesize (see Chap. 8) the excitation voltages such that maximum gain is achieved. If this were done, the resulting currents at the fed ports would be of unit magnitude, whereas the voltages needed to establish these unit magnitude currents would generally be of nonunit magnitude.

Table 10-1 Normalized Terminal Currents for a Linear Array of 12 Half-Wavelength Spaced, Parallel, Half-Wave Dipoles; $a = 0.0001\lambda$

Element Number	Zero Generator Impedance		72-Ω Generator Impedance	
	$\lvert I_A \rvert$	$\lvert Z_A \rvert$	$\lvert I_A \rvert$	$\lvert Z_A \rvert^a$
1	0.689	107.1	0.746	111.9
2	0.698	105.9	0.760	108.6
3	0.728	101.5	0.799	99.6
4	0.753	98.2	0.829	93.5
5	0.768	96.3	0.847	89.9
6	0.777	95.2	0.856	88.2
7	0.781	94.7	0.854	88.6
8	0.775	95.4	0.837	91.8
9	0.753	98.2	0.806	98.1
10	0.713	103.7	0.777	104.4
11	0.689	107.3	0.802	98.8
12	1.000	74.0	1.000	65.1

[a]Exclusive of 72-Ω generator impedance.

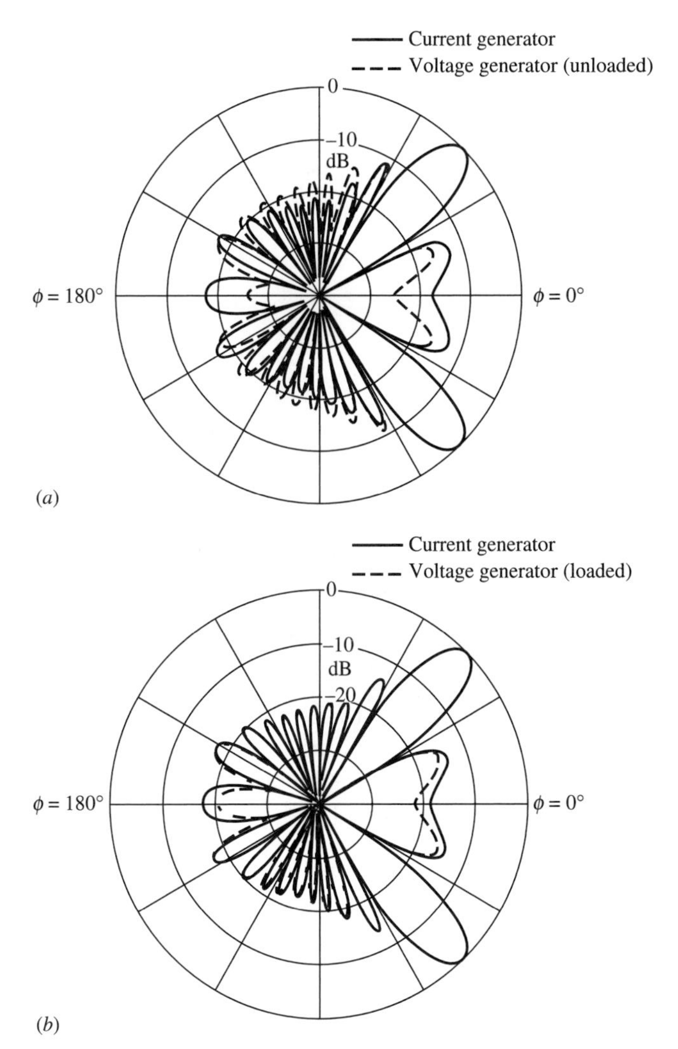

Figure 10-27 Linear array patterns with main beam steered to $\phi_o = 45°$ and ideal current generators (solid curve) compared to patterns from an array with voltage generators (see Table 10-1). (*a*) Linear array pattern for unloaded voltage generator excitations (dashed curve). (*b*) Linear array pattern for 72-Ω loaded voltage generator excitations (dashed curve).

10.11.2 The Circular Array

Consider the circular array in Fig. 10-28 that is also known as a ring array [23]. Such arrays have been used in radio direction finding, radar, sonar and in other systems applications. Usually, circular arrays are composed of identical, equally spaced elements as indicated in Fig. 10-28, and each dipole is excited at its center. If we temporarily replace each dipole with a point source at the excited dipole ports, we can write for the array factor (see Sec. 3.1):

$$\mathrm{AF}(\theta, \phi) = \sum_{n=1}^{N} I_n e^{j\alpha_n} e^{j[\beta\rho'_n \sin\theta\cos(\phi-\phi_n)]} \tag{10-97}$$

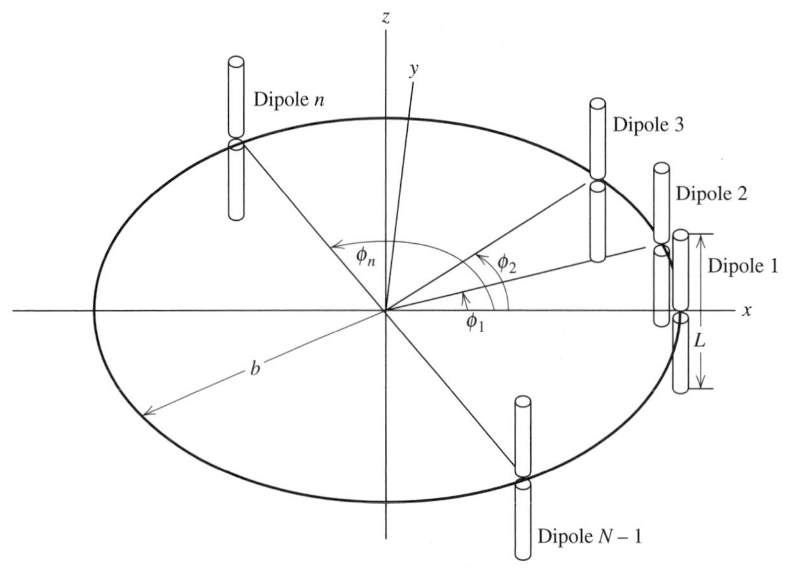

Figure 10-28 Circular array of dipoles.

where I_n is the current excitation of the nth element located at $\phi = \phi_n$, α_n is the associated phase excitation relative to the array center located at the coordinate origin, and ρ_n' is the radial distance of each element center from the origin (all of which equal b for the circular array case). For the usual case of cophasal excitation,

$$\alpha_n = -\beta\rho_n' \sin\theta_o \cos(\phi_o - \phi_n) \tag{10-98}$$

where (θ_o, ϕ_o) is the desired position of the main beam maximum.

For the half-wave dipoles, the element pattern is given approximately by (2-8). Thus, the complete pattern for the circular array of half-wave dipoles with an assumed sinusoidal current distribution can be written as

$$F(\theta, \phi) = \frac{\cos\left(\dfrac{\pi}{2}\cos\theta\right)}{\sin\theta} \frac{\sum\limits_{n=1}^{N} I_n e^{j\alpha_n} e^{j[\beta\rho_n' \sin\theta\cos(\phi-\phi_n)]}}{\sum\limits_{n=1}^{N} I_n} \tag{10-99}$$

where the assumption is made that (3-104) applies rather than (3-107).

The analysis of the circular array in (10-97) to (10-99) is, of course, based on known currents on the array elements. In practice, we usually apply voltages rather than currents to the array element ports. To determine the currents established by the voltages, we can use MoM, thereby including all mutual effects.

We will consider the circular array in Fig. 10-28 to be composed of identical, equally spaced dipoles. Thus, certain simplifications in the moment method formulation are possible. With the excitation at the centers of all dipoles, it is clear that the current distributions will have even symmetry about the $z = 0$ plane. This image symmetry can be used to compress the size of the impedance matrix $[Z_{mn}]$ of each dipole as discussed in Sec. 10.8.6. (This could also have been done for the linear array in the previous section.) In addition to this, the impedance matrix for the circular array will take the submatrix form

$$[Z]_{\text{array}} = \begin{bmatrix} [S]_{11} & [S]_{12} & \cdots & [S]_{1N} \\ [S]_{1N} & [S]_{11} & \cdots & [S]_{1(N-1)} \\ \vdots & \vdots & \ddots & \vdots \\ [S]_{12} & [S]_{13} & \dots & [S]_{11} \end{bmatrix} \tag{10-100}$$

where $[S]_{ij} = [Z_{mn}]$, and each $[Z_{mn}]$ may be compressed as described in Sec. 10.8.6. The matrix in (10-100) is not only toeplitz, but also goes by the name "block circulant." It can be shown that the inverse of a block circulant matrix is also block circulant. Thus, $[Y]_{\text{array}}$ would be block circulant. In a block circulant matrix, successive rows of blocks repeat the previous row but begin with the last block of the previous row.

If we use 1-V voltage generators in series with a 72-Ω impedance at the center of each dipole in a 12-element circular array with $\lambda/2$ spacing, the currents given in Table 10-2 resulted using phases from (10-98). The almost 2:1 variation in current magnitude is the result of mutual coupling. The corresponding pattern in the azimuthal plane is shown in Fig. 10-29. For purposes of comparison, also shown is the pattern for uniform (current) excitation calculated using (10-97). The difference between the two types of patterns is more noticeable here than in Fig. 10-27 for the linear array. Although the pattern with the voltage generator obtained using the moment method is the more realistic of the two, an advantage of the moment method is that it does yield the input impedance of the elements for any scan angle, thereby providing information for the design of the feed network (see Sec. 3.8).

Table 10-2 Normalized Terminal Currents for a Circular Array of 12 Half-Wavelength Spaced, Parallel Half-Wave Dipoles (72-Ω loaded voltage generators)

Element Number	$\lvert I_A \rvert$
1	0.735
2	0.566
3	0.628
4	0.517
5	0.547
6	0.791
7	1.000
8	0.791
9	0.547
10	0.517
11	0.628
12	0.566

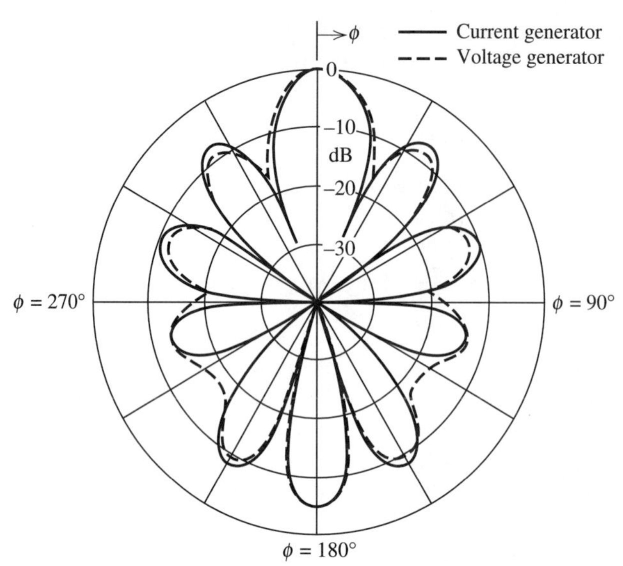

Figure 10-29 Patterns of the circular array of Fig. 10-28 with 12 elements for uniform current excitation (solid curve) and for 72-Ω loaded voltage generators with currents of Table 10-2 (dashed curve).

10.11.3 Two-Dimensional Planar Array of Dipoles

Consider a two-dimensional array of parallel dipoles located in the xz-plane as shown in Fig. 10-30. Our purpose here is to use MoM to show how the input impedance of an element in the array varies with scan angle.

Figure 10-30 shows the input impedance variation of the center element in a 7×9 array (i.e., seven collinear elements in an E-plane row by nine parallel elements in an H-plane row). Three scanning conditions are illustrated: H-plane, E-plane, and the 45° plane between the E- and H-planes. It is clear from Fig. 10-30 that the input impedance does vary considerably with scan angle and the variation depends on the plane of scan. Clearly, this variation poses a challenging design problem for the engineer responsible for designing the array feed and matching network.

Note that as the array is scanned in the E-plane (zy-plane) to 90° (i.e., "endfire"), the real part of the input impedance is tending toward zero, which in turn means the element is tending not to radiate! Indeed, although the other elements in the array will not have *exactly* the same behavior, most of them (except the edge elements) will behave similarly and the entire array will tend not to radiate! This phenomenon is known as *Wood's anomaly*, or the blindspot phenomenon, and it would seem inappropriate not to mention it in a book on antenna principles. Wood's anomaly is more likely to occur in large arrays than in the relatively small one considered here. Fortunately, Wood's anomaly can be avoided in some arrays by suitable choices in the array design parameters as well as in the element design itself. (See Sec. 3.8.3.)

An explanation for Wood's anomaly lies in an understanding of surface waves. In the E-plane, as the array is scanned farther and farther away from broadside, there is a larger and larger component of the electric field normal to the array plane. The metallic elements in the dipole array simulate a metallic surface for the currents

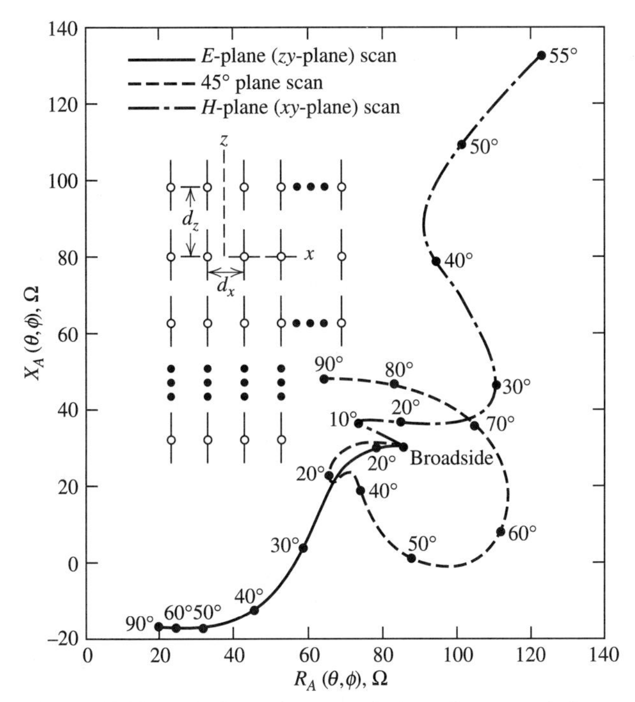

Figure 10-30 Input impedance variation of a central element in a 7 × 9 dipole array as a function of scan angle for three planes of scan ($d_x = d_z = 0.5\lambda$).

that flow in the direction established by the E-plane. Thus, a surface wave is excited in the E-plane direction that can consume the output power of the array if the array is large. This surface wave phenomenon is similar to the creeping wave discussed in Sec. 12.11.

In the H-plane, no such surface wave is possible because at large scan angles, the electric field is trying to propagate tangential to the simulated metallic surface and tends to be shorted out by it. Similarly, the creeping wave in Sec. 12.11 tends not to propagate when the electric field is parallel to a metallic surface.

The moment method analysis of a two-dimensional array such as that in Fig. 10-30 is aided by the block toeplitz nature of the problem. Much larger arrays than that in Fig. 10-30 can be analyzed and designed with the aid of MoM even if the number of unknowns is in the tens of thousands. However, for arrays of such size, other methods of analysis are available [24].

10.11.4 Summary

In this section, we illustrated, through the use of several examples, the application of the moment method to antenna arrays. The examples show us several things. First, the moment method takes into account all mutual couplings and makes it unnecessary to assume the current distribution on the elements in the array or to assume that each element has the same pattern. Second, the moment method directly provides accurate information concerning the input impedance of various elements under any scan condition. Third, the assumption of a sinusoidal current distribution on a thin half-wave dipole in an array environment is a pretty good one and, therefore, the classical methods of dipole array analysis based on this assumption are quite accurate. It is for elements other than the dipole that the moment method has an obvious additional advantage.

10.12 RADAR CROSS SECTION OF ANTENNAS

The study of antenna scattering is a combination of two electromagnetic disciplines: antennas and scattering. Usually, antenna analysis considers the antenna to be a transmitter, whereas part of the study of antenna scattering requires the antenna to be viewed as a receiver. Even in the receiving case, if we are just interested in the power delivered to a load, we can conveniently use antenna transmitting properties and reciprocity. But if we are also interested in how an antenna scatters energy into surrounding space, then a detailed knowledge of the induced currents on all parts of the antenna structure is required. In general, this is a difficult task, but one that is tractable using MoM.

To begin our discussion, consider the Thevenin equivalent circuit of Fig. 10-31 for an antenna as a function of its load impedance (see also Fig. 2-18). In this circuit, $Z_A = R_A + jX_A$ is the antenna impedance, $Z_L = R_L + jX_L$ is the load impedance, and V_A is the open-circuit voltage induced at the antenna terminals. V_A can be related to the incident electric field $\mathbf{E}^i$ as in [29] by

$$V_A = -\mathbf{h}_A^r \cdot \mathbf{E}^i \tag{10-101}$$

where $\mathbf{h}_A^r$ is the antenna vector effective length upon receiving, evaluated in the direction of reception. (Note that (10-101) from [29] differs from (9-10).) By reciprocity, this antenna vector effective length is equal to that of the antenna upon transmitting $\mathbf{h}_A^t$, evaluated in the same direction and defined such that the far field radiated by the antenna under unit current excitation is

$$\mathbf{E}^t = -j\frac{\eta}{2\lambda}\mathbf{h}_A^t\frac{e^{-j\beta r}}{r} \tag{10-102}$$

where r is the radial distance from the antenna to the observation point.

The signs associated with (10-101) and (10-102) are such that the positive terminal during reception is the terminal into which positive sense current enters during transmission. The receiving current is then given by

$$I_A(Z_L) = -\frac{V_A}{Z_A + Z_L} \tag{10-103}$$

With this terminology defined, we now state that the field scattered by an antenna as a function of its load impedance is given by

$$\mathbf{E}^s(Z_L) = \mathbf{E}^s(0) - \frac{Z_L I(0)}{Z_L + Z_A}\mathbf{E}^t \tag{10-104}$$

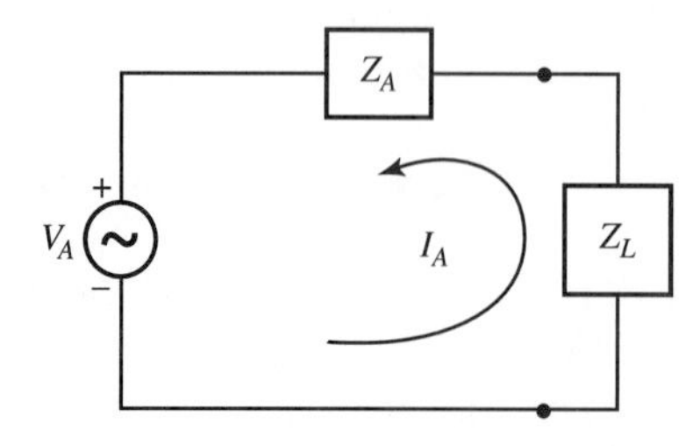

Figure 10-31 Antenna Thevenin equivalent circuit from [29].

where $\mathbf{E}^s(0)$ and $I(0)$ are, respectively, the scattered field and terminal current under short-circuit conditions, $Z_L = 0$. The frequency, polarization, and directions of incidence and reception are assumed to be fixed. This *basic equation of antenna scattering* was derived by King and Harrison [25] using compensation and superposition theorems in circuit analysis, Stevenson [26] using field theory, and Hu [27] using circuit and field theories combined. Aharoni [28] provides a textbook discussion. A derivation from the field point of view is also offered by Green [29]. Derivations employing the scattering matrix are given by Collin [30] and Hansen [31].

Equation (10-104) is not quite in the form we need for our investigation of antenna scattering. To obtain the form used by Green [29], set $Z_L = Z_A^*$, complex conjugate of the antenna impedance, and solve for $\mathbf{E}^s(0)$:

$$\mathbf{E}^s(0) = \mathbf{E}^s(Z_A^*) + \frac{Z_A^* I(0)}{2R_A} \mathbf{E}^t \tag{10-105}$$

The use of (10-103), first with $Z_L = 0$ and then with $Z_L = Z_A^*$, leads to

$$I(0) = \frac{2R_A}{Z_A} I(Z_A^*) \tag{10-106}$$

so that (10-105) becomes

$$\mathbf{E}^s(0) = \mathbf{E}^s(Z_A^*) + \frac{Z_A^*}{Z_A} I(Z_A^*)\mathbf{E}^t \tag{10-107}$$

These last two equations, substituted into (10-104), yield the fundamental equation due to Green:

$$\mathbf{E}^s(Z_L) = \mathbf{E}^s(Z_A^*) + \frac{1}{Z_A}\left[Z_A^* - \frac{2R_A Z_L}{Z_A + Z_L}\right] I(Z_A^*)\mathbf{E}^t \tag{10-108}$$

or

$$\mathbf{E}^s(Z_L) = \mathbf{E}^s(Z_A^*) - [I(Z_A^*)\mathbf{E}^t]\Gamma_m \tag{10-109}$$

where the quantity

$$\Gamma_m \equiv \frac{Z_L - Z_A^*}{Z_L + Z_A} = \frac{\dfrac{Z_L + jX_A}{R_A} - 1}{\dfrac{Z_L + jX_A}{R_A} + 1} \tag{10-110}$$

is a modified voltage reflection coefficient in contrast with the usual definition of reflection coefficient.

The quantity $[I(Z_A^*)\mathbf{E}^t]\Gamma_m$ in (10-109) is called the *antenna mode* component of the scattered field because it is completely determined by the radiation properties of the antenna. It vanishes when the antenna is conjugate-matched. This term is related to the energy absorbed in the load of a lossless antenna as well as the energy reradiated by the antenna due to load mismatch. The pattern of the energy scattered in the antenna mode is exactly that of the antenna radiation pattern. The other quantity on the right-hand side of (10-109), $E^s(Z_A^*)$, is called the *structural scattering* or *residual scattering* component. It arises from the currents induced on the antenna surface by the incident wave even when the antenna has been conjugate-matched.

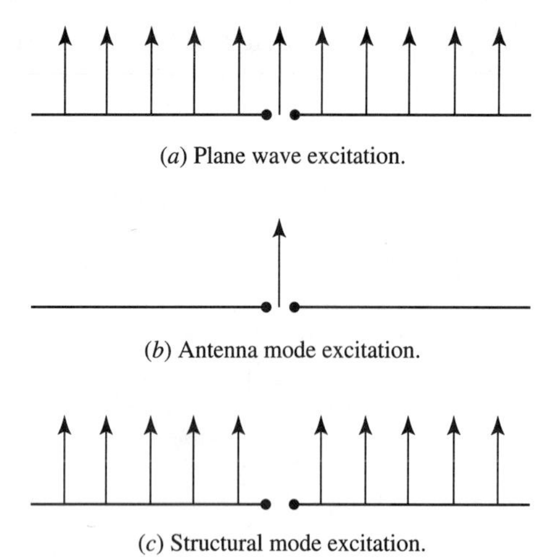

Figure 10-32 Conceptualization of antenna mode and structural mode scattering.

Hansen [31] gives an alternative formulation to that of Green in (10-109) in terms of a conventional load match. However, his formulation does not appear to be generally superior to Green's.

A conceptualization of antenna mode and residual mode scattering is possible from the point of view of point-matching [32]. Figure 10-32 shows the excitation of a dipole by a plane wave [32]. The arrows depict the amplitude of the plane wave at the match points. The (*b*) part of the figure shows the usual antenna excitation that gives rise to the antenna mode term, whereas the (*c*) part shows the difference between the plane wave excitation and the antenna mode excitation that is the residual mode excitation.

A primary reason for being interested in antenna scattering is to determine the radar cross section of an antenna. In the case of antenna mode scattering, it is possible to derive the following simple expression for the antenna mode component of the radar cross section:

$$\sigma_{\text{ant}} = \Gamma^2 G^2(\theta, \phi) \frac{\lambda^2}{4\pi} \tag{10-111}$$

However, no such simple expression is possible for the structural mode radar cross section.

Figure 10-33 shows the broadside monostatic RCS of a dipole 0.5 m in length

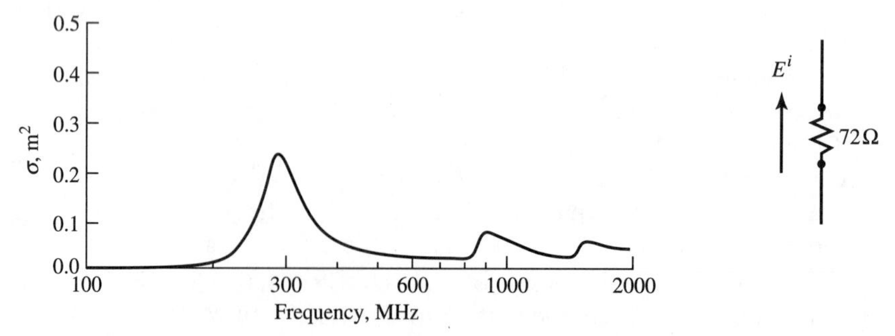

Figure 10-33 Monostatic broadside RCS vs. frequency for a 0.5-meter-long dipole terminated in a 72-Ω load.

terminated in a 72-Ω load as the frequency is varied from 100 to 2000 MHz [32]. We observe peaks in the curve at the first three dipole resonances near 300, 900, and 1500 MHz as expected. Near 300 MHz, the RCS is due to the residual mode since the dipole is conjugate-matched at its first resonance. Above about 1000 MHz, the RCS is also dominated by the residual scattering, but this is because the antenna mode radiation pattern is weak or zero in the broadside direction.

Figure 10-34 shows the bistatic scattered field for the residual mode and antenna mode when a plane wave is incident at angles of 90, 60, and 30° from the axis of a 1.723λ dipole with a load of 72 Ω. In comparing the curves in Figs. 10-34*a* and 10-34*b*, we first observe that the antenna mode bistatic electric field curves are all symmetric about $\theta_s = 90°$, whereas the structural mode curves for $\theta_s = 60°$ and 30° are not. This is a further illustration of the antenna-like behavior of the antenna

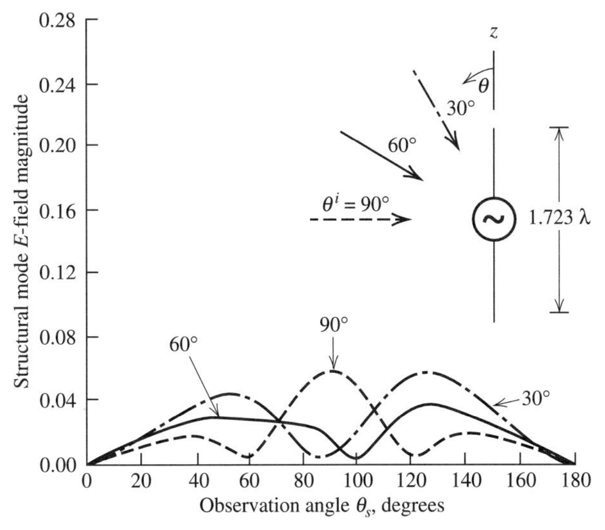

(*a*) Structural mode electric field bistatic scattering.

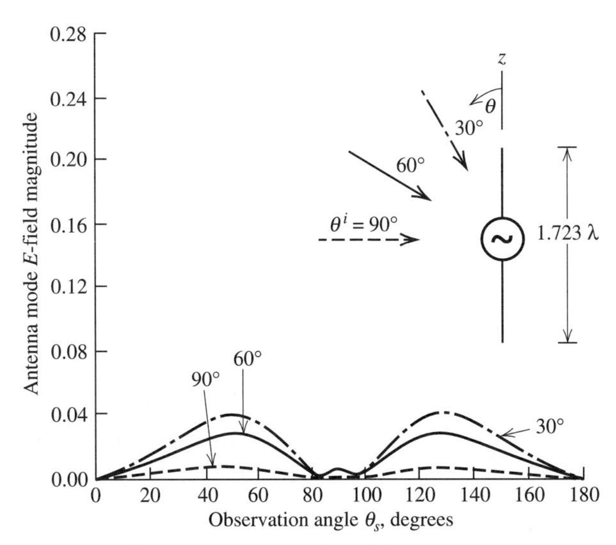

(*b*) Antenna mode electric field bistatic scattering.

Figure 10-34 Component parts of antenna scattering.

mode. To illustrate the structure-like behavior of the structural mode, note the large bistatic response around $\theta_s = 125°$ for the $\theta_i = 30°$ incidence case in Fig. 10-34*a*. This is due to the dipole structure acting like a moderately long reflecting surface. If the length of the dipole were to increase, the angle of maximum response would increase toward 150° as expected from the law of reflection.

It is worthwhile to comment on how the MoM is used to obtain the data in Fig. 10-34*a*. First, at each frequency the antenna input impedance is calculated for the dipole so that $E^s(Z_A^*)$ is known. Then the left-hand side of (10-109) is calculated for each of the three incidence angles. Next, $I(Z_A^*)$ is found so that the antenna mode scattering may be calculated. In turn, subtracting the antenna mode scattering from the total scattered field on the left-hand side of (10-109) yields the structural scattering. No other CEM technique would be as helpful as MoM for these kinds of calculations.

Figure 10-35 shows the **total** power scattered and the total power absorbed by a 0.5-m-long dipole with a 72-Ω load and a plane wave normally incident. Generally, the total power scattered is greater than the total power absorbed, the two being equal at first resonance where the dipole is conjugate-matched. Below first resonance, the dipole is not conjugate-matched and it is possible for the total power absorbed to exceed the total power scattered. The curves in Fig. 10-35 were calculated using the equivalent circuit in Fig. 10-31 by assuming that Z_A accounts for the total power scattered and Z_L for the total power absorbed by the dipole. That Z_A accounts for the total power scattered by the antenna is *not* generally true, but is approximately true when the power scattered by the open-circuited antenna is much smaller than that scattered by the terminated antenna [33] (below about 400 MHz in Fig. 10-33). Numerically integrating the scattered fields over a far-zone sphere enclosing the dipole to correctly calculate the total power scattered replicates the P_A curve up to a dipole length of about 0.6λ and provides at best only rough quantitative agreement above that length. It is sometimes thought that a matched lossless antenna cannot absorb more power than it scatters. However, this is not true theoretically [34]. In practice, it is unlikely that a matched lossless antenna will absorb more power than it scatters. In Fig. 10-35, the powers are equal just below 300 MHz where the dipole is matched.

In this section, scattering by an antenna has been examined for the purpose of understanding the radar signature of an antenna. However, this is not the only

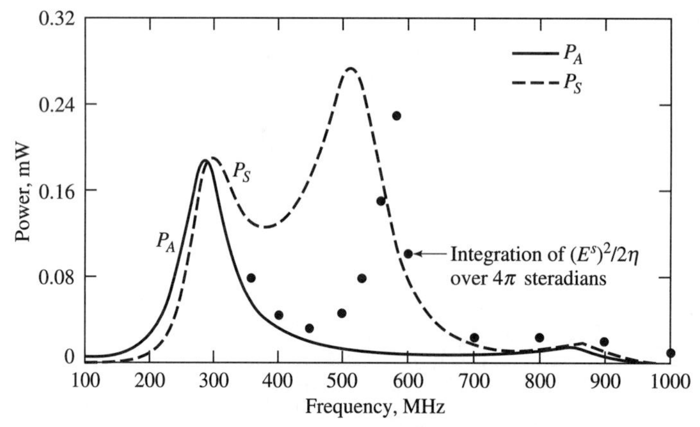

Figure 10-35 Total power absorbed and total power scattered by a 0.5-m-long dipole with a 72-Ω load.

reason to be interested in antenna scattering. For example, if an antenna did not scatter, there would be no mutual coupling in an antenna array.

10.13 MODELING OF SOLID SURFACES

There are two principal ways in which MoM can be used to model either two-dimensional or three-dimensional bodies (e.g., infinite cylinder or finite cylinder, respectively). The simplest way to model a solid surface body is with a grid of wires, the so-called *wire-grid model.* Examples of this approach are illustrated in Fig. 10-36. The other common approach is to use a magnetic field integral equation (see Prob. 10.13-5) in which the surface is broken up into patches or cells, each having a continuous metallic surface. In this section, we briefly examine both approaches.

10.13.1 Wire-Grid Model

In this subsection, we demonstrate the application of the wire segment procedure to model not just a wire antenna or wire scatterer, but also to model the metallic environment near the antenna. We can accomplish this by using a wire-grid or wire mesh to simulate an actual continuous metallic surface. The idea of using a wire mesh to simulate a continuous metallic surface precedes, of course, the time when the moment method came into widespread use. There are many practical situations where the effect of a continuous metallic surface is required, but the weight and/or wind resistance offered by a continuous surface is too large (e.g., a reflector surface).

The successful substitution of a wire grid for a continuous metallic surface (in reality or in a model) depends on the fact that as the grid size becomes smaller relative to the wavelength, the grid supports a current distribution that approximates that on the corresponding continuous surface. The current is only an approximation of the actual current, however, and as such it can be expected to reasonably predict the far fields but possibly not the near fields. This is due to the fact that the grid supports an evanescent reactive field on both sides of its surface [35]. An actual continuous conducting surface is not capable of supporting such a field.

The accuracy with which a wire-grid model simulates an actual surface depends on the computer code (i.e., expansion and weighting functions) used, the radius of the wire segments used, as well as the grid size. For example, with pulse expansion functions and point-matching, it has been found that a grid spacing of about 0.1 to 0.2λ yields good results [36]. With the piecewise sinusoidal Galerkin method, it has

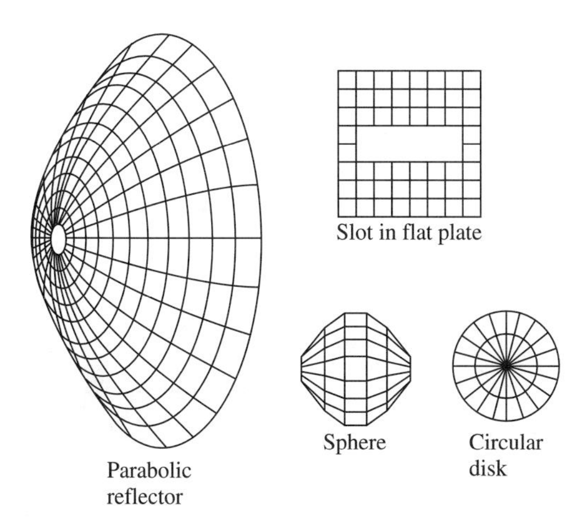

Figure 10-36 Examples of wire-grid modeling.

been found that the grid size should not exceed $\lambda/4$ and a suitable wire radius is $a = w/25$, where w denotes the width or length (whichever is greater) of the apertures [37]. A 70-segment (piecewise sinusoidal Galerkin) aircraft model is shown in Fig. 10-37. The model is a 1:200 scale. Thus, for the radar cross section results in Fig. 10-38, the actual length of the aircraft fuselage is about 15.4 m and the actual frequencies are 16.1 and 27.15 MHz, respectively. Since the radar cross section scales as the scale factor squared, the actual RCS (in square meters) at the actual frequencies would be 46 dB above that indicated in Fig. 10-38 for $f = 3.25$ and 5.43 GHz.

Let us now consider the situation where a monopole is axially mounted on the base of a cone [7] as shown in Fig. 10-39. A wire-grid representation can be used in which the cone or frustum is represented by a number of "generating lines" consisting of a number of wires joined end to end, as shown in Fig. 10-40*a*. Except for the base, no wires need to be provided in planes normal to the z-axis because of excitation symmetry.

An interesting simplification (see Sec. 10.8.6) can be obtained from the symmetry of the configuration in the case where all generating lines have the same number of segments, each segment being identical (except for the orientation on the ϕ-coordinate) to the corresponding one on each other generating line. The currents on such corresponding segments should be equal in magnitude and phase, since $I(z)$ is independent of ϕ. Let the segments be numbered in a consecutive way, starting with the line at $\phi = 0$ and proceeding in counterclockwise direction along the other lines. Let M be the number of segments on each line and L the number of generating lines. Thus, one can write

$$\sum_{j=1}^{L \cdot M} Z_{kj} I_j = E_k^i, \qquad k = 1, 2, \ldots, L \cdot M \tag{10-112}$$

Since the currents on corresponding segments are equal,

$$I_j = I_{(j+M)} = I_{(j+2M)} = I_{(j+(L-1)M)} \tag{10-113}$$

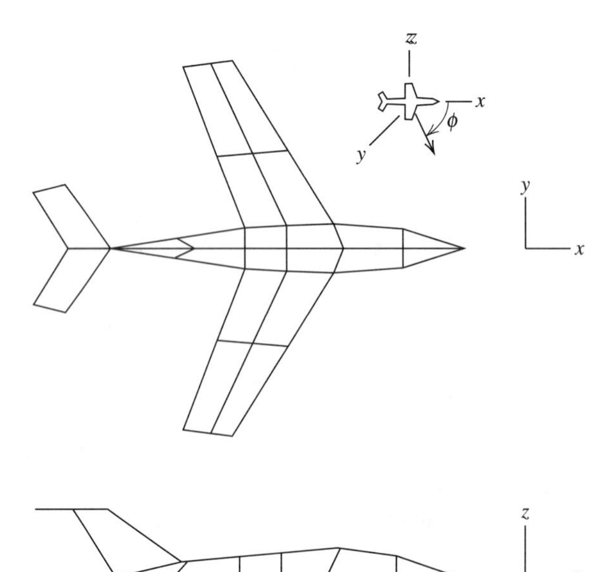

Figure 10-37 Wire-grid model for the scale model MIG 19 with 70 segments.

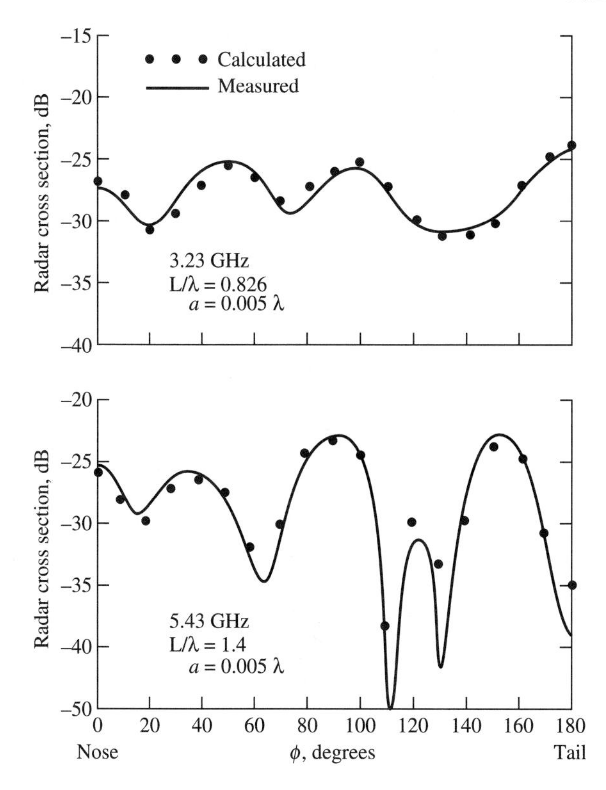

Figure 10-38 Radar cross section of a scale model MIG 19 aircraft. (L = fuselage length).

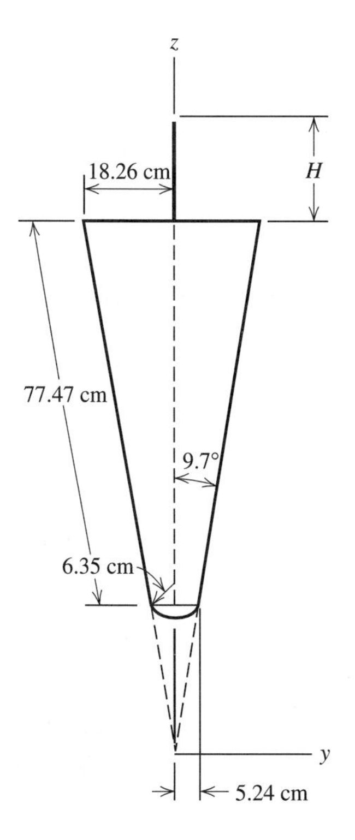

Figure 10-39 Dimensions of experimental cone model. The monopole is a quarter-wavelength at each frequency.

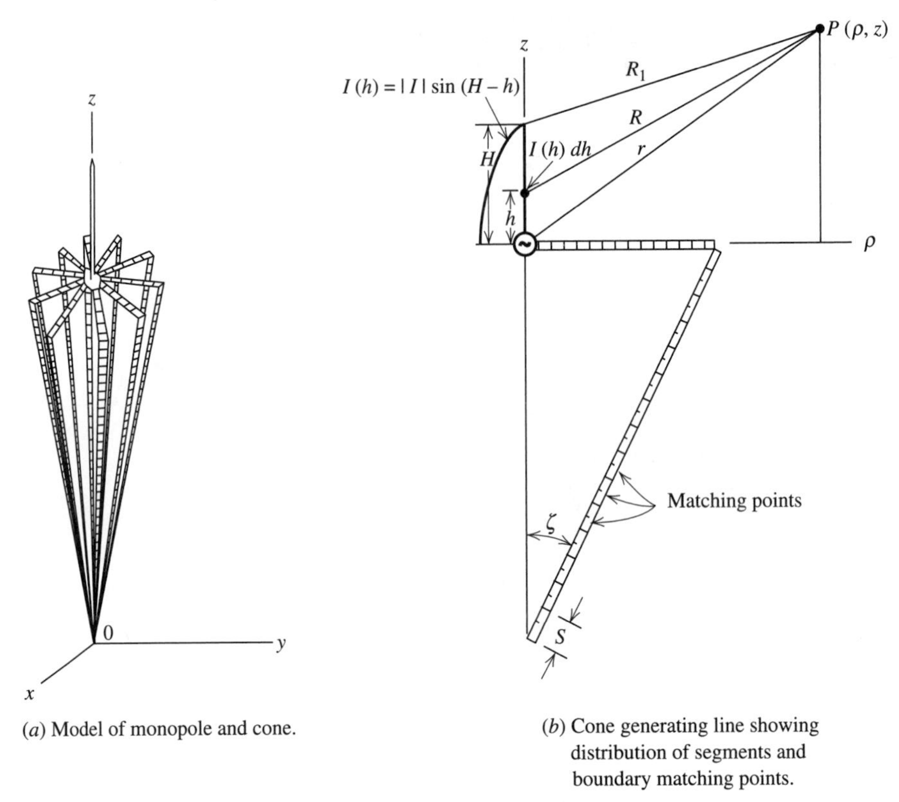

(*a*) Model of monopole and cone.

(*b*) Cone generating line showing distribution of segments and boundary matching points.

Figure 10-40 Wire-grid model of cone in Fig. 10-39.

and (10-112) can be written as (see Sec. 10.8.6)

$$\sum_{j=1}^{M} I_j \left(\sum_{n=0}^{L-1} Z_{k(j+nM)} \right) = E_k^i, \qquad k = 1, 2 \ldots, M \tag{10-114}$$

The advantage of (10-114) is that it permits us to reduce the number of unknown currents to M, while the actual number of wire segments is $L \cdot M$, where L is arbitrary. As a result, there is no limitation other than computer running time to the number of generating lines (and thus to the total number of segments represented). For the patterns calculated here, L was chosen to be 10, M to be 170, and pulse basis functions were used.

The left-hand side of (10-114) represents $-\mathbf{E}^s$ for the cone problem under consideration here. It remains to determine $\mathbf{E}^i$. For the monopole, consider the geometry depicted in Fig. 10-40*b*. If we start with the vector potential, the following near-field expressions for the monopole configuration of Fig. 10-40*b* may be derived as

$$E_z^i = -j29.975\,|I| \left[\frac{e^{-j\beta R_1}}{R_1} - \frac{e^{-j\beta r}}{r} \cos \beta H - j\frac{z}{r^2} e^{-j\beta r} \sin \beta H - \frac{z}{\beta r^3} e^{-j\beta r} \sin \beta H \right] \tag{10-115}$$

and

$$E_\rho^i = \frac{j29.975\,|I|}{\rho}\left[(z - H)\,\frac{e^{-j\beta R_1}}{R_1} - \frac{z}{r}\,e^{-j\beta r}\cos\beta H \right.$$
$$\left. - \frac{jz^2}{r^2}\,e^{-j\beta r}\sin\beta H + \frac{\rho^2}{\beta r^3}\,e^{-j\beta r}\sin\beta H\right] \qquad (10\text{-}116)$$

The right-hand side of (10-114), which constitutes the generalized voltage matrix, requires that the tangential components of the incident field be determined at each match point on one of the generating lines of the cone. The equations above for E_z^i and E_ρ^i are used to do this.

Solving for the current on the cone makes it possible to calculate the far-field pattern of the cone-monopole structure by superimposing the fields of the cone and those of the monopole. A necessary but not sufficient check on the validity of the moment method solution in this problem requires that the currents at the junction of the monopole with the wire-grid representation of the cone satisfy Kirchhoff's current law. For the formulation in (10-114) to (10-116) with $L = 10$, the current on each of the 10 wire-grid lines was found to be 0.1 A when the monopole base current $|I| = 1.0$ A. That these 11 currents satisfy Kirchhoff's current law at their common junction is a direct consequence of Maxwell's equations since Kirchhoff's current law was not explicitly built into the system of equations [i.e., a constraining equation was not one of the equations in (10-114)].

To experimentally test the validity of the wire-grid representation of a metallic surface, an actual wire-grid cone was built around a styrofoam core in a configuration similar to that shown in Fig. 10-40*a*. An experimental comparison of the solid cone and its wire-grid counterpart is shown in Fig. 10-41*a*. Some representative results showing both the results calculated for the wire-grid cone and measurements for the solid surface cone are illustrated in Fig. 10-41*b*. The results are generally quite good.

Other variations of the formulation given here are possible, of course. For example, instead of assuming the current distribution on the monopole, it can be treated as an unknown as are the currents on the metallic body. This could be done in a number of ways. The monopole terminal current value could be constrained to a particular value. This would take into account the interaction between the cone and monopole, but would not conveniently provide for the calculation of impedance. Alternatively, one could use a voltage generator at the base of the monopole such as the magnetic frill current discussed previously. Calculation of the currents on the cone and monopole would account for the cone-monopole interaction and also yield directly the monopole impedance. Note that in either case, the previously described symmetry for the cone due to the symmetrical excitation could still be used to advantage.

The accuracy of wire-grid models can be improved if the grid is reactively loaded with lumped loads [35]. The motivation for doing this is to eliminate the effects of the evanescent reactive field that is in proximity to the wire grid. Not only does this increase the accuracy of the model, but it also permits larger grid sizes to be used. Nevertheless, even without this loading, the wire-grid model is a convenient and relatively straightforward tool for engineering calculations.

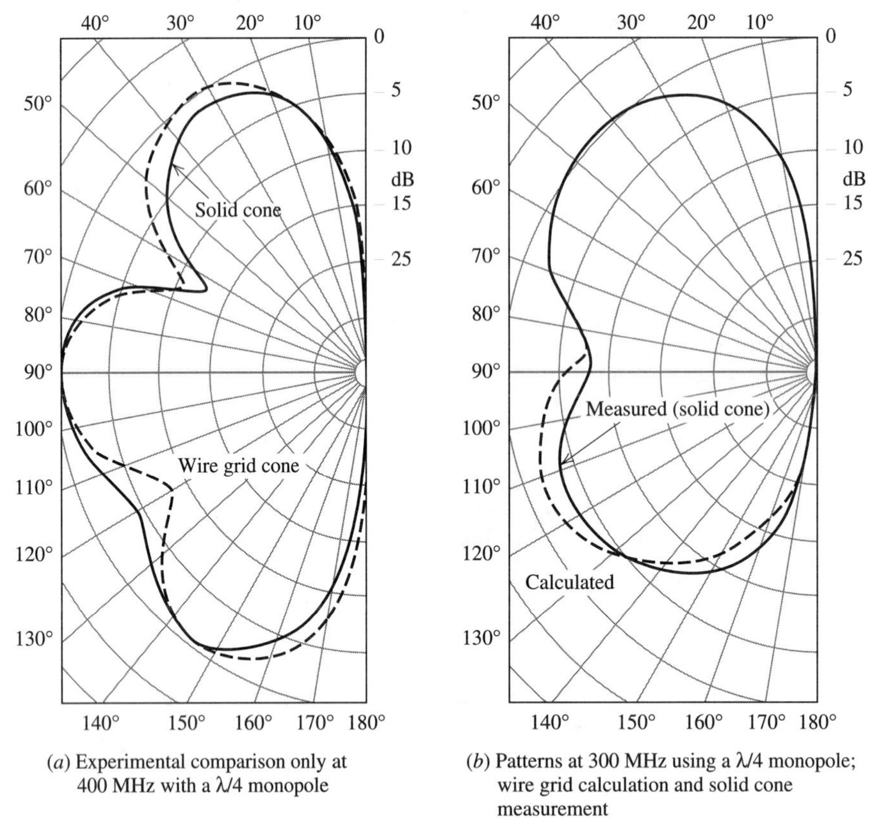

(*a*) Experimental comparison only at 400 MHz with a λ/4 monopole

(*b*) Patterns at 300 MHz using a λ/4 monopole; wire grid calculation and solid cone measurement

Figure 10-41 Far-field patterns of a monopole on the base of the cone in Fig. 10-40.

10.13.2 Continuous Surface Model

The continuous surface model of a three-dimensional body is a complex problem that is generally beyond the scope of this text. The interested reader is referred to the literature for a discussion of this topic.

On the other hand, the continuous surface model of a two-dimensional solid body follows directly from the earlier sections in this chapter. We consider two such examples here: that of a conducting cylinder with the incident electric field parallel to the axis of the cylinder (TM case) and that of a conducting cylinder with the incident magnetic field parallel to the axis of the cylinder (TE case).

In Sec. 1.5, we found the solution (Green's function) to the spherical wave equation. Here, we need a solution to the cylindrical wave equation. For the TM case, our equation is

$$\nabla^2 E_z + \beta^2 E_z = j\omega\mu J_z \tag{10-117}$$

where $E_z = E_z(x, y)$. A solution to this equation is

$$E_z = -\frac{\beta\eta}{4} I H_0^{(2)}(\beta|\boldsymbol{\rho} - \boldsymbol{\rho}'|) \tag{10-118}$$

where $H_0^{(2)}$ is the Hankel function of the second kind and zero order. It represents an outward cylindrical traveling wave just as does $e^{-j\beta r}$ for a spherical wave. The total scattered field is then the integral of (10-118) over the cross section of the cylinder or [6]

$$E_z^s(\boldsymbol{\rho}) = -\frac{\beta\eta}{4}\iint J_z(\boldsymbol{\rho}')H_0^{(2)}(\beta|\boldsymbol{\rho} - \boldsymbol{\rho}'|)\, ds' \tag{10-119}$$

where the integration is over the cross section of the cylinder of currents J_z as indicated in Fig. 10-42*a*.

A simple formulation is to require that (10-34) or (10-36) with delta weighting functions applies. Hence, the applicable integral equation is

$$E_z^i(\boldsymbol{\rho}) = \frac{\beta\eta}{4}\int_c J_z(\boldsymbol{\rho}')H_0^{(2)}(\beta|\boldsymbol{\rho} - \boldsymbol{\rho}'|)\, dc' \qquad \rho \text{ on } c \tag{10-120}$$

where $E_z^i(\boldsymbol{\rho})$ is known and J_z is the unknown to be determined. Note that (10-120) has the same form as (10-1). If pulse expansion functions are used, the impedance matrix elements are

$$Z_{mn} = \frac{\beta\eta}{4}\int_{\Delta c_n} H_0^{(2)}[\beta\sqrt{(x' - x_m)^2 + (y' - y_m)^2}]\, dc' \tag{10-121}$$

and the voltage matrix elements are

$$V_m = E_z^i(x_m, y_m) \tag{10-122}$$

Note that the incident field is present at all match points in Fig. 10-42*b* since the metallic surface has been replaced by equivalent currents in free space.

In order to calculate the elements of the generalized impedance matrix, it is necessary to evaluate (10-121). Unfortunately, there is no simple analytic expression for the integral, but it can be evaluated by one of several approximations. The simplest (and crudest) approximation is to view a current element $J_z\ \Delta c_n$ as a filament of current when the field point is not on Δc_n. Thus, when $m \neq n$,

$$Z_{mn} \approx \frac{\eta}{4}\,\beta\,\Delta c_n H_0^{(2)}[\beta\sqrt{(x_n - x_m)^2 + (y_n - y_m)^2}] \tag{10-123}$$

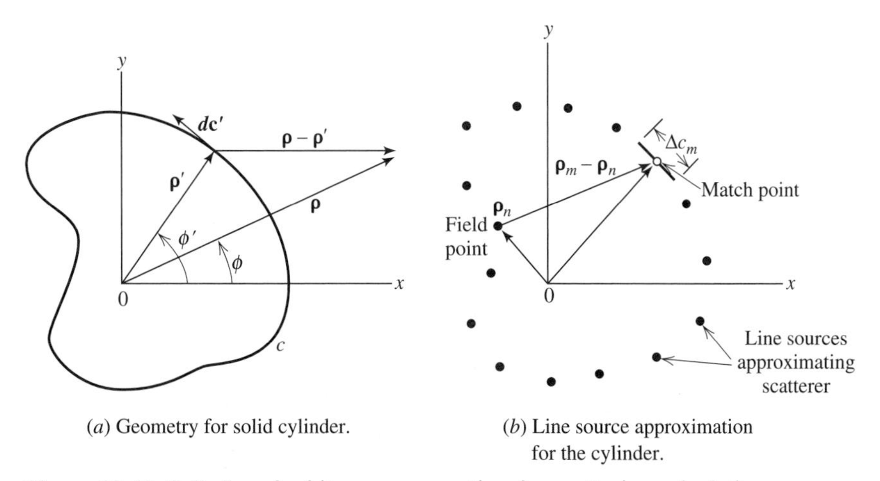

(*a*) Geometry for solid cylinder.

(*b*) Line source approximation for the cylinder.

Figure 10-42 Cylinder of arbitrary cross section for scattering calculations.

Note that although we are using a current filament approximation as shown in Fig. 10-42*b*, this is not a wire-grid model and should not be confused with that approach.

To obtain Z_{nn}, we recognize that the Hankel function has an integrable singularity and the integral must be evaluated analytically. To do this, we use the small argument formula for the Hankel function of argument $\beta\rho$:

$$H_0^{(2)}(\beta\rho) \approx 1 - j\frac{2}{\pi}\ln\left(\frac{\gamma\beta\rho}{2}\right) \tag{10-124}$$

where $\gamma = 0.5772\ldots$ is Euler's constant and obtain

$$Z_{nn} \approx \frac{\eta}{4}\beta\,\Delta c_n\left[1 - j\frac{2}{\pi}\ln\left(\frac{\gamma\beta\,\Delta c_n}{4e}\right)\right] \tag{10-125}$$

where $e = 2.718$. Better formulations (e.g., faster convergence), although somewhat more complex, are possible with the use of other expansion functions and other weighting functions.

Results for the z-directed current on a cylinder are given in Fig. 10-43 for plane wave incidence where $E_z^i = e^{-j\beta x}$. Note that the current decays to nearly zero on the shadowed side since, for this (TM) polarization, the electric field is shorted out by the metallic cylinder as it propagates from the illuminated side into the shadowed region. This will not be the case for the TE polarization.

Next consider the TE case, where the incident magnetic field is parallel to the circular cylinder axis. We will, by choice, use a magnetic field integral equation (MFIE) that has the form

$$J_\phi(\boldsymbol{\rho}) + \hat{\mathbf{z}}\cdot\nabla\times\int J_\phi(\boldsymbol{\rho})\psi(\boldsymbol{\rho},\boldsymbol{\rho}')\,d\mathbf{c}' = H_z^i(\boldsymbol{\rho}) \tag{10-126}$$

In contrast to the electric field integral equation (EFIE) in (10-1) where the unknown current only appears under the integral sign, here the unknown current appears both under and outside the integral sign. Thus, (10-126) is referred to as an integral equation of the second kind. Integral equations of the second kind are generally preferable for large smooth conducting bodies since the contribution by the integral part of the equation may be of second-order importance. However, the electric field integral equation is also useful for large conducting bodies as we have seen in the treatment of the TM case. Magnetic field integral equations are not useful for treating thin wires due to the singularity in the integral. Recall that in

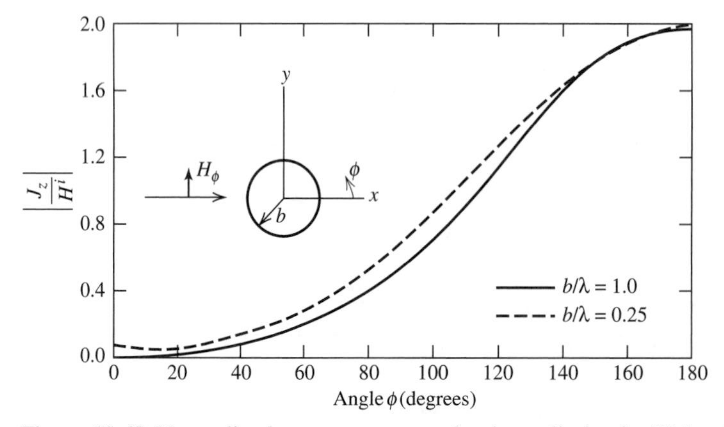

Figure 10-43 Normalized current on a conducting cylinder for TM polarization.

Sec. 10.1, we avoided the singularity in the EFIE by using the approximation that the observation points lie on the axis of the wire rather than the surface. That approach cannot be employed for the MFIE.

To derive the magnetic field integral equation for the two-dimensional problem of interest here, we note that there will only be a z-component of **H** and a transverse component of **J**, namely J_ϕ. The total magnetic field H_z at any point on or outside the surface of the conducting body is the sum of the impressed field H_z^i plus the scattered field H_z^s on the surface of the body. Thus,

$$H_z = H_z^i + H_z^s \tag{10-127}$$

Since $\mathbf{H} = \dfrac{1}{\mu} \nabla \times \mathbf{A}$, we can write

$$H_z^s(\boldsymbol{\rho}) = \hat{\mathbf{z}} \cdot \nabla \times \int_c J_\phi(\boldsymbol{\rho}) \psi(\boldsymbol{\rho}, \boldsymbol{\rho}')\, d\mathbf{c}' \tag{10-128}$$

where ψ is the two-dimensional Green's function (used for the TM case) and $d\ell'$ specifies the reference direction of **J**. The discontinuity in H_z at the conducting surface is equal to the current density J_ϕ. So,

$$J_\phi = -[H_z]_{c^+} \tag{10-120}$$

where c^+ indicates that H_z is evaluated just external to cross-sectional surface contour c. When (10-127) is applied to the contour c^+, we can use (10-128) and (10-129) to write

$$J_\phi(\boldsymbol{\rho}) = -[H_z^i(\boldsymbol{\rho}) + \hat{\mathbf{z}} \cdot \nabla \times \int_c J_\phi(\boldsymbol{\rho}) \psi(\boldsymbol{\rho}, \boldsymbol{\rho}')\, d\mathbf{c}']_{c^+} \tag{10-130}$$

This is the magnetic field integral equation for the two-dimensional problem of interest here. The current density J_ϕ is the unknown, whereas the incident field H_z^i is known. The evaluation of the integral in (10-130) must be done with care since H_z is discontinuous at c and the Green's function is singular, precluding a simple interchange of differentiation and integration.

Rewriting (10-130) as

$$J_\phi(\boldsymbol{\rho}) + [\hat{\mathbf{z}} \cdot \nabla \times \int_c J_\phi(\boldsymbol{\rho})\, H_0^{(2)}(\beta|\boldsymbol{\rho} - \boldsymbol{\rho}'|)\, d\mathbf{c}']_{c^+} = -H_z^i|_{c^+} \tag{10-131}$$

and specifying pulse expansion functions and delta weighting functions enables us to write

$$Z_{mn} = \delta_{mn} + H_z(m, n) \tag{10-132}$$

where δ_{mn} is the Kronecker delta and $H_z(m, n)$ is the magnetic field at (x_m, y_m) on c^+ due to a unit current density on Δc_n at (x_n, y_n), or

$$H_z(m, n) = [\hat{\mathbf{z}} \cdot \nabla \times \int_{\Delta c_n} H_0^{(2)}[\beta\sqrt{(x' - x_m)^2 + (y' - y_m)^2}\, d\mathbf{c}']_{c^+} \tag{10-133}$$

When the observation point and source segment coincide, $H_z(m, n)$ exhibits the singularity mentioned previously. However, Z_{mm} may be evaluated by noting that

$$H_z|_{c^+} = -H_z|_{c^-} = -\tfrac{1}{2} \tag{10-134}$$

since we are dealing with only a unit current. Thus,

$$Z_{mm} = 1 - \tfrac{1}{2} = \tfrac{1}{2} \tag{10-135}$$

To evaluate Z_{mn} for $m \neq n$, we can employ the approximation that when $\Delta c_n \ll \lambda$ and the field point at (x, y) is distant from the source Δc_n, the fields from the source appear to emanate from a magnetic line source located at the center of Δc_n. Thus,

$$H_z(m, n) = \frac{\Delta c_n}{4j} \frac{\partial}{\partial n} [H_0^{(2)}(\beta\rho)] \tag{10-136}$$

where the derivative is taken with respect to the normal to the surface and a local coordinate system is implied. If ϕ is the angle between $\hat{\boldsymbol{\rho}}$ and $\hat{\mathbf{n}}$, then

$$H_z(m, n) = \frac{j}{4} \beta \, \Delta c_n \cos \phi \, H_1^{(2)}(\beta\rho) \tag{10-137}$$

where $H_1^{(2)}$ is the Hankel function of the first order. It is necessary to translate this result from its local coordinate system to one with an arbitrary origin. This is accomplished by replacing ρ by $|\boldsymbol{\rho}_m - \boldsymbol{\rho}_n|$ and $\cos \phi$ by $\hat{\mathbf{n}} \cdot \hat{\mathbf{R}}$, where

$$\hat{\mathbf{R}} = \frac{\boldsymbol{\rho}_m - \boldsymbol{\rho}_n}{|\boldsymbol{\rho}_m - \boldsymbol{\rho}_n|} \tag{10-138}$$

is a unit vector from the source point (x_n, y_n) to the field point (x_m, y_m). Finally, for $m \neq n$, we have

$$Z_{mn} \approx \frac{j}{4} \beta \, \Delta c_n (\hat{\mathbf{n}} \cdot \hat{\mathbf{R}}) H_1^{(2)}(\beta|\boldsymbol{\rho}_m - \boldsymbol{\rho}_n|) \tag{10-139}$$

whereas for all m

$$V_m = -H_z^i(x_m, y_m) \tag{10-140}$$

Solution of the usual matrix equation $[Z_{mn}][I_n] = [V_m]$ yields the transverse currents on the conducting cylinder. A result for the current J_ϕ on a circular cylinder induced by a plane wave is shown in Fig. 10-44. The current is normalized with respect to the magnitude of the incident field. Note that the current does not go to zero on the deep shadowed side of the circular cylinder for the TE polarization. This is due to the propagation of surface waves around both sides of the cylinder that interfere in the shadowed region to produce standing waves there. These sur-

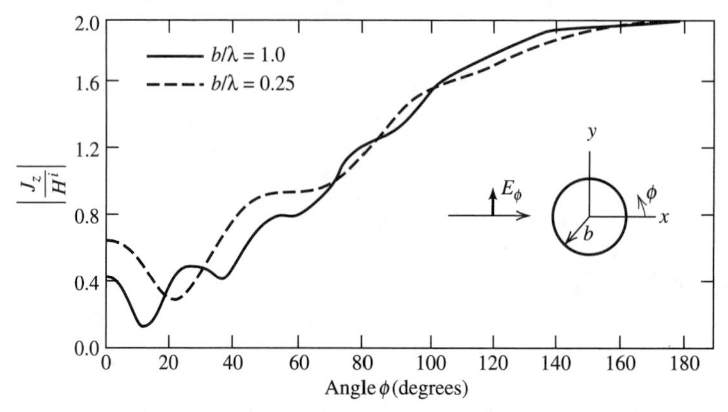

Figure 10-44 Normalized current on a conducting cylinder for TE polarization.

face waves are called creeping waves and are discussed in Chapter 12. On the other hand, the current on the most strongly illuminated portion of the cylinder can be interpreted as the physical optics current and is produced by the left-most term in (10-126). Physical optics is also discussed in Chapter 12.

We have not considered the subject of internal resonances here, but it should be pointed out that it is possible to obtain erroneous currents on the cylinder at those precise frequencies where the interior dimensions of the cylinder correspond to the resonant frequency of a waveguide type mode [38]. Such erratic behavior can be avoided if a formulation is used that combines both the EFIE and MFIE.

10.14 SUMMARY

In this chapter, we presented a very useful and powerful technique, the method of moments, for the analysis of certain types of antennas (e.g., wire antennas) and arrays of antennas (e.g., Sec. 10.10 through 10.12). Although the method has been applied primarily to z-directed wires, we have seen that it applies equally well to arbitrary configurations of wires, for example, Sec. 10.11, as well as to solid surfaces, for example, Sec. 10.13.2. Furthermore, the method of moments has been used to generate some of the data presented in Chaps. 5 and 6.

The method of moments is often thought of as a low-frequency technique because it generally cannot be applied to bodies that are arbitrarily large in terms of the wavelength (e.g., Sec. 10.8). In contrast to this, in Chapter 12 we will study high-frequency techniques that apply best to bodies which are arbitrarily large in terms of the wavelength. (See Figs. 10-1 and 10-2.)

REFERENCES

1. E. K. Miller, "A Selective Survey of Computational Electromagnetics," *IEEE Trans. Ant. & Prop.*, Vol. 36, pp. 1281–1305, Sept. 1988.
2. A. J. Poggio and E. K. Miller, "Integral Equation Solutions of Three-dimensional Scattering Problems," in *Computer Techniques for Electromagnetics*, Pergamon Press, New York, 1973, Chap. 4.
3. J. Jin, *The Finite Element Method in Electromagnetics*, Wiley, New York, 1993.
4. J. H. Richmond, "Digital Computer Solutions of the Rigorous Equations for Scattering Problems," *Proc. IEEE*, Vol. 53, Aug. 1965.
5. R. F. Harrington, "Matrix Methods for Field Problems," *Proc. IEEE*, vol. 55, Feb. 1967.
6. R. F. Harrington, *Field Computation by Moment Methods*, Macmillan, New York, 1968.
7. G. A. Thiele, "Wire Antennas," in *Computer Techniques for Electromagnetics*, Pergamon Press, New York, 1973, Chap. 2.
8. R. F. Harrington, *Time-Harmonic Electromagnetic Fields*, McGraw-Hill, New York, 1961.
9. S. A. Schelkunoff, *Advanced Antenna Theory*, Wiley, New York, 1952.
10. B. A. Finlayson, *The Method of Weighted Residuals and Variational Principles*, Academic Press, New York, 1972.
11. E. K. Miller and F. J. Deadrick, "Some Computational Aspects of Thin Wire Modeling," in *Numerical and Asymptotic Techniques in Electromagnetics*, Springer-Verlag, New York, 1975, Chap. 4.
12. V. H. Rumsey, "The Reaction Concept in Electromagnetic Theory," *Phys. Rev.*, Series 2, Vol. 94, June 1954.
13. Mininec Professional Software, EM Scientific, Inc., Carson City, NV, 1995.
14. J. H. Richmond, "Computer Program for Thin-Wire Structures in a Homogeneous Conducting Medium," National Technical Information Service, Springfield, VA, NASA Contractor Rep. CR-2399, July 1973.
15. G. J. Burke and A. J. Poggio, "Numerical Electromagnetics Code (NEC)—Method of Moments," Tech. Doc. No. 116, prepared for Naval Oceans Systems Center, San Diego, CA, NOSC/TD 116, revised Jan. 1980.
16. C. A. Klein and R. Mittra, "The Effect of Different Testing Functions in the Moment Method Solution of Thin Wire Antenna Problems," *IEEE Trans. Ant. Prop.*, Vol. AP-23, p. 258, March 1975.

17. J. Moore and R. Pizer, *Moment Methods in Electromagnetics*, Research Studies Press Ltd., Letchworth, Herefordshire, England and Wiley, New York, 1984.
18. R. Bancroft, *Understanding Electromagnetic Scattering Using the Moment Method*, Artech House, Boston, MA, 1996.
19. E. C. Jordan and K. G. Balmain, *Electromagnetic Waves and Radiating Systems*, Prentice-Hall, Englewood Cliffs, NJ, 1950, 1968.
20. R. Carrel, "Analysis and Design of the Log-periodic Dipole Antenna," Antenna Lab., University of Illinois Tech. Rep. No. 52, April 1961.
21. R. H. Kyle, "Mutual Coupling between Log-periodic Antennas," *IEEE Trans. Ant. & Prop.*, Vol. AP-18, pp. 15–22, Jan. 1970.
22. V. H. Rumsey, *Frequency Independent Antennas*, Academic Press, New York, 1966.
23. M. T. Ma, *Theory and Application of Antenna Arrays*, Wiley, New York, 1974.
24. N. Amitay, V. Galindo, and C. P. Wu, *Theory and Analysis of Phased Array Antennas*, Wiley, New York, 1972.
25. R. W. P. King and C. W. Harrison, Jr., "The Receiving Antenna," *Proc. IRE*, Vol. 32, pp. 18–49, Jan. 1944.
26. A. F. Stevenson, "Relations between the Transmitting and Receiving Properties of Antennas," *Quart. Appl. Math.*, pp. 369–384, Jan. 1948.
27. Y. Y. Hu, "Back-scattering Cross-section of a Center-loaded Antenna," *IRE Trans. Ant. & Prop.*, Vol. AP-6, No. 1, pp. 140–148, Jan. 1958.
28. J. Aharoni, *Antennae*, Oxford Univ. Press, 1946.
29. R. B. Green, "The General Theory of Antenna Scattering," Rep. No. 1223-17, ElectroScience Lab., Columbus, OH, Nov. 1963; Ph.D. dissertation, Dept. Elec. Engr., The Ohio State Univ., 1963.
30. R. E. Collin, "The Receiving Antenna," in *Antenna Theory*, Part 1, edited by R. E. Collin and F. J. Zucker, McGraw-Hill, New York, 1969.
31. R. C. Hansen, "Relationships between Antennas as Scatterers and as Radiators," *Proc. IEEE*, Vol. 77, pp. 659–662, May 1989.
32. G. A. Thiele and D. D. Richwine, "Antenna Mode and Residual Mode Scattering by a Dipole Antenna," paper presented at 1989 URSI EM Theory Symposium, Stockholm.
33. R. F. Harrington, "Electromagnetic Scattering by Antennas," *IEEE Trans. Ant. & Prop.*, Vol. AP-11, No. 5, pp. 595–596, Sept. 1963.
34. M. G. Andresen, "Airborne Jamming Antenna Study—Scattering and Absorption by a Receiving Antenna," Stanford Res. Institute, April 1960, DTIC 236342.
35. K. S. H. Lee, L. Marin, and J. P. Castillo, "Limitations of Wire-grid Modeling of a Closed Surface," *IEEE Trans. Electromag. Compatibility*, Vol. EMC-18, Aug. 1976.
36. J. H. Richmond, "A Wire-grid Model for Scattering by Conducting Bodies," *IEEE Trans. Ant. & Prop.*, Vol. AP-14, Nov. 1966.
37. Y. T. Lin and J. H. Richmond, "EM Modeling of Aircraft at Low Frequencies," *IEEE Trans. Ant. & Prop.*, Vol. AP-23, Jan. 1975.
38. M. G. Andresen, "Comments on Scattering by Conducting Rectangular Cylinders," *IEEE Trans. Ant. & Prop.*, Vol. AP-12, pp. 235–236, March 1964.
39. D. H. Werner, P. L. Werner, and J. K. Breakall, "Some Computational Aspects of Pocklington's Electric Field Integral Equation for Thin Wires," *IEEE Trans. Ant. & Prop.*, Vol. 42, No. 4, April 1994.
40. L. L. Tsai, "A Numerical Solution for the Near and Far Fields of an Annular Ring of Magnetic Current," *IEEE Trans. Ant. & Prop.*, Vol. AP-20, Sept. 1972.

PROBLEMS

10.3-1 a. Use the equivalence principle to show that the current flowing on the highly conducting wire in Fig. 10-3*a* may be replaced by the equivalent currents radiating in free space as in Fig. 10-3*b* (i.e., the wire material is replaced by free space with zero fields and zero sources inside the original wire volume) and that the equivalent currents (Fig. 10-3*b*) are the same as the currents in the original problem.

b. Why is this important?

10.3-2 Show that the left-hand side of (10-13) may be expressed as

$$-E_z^s = \frac{-1}{4\pi j\omega\varepsilon_o}\int_{-L/2}^{L/2} I(z')\,\frac{e^{-j\beta R}}{R^5}\,[(1 + j\beta R)(2R^2 - 3a^2) + \beta^2 a^2 R^2]\, dz'$$

10.3-3 Through integration by parts, show that the left-hand side of (10-13) may be written as

$$E_z^s = +\int_{-L/2}^{L/2} \left[j\omega\mu_o I(z') - \frac{1}{j\omega\varepsilon_o} \frac{\partial I(z')}{\partial z'} \frac{\partial}{\partial z} \right] \frac{e^{-j\beta R}}{4\pi R} dz'$$

This equation may be derived by using both the vector and scalar potentials [6].

10.4-1 a. What are the units of the generalized voltage, current, and impedance matrix elements in (10-26)?

b. If both sides of (10-26) are multiplied by the segment length, Δz, what are the units of the matrix elements in (10-26)?

10.4-2 Moment methods involve solving systems of linear equations. A computer technique for solving such problems is needed. This could be through a high level language (MathCAD, MatLab, Mathematica, . . .), a subroutine you write, or a canned routine such as IMSL.

a. For the problem

$$V = ZI \quad \text{where } Z = \begin{bmatrix} 1 & j \\ j & 1 \end{bmatrix}$$

obtain the solution by hand using $I = Z^{-1}V$ if

$$V = \begin{bmatrix} 2 + j \\ 1 + 2j \end{bmatrix}$$

b. Solve the problem in (a) using your chosen computer approach. Include details of the approach. Compare to your hand solution.

10.4-3 An important ingredient in moment methods is to be able to numerically integrate complex-valued functions. Select a computer approach and document it.

a. Integrate the following analytically

$$\int_0^{\pi} e^{jx}\, dx$$

b. Use your computer approach of choice to evaluate the integral in (a) and compare to results you obtained in (a).

10.5-1 In order to obtain some feeling for MoM, it is recommended that the student write a computer program to solve the following problem. Consider a straight dipole of length L (or monopole of length $L/2$) and radius a. Divide the dipole into N segments of equal length, each containing a pulse expansion function.

a. Use point-matching and the equation in Prob. 10.3-2 for the scattered field to compute the elements in the first row of the impedance matrix $[Z_{mn}]$ as given in (10-26), noting that these are the only independent matrix elements since the matrix is toeplitz (see Sec. 10.8.4). Note that the integrand tends toward singularity when $R = a$ [39], but even so one may numerically integrate through this region if reasonable care is taken.

b. Confirm the matrices in Example 10-1. Next, duplicate the curves in Fig. 10-9. (This exercise continues in Probs. 10.6-1 and 10.9-1.)

10.5-2 Starting with the electric vector potential and (10-30), derive (10-31) [40].

10.6-1 a. Extend the computer code of Prob. 10.5-1 to use pulse weighting functions in (10-40).

b. Confirm the matrices in Example 10-2 and duplicate the curves in Fig. 10-13. (This exercise continues in Prob. 10.9-1.)

10.7-1 Show that (10-47) and (10-48) follow from (10-44) and (10-45), respectively.

10.7-2 Compare (10-57) to (10-37).

10.7-3 From (10-59), derive (10-26) if delta weighting functions are used.

10.8-1 Use both vector and scalar potentials to derive (10-64).

10.8-2 Another equation for the treatment of wire antennas is Hallen's integral equation:

$$\int_{-L/2}^{L/2} I(z') \frac{e^{-j\beta R}}{4\pi R} dz' = -\frac{j}{\eta}(C_1 \cos \beta z + C_2 \sin \beta|z|)$$

where C_1 and C_2 are constants. The constant C_2 may be evaluated as $V_A/2$, where V_A is the terminal voltage of the antenna. Derive Hallen's equation for the dipole by writing a solution to the wave equation for A_z that is proportional to the right-hand side of the above equation and then equating this result to the integral form of the vector potential for A_z due to a perfectly conducting thin wire dipole.

10.8-3 In Secs. 10.4 and 10.6, we used pulse functions in the moment method. Expansion functions such as the pulse function, piecewise sinusoidal function, etc., are often called subdomain expansion functions because each expansion is generally nonzero on only a relatively small part of the radiating body. (The concept of domain relates to Sec. 10.7.2.)

There is another type of expansion function called entire-domain expansion functions. In this case, the function is generally nonzero over the entire radiating body and the concept of segments is not used. For example, if one were to treat the dipole with an entire-domain expansion function (i.e., a Fourier series), one could write for the current:

$$I(z') = \sum_{n=1}^{N} I_n F_n(z')$$

where

$$F_n(z') = \cos(2n-1)\frac{\pi z'}{L}, \qquad -\frac{L}{2} \le z' \le \frac{L}{2}$$

[Note that each term in $F_n(z')$ goes to zero at the ends of the dipole.]

a. Sketch the first two terms in the series for $F_n(z')$.

b. If there are N terms and N match points (i.e., a point-matching solution), write an expression for Z_{mn}, using the notation in Secs. 10.4 and 10.6.

c. Give a physical interpretation of Z_{25} (i.e., complete a statement similar to the following: Z_{25} represents the field from ______ at ______).

10.8-4 Verify (10-70), using the algorithm in (10-71).

10.9-1 a. After successfully completing Probs. 10.5-1 and 10.6-1, use (10-80) to compute the far-field radiation pattern of dipole antennas with lengths of 0.1, 0.5, 1.0, 1.25, and 1.5λ. Justify the value of N that you use in each case.

b. For the dipole lengths in part (a), plot the current distributions on the dipoles in magnitude and phase. Compare with the assumed sinusoidal distribution used in Sec. 5.1.

c. Using (10-81), compute the gain at broadside for the dipole lengths in part (a).

d. Consider a plane wave to be incident on a dipole short-circuited at its terminals. Use the relationship

$$E_z^i(z_m) = e^{j\beta z \cos \theta^i}$$

to compute $[V_m]$ for $\theta^i = 90°$ (i.e., the broadside case) and then compute the radar cross section, as in (10-82), when $L = \lambda/2$. Compare with Fig. 10-21 and verify several more points in Fig. 10-21.

10.9-2 Derive (10-80) by considering the dipole of length L to be comprised of N ideal collinear dipoles of length L/N.

10.9-3 Derive (10-81), starting with the Poynting theorem.

10.9-4 a. Using an available MoM code, compute, plot, and label the patterns (three planes) in polar, linear form and the current distribution for a one wavelength loop. Compare gain to that expected.

b. Find the input impedance for a loop of radius $a = 0.001$ wavelength for perimeter = 1, 1.5, and 2 wavelengths. Compare to Fig. 5-53.

10.10-1 Consider the LPDA in Fig. 10-25 to have only two dipoles. Assume each dipole is composed of three piecewise sinusoidal expansion functions numbered consecutively with the first three piecewise sinusoids on one dipole and the remaining three on the other. Thus, $[Z_{mn}]$ will be of order six, with the second and fifth piecewise sinusoids being at the centers of their respective dipoles. In accordance with (10-91) and (10-93), show that the elements of $[Y_{mn}]$ which form $[Y_A]$ are Y_{22}, Y_{25}, Y_{52}, Y_{55}.

10.10-2 a. Show that the admittance matrix for one section of the transmission line in Fig. 10-25 without dipoles attached is

$$[Y] = \begin{bmatrix} -jY_o \cot \beta d & +jY_o \csc \beta d \\ +jY_o \csc \beta d & -jY_o \cot \beta d \end{bmatrix}$$

where d is the length of one section of transmission line with propagation constant β and characteristic admittance Y_o.

b. Show that connecting N-1 of these sections using the scheme in Fig. 10-25 results in (10-94).

10.10-3 Extend the LPDA analysis in Section 10.10.2 to an array of M LPDA antennas as in [21].

10.10-4 In Sec. 10.10.2, we obtained a solution to the LPDA. One of the important points in that solution is the determination of $[Y_A]$ in the manner indicated in Eqs. (10-90) to (10-93). Had we wished to then find $[Z_A]$, we could have obtained it from $[Z_A] = [Y_A]^{-1}$. Denote this method A. Suppose instead we find $[Z_A]$ by considering the two dipole mutual impedance problem as Carrel [20] and Kyle [21] did. For example, $[Z_A]_{mn}$ is obtained by temporarily removing all dipoles except m and n from the system and then calculating $[Z_A]_{mn}$. Denote this method B.

a. Will $[Z_A]$ obtained by method A be the same as that obtained by method B? Why?

b. The following question refers to the concepts implied by part (a). When we calculate a moment method impedance matrix $[Z_{mn}]$, in what way does that calculation process relate to method B above?

10.11-1 Show that (10-100) is valid.

10.12-1 Assuming that Z_A in Fig. 10-31 can account for the power scattered by a dipole when $0 < L < 0.6\lambda$, and the load is 72 Ω purely resistive, find a relationship between R_A and R_L when:

a. More power is scattered (totally) by the dipole than is absorbed by the load.

b. The powers scattered and absorbed are equal.

c. The power absorbed by the load is greater than that scattered.

d. When the scattered and absorbed powers are equal at first resonance, what is the contribution by the antenna mode to the total RCS of the dipole?

10.12-2 Derive (10-111) in either of the following ways:

a. With a plane wave incident, define an absorption aperture of an antenna as $G\lambda^2/4\pi$, find the absorbed and scattered powers, and apply the definition of radar cross section.

b. Start with the antenna mode term in (10-109) and use the relationship that the maximum effective aperture for reradiation [28] is $A_{re} = (h_A^t)^2\eta/(4R_A)$, when $Z_A = Z_L$, and obtain an expression for the field scattered by the antenna mode, $E_{\text{ant}}^s = E^i A_{re}(-j/\lambda)\Gamma(e^{-j\beta r}/r)$, before applying the radar cross-section definition.

10.12-3 In [33] an expression for the RCS of small antennas, such as the dipole, is presented that is valid when the scattered field of the open-circuited antenna is small compared to that of the terminated antenna:

$$\sigma/\lambda^2 \approx (1/\pi)|GR_A/(Z_A + Z_L)|^2$$

a. Use this expression to compute the RCS at first resonance of the dipole used in Fig. 10-33.

b. Also use this expression to compute the RCS of the dipole at first antiresonance (i.e., about 600 MHz). Make an idealized assumption about the input impedance at first antiresonance. Compare your result to Fig. 10-33. Explain the difference.

10.12-4 A sheet of very thin conducting material (thickness, $t \lll \lambda$) has a plane wave normally incident upon it. The resistance R of the thin material is $1/\sigma t$ where σ is the conductivity of the material, and R is measured in ohms/square.

a. Find the optimum resistance of the thin material such that the sheet absorbs the maximum amount of power from the wave and that this amount is 50% of the power incident.

b. Comment on the similarity of the maximum power absorbed in (a) to that absorbed by a matched resonant antenna at first resonance (e.g., a $\lambda/2$ dipole).

c. Do the answers to (a) and (b) imply that any antenna may not absorb more power than it scatters [29, 34]?

10.13-1 Derive (10-114) from (10-112).

10.13-2 Sketch a wire-grid model for a square plate $1\lambda \times 1\lambda$. If pulse expansion functions are to be used, how many unknowns will your model have?

10.13-3 Sketch a wire-grid model for a quarter-wavelength monopole at the center of a circular ground plane of $\lambda/4$ radius. If pulse expansion functions are used, how many unknowns will your model have? If piecewise sinusoidal functions are used, how many unknowns will your model have?

10.13-4 Derive (10-120) from (10-117).

10.13-5 Derive (10-130) from (10-127).

10.13-6 Derive (10-139) from (10-130).

10.13-7 Write a computer code to solve (10-120). Verify Fig. 10-43.

10.13-8 Write a computer code to solve (10-131). Verify Fig. 10-44.

Chapter 11

CEM for Antennas: Finite Difference Time Domain Method

The computational approach of the previous chapter involved setting up and solving frequency domain integral equations for the phasor electric and/or magnetic currents induced on the surfaces of antennas or scatterers. From a computing perspective, this method of moments (MoM) procedure involves setting up and solving dense (i.e., few zero elements), complex-valued systems of linear equations. These systems can involve tens of thousands of equations in the treatment of problems of even moderate electrical size.

As powerful as MoM is, it is inadequate for some important engineering problems, particularly those involving pulsed excitations and various transient phenomena. These kinds of problems require data to be computed over a range of frequencies. This suggests the need for a solution technique in the time domain, since all of the required frequency domain data can be generated from one-time domain solution via Fourier transformation. There is an approach that directly solves Maxwell's curl equations at points on space grids in the time domain. It is the finite difference time domain (FD-TD) method (see Fig. 10-1). There are at least four reasons for the development of interest in such partial differential equation (PDE) solutions of Maxwell's equations:

1. PDE solutions are robust.
2. Time domain PDE methods usually have no matrices (frequency domain PDE methods usually have sparse matrices).
3. Complex-valued material properties are readily accommodated.
4. Computer resources are adequate to allow widespread usage of PDE methods.

The finite difference time domain (FD-TD) technique to be discussed in the following sections offers many advantages as an electromagnetic modeling, simulation, and analysis tool. Its capabilities include:

- Broadband response predictions with a single excitation
- Arbitrary three-dimensional (3-D) model geometries

- Interaction with an object of any conductivity from that of a perfect conductor, to that of low or zero conductivity
- Frequency-dependent constitutive parameters for modeling most materials:
  - Lossy dielectrics
  - Magnetic materials
  - Unconventional materials that can be anisotropic and/or nonlinear
- Any type of response such as:
  - Scattered fields
  - Antenna patterns
  - Radar cross section (RCS)
  - Surface response fields
  - Currents, power densities, charge distributions
  - Penetration/interior coupling

The basis of the FD-TD algorithm is the two Maxwell curl equations in derivative form in the time domain. These equations are expressed in linearized form by means of central finite differencing. Only nearest-neighbor interactions need be considered as the fields are advanced temporally in discrete time steps over spatial cells usually of rectangular shape as indicated in Fig. 11-1 (other cell shapes are possible, as well as two-dimensional and one-dimensional treatments).

Although FD-TD is well suited to computing responses to a continuous wave or single-frequency excitation, it is particularly well suited to computing transient responses. This is especially the case when complex geometries or difficult environments, such as an antenna that is buried in the earth or dielectrically clad, are considered. Also, interior coupling into metallic enclosures is a situation where FD-TD is a method of choice.

For problems where the modeled region must extend to infinity, absorbing boundary conditions (ABCs) are employed at the outer-grid truncation planes (grid boundary), which ideally permit all outgoing numerical waves to exit the region with negligible reflection at the grid truncation. Phenomena such as the induction of surface currents, scattering and multiple scattering, aperture penetration and cavity excitation are modeled time step by time step by the action of the numerical analog to the curl equations. The self-consistency of these modeled phenomena is

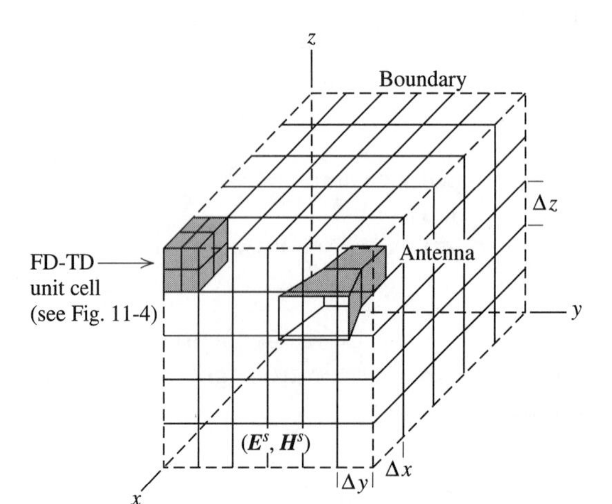

Figure 11-1 Embedding of an antenna structure in an FD-TD space lattice.

generally assured if their spatial and temporal variations are well resolved by the space and time sampling process. A self-consistent model will account for the mutual coupling of all the electrically small-volume cells constituting the structure and its near field, even if the structure spans tens of wavelengths in three dimensions and there are hundreds of millions of space cells. In contrast, MoM provides a self-consistent solution of Maxwell's equations, which includes all mutual coupling, by solving a system of simultaneous equations.

In the remainder of this chapter, we develop FD-TD for application to two classes of problems: antennas and scattering, with emphasis on the former. The theoretical development is for the three-dimensional objects in or near isotropic media, but examples are mostly specialized to two-dimensional and one-dimensional problems. The material that follows is intended as an introduction to FD-TD. For more extensive information, the reader is referred to larger works on the subject [1, 2]. The presentation here is influenced by [1–3], particularly [1] and [3].

In the following section, we examine the form of Maxwell's equations we need to solve in one, two, and three dimensions before developing the rectangular coordinate system finite difference representations for those equations in Sec. 11.2. The finite difference equations are the equations used in the Yee algorithm of the FD-TD technique and are subject to the constraints discussed in Sec. 11.3. Implementation of the finite difference equations is discussed in Sec. 11.4. Discussion of the absorbing boundary conditions follows in Sec. 11.5. Some sources used in FD-TD are presented in Sec. 11.6, whereas transformation from the near field to the far field is contained in Sec. 11.7. Sections 11.8 and 11.9 provide two-dimensional and three-dimensional examples, respectively.

11.1 MAXWELL'S EQUATIONS FOR THE FD-TD METHOD

Before developing the basis for the FD-TD method in the next section, we need to examine Maxwell's time domain curl equations in one, two, and three dimensions and put them in a form convenient for the FD-TD method. Consider a region of space that is source-free but may have lossy magnetic and/or lossy electric materials that convert energy in the electromagnetic field to heat. We define an equivalent magnetic current density $\mathbf{J_M}$ to account for the magnetic loss mechanisms:

$$\mathbf{J_M} = \rho'\mathbf{H} \tag{11-1}$$

and an equivalent electric current density $\mathbf{J}$ to account for the electric loss mechanisms:

$$\mathbf{J} = \sigma\mathbf{E} \tag{11-2}$$

Here, ρ' is an equivalent magnetic resistivity in ohms per meter and σ is the electric conductivity in siemens per meter. Thus, we can write

$$\frac{\partial \mathbf{H}}{\partial t} = -\frac{1}{\mu}\nabla \times \mathbf{E} - \frac{\rho'}{\mu}\mathbf{H} \tag{11-3}$$

$$\frac{\partial \mathbf{E}}{\partial t} = \frac{1}{\varepsilon}\nabla \times \mathbf{H} - \frac{\sigma}{\varepsilon}\mathbf{E} \tag{11-4}$$

11.1.1 Three-Dimensional Formulation

Writing out the vector components of the two curl equations above yields the following system of six coupled scalar equations in the three-dimensional rectangular coordinate system[1]:

$$\frac{\partial H_x}{\partial t} = \frac{1}{\mu}\left(\frac{\partial E_y}{\partial z} - \frac{\partial E_z}{\partial y} - \rho' H_x\right) \tag{11-5a}$$

$$\frac{\partial H_y}{\partial t} = \frac{1}{\mu}\left(\frac{\partial E_z}{\partial x} - \frac{\partial E_x}{\partial z} - \rho' H_y\right) \tag{11-5b}$$

$$\frac{\partial H_z}{\partial t} = \frac{1}{\mu}\left(\frac{\partial E_x}{\partial y} - \frac{\partial E_y}{\partial x} - \rho' H_z\right) \tag{11-5c}$$

$$\frac{\partial E_x}{\partial t} = \frac{1}{\varepsilon}\left(\frac{\partial H_z}{\partial y} - \frac{\partial H_y}{\partial z} - \sigma E_x\right) \tag{11-6a}$$

$$\frac{\partial E_y}{\partial t} = \frac{1}{\varepsilon}\left(\frac{\partial H_x}{\partial_z} - \frac{\partial H_z}{\partial x} - \sigma E_y\right) \tag{11-6b}$$

$$\frac{\partial E_z}{\partial t} = \frac{1}{\varepsilon}\left(\frac{\partial H_y}{\partial x} - \frac{\partial H_x}{\partial y} - \sigma E_z\right) \tag{11-6c}$$

This system of six coupled partial differential equations forms the basis of the FD-TD numerical algorithm to be developed in the next section. Before proceeding with the full three-dimensional FD-TD algorithm, it is useful to consider reductions to the two-dimensional and one-dimensional cases, which can yield engineering information without the computational effort required for the three-dimensional case.

11.1.2 Two-Dimensional Formulation

In the two-dimensional problem, there is no variation with respect to one of the coordinates in either the problem geometry or excitation. Here, we assume no variation with respect to z, which means that all partial derivatives of the fields with respect to z equal zero, and that the structure being modeled extends to infinity in the z-direction with no change in its geometry.

Consider grouping the previous six equations, with all partial derivatives with respect to z equal to zero, into two sets, one of which only involves magnetic field components transverse to the problem geometry axis (i.e., the z-axis) and the other in which there are only electric field components transverse to the z-axis. The first set is called the two-dimensional transverse magnetic (TM) mode and is

$$\frac{\partial H_x}{\partial t} = \frac{1}{\mu}\left(-\frac{\partial E_z}{\partial y} - \rho' H_x\right) \tag{11-7a}$$

$$\frac{\partial H_y}{\partial t} = \frac{1}{\mu}\left(\frac{\partial E_z}{\partial x} - \rho' H_y\right) \qquad \textit{two-dimensional TM mode} \tag{11-7b}$$

$$\frac{\partial E_z}{\partial t} = \frac{1}{\varepsilon}\left(\frac{\partial H_y}{\partial x} - \frac{\partial H_x}{\partial y} - \sigma E_z\right) \tag{11-7c}$$

[1]For clarity and notational convenience, script will not be used in this chapter for the time-varying field quantities as in Chap. 1.

The second set is then the two-dimensional transverse electric (TE) mode and is

$$\frac{\partial E_x}{\partial t} = \frac{1}{\varepsilon}\left(\frac{\partial H_z}{\partial y} - \sigma E_x\right) \tag{11-8a}$$

$$\frac{\partial E_y}{\partial t} = \frac{1}{\varepsilon}\left(-\frac{\partial H_z}{\partial x} - \sigma E_y\right) \qquad \textit{two-dimensional TE mode} \tag{11-8b}$$

$$\frac{\partial H_z}{\partial t} = \frac{1}{\mu}\left(\frac{\partial E_x}{\partial y} - \frac{\partial E_y}{\partial x} - \rho' H_z\right) \tag{11-8c}$$

We observe that the TM and TE modes are decoupled, that is, they contain no common field vector components. In fact, these modes are completely independent for structures comprised of isotropic materials. That is, the modes can exist simultaneously with no mutual interactions. Problems having both TM and TE excitation can be solved as a superposition of these two separate problems.

Physical phenomena associated with the TM and TE cases can be quite different. To see this, one can look at the currents on the circular cylinder in Sec. 10.13.2. In the TM case, the current goes smoothly to nearly zero on the deep shadowed side of the cylinder, whereas in the TE case, the current propagates much more readily into the shadowed region.

11.1.3 One-Dimensional Formulation

Next, assume that there is no variation with respect to two coordinates in either the problem geometry or the excitation. In this instance, assume no variation with respect to either y or z, which means that all partial derivatives with respect to either y or z equal zero. This implies that the problem is one-dimensional in nature with propagation in the x-direction, but with space infinite in the y- and z-directions. Thus, while propagating in the x-direction, a wave could encounter infinite sheets of material having thickness in the x-dimension.

The one-dimensional problem is formulated by reducing either the two-dimensional TM mode or two-dimensional TE mode and ultimately obtaining almost the same result. Reducing the two-dimensional TM mode gives:

$$\frac{\partial H_x}{\partial t} = \frac{1}{\mu}(-\rho' H_x) \tag{11-9a}$$

$$\frac{\partial H_y}{\partial t} = \frac{1}{\mu}\left(\frac{\partial E_z}{\partial x} - \rho' H_y\right) \tag{11-9b}$$

$$\frac{\partial E_z}{\partial t} = \frac{1}{\varepsilon}\left(\frac{\partial H_y}{\partial x} - \sigma E_z\right) \tag{11-9c}$$

The first of these three equations can be shown to vanish by reasoning that if the fields are all zero prior to some time, say, $t = 0$, when a source is turned on, then the time derivative of H_x is zero. This, in turn, implies that H_x remains at zero. We now have a set of just two equations involving only H_y and E_z. Designate this set the TM mode in one dimension:

$$\frac{\partial H_y}{\partial t} = \frac{1}{\mu}\left(\frac{\partial E_z}{\partial x} - \rho' H_y\right) \tag{11-10a}$$

one-dimensional TM mode

$$\frac{\partial E_z}{\partial t} = \frac{1}{\varepsilon}\left(\frac{\partial H_y}{\partial x} - \sigma E_z\right) \tag{11-10b}$$

In a similar way, we can reduce the two-dimensional TE mode to a set of two equations involving only E_y and H_z. Designate this set the TE mode in one dimension:

$$\frac{\partial E_y}{\partial t} = \frac{1}{\varepsilon}\left(-\frac{\partial H_z}{\partial x} - \sigma E_y\right) \tag{11-11a}$$

one-dimensional TE mode

$$\frac{\partial H_z}{\partial t} = \frac{1}{\mu}\left(-\frac{\partial E_y}{\partial x} - \rho' H_z\right) \tag{11-11b}$$

The only practical difference between the one-dimensional TM and TE modes is that they represent plane waves of orthogonal polarizations. This renders the TM and TE labels in the one-dimensional case uncommon since we would ordinarily identify the plane wave polarization in some other way.

From either one-dimensional set, we can easily derive the one-dimensional scalar wave equation for a component of E and that for a component of H, both of which only have for solutions plane waves traveling in the $\pm x$-direction at a speed given by $1/\sqrt{\mu\varepsilon}$. That is, in the one-dimensional case we have transverse electromagnetic (TEM) plane waves traveling at a speed determined by the constitutive parameters of the medium.

In the next section, we examine numerical solutions to the one-dimensional, two-dimensional, and three-dimensional equations developed here.

11.2 FINITE DIFFERENCES AND THE YEE ALGORITHM

In this section, we develop the Yee algorithm used in the FD-TD method. The Yee algorithm is based on finite difference approximations of the space derivatives and time derivatives in Maxwell's curl equations as shown later in this section. To begin our development, consider (11-10a) in the lossless case:

$$\frac{\partial H_y}{\partial t} = \frac{1}{\mu}\frac{\partial E_z}{\partial x} \tag{11-12}$$

Employing the classical definition of a derivative, we can write

$$\lim_{\Delta t \to 0}\frac{\Delta H_y}{\Delta t} = \frac{1}{\mu}\lim_{\Delta x \to 0}\frac{\Delta E_z}{\Delta x} \tag{11-13}$$

In Fig. 11-2 we illustrate (11-13) and note that in the limit a continuous and exact solution to (11-13) is obtained at the point (x, t). It is important to note that at this

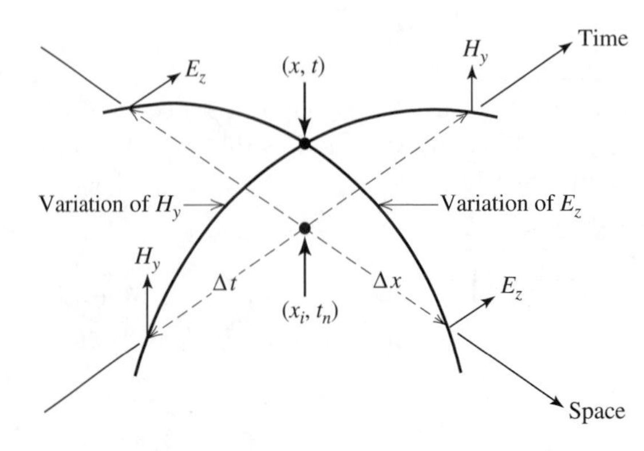

Figure 11-2 Space-time graphical interpretation of a one-dimensional component of Maxwell's equations and its discretization. (Reprinted with permission of Eric Thiele.)

point, space and time derivatives are being equated, and not the actual values of the fields. In other words, it is apparent that Maxwell's equations do not directly yield electric and magnetic field values, but rather relate the rate of change between electric and magnetic field values.

Thus the following strategy may suggest itself. Discretize space and time around the point (x, t) in such a way that Maxwell's equations hold true. That is, apply central differences to relate the derivatives of the neighboring discrete fields. Then for example, (11-13) can be expressed as

$$\left.\frac{H_y\left(t_n + \frac{\Delta t}{2}\right) - H_y\left(t_n - \frac{\Delta t}{2}\right)}{\Delta t}\right|_{x_i} = \frac{1}{\mu}\left.\frac{E_z\left(x_i + \frac{\Delta x}{2}\right) - E_z\left(x_i - \frac{\Delta x}{2}\right)}{\Delta x}\right|_{t_n} \tag{11-14}$$

This gives us the relationship between the derivatives at (x_i, t_n) which closely approximates the relationship at (x, t). However, as we shall see later, if (11-14) is solved for the field quantity at the most advanced point in time (i.e., $t_n + \Delta t/2$), then an estimate of the magnetic field value at the spatial point x_i at time $(t_n + \Delta t/2)$ can be obtained.

We could obtain (11-14) from (11-12) in a more formal way by expanding $H_y(x_i, t_n)$ in a Taylor series about the temporal point t_n to temporal point $t_n + \dfrac{\Delta t}{2}$, keeping space fixed at the point x_i. This yields an expansion for $\left.H_y\left(t_n + \dfrac{\Delta t}{2}\right)\right|_{x_i}$. Similarly, we could obtain an expansion for $\left.H_y\left(t_n - \dfrac{\Delta t}{2}\right)\right|_{x_i}$. Taking the difference would give the left-hand side of (11-14) plus remainder terms on the order of $(\Delta t)^2$. Likewise, expansions of E_z about x_i in both directions with time fixed lead to the right-hand side of (11-14) plus remainder terms on the order of $(\Delta x)^2$. In this way, we formally obtain second-order-accurate central difference approximations to the first partial derivatives in time and space.

Continuing our FD-TD solution to (11-12), we solve (11-14) for $H_y\left(t_n + \dfrac{\Delta t}{2}\right)$ and obtain [after dropping the z and y subscripts on the field components and taking (11-14) to be an equality]

$$\left.H\left(t_n + \frac{\Delta t}{2}\right)\right|_{x_i} = \left.H\left(t_n - \frac{\Delta t}{2}\right)\right|_{x_i} + \frac{\Delta t}{\mu\,\Delta x}\left[E\left(x_i + \frac{\Delta x}{2}\right) - E\left(x_i - \frac{\Delta x}{2}\right)\right]_{t_n} \tag{11-15}$$

For convenience, we notationally adopt a subscript i for the space position and a superscript n for the time observation point. If we use this shorthand notation, (11-15) can be written compactly as

$$H_i^{n+1/2} = H_i^{n-1/2} + \frac{\Delta t}{\mu \Delta x}\left[E_{i+1/2}^n - E_{i-1/2}^n\right] \tag{11-16}$$

which implies that we can solve for $H_i^{n+1/2}$ knowing the value for H at the same spatial point but at Δt earlier in time and knowing E at spatial points $\pm\Delta x/2$ removed from x_i and $\Delta t/2$ earlier in time. This is illustrated in Fig. 11-3 that shows on a time-space diagram the three quantities linked (by the various dashed lines) to the calculation of $H_i^{n+1/2}$.

How do we obtain E at spatial points $x_i \pm \Delta x/2$? The answer, of course, is to start

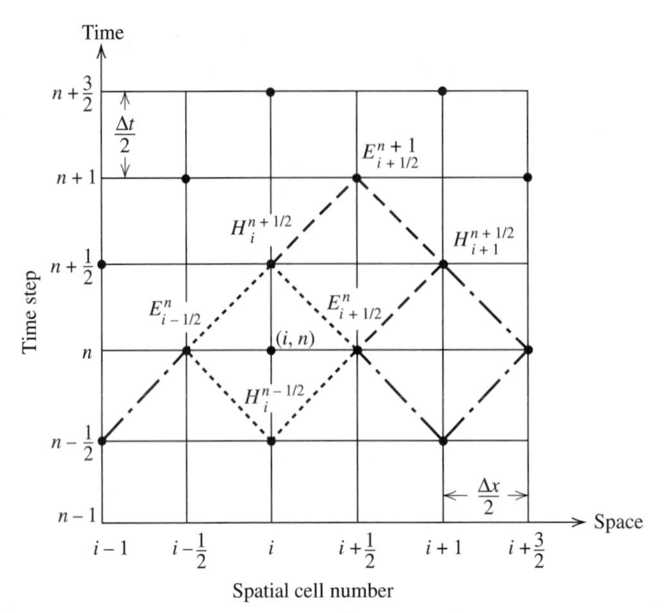

Figure 11-3 Calculation of $H_i^{n+1/2}$ and $E_{i+1/2}^{n+1}$ from its nearest neighbors in space and time [3].

with the lossless version of (11-10b) and obtain for E_z the central difference approximation

$$E_{i+1/2}^{n+1} = E_{i+1/2}^{n} + \frac{\Delta t}{\varepsilon\,\Delta x}\left[H|_{i+1}^{n+1/2} - H|_{i}^{n+1/2}\right] \tag{11-17}$$

by differencing about the temporal point $(n + 1/2)$ and the spatial point $(i + 1/2)$. This equation says that $E_{i+1/2}^{n+1}$ can be calculated with values of H and E at previous instants of time at adjacent spatial locations as Fig. 11-3 suggests. Clearly, we have the basis for a method that can march a field behavior forward in space and time through the use of difference equations like (11-16) and (11-17), more commonly called *update equations* (because they update the fields in the cells as time moves forward).

In 1966, K. S. Yee [4] originated a set of finite difference equations for the lossless three-dimensional time-dependent Maxwell's curl equations of (11-5) and (11-6) similar to that above for the lossless one-dimensional case. Yee's algorithm, introduced later in this section for the three-dimensional case, is one of great usefulness since its fundamental basis is so robust. The Yee algorithm is robust for the following reasons. First, it solves for *both* electric and magnetic fields in time and space using the coupled Maxwell's curl equations rather than solving for the electric field alone (or the magnetic field alone) as with the wave equation.

Second, the Yee algorithm interleaves its E- and H-field vector components in three-dimensional space (see Fig. 11-4), so that every E-field vector component is surrounded by H-field components, and every H-field vector component is surrounded by E-field components as suggested by Fig. 11-4. The spatial arrangement in Fig. 11-4 is not arbitrary since it must be consistent with the laws of Ampere and Faraday [1].

Third, the Yee algorithm centers its E- and H-field vector components in time in

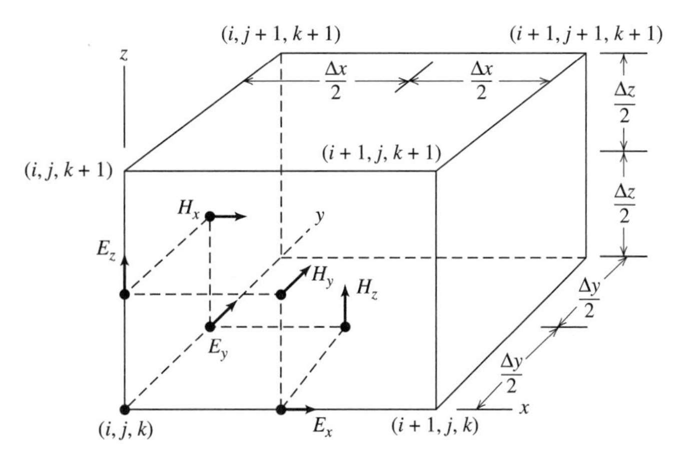

Figure 11-4 Position of the electric and magnetic field vector components in a cubic unit cell of the Yee space lattice of dimension Δx by Δy by Δz [3].

what is commonly termed a *leapfrog* arrangement. All the E-field computations in the three-dimensional space of interest are computed for a particular time point using the most recently computed H-field data stored in the computer memory (as Fig. 11-3 suggests for the one-dimensional case). Then, all the H-field computations in the three-dimensional space are computed using the E-field data just computed and stored in memory. This leapfrog arrangement is then repeated with the recomputation of the E-field based on the newly obtained H-fields. This process continues for a finite number of time steps until some desired late time response is achieved (e.g., steady state).

Fourth, no matrices are involved and no large systems of simultaneous equations need to be solved as in the method of moments.

Equation (11-17) above involved two variables: one in space and one in time. In the most general case, we have four degrees of freedom (three in space and one in time), and must choose notation carefully. Extending the notation from (11-17), denote a space point in a uniform, rectangular three-dimensional lattice as

$$(i, j, k) = (i\Delta x, j\Delta y, k\Delta z) \tag{11-18}$$

Here, Δx, Δy, and Δz are, respectively, the lattice space increments in the x-, y-, and z-coordinate directions, and i, j, and k are integers. Further, we denote any field component u as a function of space and time evaluated at a discrete point in the space lattice and at a discrete point in time as

$$u(i\Delta x, j\Delta y, k\Delta z, n\Delta t) = u^n_{i,j,k} \tag{11-19}$$

Here, Δt is the time increment, assumed uniform over the observation interval, and n is an integer. Carrying this notation to derivatives, we find, for example, Yee's expression for the first space derivative of u in the x-direction, evaluated at the fixed time $t_n = n\,\Delta t$, to be

$$\frac{\partial}{\partial x} u(i\Delta x, j\Delta y, k\Delta z, n\Delta t) = \frac{u^n_{i+1/2,j,k} - u^n_{i-1/2,j,k}}{\Delta x} + O[(\Delta x)^2] \tag{11-20a}$$

We note that the $\pm\frac{1}{2}$ increment in the i subscript (x-coordinate) of u denotes a space finite difference over $\pm\Delta x/2$ as in (11-17). The remainder term $O[(\Delta x)^2]$ is a result of the Taylor series expansions that formally lead to the second-order-accurate finite difference representation of the derivatives. The numerical approximation analogous to (11-20a) for $\partial u/\partial y$ or $\partial u/\partial z$ can be written simply by incrementing the j or k subscript of u by $\pm\frac{1}{2}\,\Delta y$ or $\pm\frac{1}{2}\,\Delta z$, respectively.

Yee's expression for the first time derivative of u, evaluated at the fixed space point (i, j, k), follows by analogy:

$$\frac{\partial}{\partial t}\,u(i\Delta x,\, j\Delta y,\, k\Delta z,\, n\Delta t) = \frac{u_{i,j,k}^{n+1/2} - u_{i,j,k}^{n-1/2}}{\Delta t} + O[(\Delta t)^2] \tag{11-20b}$$

Here, the $\pm\frac{1}{2}$ increment in the n superscript (time value) of u denotes a time finite difference over $\pm\Delta t/2$ as in the left-hand side of (11-14).

We now apply the above ideas and notation to achieve a numerical approximation of Maxwell's curl equations in three dimensions, given by (11-16) and (11-17). For example, consider (11-5a), repeated here for convenience:

$$\frac{\partial H_x}{\partial t} = \frac{1}{\mu}\left(\frac{\partial E_y}{\partial z} - \frac{\partial E_z}{\partial y} - \rho' H_x\right) \tag{11-21}$$

Substituting for the time and space derivatives at time step n and assuming the space lattice point (i, j, k), we have initially[2]

$$\frac{H_x\big|_{i,j,k}^{n+1/2} - H_x\big|_{i,j,k}^{n-1/2}}{\Delta t} = \frac{1}{\mu_{i,j,k}}\left(\begin{array}{c} \dfrac{E_y\big|_{i,j,k+1/2}^{n} - E_y\big|_{i,j,k-1/2}^{n}}{\Delta z} \\ -\dfrac{E_z\big|_{i,j+1/2,k}^{n} - E_z\big|_{i,j-1/2,k}^{n}}{\Delta y} \\ -\rho'_{i,j,k}\cdot H_x\big|_{i,j,k}^{n} \end{array}\right) \tag{11-22}$$

Note that all field quantities on the right-hand side (in a compact format) are evaluated at time step n, including the magnetic field term H_x, appearing due to the magnetic loss ρ'. Since H_x at time step n is not assumed to be stored in computer memory (only the previous values of H_x at time step $n - \frac{1}{2}$ are assumed to be in memory), we need some way to estimate this term. A very good approach is to apply a "semi-implicit" approximation to the H_x term on the right-hand side:

$$H_x\big|_{i,j,k}^{n} = \frac{H_x\big|_{i,j,k}^{n+1/2} + H_x\big|_{i,j,k}^{n-1/2}}{2} \tag{11-23}$$

H_x at time step n is assumed to be simply the arithmetic average of the stored value of H_x at the time step $n - \frac{1}{2}$ and the yet to be computed new value of H_x at time step $n + \frac{1}{2}$. Substituting into (11-22) after multiplying both sides by Δt, we obtain

$$H_x\big|_{i,j,k}^{n+1/2} - H_x\big|_{i,j,k}^{n-1/2} = \frac{\Delta t}{\mu_{i,j,k}}\left(\begin{array}{c} \dfrac{E_y\big|_{i,j,k+1/2}^{n} - E_y\big|_{i,j,k-1/2}^{n}}{\Delta z} \\ -\dfrac{E_z\big|_{i,j+1/2,k}^{n} - E_z\big|_{i,j-1/2,k}^{n}}{\Delta y} \\ -\rho'_{i,j,k}\cdot\left(\dfrac{H_x\big|_{i,j,k}^{n+1/2} + H_x\big|_{i,j,k}^{n-1/2}}{2}\right) \end{array}\right) \tag{11-24}$$

[2]For ease of presentation in some of the equations that follow, the terms inside the parentheses are "stacked" vertically rather than written horizontally.

We note that $H_x|_{i,j,k}^{n+1/2}$ and $H_x|_{i,j,k}^{n-1/2}$ appear on both sides of (11-24). Collecting all terms of these two types and isolating $H_x|_{i,j,k}^{n+1/2}$ on the left-hand side yield

$$\left(1+\frac{\Delta t}{\mu_{i,j,k}}\cdot\frac{\rho'_{i,j,k}}{2}\right)\cdot H_x|_{i,j,k}^{n+1/2}=\left(1-\frac{\Delta t}{\mu_{i,j,k}}\cdot\frac{\rho'_{i,j,k}}{2}\right)\cdot H_x|_{i,j,k}^{n-1/2}+\frac{\Delta t}{\mu_{i,j,k}}\left(\frac{E_y|_{i,j,k+1/2}^{n}-E_y|_{i,j,k-1/2}^{n}}{\Delta z}-\frac{E_z|_{i,j+1/2,k}^{n}-E_z|_{i,j-1/2,k}^{n}}{\Delta y}\right)\tag{11-25}$$

Finally, dividing both sides by $\left(1+\frac{\Delta t}{\mu_{i,j,k}}\cdot\frac{\rho'_{i,j,k}}{2}\right)$ yields an explicit expression for $H_x|_{i,j,k}^{n+1/2}$:

$$H_x|_{i,j,k}^{n+1/2}=\left(\frac{1-\frac{\rho'_{i,j,k}\,\Delta t}{2\mu_{i,j,k}}}{1+\frac{\rho'_{i,j,k}\,\Delta t}{2\mu_{i,j,k}}}\right)\cdot H_x|_{i,j,k}^{n-1/2}+\left(\frac{\frac{\Delta t}{\mu_{i,j,k}}}{1+\frac{\rho'_{i,j,k}\,\Delta t}{2\mu_{i,j,k}}}\right)\cdot\left(\frac{E_y|_{i,j,k+1/2}^{n}-E_y|_{i,j,k-1/2}^{n}}{\Delta z}-\frac{E_z|_{i,j+1/2,k}^{n}-E_z|_{i,j-1/2,k}^{n}}{\Delta y}\right)\tag{11-26}$$

In a similar manner, we can derive finite difference expressions based on Yee's algorithm for the H_y and H_z field components in the curl equations.

By analogy, we can derive finite difference expressions based on Yee's algorithm for the E_x, E_y, and E_z field components given by (11-6). Here, $\sigma E^{n+1/2}$ represents the loss term on the right-hand side of each equation that is estimated using a semi-implicit procedure similar to that of (11-23). This results in a set of three equations for E_x, E_y, and E_z. For example, the result for E_z, also at space lattice point (i, j, k), is

$$E_z|_{i,j,k}^{n+1}=\left(\frac{1-\frac{\sigma_{i,j,k}\,\Delta t}{2\varepsilon_{i,j,k}}}{1+\frac{\sigma_{i,j,k}\,\Delta t}{2\varepsilon_{i,j,k}}}\right)\cdot E_z|_{i,j,k}^{n}+\left(\frac{\frac{\Delta t}{\varepsilon_{i,j,k}}}{1+\frac{\sigma_{i,j,k}\,\Delta t}{2\sigma_{i,j,k}}}\right)\cdot\left(\frac{H_y|_{i+1/2,j,k}^{n+1/2}-H_y|_{i-1/2,j,k}^{n+1/2}}{\Delta x}-\frac{H_x|_{i,j+1/2,k}^{n+1/2}-H_x|_{i,j-1/2,k}^{n+1/2}}{\Delta y}\right)\tag{11-27}$$

With the above expressions for $H^{n+1/2}$ and E^{n+1}, the new value of a field vector component at any space lattice point depends only on its previous value and the previous values of the components of the other field vectors at adjacent points.

To implement a solution like (11-26) and (11-27) for a region having a continuous variation of isotropic material properties with spatial position, it is desirable to define and store the following constant coefficients for each field vector component before time stepping is commenced. For a cubic lattice where $\Delta x = \Delta y = \Delta z = \Delta s$, we have for the electric field algorithm coefficients at point (i, j, k)

$$C_a\big|_{i,j,k} = \frac{1 - \dfrac{\sigma_{i,j,k}\,\Delta t}{2\varepsilon_{i,j,k}}}{1 + \dfrac{\sigma_{i,j,k}\,\Delta t}{2\varepsilon_{i,j,k}}} \tag{11-28a}$$

$$C_b\big|_{i,j,k} = \frac{\dfrac{\Delta t}{\varepsilon_{i,j,k}\,\Delta s}}{1 + \dfrac{\sigma_{i,j,k}\,\Delta t}{2\varepsilon_{i,j,k}}} \tag{11-28b}$$

And for the magnetic field algorithm coefficients at point (i, j, k), we have

$$D_a\big|_{i,j,k} = \frac{1 - \dfrac{\rho'_{i,j,k}\,\Delta t}{2\mu_{i,j,k}}}{1 + \dfrac{\rho'_{i,j,k}\,\Delta t}{2\mu_{i,j,k}}} \tag{11-29a}$$

$$D_b\big|_{i,j,k} = \frac{\dfrac{\Delta t}{\mu_{i,j,k}\,\Delta s}}{1 + \dfrac{\rho'_{i,j,k}\,\Delta t}{2\mu_{i,j,k}}} \tag{11-29b}$$

Note that the lattice increment Δs is contained in C_b and D_b.

The complete set of finite difference equations suggested by (11-26) and (11-27) can now be written to conform to the spatial arrangement in Fig. 11-4 by adjusting the spatial indices appropriately. For example, to the spatial indices in (11-26), we add $\frac{1}{2}$ to both j and k to obtain the following equation for H_x and in (11-27) add $\frac{1}{2}$ to the index k to obtain the following equation for E_z. Thus, the complete set of six equations can be written as

$$\begin{aligned} H_x\big|^{n+1/2}_{i,j+1/2,k+1/2} = {} & D_{a_{HX}}\big|_{i,j+1/2,k+1/2} \cdot H_x\big|^{n-1/2}_{i,j+1/2,k+1/2} + D_{b_{HX}}\big|_{i,j+1/2,k+1/2} \\ & \cdot \left(E_y\big|^{n}_{i,j+1/2,k+1} - E_y\big|^{n}_{i,j+1/2,k} + E_z\big|^{n}_{i,j,k+1/2} - E_z\big|^{n}_{i,j+1,k+1/2}\right) \end{aligned} \tag{11-30a}$$

$$\begin{aligned} H_y\big|^{n+1/2}_{i+1/2,j,k+1/2} = {} & D_{a_{HY}}\big|_{i+1/2,j,k+1/2} \cdot H_y\big|^{n-1/2}_{i+1/2,j,k+1/2} + D_{b_{HY}}\big|_{i+1/2,j,k+1/2} \\ & \cdot \left(E_z\big|^{n}_{i+1,j,k+1/2} - E_z\big|^{n}_{i,j,k+1/2} + E_x\big|^{n}_{i+1/2,j,k} - E_x\big|^{n}_{i+1/2,j,k+1}\right) \end{aligned} \tag{11-30b}$$

$$\begin{aligned} H_z\big|^{n+1/2}_{i+1/2,j+1/2,k} = {} & D_{a_{HZ}}\big|_{i+1/2,j+1/2,k} \cdot H_z\big|^{n-1/2}_{i+1/2,j+1/2,k} + D_{b_{HZ}}\big|_{i+1/2,j+1/2,k} \\ & \cdot \left(E_x\big|^{n}_{i+1/2,j+1,k} - E_x\big|^{n}_{i+1/2,j,k} + E_y\big|^{n}_{i,j+1/2,k} - E_y\big|^{n}_{i+1,j+1/2,k}\right) \end{aligned} \tag{11-30c}$$

$$\begin{aligned} E_x\big|^{n+1}_{i+1/2,j,k} = {} & C_{a_{EX}}\big|_{i+1/2,j,k} \cdot E_z\big|^{n}_{i+1/2,j,k} + C_{b_{EX}}\big|_{i+1/2,j,k} \\ & \cdot \left(H_z\big|^{n+1/2}_{i+1/2,j+1/2,k} - H_z\big|^{n+1/2}_{i+1/2,j-1/2,k} + H_y\big|^{n+1/2}_{i+1/2,j,k-1/2} - H_y\big|^{n+1/2}_{i+1/2,j,k+1/2}\right) \end{aligned} \tag{11-31a}$$

$$\begin{aligned} E_y\big|^{n+1}_{i,j+1/2,k} = {} & C_{a_{EY}}\big|_{i,j+1/2,k} \cdot E_y\big|^{n}_{i,j+1/2,k} + C_{b_{EY}}\big|_{i,j+1/2,k} \\ & \cdot \left(H_x\big|^{n+1/2}_{i,j+1/2,k+1/2} - H_x\big|^{n+1/2}_{i,j+1/2,k-1/2} + H_z\big|^{n+1/2}_{i-1/2,j+1/2,k} - H_z\big|^{n+1/2}_{i+1/2,j+1/2,k}\right) \end{aligned} \tag{11-31b}$$

$$E_z\big|_{i,j,k+1/2}^{n+1} = C_{a_{EZ}}\big|_{i,j,k+1/2} \cdot E_z\big|_{i,j,k+1/2}^{n} + C_{b_{EZ}}\big|_{i,j,k+1/2} \cdot \left(H_y\big|_{i+1/2,j,k+1/2}^{n+1/2} - H_y\big|_{i-1/2,j,k+1/2}^{n+1/2} + H_x\big|_{i,j-1/2,k+1/2}^{n+1/2} - H_z\big|_{i,j+1/2,k+1/2}^{n+1/2}\right) \tag{11-31c}$$

The above six equations can be used for the three-dimensional case in Fig. 11-4 or reduced appropriately for the two-dimensional and one-dimensional cases. For example, the two-dimensional TE case contains the field components in the spatial arrangement given in Fig. 11-4 in the x-y plane, whereas the two-dimensional TM case contains the field components in the $k + \frac{1}{2}$ plane. A one-dimensional case may be obtained by appropriately reducing either of the two-dimensional cases.

Next, we develop bounds on the cell size and the time step used in the update equations and discuss the effects of dispersion.

11.3 CELL SIZE, NUMERICAL STABILITY, AND DISPERSION

Before we can implement the difference equations presented in the previous section, the cell size and time increment must be determined. In practice, the cell size is determined first. It is primarily influenced by numerical dispersion, which is the propagation of different frequencies with different velocities. Then after we have established the cell size, the time increment is determined such that numerical stability is achieved.

In view of our study of MoM earlier in this chapter, we can appreciate that over one FD-TD cell dimension the electromagnetic field should not change significantly. This means that for meaningful results, the grid size should be only a fraction of the wavelength of the highest significant frequency content f_u in the excitation frequency spectrum. For example, from a study of Fourier analysis we know that for a pulse of width τ, the major portion of the frequency spectrum lies between zero and $f_u = 1/\tau$. The Nyquist sampling theorem would suggest that the cell size be less than $\lambda_u/2$ in order that the spatial variation of the fields be adequately sampled. However, our pulse has frequency content higher than f_u, numerical dispersion is present in the two-dimensional and three-dimensional cases, and our difference equations are themselves approximations, so a higher spatial sampling rate (i.e., smaller cell size) is required. Depending on the accuracy of desired results, it has been found that the cell size should be smaller than approximately $\lambda_u/10$ in the material medium (e.g., $\lambda_u/20$ if computational resources allow), primarily to minimize the effects of numerical dispersion. Figure 11-5 shows, for the one-dimensional case, the effects of cell size on phase velocity and suggests a cell size at least as small as $\lambda_u/20$. Details of the geometry may dictate a still smaller cell size. For example, in Sec. 11.9.3 a cell size of $\lambda/99$ was required to model certain fine geometrical details of the Vivaldi antenna.

Now that we have established cell size, the time step Δt can be determined. Let us first consider the one-dimensional case. In one time step, any point on the wave must not travel more than one cell because during one time step the FD-TD algorithm can propagate the wave only from one cell to its nearest neighbors. Any attempt to use even a slightly larger time step will quickly lead to numerical instability. We can do less than one cell in one time step, but it is not an optimum situation and will not lead to increased accuracy. Thus, the condition in the one-dimensional case is

$$\Delta t \leq \frac{\Delta x}{c} \tag{11-32}$$

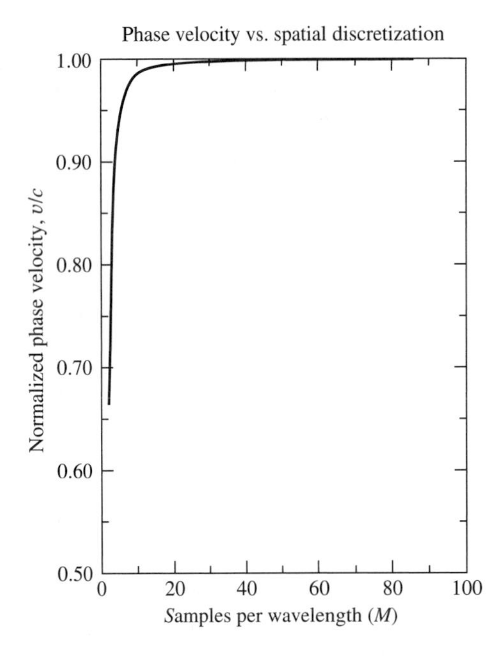

Figure 11-5 Variation of numerical phase velocity with samples (cells) per wavelength in a one-dimensional FD-TD grid [3].

If the equality sign is used, we have what is referred to by Taflove [1] as the *magic time step*, $c\,\Delta t = \Delta x$. It can be shown that the one-dimensional central difference equations produce an exact solution when the magic time step is used. This is an interesting result given that the difference equations are themselves approximations (see Fig. 11-2). Unfortunately, a similar condition does not exist in the two-dimensional and three-dimensional cases.

To guarantee numerical stability in the general case, it has been shown that

$$\Delta t \leq \frac{1}{c\sqrt{\dfrac{1}{(\Delta x)^2} + \dfrac{1}{(\Delta y)^2} + \dfrac{1}{(\Delta z)^2}}} \tag{11-33}$$

The above condition was obtained using the classical approach first suggested by Courant et al. in [5]. In this approach, a time eigenvalue problem is first solved and then a space eigenvalue problem is solved. Next, a stable range of space eigenvalues is forced to lie within the stable range of the time eigenvalues, resulting in the general relation above.

In the two-dimensional case, if $\Delta x = \Delta y = \Delta z = \Delta s$, (11-33) reduces to

$$\Delta t \leq \frac{\Delta s}{c\sqrt{2}} \tag{11-34}$$

whereas in the three-dimensional case, (11-33) reduces to

$$\Delta t \leq \frac{\Delta s}{c\sqrt{3}} \tag{11-35}$$

Examination of the above results shows that the minimum number of time steps required to travel the maximum dimension of a unit cell is equal to the dimensionality of the cell. Thus, at least two time steps are required to traverse the diagonal of a two-dimensional square cell and at least three time steps to traverse the diagonal of a three-dimensional cubic cell.

Before we leave this section, it is necessary to mention dispersion. Dispersion is the propagation of different numerical wavelengths with different velocities within the grid. Dispersion, for example, can cause the distortion of a pulse shape. In the one-dimensional case, dispersion is zero if the magic time step is used.

In the two-dimensional case, dispersion is zero if the equality is used in (11-33) and propagation is along the square cell diagonal. In any other direction of propagation, there will be dispersion. The situation is similar in the three-dimensional case. If the equality sign in (11-33) is used and propagation is along the cube diagonal, dispersion will be zero, otherwise not. Generally, numerical dispersion can be reduced, but not eliminated, by reducing the cell size.

Dispersion is illustrated in Fig. 11-6, which shows the variation of the normalized numerical phase velocity with the propagation angle in a two-dimensional FD-TD grid where the inequality of (11-33) was used. The time step $c\,\Delta t = \Delta s/2$ was employed; it is an example of a time step commonly used in two-dimensional (and three-dimensional) grids to satisfy the stability criterion in (11-33) with a margin of error. The figure shows that the phase velocity is a minimum along the Cartesian axes ($\alpha = 0°$ and $\alpha = 90°$) and is a maximum at $\alpha = 45°$ (along the square cell diagonal), but is slightly less than c even there since the equality of (11-33) was not used. The general behavior in Fig. 11-6 represents a numerical anisotropy that is inherent in the Yee algorithm.

Figure 11-7 shows the variation of the numerical phase velocity versus cell size for the same incidence angles and time step. The beneficial effect of small cell size is apparent. If too large of a cell size (i.e., too close to the Nyquist limit) is used, the wave will actually stop propagating.

Both figures imply that different frequency components of the excitation will propagate with different speeds, resulting in pulse distortion that will increase with distance. On the other hand, for a sinusoidal wave, the effect of an incorrect phase velocity would be to develop a lagging phase error that increases with propagation distance.

Now that we have bounds on the cell size and the time step, and understand the

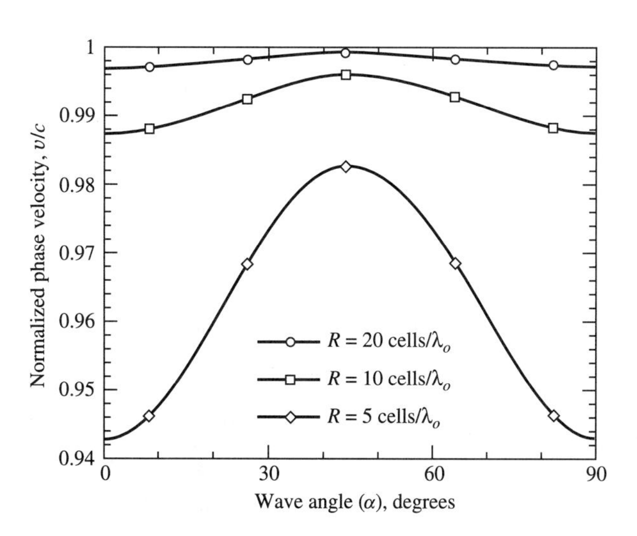

Figure 11-6 Variation of the numerical phase velocity with wave propagation angle in a two-dimensional FD-TD grid for three grid resolutions. At 0 and 90°, incidence is along either Cartesian grid axis. (From Taflove [1] © 1995. Reprinted by permission of Artech House, Inc., Boston, MA.)

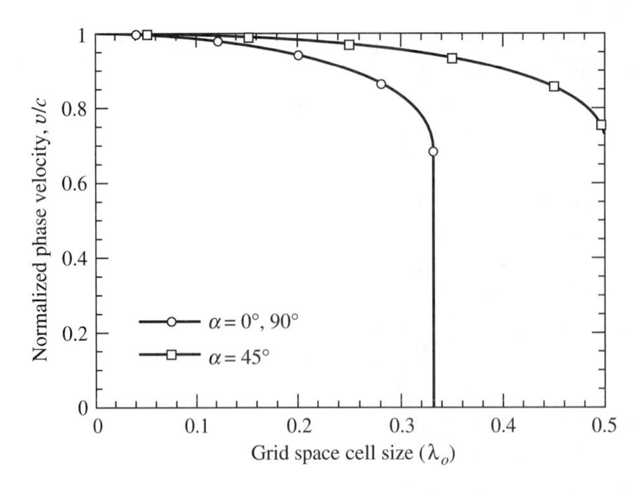

Figure 11-7 Variation of the numerical phase velocity with grid cell size in a two-dimensional FD-TD grid for three wave propagation angles relative to Cartesian grid axes at 0 and 90°. (From Taflove [1]. Reprinted with permission.)

effects of dispersion, we are in a position to implement the central difference update equations. The next section examines how to do this.

11.4 COMPUTER ALGORITHM AND FD-TD IMPLEMENTATION

In previous sections, we developed the Yee algorithm and explained some of the basic concepts in the FD-TD approach. Although there are other fundamental issues to be discussed, it is helpful at this point to view the generalities of the computer architecture and how the Yee algorithm is implemented. Some specific points will be illustrated with a one-dimensional example.

The primary computational feature of an FD-TD code is the time stepping process. This is a small part of the code, but the most heavily used part. Prior to time stepping, the FD-TD grid must be defined as well as parameters such as cell size, time step, and the source condition. Constant multipliers that are not computed at each time step, such as the C and D coefficients in (11-28) and (11-29), should also be evaluated and stored before time stepping begins. There must be a geometrical definition of the antenna or scatterer that consists of identifying those cell locations containing material other than free space. This is done via the C_a, C_b, D_a, and D_b coefficients. In addition, desired responses must be specified so that they will be available for output after time stepping is completed (or perhaps during the time stepping if transient information is desired).

The code requirements consist of the following major steps:

Preprocessing

- Define the FD-TD grid (sets the number of cells in each dimension and the cell size).
- Calculate the time step according to the Courant stability condition presented in the previous section.
- Calculate constant multipliers, including the C and D coefficients from Sec. 11.2, which serves to define the antenna or scatterer geometry in the FD-TD grid.

Time stepping

- Update the source conditions (to be discussed in Sec. 11.6).
- Calculate the response of an E-field component from that of the nearest-

neighbor field quantities according to the type of material present at the nearest-neighbor locations.
- Update the absorbing boundary condition (ABC), also called the outer radiation boundary condition. The purpose of the ABC (discussed in Sec. 11.5) is to absorb, at the extremities of the FD-TD grid, as much of the radiation field as possible to prevent nonphysical reflections within the FD-TD grid.
- Update *H*-field components.

Postprocessing
- In software arrays save response data such as *E*- and *H*-field components, currents, voltages, etc., at desired time steps.
- Determine the tangential electric and magnetic fields on a closed fictitious surface surrounding the antenna or scatterer and compute the corresponding scattered or radiated fields in the far zone (see Sec. 11.7).

A code structure that will implement the above requirements is suggested by the simplified flowchart shown in Fig. 11-8.

To illustrate how some of the calculations are done, a simplified one-dimensional model is used. The model is along the x-axis and in free space. We use the one-dimensional equations of (11-10a) and (11-10b), but the FD-TD equations are taken from (11-30b) with $E_x = 0$ and from (11-31c) with $H_x = H_z = 0$. Note that $C_a = 1$

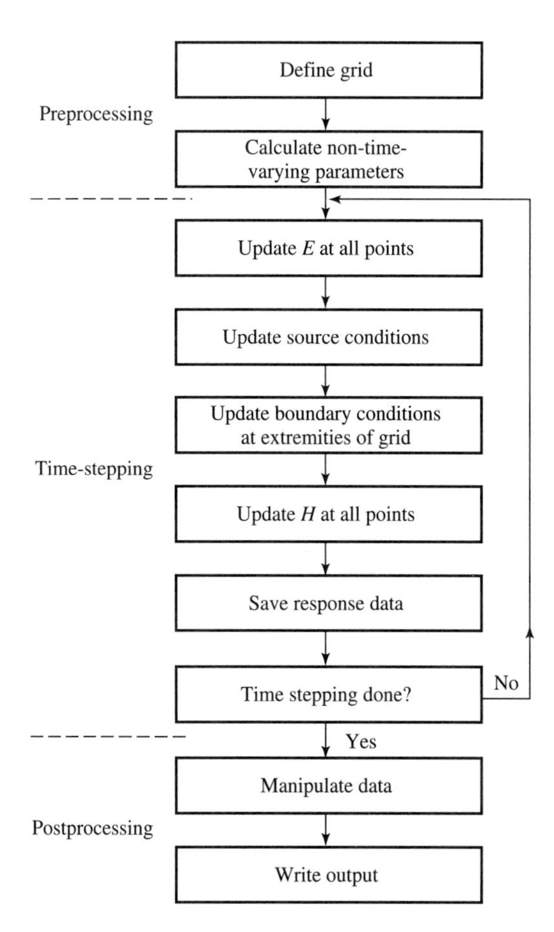

Figure 11-8 FD-TD flowchart.

and $C_b = \dfrac{\Delta t}{\varepsilon_o \Delta x}$ and that $D_a = 1$ and $D_b = \dfrac{\Delta t}{\mu_o \Delta x}$ using the magic time step condition. Thus,

$$H_y\big|_{i+1/2}^{n+1/2} = H_y\big|_{i+1/2}^{n-1/2} + \frac{\Delta t}{\mu_o \Delta x}\left(E_z\big|_{i+1}^{n} - E_z\big|_{i}^{n}\right) \tag{11-36}$$

$$E_z\big|_{i}^{n+1} = E_z\big|_{i}^{n} + \frac{\Delta t}{\varepsilon_o \Delta x}\left(H_y\big|_{i+1/2}^{n+1/2} - H_y\big|_{i-1/2}^{n+1/2}\right) \tag{11-37}$$

We note that $\dfrac{\Delta t}{\varepsilon_o \Delta x} = \dfrac{1}{\varepsilon_o c} = \eta$ and that $\dfrac{\Delta t}{\mu_o \Delta x} = \dfrac{1}{\mu_o c} = \dfrac{1}{\eta}$, where $c = (\mu_o \varepsilon_o)^{-1/2}$.

Next, an FD-TD grid is set up along the x-axis starting with cell #1, ending with cell #53, and a source at cell 50 as indicated in Fig. 11-9. The source is a delta function with amplitude η. As an initial condition, for $n < 1$ all fields in the FD-TD grid are taken to be zero. The source turns on at $n = 1$. Thus, dropping the coordinate subscripts on H and E in (11-36) and (11-37), we can write (for $i \geq 50$)

$$E_{50}^{1.0} = \eta \tag{11-38a}$$

$$H_{50.5}^{1.5} = 0 + \frac{1}{\eta}\left(E_{51}^{1.0} - E_{50}^{1.0}\right) = -1 \tag{11-38b}$$

All other $E_i^{1.0}$ and $H_{i+1/2}^{1.5}$ for $i > 50$ are zero since the delta function has not yet propagated to those other locations in the FD-TD grid. A word on the notation is in order here. Equation (11-36) shows $H_y\big|_{i+1/2}^{n+1/2}$ on the left side, whereas (11-38b) shows $H_{50.5}^{1.5}$. In a computer array, there is no location 50.5. There are only the data locations for E_{50} and H_{50}. It is important to understand that $H_{i+\Delta x/2}$ is H_i in the computer. At the next time step, we write (for $i \geq 50$)

$$E_{50}^{2.0} \equiv 0 \tag{11-39a}$$

$$E_{51}^{2.0} = E_{51}^{1.0} + \eta\left(H_{51.5}^{1.5} - H_{50.5}^{1.5}\right) = 0 + \eta(0 + 1) = \eta \tag{11-39b}$$

$$E_{52}^{2.0} = E_{52}^{1.0} + \eta\left(H_{52.5}^{1.5} - H_{51.5}^{1.5}\right) = 0 + \eta(0 - 0) = 0 \tag{11-39c}$$

$$E_{53}^{2.0} = 0 \tag{11-39d}$$

$$H_{50.5}^{2.5} = H_{50.5}^{1.5} + \frac{1}{\eta}\left(E_{51}^{2.0} - E_{50}^{2.0}\right) = -1 + \frac{1}{\eta}(\eta - 0) = 0 \tag{11-39e}$$

$$H_{51.5}^{2.5} = H_{51.5}^{1.5} + \frac{1}{\eta}\left(E_{52}^{2.0} - E_{51}^{2.0}\right) = 0 + \frac{1}{\eta}(0 - \eta) = -1 \tag{11-39f}$$

$$H_{52.5}^{2.5} = H_{52.5}^{1.5} + \frac{1}{\eta}\left(E_{53}^{2.0} - E_{52}^{2.0}\right) = 0 + \frac{1}{\eta}(0 - 0) = 0 \tag{11-39g}$$

$$H_{53.5}^{2.5} = 0 \tag{11-39h}$$

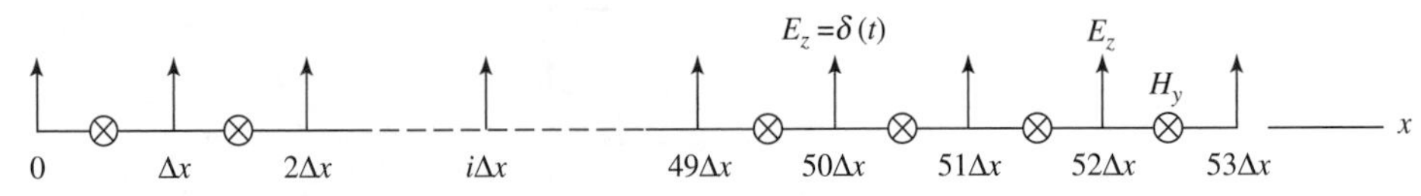

Figure 11-9 A one-dimensional 53-cell model.

At the third time step, we find that, for $i \geq 50$,

$$E_{50}^{3.0} \equiv 0, \qquad E_{51}^{3.0} = 0 \tag{11-40a}$$

$$E_{52}^{3.0} = E_{52}^{2.0} + \eta(H_{52.5}^{2.5} - H_{51.5}^{2.5}) = 0 + \eta(0 + 1) = \eta \tag{11-40b}$$

$$E_{53}^{3.0} = 0, \qquad H_{50.5}^{3.5} = 0, \qquad H_{51.5}^{3.5} = 0 \tag{11-40c}$$

$$H_{52.5}^{3.5} = H_{52.5}^{2.5} + \frac{1}{\eta}(E_{53}^{3.0} - E_{52}^{3.0}) = 0 + \frac{1}{\eta}(0 - \eta) = -1 \tag{11-40d}$$

$$H_{53.5}^{3.5} = 0 \tag{11-40e}$$

At the fourth time step, for $i \geq 50$, we encounter the following:

$$E_{50}^{4.0} \equiv 0, \qquad E_{51}^{4.0} = 0, \qquad E_{52}^{4.0} = 0 \tag{11-41a}$$

$$E_{53}^{4.0} = E_{53}^{3.0} + \eta(H_{53.5}^{3.5} - H_{52.5}^{3.5}) = 0 + \eta(0 + 1) = \eta \tag{11-41b}$$

$$H_{50.5}^{4.5} = 0, \qquad H_{51.5}^{4.5} = 0, \qquad H_{52.5}^{4.5} = 0 \tag{11-41c}$$

$$H_{53.5}^{4.5} = H_{53.5}^{3.5} + \frac{1}{\eta}(E_{54}^{4.0} - E_{53}^{4.0}) \tag{11-41d}$$

Here, we have a problem in that $E_{54}^{4.0}$ is undefined. If a computer software could take $E_{54}^{4.0}$ to be zero, then in the computer we would obtain the "correct" value for $H_{53.5}^{4.5}$. However, since the grid was specified to extend only to $i = 53$, we have no reason to expect this will happen. The difficulty can be overcome with an absorbing boundary condition, as discussed in the next section.

Before proceeding to the next section, we observe that in this section we applied the leapfrog time-marching finite difference algorithm equations, and that for our impulsive source, propagation at the speed of light is predicted in the positive x-direction. It is left as an exercise for the reader to show that the same equations will predict propagation in the negative x-direction, and that the right-hand rule for power flow is automatically obeyed.

11.5 ABSORBING BOUNDARY CONDITIONS

At the end of the previous section, it was seen that there was a problem in computing the fields at the edge of the FD-TD grid. Without some means of absorbing the outward propagating waves at the extremities of the FD-TD grid, nonphysical reflections at the edge of the grid will contaminate the fields inside the grid. Of course, we could terminate the time-stepping procedure before such a reflection reaches the observation area of interest or make the grid very, very large, but these are not computationally viable alternatives. Therefore, some special attention must be given to the problem of updating field components at the edge of the grid.

The most practical solution to updating at the edge of the grid is to employ an *absorbing boundary condition* (ABC), sometimes referred to as a radiation boundary condition (RBC). In the one-dimensional case, the required condition is simple and exact because there is a plane wave normally incident on the edges of the grid. Thus, simple propagation delay can be used. In the two-dimensional and three-dimensional cases, the problem is considerably more difficult because the wave is not likely to be normally incident on the edges of the grid and the waves are not likely to be planar as indicated in Fig. 11-10.

Numerous ABCs have been developed over the past several decades. It is beyond

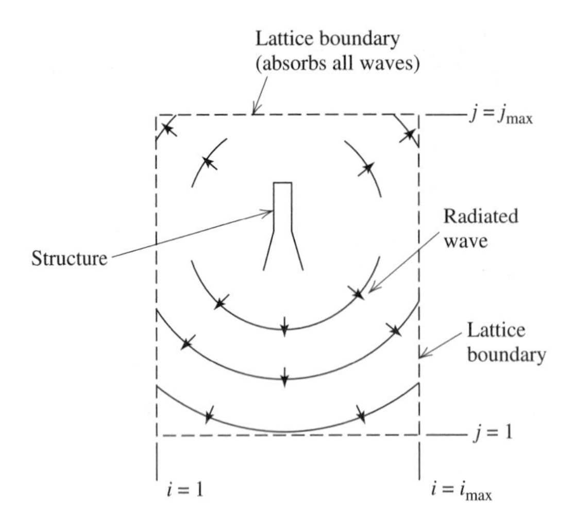

Figure 11-10 FD-TD electromagnetic wave interaction region with (ideally) no reflection from the lattice boundary.

the scope of this book to derive them or even to present more than two of them. Thus, we will present only the Mur ABC [6] and the more recently developed Berenger perfectly matched layer ABC [7].

There exist two Mur estimates for the fields on the boundary which are first-order- and second-order-accurate. Consider the E_z component located at $x = i\Delta x$, $y = j\Delta y$ for the two-dimensional case. The first-order Mur estimate of this E_z field component is [2]

$$E_{i,j}^{n+1} = E_{i-1,j}^{n} + \frac{c\Delta t - \Delta x}{c\Delta t + \Delta x}\left(E_{i-1,j}^{n+1} - E_{i,j}^{n}\right) \tag{11-42}$$

In the one-dimensional case if $\Delta x = c\Delta t$, $E_i^{n+1} = E_{i-1}^{n}$, which says that the estimate at location i at the $n + 1$ time step is the field from the previous location and previous time step. This is an exact result in one dimension only.

The second-order Mur estimate for E_z in the two-dimensional case, if we assume $\Delta x = \Delta y$, is [2]

$$\begin{aligned} E_{i,j}^{n+1} = {} & -E_{i-1,j}^{n-1} + \frac{c\Delta t - \Delta x}{c\Delta t + \Delta x}\left(E_{i-1,j}^{n+1} + E_{i,j}^{n-1}\right) \\ & + \frac{2\Delta x}{c\Delta t + \Delta x}\left(E_{i-1,j}^{n} + E_{i,j}^{n}\right) \\ & + \frac{(c\Delta t)^2}{2(\Delta x)(c\Delta t + \Delta x)}\left(E_{i,j+1}^{n} - 2E_{i,j}^{n} + E_{i,j-1}^{n}\right. \\ & \left. + E_{i-1,j+1}^{n} - 2E_{i-1,j}^{n} + E_{i-1,j-1}^{n}\right) \end{aligned} \tag{11-43}$$

In the case of the first-order Mur estimate, we see that the current value of E_z at $x = i\,\Delta x$ is estimated from the previous E_z value at $x = i\,\Delta x$ and the value of E_z at $x = (i - 1)\,\Delta x$ at the current time step, both at the same y-position. The second-order Mur estimate uses values from the preceding two time steps, and values at the adjacent x- and y-positions. The equations needed at the $y = j\Delta y$ surface (where the index j is not to be confused with $\sqrt{-1}$) are appropriate permutations of the positional coordinates given in (11-43) above. The second-order Mur estimate is an exact solution for waves impinging normal to a grid boundary. At the intersection of the xz- and yz-planes, some type of first-order-accurate approximation may be employed based on propagation delay as suggested by (11-42).

In 1994, Berenger [7] published a technique that lowered the reflection from the outer grid boundary by several orders of magnitude over other approaches. He called his approach the "perfectly matched layer (PML) for the absorption of electromagnetic waves" in his paper that treated the two-dimensional TE and TM cases. Ingenuously, he artificially split the fields at the boundaries into two components, creating four coupled equations rather than the usual three. This extra degree of freedom permitted Berenger to derive a nonphysical anisotropic absorbing medium, adjacent to the outer boundary (see Fig. 11-11), with a remarkable wave impedance that is independent of the angle of incidence and frequency of the outgoing waves. For the TM case, except in the interface for H_y, the applicable FD-TD equations for H_y and E_z are

$$\begin{aligned} H_y\big|_{i+1/2,j}^{n+1/2} &= e^{-\sigma_x^*(i+1/2)\Delta t/\mu_0}\, H_y\big|_{i+1/2,j}^{n-1/2} - \frac{1 - e^{\sigma_x^*(i+1/2)\Delta t/\mu_0}}{\sigma_x(i+\frac{1}{2})\,\Delta x} \\ &\quad \times \left[E_{zx}\big|_{i+1,j}^{n} + E_{zy}\big|_{i+1,j}^{n} - E_{zx}\big|_{i,j}^{n} - E_{zy}\big|_{i,j}^{n}\right] \end{aligned} \tag{11-44}$$

$$\begin{aligned} E_{zx}\big|_{i,j}^{n+1} &= e^{-\sigma_x(i)\Delta t/\varepsilon_0}\, E_{zx}\big|_{i,j}^{n} - \frac{1 - e^{\sigma_x(i)\Delta t/\varepsilon_0}}{\sigma(i)\,\Delta x} \\ &\quad \times \left[H_y\big|_{i+1/2,j}^{n+1/2} - H_y\big|_{i-1/2,j}^{n+1/2}\right] \end{aligned} \tag{11-45}$$

where the electric and magnetic conductivities σ_x and σ_x^* are functions of $x(i)$ in the left, right, and corner layers. In the upper and lower PML layers, σ_x and σ_x^* are equal to zero for all $x(i)$, where in fact, the medium behaves as a vacuum for the equations dependent on $x(i)$. Note that E_{zx} and E_{zy} are colocated at the same point.

For an H_y component lying on the interface, the update equation is based on the values of three adjacent E-field components: one E_z component in the regular

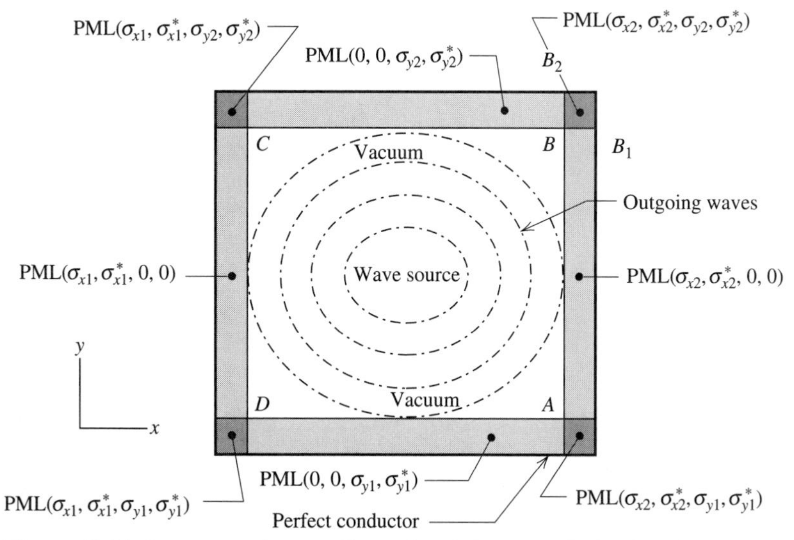

Figure 11-11 Structure of a two-dimensional FD-TD grid having the Berenger PML ABC. (J. Berenger, *Comput. Phys.*, Vol. 114, 1994, pp. 185–200. Reprinted with permission.)

FD-TD grid, and two components, E_{zx} and E_{zy}, in the PML region. Through the application of a normal Maxwell-based FD-TD update equation to this H_y component, an interface is established between the regular FD-TD grid and PML grid. Here, the assumption is made that the E_{zx} and E_{zy} components are summed to yield an E_z component that is effectively a regular FD-TD field component. This is seen to be valid by simply letting $\sigma_x = \sigma_y$ and reducing the four PML update equations to the usual three FD-TD governing equations. Thus, the finite difference equations have to be modified. So, in the right side interface normal to x, the equation for H_y above becomes

$$H_y\big|_{il+1/2,j}^{n+1/2} = e^{-\sigma_x^*(il)\Delta t/\mu_0} H_y\big|_{il+1/2,j}^{n-1/2} - \frac{1 - e^{\sigma_x^*(il)\Delta t/\mu_0}}{\sigma_x^*(il)\,\Delta x} \times \left[E_{zx}\big|_{il+1,j}^{n} + E_{zy}\big|_{il+1,j}^{n} - E_z\big|_{il,j}^{n}\right] \tag{11-46}$$

This particular update is applied at the right-side interface normal to x, where il indicates the inner boundary of the PML.

The discretized equations needed for the components H_x and E_{zy} are left as an exercise for the reader. Not surprisingly, the two-dimensional PML equations reduce to the exact one-dimensional result. This is also left as an exercise for the reader. A Maxwellian derivation of the PML technique may be found in [8].

Of the two ABCs presented here, Mur and Berenger, the former is somewhat simpler to implement but the latter offers substantially lower reflection characteristics, perhaps more than is really necessary for many applications. One general consideration with ABCs is that of determining the necessary distance from the antenna or scatterer to the outer boundary where the ABC is applied. The greater this distance, the more effective the ABC tends to be. (This is particularly true with the Mur ABC but not necessarily so with the PML.) The increased effectiveness of the ABC is due to the more plane-like nature of the outward traveling wavefront as the distance becomes large. A common criterion is a minimum of 10 cells between the antenna or scatterer and the outer boundary, with 15 to 20 being preferred for the Mur ABC and as few as 4 or 5 for the PML.

Before we leave this section, let us apply the one-dimensional exact result to (11-41d). In this case, the exact result is $E_{54}^{4.0} = E_{53}^{3.0} = 0$. Thus,

$$H_{53.5}^{4.5} = 0 + \frac{1}{\eta}(0 - \eta) = -1 \tag{11-47}$$

Then, to evaluate the reflection from the edge of the grid,

$$E_{53}^{5.0} = E_{53}^{4.0} + \eta(H_{53.5}^{4.5} - H_{52.5}^{4.5}) = \eta + \eta(-1 - 0) = 0 \tag{11-48a}$$

$$H_{52.5}^{5.5} = H_{52.5}^{4.5} + \frac{1}{\eta}(E_{53}^{5.0} - E_{52}^{5.0}) = 0 + \frac{1}{\eta}(0 - 0) = 0 \tag{11-48b}$$

We see that there is perfect absorption. Table 11-1 summarizes the situation for $1 \le n < 6$ and $i \ge 50$. Note that all points in the grid are calculated at all time steps.

The grid used for Table 11-1 ends on an H-field calculation. The example could also be developed to end on an E-field calculation at E_{53} by defining the H-field only out to $H_{52.5}$. In this case, $H_{53.5}^{3.5}$ would be undefined in (11-41b) and the one-dimensional exact absorption condition would be applied to E_{53}. For simplicity, an ABC is typically applied to only E- or only H-fields. In this way, regular FD-TD

Table 11-1 Pulse Propagation with Perfect Absorption ($|E| = \eta$, $|H| = 1$)
Spatial Cell Location ($i \geq 50$)

Time Step	50	50.5	51	51.5	52	52.5	53	53.5	
1	η		0		0		0		E
1.5		−1		0		0		0	H
2	0		η		0		0		E
2.5		0		−1		0		0	H
3	0		0		η		0		E
3.5		0		0		−1		0	H
4	0		0		0		η		E
4.5		0		0		0		−1	H
5	0		0		0		0		E
5.5		0		0		0		0	H

update equations will govern all behavior for one type of field (say, H) and special updates then need only be applied at the grid boundary to the other type of field (say, E). Note that the Mur equations are given here for the electric field.

11.6 SOURCE CONDITIONS

In this section, we introduce into the FD-TD lattice several electromagnetic wave excitations appropriate for modeling engineering problems. An excitation of interest is the linearly polarized plane wave propagating in free space for use in scattering analysis, but we are also interested in waves radiated by antennas. With FD-TD, we usually study antennas in the transmitting mode since it is not computationally efficient to do so in the receiving mode.

This section covers source conditions for antennas and scatterers. Following the designations of Taflove [1], the sources will be classified as either "hard" or "soft." A hard source forces a field quantity to a value independent of neighboring fields, which means that the update equations are not allowed to update the field(s) at the source location (e.g., a metallic monopole near a scatterer). A soft source does permit the fields to be updated at the source location(s) (e.g., a plane wave injected into the grid).

11.6.1 Source Functionality

A common source is one that generates a continuous sinusoidal wave of frequency f_o that is switched on at $n = 0$:

$$f(t) = E_o \sin(2\pi f_o n\Delta t) \tag{11-49a}$$

A second source provides a wideband Gaussian pulse with finite dc spectral content that is centered at time step n_o and has a $1/e$ characteristic decay of n_{decay} time steps:

$$f(t) = E_o e^{-[(n-n_o)/n_{\text{decay}}]^2} \tag{11-49b}$$

Note that (11-49b) has a nonzero value at $n = 0$, so that if a smooth transition from zero into the Gaussian pulse is required, n_o should be taken as at least $3n_{\text{decay}}$. A third source that provides a zero-dc content is a sine modulated (bandpass) Gaussian pulse with a Fourier spectrum symmetrical about f_o. The pulse is again centered at time step n_o and has a $1/e$ characteristic decay of n_{decay} time steps:

$$f(t) = E_o e^{-[(n-n_o)/n_{\text{decay}}]^2} \sin[2\pi f_o(n - n_o)\,\Delta t] \tag{11-49c}$$

Each source of (11-49) radiates a numerical wave having a time waveform corresponding to the source function $f(t)$. The numerical wave propagates symmetrically in all directions from the source point at i_s. If a material structure is specified at some distance from the source point, the radiated numerical wave eventually propagates to this structure and undergoes partial transmission and partial reflection. In principle, time-stepping can be continued until all transients decay. For the source of (11-49a), this would mean the attainment of the sinusoidal steady state for the transmitted and reflected fields. For the sources of (11-49b) and (11-49c), this would mean the evolution of the complete time histories of the transmitted and reflected waves. Discrete Fourier analysis of these time histories obtained in a single FD-TD run can provide the magnitude and phase of the transmission and reflection coefficients over a potentially wide frequency band starting at dc.

Thus far, we have discussed three time functions used in FD-TD work. The delta function used in Secs. 11.4 and 11.5 is not a generally useful time function for FD-TD calculations because of its theoretically infinite bandwidth. The delta function was used in Secs. 11.4 and 11.5 because it provided a simple way of illustrating how the update equations worked, how the one-dimensional absorbing boundary condition worked, and even permitted a small number of calculations to be easily done by hand. Unfortunately, the delta function hard source will only work in one dimension, and then only when propagation is exactly one cell per time step (e.g., the magic time step). Figure 11-12 illustrates the consequences of violating this condition. Consider Fig. 11-9 but with the grid extending from zero to $i\,\Delta x = 1200\,\Delta x$,

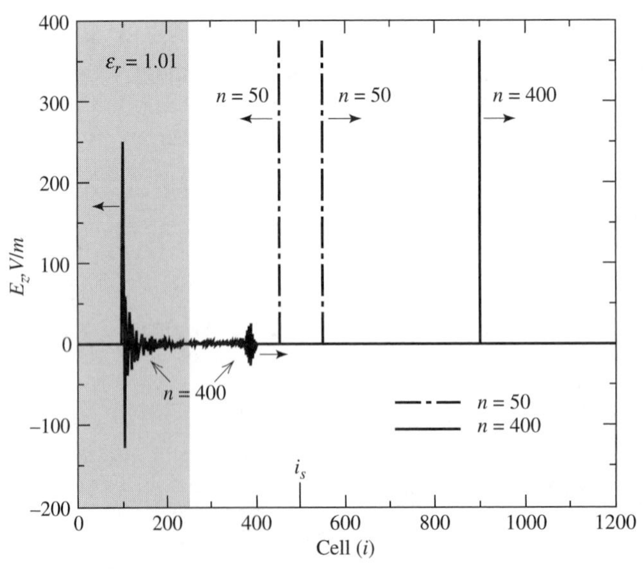

Figure 11-12 Delta wave function at $n = 50$ and 400 with $\varepsilon_r = 1.01$ for $0 \le i \le 250$ and $i_s = 500$. Source magnitude is 377 V/m.

with a delta function source of amplitude η at $i = 500$ and the region from $i = 0$ to $i = 250$ filled with a relative permitivity of only 1.01. Figure 11-12 shows the electric field in the grid after 50 and 400 time steps. Note that the wave encounters the small change in dielectric constant at $n = 250$ and there is obvious evidence of numerical dispersion for $n = 400$ in the region $i < 400$ as a consequence of the (slight) violation of the condition one cell in one time step for $i < 250$ and $n > 250$. On the other hand, at $n = 400$, $i = 900$ the delta function propagates to the right in a dispersionless manner since no violation of the one cell in one time step condition has occurred.

11.6.2 The Hard Source

The hard source is set up simply by assigning a desired time function to specific electric or magnetic field components in the FD-TD lattice as we did in the example in Sec. 11.4. For example, in a one-dimensional grid, the following hard source on E_z could be established at the grid source point i_s to generate a continuous sinusoid that is switched on at $n = 0$:

$$E_z|_{i_s}^{n} = f(t) = E_o \sin(2\pi f_o n\Delta t) \tag{11-50}$$

Note that the electric field at i_s is forced to have a value determined entirely by the source and it is independent of the update equation.

There are some difficulties with the hard source scenario. As time-stepping is continued to obtain either the sinusoidal steady state or the late-time pulse response, we note that the reflected numerical wave eventually returns to the source grid location i_s. Because the total electric field is specific at i_s without regard to any possible reflected waves in the grid (hence the terminology, "hard source"), the hard source causes a retro-reflection of these waves at i_s back toward the material structure of interest. In effect, it prevents the movement of reflected wave energy through its position toward infinity, and thereby may fail to properly simulate the true physical situation.

11.6.3 The Soft Source

A simple way to mitigate the reflective nature of a hard source is to allow a new value of the electric field at the source location i_s to equal the update value plus the value of an impressed electric field described by the time function $f(t)$. For our one dimensional example, this means that

$$E_z|_{i_s}^{n} = E_z|_{i_s}^{n-1} + \frac{\Delta t}{\varepsilon_o \Delta x}\left(H_y|_{i_s+1/2}^{n-1/2} - H_y|_{i_s-1/2}^{n-1/2}\right) + f(t) \tag{11-51}$$

where $f(t)$ can be obtained, for example, from (11-49). The relationship in (11-51) is conceptually similar to that of the resistive voltage source in Taflove [1, pg. 459].

Figure 11-13 illustrates the difference between one-dimensional hard and soft sources. The FD-TD model has cells from $i = 0$ to $i = 1200$ with a source at $i = 500$ and a dielectric with $\varepsilon_r = 9$ in cells 1 to 200. The time function is a Gaussian pulse as in (11-49b). The (*a*) and (*b*) parts of the figure apply whether the source is hard or soft, whereas the (*c*) and (*d*) parts only apply to the soft source and the (*e*) and (*f*) to the hard source. For $n = 600$, 700, and 800, the differences between the hard and soft sources are apparent. Note the effect of the ABC at $i = 1200$.

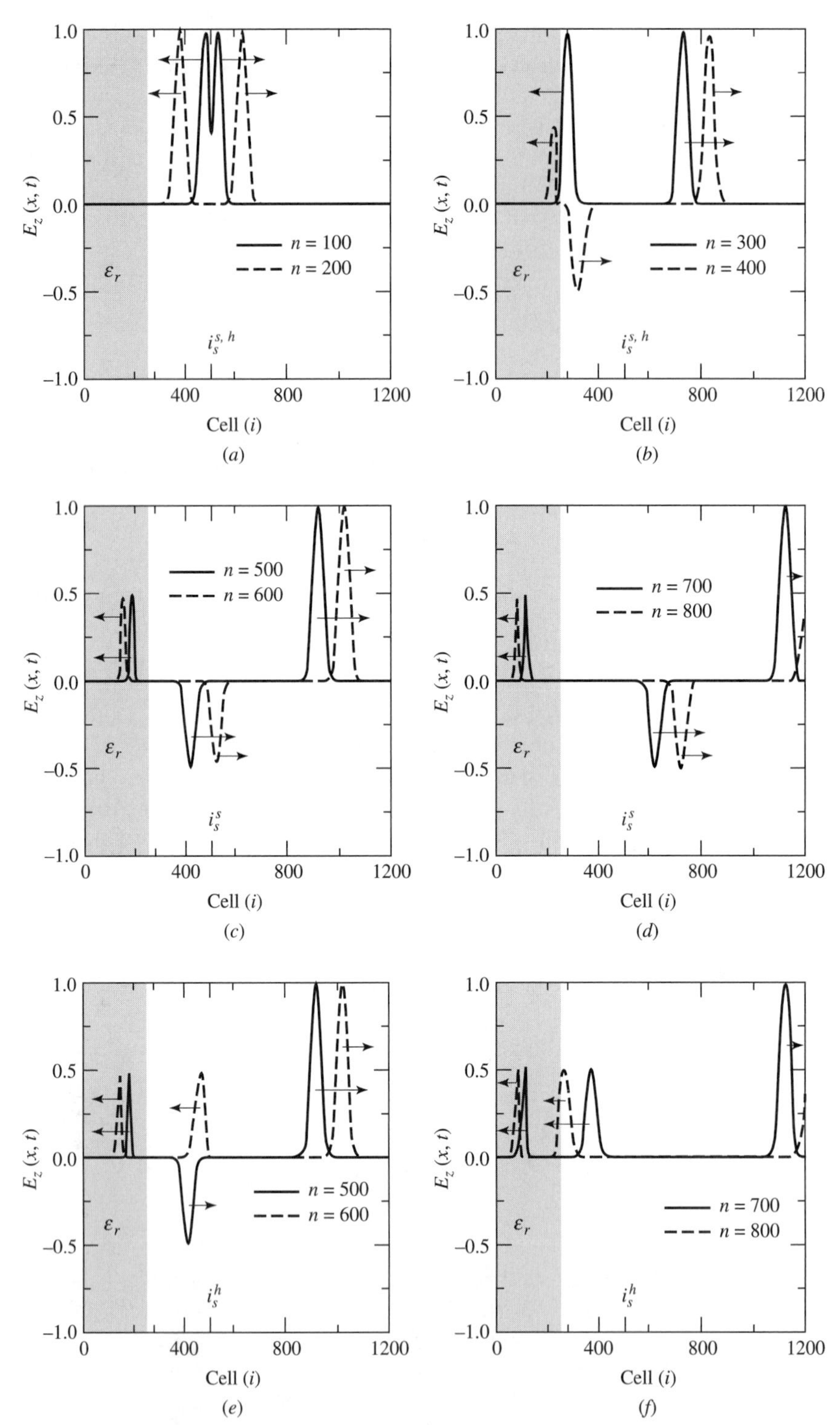

Figure 11-13 Gaussian pulse electric field at n = 100, 200, 300, 400, 500, 600, 700, and 800. Parts (a–d) are for a soft source at $i_s = 500$. Parts (a), (b), (e), and (f) are for a hard source at $i_s = 500$. Cells from $i = 0$ to $i = 200$ contain $\varepsilon_r = 9$.

11.6.4 Total-Field/Scattered-Field Formulation

The total-field/scattered-field FD-TD formulation [1] (see Fig. 11-14) finds its greatest use in simulating plane wave illumination. This approach is based on the linearity of Maxwell's equations and the usual decomposition of the total electric and magnetic fields into incident fields ($\mathbf{E}^i$, $\mathbf{H}^i$) and scattered fields ($\mathbf{E}^s$, $\mathbf{H}^s$). $\mathbf{E}^i$ and $\mathbf{H}^i$ are the values of the incident fields that are *known* at all points of an FD-TD grid at all time steps as Fig. 11-15 indicates. These are the field values that would exist if there were no materials of any sort in the modeling space. Figure 11-15 actually shows the fields on two separate grids, one being used for the incident field and the second for the total field-scattered field. In practice, the data in the incident field grid are used to inject the incident field into the total-field region of the second grid at the total-field/scattered-field boundary. Since the total-field/scattered-field grid in Fig. 11-15 has no material objects, the scattered fields for $i < 100$ and $i > 1100$ are always zero. $\mathbf{E}^s$ and $\mathbf{H}^s$ are the values of the scattered wave fields that are initially *unknown* as indicated in Figs. 11-15*a* and 11-15*b*, and are the fields that result from the interaction of the incident wave with any materials in the grid.

The finite difference approximations of the Yee algorithm can be applied with equal validity to either the incident-field vector components, the scattered-field vector components, or the total-field vector components. FD-TD codes can utilize this property to zone the numerical space lattice into two distinct regions separated by a nonphysical surface that serves to connect the fields in each region as shown in Fig. 11-14.

Region 1, the inner zone of the lattice, is denoted as the *total-field* region where the Yee algorithm operates on total-field vector components. The interacting structure of interest is embedded within this region.

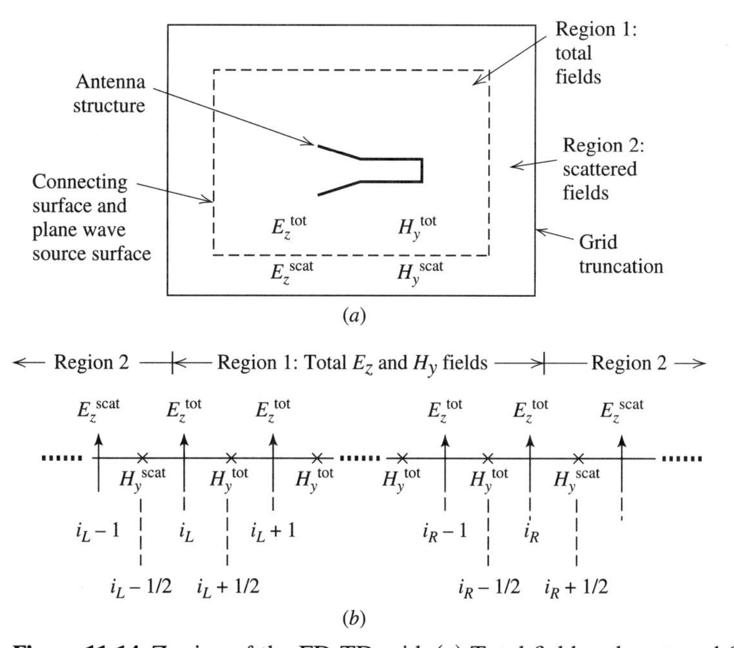

Figure 11-14 Zoning of the FD-TD grid. (*a*) Total-field and scattered-field regions, connecting surface/plane wave source, and lattice truncation (ABC). (*b*) Detail of field component locations in a one-dimensional horizontal cut through the grid of (*a*).

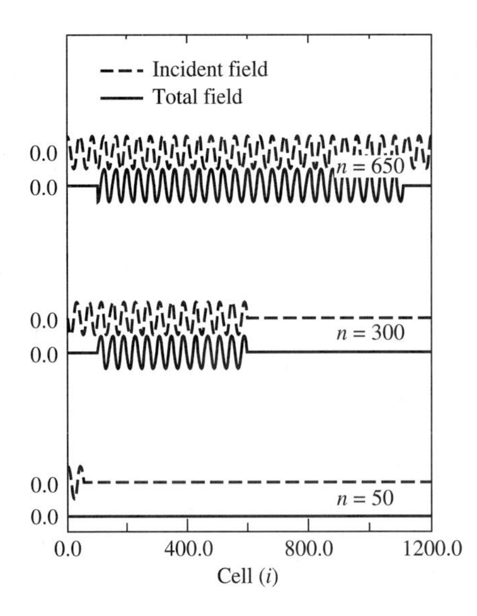

Figure 11-15 One-dimensional free-space grid from $i = 0$ to $i = 1200$. Total field region is from $i = 100$ to $i = 1100$. 40 cells per wavelength with a 10-GHz sine wave propagating from left to right. E-fields shown at $n = 50$, $n = 300$, and $n = 650$.

Region 2, the outer zone of the lattice, is denoted as the *scattered-field* region where the Yee algorithm operates only on scattered-field vector components. This implies that there is no incident wave in Region 2 as illustrated by Fig. 11-16 for $i < 100$ and $n \leq 330$. For $n = 200$, the total field is the incident field (E and H considered to be 180° out of time phase), whereas for $n = 330$ reflection has occurred from the perfect electric conductor (PEC) and some of the total field is just the incident field and some is a standing wave of the incident and scattered (reflected) fields (E and H 90° out of time phase). In Fig. 11-16, for $n = 650$ and $n = 671$ there is a scattered field propagating to the left for $i < 100$ (E and H are in time phase) with a standing wave for $100 < i < 250$, with E always zero at $i = 250$ and H at times reaching a maximum value of 2 at the PEC.

To illustrate how the total-field/scattered-field formulation is implemented, consider the one-dimensional case. The nonphysical surface constituting the interface of Regions 1 and 2 contains **E** and **H** components that obviously require the formulation of various field component spatial differences in the update equations. When a spatial difference is taken across the interface plane, a problem of consistency arises. That is, on the Region 1 side of the interface, the field in the difference expression is assumed to be a total field, whereas on the Region 2 side of the interface, the field in the difference expression is assumed to be a scattered field. It is inconsistent to perform an arithmetic difference between scattered- and total-field values.

This problem of consistency can be solved by using the values of the components of the incident-field vectors[3] $\mathbf{E}^{\text{inc}}$ and $\mathbf{H}^{\text{inc}}$, which are assumed to be known or calculable at each space lattice point. As illustrated in Fig. 11-14*b*, let the left interface between scattered-field and total-field zones be positioned between E_z at i_L and H_y

[3]To avoid confusion with the index i, "inc" is used to denote the incident field. For further consistency, "scat" denotes the scattered field and "tot" the total field.

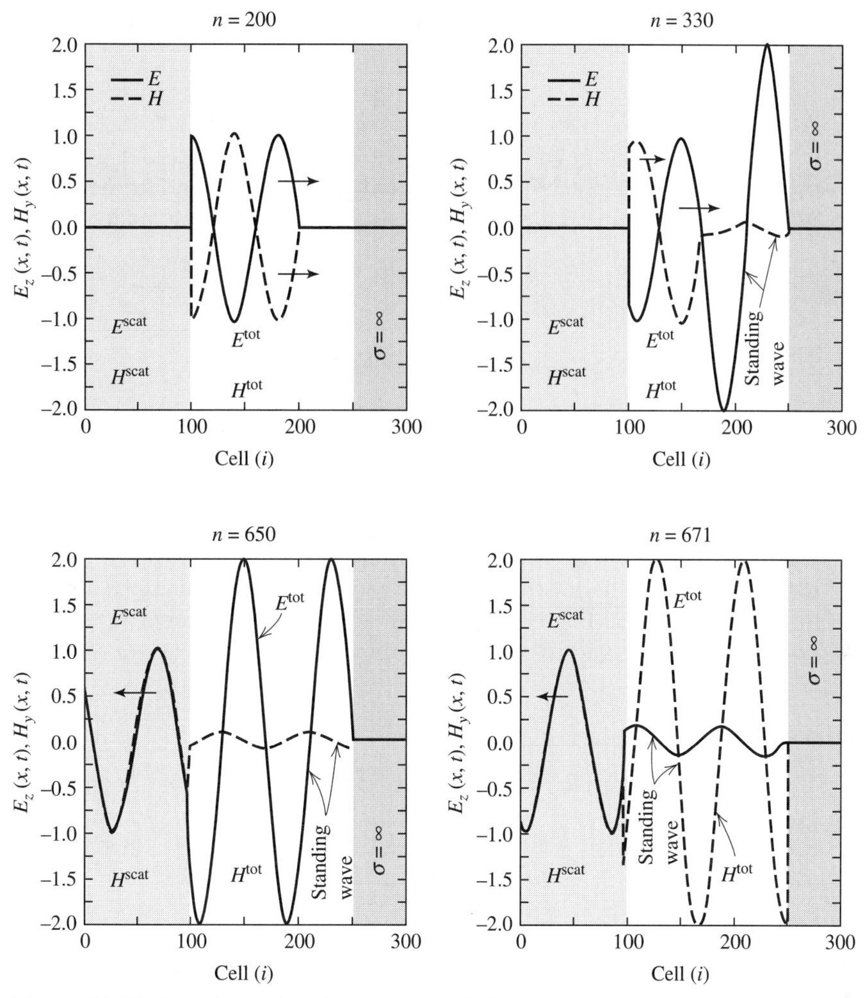

Figure 11-16 One-dimensional free-space grid from $i = 0$ to $i = 249$ with a PEC from $i = 250$ to $i = 300$. Total field region is from $i = 100$ to $i = 300$. 40 cells per wavelength with a 10-GHz sine wave incident from the left. Normalized E- and H-fields shown at $n = 200$, 330, 650, and 671.

at $i_{L-1/2}$. From this arrangement, it is clear that E_z is a total-field component. We then write

$$E_z^{\text{tot}}\Big|_{i_L}^{n+1} = E_z^{\text{tot}}\Big|_{i_L}^{n} + \frac{\Delta t}{\varepsilon_o \Delta x}\left(H_y^{\text{tot}}\Big|_{i_L+1/2}^{n+1/2} - H_y^{\text{scat}}\Big|_{i_L-1/2}^{n+1/2}\right) - \frac{\Delta t}{\varepsilon_o \Delta x} H_y^{\text{inc}}\Big|_{i_L-1/2}^{n+1/2} \tag{11-52}$$

where the right-most term corrects the problem of inconsistency since

$$-H_y^{\text{scat}}\Big|_{i_L-1/2}^{n+1/2} - H_y^{\text{inc}}\Big|_{i_L-1/2}^{n+1/2} = -H_y^{\text{tot}}\Big|_{i_L-1/2}^{n+1/2} \tag{11-53}$$

Similarly at grid point $(i_L - \frac{1}{2})$, we write

$$H_y^{\text{scat}}\Big|_{i_L-1/2}^{n+1/2} = H_y^{\text{scat}}\Big|_{i_L-1/2}^{n-1/2} + \frac{\Delta t}{\mu_o \Delta x}\left(E_z^{\text{tot}}\Big|_{i_L}^{n} - E_z^{\text{scat}}\Big|_{i_L-1}^{n}\right) - \frac{\Delta t}{\mu_o \Delta x} E_z^{\text{inc}}\Big|_{i_L}^{n} \tag{11-54}$$

where the right-most term corrects the problem of inconsistency since

$$E_z^{\text{scat}}\Big|_{i_L}^{n} = E_z^{\text{tot}}\Big|_{i_L}^{n} - E_z^{\text{inc}}\Big|_{i_L}^{n} \tag{11-55}$$

A similar procedure is carried out at the right-hand-side total-field/scattered-field interface. Let the right interface between scattered-field and total-field zones be positioned exactly at an E_z component at grid point i_R, and further assume that this E_z is a total-field component. The electric field expression analogous to (11-52) is

$$E_z^{\text{tot}}\Big|_{i_R}^{n+1} = E_z^{\text{tot}}\Big|_{i_R}^{n} + \frac{\Delta t}{\varepsilon_o \Delta x}\left(H_y^{\text{scat}}\Big|_{i_R+1/2}^{n+1/2} - H_y^{\text{tot}}\Big|_{i_R-1/2}^{n+1/2}\right) + \frac{\Delta t}{\varepsilon_o \Delta x} H_y^{\text{inc}}\Big|_{i_R+1/2}^{n+1/2} \tag{11-56}$$

The magnetic field expression analogous to (11-54) is

$$H_y^{\text{scat}}\Big|_{i_R-1/2}^{n+1/2} = H_y^{\text{scat}}\Big|_{i_R+1/2}^{n-1/2} + \frac{\Delta t}{\mu_o \Delta x}\left(E_z^{\text{scat}}\Big|_{i_R+1}^{n} - E_z^{\text{tot}}\Big|_{i_R}^{n}\right) + \frac{\Delta t}{\mu_o \Delta x} E_z^{\text{inc}}\Big|_{i_R}^{n} \tag{11-57}$$

The important effect of this procedure is to generate a plane wave at the left-hand scattered-field/total-field interface point i_L, propagate it through the total-field region to the right-hand total-field/scattered-field interface point i_R, and then cancel it out in the right-hand scattered-field region. In the absence of a scattering object in the central total-field zone, there are zero fields present in the scattered-field regions to the left and right of the center zone as is the case in Fig. 11-15.

11.6.5 Pure Scattered-Field Formulation

The pure scattered-field formulation borrows from a method popular with the frequency domain integral equation (i.e., MoM) community. Again, the concept evolves from the linearity of Maxwell's equations and the decomposition of the total electric and magnetic fields into a known incident field and an unknown scattered field. Here, however, the FD-TD method is used to time step only the scattered electric and magnetic fields. That is, the FD-TD grid is not segmented into total-field and scattered-field regions, but instead assumes scattered-field quantities everywhere. This is the case for (transmitting) antenna analysis where the scattered field is thought of as the radiation field. The scattered (radiation) field is, however, a near field since it is not practical to extend the grid to the far field. To obtain a far-field radiation pattern, it is necessary to transform data in the near field to the far field as discussed in the next section.

11.7 NEAR FIELDS AND FAR FIELDS

As we have implied earlier, it is not practical to directly calculate far-field FD-TD data within the FD-TD grid because for most problems, the grid space cannot be made large enough to include the far field. Thus, near-field data must be transformed into far-field data. The existence of a well-defined scattered-field region in the FD-TD lattice, as described in the previous section, facilitates a near-to-far-field transformation that is discussed here. If we use the near-field data stored in a single FD-TD modeling run, this transformation efficiently and accurately calculates the complete radiation pattern of an antenna or the complete far-field bistatic scattering response of an illuminated structure for a single illumination angle.

To begin developing the near-to-far-field transformation, refer to Fig. 11-17, where a rectangular virtual surface S_{ab} fully enclosing the scatterer (region B) is

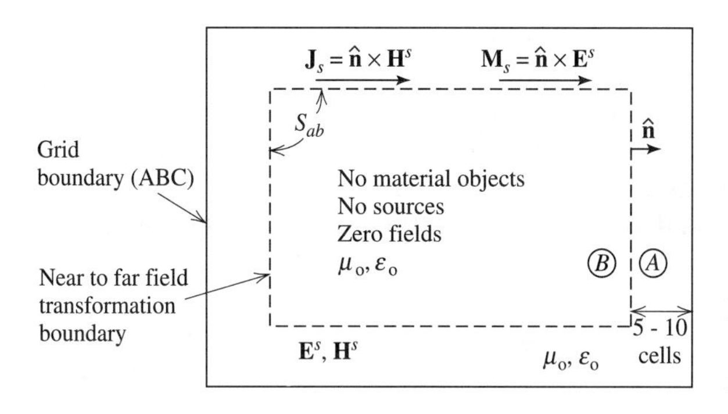

Figure 11-17 Electromagnetic equivalence to transform near fields to far fields.

located in the scattered-field region (region 2 of Fig. 11-17) near the lattice boundary. The tangential components of the scattered fields $\mathbf{E}^s$ and $\mathbf{H}^s$ are first obtained at S_{ab} using FD-TD. Then, as indicated in Fig. 11-17, an equivalent problem is set up that is completely valid for Region A, external to S_{ab}. The new excitation data are $\mathbf{J}_s$ and $\mathbf{M}_s$, the equivalent surface electric and magnetic currents, respectively, on S_{ab} that are obtained according to (see Sec. 7.1)

$$\mathbf{J}_s(\mathbf{r}) = \hat{\mathbf{n}} \times \mathbf{H}^s(\mathbf{r}) \tag{11-58a}$$

$$\mathbf{M}_s(\mathbf{r}) = -\hat{\mathbf{n}} \times \mathbf{E}^s(\mathbf{r}) \tag{11-58b}$$

where $\hat{\mathbf{n}}$ is the outward unit normal vector at the surface S_{ab}.

The scattered far fields are then given by the integration of the equivalent currents of (11-58a) and (11-58b). If (μ_o, ε_o) are the region A medium characteristics, then the following scattered far-field expressions for θ and ϕ polarizations are obtained:

$$E_\theta = -j\omega\,[A_\theta - \eta F_\phi] \tag{11-59a}$$

$$E_\phi = -j\omega\,[A_\phi + \eta F_\theta] \tag{11-59b}$$

where

$$A_\theta = A_x \cos\theta\cos\phi + A_y \sin\theta\sin\phi - A_z \sin\theta \tag{11-59c}$$

$$F_\theta = F_x \cos\theta\cos\phi + F_y \cos\theta\sin\phi - F_z \sin\theta \tag{11-59d}$$

$$A_\phi = -A_x \sin\phi + A_y \cos\phi \tag{11-59e}$$

$$F_\phi = -F_x \sin\phi + F_y \cos\phi \tag{11-59f}$$

and the potentials in the far-field region are given by

$$\begin{bmatrix} \mathbf{A} \\ \mathbf{F} \end{bmatrix} = \left(\frac{e^{-j\beta r}}{4\pi r}\right) \iint \begin{bmatrix} \mu_o \mathbf{J}_s \\ \varepsilon_o \mathbf{M}_s \end{bmatrix} e^{j\beta r' \cos\xi}\, dS'_{ab} \tag{11-60a}$$

$$r' \cos\xi = (x' \cos\phi + y' \sin\phi)\sin\theta + z' \cos\theta \tag{11-60b}$$

This approach to computing the far scattered fields is straightforward because (1) the near-field data for arbitrary antennas or scatterers can be obtained from the FD-TD calculations themselves and (2) the transformation of the near-field data to the far field is independent of the nature of the scatterer that resides within the integration surface S_{ab}.

Early FD-TD calculations of far-zone scattered fields used sinusoidal excitation. Because of this, the FD-TD far-zone results were obtained at only one frequency per FD-TD calculation run. The procedure for such single-frequency far-zone calculations is straightforward. First, the FD-TD calculations are stepped through time until steady-state conditions are reached. Then the complex time-harmonic electric and magnetic currents flowing on a closed surface surrounding the object are obtained. If these time-harmonic fields or currents are stored, then during postprocessing the far-zone radiated or scattered fields can be calculated in any desired direction. This approach is particularly suited to far-zone radiation or scattering patterns at only a single frequency.

To obtain far-zone results at multiple frequencies, the approach is to use pulsed excitations for the FD-TD calculations. For each frequency of interest, a running discrete Fourier transform (DFT) of the time-harmonic surface currents on a closed surface surrounding the FD-TD geometry is updated at each time step. The running DFT provides the complex frequency domain currents for any number of frequencies when using pulse excitation for the FD-TD calculation. This is much more efficient than using a time-harmonic excitation for each frequency of interest. It requires no more computer storage per frequency for the surface currents than when sinusoidal excitation is used and provides frequency domain far-zone fields at any far-zone angle. If far-zone results are desired at several frequencies, then the running DFT approach is the better choice.

Before leaving this section, we should mention that the near-field data itself may be of interest. Near-field data are readily obtained by selecting appropriate field values directly from the FD-TD grid. Data can include instantaneous fields, phasor fields obtained via Fourier transformation of the instantaneous fields, scalar or vector-interpolated field maps. The near-field radiation pattern of an antenna is simply the spatial distribution of the FD-TD computed radiated fields in the vicinity of the antenna. Near-fields provide insight into basic physical interactions such as reflection and diffraction. Near-field data can also be used to determine, for example, magnitude and phase data across an antenna aperture (as in the next section), surface current densities on an antenna, and current or field distributions in or along an antenna feed.

11.8 A TWO-DIMENSIONAL EXAMPLE: AN *E*-PLANE SECTORAL HORN ANTENNA

The previous sections have presented all the basics needed to do a problem from grid layout to computation of the far field. This section considers a two-dimensional TE example problem in detail. The problem is a two-dimensional model of an *E*-plane sectoral horn as illustrated in Fig. 7-15. The horn was chosen according to the optimum condition $B = \sqrt{2\lambda R_2}$ with $R_2 = 8\lambda$ so that $B = 4\lambda$. (See Sec. 7.4.2.) The resulting FD-TD model of the horn is illustrated in Fig. 11-18. Notice that the walls of the horn are "stepped." To see why this stepping comes about, examine the two-dimensional grid in Fig. 11-19. At the walls of the horn and waveguide, the coefficients C_a and C_b are calculated with a high value of conductivity (e.g., 5.7×10^7 siemens/m). Since the cells are square, a stepped contour naturally results.

The two-dimensional grid used for this example was 260×200 cells as indicated in Fig. 11-18. The cells are $\lambda/20$ by $\lambda/20$ at the center frequency. The boundary for the near-to-far-field transformation is taken to be 12 cells inside the extremities of the grid. The time function chosen is the sine modulated Gaussian pulse of (11-49c) expressed as

$$f(t) = 1.484e^{[(t-3\tau_o)/\tau_o]^2} \sin[2\pi f_c t] \quad (11\text{-}61)$$

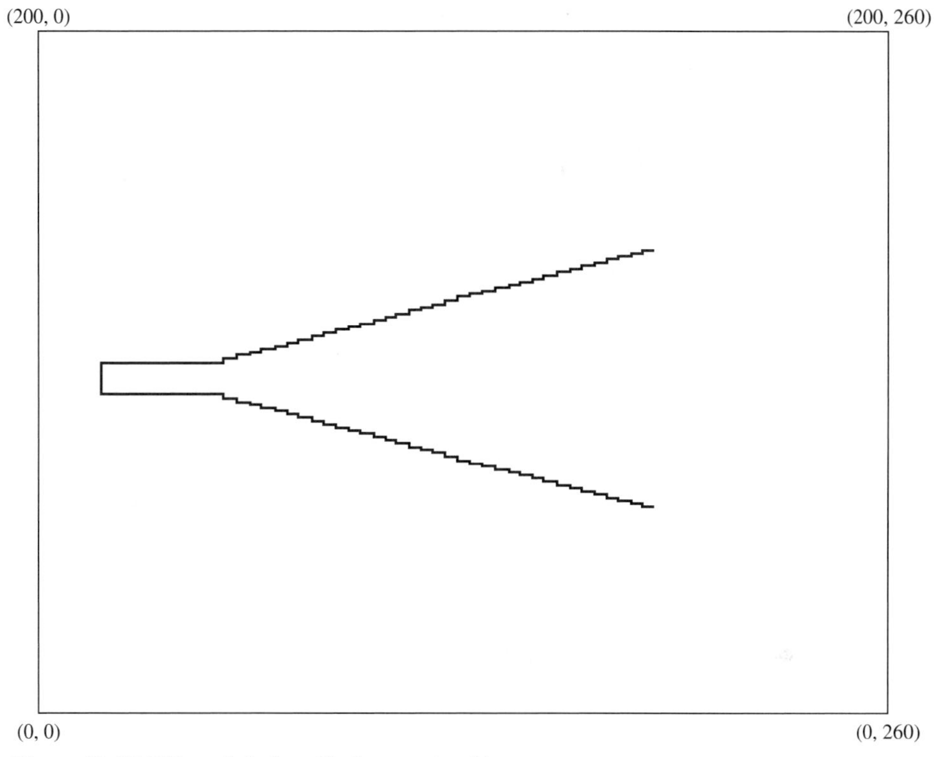

Figure 11-18 TE model of an E-plane sectoral horn.

where

$$\tau_o = 1.0/[\pi(f_h - f_c)]$$
$$f_h = 15 \text{ GHz}$$
$$f_c = 10 \text{ GHz}$$
$$\Delta x = \lambda/20 \text{ at the center frequency}$$
$$2\Delta t = \Delta x/2.99792458 \times 10^8$$
$$t = n\Delta t, \qquad \Delta t = 2.5 \times 10^{-12}$$

The plot of this function is shown in Fig. 11-20. A soft source with this time function is located $\lambda/4$ from the back wall of the waveguide at the center frequency.

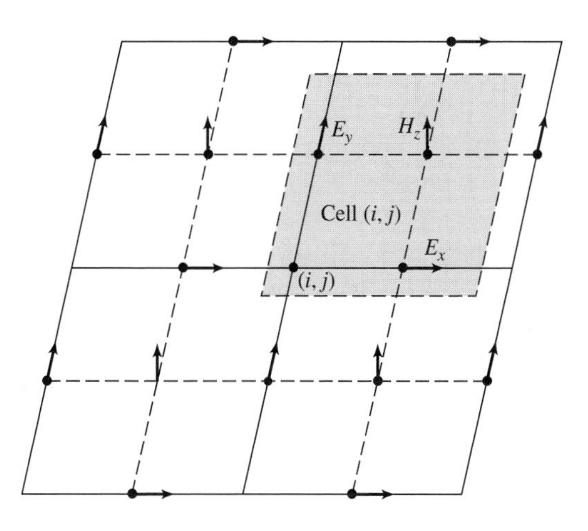

Figure 11-19 Perspective view of a two-dimensional TE grid.

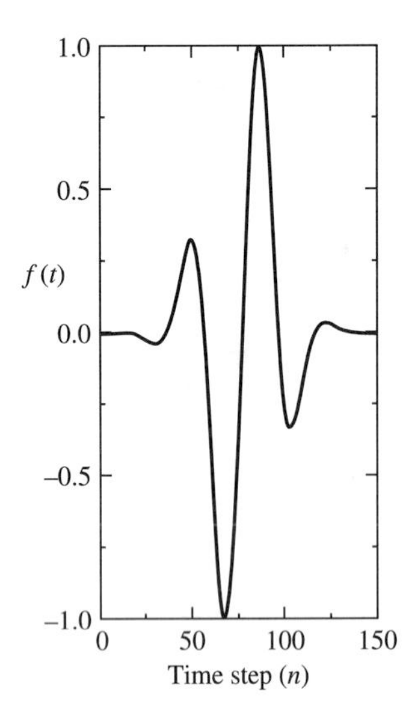

Figure 11-20 Sine modulated Gaussian waveform.

Figure 11-21*a* shows the fields in the horn at the 250th time step. The pulse has not yet reached the aperture and there is evidence of some reflection at the discontinuity where the walls of the horn join with the waveguide walls. In Fig. 11-21*b*, where $n = 475$, the peak energy content of the pulse has passed just beyond the aperture and there is some evidence of energy reflected from the aperture traveling back toward the waveguide. Diffraction at the edges of the horn is in clear evidence. In Fig. 11-21*c*, where $n = 600$, the peak energy of the pulse has moved well beyond the aperture by the 600th time step. By $n = 800$ in Fig. 11-21*d*, the pulse has reached the rear-most (left) portion of the horn geometry. The trailing weaker pulse is due to double diffraction.

In Figs. 11-21 a–d, the darkest areas indicate highest levels of field intensity and the white areas indicate zero field intensity. Even within the pulse itself there are several instants in time when the pulse is zero (see Fig. 11-20). These times of zero field intensity are the thin white lines within the main pulse and its diffractions or reflections in Fig. 11-21. Large almost-white areas are evidence of numerical noise.

Figure 11-22 shows the amplitude and phase distributions across the aperture at 9, 10, and 11 GHz. These are obtained from a Fourier transform of the fields at the aperture. From our design condition for this horn, $B = \sqrt{2\lambda R_2}$, a 90° phase change is expected from the center of the aperture to the edge. The FD-TD results are nearly in agreement with this if allowance is made for the diffraction effects near the edges of the aperture that are not included in the classical analysis of Chap. 7, but are included in the nearly exact FD-TD results. From the amplitude distribution near the edges of the aperture in Fig. 11-22 as well as the more rapid change in phase there, it is apparent that the electromagnetic wave has a strong interaction with the edges of the horn.

Figure 11-23 shows the magnitude of the electric field along the transformation boundary that partly contributes to the near-to-far-field transformation; the magnetic field accounts for the remainder. The field is evaluated at $n = 800$. Not surprisingly, the field is strongest on the side in front of the aperture and is zero on the side opposite the aperture (i.e., the back side) since the pulse has not yet had time to reach the back side. Along the sides (i.e., the top or bottom of Fig. 11-18), there is some field present, particularly near the horn aperture.

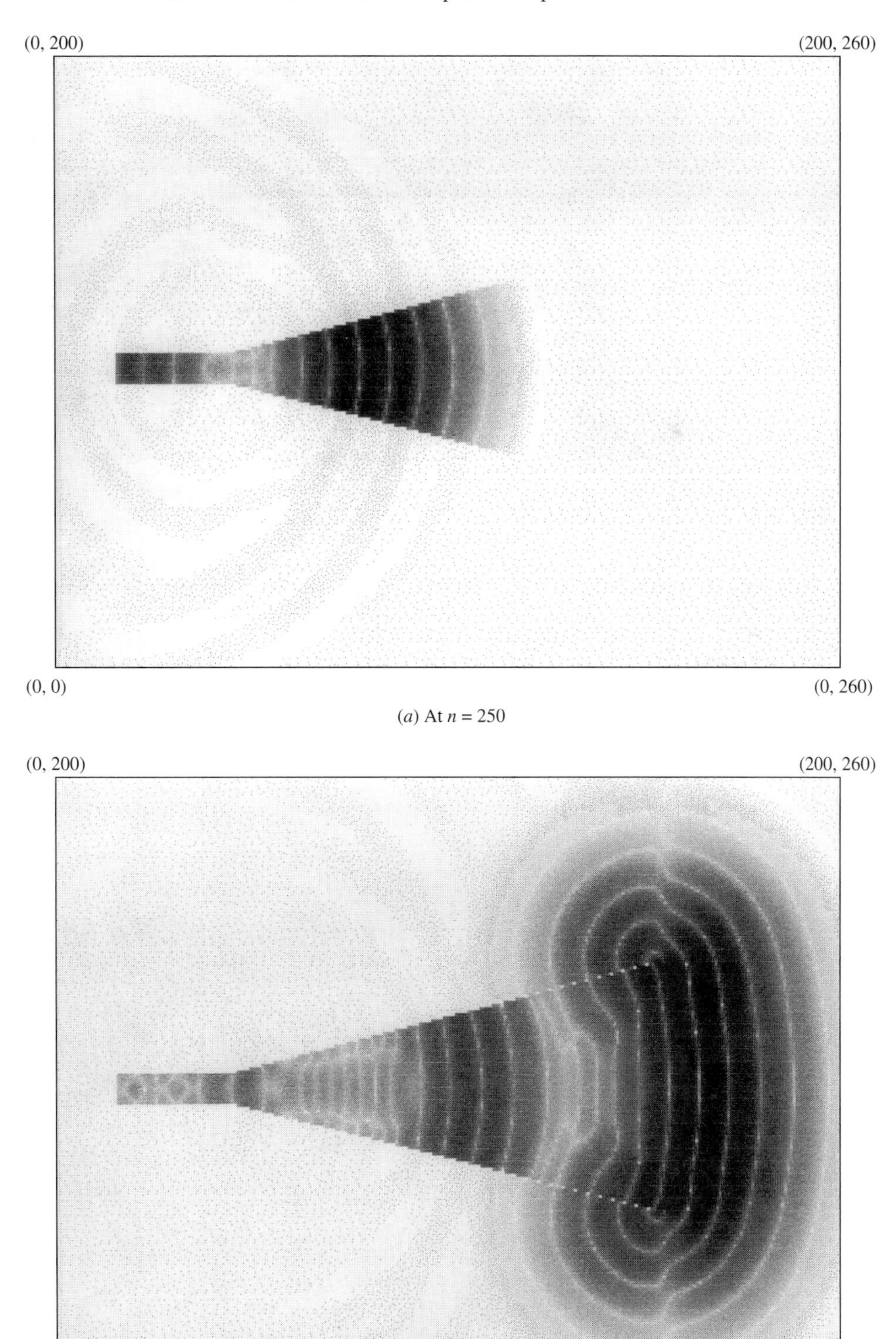

Figure 11-21 Pulse propagation at $n = 250, 475, 600, 800$.

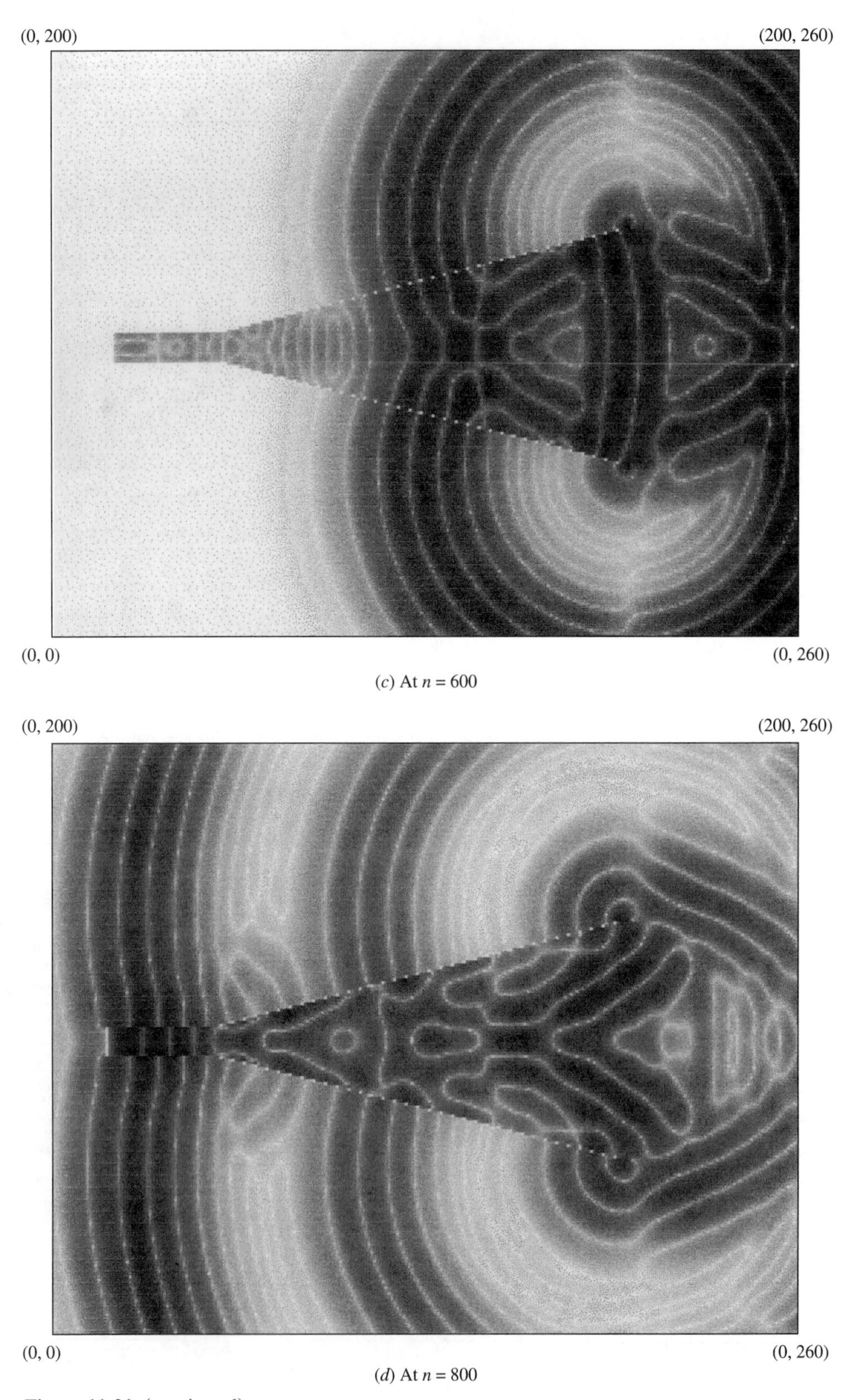

(*c*) At $n = 600$

(*d*) At $n = 800$

Figure 11-21 (continued)

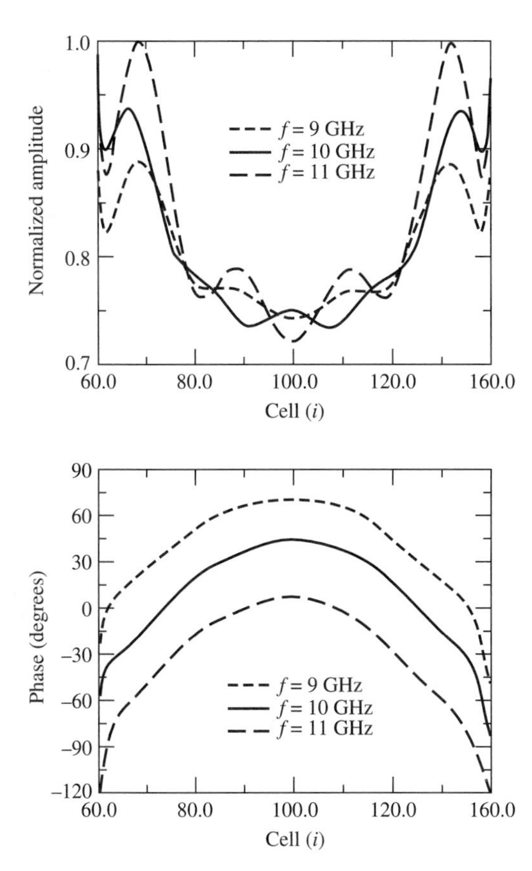

Figure 11-22 Horn aperture field distribution at $n = 800$.

Figure 11-24 shows the far-field pattern at 10 GHz computed from the near-to-far-field transformation using data at a boundary 12 cells inside the extremities of the grid in Fig. 11-18. The pattern is in good agreement with the classical pattern shown in Fig. 7-16 for the case $s = \frac{1}{4}$ when the E-plane scale on the abscissa is 4 $\sin\theta$ since $B/\lambda\ \sin\theta = 4\ \sin\theta$ here.

The forward-region far-field pattern was calculated at $n = 800$ because data in Fig. 11-25 show that steady state has been achieved on the front (right) face of the near-to-far-field transformation boundary. Near-field convergence guarantees far-

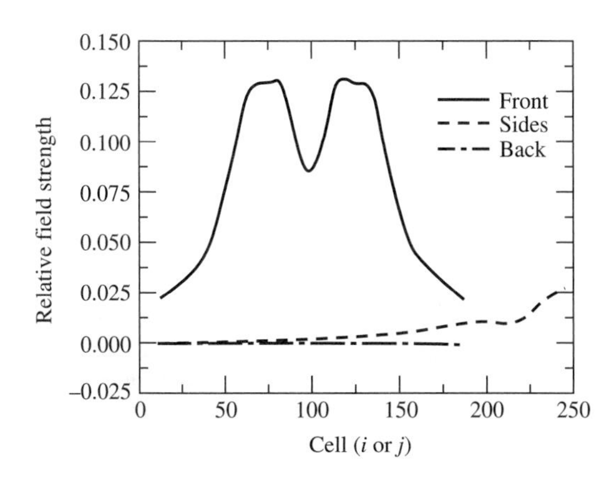

Figure 11-23 Amplitude of the 10-GHz electric field on the near-to-far-field transformation boundary at $n = 800$.

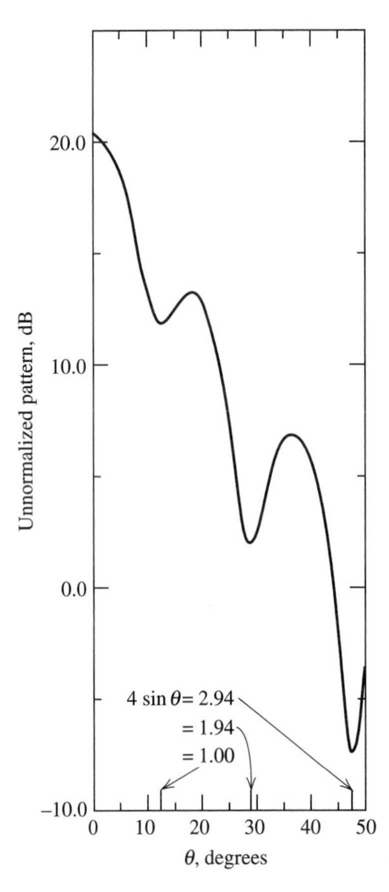

Figure 11-24 Far-field pattern at 10 GHz after 800 time steps. $B/\lambda \sin\theta = 4 \sin\theta$.

field convergence, but the reverse is not true. One could calculate the far-field pattern after more time steps, but unwanted effects such as the interaction of the wave with the exterior of the waveguide may appear in the data and may not be wanted. Figure 11-21*d* shows the start of such interaction before the disturbance has had time to reach the back face of the near-to-far-field transformation boundary. In other words, it is important to march out in time far enough to achieve steady state,

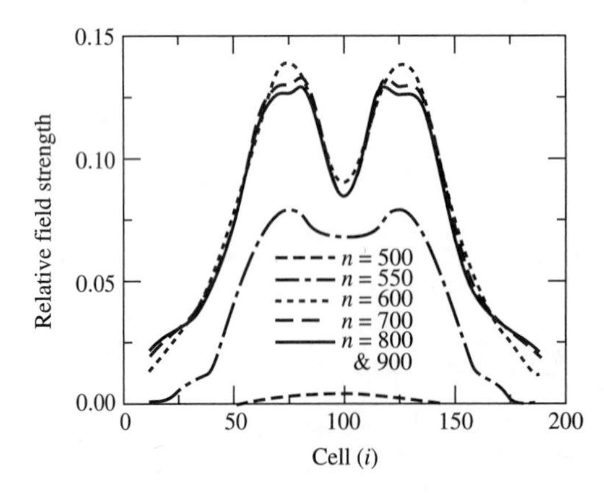

Figure 11-25 Amplitude of the 10-GHz electric field on the front part of the near-to-far-field boundary at six different time steps.

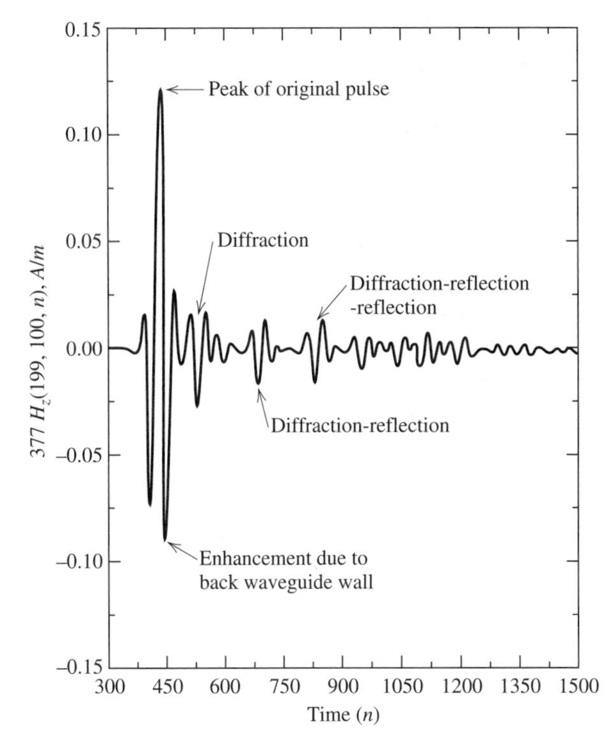

Figure 11-26 *H*-field at the center of the aperture (i = 199, j = 100) vs. n.

but not so far out in time that the desired data become contaminated with unwanted effects.

FD-TD has the potential to provide other data as well. Consider Figure 11-26 that shows the magnetic field at the center of the aperture ($i = 199, j = 100$) from $n = 300$ to $n = 1500$. The first major feature to appear is the main pulse at approximately $n = 420$ (e.g., about 210 cells traveled with two time steps per cell). About 80 time steps later (i.e., 40 cells from an edge to the aperture center), the diffraction from the edges of the horn arrives. At approximately $n = 700$ and $n = 850$, respectively, diffraction that has reflected from the opposite wall arrives, followed by diffraction that has again reflected from the other wall. These last two effects at around $n = 700$ and $n = 825$ do not affect the main-lobe region of the far-field pattern because they are propagating in directions that are not in the main-lobe direction.

The next section will consider two three-dimensional antenna problems, although not in as much detail as was done here for the two-dimensional model of the E-plane sectoral horn.

11.9 ANTENNA ANALYSIS AND APPLICATIONS

Application of FD-TD to antennas has occurred only recently relative to other applications such as shielding and radar cross section. One reason for this is that MoM can provide results for small, relatively simple antennas with much less computer time and memory than FD-TD, since MoM finds only the currents flowing on the wire or conducting surface, whereas FD-TD must calculate the fields in the entire computational region. This region should contain enough cells to allow some near-field decay between the antenna and the absorbing boundaries. If the antenna is small and geometrically simple, computing the fields in all the surrounding free-space cells makes FD-TD much less efficient than MoM. For medium-sized antennas, or antennas with geometries and/or materials [9] that are not easily included in MoM formulations, or where data are needed at many frequencies, FD-TD becomes a superior method.

When FD-TD is applied to receiving antenna calculations, it loses one of its advantages relative to the MoM in applications that require results at multiple far-zone angles. For example, in scattering applications the MoM produces results for different plane wave incidence angles efficiently from a single impedance matrix, whereas FD-TD requires a complete recalculation for each different incidence angle. However, for antenna transmitting problems, FD-TD can produce far-zone fields in any number of different directions efficiently during one computation, as can MoM. Because FD-TD also provides wide frequency band results with pulse excitation, it is extremely efficient in antenna applications, since results for impedance and radiation patterns over a wide frequency band can be obtained from one FD-TD computation. We shall see an example of this for the Vivaldi antenna at the end of this section.

11.9.1 Impedance, Efficiency, and Gain

It should not be forgotten that antenna descriptors we have become so comfortable with elsewhere in this book, such as impedance, gain, far-field patterns, and radar cross section, are frequency domain concepts. To obtain them from the FD-TD calculation process, it is necessary to Fourier-transform the appropriate voltages, currents, and fields from the time domain to the frequency domain.

In order to capitalize on the advantages of FD-TD (i.e., wide bandwidth data), it is common to utilize a Gaussian voltage pulse to excite an antenna. The Fourier transform of the voltage excitation pulse at the feed point is denoted as $V_A(\omega)$ and the Fourier transform of the current at the feed point is denoted $I_A(\omega)$. Then the input impedance is given by

$$Z_A(\omega) = \frac{V_A(\omega)}{I_A(\omega)} \tag{11-62}$$

To determine $v_A(t)$ and $i_A(t)$ in the FD-TD grid, from which $V_A(\omega)$ and $I_A(\omega)$ are derived, we employ the line integrals of **E** and **H**, respectively.

Consider a situation in which an antenna is fed with one voltage source modeled as an electric field $E_z|_{i,j,k}^n$ with corresponding voltage $v_i(t)$ across the cell at the antenna feed gap, and this source supplies a time domain current $i_i(t)$. After all transients are dissipated and the time domain results for these two quantities are Fourier-transformed, the equivalent steady-state input power *at each frequency* is given quite simply by

$$P_{\text{in}}(\omega) = \frac{1}{2} Re[V_A(\omega) I_A^*(\omega)] \tag{11-63}$$

Dissipated power due to ohmic losses is computed as follows. Suppose that an FD-TD electric field component $E_z(t)$ is in a region with conductivity σ. If we assume that the electric field is uniform within a single FD-TD cell, then at each frequency the equivalent steady-state power dissipated in this region is given by

$$\begin{aligned} P_{\text{ohmic}} &= \frac{1}{2}\iiint \sigma|E_z(\omega)|^2\,dv = \frac{1}{2}\sigma|E_z(\omega)|^2\,\Delta x\,\Delta y\,\Delta z \\ &= \frac{1}{2}\frac{\sigma\,\Delta x\,\Delta y}{\Delta z}|E_z(\omega)\,\Delta z|^2 = \frac{|V(\omega)|^2}{2R} \end{aligned} \tag{11-64}$$

where $E_z(\omega)$ is the Fourier transform of $E_z(t)$ and R is a lumped resistance across the cell in the z-direction. Knowing P_{in} and P_{ohmic} leads to a determination of radiation efficiency from (1-173).

To determine the antenna gain, the far-zone electric field in the desired direction must be determined at specified frequencies. If we use the approach given in the previous section, this can be done for pulsed far-zone fields. Since the far-zone electric field is computed so that the $1/r$ amplitude factor and the propagation delay are suppressed, the antenna gain relative to a lossless isotropic antenna in the (θ, ϕ) direction is given by

$$G(\omega, \theta, \phi) = \frac{1}{2}\frac{|E(\omega, \theta, \phi)|^2/\eta}{P_{\text{in}}/4\pi} \tag{11-65}$$

where $E(\omega, \theta, \phi)$ is the peak value of the Fourier transform of the pulsed far-zone time domain electric field radiated in the (θ, ϕ) direction.

11.9.2 The Monopole over a PEC Ground Plane

Maloney et al. [10] used FD-TD to model the radiation from two simple antennas: the cylindrical monopole and the conical monopole. Here, we shall consider only the former, the cylindrical monopole of height h over a PEC ground plane. The FD-TD grid used to model this antenna is shown in Fig. 11-27. The grid used a two-

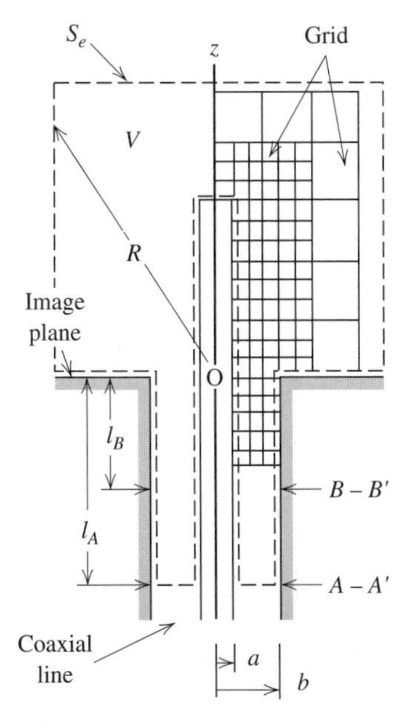

Figure 11-27 Geometry of the two-dimensional cylindrical-coordinate FD-TD grid used to model the transient excitation of a coaxial-fed monopole over a PEC ground plane. (Maloney et al., *IEEE Trans. Ant. & Prop.*, Vol. 38, 1990, pp. 1059–1068. Reprinted with permission.)

dimensional cylindrical-coordinate algorithm, exploiting the rotational symmetry of both the antenna and feeding coaxial line about the z-axis. In the cylindrical system, the TE mode is composed of E_ϕ, H_ρ, and H_z, whereas the TM mode has the components E_ρ, E_z, and H_ϕ. Since the coaxial line was excited with a TEM mode consisting of just E_ρ and H_ϕ, only the TM cylindrical mode was modeled.

A Gaussian pulse voltage excitation within the coaxial line $v(t) = v_0 \exp(-t^2/2\tau_p^2)$ was used at source plane $A - A'$ in combination with an exact ABC to emulate a matched source at that location. The following normalized parameters apply to the model: $b/a = 2.30$ (50-Ω coaxial line); $h/a = 65.8$; $\tau_p/\tau_a = 8.04 \times 10^{-2}$. Here τ_p is the $1/e$ width of the excitation pulse and $\tau_a = h/c =$ characteristic antenna height. In other words, τ_a represents the time required for an electromagnetic wave to travel the length of the monopole. Finally, an ABC of only first-order accuracy was used to terminate the grid at its outer boundary S_e.

The cylindrical monopole represents a two-dimensional electromagnetic problem. For example, the radiator in Fig. 11-27 is rotationally symmetric and is excited by a rotationally symmetric source

$$\mathbf{E}^i(t) = \frac{v^i(t)}{\ln(b/a)\rho}\,\hat{\boldsymbol{\rho}} \tag{11-66}$$

The applicable components of Maxwell's curl equations are

$$\frac{\partial E_\rho}{\partial z} - \frac{\partial E_z}{\partial \rho} = -\mu_o \frac{\partial H_\phi}{\partial t} \tag{11-67a}$$

$$-\frac{\partial H_\phi}{\partial z} = \varepsilon_o \frac{\partial E_\rho}{\partial t} \tag{11-67b}$$

$$\frac{1}{\rho}\frac{\partial(\rho H_\phi)}{\partial \rho} = \varepsilon_o \frac{\partial E_z}{\partial t} \tag{11-67c}$$

After discretization of the above, we have

$$\begin{aligned} H_\phi\Big|_{i,j}^{n+1/2} = H_\phi\Big|_{i,j}^{n-1/2} &+ \frac{\Delta t}{\mu_o \Delta \rho}\left[E_z\Big|_{i+1/2,j}^{n} - E_z\Big|_{i-1/2,j}^{n}\right] \\ &- \frac{\Delta t}{\mu_o \Delta z}\left[E_\rho\Big|_{i,j+1/2}^{n} - E_\rho\Big|_{i,j-1/2}^{n}\right\} \end{aligned} \tag{11-68a}$$

$$E_\rho\Big|_{i,j-1/2}^{n+1} = E_\rho\Big|_{i,j-1/2}^{n} - \frac{\Delta t}{\varepsilon_o \Delta z}\left[H_\phi\Big|_{i,j}^{n+1/2} - H_\phi\Big|_{i,j-1}^{n+1/2}\right] \tag{11-68b}$$

$$E_z\Big|_{i+1/2,j}^{n+1} = E_z\Big|_{i+1/2,j}^{n} + \frac{\Delta t}{\varepsilon_o \Delta \rho}\frac{1}{\rho_{i+1/2}}\left[\rho_{i+1}\, H_\phi\Big|_{i+1,j}^{n+1/2} - \rho_i\, H_\phi\Big|_{i,j}^{n+1/2}\right] \tag{11-68c}$$

Note that the grid in Fig. 11-28 is arranged so that the electric field component tangential to the surface of a perfect conductor is evaluated at the surface.

An absorbing boundary condition is used at the surface S_e; this allows the observation period to be extended beyond $t = t_o$. If we look in the opposite direction, the TEM field within the coaxial line behaves like the one-dimensional case examined in previous sections. Thus, an exact absorbing boundary condition can be constructed within the coaxial line. The incident field is additively injected at a plane $z = -\ell$, and the absorbing boundary condition, placed at $z = -(\ell + \Delta z)$, exactly absorbs the field of a TEM mode propagating in the $-z$-direction. This allows the

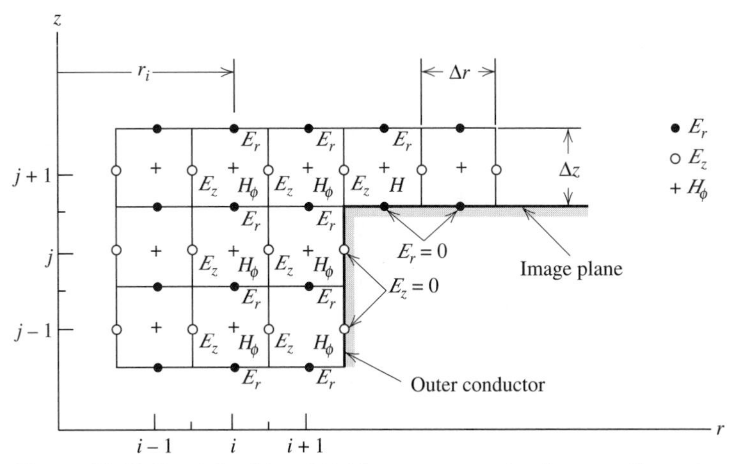

Figure 11-28 Spatial grid and field components for the two-dimensional problem with cylindrical symmetry. (Maloney et al., *IEEE Trans. Antennas and Propagation*, 1990, pp. 1059–1068, 1990 IEEE. Reprinted with permission.)

cross section at which the incident field is specified to be moved closer to the ground plane; namely, in Fig. 11-27, $B - B'(z = \ell_B)$ is used instead of $A - A'(z = -\ell_A)$. This reduces both the time required for observation and the size of the grid.

The spatial and temporal increments ($\Delta\rho$, Δz, and Δt) are chosen to satisfy the Courant–Friedrichs–Levy condition in the cylindrical system [5]:

$$c\Delta t \le \sqrt{\frac{\Delta\rho^2\,\Delta z^2}{\Delta\rho^2 + \Delta z^2}} \tag{11-69}$$

In this work, two spatial grid spacings are used: a fine spacing ($\Delta\rho_1 \approx \Delta z_1$) within the coaxial line and close to the antenna where the field is varying rapidly with spatial position, and a coarse grid [$\Delta\rho_2 = (3 - 5)\,\Delta\rho_1$, $\Delta z_2 = (3 - 5)\,\Delta z_1$] in the remainder of the space. The use of the dual grid reduces computer storage. Note when (11-69) is satisfied for the fine grid, it is automatically satisfied for the course grid. In the example that follows,

$$c\Delta t = \frac{\min(\Delta\rho_1, \Delta z_1)}{2} \tag{11-70}$$

and the increments $\Delta\rho_1$, Δz_1 are chosen small enough to resolve the spatial variation of the field.

Figure 11-29 is a space-time plot of the FD-TD calculated surface charge density on the monopole antenna and its feeding coaxial line. At point A in this figure, the incident pulse has reached the antenna. An impedance mismatch between the feedline and the antenna causes some of the energy to reflect back down the line. The remaining energy then propagates along the length of the antenna until the end of the antenna is reached at point B. Here, some energy radiates while the remaining energy reflects back down the antenna. This represents the (imperfect) transition from the antenna to free space.

At point C, the antenna/feedline junction causes a partial retroreflection, with some energy continuing down the coax and the remainder going back up the monopole. This process repeats itself until all transients have decayed. It is significant to note that at no time did energy enter the antenna from the coax once the incident

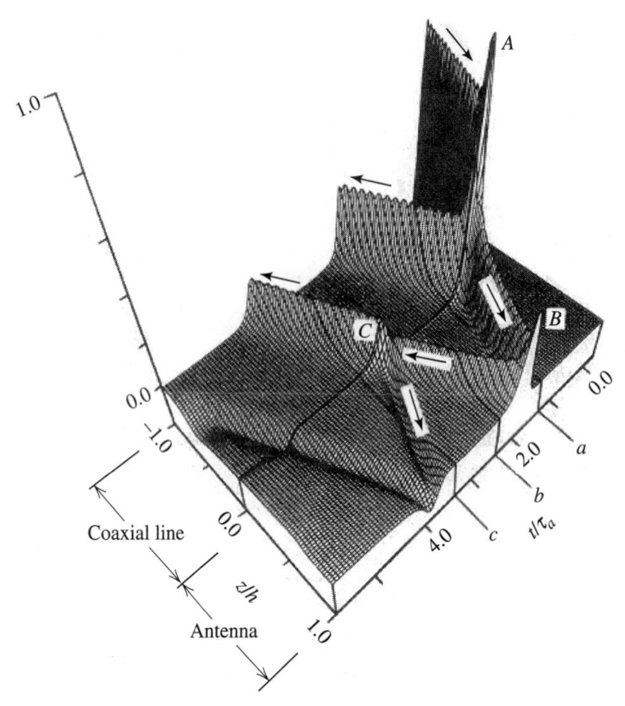

Figure 11-29 Normalized surface charge density on the cylindrical monopole antenna as a function of the normalized position z/h and the normalized time t/τ_a. (Maloney et al., *IEEE Trans. Ant. & Prop.*, Vol. 38, July 1990, pp. 1059–1068. Reprinted with permission.)

pulse had propagated through. This verified that the ABC at the line feedpoint was working properly.

Figure 11-30 shows the FD-TD-computed radiated fields for three snapshots in time. In Fig. 11-30*a*, the initial outgoing wavefront W_1 was produced after the exciting pulse passed the feedline/antenna transition. Note the reflected energy traveling back down the coaxial feedline. In Fig. 11-30*b*, a second outgoing wavefront W_2 was produced when the energy traveling up the antenna was reflected from its top end. In Fig. 11-30*c*, both W_1 and W_2 have propagated away from the antenna, but a third wavefront W_3 was generated when energy retroreflected from the feedline/antenna transition. Also in Fig. 11-30*c*, the wavefront W_{2R} arises from the reflection of the W_2 wavefront from the ground plane. This process repeats until the surface charge density decays to zero.

The far-zone electric field E_θ for the cylindrical monopole antenna is shown in Fig. 11-31. The surface used for these calculations was the cylindrical boundary separating the fine and coarse grids in Fig. 11-27. Each trace in this figure shows the electric field at a fixed polar angle θ as a function of the normalized time t/τ_a. Notice that the shape of each time domain trace is different for each polar angle because each trace has a different frequency content. This is due to the radiation patterns in the frequency domain being different at each frequency. Also notice that wavefronts from the same point on the antenna are always separated by a time interval that is a multiple of $2\tau_a$, the round-trip transit time for the pulse on the antenna. For example, wavefronts W_1 and W_3, which are centered on the drive point, are separated by the time $2\tau_a$, as are wavefronts W_2 and W_4, which are centered on the end.

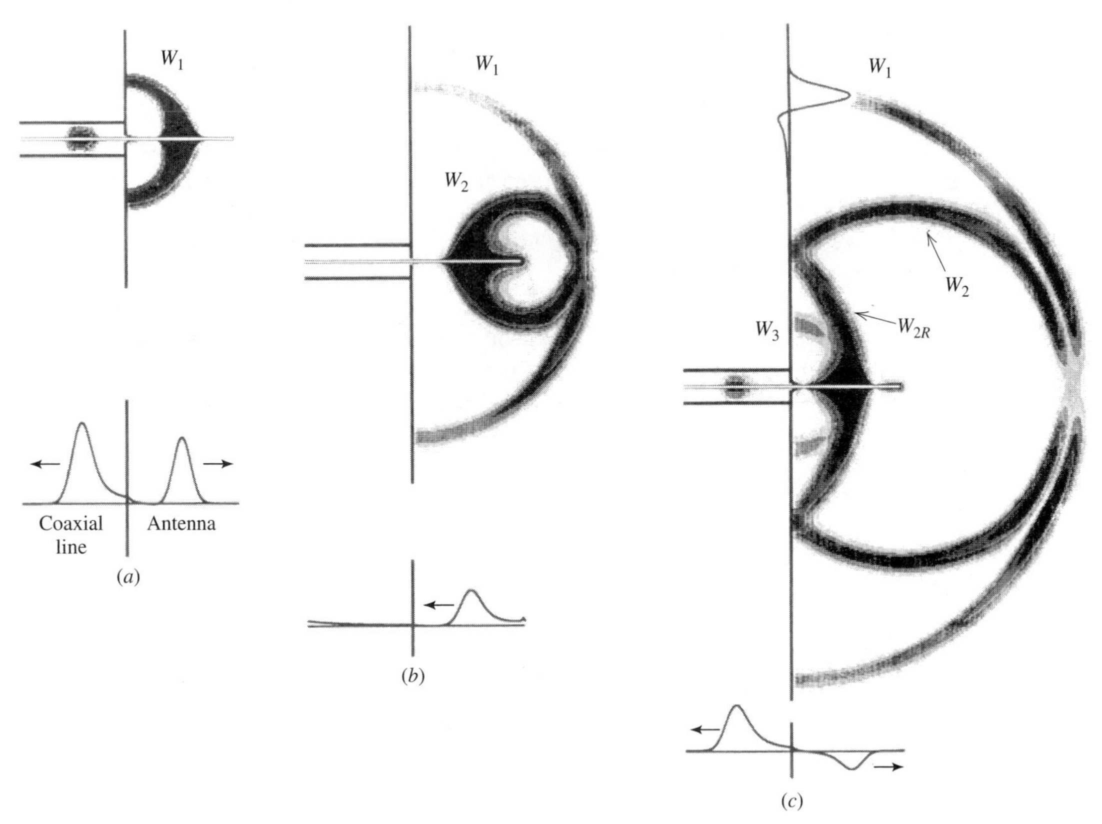

Figure 11-30 FD-TD-computed radiation of a Gaussian pulse from the cylindrical monopole antenna. The gray scale plots show the magnitude of the electric field, whereas the line drawings show the surface charge density on the antenna and the feeding coaxial life. (Maloney et al., *IEEE Trans. Ant. & Prop.*, Vol. 38, July 1990, pp. 1059–1068. Reprinted with permission.)

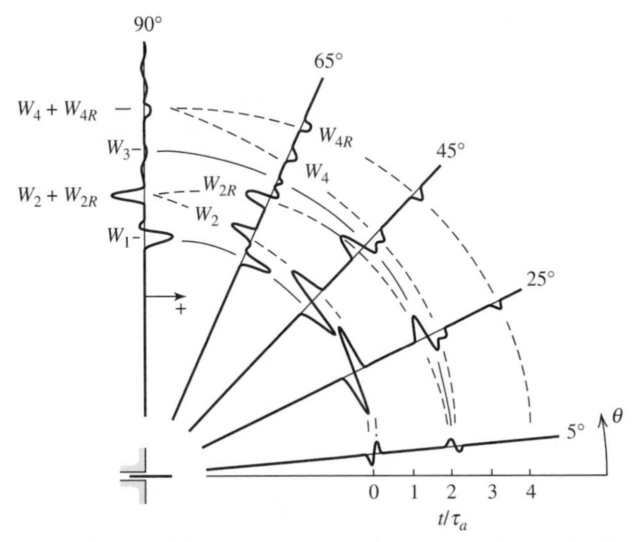

Figure 11-31 Radiation of a Gaussian pulse from a cylindrical monopole antenna. Each trace shows the far-zone electric field ε_θ at a fixed polar angle θ as a function of the normalized time t/τ_a. $b/a = 2.30$, $h/a = 65.8$, and $\tau_p/\tau_a = 8.04 \times 10^{-2}$. (Maloney et al., *IEEE Trans. Antennas and Propagation*, 1990, pp. 1059–1068, 1990 IEEE. Reprinted with permission.)

11.9.3 The Vivaldi Slotline Array

Slotline antennas, such as the Vivaldi, are traveling-wave antennas that produce broadband endfire radiation. In this structure, a microstrip slotline is flared outward to an aperture where the wave impedance matches free space. This is often referred to as a *tapered slot antenna* (TSA). It is this type of slot antenna that was modeled by Thiele [11, 13] using FD-TD.

Figure 11-32 illustrates the geometry of a planar Vivaldi element. The FD-TD cell size was set to $\Delta s = 0.5$ mm, based on the smallest physical dimension of this model that is the throat of the horn. This corresponds to a resolution of $\lambda/33$ to $\lambda/99$ over the 6 to 18 GHz bandwidth of the element. This high resolution permitted a simple stepped-edge model to simulate the antenna radiation characteristics nearly as accurately as a more elaborate conformal contour-path model [12]. Therefore, all subsequent modeling was performed using stepped edges (see Fig. 11-18). The

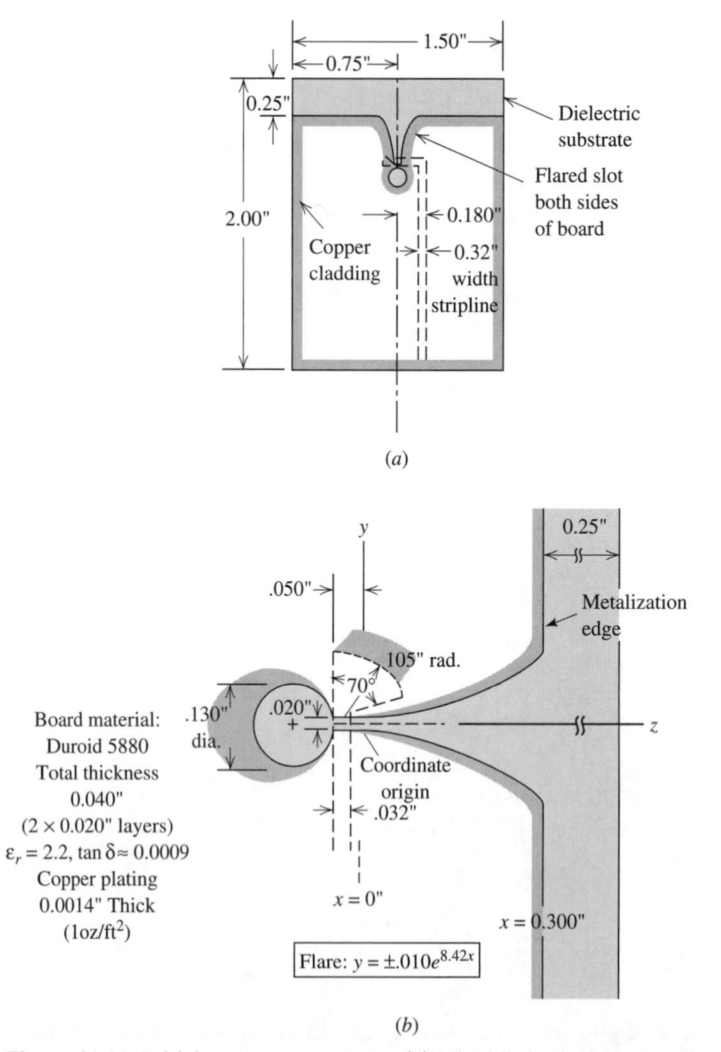

Figure 11-32 Initial antenna geometry. (*a*) Vivaldi single-flare baseline element with 0.25″ substrate protrusion. (*b*) Detail of the strip line feed and the slot element. (E. Thiele and A. Taflove, *IEEE Trans. Ant. & Prop.*, Vol. 42, 1994, pp. 633–641. Reprinted with permission.)

resulting grid size for the individual flare Vivaldi models was $42 \times 116 \times 142$ cells, corresponding to 4.2 million field unknowns. CPU times were on the order of 800 sec for a single-processor Cray Y-MP.

We examine FD-TD and measurement data for the eight-element array of Vivaldi quads depicted in Fig. 11-33*b*. This linear array, modeled with 32 individual feeds, can be excited with varying phase and amplitude distribution to steer the beam and select polarization. Co-pol and cross-pol gain patterns were calculated in the plane of the array (the E-plane) between 6 and 18 GHz for beam-steer angles of 0°, 20°, 45°, and 60°. Initially, sinusoidal excitations at selected frequencies were used, with an appropriate phase taper across the array provided for beam-steering. This excitation method was later dropped in favor of using pulsed array excitations coupled with on-the-fly DFTs of the fields at the near-to-far-field observation locus, thereby reducing computer time requirements by about two orders of magnitude. A single pulsed excitation run could cover the complete 6- to 18-GHz band, with an appropriate time-delay taper across the array provided for the desired beam-steering angle.

The eight-element array increased the FD-TD grid size to $222 \times 222 \times 140$ cells containing 41.4 million vector-field components. Run times were about one CPU hour using automatic multiprocessor tasking on a dedicated eight-processor Cray Y-MP/8. It is probable that this intensive use of supercomputing resources for an FD-TD antenna model was without precedent at the time of these runs.

Figure 11-34 graphs the FD-TD results for the E-plane co-pol and cross-pol radiation patterns for the eight-element array, if we assume a nominal 45° beam steer. It is clear that grating lobes evolved as the operating frequency increased. In fact, the principal grating lobe equaled or exceeded the nominal main beam for frequencies greater than 15 GHz. The cross-pol levels were quite high, rising to within 10 dB of the co-pol levels in the main beam at all the frequencies modeled.

Simple array theory can be used to qualitatively assess the results of this section. We note that in *both* the computed and measured patterns, the nominal beam-

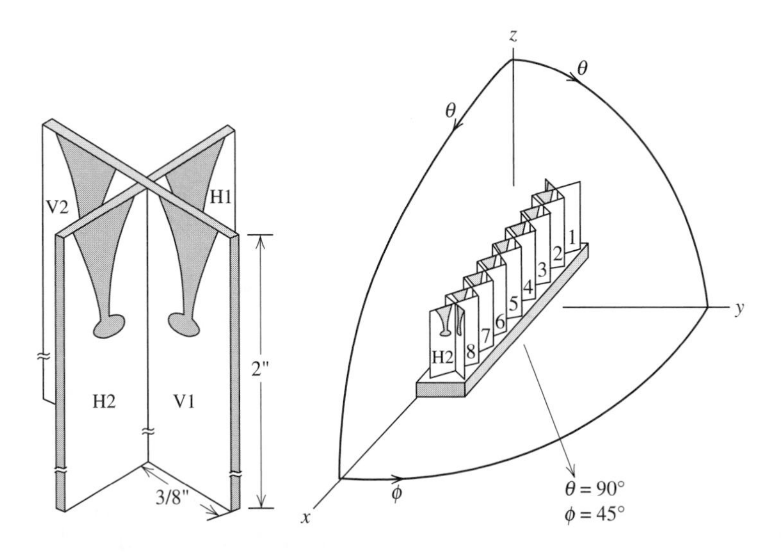

(*a*) Single quad element (*b*) Array of eight quad elements

Figure 11-33 Final geometry of the quad Vivaldi element and the eight-element array (without 0.25″ protrusion of the dielectric substrate). (E. Thiele and A. Taflove, *IEEE Trans. Ant. & Prop.*, Vol. 42, 1994, pp. 633–641. Reprinted with permission.)

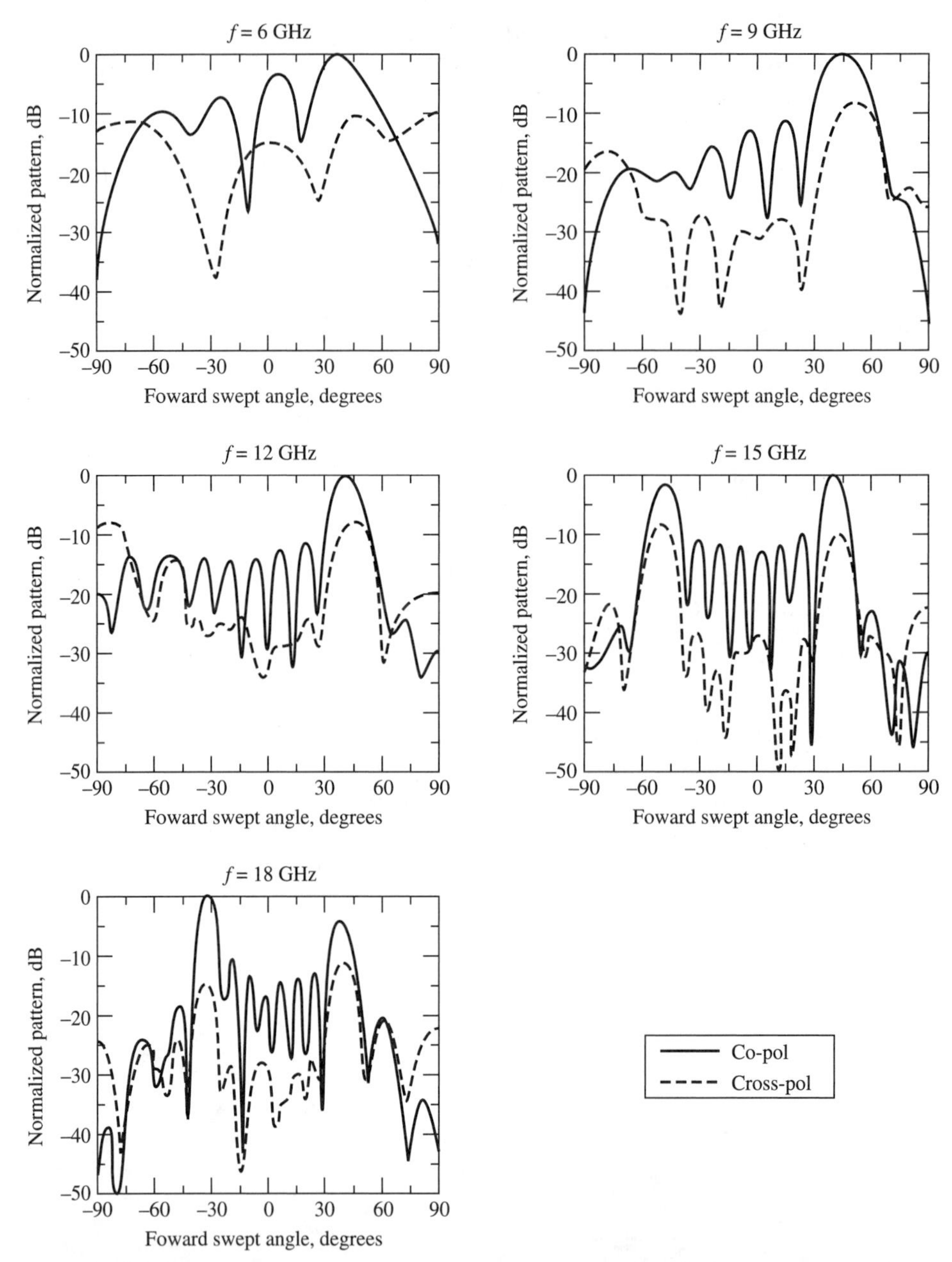

Figure 11-34 FD-TD-computed E-plane co-pol and cross-pol radiation patterns for the eight-element linear array of Vivaldi quads of Fig. 11-33 phased for a 45° beam-steer. Note the evolution of grating lobes as the operating frequency increases. (From [13], reprinted with permission.)

steering angle was not quite reached. For example, in Figure 11-34 (bottom), a desired beam-steer angle of 45° resulted in an actual beam-steer angle of approximately 40°. Array theory predicts the overall radiation pattern to be the product of the element pattern and the array factor. The deviation of the beam-steer angle from this pattern multiplication prediction can be attributed to mutual coupling.

Active impedance was defined in Chap. 3 as the driving-point impedance of a given array element when all the elements of the array are excited. When antenna elements are near one another, as in the case of most arrays, complex interactions occur between all of the elements, changing the current distribution of any particular element relative to its distribution when isolated in free space. Since beam-steering is implemented via changes in the excitation of a given element, the current distributions vary on all the elements due to their mutual coupling, in turn varying the driving-point impedance of each element. In practice, array active impedance has been difficult to predict and measure due to the complexity of the mutual coupling and its sensitivity to the test setup. A direct FD-TD approach used to calculate the driving-point impedance is discussed in [1].

FD-TD results are presented for the active impedance of the eight-element quad-Vivaldi array from 6 to 18 GHz for a beam-steer angle of 45°. In the presentation of these data in [13], the driving-point impedance of each of the four feeds of each of the eight quad elements is depicted separately. This is because geometrical asymmetries arising in the construction of each quad element caused corresponding electrical asymmetries of the driving-point impedance noted for each of the four feeds of each element. Here we present data for only two (orthogonal) feeds.

Figure 11-35 graphs the driving-point impedance data in a three-dimensional per-

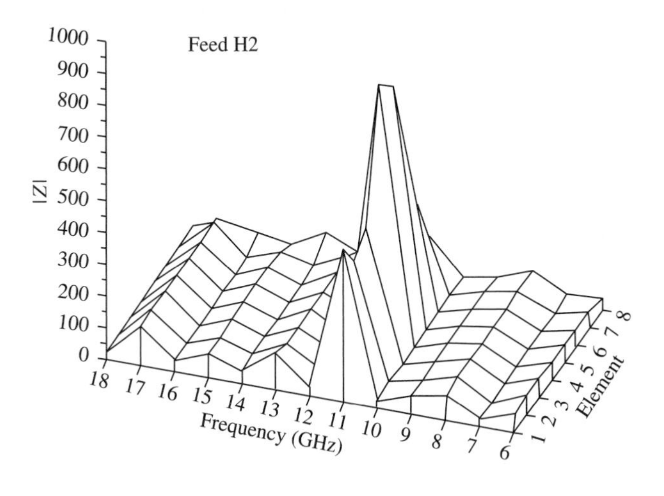

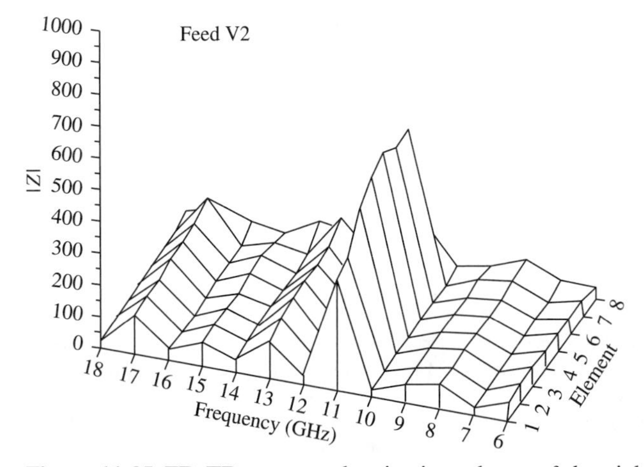

Figure 11-35 FD-TD-computed active impedance of the eight-element array of Fig. 11-33 for a 45° beam-steer and the type 2 strip line feeds. (From [13], reprinted with permission.)

spective view. The frequency (in 1-GHz increments) and the quad-element number (1 through 8) are shown as independent variables, and the magnitude of the driving-point impedance is depicted as a "height" above the frequency/element-number plane. In Fig. 11-35, the FD-TD calculated impedance for the feeds classified in the groups "H2" and "V2" was high at 11 GHz, with lesser peaks at 13 and 17 GHz.

Overall, FD-TD calculations indicated that the impedance behavior of the array was within desired voltage standing wave ratio (VSWR) specifications of 2:1 or less (25 to 100 Ω) for much of the 6- to 18-GHz design bandwidth. However, the case shown here had very high VSWR at 11 GHz. Therefore, the array would likely fail at this frequency. Knowledge of this problem would be sufficient to permit (it is hoped) modest changes in the feeds to meet the VSWR specification throughout the entire bandwidth.

11.10 SUMMARY

In this chapter, the finite difference time domain, or FD-TD, method has been presented. FD-TD is a differential-equation-based method in the time domain that employs approximations to derivatives in a solution of Maxwell's equations that "marches on in time and space." The basic features of FD-TD were presented in Secs. 11.2 through 11.6. The one-dimensional case was used to illustrate fundamental principles, but the two-dimensional and three-dimensional cases were also considered (e.g., Secs. 11.8 and 11.9). Only simple media were employed, but FD-TD is well suited to complex media as well, such as inhomogeneous and anisotropic media [1, 2, 12].

Both MoM (in the previous chapter) and FD-TD are usually thought of as intermediate frequency techniques because they cannot easily accommodate bodies that are arbitrarily large in terms of the wavelength. In contrast to this, Chap. 12 presents high-frequency or asymptotic methods that apply best to material structures (i.e., antennas or scatterers) arbitrarily large in terms of the wavelength.

REFERENCES

1. A. Taflove, *Computational Electrodynamics—The Finite-Difference Time-Domain Method*, Artech House, Boston, MA, 1995.
2. K. S, Kunz and R. J. Luebbers, *The Finite Difference Time Domain Method*, CRC Press, Boca Raton, FL, 1993.
3. E. T. Thiele, private communications, 1995.
4. K. S. Yee, "Numerical Solution of Initial Boundary Value Problems Involving Maxwell's Equations in Isotropic Media," *IEEE Trans. Ant. & Prop.*, Vol. 14, April 1966, pp. 302–307.
5. E. Isaacson and H. B. Keller, *Analysis of Numerical Methods*, Wiley, New York, 1967.
6. G. Mur, "Absorbing Boundary Conditions for the Finite-Difference Approximation of the Time-Domain Electromagnetic Field Equations," *IEEE Trans. Electromag. Compatibility*, Vol. 23, 1981, pp. 377–382.
7. J. P. Berenger, "A Perfectly Matched Layer for the Absorption of Electromagnetic Waves," *J. Computational Phys.*, Vol. 114, 1994, pp. 185–200.
8. S. D. Gedney, "An Anistropic Perfectly Matched Layer-absorbing Medium for the Truncation of FDTD Lattices," *IEEE Trans. Ant. & Prop.*, Vol. 44, December 1996, pp. 1630–1639.
9. M. A. Jensen and Y. Rahmat-Samii, "EM Interaction of Handset Antennas and a Human in Personal Communications," *Proc. IEEE*, Vol. 83, January 1995, pp. 7–17.
10. J. G. Maloney, G. S. Smith, and W. R. Scott, "Accurate Computation of the Radiation from Simple Antennas Using the Finite-Difference Time-Domain Method," *IEEE Trans. Ant. & Prop.*, Vol. 38, July 1990, pp. 1059–1068.

11. E. Thiele and A. Taflove, "FD-TD Analysis of Vivaldi Flared Horn Antennas and Arrays," *IEEE Trans. Ant. & Prop.*, Vol. 42, May 1994, pp. 633–640.
12. K. L. Shlager and J. B. Schneider, "A Selective Survey of the Finite-Difference Time-Domain Literature," *IEEE Ant. & Prop. Magazine*, Vol. 37, January 1995, pp. 39–56.
13. E. T. Thiele, "FD-TD Analysis of Vivaldi Flared Horn Antennas and Arrays," Ph.D. dissertation, Northwestern University, Evanston, IL, 1994.

PROBLEMS

11.1-1 Reduce (11-5) to (11-7) and (11-8).
11.1-2 Reduce (11-7) to (11-10) and reduce (11-8) to (11-11).
11.2-1 a. Obtain (11-14) from (11-12) by expanding H_y in a Taylor series about the temporal point t_n and by expanding E_z about the spatial point x_i.
b. Show that the central difference approximations are second-order accurate. That is, error $= O(\Delta z^2) + O(\Delta t^2)$.
c. Does this tell how much error is in the solution?
11.2-2 Derive (11-17) from (11-10b).
11.2-3 Derive (11-22) from (11-21).
11.2-4 Derive (11-27) from (11-6c).
11.2-5 Derive from (11-5b) an expression for H_y similar to (11-26).
11.2-6 Derive from (11-6b) an expression for E_y similar to (11-27).
11.2-7 Derive (11-30a) from (11-26).
11.3-1 Show that the one-dimensional central difference equations (11-14) and (11-17) produce an exact solution when $c\Delta t = \Delta x$.
11.3-2 Can numerical dispersion occur in a non-dispersive medium?
11.3-3 Consider a plane wave in free space:

$$H_y = H_o e^{j(\omega t - \beta x)}$$

If the plane wave is discretized in time and space

$$H_y(x_i, t_n) = H_o e^{j(\omega n \Delta t - \beta_{\text{num}} i \Delta x)}$$

where $t = n\Delta t$, $x = i\Delta x$, $\beta_{\text{num}} = \dfrac{\omega}{v}$. v is the numerical phase velocity.

a. Write an expression for the phase error of the discretized wave relative to the actual wave and comment on the amount of phase error as the propagation distance increases.
b. How can the error in phase be overcome?
11.4-1 With reference to Fig. 11-9, calculate by hand E and H for the fourth time step when $i < 50$. Then, from your results verify that there is power flow in the negative x-direction.
11.5-1 Recompute by hand Table 11-1 if there is a perfect electrical conductor in cell 53.
11.5-2 Show that the second order Mur estimate in (11-43) reduces to an exact 1-D result.
11.6-1 Write a one-dimensional computer code and verify the results presented in Fig. 11-12 for the delta function hard source.
11.6-2 Extend the computer code in Prob. 11.6-1 to accommodate soft sources and verify Fig. 11-13.
11.6-3 With reference to Fig. 11-13, calculate by classical means the reflection and transmission coefficients at the dielectric to air interface and compare with the magnitude of the reflected and transmitted fields in the figure when $n = 400$.
11.6-4 At what time step in Fig. 11-13 does the peak value of the reflected Gaussian pulse arrive back at the source? Arrive at your answer by assuming the magic time step and then determining how many time steps are required.
11.6-5 In terms of the update equations, explain why the soft source in Fig. 11-13 allows the wave reflected from the dielectric to pass onward to the right and the hard source does not.
11.6-6 Verify (11-52) using (11-53).
11.6-7 Verify (11-54) using (11-55).

11.6-8 Explain why it is more computationally efficient to use FD-TD for antennas when the antenna is transmitting than when it is receiving.
11.6-9 Extend Prob. 11.5-1 out to $n = 8$ if the source at $i = 50$ is a hard source. Repeat if the source is a soft source.
11.7-1 Using the surface equivalence theorem, show that the far-field pattern of an antenna may be computed using (11-59) and (11-60) applied to a surface surrounding the antenna.
11.8-1 In Fig. 11-22, exclude edge effects on the phase and show that the phase change from the center of the aperture to the edges is 90° at 10 GHz as required by the optimum condition under which this horn was designed. (Note the amplitude distribution near the edges and use that to estimate the region over which the edges are having a strong effect on both the amplitude and phase distributions.)
11.8-2 Compare Fig. 11-24 to Fig. 7-16.
11.8-3 By making measurements on Fig. 11-18 with a ruler, verify the time at which the various physical phenomena depicted in Fig. 11-26 occur.
11.8-4 Write a two-dimensional computer code (or use an existing one) and verify Fig. 11-22.
11.8-5 Make a photocopy of Figs. 11-21 a–d and indicate on the copy the various physical phenomena that you see there.
11.9-1 Verify (11-64).
11.9-2 Derive (11-65).
11.9-3 Derive (11-68) from (11-67).

Chapter 12

CEM for Antennas: High-Frequency Methods

Optics is a well-understood area of physics that deals with the characteristics of light wave propagation. It was Maxwell who showed before 1873 that the propagation of light could be viewed as an electromagnetic phenomenon. Since the wavelength of light waves is usually small compared to objects with which it interacts, the analytical treatment of light wave propagation is much different than that employed to analyze lower-frequency propagation where the size of a scattering surface is comparable to the wavelength.

A very useful and easily understood method for analyzing optical problems is the ray concept. The relationship between ray optics and wave propagation is apparent from the famous works of Huygens in 1690 and Fresnel in 1818, but was not formally shown until the works of Luneberg in 1944 and Kline in 1951 [1]. Since that time the well-known methods of optics have found increasing use in the treatment of many electromagnetic problems in the radio frequency portion of the spectrum for situations where the wavelength is small compared to the geometrical dimensions of the scatterer or antenna. In these cases, asymptotic high-frequency methods must be employed since it is not practical to use moment methods (Chap. 10) or eigen-function expansions. This is because the rate of convergence of both of these techniques is generally quite poor when dealing with an electrically large antenna or scatterer.

In this chapter, we will first examine the principles of geometrical optics. We will then see that in many situations geometrical optics is inadequate to completely describe the behavior of the electromagnetic field and it is necessary to include another field called the diffracted field. The diffracted field, when added to the geometrical optics field, permits us to solve many practical radiation and scattering problems in a moderately straightforward manner that could not be solved any other way.

Geometrical optics and its extension to include diffracted fields is a field-based method (see Fig. 10-1) and does not require the calculation of currents. Later in this chapter, current-based methods will be discussed wherein currents are used to ultimately determine the field quantities of interest. These methods are physical optics and its extension to include diffraction. In many situations, a physical optics current is inadequate to produce accurate fields from a radiating object and it is necessary to include another current called the non-uniform current. The non-uniform current, when added to the physical optics current, permits an accurate representation of the fields to be obtained. Whether a field-based or a current-based method is to be used depends on the specific application, as we shall see in the sections that follow.

12.1 GEOMETRICAL OPTICS

Geometrical optics, or *ray optics* as it is often called, was originally developed to analyze the propagation of light where the frequency is sufficiently high that the wave nature of light need not be considered. Indeed, geometrical optics can be developed by simply considering the transport of energy from one point to another without any reference to whether the transport mechanism is particle or wave in nature.

Classical geometrical optics applies to isotropic lossless media that may or may not be homogeneous. In this chapter, we will only consider homogeneous media where the index of refraction n is assumed to be real and is given by

$$n = \frac{c}{v} \tag{12-1}$$

and is not a function of position within a given medium. Here, c is approximately 3×10^8 m/s and v is the velocity of propagation in the medium. In a homogeneous medium, energy moves along ray paths that are straight lines. Normal to these ray paths are a family of surfaces called the *eikonal* of the ray system. In applying geometrical optics, it is only necessary that we know either the eikonal of the ray system or the ray paths, since the two are uniquely related.

For a plane wave in homogeneous media, the eikonal surfaces are planes perpendicular to the ray paths as shown in Fig. 12-1*a*. For a spherical source, the eikonal surfaces are spherical surfaces perpendicular to the ray paths as shown in Fig. 12-1*b*.

The variation of the amplitude of the geometrical optics field within a ray tube is determined by the law of energy conservation since the rays are lines of energy flow. Consider two surfaces ρ_o and $\rho_o + \Delta\rho$ as shown in Fig. 12-2. Between the two surfaces, we can construct a tube of constant energy flux by using the rays. Thus, the energy through cross section $d\sigma_o$ at P_o must equal the energy flux through cross

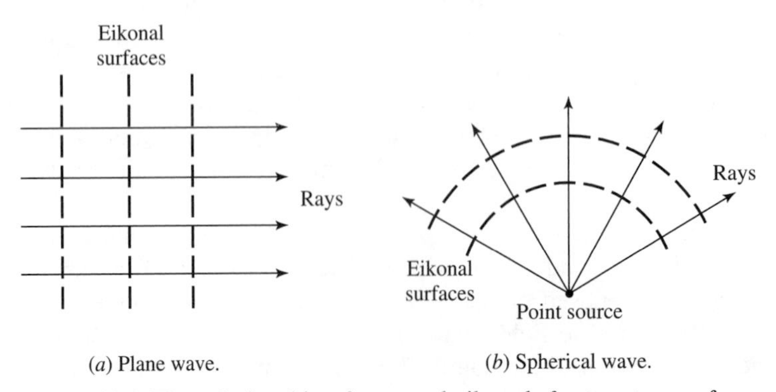

(*a*) Plane wave. (*b*) Spherical wave.

Figure 12-1 The relationship of rays and eikonals for two types of sources.

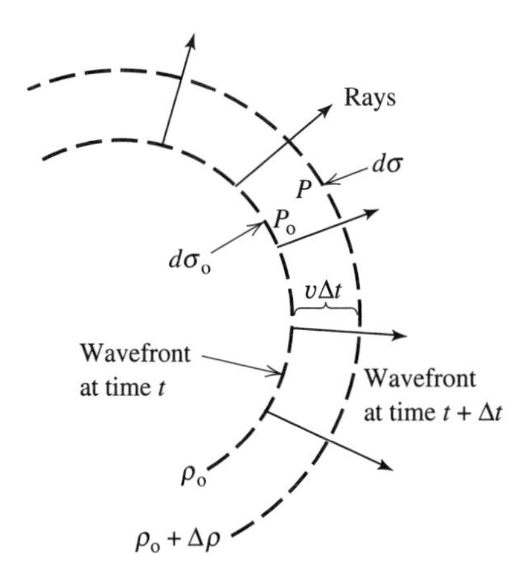

Figure 12-2 The relationship of rays and wavefronts.

section $d\sigma$ at P. If S is the power per unit area, the condition of constant energy flow through the flux tube is

$$S_o \, d\sigma_o = S \, d\sigma \tag{12-2}$$

In the case of electromagnetic waves, the quantity S is the real part of the complex Poynting vector and we can assume that

$$S = \frac{1}{2}\sqrt{\frac{\varepsilon}{\mu}}\,|E|^2 \tag{12-3}$$

Substituting (12-3) into (12-2) yields

$$|E_o|^2 \, d\sigma_o = |E|^2 \, d\sigma \tag{12-4}$$

Solving for $|E|$, we obtain

$$|E| = |E_o|\sqrt{\frac{d\sigma_o}{d\sigma}} \tag{12-5}$$

Therefore, we have obtained a relationship between the amplitude of the geometrical optics field at one point in terms of the amplitude at another.

The relationship in (12-5) would be more useful if the radii of curvature of the wavefront surfaces $d\sigma$ and $d\sigma_o$ were used. Consider the astigmatic ray tube picture in Fig. 12-3. The principal radii of curvature of $d\sigma_o$ are ρ_1 and ρ_2, whereas the

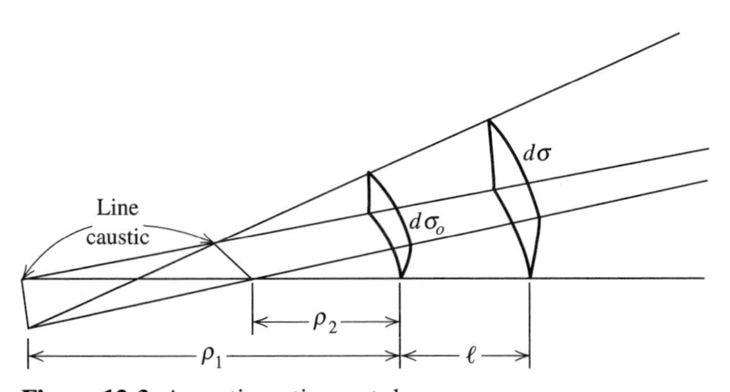

Figure 12-3 An astigmatic ray tube.

principal radii of curvature of $d\sigma$ are $(\rho_1 + \ell)$ and $(\rho_2 + \ell)$. We can write for the ratios

$$\frac{d\sigma_o}{\rho_1\rho_2} = \frac{d\sigma}{(\rho_1 + \ell)(\rho_2 + \ell)} \tag{12-6}$$

and thus

$$\frac{d\sigma_o}{d\sigma} = \frac{\rho_1\rho_2}{(\rho_1 + \ell)(\rho_2 + \ell)} \tag{12-7}$$

From (12-5), we have

$$|E| = |E_o|\sqrt{\frac{\rho_1\rho_2}{(\rho_1 + \ell)(\rho_2 + \ell)}} \tag{12-8}$$

Note that the tube of rays converge to a line at $\rho_1 = 0$ and $\rho_2 = 0$ where the cross section of the ray tube goes to zero. Therefore, the amplitude of the geometrical optics field description becomes infinite there although the actual field does not. The locus of points where the ray tube cross section exhibits such behavior is called a *caustic*. Caustics may be a point, line, or surface. For example, consider a point source as shown in Fig. 12-4. We can construct a ray tube from four rays and write

$$\frac{d\sigma_o}{\rho^2} = \frac{d\sigma}{(\rho + \ell)^2} \tag{12-9}$$

Thus,

$$|E| = |E_o|\sqrt{\frac{\rho^2}{(\rho + \ell)^2}} = |E_o|\frac{\rho}{\rho + \ell} \tag{12-10}$$

The caustic would be located at the point source in this case.

In both (12-8) and (12-10), we note that as ℓ becomes large, we have the usual inverse distance-type field dependence found in the far zone of a three-dimensional source. Often, however, one is concerned with two-dimensional problems where one of the radii of curvature, say, ρ_2, becomes infinite. In such problems,

$$|E| = |E_o|\sqrt{\frac{\rho_1}{\rho_1 + \ell}} \tag{12-11}$$

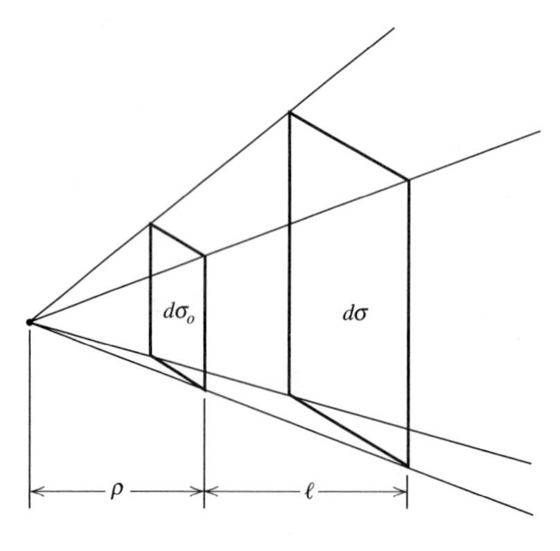

Figure 12-4 A tube of rays from a point source.

Here, the eikonal surfaces are cylindrical and, as $\ell \to \infty$, we have an amplitude dependence of the field at large distances of the form $1/\sqrt{\ell}$. Obviously, if both ρ_1 and ρ_2 are infinite, the eikonal surfaces are planes and $|E|$ is a constant for all values of ℓ (e.g., a plane wave).

The results of (12-8), (12-10), and (12-11) are extremely important for they permit us to easily compute the amplitude of the geometrical optics field at one point in terms of its known value at another. In electromagnetic field problems, however, we must also include the phase. Phase can be introduced into (12-8) artificially. First, we take our phase reference to coincide with the amplitude reference. Thus, the electrical phase of the ray tube is given by $e^{-j\beta\ell}$ and we may write for the amplitude and phase of the field in the ray tube of Fig. 12-3

$$E = E_o e^{j\phi_o} \sqrt{\frac{\rho_1 \rho_2}{(\rho_1 + \ell)(\rho_2 + \ell)}}\, e^{-j\beta\ell} \tag{12-12}$$

or

$$E = E_o e^{j\phi_o} A(\rho_1, \rho_2, \ell) e^{-j\beta\ell} \tag{12-13}$$

where E_o is the reference amplitude at $\ell = 0$, ϕ_o is the reference phase at $\ell = 0$, $A(\rho_1, \rho_2, \ell)$ is the general spatial attenuation factor, and $e^{-j\beta\ell}$ is the spatial phase delay factor.

Note that when ℓ becomes less than $-\rho_2$, the quantity under the radical sign in $A(\rho_1, \rho_2, \ell)$ becomes negative and a phase jump of $\pi/2$ occurs when the observer passes through the caustic. Although we cannot predict the amplitude or the phase of the geometrical optics field at the caustic, we can determine the fields on either side of the caustic.

Equation (12-12) or (12-13) permits us to approximately express the field at a point (i.e., ℓ) in terms of the value at a known point (i.e., $\ell = 0$). Rigorously, the result is only approximate, becoming more accurate as the wavelength tends toward zero. In practice, however, we will find the geometrical optics expression above to be highly accurate for engineering purposes where the assumptions of geometrical optics are valid.

To finish our initial discussion of geometrical optics, we illustrate its use by considering the problem of reflection at a curved smooth surface and the subsequent calculation of the radar cross section of a sphere. From (12-12), it is apparent that we need an expression for the radii of curvature of the wavefront in terms of the geometrical radii of curvature of the surface. Consider Fig. 12-5 that depicts a line source parallel with the axis of a convex cylinder of arbitrary cross section. From Fig. 12-5*a*,

$$\gamma_1 = \pi - \alpha - (\pi - \theta_o) = \theta_o - \alpha \tag{12-14}$$

The element of arc length in Fig. 12-5*b* is equal to $r_1^c\, \Delta\alpha$ and

$$r_1^c\, \Delta\alpha = \frac{\Delta\gamma_1 \ell_o}{\cos\theta_o} = \frac{(\Delta\theta_o - \Delta\alpha)\ell_o}{\cos\theta_o} \tag{12-15}$$

Since $\Delta\gamma_2 = \Delta\theta_o + \Delta\alpha$, we have

$$r_1^c\, \Delta\alpha = \frac{\rho_1\, \Delta\gamma_2}{\cos\theta_o} = \rho_1 \frac{\Delta\theta_o + \Delta\alpha}{\cos\theta_o} \tag{12-16}$$

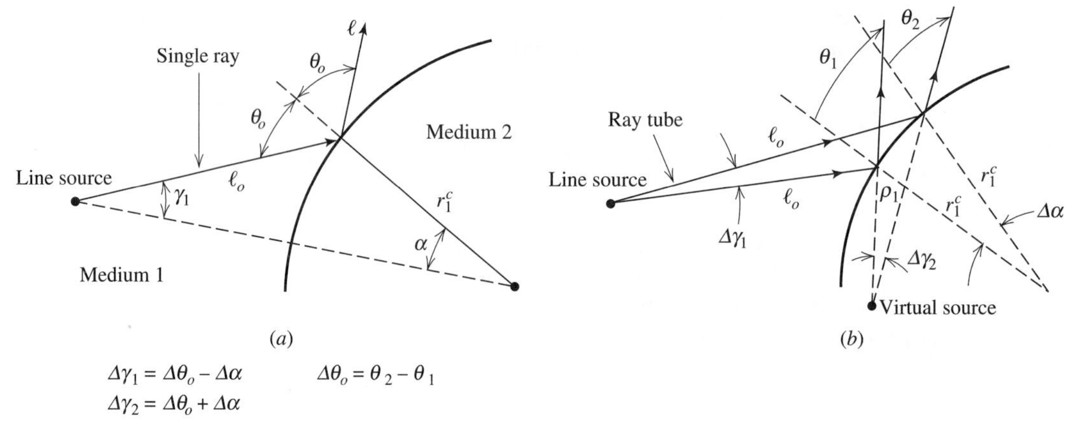

Figure 12-5 Ray geometry for reflection by a curved conducting surface.

Solving both (12-15) and (12-16) for $r_1^c\,\Delta\alpha\cos\theta_o$, we have, respectively,

$$r_1^c\,\Delta\alpha\cos\theta_o = \ell_o\,\Delta\theta_o - \ell_o\,\Delta\alpha \tag{12-17}$$

and

$$r_1^c\,\Delta\alpha\cos\theta_o = \rho_1\,\Delta\theta_o + \rho_1\,\Delta\alpha \tag{12-18}$$

Solving both these equations for $\Delta\alpha$ and equating the two results yield

$$\frac{\ell_o\,\Delta\theta_o}{r_1^c\cos\theta_o + \ell_o} = \frac{\rho_1\,\Delta\theta_o}{r_1^c\cos\theta_o - \rho_1} \tag{12-19}$$

which after some manipulation gives us the desired result

$$\frac{1}{\rho_1} = \frac{1}{\ell_o} + \frac{2}{r_1^c\cos\theta_o} \tag{12-20}$$

This equation[1] relates a principal radii of curvature of the reflected wavefront to the geometrical radius of curvature of the surface at the point where the ray strikes the surface.

As a simple example of the application of (12-20), consider the situation shown in Fig. 12-6 where a plane wave is incident on a sphere. We wish to find the field

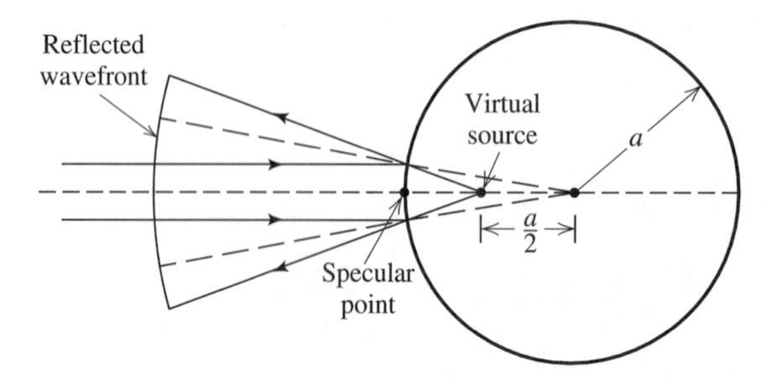

Figure 12-6 Geometrical optics scattering by a sphere.

[1]Even though this result is based on a two-dimensional configuration, the result is somewhat more general than this in that it holds true in the plane of incidence (see Sec. 12-1) whenever the plane of incidence coincides with the principal planes of the surface.

scattered back in the direction of the transmitter and from this *back-scattered field* find the radar cross section of the sphere. Thus, the only ray we need consider is that reflected from what is called the *specular point.* In this situation, then, $\ell_o = \infty$, $\theta_o = 0°$, and $r_c = a$ in (12-20) and we have the following result:

$$\rho_1 = \frac{a}{2} = \rho_2 \tag{12-21}$$

where ρ_2 is the radius of curvature of the reflected wavefront orthogonal to ρ_1. (See Prob. 12-1 for an expression for ρ_2.)

If the incident field has a value of E_o at the specular point, then in the backscattered direction,

$$E^s = -E_o \frac{\rho_1}{\rho_1 + \ell} e^{-j\beta\ell} \tag{12-22}$$

with ρ_2 having the same value as ρ_1 in this example. Therefore, if we use (9-35) the radar cross section is (at high frequencies)

$$\sigma = \lim_{\ell\to\infty} 4\pi\ell^2 \left[\frac{a/2}{a/2 + \ell}\right]^2 = \pi a^2 \tag{12-23}$$

The exact value for $\sigma/\pi a^2$ is shown in Fig. 12-7. We note that as the radius of the sphere becomes larger, the more closely the geometric optics cross section approaches the exact result. That is what one would expect since geometrical optics assumes the wavelength is small when compared to the geometrical dimensions of the scattering surface. Furthermore, the result in (12-23) is frequency-independent, which is typical of geometrical optics calculations [2–4].

We can extrapolate from (12-22) to write a general expression for the geometrical optics field due to a plane wave reflected from a smooth surface. Let a plane of incidence be defined by the incident ray and the normal to the surface. Let $E^i_\parallel(Q_r)$ and $E^i_\perp(Q_r)$ be the components of the incident field that are parallel and perpendicular, respectively, to the plane of incidence at the point of reflection Q_r, and let $E^r_\parallel(\ell)$ and $E^r_\perp(\ell)$ be the components of the reflected field that are parallel and perpendicular to the plane of incidence, respectively. Then, in matrix form,

$$\begin{bmatrix} E^r_\parallel(\ell) \\ E^r_\perp(\ell) \end{bmatrix} = \begin{bmatrix} E^i_\parallel(Q_r) \\ E^i_\perp(Q_r) \end{bmatrix} \cdot [R]\sqrt{\frac{\rho_1\rho_2}{(\rho_1 + \ell)(\rho_2 + \ell)}}\, e^{-j\beta\ell} \tag{12-24}$$

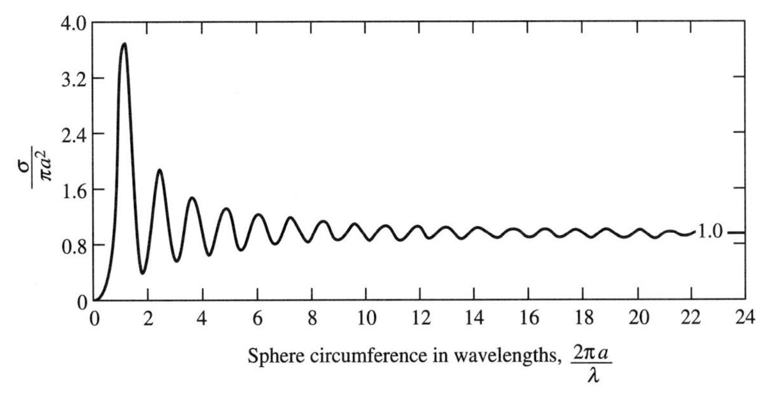

Figure 12-7 Radar cross section σ of a sphere versus the electrical size of the sphere.

where R is a reflection coefficient matrix, and for perfectly conducting surfaces appears as

$$R = \begin{bmatrix} R_{\parallel} & 0 \\ 0 & R_{\perp} \end{bmatrix} \tag{12-25}$$

where $R_{\parallel} = +1$ and $R_{\perp} = -1$ and are recognized as representing the Fresnel reflection coefficients for parallel and perpendicular polarization reflection from plane perfectly conducting surfaces.[2] The Fresnel reflection coefficients imply that the incident wave is a plane wave and the reflecting surface is also planar. We can deviate from these restrictions at high frequencies (short wavelengths) by noting that geometrical optics reflection is a local phenomenon and, therefore, the incident field need only be locally plane at the reflecting point and the surface need only be adequately approximated by a plane tangent to the surface at the point of reflection.

If we apply geometrical optics to reflection from a surface when the source point and reflection point are fixed, then the observation point is determined for us by the law of reflection. That is, we obtain information about the reflected field in one direction only (the specular direction), whereas the reflection typically spreads out over some angular region. To obtain information about the reflected field in non-specular directions, it is necessary to first consider what the current is on the reflecting surface and then integrate that current to get the reflected field (e.g., the aperture integration of Chap. 7). In the Sec. 12.13, we will examine the physical optics method of doing this.

12.2 WEDGE DIFFRACTION THEORY

In the previous section, we introduced the ray-optical concept of geometrical optics. The theory was applied to the calculation of the backscattered field from a sphere, but no attempt was made to determine the field in the forward scattering direction, in particular, the shadowed region in Figs. 12-6 or 12-8. By simple ray tracing, it is

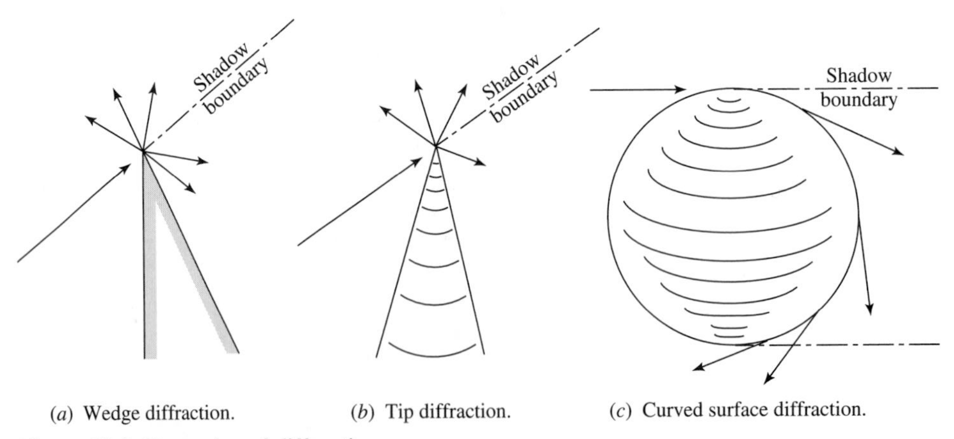

(*a*) Wedge diffraction. (*b*) Tip diffraction. (*c*) Curved surface diffraction.

Figure 12-8 Examples of diffraction.

[2]The signs on the entries $R_{\parallel}$ and $R_{\perp}$ depend on the reference directions used for the incident and reflected electric field vectors. The treatment here is consistent with that in Fig. 5-47, but others are in common use; see [35].

quite apparent that geometrical optics is incapable of correctly predicting a nonzero field in the shadow region. However, geometrical optics may be extended to include a class of rays, called *diffracted rays* [5, 6], that permit the calculation of fields in the shadow region of a scatterer. Diffracted rays are produced, for example, when a ray strikes an edge, a vertex, or is incident tangentially to a curve surface as illustrated in Fig. 12-8. It is these rays that account for a nonzero field in the shadow region. In addition, they also modify the geometrical optics field in the illuminated region. It is the purpose of this section to examine in some detail one type of diffracted ray, the wedge diffracted ray of Fig. 12-8*a*.

Consider the wedge diffraction situation shown in Fig. 12-8*a*. Geometrical optics would predict a sharp discontinuity in the field at a *shadow boundary* as shown in Fig. 12-9. Since physical phenomena in nature are not perfectly discontinuous, it is apparent that geometrical optics by itself constitutes an incomplete treatment of problems such as those in Fig. 12-8. It will be shown that the wedge diffracted rays will make the total electric field continuous across the shadow boundary in Fig. 12-8*a*.

Because diffraction is a local phenomena at high frequencies, the value of the field of a diffracted ray is proportional to the field value of the incident ray at the point of diffraction multiplied by a coefficient called the *diffraction coefficient*. That is, the diffraction coefficient is determined largely by the local properties of the field and the boundary in the immediate neighborhood of the point of diffraction. Since it is only the local conditions near the point of diffraction that are important, the diffracted ray amplitude may be determined from the solution of the appropriate boundary value problem having these local properties. Such a problem is called a canonical problem and wedge diffraction is one such canonical problem. Wedge diffraction is perhaps the most important canonical problem in the extension of geometrical optics as originally proposed by Joseph Keller in 1953. Keller's theory is known as the *geometrical theory of diffraction*, or GTD [5–7].

Through the use of geometrical optics and the solution to a number of canonical problems, such a those in Fig. 12-8, we can construct solutions to more complex problems via the principle of superposition. Let us now consider the canonical problem of wedge diffraction. To start, we will consider scalar diffraction by an infinitely conducting and infinitesimally thin half-plane sheet as shown in Fig. 12-10. The half-plane is a wedge of zero included angle. To calculate the field in the region $z > 0$, we will use Huygens' principle in two dimensions. Thus, each point on the primary wavefront along $z = 0$ is considered to be a new source for a secondary cylindrical wave, the envelope of these secondary cylindrical waves being the secondary wavefront. Thus,

$$E(P) = \int_{x=a}^{x=\infty} dE \tag{12-26}$$

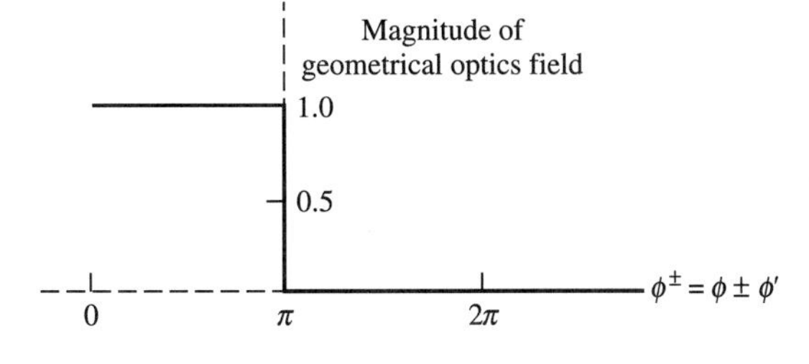

Figure 12-9 Magnitude of the geometrical optics field near either a reflected field shadow boundary ($\phi^+ = \phi + \phi' = \pi$) or an incident field shadow boundary ($\phi^- = \phi - \phi' = \pi$).

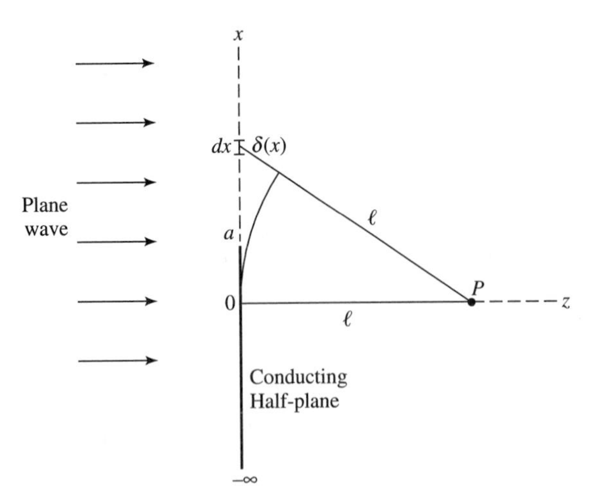

Figure 12-10 Plane wave diffraction by a conducting half-plane.

where dE is the electric field at P due to a magnetic line source parallel to the y-axis in the $z = 0$ plane, or

$$dE = \frac{C_1}{\sqrt{\ell + \delta(x)}} e^{-j\beta[\ell+\delta(x)]}\, dx \tag{12-27}$$

where C_1 is a constant. If $(\ell + \delta) \gg \lambda$ and $\ell \gg \delta$, we may write for the contribution to $E(P)$ from those two-dimensional Huygens' sources between $x = a$ and $x = x_o$

$$E(P) = \frac{C_1}{\sqrt{\ell}} e^{-j\beta\ell} \int_a^{x_o} e^{-j\beta\delta(x)}\, dx \tag{12-28}$$

We will consider the contribution from x_o to ∞ later. When $\delta \ll \ell$, we can follow the same reasoning as in (1-84) to show that $\ell + \delta \approx \ell + x^2/2\ell$. Making the substitutions $\gamma^2 = 2/\lambda\ell$ and $u = \gamma x$ gives

$$E(P) = C_1\sqrt{\lambda/2}e^{-j\beta\ell} \int_{\gamma a}^{\gamma x_o} e^{-j(\pi/2)u^2}\, du \tag{12-29}$$

If the upper limit in (12-29) is allowed to go to infinity, the integral will be in the standard form of a Fresnel integral [7, 8]. The Fresnel integral may be easily evaluated on a computer or from a graph known as Cornu's spiral, which is shown in Fig. 12-11*a*. A vector drawn from the origin to any point on the curve represents the magnitude of a Fresnel integral with lower limit zero and upper limit u_o. As u_o approaches infinity, the tip of the vector will circle the point $(\frac{1}{2}, \frac{1}{2})$ an infinite number of times, which suggests that the contribution to the value of the integral comes primarily between the limits zero and u_o provided $u_o > 1.26$. For this reason, we can argue that allowing $\gamma x_o \to \infty$ in (12-29) has little effect on the value of the integral. Thus,

$$E(P) \approx C_1\sqrt{\lambda/2}e^{-j\beta\ell} \int_{\gamma a}^{\infty} e^{-j(\pi/2)u^2}\, du \tag{12-30}$$

The value of the integral in (12-30) can be represented by a vector drawn from any point on the Cornu spiral to the point $(\frac{1}{2}, \frac{1}{2})$ (e.g., see Prob. 12.2-2).

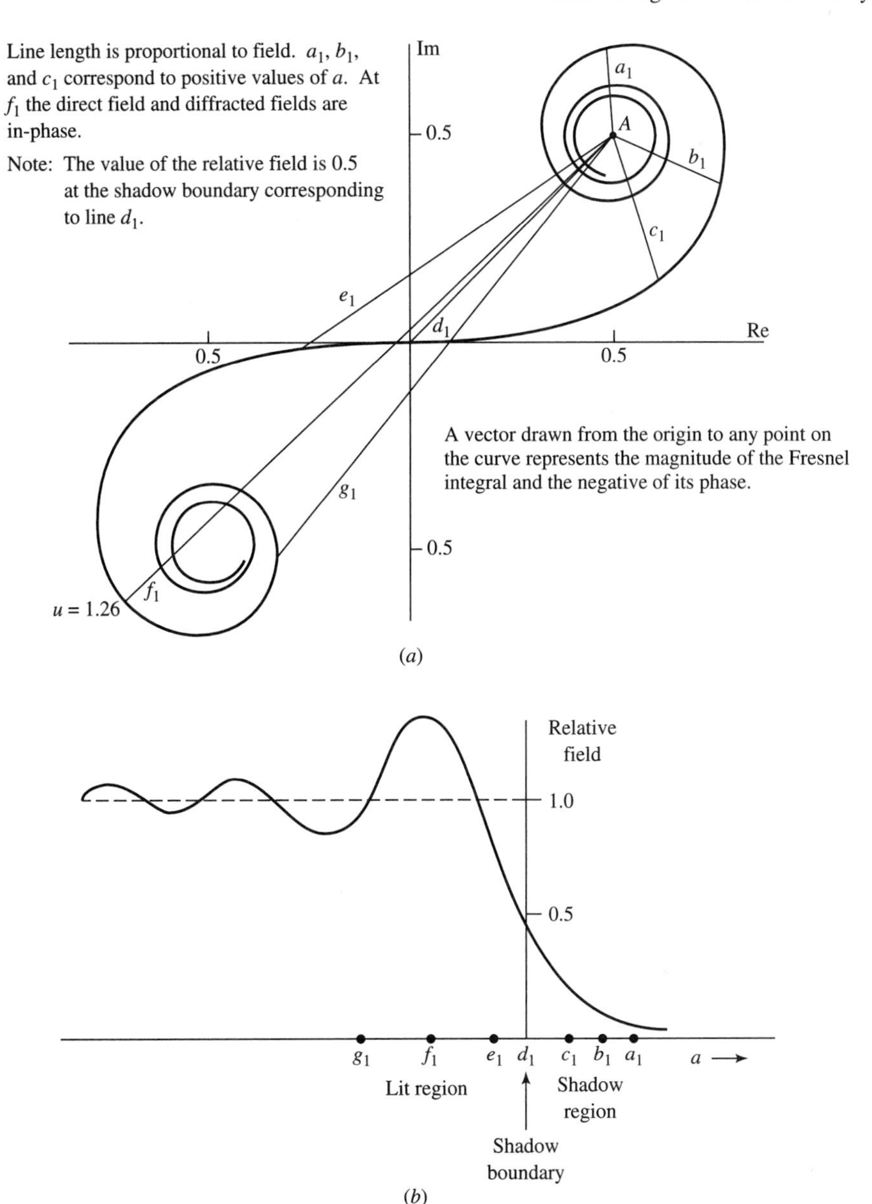

Figure 12-11 (*a*) Use of the Cornu spiral in evaluating the Fresnel integral as a function of the parameter *a*. (*b*) Relative magnitude of the diffracted field in the vicinity of a shadow boundary. Refer to Fig. 12-10 for values of *a*.

If the lower limit in (12-30) is allowed to go to minus infinity, $E(P)$ will equal the field strength without the half-plane present [8]. Thus,

$$E(P)|_{a=-\infty} = C_1\sqrt{\lambda/2}(1 - j)e^{-j\beta\ell} = E_o e^{-j\beta\ell} \tag{12-31}$$

Solving for C_1 and substituting into (12-30) give the value of $E(P)$ in terms of the free-space field E_o:

$$E(P) \approx \frac{E_o e^{+j(\pi/4)}}{\sqrt{2}} e^{-j\beta\ell} \int_{\gamma a}^{\infty} e^{-j(\pi/2)u^2}\, du \tag{12-32}$$

where for this approximate analysis to hold, it is necessary that $\ell \gg \lambda$ and the point $x = a$ not be far removed from the z-axis so that the assumption $\ell \gg \delta$ holds. A more exact (and complicated) analysis of this problem is possible, but it has not been presented here for we wish simply to show how the Fresnel integral arises naturally in the study of wedge diffraction.

Equation (12-32) and the Cornu spiral make it possible to visualize the variation of the electric field as the point a moves along the x-axis, causing the observation point to change from the lit region to the shadow region. The corresponding plot of the relative electric field in the vicinity of the shadow boundary is shown in Fig. 12-11b. We note that on the shadow boundary the value of the relative field is $\frac{1}{2}$ and in the lit region the value of the field oscillates about the value of unity. This oscillation can be interpreted as being caused by interference between the diffracted field and the direct field. Since there is no direct field in the shadow region, we observe that no such oscillation occurs. Unfortunately, it is not convenient to explicitly distinguish between the direct and diffracted field in (12-32). In many applications of diffraction theory, it is essential that we be able to mathematically distinguish between the direct and diffracted fields, as well as the reflected field that we have yet to consider.

Referring to Fig. 12-12, we can identify two shadow boundaries: the incident or direct field shadow boundary and the reflected field shadow boundary. These two shadow boundaries serve to divide space into three regions where region I contains direct and diffracted rays as well as reflected rays, region II direct and diffracted rays but no reflected rays, and region III only diffracted rays.

For a field in any one of the three regions, let us write $E(\rho, \phi)$ as consisting of a reflected field $v^r(\rho, \phi + \phi')$ and an incident field $v^i(\rho, \phi - \phi')$. Thus,

$$E(\rho, \phi) = \pm v^r(\rho, \phi + \phi') + v^i(\rho, \phi - \phi') \tag{12-33}$$

The choice of sign depends on the polarization of the incident field. If the electric field is perpendicular (parallel) to the diffracting edge, the plus (minus) sign is used. The field E at the point P must be a solution to the scalar wave equation with the appropriate boundary conditions. The boundary value problem depicted in Fig. 12-12 was first solved by Sommerfeld in 1896. We will first consider his solution. To do so, we must examine (12-33) more fully.

The first term in (12-33) gives the reflected fields, whereas the term $v^i(r, \phi - \phi')$ represents the incident field. If the ground plane were infinite in extent, the reflected field term would simply be the geometrical optics reflected field. However, in the case of the half-plane in Fig. 12-12, the reflected field will consist of two parts:

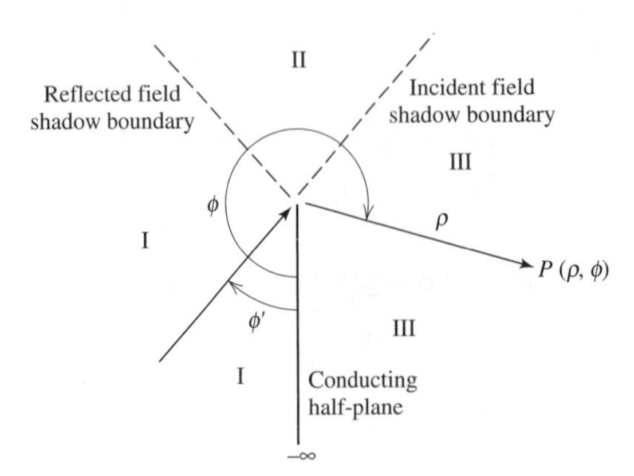

Figure 12-12 Diffraction by a conducting half-plane showing the location of shadow boundaries.

namely, a geometrical optics reflected field and a diffracted field. Both parts of the reflected field will appear to originate from an image source behind the half-plane. Similarly, the incident field can be thought to consist of two parts: a geometrical optics incident field and a diffracted field. Thus, for the reflected field,

$$\pm v^r(\rho, \phi + \phi') = \pm[v^r_*(\rho, \phi + \phi') + v^r_B(\rho, \phi + \phi')] \tag{12-34}$$

and for the incident field,

$$v^i(\rho, \phi - \phi') = v^i_*(\rho, \phi - \phi') + v^i_B(\rho, \phi - \phi') \tag{12-35}$$

where v_* denotes the geometrical optics field and v_B the diffracted field. Thus, (12-33) may be thought of as being composed of four parts. Each of the terms on the right-hand side of (12-34) and (12-35) satisfies the wave equation individually except at the reflected field and incident field shadow boundaries, respectively. However, the sum of v^r_* and v^r_B makes v^r continuous across the reflected field shadow boundary and thus v^r satisfies the wave equation there. (Similar comments apply to v^i.) But, neither v^r nor v^i alone satisfies the boundary conditions at the wedge. However, the sum of v^r and v^i in (12-33) does satisfy the boundary conditions as well as the wave equation.

From simple geometrical considerations, we can see that for reflected geometrical optics rays, all points on a constant phase wavefront are given by

$$v^r_*(\rho, \phi + \phi') = e^{j\beta\rho\cos(\phi+\phi')}, \qquad 0 < \phi < \pi - \phi' \text{ in region I} \tag{12-36}$$

where the phase reference is taken to be at the edge of the half-plane in Fig. 12-13 since we are using a cylindrical coordinate system whose origin is on the edge of

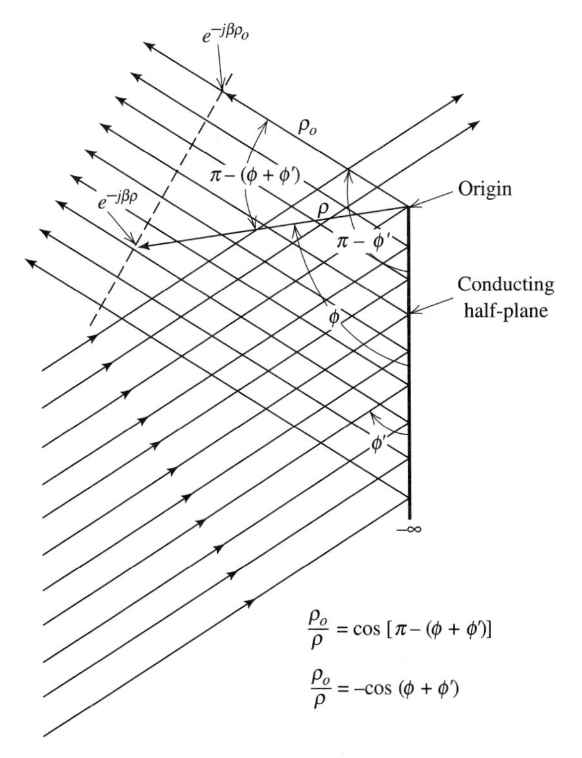

Figure 12-13 Geometry for the reflected field wavefront from a conducting half-plane.

the wedge. By similar considerations for direct incident rays, points on a constant phase wavefront are given by

$$v_*^i(\rho, \phi - \phi') = e^{j\beta\rho\cos(\phi-\phi')}, \qquad 0 < \phi < \pi + \phi' \text{ in regions I and II} \tag{12-37}$$

For other values of ϕ,

$$v_*^i = 0 = v_*^r \tag{12-38}$$

In other words, v_* is zero in regions II and III for reflected rays and is zero in region III for direct or incident rays. It is the diffracted field that compensates for this discontinuity in the geometrical optics field as shown in Fig. 12-14. We see in Fig. 12-14*a* that for $\phi > 255°$, the total field is just the diffracted field and the total field is continuous across the incident field shadow boundary at $\phi = 255°$, where the value of the diffracted field is 0.5. For $105° < \phi < 255°$, the total field oscillates due to the interference between the incident field and the diffracted field. At $\phi = 105°$, the diffracted field again rises to 0.5 and the total field is continuous across the field shadow boundary. For $\phi < 105°$, the total field oscillates almost between zero and 2 due mainly to the standing wave produced in region I by the incident and reflected fields and the fact that the field is observed at a constant distance ($\rho = 3\lambda$) from the edge of the half-plane, requiring the observation point to sweep through the standing wave field. The field is normal to the half-plane at $\phi = 0$ and is nonzero there. Figure 12-14b shows a time domain representation of the total electric field in the vicinity of the edge of the half-plane when a sinusoidal plane wave is incident at $\phi' = 75°$. Since the presentation is essentially a "snap shot" in time, the almost white areas indicate zero field at an instant of time. (See Fig. 11-21 and associated text.) Note evidence of the reflection and shadow boundaries, the weak field when $\phi > 255°$, and the interference pattern when $\phi < 105°$. In the interference pattern for $\phi < 105°$ there is a standing wave in directions both normal and tangential to the half plane since $\phi' \neq 90°$.

Mathematical expressions for the diffracted field v_B have been a subject of considerable research in the past several decades in an effort to improve on the early classical work of Sommerfeld [9]. For the half-plane problem of Fig. 12-12, Sommerfeld obtained an expression for the diffracted field due to an incident plane wave in terms of the Fresnel integral. This expression is[3]

$$v_B(\rho, \phi^\pm) = -e^{j(\pi/4)}\sqrt{\frac{2}{\pi\alpha}} \cdot e^{j\beta\rho\cos\phi^\pm} \cos\frac{\phi^\pm}{2} \int_{\sqrt{\alpha\beta\rho}}^{\infty} e^{j\tau^2}\, d\tau \tag{12-39}$$

where

$$\phi^\pm = \phi \pm \phi' \tag{12-40}$$

and

$$\alpha = 1 + \cos\phi^\pm \tag{12-41}$$

We note that this solution is in a form somewhat similar to that of (12-32). The mathematical details of deriving the above are beyond the scope of this text.

Sommerfeld's work was more general than that of just a half-plane. He also con-

---

[3]Note that in (12-39), we are really writing two equations, one for $v_B^r(r, \phi^+)$ and the other for $v_B^i(r, \phi^-)$. The use of the notation $\phi^\pm$ is for convenience and the reader should keep in mind that wherever it appears there are two separate equations implied, one associated with the reflected field and one associated with the incident field.

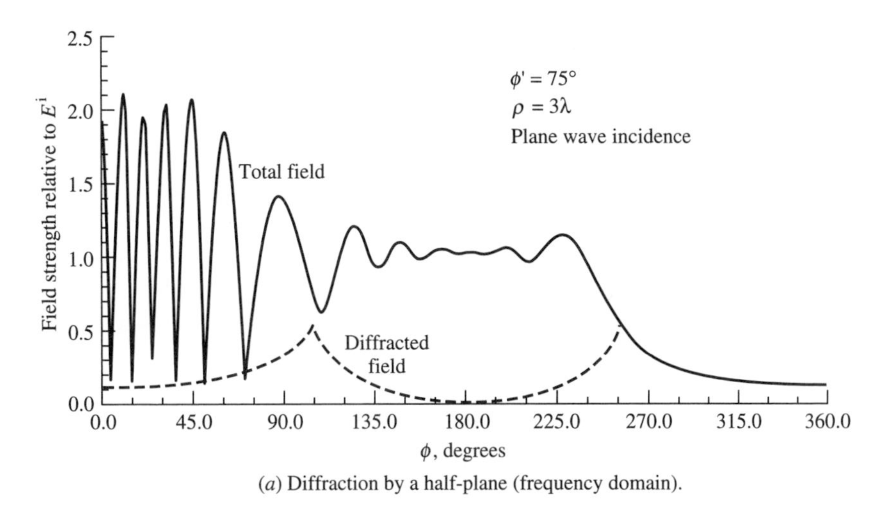

(a) Diffraction by a half-plane (frequency domain).

180°

105°

255°

75°

(b) FD-TD calculated field distribution for a fixed instant of time in the vicinity of a half-plane. Sinusoidal plane wave incident at $\phi' = 75°$.

Figure 12-14 Diffraction by a half-plane. The incident field is polarized perpendicular to the edge of the half-plane. Refer to Figs. 12-12 and 12-13.

sidered the more general case of a plane wave illuminating a conducting wedge of interior angle $(2 - n)\pi$, where $0 < n \leq 2$.[4] For this case, he obtained an asymptotic evaluation of a contour integral representation for the diffracted field that is given by

$$v_B(\rho, \phi^{\pm}) = \frac{e^{-j(\beta\rho + \pi/4)}}{\sqrt{2\pi\beta\rho}} \frac{(1/n)\sin(\pi/n)}{\cos(\pi/n) - \cos(\phi^{\pm}/n)} \tag{12-42}$$

[4]Refer forward to Fig. 12-15.

Unfortunately, this asymptotic form yields infinite fields in the immediate vicinity of the shadow boundary [10]. The region near a shadow boundary is usually referred to as a *transition region*. Equation (12-42) is only valid outside a transition region where the condition

$$\beta\rho\left(\cos\frac{\pi}{n} - \cos\frac{\phi^{\pm}}{n}\right)^2 \gg 1 \tag{12-43}$$

is satisfied. This condition is always met if the quantity $\beta\rho(1 + \cos\phi^{\pm})$ is large, which means that the observation point at $P(\rho, \phi, z)$ must be at a large electrical distance from the diffracting edge. Nevertheless, (12-42) is a useful one if the observation point is not near a shadow boundary and the above conditions are met.

In 1938, Pauli [11] improved on the work of Sommerfeld by obtaining a series form for Sommerfeld's contour integral solution. Pauli's result is given by

$$v_B(\rho, \phi^{\pm}) = \frac{2e^{j(\pi/4)}}{n\sqrt{\pi}} \frac{\sin(\pi/n)}{\cos(\pi/n) - \cos(\phi^{\pm}/n)} \left|\cos\frac{\phi^{\pm}}{2}\right| \cdot e^{j\beta\rho\cos\phi^{\pm}} \int_{\sqrt{\alpha\beta\rho}}^{\infty} e^{-j\tau^2}\,d\tau + [\text{higher-order terms}] \tag{12-44}$$

This expression is far more accurate, particularly near the shadow boundaries, than (12-42) while being only slightly more difficult to evaluate. It is valid for $0 < n \leq 2$. In the case of the half-plane ($n = 2$), the higher-order terms are identically zero and Pauli's result in (12-44) reduces to that of Sommerfeld in (12-39). Pauli's expression was the first practical formulation of Sommerfeld's original solution that included a finite observation distance.

EXAMPLE 12-1 *Sample Wedge Diffraction Calculations*

The use of Eqs. (12-39), (12-42), and (12-44) is best illustrated by an example. Let us calculate the diffracted field in Fig. 12-14 for $\phi = 250°$. Using (12-39), we obtain

$$\begin{aligned} v_B(3, \phi^-) &= (-9.146 - j9.146)(0.0436) \\ &\quad \cdot (0.997 + j0.0717)(0.359 - j0.620) \\ &= -0.397 + j0.0760 \end{aligned}$$

$$\begin{aligned} v_B(3, \phi^+) &= (-0.418 - j0.418)(-0.954) \\ &\quad \cdot (-0.964 + j0.264)(-0.0237 + j0.0820) \\ &= 0.0345 - j0.0335 \end{aligned}$$

Thus, the exact Sommerfeld solution gives for the diffracted field magnitude $|-0.3625 + j0.0435| = 0.365$, which agrees with Fig. 12-14. Using (12-44), we should obtain the same result for the half-plane case since Pauli's equation reduces to Sommerfeld's. Thus,

$$\begin{aligned} v_B(3, \phi^-) &= (0.798 + j0.798)(-11.46)(0.0436) \\ &\quad \cdot (0.997 + j0.0717)(0.359 - j0.620) \\ &= -0.397 + j0.0760 \end{aligned}$$

$$\begin{aligned} v_B(3, \phi^+) &= (0.798 + j0.798)(0.524) \\ &\quad \cdot (0.954)(-0.964 + j0.264) \\ &\quad \cdot (-0.0237 + j0.0820) \\ &= 0.0345 - j0.0335 \end{aligned}$$

and the diffracted field magnitude is once again 0.365. We note that since $\phi = 250°$ is near the incident field shadow boundary, $v_B(3, \phi^-)$ is the major contributor to the diffracted field and $v_B(3, \phi^+)$, which is associated with the reflected field shadow boundary, makes only a minor contribution. Both (12-39) and (12-44) would go to infinity precisely at the shadow boundary $\phi = 255°$ (or $\phi = 105°$). For this reason, we have elected to use $\phi = 250°$ in this example. Finally, let us use the asymptotic form in (12-42). Thus,

$$v_B(3, \phi^-) = (0.065 - j0.065)(-11.46)$$
$$= -0.745 + j0.745$$

$$v_B(3, \phi^+) = (0.065 - j0.065)(0.524)$$
$$= 0.034 - j0.034$$

and the magnitude of the diffracted field alone exceeds unity or that of the incident field. This result is in error because the condition in (12-43) has been violated. The result would be only 10% in error at $\rho = 10\lambda$ if $\phi = 255° \pm 12°$, at 20λ if $\phi = 255° \pm 5°$, at 30λ if $\phi = 255° \pm 4°$, and at 100λ if $\phi = 255° \pm 3°$. However, no matter how large ρ is, the asymptotic form will be singular right at the shadow boundary.

Starting in 1953, it was Keller [5, 6] who systematically developed the geometrical theory of diffraction, or GTD as it is often referred to. In his work, he has called the quantities $D(\phi^+)$ and $D(\phi^-)$ diffraction coefficients, where

$$[v_B^i(\rho, \phi^-) \mp v_B^r(\rho, \phi^+)] = [D(\phi^-) \mp D(\phi^+)] \frac{e^{-j\beta\rho}}{\sqrt{\rho}} \tag{12-45}$$

and used the asymptotic expression of Sommerfeld in (12-42) to calculate the diffracted field due to plane wave incidence. The postulates of Keller's theory are:

1. The diffracted field propagates along ray paths that include points on the boundary surface. These ray paths obey the principle of Fermat, known also as the principle of shortest optical path.
2. Diffraction, like reflection and transmission, is a local phenomenon at high frequencies. That is, it depends only on the nature of the boundary surface and the incident field in the immediate neighborhood of the point of diffraction.
3. A diffracted wave propagates along its ray path so that:
   a. power is conserved in a tube of rays, and
   b. phase delay equals the wave number times the distance along the ray path.

A consequence of the second postulate is that the diffracted fields caused by the edge of the infinite wedge in Fig. 12-13, for example, appear to be cylindrical wave fields that originate at the wedge edge. This is consistent with the $(\rho)^{-1/2}$ dependence in (12-45).

The simple ray formulation of Keller's geometrical theory of diffraction is restricted to the calculation of fields in regions of space that exclude transition regions adjacent to shadow boundaries, caustics, and focal points. To calculate the field at such points, additions and modifications to the geometrical theory of diffraction are required. Further, if the incident field is not a plane wave, but a cylindrical or spherical wave, GTD must be modified to accept these incident fields as well. These various modifications will be considered in later sections.

12.3 THE RAY-FIXED COORDINATE SYSTEM

In the previous section, we considered the scalar diffracted field due to a plane wave normally incident (i.e., traveling in the negative ρ-direction) on a perfectly conducting infinite wedge whose edge was along the z-axis. Such a coordinate system is said to be an edge-fixed coordinate system. On the other hand, the obliquely incident and diffracted rays associated with the point Q in Fig. 12-15 are more conveniently described in terms of spherical coordinates centered at Q. Such a coordinate system is said to be ray-fixed [1]. Let the position of the source of the incident ray be defined by the spherical coordinates (s', γ_o', ϕ'), and the observation point by the coordinates (s, γ_o, ϕ) as indicated in Fig. 12-15. Note that the point Q is a unique point on the edge for a given source location and observation point.

The plane containing the incident ray and the edge of the wedge will be referred to as the plane of incidence, whereas that plane containing the diffracted ray and the edge of the wedge will be referred to as the plane of diffraction. The unit vector $\hat{\mathbf{s}}'$ is in the direction of incidence and the unit vector $\hat{\mathbf{s}}$ in the direction of diffraction. It is then apparent that the unit vectors $\hat{\boldsymbol{\gamma}}_o'$ and $\hat{\boldsymbol{\phi}}'$ are parallel and perpendicular, respectively, to the plane of incidence, and that the unit vectors $\hat{\boldsymbol{\gamma}}_o$ and $\hat{\boldsymbol{\phi}}$ are parallel and perpendicular, respectively, to the plane of diffraction as shown in Fig. 12-16. γ_o' and γ_o are angles less than $\pi/2$ measured from the edge to the incident and diffracted rays, respectively, whereas $\hat{\boldsymbol{\gamma}}_o'$ and $\hat{\boldsymbol{\gamma}}_o$ are the implied unit vectors. Further, ϕ' and ϕ are angles measured from one face of the wedge to the plane of incidence and diffraction, respectively, whereas $\hat{\boldsymbol{\phi}}'$ and $\hat{\boldsymbol{\phi}}$ are the implied unit vectors. Note that ϕ' and ϕ are measured from the same face of the wedge.

Let us write a symbolic expression for the diffracted field in matrix form as

$$[\mathbf{E}^d] = [D][\mathbf{E}^i]A(\rho)e^{-j\beta\rho} \tag{12-46}$$

where $[\mathbf{E}^d]$ and $[\mathbf{E}^i]$ are column matrices consisting of the scalar components of the diffracted and incident fields respectively, $[D]$ is a square matrix of the appropriate scalar diffraction coefficients, and ρ is the distance from the wedge edge to the observation point, and $A(\rho)$ is a spreading factor. Now if the edge-fixed coordinate

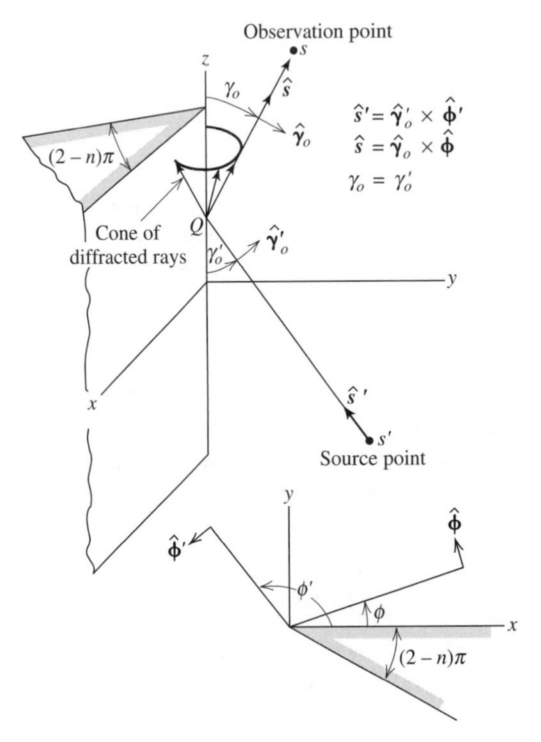

Figure 12-15 Geometry for three-dimensional wedge diffraction problem.

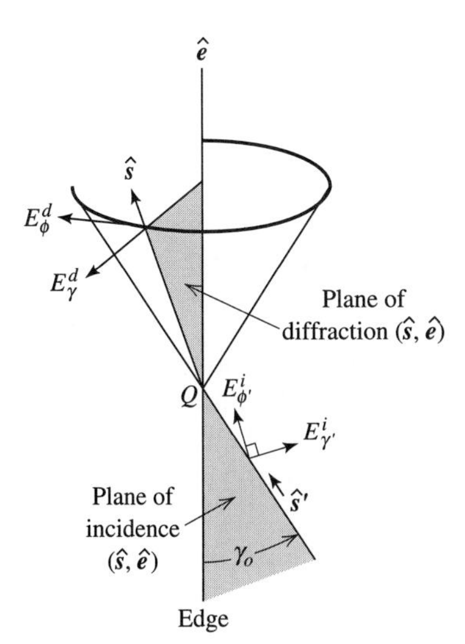

Figure 12-16 Ray-fixed coordinate system.

system is used, it is clear that $[\mathbf{E}^d]$ will have, in general, three scalar components E^d_ρ, E^d_ϕ, and E^d_z and $[D]$ will be a 3×3 matrix. It can be shown that in such a situation seven of the nine terms in $[D]$ are nonvanishing. However, when the ray-fixed coordinate system is used, there is no (radial) component of the diffracted field in the direction of the diffracted ray tube since the incident field is not allowed to have a component in the direction of the incident ray tube. It follows that there are then only two possible components of the diffracted field, E^d_γ and E^d_ϕ, and only two components of the incident field, $E^i_{\gamma'}$ and $E^i_{\phi'}$. Clearly, $[D]$ is then a 2×2 matrix. In this case, $[D]$ has nonvanishing terms on the main diagonal. Thus, for plane wave incidence in the ray-fixed system, (12-46) can be written as

$$\begin{bmatrix} E^d_\gamma(s) \\ E^d_\phi(s) \end{bmatrix} = \begin{bmatrix} -D_\| & 0 \\ 0 & -D_\perp \end{bmatrix} \begin{bmatrix} E^i_{\gamma'}(Q) \\ E^i_{\phi'}(Q) \end{bmatrix} A(s)e^{-j\beta s} \tag{12-47}$$

where the scalar diffraction coefficients $D_\|$ and $D_\perp$ are momentarily undefined and will be given in the following section.

It is apparent that the use of the ray-fixed coordinate system instead of the edge-fixed system reduces the diffraction matrix from a 3×3 matrix with seven nonvanishing terms to a 2×2 matrix with but two nonvanishing terms. Thus, the ray-fixed system is the natural coordinate system to be used for wedge diffraction and the importance of using it can hardly be overemphasized.

We have chosen to use the notation $D_\|$ in association with $E^i_{\gamma'}(Q)$ and $D_\perp$ in association with $E^i_{\phi'}(Q)$ not because $E^i_{\gamma'}$ and $E^i_{\phi'}$ are parallel and perpendicular, respectively, to the diffracting edge (which they are at normal incidence when $\gamma'_o = 90°$), but because $E^i_{\gamma'}$ and $E^i_{\phi'}$ are parallel and perpendicular, respectively, to the plane of incidence as shown in Fig. 12-16.

Since $E^i_{\gamma'}$ and $E^i_{\phi'}$ are parallel and perpendicular, respectively, to the plane of incidence, we will let $E^i_{\gamma'}$ be written as $E^i_\|$ and $E^i_{\phi'}$ as $E^i_\perp$. Similarly, $E^d_\gamma(s)$ and $E^d_\phi(s)$ are parallel and perpendicular, respectively, to the plane of diffraction as shown in Fig. 12-16. Thus, we let E^d_γ be written as $E^d_\|$ and E^d_ϕ as $E^d_\perp$. With these notational changes, (12-47) may be rewritten as

$$\begin{bmatrix} E^d_\|(s) \\ E^d_\perp(s) \end{bmatrix} = \begin{bmatrix} -D_\| & 0 \\ 0 & -D_\perp \end{bmatrix} \begin{bmatrix} E^i_\|(Q) \\ E^i_\perp(Q) \end{bmatrix} A(s)e^{-j\beta s} \tag{12-48}$$

We will use this notation throughout the remainder of the chapter, keeping in mind that when the $\parallel$ and $\perp$ symbols are associated with E^i, reference to the plane of incidence is implied. When the $\parallel$ and $\perp$ symbols are associated with E^d, reference to the plane of diffraction is implied.

12.4 A UNIFORM THEORY OF WEDGE DIFFRACTION

The modern version of GTD can be divided into the two basic canonical problems of wedge diffraction and curved surface diffraction plus the lesser but more complex problems of vertex diffraction, tip diffraction, and other higher-order phenomena. In the application of wedge diffraction to antenna problems, the important features of antennas are modeled by perfectly conducting wedges. For example, the sectoral horn antenna can be modeled by two half-planes as shown in Fig. 12-18 for the purpose of analyzing the E-plane pattern [10]. In such a problem, however, it is necessary to use cylindrical wave diffraction coefficients instead of plane wave diffraction coefficients as in Sec. 12.2. The first use of cylindrical wave diffraction in the treatment of antenna problems, such as in Sec. 12.5, was by Rudduck [10] who used Pauli's formulation together with the principle of reciprocity to calculate the necessary cylindrical wave diffraction. Problems involving spherical wave diffraction are also common.

In Sec. 12.2, some early developments in the study of diffraction by a conducting wedge were presented. We saw that although some of the formulas presented are certainly useful for some engineering calculations, they are limited in their accuracy in a transition (shadow boundary) region [e.g., (12-42)], or when the observation point is near ($r < \lambda$) the diffracting edge [e.g., (12-44)]. It would obviously be useful and convenient if there were available to us a theory of wedge diffraction having the property that it could accurately predict the diffracted field in such places as the transition regions or near the diffracting edge without the necessity for considering each type of incident field separately. Such a theory is available and is known as a *uniform theory* of wedge diffraction because it applies in all situations consistent with the postulates of the geometrical theory of diffraction given in Sec. 12.2. It is the purpose of this section to present the important results in this theory, known as UTD, which is based on the numerous works of Kouyoumjian and Pathak [12–14].

In 1967, Kouyoumjian and co-workers obtained a generalized version of Pauli's result [i.e., (12-44)] with the resultant diffraction function v_B expressed as $v_B(L, \phi^{\pm})$, where L is a distance parameter more general than just the distance ρ used in Sec. 12.2, whereas $\phi^{\pm}$ retains the meaning used previously. In their work, the distance parameter is given by

$$L = \begin{cases} s \sin^2 \gamma_o' & \text{for plane waves} \\ \dfrac{\rho' \rho}{\rho + \rho'} & \text{for cylindrical waves} \\ \dfrac{s' s \sin^2 \gamma_o'}{s + s'} & \text{for conical and spherical waves.} \end{cases} \qquad (12\text{-}49)$$

We note immediately that L is dependent on the type of incident wave and the angle of incidence γ_o' (which equals the angle of reflection γ_o) as well as the distances involved. The distance parameter L in (12-49) can be found by imposing the condition that the total field, which is the sum of the geometrical-optics field and the diffracted field, be continuous at shadow or reflection boundaries.

When the work of Kouyoumjian and co-workers is expressed in terms of the scalar diffraction coefficients $D_{\|}$ and $D_{\perp}$ where

$$D_{\|}(L, \phi, \phi') = [v_B(L, \phi^-) - v_B(L, \phi^+)] \frac{\sqrt{L}e^{j\beta L}}{\sin \gamma_o} \tag{12-50}$$

$$D_{\perp}(L, \phi, \phi') = [v_B(L, \phi^-) + v_B(L, \phi^+)] \frac{\sqrt{L}e^{j\beta L}}{\sin \gamma_o} \tag{12-51}$$

we have (without proof) [12, 14]

$$\begin{aligned} D_{\substack{\| \\ \perp}}(L, \phi, \phi') = & \frac{-e^{-j(\pi/4)}}{2n\sqrt{2\pi\beta} \sin \gamma_o'} \\ & \times \left[\cot\left(\frac{\pi + (\phi - \phi')}{2n}\right) F[\beta L a^+(\phi - \phi')] \right. \\ & + \cot\left(\frac{\pi - (\phi - \phi')}{2n}\right) F[\beta L a^-(\phi - \phi')] \\ & \mp \left\{ \cot\left(\frac{\pi + (\phi + \phi')}{2n}\right) F[\beta L a^+(\phi + \phi')] \right. \\ & \left.\left. + \cot\left(\frac{\pi - (\phi + \phi')}{2n}\right) F[\beta L a^-(\phi + \phi')] \right\} \right] \end{aligned} \tag{12-52}$$

where, if the argument of F is represented by X,

$$F(X) = 2j|\sqrt{X}|e^{jX} \int_{|\sqrt{X}|}^{\infty} e^{-j\tau^2}\, d\tau \tag{12-53}$$

Again, we see that a Fresnel integral appears in the expression for the diffraction coefficient. The factor $F(X)$ may be regarded as a correction factor to be used in the transition regions of the shadow and reflection boundaries. Outside of the transition regions where the argument of F exceeds about 3, the magnitude of F is approximately equal to 1 as Fig. 12-17 shows. Even within a given transition region, usually only one of the four terms in (12-52) is significantly different from unity. The transition function that is significantly different from unity goes to zero at the

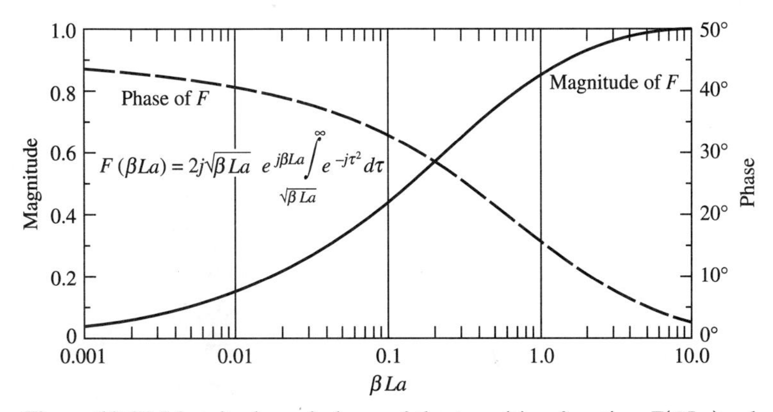

Figure 12-17 Magnitude and phase of the transition function $F(\beta La)$, where $a = a^+$ or a^-.

same rate that its cotangent multiplier goes to infinity. Thus the transition function prevents the singularity in (12-42) from occurring in (12-52) (see Prob. 12.4-7).

The argument of the transition function, which is $X = \beta L a^{\pm}(\phi \pm \phi')$, may be calculated for a known value of βL if $a^{\pm}$ as a function of $(\phi \pm \phi')$ is known. To determine $a^{+}(\phi \pm \phi')$ and $a^{-}(\phi \pm \phi')$, we use

$$a^{\pm}(\phi \pm \phi') = 2\cos^2\left[\frac{2n\pi N^{\pm} - (\phi \pm \phi')}{2}\right] \tag{12-54}$$

in which $N^{\pm}$ are the integers that most nearly satisfy the four equations

$$2\pi n N^{+} - (\phi \pm \phi') = \pi \tag{12-55}$$

and

$$2\pi n N^{-} - (\phi \pm \phi') = -\pi \tag{12-56}$$

We note that N^{+} and N^{-} may each have two separate values in a given problem. For exterior wedge diffraction where $1 < n \leq 2$, $N^{+} = 0$ or 1 but $N^{-} = -1, 0$ or 1. The factor $a^{\pm}(\phi \pm \phi')$ may be interpreted physically as a measure of the angular separation between the field point and a shadow or reflection boundary.

Now that we have all the necessary relationships to calculate $D_{\parallel}$ and $D_{\perp}$, we repeat (12-47) in the format of UTD as

$$\begin{bmatrix} E^d_{\parallel}(s) \\ E^d_{\perp}(s) \end{bmatrix} = \begin{bmatrix} -D_{\parallel} & 0 \\ 0 & -D_{\perp} \end{bmatrix} \begin{bmatrix} E^i_{\parallel}(Q) \\ E^i_{\perp}(Q) \end{bmatrix} A(s) e^{-j\beta s} \tag{12-57}$$

where the spatial attenuation factor $A(s)$ is defined as

$$A(s) = \begin{cases} \dfrac{1}{\sqrt{s}} & \text{for plane, cylindrical, and conical wave incidence} \\ \left[\dfrac{s'}{s(s'+s)}\right]^{1/2} & \text{for spherical wave incidence} \end{cases} \tag{12-58}$$

It should also be mentioned that, since diffraction concepts apply to acoustical problems, the diffraction coefficients $D_{\parallel}$ and $D_{\perp}$ in (12-57) are sometimes written D_s and D_h, respectively, which correspond to the acoustic soft and hard boundary conditions [14]. Software code based upon (12-52) for UTD is available in the public domain (see Appendix G.5) and in the previous edition of this text.

EXAMPLE 12-2 *Sample UTD Calculation*

The use of (12-49) to (12-58) is best illustrated by an example. Suppose we wish to calculate the diffracted field in Fig. 12-14 when $\phi = 250°$. We have in this case: $\phi + \phi' = 325°$; $\phi - \phi' = 175°$; $L = 3\lambda$; $\beta L = 6\pi$; n = 2. Thus, from (12-54) to (12-56)

$$\begin{aligned} a^{+}(\phi + \phi') &= 2\cos^2(197.5°), &\quad \text{where } N^{+} = 1 \\ a^{+}(\phi - \phi') &= 2\cos^2(87.5°), &\quad \text{where } N^{+} = 0 \\ a^{-}(\phi + \phi') &= 2\cos^2(162.5°), &\quad \text{where } N^{-} = 0 \\ a^{-}(\phi - \phi') &= 2\cos^2(87.5°), &\quad \text{where } N^{-} = 0 \end{aligned}$$

From Fig. 12-17, using the respective values of a^+ and a^- above, we obtain

$$F(6\pi \cdot 1.819) = 0.999 + j0.0146$$
$$F(6\pi \cdot 0.0038) = 0.318 + j0.216$$
$$F(6\pi \cdot 1.819) = 0.999 + j0.0146$$
$$F(6\pi \cdot 0.0038) = 0.318 + j0.216$$

Using (12-52) and (12-58), we obtain

$$D_\perp(L, \phi, \phi') = -0.628 + j\,0.0735$$
$$A(s)e^{-j\beta s} = 0.577$$

From (12-57),

$$E_\perp^d(s) = -0.363 + j0.0424$$

or

$$|E_\perp^d(s)| = 0.365$$

which agrees with Fig. 12-14. It is worth noting that when the four correction factors F above are multiplied by their associated cotangent factor, it is the fourth term above that is much larger than the others. As mentioned earlier, usually just one of the terms in (12-52) turns out to be large, even close to a shadow boundary. Equation (12-52) will not exhibit a singular behavior at a shadow boundary as was the case in Sec. 12.2 with (12-39) and (12-42).

If the field point is not close to a shadow or reflection boundary and $\phi' \neq 0$ or $n\pi$ (grazing incidence), the scalar diffraction coefficients $D_\parallel$ and $D_\perp$ reduce to Keller's diffraction coefficients [see (12-42) and (12-45)] that may be written as

$$D_{\parallel \atop \perp}(\phi, \phi'; \gamma_o') = \frac{e^{-j(\pi/4)}\sin(\pi/n)}{n\sqrt{2\pi\beta}\sin\gamma_o'} \cdot \left[\frac{1}{\cos\dfrac{\pi}{n} - \cos\dfrac{\phi - \phi'}{n}} \mp \frac{1}{\cos\dfrac{\pi}{n} - \cos\dfrac{\phi + \phi'}{n}}\right] \tag{12-59}$$

This expression is valid for all four types of incident waves given in (12-49), which is important because the diffraction coefficient should be independent of the edge illumination away from shadow and reflection boundaries. However, from Sec. 12.2, we know that (12-59) will become singular as a shadow or reflection boundary is approached.

Grazing incidence, where $\phi' = 0$ or $n\pi$, is a special case that must be considered separately. In this case, $D_\parallel \approx 0$, and the expression for $E_\perp^d$ must be multiplied by a factor of $\frac{1}{2}$. If we consider grazing incidence to be the limit of oblique incidence, we can see how the need for the factor of $\frac{1}{2}$ arises, because at grazing incidence the incident and reflected fields merge. When they merge, one-half of the total field propagating along the face of the wedge toward the edge is the incident field and the other half is the reflected field. The merged field is then regarded as being the "incident" field, but it is too large by a factor of 2 and the factor of $\frac{1}{2}$ becomes necessary. That is, (12-57) requires the use of the free-space incident field and not the merged field.

The uniform theory of wedge diffraction described in this section permits us to consider diffraction problems wherein both the source and observation points are quite close to the diffracting edge (i.e., a wavelength or even less). It also permits us to consider any type of TEM incident field. A more general expression for L, valid for an arbitrary wavefront incident on the straight edge of a wedge, appears in the literature [13, 14].

Unlike the edge diffraction formulas presented in Sec. 12.2, (12-52) is valid in the transition regions of the incident field shadow boundary and the reflected field shadow boundary. Equation (12-52) cannot be used to calculate the field at a caustic of the diffracted ray. This does not conflict with the concept of a uniform theory of wedge diffraction because geometrical optics itself is incapable of determining the field at a caustic. The field at a caustic may, however, be found through the use of a supplementary solution in the form of an integral representation of the field. The equivalent sources in the integral representation are determined from a suitable high-frequency approximation such as geometrical optics or the geometrical theory of diffraction. The calculation of the field at a caustic by such methods will be considered in Sec. 12.9.

12.5 *E*-PLANE ANALYSIS OF HORN ANTENNAS

To illustrate the application of the uniform theory of diffraction presented in the previous section, consider the E-plane horn antenna shown in Fig. 12-18*a*. In this section, we use the model shown in Fig. 12-18*b* to compute the complete E-plane pattern of the horn antenna. The model is simple and therefore particularly well-

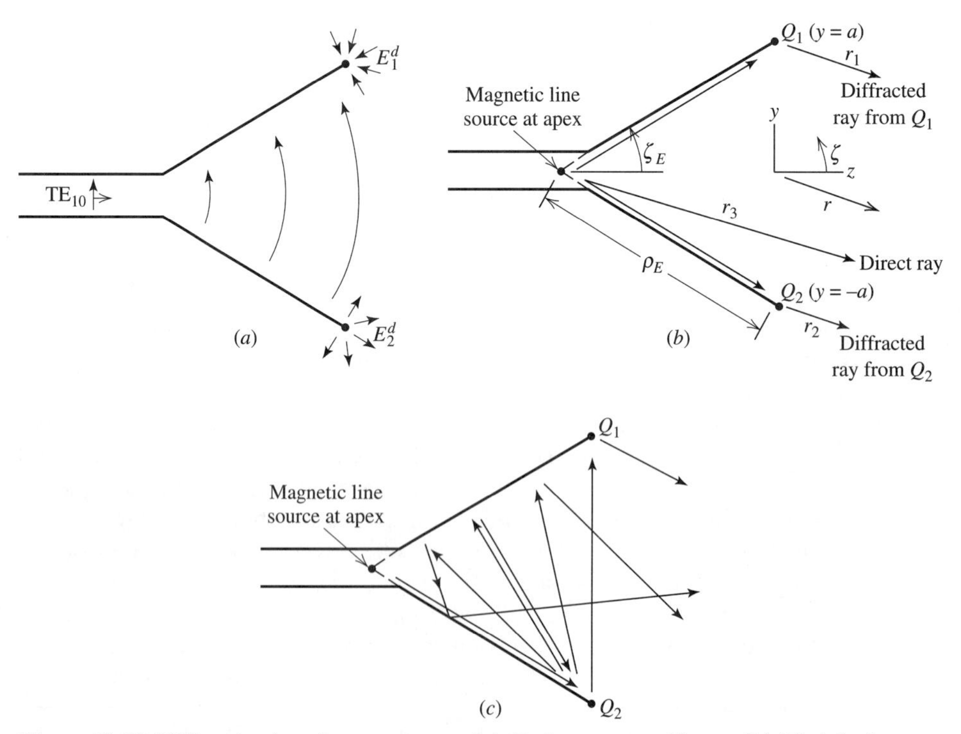

Figure 12-18 Diffraction by a horn antenna. (*a*) E-plane sectoral horn. (*b*) Model of E-plane sectoral horn. (*c*) Neglected rays.

suited to use as a first example of the application of UTD. The model has only three sources of radiation and is two-dimensional in nature (i.e., infinite in the $\pm x$-directions), which in the E-plane well-represents a three-dimensional horn antenna.

The equations applicable to the analysis are as follows. Note that the angle ζ $(0 \leq \zeta \leq 2\pi)$ is used instead of θ $(0 \leq \theta \leq \pi)$ so that positions in the yz-plane may be defined unambiguously. In the far field, we have (in the yz-plane)

$$r_1 = r - a \sin \zeta \tag{12-60}$$

$$r_2 = r + a \sin \zeta \tag{12-61}$$

$$r_3 = r + \rho_E \cos \zeta \cos \zeta_E \tag{12-62}$$

where r_1 and r_2 are distances to the far-field observation point $P(r, \zeta)$ from diffracting edges 1 and 2, respectively, and r_3 is the distance from the line source to the far-field observation point as shown in Fig. 12-18*b*. Thus, the incident field along the direct ray can be expressed by

$$E^i(P) = \frac{e^{-j\beta r_3}}{\sqrt{r_3}} \approx \frac{e^{-j\beta r}}{\sqrt{r}} e^{j\beta\rho_E \cos \zeta \cos \zeta_E}, \qquad -\zeta_E \leq \zeta \leq \zeta_E \tag{12-63}$$

and

$$E^i(P) = 0, \qquad \zeta_E < \zeta < 2\pi - \zeta_E \tag{12-64}$$

Note that in applying UTD, we do not replace the conducting surfaces with equivalent currents radiating in free space as in the preceding chapters of this book. Instead, the conducting surfaces are retained. As a consequence, for example, $E^i(P) = 0$ when $\zeta > \zeta_E$.

The edge diffracted field at $P(r, \zeta)$ from a diffraction point Q_1 on the "top" edge may be written as

$$E_1^d(P) = \frac{1}{2} E_\perp^i(Q_1) D_\perp(L, \phi, \phi') \frac{e^{-j\beta r_1}}{\sqrt{r_1}} = \frac{1}{2} E_\perp^i D_\perp(L, \phi, \phi') \frac{e^{-j\beta r}}{\sqrt{r}} e^{j\beta a \sin \zeta}, \quad -\frac{\pi}{2} \leq \zeta \leq \pi + \zeta_E \tag{12-65}$$

and

$$E_1^d(P) = 0, \qquad \pi + \zeta_E < \zeta < \frac{3\pi}{2} \tag{12-66}$$

Similarly, the diffracted field at $P(r, \zeta)$ from a diffraction point Q_2 on the "bottom" edge may be written

$$E_2^d(P) = \frac{1}{2} E_\perp^i(Q_2) D_\perp(L, \phi, \phi') \frac{e^{-j\beta r_2}}{\sqrt{r_2}} = \frac{1}{2} E_\perp^i D_\perp(L, \phi, \phi') \frac{e^{-j\beta r}}{\sqrt{r}} e^{-j\beta a \sin \zeta}, \quad -\pi - \zeta_E \leq \zeta \leq \frac{\pi}{2} \tag{12-67}$$

and

$$E_2^d(P) = 0, \qquad \frac{\pi}{2} < \zeta < \pi - \zeta_E \tag{12-68}$$

where

$$E_{\perp}^{i}(Q_1) = E_{\perp}^{i}(Q_2) = \frac{e^{-j\beta\rho_E}}{\sqrt{\rho_E}} \tag{12-69}$$

Thus, the total field at an observation point $P(r, \zeta)$ may be written as the scalar sum.

$$E(P) = E^{i}(P) + E_{1}^{d}(P) + E_{2}^{d}(P) \tag{12-70}$$

In the above equations, scalar $D_{\perp}$ denotes the diffraction coefficient at the point of diffraction Q_m for the case where the incident electric field is normal to the edge. The diffraction coefficient at Q_m depends on the geometry of the incident and diffracted rays at Q_m and is most accurately given by (12-49) and (12-52). Here, of course, we consider the incident field to be cylindrical and use the cylindrical wave form for the distance parameter L. $E^{i}(Q_m)$ is the incident field that is perpendicular to both the edge and incident ray.

At first glance, the factor of $\frac{1}{2}$ in (12-65) and (12-67) might appear to be incorrect. However, in this problem, the rays from the line source are incident at a grazing angle with the surface of the horn walls and therefore deserve special consideration. Grazing incidence, where $\phi' = 0$ or $n\pi$, requires that $D_{\perp}$ in (12-57) be multiplied by a factor of $\frac{1}{2}$ as discussed in the preceding section below (12-59).

Figure 12-19 shows results calculated with the model shown in Fig. 12-18*b* and also experimental data. The agreement between the calculated results without using double diffractions (dashed curve) and the experimental results is seen to be very good. Note that there is a discontinuity in the calculated results when $\zeta = 90°$ (or 270°). This discontinuity may be removed simply by including rays that diffract from Q_2 (or Q_1) and travel across the horn aperture to Q_1 (or Q_2) and are diffracted a second time as indicated in Fig. 12-18*c*.

Also shown in Fig. 12-18*c* are several other rays that have not been included in the calculated results because in this problem they provide a relatively weak numerical contribution. Strictly speaking, those rays shown in Fig. 12-18*c* that do not involve double diffractions should be included in the analysis. These are the two

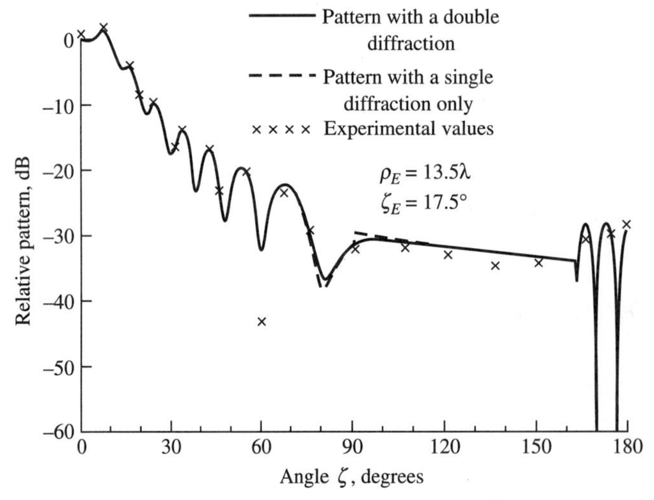

Figure 12-19 Calculated and experimental E-plane patterns of an E-plane sectoral horn.

rays that experience a reflection after undergoing diffraction at Q_2 (or Q_1, which are not shown). Of the two doubly diffracted rays shown, here only the one from Q_2 to Q_1 is important because it compensates for the shadowing of Q_2 when $\zeta > 90°$. There is no similar compensation needed in the case of the other doubly diffracted ray that goes from Q_2 to the "top" wall and back to Q_2.

In conclusion, we have used a simple model (i.e., Fig. 12-18*b*) to calculate the E-plane pattern of horn antennas with good results. Strictly speaking, we should have included some of the rays in Fig. 12-18*c*, but did not do so for the sake of simplicity without a loss of accuracy. It is a fundamental fact that in applying UTD (or GTD), one must be careful to identify and include all rays that arise in the problem. In the horn problem here, we were able to omit some of the rays only because they were not in or near a transition region and also because the rays in Fig. 12-18*b* are one or more orders of magnitude stronger than those in Fig. 12-18*c*.

12.6 CYLINDRICAL PARABOLIC ANTENNA

As a second example of the application of UTD, we consider the cylindrical parabolic antenna shown in Fig. 12-20. We use the aperture integration procedure given in Chap. 7 to obtain the pattern in and near the main beam, but use UTD to compute the pattern everywhere else. As in the study of the horn antenna in the previous section, the model here is two-dimensional. We consider only the diffractions that occur at the edges of the parabolic surface and ignore any higher-order rays associated with the curved surface (e.g., see Sec. 12.11).

First, let us consider the equation for obtaining the main beam and first few side lobes. From Sec. 7.1, we may write for the far field E^A obtained by aperture integration

$$E^A(P) = \sqrt{\frac{j\beta}{2\pi r}}\, e^{-j\beta r} \cos\zeta \int_{-a}^{a} \frac{F_f(\theta_s)}{\sqrt{\rho}}\, e^{j\beta y' \sin\zeta}\, dy' \tag{12-71}$$

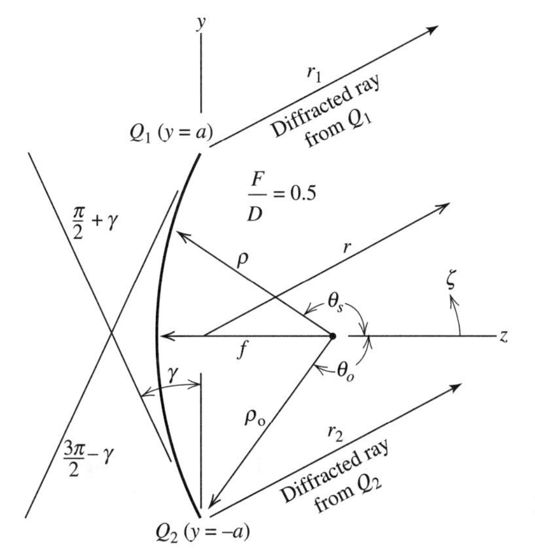

Figure 12-20 Cylindrical parabolic antenna geometry.

where $F_f(\theta_s)$ is the pattern of the electric line source current I^e that serves as the feed for the cylindrical parabolic reflector antenna. (If the line source pattern is isotropic, $F_f(\theta_s) = 1$.) Equation (12-71) is a two-dimensional specialization of the equations in Sec. 7.1.

Equation (12-71) can, of course, give us the pattern for $90° \geq \zeta \geq -90°$. However, since we must perform the aperture integration anew for each value of ζ, it is more efficient in the computational sense to use (12-71) for $\zeta_o \geq \zeta \geq -\zeta_o$, where ζ_o is the angular extent of the main beam and the first side lobe or two, and then to use UTD for the remainder of the pattern. Clearly, we do not use aperture integration and UTD simultaneously in the *same* angular sector.

For the UTD model of the antenna, the following equations apply. For the singly diffracted field from Q_1, we have at the far-field observation point $P(r, \zeta)$

$$E_1^d(P) = 0, \qquad \frac{3\pi}{2} - \gamma \leq \zeta \leq \frac{3\pi}{2} \tag{12-72}$$

and elsewhere

$$\begin{aligned} E_1^d(P) &= E_\parallel^i(Q_1) D_\parallel(L, \phi, \phi') \frac{e^{-j\beta r_1}}{\sqrt{r_1}} \\ &\approx E_\parallel^i(Q_1) D_\parallel(L, \phi, \phi') \frac{e^{-j\beta r}}{\sqrt{r}} e^{j\beta a \sin\zeta} \end{aligned} \tag{12-73}$$

where (12-60) has been used in (12-73). Similarly, the diffracted field at $P(r, \zeta)$ from Q_2 may be written

$$E_2^d(P) = 0, \qquad \frac{\pi}{2} \leq \zeta \leq \frac{\pi}{2} + \gamma \tag{12-74}$$

and elsewhere

$$\begin{aligned} E_2^d(P) &= E_\parallel^i(Q_2) D_\parallel(L, \phi, \phi') \frac{e^{-j\beta r_2}}{\sqrt{r_2}} \\ &\approx E_\parallel^i(Q_2) D_\parallel(L, \phi, \phi') \frac{e^{-j\beta r}}{\sqrt{r}} e^{-j\beta a \sin\zeta} \end{aligned} \tag{12-75}$$

where (12-61) has been used in (12-75). In both (12-73) and (12-75),

$$E_\parallel^i(Q_1) = E_\parallel^i(Q_2) = \frac{e^{-j\beta\rho_o}}{\sqrt{\rho_o}} F_f(\theta_o) \tag{12-76}$$

The total field at an observation point $P(r, \zeta)$ may be written as either

$$E(P) = E^i(P) + E^A(P) \tag{12-77}$$

or

$$E(P) = E^i(P) + E_1^d(P) + E_2^d(P) \tag{12-78}$$

depending on the angle ζ as mentioned earlier.

Figure 12-21 shows a calculated pattern for a cylindrical parabolic reflector having a 10λ aperture (i.e., $2a = 10\lambda$) and a focal length-to-diameter ratio of 0.5. The electric line source that models the feed has a pattern of $F_f(\theta_s) = \cos^2\theta_s$ for $\theta_s \geq 90°$ and $F_f(\theta_s) = 0$ in the forward half-space where $\theta_s < 90°$. We note that the pattern

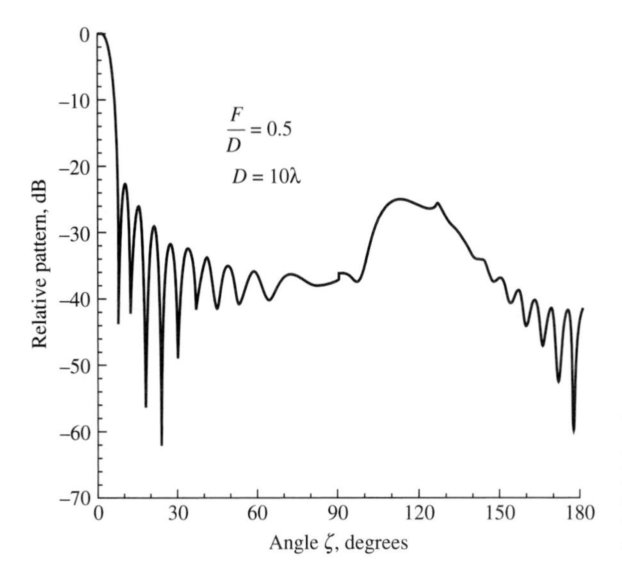

Figure 12-21 Calculated H-plane pattern of a cylindrical parabolic reflector with $D = 10\lambda$ having a focal-length-to-diameter ratio of 0.5.

has a small discontinuity at $\zeta = 90°$ (and 270°) and this discontinuity can be removed by including double diffracted rays between Q_1 and Q_2 as was done for the horn in the previous section. We also note that there is a small discontinuity at about $\zeta =$ 127° (and 233°) that is a result of the shadowing of Q_2 (or Q_1 when $\zeta \approx 233°$). The relatively high level of the pattern in the vicinity of $\zeta = 120°$ is due to the spillover caused by the feed pattern.

It is interesting to note that for the chosen feed pattern of $\cos^2 \theta_s$, the aperture electric field distribution is almost that of a cosine on a pedestal with a -15-dB edge illumination as shown in Fig. 12-22. Referring to Table 4-2, we see that such a distribution should produce a pattern with a side-lobe level of -22 dB. Examination of the pattern in Fig. 12-21 shows that indeed the side-lobe level is -22 dB. Thus, the pattern in the forward half-space could be well represented by a line source, as discussed in Chap. 4, once the aperture field distribution is known.

In this section, we have examined the H-plane pattern of a cylindrical parabolic antenna (i.e., an electric line source was used to model the feed). We could also analyze the E-plane pattern when a magnetic line source is used to model the feed. This is left as an exercise for the student.

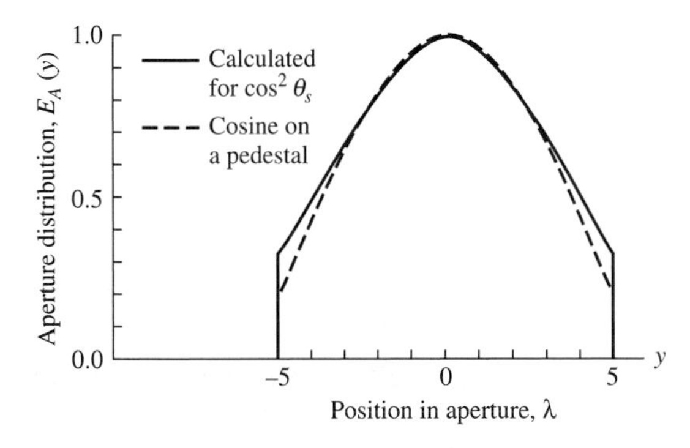

Figure 12-22 Aperture distribution for the parabola of Fig. 12-21 with a feed pattern of $\cos^2 \theta_s$.

12.7 RADIATION BY A SLOT ON A FINITE GROUND PLANE

To illustrate further the application of the uniform theory of diffraction, consider the situation in Fig. 12-23 where a radiating slot is asymmetrically located along the x-axis of the rectangular plate. We desire pattern information in both principal planes to determine the amount of ripple in the pattern caused by edge diffraction. In general, the edges denoted Q_1 and Q_2 will be illuminated unequally unless $d_1 = d_2$ and thus the pattern in the xz-plane will not be symmetrical about the z-axis.

The geometry of the problem to be investigated, as depicted in Fig. 12-23, is a narrow aperture (or slot) with length T on a finite ground plane of dimensions A and B. The narrow slot has an electric field polarized in the x-direction and has a cosine-distribution in the y-direction. The length of the slot is taken to be one-half wavelength at the operating frequency.

For radiation in the xz-plane above the ground plane, the problem is represented, to a first degree approximation, by an infinitely long slot. According to UTD, there exist two edge-diffracted rays originating from edge points Q_1 and Q_2 due to the finiteness of the ground plane. Therefore, for a far-field observation point $P(r, \theta, \phi = 0)$ in the region of interest, the total field is the sum of the contributions from the direct ray and two edge-diffracted rays as shown in Fig. 12-24. Doubly diffracted rays exist but are small compared to the singly diffracted rays shown in Fig. 12-24 and are not included in the present analysis.

For radiation in the yz-plane above the ground plane, a sampling of $N + 1$ ideal sources with cosine distribution is performed. There exist no first-order edge-diffracted rays because the incident ray is zero in the yz-plane. A geometry of five samplings is shown in Fig. 12-25. The end sources are of zero amplitude since tangential E is zero at the ends of the slot.

First, let us consider the radiation pattern in the xz-plane. The direct ray from the narrow slot at an observation point $P(r, \theta, \phi = 0)$ is

$$\mathbf{E}^i(P) = \hat{\boldsymbol{\theta}} E_o \frac{e^{-j\beta r}}{\sqrt{r}} \tag{12-79}$$

The edge-diffracted ray from Q_1 at $P(r, \theta, \phi = 0)$ becomes

$$\mathbf{E}_1^d(P) = \hat{\boldsymbol{\theta}} \frac{1}{2} E_\perp^i(Q_1) D_\perp(L, \phi, \phi') \frac{e^{-j\beta r_1}}{\sqrt{r_1}} \tag{12-80}$$

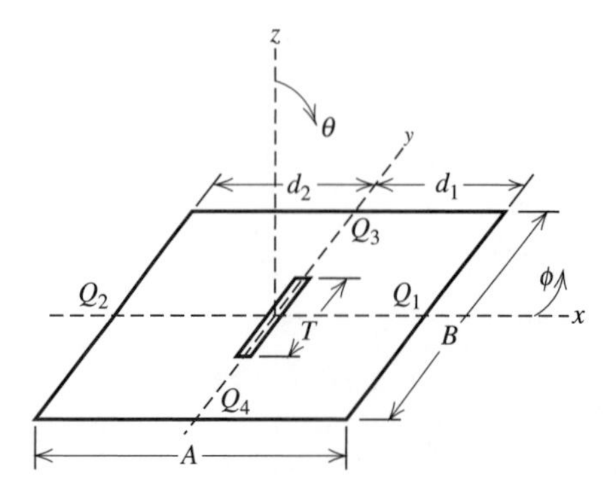

Figure 12-23 Geometry of a slot on a rectangular conducting plate.

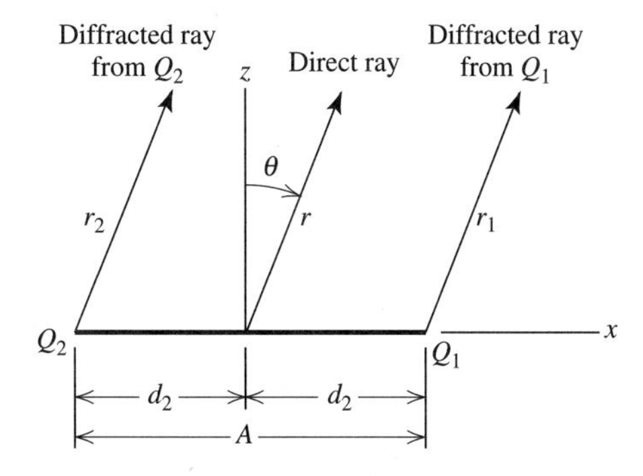

Figure 12-24 Direct and diffracted rays contributing to the xz-plane pattern.

with

$$\mathbf{E}_\perp^i(Q_1) = \hat{\mathbf{z}} E_o \frac{e^{-j\beta d_1}}{\sqrt{d_1}} = \hat{\mathbf{z}} E_\perp^i(Q_1) \tag{12-81}$$

The edge-diffracted ray from Q_2 at $P(r, \theta, \phi = 0)$ yields

$$\mathbf{E}_2^d(P) = \hat{\boldsymbol{\theta}} \frac{1}{2} E_\perp^i(Q_2) D_\perp(L, \phi, \phi') \frac{e^{-j\beta r_2}}{\sqrt{r_2}} \tag{12-82}$$

with

$$\mathbf{E}_\perp^i(Q_2) = \hat{\mathbf{z}} E_o \frac{e^{-j\beta d_2}}{\sqrt{d_2}} = \hat{\mathbf{z}} E_\perp^i(Q_2) \tag{12-83}$$

The total field at an observation point $P(r, \theta, \phi = 0)$ then becomes (in the symmetrical case)

$$\mathbf{E}(P) = \mathbf{E}^i(P) + \mathbf{E}_1^d(P) + \mathbf{E}_2^d(P) \tag{12-84}$$

The parameters r, r_1, d_1, r_2, and d_2 are shown in Fig. 12-24. The parameter E_o represents the magnitude of the electric field at the narrow slot in the xz-plane. $E_\perp^i(Q_m)$ is that component of the incident field which is perpendicular to both the edge and the incident ray. To first order $D_\parallel$ is zero. However, there is a small amount of diffraction that does take place and this is called the slope diffraction (see Prob. 12.7-1). The addition of slope diffraction to the diffracted field ensures that not only is the total field continuous across a shadow boundary, but also the derivative of the total field is continuous.

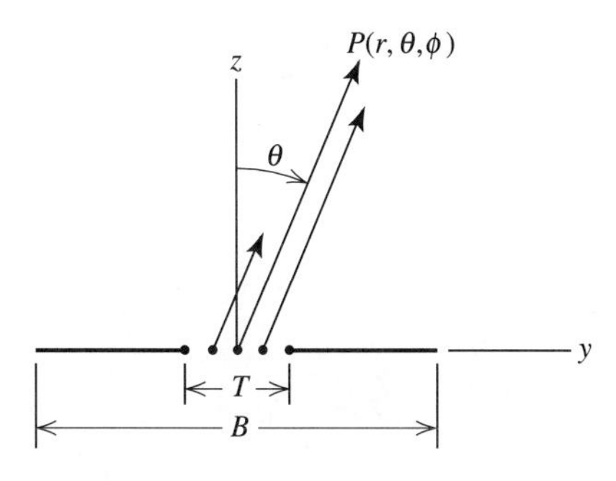

Figure 12-25 Direct rays from the weighted sources representing the slot contribution to the yz-plane pattern. See Fig. 12-23.

For the slot problem of Fig. 12-23, the radiation in the yz-plane may be analyzed in the region above the ground plane, to a first degree approximation, by an array of dipole sources with a cosine-distributed amplitude across the array. Let the total number of dipoles in the array be $N + 1$; then the separation between dipoles is

$$s = \frac{T}{N} \tag{12-85}$$

The total field at an observation point $P(r, \theta, \phi = \pi/2)$ then becomes

$$\mathbf{E}(P) = \hat{\boldsymbol{\theta}} E_o \sin(90^\circ - \theta) \sum_{n=-N/2}^{N/2} \cos\left(\frac{ns\pi}{T}\right) e^{jn\beta s \sin\theta} \tag{12-86}$$

In Fig. 12-25 is shown the geometry of the yz-plane with five dipoles ($N = 4$) in the array.

Figure 12-26 shows the far-field pattern results in the xz- and yz-planes at both 1 and 3 GHz. The ground plane is 61 × 61 cm but the slot is taken to be one-half wavelength at each frequency and diffraction in the yz-plane has been assumed to be negligible, and under this assumption the pattern in the yz-plane is the same at each frequency as indicated in Fig. 12-26. However, due to diffraction, the two patterns in the xz-plane are different, the "ripple" in the patterns being the result of the diffracted energy. Since the slot is located symmetrically on the ground plane, we see that the pattern is symmetric about the z-axis. For completeness, the slope diffraction contribution to the yz-plane pattern at 1 and 3 GHz is also shown. Slope diffraction is proportional to the spatial derivative of the incident field with respect to the direction that is normal to both the incident ray and the edge at Q (see Prob. 12.7-1). In contrast, the edge diffraction we have considered thus far is proportional to the incident field at Q. In the problem considered in this section, the incident field at Q_3 and Q_4 in Fig. 12-23 is zero. However, the derivative of the incident field with respect to the normal (z in this case) is not zero at either Q_3 or Q_4.

Although an experimental comparison is not shown here, such comparisons have been made with excellent results [14], even though diffraction from the four corners or vertices of the ground plane has been neglected. We know from experimental measurements that vertex diffraction is generally much weaker than wedge diffrac-

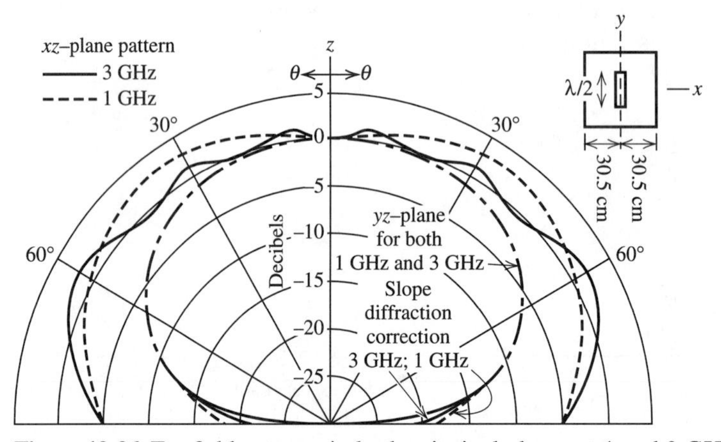

Figure 12-26 Far-field patterns in both principal planes at 1 and 3 GHz.

tion. Thus, the total far field is given to a good approximation by (12-84). In the problem considered here, vertex diffraction is weak in the xz- and yz-planes and somewhat stronger in the $\phi = 45°$ and $\phi = 135°$ planes.

12.8 RADIATION BY A MONOPOLE ON A FINITE GROUND PLANE

As another application of the uniform theory of diffraction and also as an example of a problem with a caustic, consider the two situations depicted in Fig. 12-27. First, consider the situation in Fig. 12-27*a* of a $\lambda/4$ monopole on a square plate and suppose we wish to obtain the pattern in the xz-plane. For purposes of far-field calculation and conceptual simplicity, a suitable approximation to the $\lambda/4$ monopole is the ideal dipole of Chap. 1. We will consider the ideal dipole to be resting on the surface of the ground plane. Thus, following the development of the previous section for the slot, we have for the direct ray from the ideal dipole at $P(r, \theta, \phi = 0)$

$$\mathbf{E}^i(P) = \hat{\boldsymbol{\theta}} E_o \frac{e^{-j\beta r}}{r} \sin\theta \tag{12-87}$$

which now must obviously be considered a spherical wave. The edge-diffracted ray from Q_1 at $P(r, \theta, \phi = 0)$ appears to emanate from a single point and is therefore

$$\mathbf{E}_1^d(P) = \hat{\boldsymbol{\theta}} \frac{1}{2} E_\perp^i(Q_1) D_\perp(L, \phi, \phi') \sqrt{d_1} \frac{e^{-j\beta r_1}}{r_1} \tag{12-88}$$

with

$$\mathbf{E}_\perp^i(Q_1) = \hat{\mathbf{z}} E_o \frac{e^{-j\beta d_1}}{d_1} = \hat{\mathbf{z}} E_\perp^i(Q_1) \tag{12-89}$$

Similarly, the edge-diffracted ray from Q_2 at $P(r, \theta, \phi = 0)$ is

$$\mathbf{E}_2^d(P) = \hat{\boldsymbol{\theta}} \frac{1}{2} E_\perp^i(Q_2) D_\perp(L, \phi, \phi') \sqrt{d_2} \frac{e^{-j\beta r_2}}{r_2} \tag{12-90}$$

with $E_\perp^i(Q_2)$ given by (12-89) since the source is located at the center of the ground plane. Diffraction from the sides containing Q_3 and Q_4 does not contribute to the

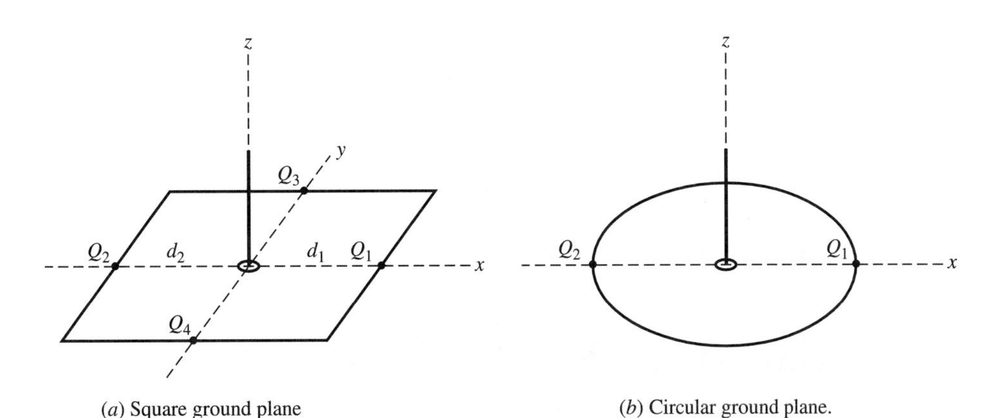

(*a*) Square ground plane (*b*) Circular ground plane.

Figure 12-27 Monopole on a finite ground plane.

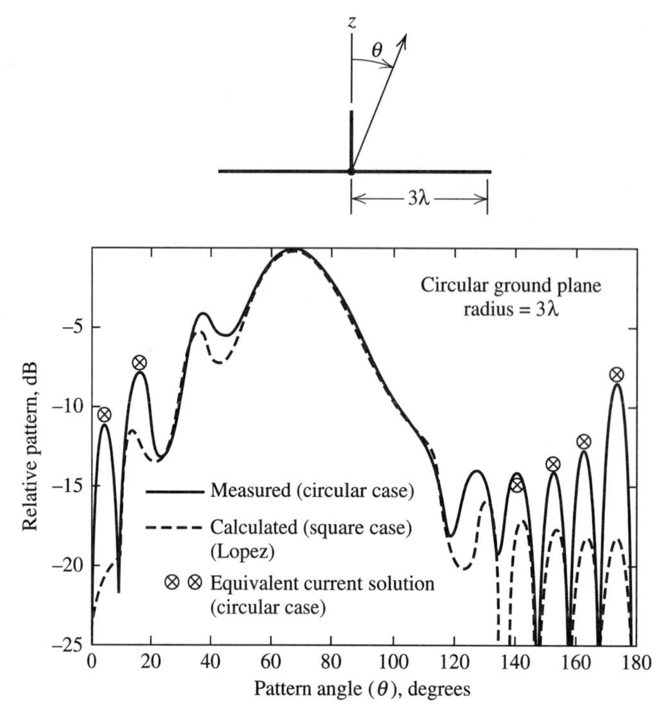

Figure 12-28 Radiation pattern of an ideal monopole on a circular ground plane 6λ in diameter. The calculated curve by Lopez [15] is for a $6\lambda \times 6\lambda$ square ground plane.

far field in the xz-plane since the monopole is positioned at the center of the ground plane and the diffracted fields from these two sides will cancel one another. As in the case of the slot of the previous section, we neglect diffraction from the four corners or vertices of the ground plane. The corresponding pattern for a 6λ square ground plane is given in Fig. 12-28 (dashed curve).

If we now consider the geometry of Fig. 12-27*b*, we note that in the xz-plane, the diffracted radiation will also appear (due to Fermat's principle) to come from two points that are called stationary points. We note also that the z-axis is a caustic in this problem because all rays from the circular edge of the ground plane intersect along the z-axis. Therefore, although we can expect to use the two stationary points to calculate the diffracted field contribution to the pattern in regions not near the caustic [15], we can likewise expect the "two-point approximation" to be increasingly in error as the observation point P moves nearer the caustic. Figure 12-28 shows that indeed this is the case since the measured and two-point calculated patterns diverge as both $\theta \to 0$ and $\theta \to \pi$, which is also a caustic. The apparent difficulty in the vicinity of the caustic can be overcome, as suggested in Fig. 12-28, by the use of a fictitious equivalent edge current. As will be seen in the next section, the so-called equivalent current is not a physical current at all, but rather a mathematical artifice for predicting the correct diffracted field at or near a caustic.

12.9 EQUIVALENT CURRENT CONCEPTS

In the previous section, we saw that, in the treatment of the circular ground plane, we could obtain the diffracted field using ordinary wedge diffraction theory if the point of observation was not near a caustic. In essence, we treated the problem as a two-dimensional one with the diffraction taken as that from an infinite two-dimensional wedge, whereas in fact we had a finite edge that was not straight, but curved.

To properly treat the diffraction by a curved edge or finite wedge (i.e., finite length of the edge), it is necessary to consider the concept of equivalent currents [16]. As we shall see, the strengths (i.e., amplitude and phase) of these so-called equivalent currents will be determined by the canonical problem of wedge diffraction.

Consider the wedge of Fig. 12-15 to be of finite extent, $-\ell/2 \leq z(Q) \leq \ell/2$. To start, we assume the currents are the same as those on an infinite wedge. Let us determine the current flowing on the edge of the infinite wedge that would produce the scattered field predicted by wedge diffraction analysis. Thus, we specify an infinite line source whose current is determined by the diffraction coefficient. If the z-directed line source is an electric current, it can be shown in a manner similar to that used in Secs. 1.5 and 10.13.2, that the solution to the scalar wave equation is [17]

$$\psi = \frac{I^e}{4j} H_0^{(2)}(\beta\rho) \tag{12-91}$$

and therefore that the z-component of the electric field is

$$E_z = \frac{-\beta^2 I^e}{4\omega\varepsilon} H_0^{(2)}(\beta\rho) \tag{12-92}$$

where I^e denotes an electric current. If the argument of the Hankel function $H_0^{(2)}(\beta\rho)$ is large, then using the asymptotic representation of the Hankel function, we obtain

$$E_z = \eta\beta I^e \frac{e^{j(\pi/4)}}{2\sqrt{2\pi\beta\rho}} e^{-j\beta\rho} \tag{12-93}$$

We note that (12-93) represents an outward traveling wave in the cylindrical coordinate system with the proper $\rho^{-1/2}$ dependence for a two-dimensional problem. If instead the line source is a magnetic current I^m, then we have

$$H_z = -\frac{\beta}{\eta} I^m \frac{e^{j(\pi/4)}}{2\sqrt{2\pi\beta\rho}} e^{-j\beta\rho} \tag{12-94}$$

Since we are considering a two-dimensional problem, we can also apply wedge diffraction theory to obtain the diffracted field from the edge for the two orthogonal polarizations. Thus,

$$E_z = D_{\parallel}(L, \phi, \phi')E_z^i \frac{e^{-j\beta\rho}}{\sqrt{\rho}} \tag{12-95}$$

and

$$H_z = D_{\perp}(L, \phi, \phi')H_z^i \frac{e^{-j\beta\rho}}{\sqrt{\rho}} \tag{12-96}$$

where $D_{\parallel}$ and $D_{\perp}$ are given in Sec. 12.4. Usually, however, we find that the use of equivalent currents involves the calculation of diffracted fields in regions away from an incident field or a reflected field shadow boundary or their associated transition regions. Thus, the asymptotic form in (12-59) for arbitrary incidence angle γ_o is usually sufficient.

From (12-93) with (12-95) and also (12-94) with (12-96), we can solve for the electric and magnetic currents of an infinite line source that will produce the same far fields predicted using the diffraction coefficients. Thus,[5]

$$I^e = \frac{-2j}{\eta\beta} E_z^i D_{\parallel}\left(\phi, \phi'; \frac{\pi}{2}\right)\sqrt{2\pi\beta}\, e^{j(\pi/4)} \tag{12-97}$$

and

$$I^m = \frac{2j\eta}{\beta} H_z^i D_{\perp}\left(\phi, \phi'; \frac{\pi}{2}\right)\sqrt{2\pi\beta}\, e^{j(\pi/4)} \tag{12-98}$$

We note that (12-97) and (12-98) give the equivalent currents I^e and I^m, but they are numerically different for each value of ϕ and ϕ'. The fact that these currents are different for different observation points (i.e., values of ϕ) serves to emphasize the fact that these equivalent currents are not true currents, but fictitious currents that simply aid us in calculating diffracted fields.

Considering Fig. 12-15 with the ray incident normally on the edge ($\gamma_o = \pi/2$), we have, respectively, for the far-zone diffracted fields

$$E_\theta^e = \frac{j\omega\mu \sin\theta}{4\pi r} e^{-j\beta r} \int_{-\ell/2}^{\ell/2} I_z^e(z') e^{j\beta z' \cos\theta}\, dz' \tag{12-99}$$

and

$$H_\theta^m = \frac{j\omega\mu \sin\theta}{4\pi r} e^{-j\beta r} \int_{-\ell/2}^{\ell/2} I_z^m(z') e^{j\beta z' \cos\theta}\, dz' \tag{12-100}$$

As in Chap. 4, we see that since the currents are constant with respect to z', (12-99) and (12-100) reduce to results in the general form of $\sin(x)/x$ with respect to the θ-coordinate.

For the case of nonnormal incidence (i.e., $\gamma_o \neq \pi/2$), we can proceed in the same manner and show that

$$I^e = \frac{-2j}{\eta\beta} E_z^i D_{\parallel}(\phi, \phi'; \gamma_o)\sqrt{2\pi\beta}\, e^{j(\pi/4)} e^{j\beta\ell \cos\gamma_o} \tag{12-101}$$

and

$$I^m = \frac{2j\eta}{\beta} H_z^i D_{\perp}(\phi, \phi'; \gamma_o)\sqrt{2\pi\beta} e^{j(\pi/4)} e^{j\beta\ell \cos\gamma_o} \tag{12-102}$$

which includes the phase term to account for the traveling-wave-type current due to the oblique angle of incidence. In obtaining (12-101) and (12-102), we have neglected the effects of the terminations at $z = \pm\ell/2$. If the effect of the termination could be specified, an alternative equivalent current could be composed of the currents given above plus a reflected current due to the termination. These reflection effects would be expected to be of most concern in the backscatter direction, rather

[5]Note that we denote the diffraction coefficient to be a function of L, ϕ, and ϕ' to imply the Fresnel integral form of the uniform theory in (12-52) and use ϕ, ϕ', and γ_o when the asymptotic form in (12-59) is intended.

than in the direction of the bistatic scattered field. Even so, as the edge becomes long in terms of the wavelength, termination effects diminish. In addition, usually the above currents find their application in the angular region near the plane normal to the edge, further diminishing any possible termination effects.

When we obtain equivalent currents, we invoke the postulate of diffraction theory that diffraction is a local phenomena. For curved edges, we stretch this postulate even further than for the straight edge. That is, we assume that each point on a curved edge acts as an incremental section of an infinite straight edge and thereby determine the equivalent current. Thus, for example, the equivalent current that would enable us to calculate the diffracted field at the caustic of the problem in Fig. 12-27*b* would be [14]

$$I^m = -(\hat{\boldsymbol{\phi}} \times \hat{\mathbf{s}}') \cdot \mathbf{E}^i D_\perp\left(\phi, \phi'; \frac{\pi}{2}\right)\sqrt{\frac{8\pi}{\beta}}\, e^{-j(\pi/4)} \tag{12-103}$$

where we have used the result of (12-98) and the fact that $(\hat{\boldsymbol{\phi}} \times \hat{\mathbf{s}}')$ gives us the unit vector perpendicular to the ray from the sources to the edge. The use of (12-103) gives the calculated results in Fig. 12-28, which agree with experimental measurements in the caustic region.

If, on the other hand, the source in Fig. 12-27*b* were a magnetic dipole, then the required equivalent current would be [14]

$$I^e = -\frac{\hat{\boldsymbol{\phi}} \cdot \mathbf{E}^i}{\eta} D_\parallel\left(\phi, \phi'; \frac{\pi}{2}\right)\sqrt{\frac{8\pi}{\beta}}\, e^{-j(\pi/4)} \tag{12-104}$$

For an arbitrary polarization of the incident wave, both electric and magnetic currents are necessary to obtain the total diffracted field. Such a situation would occur, for example, in the calculation of the fields at or near the rear axis (caustic region) of a circular parabolic reflector antenna. At the rim of the parabolic dish, the polarization of the field incident from the feed is generally neither perpendicular nor parallel to the edge. Thus, both electric and magnetic equivalent currents at the rim would be required to obtain the total diffracted field in the rear axial region.

12.10 A MULTIPLE DIFFRACTION FORMULATION

In the previous two sections, we considered radiating elements on infinitely thin ground planes (i.e., $n = 2$). If, instead, the ground plane were "thick" such that one side could be represented by two 90° wedges as shown in Fig. 12-29, then it would

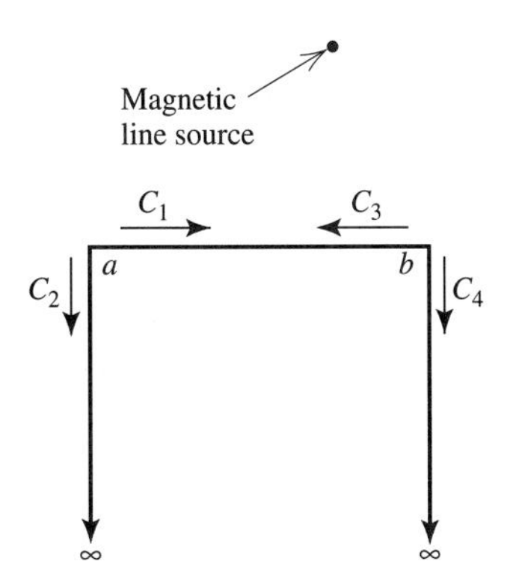

Figure 12-29 Magnetic line source exciting surface rays on a half-plane of finite thickness.

have been necessary to consider the multiple diffractions that occur between the two closely spaced edges. In such a situation, some of the energy diffracted by one edge is, in turn, diffracted by the other, giving rise to second-order diffraction or double diffracted rays. Clearly, some of these double diffracted rays give rise to still higher-order multiple diffractions.

If, to reasonably approximate the total diffracted energy, it is necessary to include doubly diffracted rays, then it is usually simplest to include them in the same manner used to account for the first-order diffraction in the previous two sections. On the other hand, if it is necessary to account for diffractions higher than second-order, it is advantageous to use a procedure known as the *method of self-consistency*.

Briefly, the method of self-consistency incorporates all the diffracted rays (i.e., single as well as all higher-order multiple ones) into a *total* (or net) diffracted field from each diffracting edge. Thus, each of these total edge diffracted fields is excited by a surface ray. Between the two diffracting edges there are, therefore, two surface rays traveling in opposite directions. The amplitudes and phases of these two surface rays are treated as unknowns. To solve for the two unknowns, two equations are generated by applying single diffraction conditions at each of the two diffracting edges.

To illustrate, consider Fig. 12-29. The coefficients C_1 and C_3 are the unknown amplitudes (i.e., magnitude and phase) of the two surface rays on the surface ab. The coefficients C_2 and C_4 are known once C_1 and C_3 are known. Thus, we may write the following equations. At edge a,

$$C_1 = C_3 R_{ba} + V_1 \tag{12-105}$$

and at edge b,

$$C_3 = C_1 R_{ab} + V_3 \tag{12-106}$$

and knowing C_1 and C_3, we have

$$C_2 = T_{ba} C_3 + V_2 \tag{12-107}$$

$$C_4 = T_{ab} C_1 + V_4 \tag{12-108}$$

where R and T are reflection and transmission coefficients, respectively, and V is the direct source contribution to the corresponding surface ray.

Equations (12-105) and (12-106) may be written in matrix form as

$$\begin{bmatrix} 1 & -R_{ba} \\ -R_{ab} & 1 \end{bmatrix} \begin{bmatrix} C_1 \\ C_3 \end{bmatrix} = \begin{bmatrix} V_1 \\ V_3 \end{bmatrix} \tag{12-109}$$

or compactly as

$$[Z][C] = [V] \tag{12-110}$$

where $[Z]$ is taken to be a coupling matrix and $[V]$ is the excitation matrix. The elements of the coupling matrix specify the interactions between the two surface rays. In general, two surface rays can couple only if they travel on the same or adjacent faces of a polygon as shown in Fig. 12-30. This, in general, leads to a sparse $[Z]$ matrix.

For the situation in Fig. 12-29, the reflection and transmission coefficients are

$$R_{ab} = \frac{e^{-j\beta\rho_{ab}}}{\sqrt{\rho_{ab}}} \frac{1}{2} D_{\perp}(L, \phi, \phi') \tag{12-111}$$

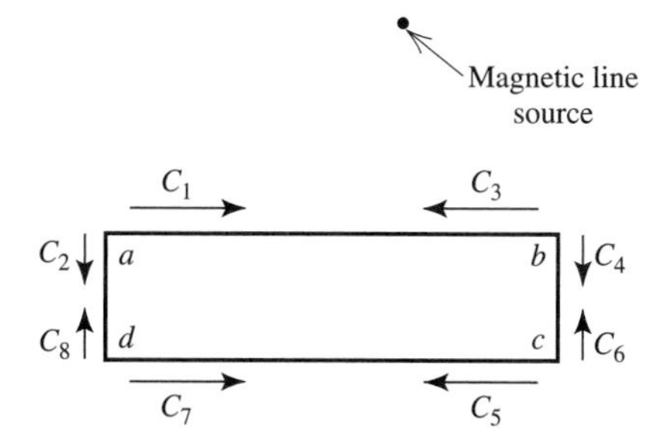

Figure 12-30 Magnetic line source exciting surface rays on an infinite four-sided polygon cylinder.

where $\phi^+ = \phi^- = 0$ and $\gamma_o = 90°$, and

$$T_{ab} = \frac{e^{-j\beta\rho_{ab}}}{\sqrt{\rho_{ab}}} \frac{1}{2} D_{\perp}(L, \phi, \phi') \tag{12-112}$$

where $\phi^+ = 2\pi - \pi/2$, $\phi^- = 0$, and $\gamma_o = 90°$. For T_{ab}, ϕ^+ is 2π less the interior wedge angle of $\pi/2$. In both cases, the distance parameter L used is that for cylindrical waves. For the special situation depicted in Fig. 12-29, $R_{ab} = R_{ba}$ and $T_{ab} = T_{ba}$. This is not true in general. For example, it would not be true for the situation depicted in Fig. 12-31.

For the two excitation matrix elements, we have

$$V_1 = \frac{e^{-j\beta\rho_{sa}}}{\sqrt{\rho_{sa}}} D_{\perp}(L, \phi, \phi') \tag{12-113}$$

$$V_3 = \frac{e^{-j\beta\rho_{sb}}}{\sqrt{\rho_{sb}}} D_{\perp}(L, \phi, \phi') \tag{12-114}$$

where ρ_{sa} is the distance from the line source to edge a and ρ_{sb} is the distance from the line source to edge b. If the line source did not directly illuminate, say, edge b, then V_3 would be zero. Here, we have considered only the TE case. A consideration of the TM case requires a knowledge of slope diffraction (see Sec. 12.7).

As stated earlier, it may be sufficient in many problems to only take into account second-order diffraction, thereby neglecting all higher-order multiply-diffracted rays. An example of a situation where the method of self-consistency greatly simplifies the amount of work required for solution is suggested by Fig. 12-29. It is possible and practical to approximate the curved surface of Fig. 12-29*a* with a polygon such as that in Fig. 12-31*b*. For an accurate approximation to the curved surfaces, the edges in Fig. 12-31*b* may be sufficiently close together that higher-order multiple diffractions should be taken into account. The easiest way of doing that is via the method of self-consistency. However, the self-consistent field method only works well provided an edge is not in the transition region of a diffraction from another edge. This is a possibility if adjacent edges of the polygon are closely aligned and this limits the degree to which the curved surface may be approximated.

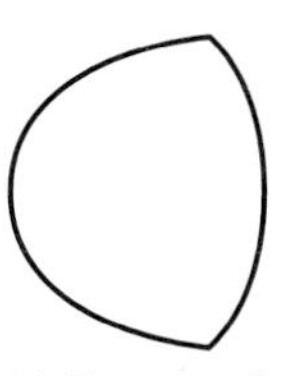

(*a*) Curved surface.

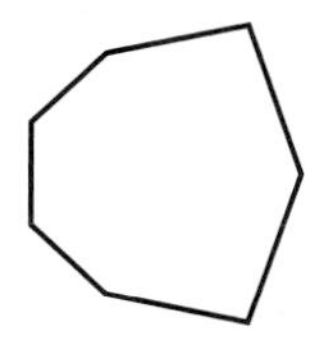

(*b*) Polygon approximation.

Figure 12-31 Polygon approximation of a curved surface cylinder.

12.11 DIFFRACTION BY CURVED SURFACES

In previous sections, we saw how a perfectly conducting wedge diffracts energy into the shadow region. Curved surfaces can also diffract energy. That is, when an incident ray strikes a smooth, convex-curved perfectly conducting surface at grazing incidence, a part of its energy is diffracted into the shadow region as illustrated by Fig. 12-32.

In Fig. 12-32, the incident plane wave undergoes diffraction at the shadow boundary at point Q_1 that is a point of tangency for the incident ray. At this point, a portion of the energy is trapped, resulting in a wave that propagates on the surface of the scatterer, shedding energy by radiation in directions tangent to the surface (e.g., point Q_2) as it progresses around the curved surface.

This wave that propagates along the surface in the shadow region is known as a *creeping wave*. The creeping wave can be described by an attachment (diffraction) coefficient at the point of capture, a launching (diffraction) coefficient at the point of radiation, an attenuation factor to account for the rate of radiation, and a description of the path on the scatterer transversed by the creeping wave in order to account for phase delay and total attenuation. Thus, the creeping wave field $E^{cw}(s'')$ along the path s'', in the case of a two-dimensional problem, can be written as

$$E^{cw}(s'') = E^i(Q_1)D^s(Q_1)G(s'')\, e^{-\int \gamma(s'')\, ds''} \qquad (12\text{-}115)$$

where

$E^{cw}(s'')$ = creeping wave field along s''
$D^s(Q_1)$ = the surface diffraction coefficient (attachment coefficient) at point A
$\gamma(s'') = \alpha(s'') + j\beta(s'')$ = creeping wave propagation factor
s'' = arc length along the creeping wave path
$G(s'')$ = the ray divergence factor determined by the geometry of the ray

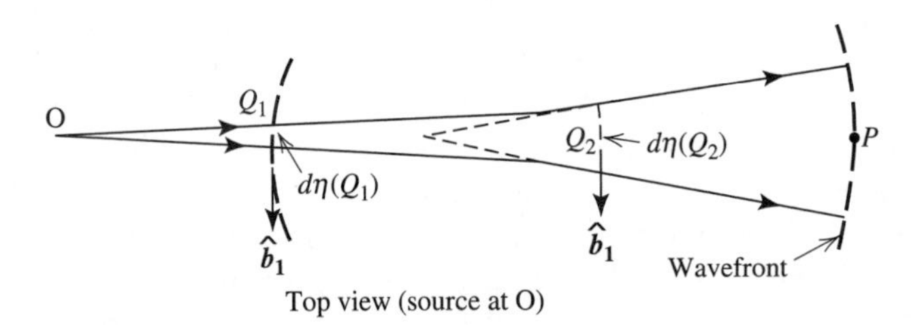

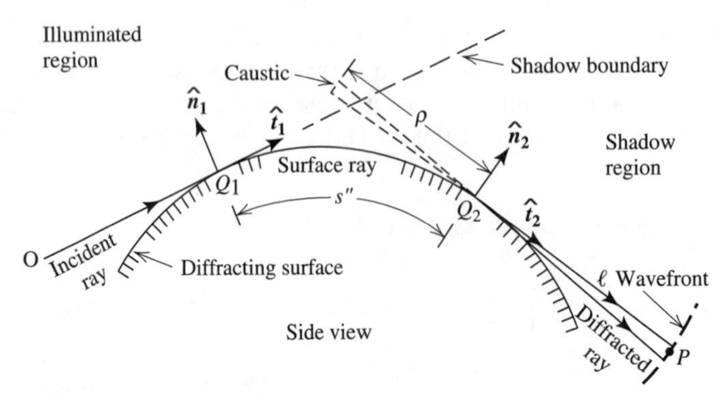

Figure 12-32 Diffraction by a smooth convex surface.

Table 12-1 Diffraction and Attenuation Coefficients for a Convex Cylindrical Surface

Case	$(D^s)^2$	α	Airy Function and its Zeros
$E_\parallel$	$\dfrac{\rho_g^{1/3}\, e^{-j\pi/12}}{\pi^{1/2}\, 2^{5/6}\, \beta^{1/6}\, (Ai'(-q))^2}$	$\dfrac{q}{\rho_g}\, e^{j\pi/6}\left(\dfrac{\beta\rho_g}{2}\right)^{1/3}$	$q = 2.33811$ $A_i'(-q) = 0.7012$
$E_\perp$	$\dfrac{\rho_g^{1/3}\, e^{-j\pi/12}}{\pi^{1/2}\, 2^{5/6}\, \beta^{1/6}(\bar{q})\, (Ai(-\bar{q}))^2}$	$\dfrac{\bar{q}}{\rho_g}\, e^{j\pi/6}\left(\dfrac{\beta\rho_g}{2}\right)^{1/3}$	$\bar{q} = 1.01879$ $Ai(-\bar{q}) = 0.5356$

As already stated, the surface ray sheds energy tangentially as it propagates along a geodesic on the curved surface, with the result that energy is continuously lost, resulting in attenuation. As in geometrical optics, we assume that energy in the flux tube between adjacent rays is conserved, which gives the two-dimensional geometrical optics spreading factor as

$$G(s'') = \sqrt{\frac{d\eta(Q_1)}{d\eta(s'')}} \tag{12-116}$$

where $d\eta(s'')$ is the transverse dimension of the surface ray tube as shown in Fig. 12-32. For a plane wave at normal incidence, $G(s'')$ is unity.

Keller and Levy [21, 22] have given the first-order terms in the expressions for the curved surface diffraction coefficients and attenuation constants. Kouyoumjian [14] gives higher-order terms and uses the notation of a soft surface for the case where E is tangential to the surface and hard surface for the case where E is normal to the surface. To be consistent with our earlier notation, we will use the perpendicular and parallel notation of earlier parts of this chapter.

At point Q_2, there will be a second surface diffraction coefficient $D^s(Q_2)$ that will account for the tangential detachment (launch) of a ray from the surface toward the observation point at a distance ℓ from point Q_2. Thus, we can write for the creeping wave field shed at Q_2 and observed at the observation point P

$$E^{cw}(P) = E^i(Q_1)D^s(Q_1)D^s(Q_2)\sqrt{\frac{d\eta(Q_1)}{d\eta(Q_2)}}\, e^{-\int_{Q_1}^{Q_2}\alpha(s'')\, ds''} e^{-j\beta(Q_2-Q_1)}\, \frac{e^{-j\beta\ell}}{\sqrt{\ell}} \tag{12-117}$$

For a circular cylinder, the diffraction coefficients and attenuation constants for the asymptotic approximation to the exact solution are given by the quantities in Table 12-1, where ρ_g is the radius of curvature along a geodesic. For a normally incident plane wave on a circular cylinder, $\rho_g = a$. $\mathrm{Ai}(-x)$ is the Airy function [7, 14]. The creeping wave surface field is more accurately represented by a series of modes, but only the first such mode is given in Table 12-1 since the higher-order modes are not numerically significant for the circular cylinder treated here.

EXAMPLE 12-3 *Creeping Wave on a Circular Cylinder*

Consider the two-dimensional problem of calculating the radar echo width of a right circular cylinder normal to the axis of the cylinder. The echo width σ_w is the two-dimensional counterpart to the three-dimensional echo area or radar cross section. σ_w is defined to be

$$\sigma_w = \lim_{\ell\to\infty} 2\pi\ell\, \frac{|E^s|^2}{|E^i|^2} \tag{12-118}$$

where ℓ is the range to the target.

Let the cylinder be of radius a and centered about the z-axis as in Fig. 12-33. Assume the incident wave is given by

$$\begin{pmatrix} E^i_\parallel \\ E^i_\perp \end{pmatrix} = e^{+j\beta\ell} \tag{12-119}$$

The two-dimensional geometrical optics field reflected in the backscatter direction may be found from (12-21) and (12-24) to be

$$\begin{pmatrix} E^r_\parallel \\ E^r_\perp \end{pmatrix} = \begin{pmatrix} E^i_\parallel \\ E^i_\perp \end{pmatrix} [R] \sqrt{\frac{a}{2\ell}} \tag{12-120}$$

where $\ell \gg a$ and where R is given by (12-25). Applying the echo width definition above gives

$$\sigma_w = \lim_{\ell\to\infty} 2\pi\ell \frac{|E^s|^2}{|E^i|^2} = \lim_{\ell\to\infty} 2\pi\ell\left(\frac{a}{2\ell}\right) = \pi a \tag{12-121}$$

Next, the effects of creeping waves around the cylinder are included. At the attachment point, write

$$E^{cw}(Q_1) = E^i(Q_1)D^s(Q_1) \tag{12-122}$$

and at the detachment point, write

$$E^{cw}(Q_2) = E^i(Q_1)D^s(Q_1)\, e^{-j\beta\pi a} e^{-\int_0^{\pi a} \alpha\, ds''} = E^i(Q_1)D^s(Q_1)\, e^{-j\beta\pi a} e^{-\alpha\pi a} \tag{12-123}$$

$G(s'')$ is unity since there is no transverse spreading of the rays on the surface of the cylinder.

When $E^{cw}(Q_2)$ is multiplied by the detachment coefficient $D(Q_2)$, we have the radiated creeping wave field. By a reciprocity argument, we can see that for the circular cylinder, the launching and attachment coefficients are the same. Thus in the following expression for the radiated field in the backscatter direction, the surface diffraction coefficient is squared:

$$\begin{pmatrix} E^{cw}_\parallel \\ E^{cw}_\perp \end{pmatrix} = \begin{pmatrix} E^i_\parallel \\ E^i_\perp \end{pmatrix} \cdot \begin{pmatrix} D^s_\parallel \\ D^s_\perp \end{pmatrix}^2 e^{-j\beta\pi a} e^{-\pi a \alpha_{\parallel\perp}} \sqrt{\frac{a}{2\ell}} \tag{12-124}$$

We must consider the fact that there are attachment points at both the top and bottom of the cylinder requiring us to double the creeping wave field strength. Thus, the total back-scattered field is

$$\begin{pmatrix} E^s_\parallel \\ E^s_\perp \end{pmatrix} = \begin{pmatrix} E^r_\parallel \\ E^r_\perp \end{pmatrix} + 2\begin{pmatrix} E^{cw}_\parallel \\ E^{cw}_\perp \end{pmatrix} \tag{12-125}$$

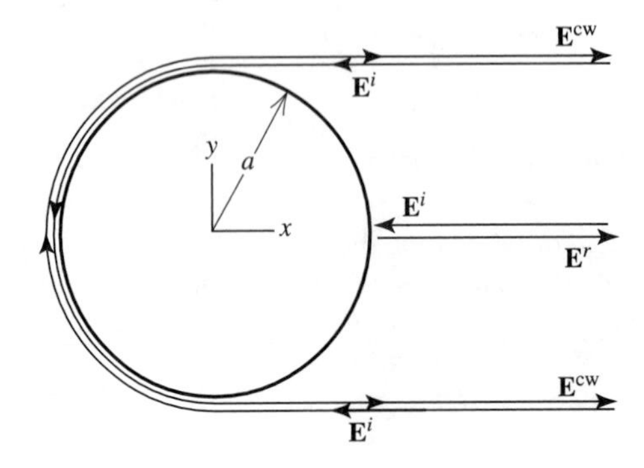

Figure 12-33 Backscatter from a circular cylinder.

which for the cylinder is

$$\begin{pmatrix} E_{\|}^{s} \\ E_{\perp}^{s} \end{pmatrix} = \begin{pmatrix} E_{\|}^{i} \\ E_{\perp}^{i} \end{pmatrix} \cdot \begin{pmatrix} -\sqrt{\dfrac{a}{2}}\, e^{j2\beta a} + 2(D_{\|}^{s})^{2} e^{-\pi a \alpha_{\|}} e^{-j\beta\pi a} \\ +\sqrt{\dfrac{a}{2}}\, e^{j2\beta a} + 2(D_{\perp}^{s})^{2} e^{-\pi a \alpha_{\perp}} e^{-j\beta\pi a} \end{pmatrix} \frac{e^{-j\beta\ell}}{\sqrt{\ell}} \tag{12-126}$$

The echo width then becomes

$$\sigma_w = \pi a \left| \mp e^{j2\beta a} + 2(D_{\|,\perp}^{s})^{2} \left(\frac{2}{a}\right)^{1/2} e^{-\pi a \alpha_{\|,\perp}} e^{-j\beta\pi a} \right|^{2} \xrightarrow[\beta a \to \infty]{} \pi a^{2} \tag{12-127}$$

Expressions for D and α can be obtained from Table 12-1 with $\rho_g = a$.

For the problem considered in the above example, the exact result for the echo width for both principal polarizations appears in Fig. 12-34*a*. The echo width is normalized with respect to πa just as the sphere echo area was normalized with respect to its specular contribution. Note that the creeping wave contribution to the echo width for the cylinder is quite visible for the perpendicular polarization and does not appear for the parallel polarization. This is due to the tangential electric field boundary condition that tends to short out the creeping wave contribution in the parallel case but not the perpendicular case.

The above GTD-based expression in (12-127) for the echo width of a circular cylinder cannot be expected to reproduce the curves in Fig. 12-34*a*, which are based on the exact eigenfunction solution, when the diameter is on the order of the wavelength or less. This, of course, is because GTD is an asymptotic theory valid most when the wavelength is small compared to the scatterer. In Fig. 12-34*a*, the wavelength is not small compared to the scatterer. Results using (12-127) are shown in Fig. 12-34*b*. The differences between the results from (12-127) and the exact solution are apparent. As a point of fact, there is no theoretical reason to expect them to agree. Interestingly, most of the error in attempting to reproduce Fig. 12-34*a* with (12-127) arises from the geometrical optics term and not the creeping wave term.

To improve the accuracy of the geometrical optics contribution, it is necessary to include correction terms [23] (see Prob. 12.11-1). The application of correction terms to the geometrical optics contribution is shown in Fig. 12-34*c*, and the improvement over Fig. 12-34*b* is substantial. In fact, the agreement between Fig. 12-34*a* and Fig. 12-34*c* when $\beta a > 2$ (i.e., $a > \lambda/3$) is surprisingly good for a high-frequency method in a lower portion of the intermediate frequency region. In this region, one would normally employ the method of moments or MoM (see Sec. 10.13.2) to produce results such as those in Fig. 12-34*a*.

The concept of a creeping wave is valuable in that it helps one visualize the physical process involved in diffraction by curved surfaces. For example, the RCS of a sphere as a function of the sphere radius is presented in Fig. 12-7. We can interpret the oscillatory feature of the curve as being caused by two creeping waves traveling around the sphere in opposite directions. A similar explanation applies to the cylinder in Fig. 12-34. Depending on the electrical size of the sphere or cylinder, these two creeping waves tend to either constructively or destructively interfere with each other, causing the RCS to oscillate about the value contributed by the specular scattering. As the sphere or cylinder becomes larger, the amount of oscillation decreases, which may be attributed to the decreasing amplitudes of the two creeping waves due to the product α times the total path length s''. In the backscatter case, as the radius becomes large, α becomes small, but the product $\alpha s''$ becomes

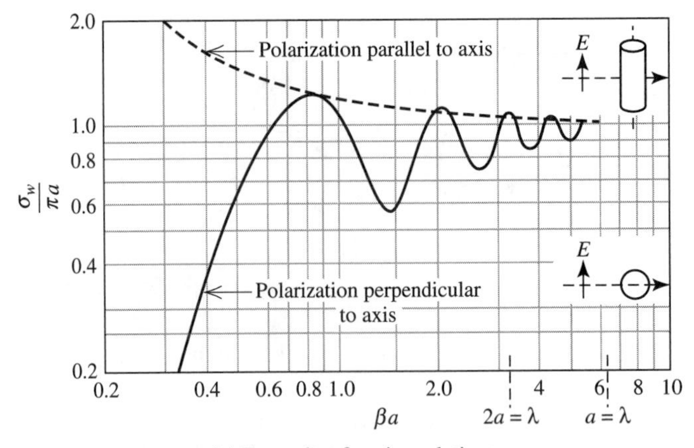

(*a*) Exact eigenfunction solution.

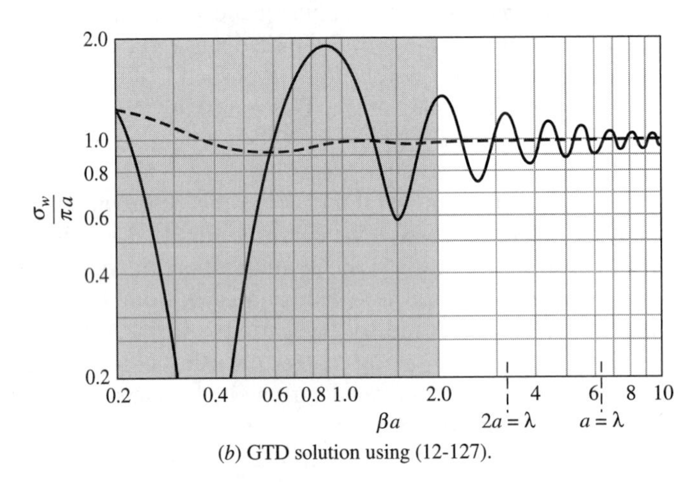

(*b*) GTD solution using (12-127).

(*c*) GTD solution with GO correction.

Figure 12-34 Echo width of an infinitely long circular cylinder calculated by three methods.

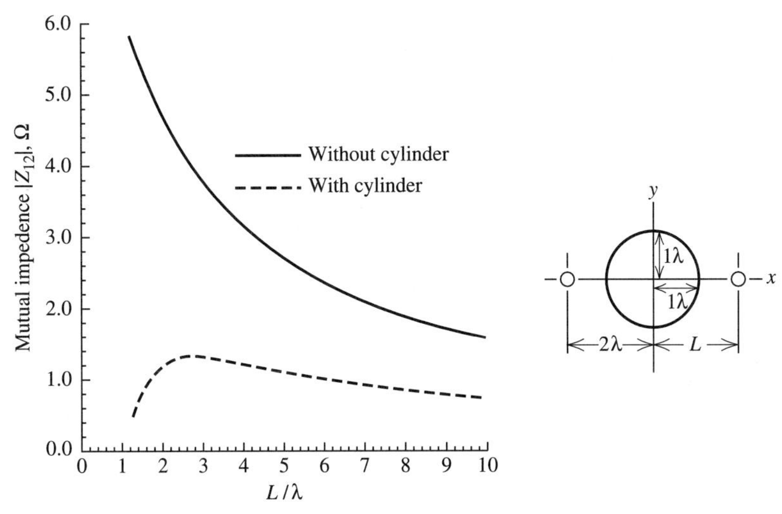

Figure 12-35 Mutual impedance Z_{12} between two $\lambda/2$ dipoles in the deep shadow region of each other compared to Z_{12} without cylinder present.

large, causing the creeping wave to become small compared to the specular contribution as evidenced by Figs. 12-7 and 12-34.

In the case of backscatter for the cylinder or sphere, the creeping wave travels a 180° geodesic path. A 180° path is not the only path of possible interest. Consider Fig. 12-35 that shows the mutual coupling between two half-wave dipoles on opposite sides of a circular cylinder [24]. The coupling with the cylinder present can be approximately calculated by the creeping wave formulation in this section. A more robust UTD formulation can be found in [25].

12.12 EXTENSION OF MOMENT METHOD USING THE GTD

In Chap. 10, we saw how the MoM could be applied to many antenna and scattering problems where the antenna or scatterer was not excessively large in terms of the wavelength. In this chapter, we have seen how geometrical optics and GTD can be applied to problems that are large in terms of wavelength. The purpose of this section is to show how the class of problems solvable by moment methods can be enlarged by incorporating GTD into the moment method solution [26]. In studying this section, the student will have an opportunity to test his or her understanding of the fundamental concepts developed in Chap. 10 and the previous sections of this chapter.

Recall from Chap. 10 the elements of the generalized impedance matrix can be given in inner product notation as

$$Z_{mn} = \langle \mathbf{J}_m, \mathbf{E}_n \rangle \tag{12-128}$$

This states that the Z_{mn}th element of the impedance matrix is found by reacting the mth test function (weight function) with the electric field from the nth basis function. Similarly, the mth element in the generalized voltage matrix is found by reacting the mth test function with the incident field.

In a strictly moment method formulation of a given problem, all material bodies are removed and replaced with equivalent currents radiating in free space. Thus, when one reacts the mth test function with the field from the nth basis function, it is only that field which directly arrives at the mth test function via the shortest free-space path that one needs to consider since it is the only possible field. However, suppose there exists in a given situation a portion of the structure that is not represented by equivalent currents (i.e., a material body remains as shown in Fig. 12-36). In this case, the calculation of the impedance matrix elements is more complex but not unduly so. Let these new impedance matrix elements be denoted Z'_{mn}. In terms of (12-128), the reaction of $\mathbf{J}_m$ with $\mathbf{E}_n$ may be interpreted to mean the reaction of the test source with not only the field from the true source arriving at the test source directly, but also the reaction of the test source with fields from the true source that arrive by other means as suggested by Fig. 12-36. Therefore, one can write

$$Z'_{mn} = \langle \mathbf{J}_m, a\mathbf{E}_n + b\mathbf{E}_n \rangle \tag{12-129}$$

where a may be set to unity and $b = b(m, n)$ is different for each m and n. The quantity $b\mathbf{E}_n$ also represents the field due to $\mathbf{J}_n$, but arriving at the mth observation point or region due to a physical process, such as a geometrical optics or diffraction mechanism, which is not accounted for in that portion of the problem formulated by the moment method. Thus,

$$Z'_{mn} = \langle \mathbf{J}_m, \mathbf{E}_n \rangle + \langle \mathbf{J}_m, b\mathbf{E}_n \rangle \tag{12-130}$$

or

$$Z'_{mn} = Z_{mn} + Z^g_{mn} \tag{12-131}$$

where the superscript g denotes that Z^g_{mn} is an additional term added to, in general, each impedance matrix element due to a physical process g that redirects energy from the nth basis current function to the location of the mth test source.

As implied by Fig. 12-36, there is also a modification of the usual generalized voltage matrix terms. That is,

$$V'_m = \langle \mathbf{J}_m, \mathbf{E}^i + c\mathbf{E}^i \rangle \tag{12-132}$$

where $\mathbf{E}^i$ is the incident field arriving directly at region m and $c\mathbf{E}_i$ is that field from the source redirected to region m by a physical process g. We note that $c = c(m)$ is different for each m:

$$V'_m = \langle \mathbf{J}_m, \mathbf{E}^i \rangle + \langle \mathbf{J}_m, c\mathbf{E}^i \rangle \tag{12-133}$$

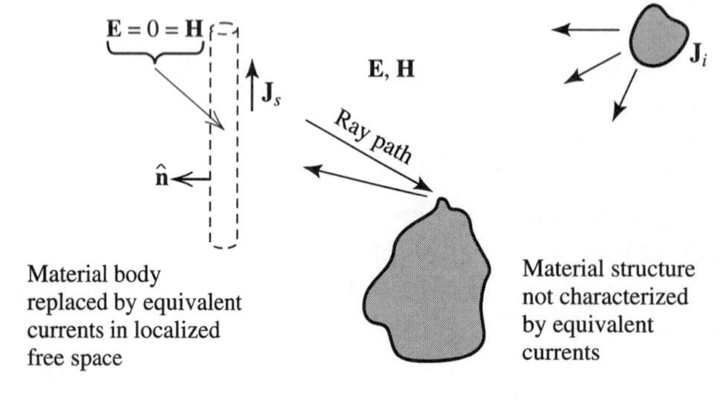

Figure 12-36 A source radiating in free space with one scatterer replaced by an equivalent current and the other remaining as a material body.

or

$$V'_m = V_m + V^g_m \tag{12-134}$$

As a direct consequence of the foregoing discussion, we have

$$[Z'][I'] = [V'] \tag{12-135}$$

and its solution as

$$[I'] = [Z']^{-1}[V'] \tag{12-136}$$

where $[I']$ is the current on, for example, an antenna operating in the presence of scattering mechanisms that may be accounted for by either geometrical optics techniques or GTD.

Initially, to combine the method of moments and GTD into a hybrid technique, consider the problem of a monopole near a perfectly conducting wedge as shown in Fig. 12-37. If we describe the monopole on an infinite ground plane strictly by the moment method matrix representation given in (10-43), then for the monopole near the conducting wedge, we utilize (12-135), where in (12-131) the term Z^g_{mn} is obtained by considering that energy radiated by the nth basis function on the monopole that is diffracted by the wedge to the mth observation point or region. In the work here, we employ pulse basis functions and point-matching where the testing functions are delta functions. However, the choice of basis and testing functions is not restricted to these functions.

To calculate Z^g_{mn}, we compute the electric field from the nth pulse basis function incident on the edge of the wedge at the stationary point. Taking that component of the electric field perpendicular to the edge and to the direction of propagation of the incident field, we then compute the energy diffracted to the observation point at the center of the mth segment on the monopole. The component of this field tangential to segment m is the term Z^g_{mn} of (12-131) since we are employing delta-weighting functions. To compute the diffracted field, we use the formulation in Sec. 12.4 for the case of spherical wave incidence.

Shown in Fig. 12-38*a* is a calculated curve for the input resistance of a quarter-wavelength monopole a distance d from the edge of a perfectly conducting wedge (see Fig. 12-37). We note that the resistance oscillates about the value for a quarter-wavelength monopole on an infinite ground plane and also the amount of variation is relatively small, being only a few ohms. A similar curve is shown in Fig. 12-38*b* for the input reactance. Data for both curves were obtained directly from (12-136) without the need for any apriori knowledge of the current distribution or terminal current value.

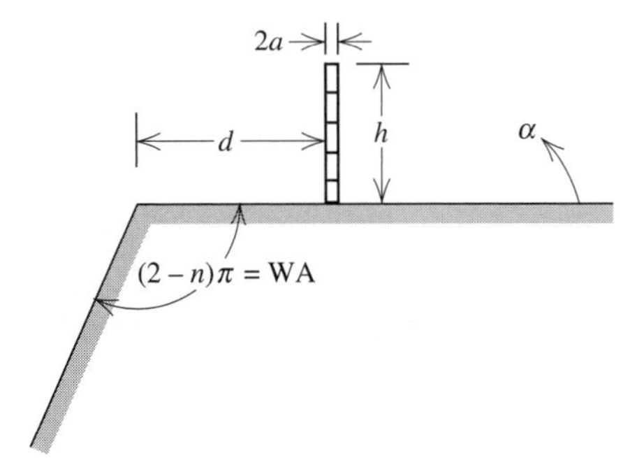

Figure 12-37 Monopole on a conducting wedge.

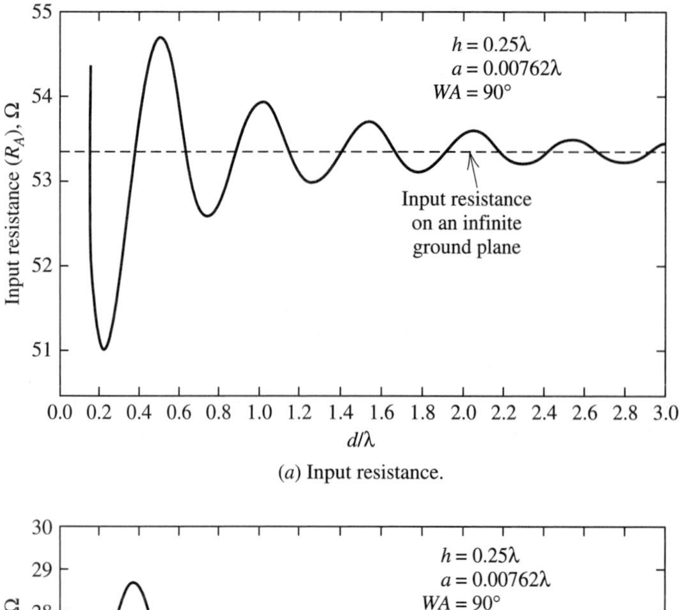

(*a*) Input resistance.

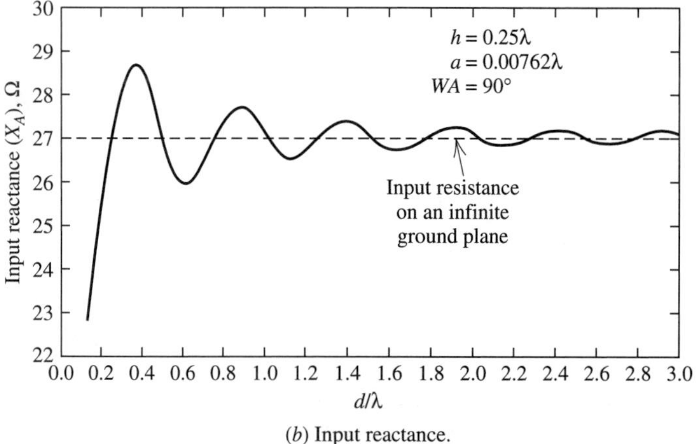

(*b*) Input reactance.

Figure 12-38 Input impedance of a monopole on a conducting wedge as a function of the distance *d* from the edge as shown in Fig. 12-37.

Thus far, the discussion has centered on the calculation of input impedance. Obviously if one can accurately compute the input impedance, then quite accurate far-field information can readily be obtained also. For example, in the case of a monopole near a single wedge, as in Fig. 12-37, there may be as many as three contributors to the far field. First, there is direct source radiation except in the shadow region. Second, there is the reflected field, which is most conveniently accounted for by using the image in the horizontal surface. Third, there is the diffracted field that contributes in all regions and, of course, is the only source of radiation in the shadow region. A typical far-field pattern is shown in Fig. 12-39. Note that for $\alpha = 90°$, the field does not go to zero as would be the case if the ground plane was infinite in extent.

If we wish to investigate a circular ground plane as in Fig. 12-40, we must use the equivalent edge currents described in Sec. 12.9. Thus, we replace the edge of the disk with an equivalent magnetic current M given by

$$M = -2E_\theta e^{-j(\pi/4)} D_\perp(L, \phi, \phi')\sqrt{\lambda} \tag{12-137}$$

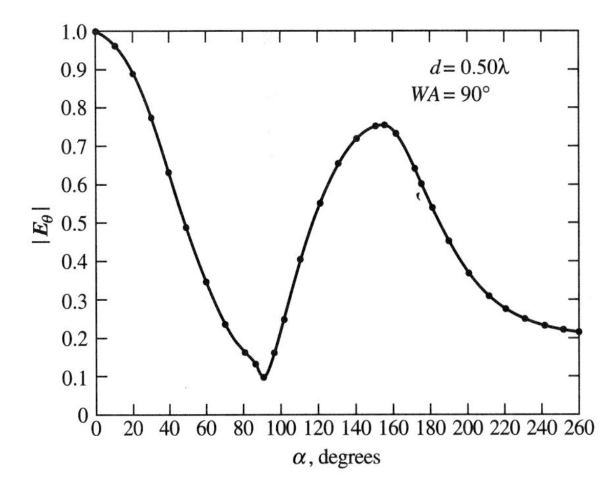

Figure 12-39 Normalized far-field pattern of a quarter-wave monopole on a conducting wedge as shown in Fig. 12-37.

This equivalent magnetic current is used to calculate the field at the segment at s due to the current at s' as indicated in Fig. 12-40. Note that an equivalent magnetic ring current must be calculated for each choice of s and s'.

It is useful for us to break up the equivalent magnetic ring current of Fig. 12-40 into differential elements dc' so that the observation point is in the far field of each element even though it may be in the near field of the total ring current.

The electric field in a plane perpendicular to an element dc' is given by

$$dE_z = \frac{M\,dc'}{4\pi}\left(\frac{j\omega}{cr} + \frac{1}{r^2}\right)e^{-j\beta r} \tag{12-138}$$

where $r = (a^2 + z^2)^{1/2}$. Letting $dc' = a\,d\psi$ where ψ is the azimuth angle, taking only the z-component at the monopole, and integrating over the range $\psi = 0$ to $\psi = 2\pi$ yields

$$E_z = \frac{Ma^2}{2r}\left(\frac{j\beta}{r} + \frac{1}{r^2}\right)e^{-j\beta r} \tag{12-139}$$

The value for E_z is the term Z^g_{mn} that is added to the impedance element obtained for a monopole on an infinite ground plane. This process gives the modified imped-

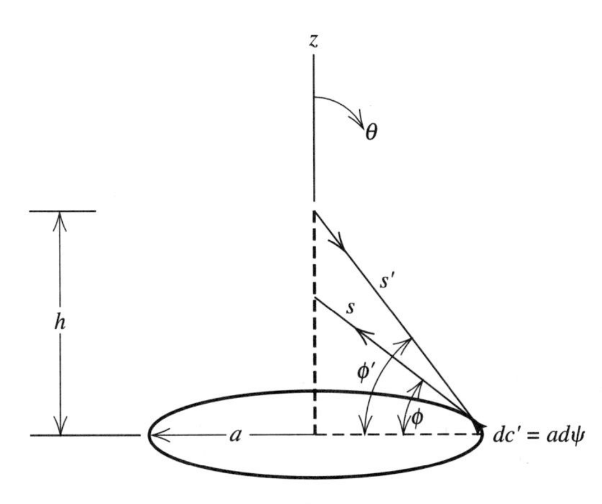

Figure 12-40 Segmented monopole encircled by a magnetic ring current for analysis of a monopole on a circular ground plane.

ance element needed, Z'_{mn}, to calculate the modified currents (and hence input impedance) of a monopole on the finite circular ground plane.

Figures 12-41*a* and 12-41*b* show a comparison between calculations made with the equivalent magnetic ring current and measurements for a monopole of length 0.224λ and radius 0.003λ on a circular ground plane for varying radius. It is apparent that the correct variation is accurately predicted for both the real and imaginary parts of the input impedance. For the input resistance, the agreement between the measurements and the theory is excellent. For the input reactance, the agreement is very good, but there is a slight shift in the calculated curve when compared to the measurements. The amount of this shift is sufficiently small that it can be attributed to the usual problems associated with modeling the region in proximity to the driving point.

Next, consider the situation shown in Fig. 12-42 where a monopole of height h is a distance d_1 away from a vertical conducting step. To properly determine the Z^g_{mn} term in (12-131), it is necessary to determine all the various combinations of

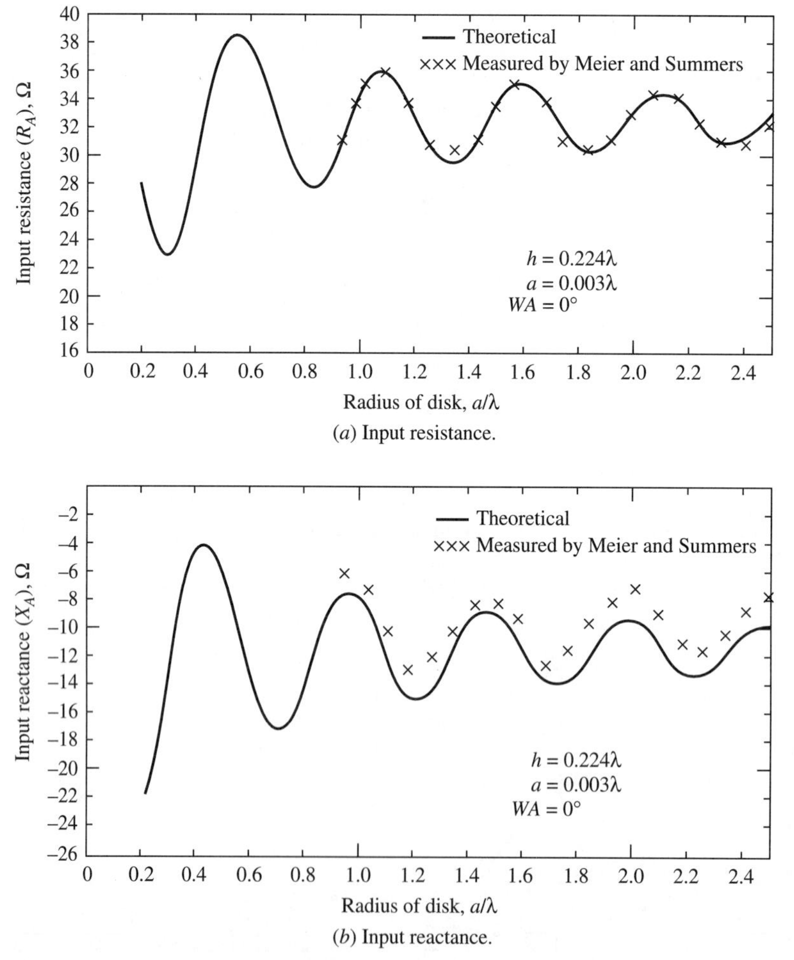

Figure 12-41 Theoretical and experimental input impedance of a monopole of radius 0.003λ at the center of a circular disk as shown in Fig. 12-40.

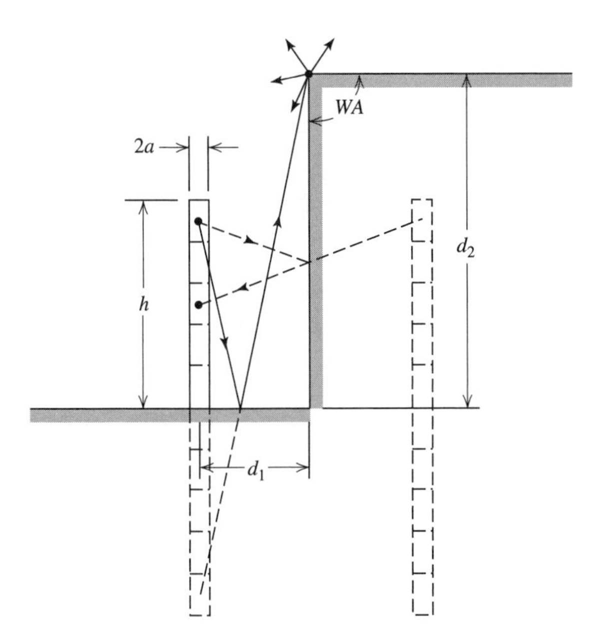

Figure 12-42 Monopole near a conducting step showing the partial use of images.

reflections that can occur for rays emanating from the monopole and reflecting back to it as well as the diffraction from the top edge of the step. Since the vertical wall is at a right angle to the lower horizontal surface, there will be no diffraction from the interior wedge and all the reflections can most conveniently be accounted for by imaging the monopole into the horizontal ground plane and then imaging the resulting dipole into the plane of the vertical wall.

Shown in Fig. 12-42 are two example situations that depict the utilization of the images. If we consider the uppermost segment of the monopole to be the source segment, one set of rays shows the use of the image in the horizontal surface to calculate reflected-diffracted energy reaching the segments of the monopole. The other set of rays shows the use of the image in the vertical wall to calculate singly reflected energy. In the calculated results that follow, all combinations of singly reflected, doubly reflected, diffracted, diffracted-reflected, reflected-diffracted, and reflected-diffracted-reflected rays are taken into account. All rays that involve combinations of double (or higher-order) diffractions are negligible.

Figure 12-43 shows the calculated input impedance for a quarter-wavelength monopole a quarter-wavelength away from a vertical wall whose height is $d_2 > 0.25\lambda$. As d_2 increases, the impedance oscillates about the value for the case where $d_2 = \infty$. The results of Figs. 12-43*a* and 12-43*b* show that as the diffracting edge recedes from the vicinity of the monopole, its effect on the input impedance rapidly diminishes. Although we have not shown results for the case where the step height is less than the height of the monopole, the same method could be used to investigate such situations.

In combining moment methods with GTD, we have proceeded from the philosophical viewpoint of extending the method of moments via GTD. In so doing, we have shown that modifying the impedance matrix to account for diffraction effects (or geometrical optics effects) enables one to accurately treat a larger class of problems than could be treated by moments methods alone. An alternative interpretation of the hybrid method is also possible. That is, the procedure employed can be

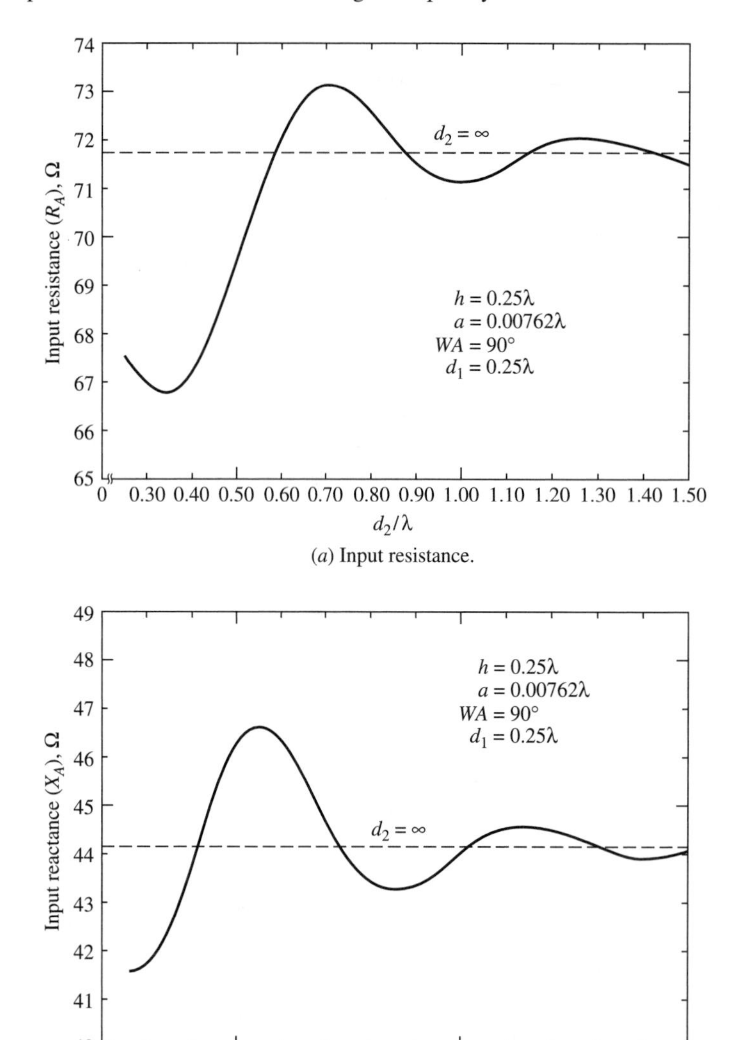

(*a*) Input resistance.

(*b*) Input reactance.

Figure 12-43 Input impedance of a quarter-wave monopole as a function of step height for the geometry of Fig. 12-42.

viewed as using GTD to obtain an approximation to the exact Green's function needed. Other hybrid methods are discussed in [27].

Although this hybrid method possesses many of the advantages inherent in both the moments method (MM) and GTD, it also has some of the limitations peculiar to each. For example, as in the usual moment method, one can treat arbitrary configurations of wire antennas (or slot antennas), taking into account lumped loading, finite conductivity, and so forth, and obtain accurate impedance data and current distributions. Naturally, one still must take the usual precaution of using a sufficient number of basis functions to assure convergence. On the other hand, as in the usual GTD problem, one must take care that the antenna is not too close to a source of diffraction (e.g., $d > 0.2\lambda$).

12.13 PHYSICAL OPTICS

In Sec. 12.1, we calculated the scattered field from the sphere by geometrical optics. Often, we can calculate these same scattered fields by physical optics. The concept of *physical optics* can be considered to be somewhat more general than geometrical optics since the equations obtained from physical optics for the scattered field from a conducting body often reduce to the equations of geometrical optics in the high-frequency limit. In fact, it is assumed in physical optics that the field at the surface of the scattering body is the geometrical optics surface field. This implies that, at each point on the illuminated side of the scatterer, the scattering takes place as if there were an infinite tangent plane at that point, while over the shadowed regions of the scatterer the field at the surface is zero [2].

For a perfectly conducting body, the assumed physical optics surface current is

$$\mathbf{J}_{\text{PO}} = \begin{cases} \hat{\mathbf{n}} \times \mathbf{H}_{total} & \textit{in the illuminated region} \\ 0 & \textit{in the shadowed region} \end{cases} \tag{12-140}$$

where $\hat{\mathbf{n}}$ is a unit normal vector outward from the surface of interest as shown in Fig. 12-44.

From image theory, the tangential components of $\mathbf{H}$ at a perfect conductor are just twice those from the same source when the conducting scatterer is replaced by equivalent currents in free space. Thus, the physical optics current is given by

$$\mathbf{J}_{\text{PO}} = 2(\hat{\mathbf{n}} \times \mathbf{H}^i) \tag{12-141}$$

From Chap. 1, we know that in the far field

$$\mathbf{E}^s = -j\omega\mathbf{A} \tag{12-142}$$

having neglected any radial terms. For the purposes of simplification, let ψ represent the free-space Green's function as in (1-56). Then

$$\mathbf{E}^s = -j\omega\mu \iint \mathbf{J}\psi \, ds' \tag{12-143}$$

Using the curl E equation, we can write for $\mathbf{H}^s$

$$\mathbf{H}^s = \nabla \times \iint \mathbf{J}\psi \, ds' \tag{12-144}$$

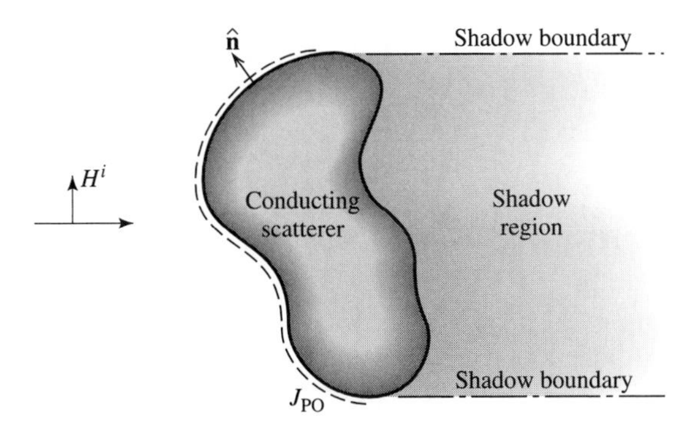

Figure 12-44 Physical optics current on a conducting scatterer.

Since the del operation is in the unprimed coordinate system and the integration is in the primed coordinate system, we can write

$$\mathbf{H}^s = \iint (\nabla \times \mathbf{J}\psi)\, ds' \tag{12-145}$$

Since $\nabla \times \mathbf{J}\psi = \nabla\psi \times \mathbf{J} + \psi\nabla \times \mathbf{J}$ using (C-16), and the last term on the right is zero,

$$\mathbf{H}^s = \iint (\nabla\psi \times \mathbf{J})\, ds' \tag{12-146}$$

Since $R = r - \hat{\mathbf{r}} \cdot \mathbf{r}'$, as in (1-96), we can express $\nabla\psi$ in the far field as

$$\nabla\psi = -\hat{\mathbf{r}}\, \frac{1 + j\beta r}{4\pi r^2}\, e^{-j\beta r} e^{j\beta \hat{\mathbf{r}}\cdot\mathbf{r}'} \tag{12-147}$$

giving

$$\mathbf{H}^s = e^{-j\beta r} \iint (\mathbf{J} \times \hat{\mathbf{r}})\, \frac{1 + j\beta r}{4\pi r^2}\, e^{j\beta \hat{\mathbf{r}}\cdot\mathbf{r}'}\, ds' \tag{12-148}$$

which is approximately equal to

$$\mathbf{H}^s \approx \frac{j\beta}{4\pi r}\, e^{-j\beta r} \iint (\mathbf{J} \times \hat{\mathbf{r}})\, e^{j\beta \hat{\mathbf{r}}\cdot\mathbf{r}'}\, ds' \tag{12-149}$$

since $\beta r \gg 1$ and $r \approx R$ in the denominator. It should be noted that this expression for the scattered field is frequency-dependent in contrast to the geometrical optics expression that is frequency-independent. It might, therefore, be intuitively inferred that physical optics provides a more accurate approximation to the scattered field. Although this may be so in certain cases, a general conclusion cannot be reached since necessary and sufficient conditions for the valid application of physical optics are not known [2]. It is fortunate for the engineer that physical optics works in many practical problems, even though in some problems prior justification of its application would be difficult to make.

Next, let us develop a general expression for the radar cross section using our expression in (12-149) for $\mathbf{H}^s$. Writing the radar cross-section definition as

$$\sigma = \lim_{r\to\infty} 4\pi r^2 \frac{|\mathbf{H}^s|^2}{|\mathbf{H}^i|^2} \tag{12-150}$$

and inserting the physical optics current $\mathbf{J} = 2\hat{\mathbf{n}} \times \mathbf{H}^i$ gives

$$\sigma = \lim_{r\to\infty} 4\pi r^2 \left(\frac{\beta}{4\pi}\right)^2 \left|\frac{1}{\mathbf{H}^i}\right|^2 \left|\iint ((2\hat{\mathbf{n}} \times \mathbf{H}^i) \times \hat{\mathbf{r}})\, \frac{e^{j\beta \hat{\mathbf{r}}\cdot\mathbf{r}'}}{r}\, ds'\right|^2 \tag{12-151}$$

which reduces to

$$\sigma = \frac{\pi}{\lambda^2} \left|\frac{1}{\mathbf{H}^i}\right|^2 \left|\iint ((2\hat{\mathbf{n}} \times \mathbf{H}^i) \times \hat{\mathbf{r}}) e^{j\beta \hat{\mathbf{r}}\cdot\mathbf{r}'}\, ds'\right|^2 \tag{12-152}$$

Using the vector identity (C-8), we can write

$$(\hat{\mathbf{n}} \times \mathbf{H}^i) \times \hat{\mathbf{r}} = (\hat{\mathbf{r}} \cdot \hat{\mathbf{n}})\mathbf{H}^i - (\hat{\mathbf{r}} \cdot \mathbf{H}^i)\hat{\mathbf{n}} \tag{12-153}$$

At this point, consider the backscattered or monostatic radar cross section where (12-153) reduces to $(\hat{\mathbf{r}} \cdot \hat{\mathbf{n}})\mathbf{H}^i$ and the phase of $\mathbf{H}^i$ is $e^{j\beta\hat{\mathbf{r}}\cdot\mathbf{r}'}$ on the illuminated surface. Thus,

$$\sigma = \frac{4\pi}{\lambda^2}\left|\iint (\hat{\mathbf{r}} \cdot \hat{\mathbf{n}})e^{j2\beta\hat{\mathbf{r}}\cdot\mathbf{r}'}\, ds'\right|^2 \tag{12-154}$$

Next, take $\hat{\mathbf{r}} = \hat{\mathbf{z}}$ for the purpose of monostatic illustration (i.e., the radar is on $+z$ axis), giving us a final result of

$$\sigma = \frac{4\pi}{\lambda^2}\left|\iint (\hat{\mathbf{z}} \cdot \hat{\mathbf{n}})e^{j2\beta\hat{\mathbf{z}}\cdot\mathbf{r}'}\, ds'\right|^2 \tag{12-155}$$

where the factor of 2 in the exponent represents the phase advance of the backscattered field relative to the origin due to the two-way path.

EXAMPLE 12-4 *RCS of a Sphere by Physical Optics*

Here we apply the result in (12-155) to obtain the physical optics expression for the monostatic RCS of a sphere and compare it to the result obtained by geometrical optics. First, we note that from Fig. 12-45

$$\hat{\mathbf{z}} \cdot \hat{\mathbf{n}} = \cos\theta' = \frac{a - \ell'}{a} \tag{12-156}$$

where ℓ' is the distance from the reference plane to the spherical surface, and that an element of surface area is $a^2 \sin\theta'\, d\theta'\, d\phi'$ on the spherical surface. Since $a - \ell' = a\cos\theta'$ and $d\ell' = a\sin\theta'\, d\theta'$, we find that on the projected area of the sphere onto the reference plane is $ds' = a^2 \sin\theta'\, d\theta'\, d\phi' = a\, d\phi'\, d\ell'$. Noting that $\hat{\mathbf{z}} \cdot \mathbf{r}' = (a - \ell')$, substituting into our general expression for monostatic RCS, and performing the ϕ integration, we obtain

$$\sigma_{sp} = \frac{4\pi}{\lambda^2}\left|2\pi\, e^{j2\beta a}\int_0^a e^{-j2\beta\ell'}(a - \ell')\, d\ell'\right|^2 \tag{12-157}$$

Performing the remaining integration yields

$$\sigma_{sp} = \frac{4\pi}{\lambda^2}\left|2\pi\frac{e^{j2\beta a}}{j2\beta}\left[a - \frac{1 - e^{-j2\beta a}}{j2\beta}\right]\right|^2 \tag{12-158}$$

which can be put in the form

$$\sigma_{sp} = \frac{4\pi}{\lambda^2}\left|\frac{a\lambda}{j2}\left[\left(1 + \frac{j}{2\beta a}\right)e^{j2\beta a} - \frac{j}{2\beta a}\right]\right|^2 \tag{12-159}$$

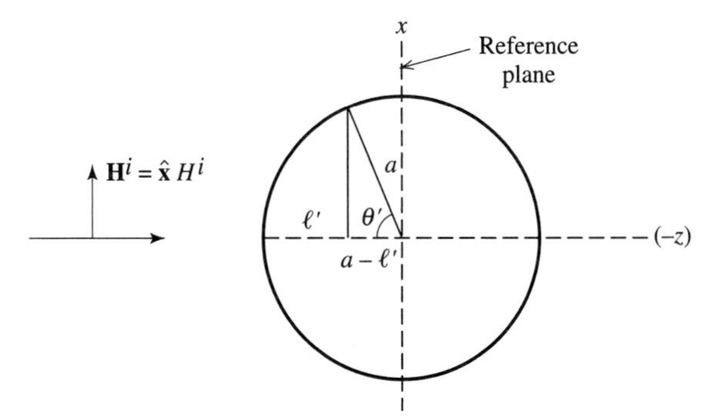

Figure 12-45 Physical optics scattering by a sphere.

The terms associated with the exponential $e^{j2\beta a}$ are due to the front face (i.e., specular) reflection, whereas the right-most term $(j/2\beta a)$ is due to the contribution from the artificially imposed discontinuity in the current at the $\theta = \pi/2$ location on the sphere (i.e., shadow boundary). Since this discontinuity is nonphysical, so too is the right-most term $(j/2\beta a)$ and we must disregard it. Thus,

$$\sigma = \pi a^2 \left| \frac{1}{j}\left(1 + \frac{j}{2\beta a}\right) e^{j2\beta a} \right|^2 \xrightarrow[\beta a \to \infty]{} \pi a^2 \tag{12-160}$$

We see then that the radar cross section of the sphere obtained via physical optics reduces to the geometrical optics result of (12-23) in the high-frequency limit.

The fact that we have had to eliminate the right-most term $(j/2\beta a)$ in (12-159) is not a peculiarity of the sphere, but is common to any problem employing physical optics where a nonphysical discontinuity in current gives rise to an erroneous contribution to the scattered field that can be numerically significant when compared to the geometrical optics contribution.

The second term in (12-160) may be taken to be the second term in a high-frequency asymptotic expansion of the scattered field. Such an expansion is in inverse powers of the frequency and is known as a Luneburg–Kline expansion [1]. The Luneburg–Kline expansion satisfies the wave equation and is a formal way of showing the correspondence between optics and electromagnetics in the high-frequency limit. The leading term in the Luneburg–Kline expansion is, in fact, the geometrical optics term that is also the first term in (12-160).

Physical optics is more useful to us than just finding radar cross sections. For example, if we wish to find the far-field pattern of a parabolic reflector antenna, physical optics is one way of doing so. In fact, it is probably the easiest way of finding the radiated field on the forward axis of the reflector antenna. In directions other than on the forward axis of the reflector antenna, physical optics provides us with a nonzero estimate of the radiation pattern. This should be contrasted with geometrical optics that can only provide information in a specular direction (see Fig. 12-46), but does so in a straightforward manner. Figure 12-46 shows a ray normally incident on a flat plate and the reflected or scattered field coming back only in one direction, whereas the figure indicates that the physical optics current produces a scattered field in all directions for the same incident field.

In summary, physical optics is an approximate high-frequency method of considerable usefulness that can be expected to provide an accurate representation of the scattered field arising from a surface where the postulated physical optics current is reasonably close to the true current distribution. We recall from the discussion at the beginning of this section that the physical optics current will be a reasonable representation of the true current if the field at the scatterer surface is correctly given by the geometrical optics surface field.

An example of a situation where the geometrical optics surface field does not give us the true current is in the vicinity of an edge (where a plane tangent to the

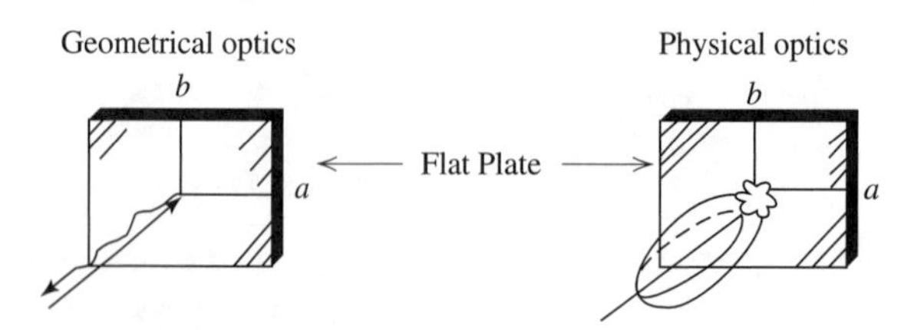

Figure 12-46 Geometrical optics compared to physical optics.

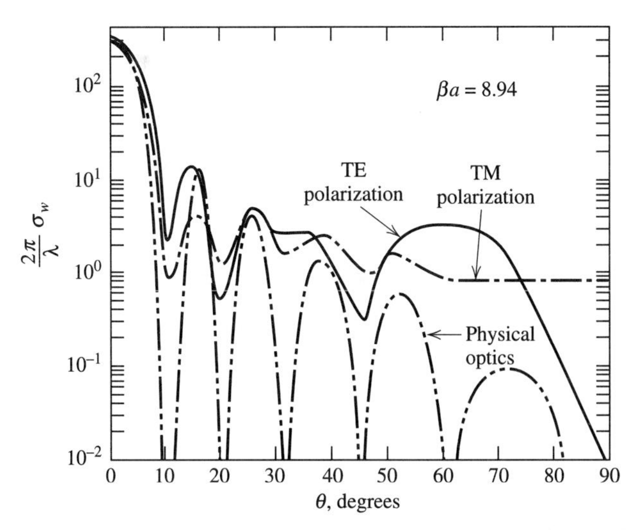

Figure 12-47 TE and TM physical optics scattering by a strip of width a.

surface is not defined). Consider Fig. 12-47 that shows $\beta\sigma_w$ for an infinite strip when physical optics is used compared to an exact solution [2]. Both TE (E perpendicular to the edges) and TM (E parallel to the edges) are shown. As the backscatter angle θ moves away from the normal to the strip ($\theta = 0°$), the difference between physical optics and the correct solution for the two polarizations becomes increasingly larger. This difference may be eliminated by adding to the physical optics current an additional current to account for each edge. This is the subject of Sec. 12.15. Before doing this, it is necessary to consider the principle of stationary phase in the next section.

12.14 METHOD OF STATIONARY PHASE

As we have seen many times earlier in this book, integrals describing radiation have integrands that consist of an amplitude function times a phase function. In many cases, an asymptotic evaluation is possible if the amplitude function is slowly varying and the exponential function is rapidly varying.

Consider the integral

$$I = \int_a^b f(x)e^{j\beta\gamma(x)}\,dx \tag{12-161}$$

in which $f(x)$ and $\gamma(x)$ are real functions. If $f(x)$ is slowly varying and $\beta\gamma(x)$ is a rapidly varying function over the interval of integration due to β being large, the major contribution from the integral comes from the point or points of stationary phase [22]. A point of stationary phase is defined as a point where the first derivative of the phase function γ vanishes:

$$\frac{d\gamma}{dx} = 0, \qquad \text{at } x = x_o \tag{12-162}$$

To expand the phase function in a Taylor series about the point of stationary phase, write

$$\gamma(x) \simeq \gamma_o + (x - x_o)\gamma_o' + (x - x_o)^2\,\gamma_o''/2 + \dots \tag{12-163}$$

where γ_o' and γ_o'' represent the derivatives of γ with respect to x, evaluated at x_o. Now γ_o' is zero by definition [see (12-162)], and in the neighborhood of the point of stationary phase the quantity $(x - x_o)$ is small so that the high-order terms (i.e., order 3 and higher) indicated in (12-163) may be neglected. If there is one and only one stationary point x_o in the interval from a to b, and x_o is not near either a or b, (12-161) thus becomes

$$I = \int_{x_o-\delta}^{x_o+\delta} f(x) e^{j\beta\gamma_o} e^{j\beta(x-x_o)^2\gamma_o''/2}\, dx \quad (\text{if } \gamma_o'' \neq 0) \tag{12-164}$$

where δ represents a small number. Thus, the range of integration has been reduced to a small neighborhood about the point of stationary phase. If $f(x)$ is slowly varying, it may be approximated by $f(x_o)$ over this small interval. Thus, (12-164) becomes

$$I_o = f(x_o) e^{j\beta\gamma_o} \int_{-\infty}^{\infty} e^{j\beta(x-x_o)^2\gamma_o''/2}\, dx = f(x_o) e^{j\beta\gamma_o} \int_{-\infty}^{\infty} e^{j\beta z^2\gamma_o''/2}\, dz \tag{12-165}$$

where $(x - x_o) = z$. For convenience, the limits of integration have been changed again, this time to infinity. This introduces a little error if the chief contribution to the integral comes from the neighborhood of the point of stationary phase. In other regions, the rapid phase variations cause the contribution from one half-period to be nearly canceled by that from the next half-period if the amplitude $f(x)$ is constant or varies slowly.

Now consider the integral

$$\int_{-\infty}^{\infty} e^{jaz^2}\, dz = \int_{-\infty}^{\infty} (\cos az^2 + j \sin az^2)\, dz = \sqrt{\frac{\pi}{|a|}}\, e^{j(\pi/4)\,\mathrm{sgn}(a)} \tag{12-166}$$

where

$$\mathrm{sgn}(a) = \begin{cases} 1 & \text{if } a > 0 \\ -1 & \text{if } a < 0 \end{cases}$$

If we use (12-166) to evaluate (12-165), the stationary phase approximation is

$$\int_{x_o-\delta}^{x_o+\delta} f(x)\, e^{j\beta\gamma(x)}\, dx \approx f(x_o)\, e^{j\beta\gamma(x_o)} \sqrt{\frac{2\pi}{\beta|\gamma_o''|}}\, e^{j(\pi/4)\,\mathrm{sgn}(\gamma_o'')} \tag{12-167}$$

If two or more points of stationary phase exist in the interval of integration (a to b) and there is no coupling between them, the total value of the integral is obtained by summing the contributions from each such point as given in (12-167).

Equation (12-167) is not valid if the second derivative of the phase function vanishes at the point of stationary phase. In this event, $\gamma_o'' = 0$ and it is necessary to retain the third-order term in the Taylor series in (12-163).

Equation (12-167) also becomes invalid if one of the limits of integration, a or b, is close to the point of stationary phase x_o. In this event, however, it is possible to express the integral in the form of a Fresnel integral as discussed later. A problem also arises if there exist two or more stationary points close together in the range of integration.

To obtain the endpoint contribution, it is best to write (12-161) as

$$I = \int_{-\infty}^{\infty} f(x) e^{j\beta\gamma(x)}\, dx - \int_{b}^{\infty} f(x) e^{j\beta\gamma(x)}\, dx - \int_{-a}^{\infty} f(-x) e^{j\beta\gamma(-x)}\, dx \tag{12-168}$$

or

$$I = I_o - I_b - I_{-a} \tag{12-169}$$

The evaluation of I_o has been done in (12-167). I_b, for instance, can be evaluated via integration by parts (see Prob. 12.14-1) by allowing the wave number to be complex and to have a small amount of loss (i.e., small α) so that the contribution to the integral by the upper limit at infinity vanishes, and then letting the wave number be approximated by β, as before. Thus,

$$I_b \cong -\frac{1}{j\beta}\frac{f(b)}{\gamma'(b)}e^{j\beta\gamma(b)} \tag{12-170}$$

A similar expression can be found for I_{-a}. Equation (12-170) is valid when b is not near (or coupled) to x_o. When the stationary point is coupled to the endpoint, we have [22]

$$I_b \cong U(-\epsilon_1)I_o + \epsilon_1 f(b)e^{j\beta\gamma(b)\mp j\nu^2}\sqrt{\frac{2}{\beta|\gamma''(b)|}}F_{\pm}(\nu) \tag{12-171}$$

where $\gamma''(b) \neq 0$, $\epsilon_1 = \text{sgn}(b - x_o)$, $\nu = \sqrt{\dfrac{\beta}{2|\gamma''(b)|}}|\gamma'(b)|$, U = unit step function, and $F_{\pm}(\nu)$ is the Fresnel integral.

There also are formulas for the stationary phase evaluation of double integrals [2]. However, (12-167) and (12-170) are sufficient to develop the physical theory of diffraction in the next section.

EXAMPLE 12-5 *Echo Width of a Circular Cylinder*

Consider the radar echo width of an infinite circular cylinder about the z-axis (see Fig. 12-48). We employ (12-167) to do this; (12-170) is used in the next section. Starting with the two-dimensional counterpart to (12-167), write for ψ in the cylindrical system:

$$\psi = \frac{1}{4j}H_0^{(2)}(\beta\rho) \cong \frac{1}{4j}\sqrt{\frac{2j}{\pi\beta|\boldsymbol{\rho} - \boldsymbol{\rho}'|}}e^{-j\beta|\boldsymbol{\rho}-\boldsymbol{\rho}'|} \tag{12-172}$$

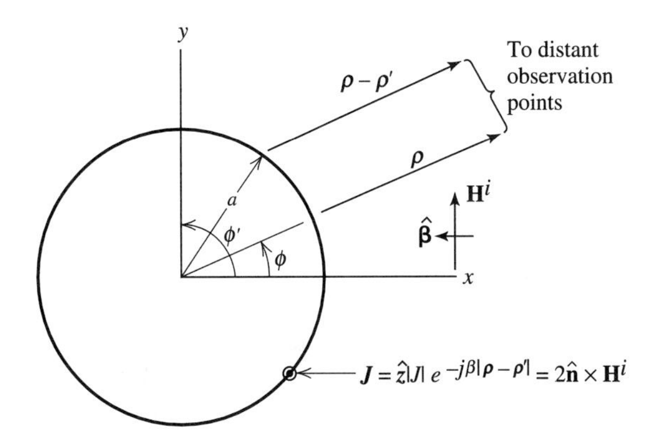

Figure 12-48 Geometry for radiation by a circular cylinder.

where $|\boldsymbol{\rho} - \boldsymbol{\rho}'| = \rho - \rho' \cos(\phi - \phi')$ and the asymptotic or large argument approximation for the Hankel function, $H_0^{(2)}(\beta\rho)$, of the second kind and zeroth order has been used. For amplitude purposes in the far field, $|\boldsymbol{\rho} - \boldsymbol{\rho}'| \approx \rho$. Thus, similar to (12-168), we obtain

$$\nabla\psi = \hat{\boldsymbol{\rho}}\,\frac{1}{4j}\sqrt{\frac{2j}{\pi\beta}}\left[-\frac{1/2 + j\beta\rho}{\rho^{3/2}}\right]e^{-j\beta|\boldsymbol{\rho}-\boldsymbol{\rho}'|} \tag{12-173}$$

In the far field, the $\frac{1}{2}$ is insignificant compared to $\beta\rho$. Thus,

$$\mathbf{H}^s \approx \sqrt{\frac{j}{8\pi\beta}}\int (\mathbf{J} \times \hat{\boldsymbol{\rho}})\left[j\,\frac{\beta}{\rho^{1/2}}\right]e^{-j\beta|\boldsymbol{\rho}-\boldsymbol{\rho}'|}\,dc' \tag{12-174}$$

where dc' is an incremental element on a circumferential line of the cylinder, $dc' = a\,d\phi'$. Applying the definition of echo width (see Sec. 12.11) and taking $\mathbf{J} = 2\hat{\mathbf{n}} \times \mathbf{H}^i$ give

$$\sigma_w = \frac{\beta}{4}\frac{1}{|\mathbf{H}^i|^2}\left|\int ((2\hat{\mathbf{n}} \times \mathbf{H}^i) \times \hat{\boldsymbol{\rho}})e^{-j\beta|\boldsymbol{\rho}-\boldsymbol{\rho}'|}\,dc'\right|^2 \tag{12-175}$$

The phase of the current on the cylinder is $e^{+j\beta a\cos(\phi-\phi')}$ (i.e., advanced relative to the origin as indicated in Fig. 12-45). Thus,

$$\sigma_w = \beta\left|\int e^{j\beta a\cos(\phi-\phi')}(\hat{\boldsymbol{\rho}} \cdot \hat{\mathbf{n}})e^{-j\beta|\boldsymbol{\rho}-\boldsymbol{\rho}'|}\,dc'\right|^2 \tag{12-176}$$

Noting that $(\hat{\boldsymbol{\rho}} \cdot \hat{\mathbf{n}}) = \cos\phi'$ and $|\boldsymbol{\rho} - \boldsymbol{\rho}'| = \rho - a\cos\phi'$ when $\rho' = a$ and $\phi = 0$ yields

$$\sigma_w = \beta\left|e^{-j\beta\rho}\int_{-\pi/2}^{\pi/2}\cos\phi'\;e^{+j2\beta a\cos\phi'}\,a\,d\phi'\right|^2 \tag{12-177}$$

The integral can be evaluated in a straightforward manner by the method of stationary phase. Using (12-167), identify $f(\phi') = a\cos\phi'$ and $\gamma(\phi') = -2a\cos\phi'$. To find the stationary point(s) ϕ_o, use $f'(\phi') = 0 = -a\sin\phi'$ and determine that $\phi_o' = 0, \pi$. Due to the physical optics assumption of no current at $\phi_o' = \pi$, the value of π is discarded. Therefore, $f(\phi_o') = a\cos(0°) = a$. Since $\gamma''(\phi') = 2a\cos\phi'$, then $\gamma''(\phi_o') = 2a$. Therefore,

$$\sigma_w = \beta\left|ae^{-j\beta\rho}\,e^{j\beta a}\sqrt{\frac{2\pi}{\beta 2a}}\,e^{j\pi/4\,\mathrm{sgn}(\gamma_o'')}\right|^2 = \pi a \tag{12-178}$$

Thus, our stationary phase evaluation has produced the same result for the echo width as we obtained in Sec. 12.13 using geometrical optics.

It is interesting to compare the treatment here for the cylinder with the treatment of the sphere in the previous section. In the case of the sphere, the projection of the currents onto a plane was integrated. This produced an integral that could be evaluated in closed form. Had the sphere problem been formulated in a manner similar to that used here for the cylinder by integrating on the actual surface, it would have been necessary to use stationary phase for double integrals. The single integral treatment is, however, sufficient for our development of PTD in the next section.

12.15 PHYSICAL THEORY OF DIFFRACTION

The physical theory of diffraction (PTD) is an extension of physical optics (PO) that refines the PO surface field approximation just as GTD or UTD refines the geometrical optics surface-field approximation. The original PTD formulation was developed by Ufimtsev [28] for surfaces with perfectly conducting edges. Ufimtsev's original work was done at about the same time as the ray-optical work of Keller and independently of Keller. However, Ufimtsev was aware of the work of Sommerfeld [9] and Pauli [11], and used the asymptotic form in (12-42) as did Keller.

In his work, Ufimtsev postulated that there was a nonuniform component of the current that would include effects not accounted for in the physical optics current, called the uniform current in his work (see Fig. 12-49). Ufimtsev did not actually find the nonuniform current for the wedge, but instead found the field due to the nonuniform component of the current by indirect means. (More recently, expressions for these nonuniform currents have been found [29, 30].) He found the nonuniform current contribution to the field by subtracting the PO field from the known total field for the wedge. The result was the field due to what was left, that from the nonuniform current. To see this, write

$$\mathbf{E}^s_{\text{total}} = \mathbf{E}^r_{\text{GO}} + \mathbf{E}^d_K \tag{12-179}$$

where

$\mathbf{E}^s_{\text{total}}$ = the total scattered field

$\mathbf{E}^r_{\text{GO}}$ = the reflected field obtained by geometrical optics

$\mathbf{E}^d_K$ = the diffracted field found using the Keller diffraction coefficient in (12-59)

Then write $\mathbf{E}_{\text{PO}}$ for the field due to the physical optics current, as

$$\mathbf{E}_{\text{PO}} \simeq \mathbf{E}^r_{\text{PO}} + \mathbf{E}^d_{\text{PO}} \tag{12-180}$$

where $\mathbf{E}^d_{\text{PO}}$ is called the physical optics diffracted field (which is not the total diffracted field) and is due to the abrupt termination of the physical optics current at, for example, the edge of the half-plane in Fig. 12-49. $\mathbf{E}^r_{\text{PO}}$ is the reflected field obtained by integration of the currents and is theoretically equal to $\mathbf{E}^r_{\text{GO}}$ since both represent the reflected field, although by different means.

Subtracting (12-180) from (12-179) or ($\mathbf{E}^s_{\text{total}} - \mathbf{E}_{\text{PO}}$) yields the field due to the nonuniform current $\mathbf{E}^{\text{nu}}$ as

$$\mathbf{E}^{\text{nu}} = \mathbf{E}^d_K - \mathbf{E}^d_{\text{PO}} \tag{12-181}$$

The field due to the nonuniform current, when added to the field due to the uniform current (e.g., the specular contribution), gives the total scattered field:

$$\mathbf{E}^s_{\text{total}} = \mathbf{E}^{\text{unif}} + \mathbf{E}^{\text{nu}} \tag{12-182}$$

If we take $\mathbf{E}^r_{\text{GO}} = \mathbf{E}^r_{\text{PO}}$, it follows from (12-180) that

$$\mathbf{E}^r_{\text{GO}} = \mathbf{E}_{\text{PO}} - \mathbf{E}^d_{\text{PO}} \tag{12-183}$$

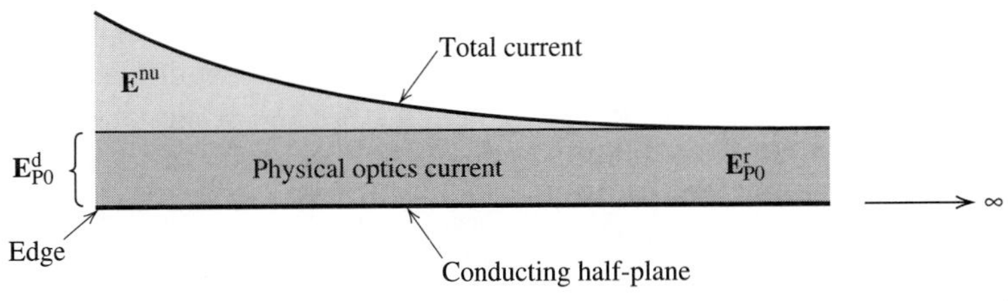

Figure 12-49 Conceptualization of PTD currents. Also indicated are the fields generated by those currents.

Substituting (12-183) into (12-179) gives

$$\mathbf{E}^s_{\text{total}} = \mathbf{E}_{\text{PO}} + (\mathbf{E}^d_K - \mathbf{E}^d_{\text{PO}}) = \mathbf{E}^{\text{unif}} + (\mathbf{E}^{\text{nu}}) \tag{12-184}$$

Next, we present expressions for the fields due to the nonuniform current. As in GTD, each polarization is considered separately. Thus for the nonuniform contribution in the two-dimensional case, the parallel and perpendicular polarizations (where $\|$ and $\perp$ refer to the orientation of the *electric* field) are

$$E^{\text{nu}}_{\|} = E^i f' \frac{e^{j(\beta\rho + \pi/4)}}{\sqrt{2\pi\beta\rho}} \tag{12-185}$$

$$H^{\text{nu}}_{\perp} = H^i g' \frac{e^{j(\beta\rho + \pi/4)}}{\sqrt{2\pi\beta\rho}} \tag{12-186}$$

Thus, $E^{\text{nu}}_{\|}$ and $H^{\text{nu}}_{\perp}$ are known in terms of two simple functions f' and g', where $f' = f - f_o$ and $g' = g - g_o$ and

$$\begin{Bmatrix} f \\ g \end{Bmatrix} = \frac{1}{n}\sin\frac{\pi}{n}\left[\frac{1}{\cos\frac{\pi}{n} - \cos\frac{\phi - \phi'}{n}}\right] \mp \left[\frac{1}{\cos\frac{\pi}{n} - \cos\frac{\phi + \phi'}{n}}\right] \tag{12-187}$$

$$f_o = \begin{cases} f_a & 0 < \phi' \le \pi - \phi_{\text{int}} \\ f_a + f_b & \pi - \phi_{\text{int}} \le \phi' \le \pi \\ f_b & \pi < \phi' < 2\pi - \phi_{\text{int}} \end{cases} \tag{12-188}$$

$$g_o = \begin{cases} g_a & 0 < \phi' \le \pi - \phi_{\text{int}} \\ g_a + g_b & \pi - \phi_{\text{int}} \le \phi' \le \pi \\ g_b & \pi < \phi' < 2\pi - \phi_{\text{int}} \end{cases} \tag{12-189}$$

$$\begin{Bmatrix} f_a \\ g_a \end{Bmatrix} = \begin{Bmatrix} \sin\phi' \\ -\sin\phi \end{Bmatrix} \frac{1}{\cos\phi + \cos\phi'} \tag{12-190}$$

$$\begin{Bmatrix} f_b \\ g_b \end{Bmatrix} = \begin{Bmatrix} \sin(2\pi - \phi_{\text{int}} - \phi') \\ -\sin(2\pi - \phi_{\text{int}} - \phi \end{Bmatrix} \tag{12-191}$$

$$\times \frac{1}{\cos(2\pi - \phi_{\text{int}} - \phi) + \cos(2\pi - \phi_{\text{int}} - \phi')}$$

with $\phi_{\text{int}} = (2 - n)\pi$, the interior wedge angle. The a subscript denotes that the A face is illuminated and b denotes that the B face is illuminated as in Fig. 12-50.

Clearly, f and g correspond to the Keller GTD diffraction coefficients. Even though we know that f and g tend to infinity at the reflection and shadow boundaries, f' and g' do not because the singularities in f and g are canceled by identical singularities in f_o and g_o. The quantities f_o and g_o are the PO diffraction coefficients.

The expressions for f_o and g_o are obtained from the stationary phase endpoint contribution in (12-170). To demonstrate this [31], consider the wedge in Fig. 12-50 for the perpendicular (TE) case with only face A illuminated. Write for the incident magnetic field [see Eq. (12-37)]

$$\mathbf{H}^i = \hat{\mathbf{z}}\, H_o e^{j\beta\rho\cos(\phi - \phi')} \tag{12-192}$$

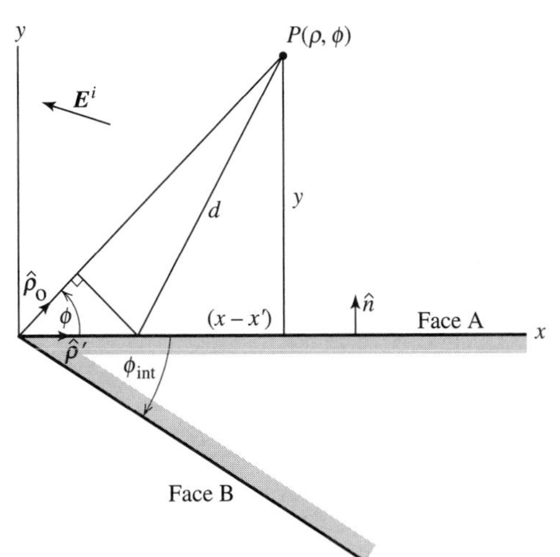

Figure 12-50 TE plane wave incidence on the A face of a wedge.

The physical optics current on the illuminated face of the wedge, face A in Fig. 12-50, is given by

$$\begin{aligned}\mathbf{J}_{\mathrm{PO}} &= 2\hat{\mathbf{n}} \times \mathbf{H}^i|_{\rho=x',\phi=0} \\ &= 2(\hat{\mathbf{y}} \times \hat{\mathbf{z}})H^i e^{j\beta x' \cos\phi'} = 2\hat{\mathbf{x}}\, H_o\, e^{j\beta x' \cos\phi'}\end{aligned} \tag{12-193}$$

The vector potential for the two-dimensional wedge is

$$\mathbf{A} = \mu \int_0^\infty \mathbf{J}_{\mathrm{PO}} \frac{1}{4j} H_0^{(2)}(\beta|\boldsymbol{\rho}_o - \boldsymbol{\rho}'|)\, dx' \tag{12-194}$$

Using the asymptotic expression of the Hankel function gives

$$\mathbf{A} \simeq \mu \frac{2H^i}{4j} \hat{\mathbf{x}} \int_0^\infty e^{j\beta x' \cos\phi'} \sqrt{\frac{2j}{\pi\beta|\boldsymbol{\rho}_o - \boldsymbol{\rho}'|}}\, e^{-j\beta|\boldsymbol{\rho}_o - \boldsymbol{\rho}'|}\, dx' \tag{12-195}$$

From Fig. 12-50, note that $|\boldsymbol{\rho}_o - \boldsymbol{\rho}'| \cong \rho_o - \hat{\boldsymbol{\rho}} \cdot \boldsymbol{\rho}' = \rho_o - x' \cos\phi$. Thus, $\mathbf{A}$ becomes, with the usual far-field approximations,

$$\mathbf{A} \cong \mu \frac{H^i}{2j} \hat{\mathbf{x}} \sqrt{\frac{2j}{\pi\beta\rho_o}} \int_o^\infty e^{-j\beta\rho_o} e^{j\beta x'(\cos\phi' + \cos\phi)}\, dx' \tag{12-196}$$

Considering just the integral with $d \approx \rho_o - x' \cos\phi$, we get

$$I = \int_0^\infty e^{j\beta(x' \cos\phi' - d)}\, dx' \cong I_o + I_b \tag{12-197}$$

where I_o is the stationary contribution discussed in the previous section and I_b is the endpoint contribution in Eq. (12-170). Evaluation of I_o is not of immediate interest to us here and is left as an exercise for the reader. Our attention is turned to I_b, where

$$I_b = -\frac{1}{j\beta} \frac{f(0)}{\gamma'(0)} e^{j\beta\gamma(0)} \tag{12-198}$$

Here,

$$f(0) = 1 \tag{12-199a}$$

$$\gamma(x') = x' \cos \phi' - d \tag{12-199b}$$

$$d = \{(x - x')^2 + (y)^2\}^{1/2} \tag{12-199c}$$

$$\frac{d}{dx'}(d) = -\frac{(x - x)'}{d} \tag{12-199d}$$

$$\gamma'(x') = \cos \phi' - \frac{d}{dx'}(d) = \cos \phi' + \frac{x - x'}{d} \tag{12-199e}$$

$$\gamma'(0) = \cos \phi' + \frac{x}{\rho_o} = \cos \phi' + \cos \phi \tag{12-199f}$$

$$\gamma(0) = -d(0) = -\rho_o \tag{12-199g}$$

Thus, the endpoint contribution to the integral is

$$I_b = -\frac{1}{j\beta} \frac{1}{\cos \phi' + \cos \phi} e^{-j\beta\rho_o} \tag{12-200}$$

which means that the endpoint contribution $\mathbf{A}_\perp^{\text{ep}}$ is

$$\mathbf{A}_\perp^{\text{ep}} = -\mu \frac{H^i \hat{\mathbf{x}}}{\sqrt{2\pi j\beta\rho_o}} e^{j\beta\rho_o} \frac{-1}{j\beta} \frac{1}{\cos \phi' + \cos \phi} \tag{12-201}$$

Since $\mathbf{E}_\perp^{\text{ep}} = -j\omega \mathbf{A}_\perp^{\text{ep}}$ and $\eta H^i = E^i$,

$$\mathbf{E}_\perp^{\text{ep}} = E^i \frac{1}{\sqrt{2\pi j\beta\rho_o}} \frac{1}{\cos \phi + \cos \phi'} e^{-j\beta\rho_o} \hat{\mathbf{x}} \tag{12-202}$$

And finally for the magnetic field, we have

$$\begin{aligned} \mathbf{H}_\perp^{\text{ep}} &= (\hat{\boldsymbol{\rho}}_o \times \hat{\mathbf{x}}) H^i \frac{1}{\sqrt{2\pi j\beta\rho_o}} \frac{1}{\cos \phi + \cos \phi'} e^{-j\beta\rho_o} \\ &= -H^i \frac{1}{\sqrt{2\pi j\beta\rho_o}} \frac{\sin \phi}{\cos \phi + \cos \phi'} e^{-j\beta\rho_o} \hat{\mathbf{z}} = H_{\text{PO}}^d \hat{\mathbf{z}} \end{aligned} \tag{12-203}$$

which is the postulated result given in (12-185) through (12-191).

Suppose that PTD is to be used to calculate the results in Fig. 12-14a. First, it must be kept in mind that PTD uses the equivalence principle, as does physical optics, where all conducting media are replaced with equivalent currents radiating in free space. This should be contrasted with GTD, where all material media are retained. Starting with region I (Fig. 12-12), one can write

$$\mathbf{E}_{\text{total}} \cong \mathbf{E}^i + \mathbf{E}^{\text{unif}} + \mathbf{E}^{\text{nu}} \tag{12-204a}$$

or

$$\begin{aligned} \mathbf{E}_{\text{total}} &= \mathbf{E}^i + (\mathbf{E}_{\text{PO}}) + (\mathbf{E}_K^d - \mathbf{E}_{\text{PO}}^d) \\ &= \mathbf{E}^i + (\mathbf{E}_{\text{PO}}^r + \mathbf{E}_{\text{PO}}^d) + (\mathbf{E}_K^d - \mathbf{E}_{\text{PO}}^d) \end{aligned} \tag{12-204b}$$

which in terms of GTD is

$$\mathbf{E}_{\text{total}} = \mathbf{E}^i + \mathbf{E}_{\text{GO}}^r + \mathbf{E}_K^d \quad \text{in region I} \tag{12-204c}$$

To apply PTD, (12-204a) is used, whereas the use of (12-204c) is the application of GTD. In the former, fields from currents are used, whereas in the latter, ray-optical fields are used. More specifically, to apply PTD as given in (12-204a) to the half-plane problem of Fig. 12-14, $\mathbf{E}^i$ is represented by (12-37), $\mathbf{E}^r_{\text{PO}}$ by (12-167), $\mathbf{E}^d_{\text{PO}}$ by (12-170), and $(\mathbf{E}^d_k - \mathbf{E}^d_{\text{PO}})$ by (12-186). Figure 12-51 is helpful in understanding the PTD calculation in region I. As the reflection boundary is approached in Fig. 12-51, (12-171) must be used since the stationary point and endpoint become coupled.

For region II, as in Fig. 12-51, all the quantities in (12-204b) are also present, except $\mathbf{E}^r_{\text{PO}}$ that is absent outside the two transition regions. Here with PTD, $\mathbf{E}^{\text{unif}}$ is continuous across the reflected field shadow boundary. $\mathbf{E}^{\text{nu}}$ is also continuous. Thus, $\mathbf{E}_{\text{total}}$ is continuous across the reflected field shadow boundary. (See Prob. 12.15-2.)

With regard to Fig. 12-51, in region III all PTD quantities in (12-204a) are present just as in region II. Moving across the incident field shadow boundary, $\mathbf{E}^{\text{nu}}$ will, of course, be continuous. Both $\mathbf{E}^i$ and $\mathbf{E}^{\text{unif}}$ are also continuous and, therefore, $\mathbf{E}_{\text{total}}$ is continuous. Moving deeper into the shadow region, $\mathbf{E}^i$ will (theoretically) be canceled by $\mathbf{E}^r_{\text{PO}}$, leaving $\mathbf{E}^d_K$. This is not surprising because it is known from Sec. 12.2 that $\mathbf{E}^d_K$ gives the correct field in the deep shadow region. To calculate the field by PTD in region III, note from Fig. 12-51 that $\mathbf{E}^r_{\text{PO}}$ appears in the shadow region and (12-170) is used in its evaluation away from the shadow boundary, just as it was used in region I.

PTD, just like GTD, applies only to scattering directions lying on the cone of diffracted rays shown in Fig. 12-15. This restriction for PTD may be overcome by

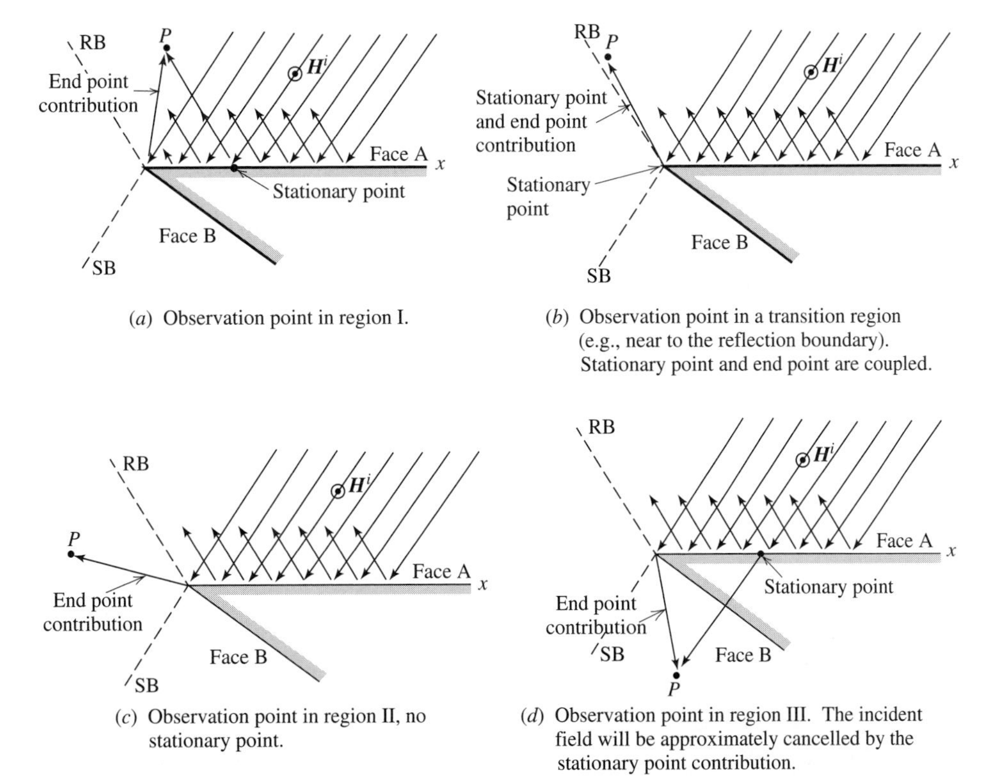

Figure 12-51 The relationship between stationary phase and geometrical optics for a conducting wedge.

using incremental length diffraction coefficients (ILDC) devised by Mitzner [32]. The ILDCs do for PTD what equivalent currents in Sec. 12.9 do for GTD [33].

12.16 CYLINDRICAL PARABOLIC REFLECTOR ANTENNA—PTD

As a second example of the application of the PTD, consider again the cylindrical parabolic reflector antenna shown in Fig. 12-20. In the half-space where z is positive (region A in Fig. 12-52), we use aperture integration for $0 < \zeta < \pi/2$. This gives the contribution from $\mathbf{E}^{\text{unif}}$. The contribution from $\mathbf{E}^{\text{unif}}$ will increasingly disagree with the UTD result in Fig. 12-53 as ζ moves from 30 to 90°. This discrepancy may be removed by including $\mathbf{E}^{\text{nu}}$ in the calculation as Fig. 12-53 shows for the PTD case.

To obtain the field or pattern in the deep shadow region (the shaded part of region C in Fig. 12-52), we cannot use the field from the aperture integration. An examination of the equivalence principle shows that this is so. For example, from an examination of the half-plane example in the previous section, we know that in the deep shadow region the field may be found from just $\mathbf{E}_k^d$ if $\mathbf{E}^i$ is canceled by $\mathbf{E}_{\text{PO}}^r$. However, if $\mathbf{E}_{\text{PO}}^r$ is taken to be from the equivalent currents in the aperture, $\mathbf{E}_{\text{PO}}^r$ gives a collimated beam in the negative z-direction that clearly cannot be canceled by the field from the feed. However, if $\mathbf{E}_{\text{PO}}^r$ is taken to be the field from the currents on the parabolic surface, it will cancel $\mathbf{E}^i$ in the shaded region.

To get the field in the unshaded part of regions C and B, we need to use $\mathbf{E}^{\text{unif}}$ and $\mathbf{E}^{\text{nu}}$, which means the integration over the parabolic surface itself must be done. Generally, it is easier to integrate over the aperture than over the parabolic surface, but that is not a valid option here. In obtaining the field in regions B and C, the need to integrate over the parabolic surface can be avoided by simply using the UTD diffraction coefficients in a GTD model rather than the Keller coefficients in a PTD model, realizing that UTD gives the correct fields without the singularity at the incident field shadow boundary. That is to say, in this problem the simpler model is a GTD model with UTD coefficients for regions C, B, and part of A.

Figure 12-53 shows the E-plane radiation pattern for the cylindrical parabolic reflector antenna of Fig. 12-20. Figure 12-53 shows a comparison of aperture integration, PTD, and UTD for the TE (or perpendicular) case. The agreement between classical aperture integration and the two asymptotic theories is excellent in the region of the main beam and the first few side lobes; thereafter, there is increasing disagreement because aperture integration does not fully account for edge diffraction effects that are increasingly important at larger angles.

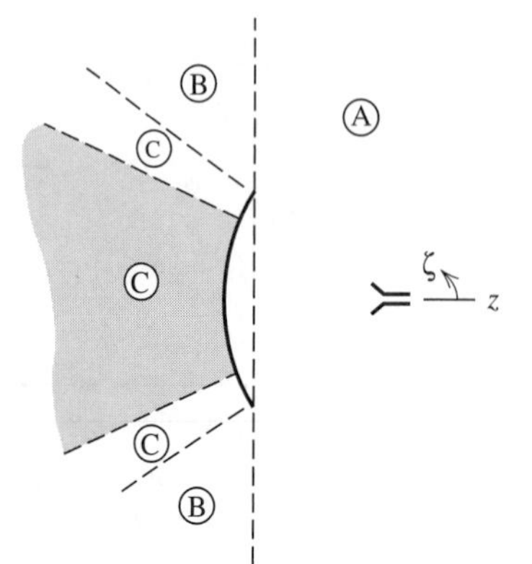

Figure 12-52 Regions of a cylindrical parabolic reflector antenna (see Fig. 12-20).

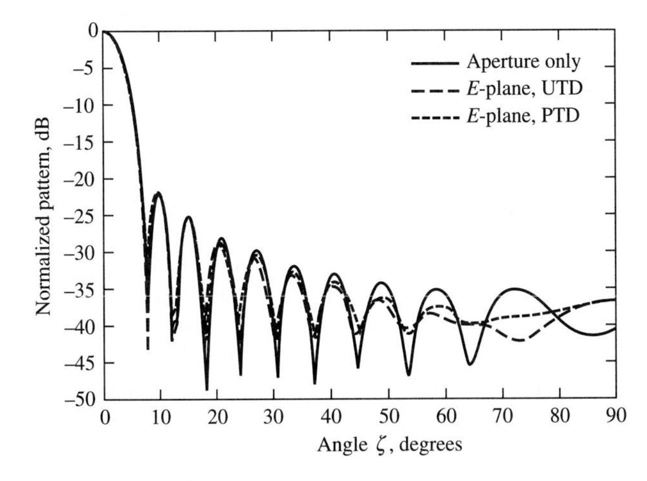

Figure 12-53 E-plane comparison of physical optics aperture integration with UTD and PTD in the side-lobe region for the geometry of Figure 12-20.

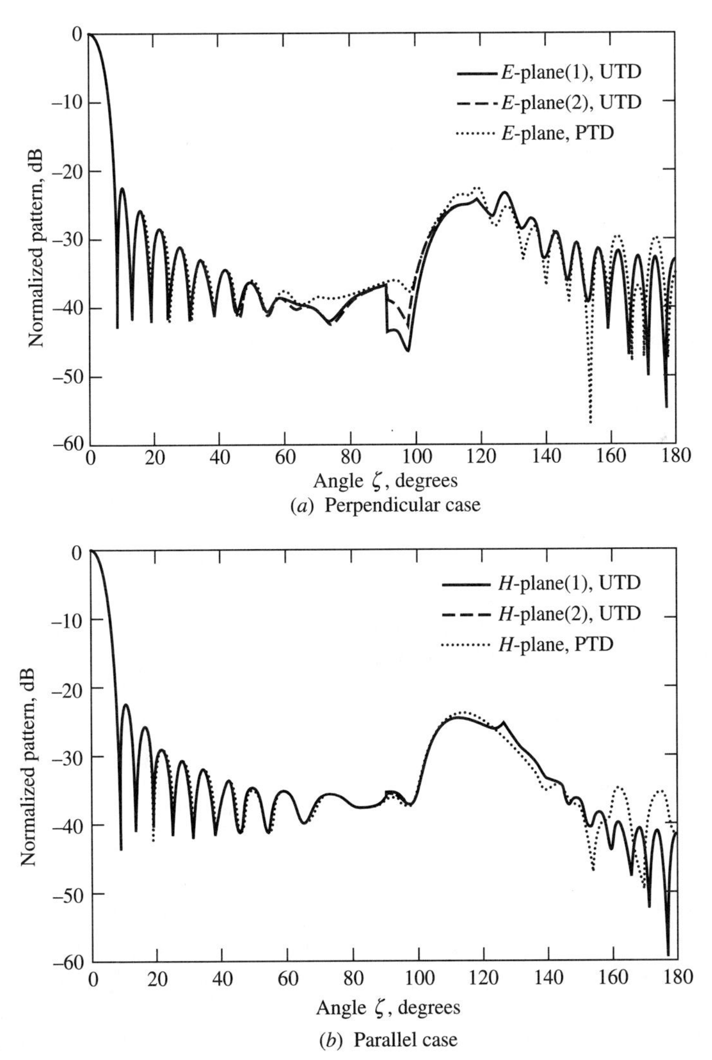

Figure 12-54 A comparison of PTD with UTD single (1) and double (2) diffraction calculations for the geometry of Figure 12-20.

Figure 12-54*a* shows a comparison between UTD single diffraction, UTD double diffraction, and PTD for the full E-plane pattern. Aperture integration is used for the main beam and the first side lobe in the UTD calculations; thereafter, starting in the second null, UTD alone is used. With two exceptions, the agreement is quite good between UTD and PTD. First, in the vicinity of 90°, there is some disagreement since edge diffraction effects are particularly strong for this polarization and one edge is in the transition region of the other (see Fig. 12-18*c*). Note, however, that the pattern is near the −40-dB level. Second, there is some disagreement in and near the back-lobe region. This disagreement is not due to a deficiency in the diffraction calculations, but the inability of $\mathbf{E}^{\text{unif}}$ from the integration over the currents on the parabolic surface to exactly cancel $\mathbf{E}^i$ from the feed in the PTD calculation. If the feed had less taper (e.g., $\cos\theta_s$ instead of $\cos^2\theta_s$), the diffracted field would be stronger and the incomplete cancellation of $\mathbf{E}^{\text{unif}}$ by $\mathbf{E}^i$ less apparent at the higher back-lobe level. For this geometry, disagreement in the back-lobe region is greater in the H-plane case than the E-plane case because the diffracted field in the former is weaker than in the latter, resulting in a more noticeable incomplete cancellation effect.

12.17 SUMMARY

In this chapter on high-frequency methods, a variety of techniques have been presented for predicting both the near- and far-zone fields from perfectly conducting bodies whose dimensions are large in terms of the wavelength. The GTD approach is ray-based and relatively simple when the number of rays is not large. The PTD approach is current-based and requires the integration of currents. Since integration is a smoothing process, a geometrical surface description in PTD does not have to be as accurate as in GTD. In both GTD and PTD, the most difficult part to calculate is usually the most basic part: GO in the case of GTD and PO in the case of PTD. An example of this is the calculation of scattering by an infinite cylinder in Sec. 12.11 where, for small radii, the creeping wave contribution was more accurate than that of GO.

The importance of the GTD method in antenna and scattering problems stems from the significant advantages to be gained from its use, namely (1) it is simple to use and yields accurate results; (2) it provides some physical insight into the radiation and scattering mechanisms involved; (3) it can be used to treat problems for which exact analytical solutions are not available. GTD is also used in acoustic problems such as SONAR and problems involving inhomogeneous or anistropic media [34].

The importance of the PTD method is mostly in scattering problems. An advantage of the PO is that it provides scattering information in directions that are not necessarily in the specular direction or on the cone of diffracted rays. PTD is thought to have played a key part in the development of the B-2 stealth bomber.

The methods of this chapter tend to complement the intermediate frequency moment method techniques presented in Chap. 10 and the FD-TD technique in Chap. 11. And, as was seen in Sec. 12.12, the moment method can be formally combined with GTD into a hybrid technique that extends the class of problems to which moment methods can be applied. This can be done not only because both MoM and GTD are highly practical techniques, but also because they are inherently flexible in their application to analysis and design problems. Hybrid methods involving FD-TD are being developed.

REFERENCES

1. M. Born and E. Wolf, *Principles of Optics*, Pergamon, New York, 1959.
2. G. T. Ruck, Ed., *Radar Cross Section Handbook*, Plenum, New York and London, 1970.
3. J. J. Bowman, T. B. A. Senior, and P. L. E. Uslenghi, Eds., *Electromagnetic and Acoustic Scattering by Simple Shapes*, North-Holland, Amsterdam, 1969.
4. J. W. Crispin and K. M. Siegel, Eds., *Methods of Radar Cross-Section Analysis*, Academic Press, New York and London, 1968.
5. J. B. Keller, "Geometrical Theory of Diffraction," *J. Opt. Soc. Amer.*, Vol. 52, pp. 116–130, 1962.
6. L. M. Graves, Ed., "A Geometrical Theory of Diffraction," in *Calculus of Variations and Its Applications*, McGraw Hill, New York, pp. 27–52, 1958.
7. G. L. James, *Geometrical Theory of Diffraction*, Peter Peregrinus, England, 1976.
8. E. C. Jordan and K. G. Balmain, *Electromagnetic Waves and Radiating Systems*, Prentice-Hall, Englewood Cliffs, NJ, 1968.
9. A. Sommerfeld, *Optics*, Academic Press, New York, 1954.
10. P. M. Russo, R. C. Rudduck, and L. Peters, Jr., "A Method for Computing *E*-plane Patterns of Horn Antennas," *IEEE Trans. Ant. & Prop.*, Vol. AP-13, pp. 219–224, March 1965.
11. W. Pauli, "On Asymptotic Series for Functions in the Theory of Diffraction of Light," *Phys. Rev.*, Vol. 54, pp. 924–931, 1938.
12. R. G. Kouyoumjian and P. H. Pathak, "A Uniform Geometrical Theory of Diffraction for an Edge in a Perfectly Conducting Surface," *Proc. IEEE*, Vol. 62, pp. 1448–1461, 1974.
13. R. G. Kouyoumjian, "Asymptotic High Frequency Methods," *Proc. IEEE*, Vol. 53, pp. 864–876, 1965.
14. R. G. Kouyoumjian, "The Geometrical Theory of Diffraction and Its Application," in *Numerical and Asymptotic Techniques in Electromagnetics*, Springer-Verlag, New York, 1975.
15. A. R. Lopez, "The Geometrical Theory of Diffraction Applied to Antenna and Impedance Calculations," *IEEE Trans. Ant. & Prop.*, Vol. 14, pp. 40–45, Jan. 1966.
16. C. E. Ryan, Jr. and L. Peters, Jr., "Evaluation of Edge Diffracted Fields Including Equivalent Currents for the Caustic Regions," *IEEE Trans. Ant. & Prop.*, Vol. AP-17, pp. 292–299, March 1969. (See also correction in Vol. AP-18, pp. 275, 1970.)
17. R. F. Harrington, *Time-Harmonic Electromagnetic Fields*, McGraw-Hill, New York, 1961.
18. N. Wang, "Self-consistent GTD Formulation for Conducting Cylinders with Arbitrary Convex Cross Section," *IEEE Trans. Ant. & Prop.*, Vol. AP-24, pp. 463–468, July 1976.
19. P. H. Pathak and R. G. Kouyoumjian, "An Analysis of the Radiation from Apertures in Curved Surfaces by the Geometrical Theory of Diffraction," *Proc. IEEE*, Vol. 62, pp. 1438–1447, 1974.
20. G. A. Thiele and T. H. Newhouse, "A Hybrid Technique for Combining Moment Methods with the Geometrical Theory of Diffraction," *IEEE Trans. Ant. & Prop.*, Vol. AP-23, pp. 62–69, Jan. 1975.
21. M. Levy and J. B. Keller, "Diffraction by a Smooth Object," *Comm. Pure Appl. Math.*, Vol. 12, pp. 159–209, 1959.
22. G. L. James, *Geometrical Theory of Diffraction for Electromagnetic Waves*, 3rd ed., Peter Peregrinus, London, 1986.
23. D. R. Voltmer, "Diffraction of Doubly Curved Convex Surfaces," Ph.D. diss., Ohio State University, 1970.
24. L. W. Henderson and G. A. Thiele, "A Hybrid MM-GTD Technique for the Treatment of Wire Antennas Near a Curved Surface," *Radio Sci.*, Vol. 16, pp. 1125–1130, Nov.–Dec. 1981.
25. D. A. McNamara, C. W. I. Pistorius, and J. A. G. Malherbe, *Introduction to the Uniform Theory of Diffraction*, Artech House, Boston, MA, 1990.
26. G. A. Thiele and T. H. Newhouse, "A Hybrid Technique for Combining Moment Methods with the Geometrical Theory of Diffraction," *IEEE Trans. Ant. & Prop.*, Vol. AP-23, pp. 62–69, Jan. 1975.
27. G. A. Thiele, "Hybrid Methods in Antenna Analysis," *IEEE Proc.*, Vol. 80, No. 1, pp. 66–78, Jan. 1992.
28. P. Ia. Ufimtsev, "Approximate Computation of the Diffraction of Plane Electromagnetic Waves at Certain Metal Bodies. Part I. Diffraction Patterns at a Wedge and a Ribbon," *Zh. Tekhn. Fiz.* (USSR), Vol. 27, No. 8, pp. 1708–1718, 1957.
29. P. K. Murthy and G. A. Thiele, "Non-Uniform Currents on a Wedge Illuminated by a TE Plane Wave," *IEEE Trans. Ant. & Prop.*, Vol. AP-34, pp. 1038–1045, Aug. 1986.
30. K. M. Pasala, "Closed-form Expressions for Nonuniform Currents on a Wedge Illuminated by TM Plane Wave," *IEEE Trans. Ant. & Prop.*, Vol. 36, pp. 1753–1759, Dec. 1988.
31. K. M. Pasala, unpublished notes.

32. K. M. Mitzner, "Incremental Length Diffraction Coefficients," Tech. Rep. No. AFAL-TR-73-296, Northrop Corporation.
33. E. F. Knott, J. F. Shaeffer, and M. T. Tuley, *Radar Cross Section*, Artech House, Boston, MA, 1985.
34. L. B. Felsen and N. Marcuvitz, *Radiation and Scattering of Waves*, Prentice-Hall, Englewood Cliffs, NJ, 1973.
35. W. L. Stutzman, *Polarization in Electromagnetic Systems,* Artech House, Inc., Norwood, MA, 1993, pg. 189.

PROBLEMS

12.1-1 It can be shown [12] that the principal radii of curvature of the geometrical optics reflected wavefront are given by

$$\frac{1}{\rho_1} = \frac{1}{2}\left(\frac{1}{\rho_1^i} + \frac{1}{\rho_2^i}\right) + \frac{1}{f_1} \quad \text{and} \quad \frac{1}{\rho_2} = \frac{1}{2}\left(\frac{1}{\rho_1^i} + \frac{1}{\rho_2^i}\right) + \frac{1}{f_2}$$

where ρ_1^i and ρ_2^i are the principal radii of curvature of the incident wavefront and ρ_1 and ρ_2 are the principal radii of curvature of the reflected wavefront. General expressions for f_1 and f_2 are given in the literature [12]. However, for an incident spherical wave,

$$\frac{1}{f_{1,2}} = \frac{1}{\cos\theta_i}\left(\frac{\sin^2\theta_2}{r_1^c} + \frac{\sin^2\theta_1}{r_2^c}\right) \pm \sqrt{\frac{1}{\cos^2\theta_i}\left(\frac{\sin^2\theta_2}{r_1^c} + \frac{\sin^2\theta_1}{r_2^c}\right)^2 - \frac{4}{r_1^c r_2^c}}$$

where θ_1 and θ_2 are the angles between the incident ray and principal directions (i.e., tangent unit vectors) associated with the principal radii of curvature of the surface r_1^c and r_2^c, respectively.

a. Show that for $\theta_1 = \theta_o$ and $\theta_2 = 90°$, the first equation reduces to (12-20) and the second to

$$\frac{1}{\rho_2} = \frac{1}{\ell_o} + \frac{2\cos\theta_o}{r_2^c}$$

b. Without using (12-20) or the expression for ρ_2 immediately above, show that in the case of plane wave illumination

$$\sqrt{\rho_1\rho_2} = \tfrac{1}{2}\sqrt{r_1^c r_2^c}$$

12.1-2 An infinite elliptical paraboloid is described by the equation

$$\frac{x^2}{2r_1^c} + \frac{y^2}{2r_2^c} = -z$$

where r_1 and r_2 are the principal radii of curvature at the specular point. Using geometrical optics, show that the radar cross section for axial incidence is

$$\sigma = \pi r_1^c r_2^c$$

Actually, this result applies to any surface expressible in terms of a second-degree polynomial, where r_1^c and r_2^c are the principal radii of curvature at the reflection point [2, 3]. Is the above result valid for a cylindrical surface or flat plate? Why not?

12.1-3 A plane wave is incident on a smooth three-dimensional conducting convex body. The two principal radii of curvature of the body at the specular point are $r_1^c = 5\lambda$ and $r_2^c = 10\lambda$. Write expressions for the electric and magnetic backscattered fields if the incident plane wave fields are

$$\mathbf{E}^i = \hat{\mathbf{y}}e^{-j\beta x} \quad \text{and} \quad \mathbf{H}^i = \hat{\mathbf{z}}\frac{e^{-j\beta x}}{\eta}$$

12.2-1 A cylindrical wave is incident on a cylindrical parabolic reflector as shown. To obtain the diffracted field from the top edge (only) at any point in space, the edge may be analyzed as if a half-plane were tangent to the uppermost portion of the parabolic surface. Divide the space around the top edge into three separate regions and write general expressions (with numerical values for ϕ') for the total electric field from the top edge in those three regions of space. In which of the three regions is the *total* geometrical optics field zero?

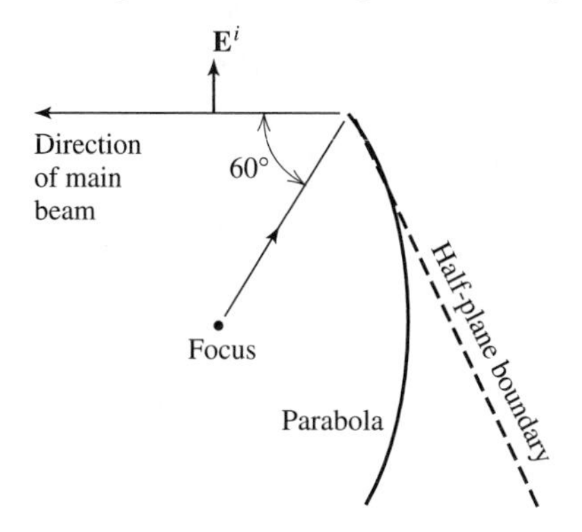

12.2-2 Evaluate the following Fresnel integrals:

a. $\int_0^\infty e^{-j\tau^2}\, d\tau$

b. $\int_0^5 e^{-j\tau^2}\, d\tau$

c. $\int_5^\infty e^{-j\tau^2}\, d\tau$

12.2-3 Find $v_B(\rho, \phi^\pm)$, using both (12-42) and (12-44) for a 90° interior angle wedge when:

a. $\phi' = 45°$, $\rho = 10\lambda$, $\phi = 220°$

b. $\phi' = 45°$, $\rho = 10\lambda$, $\phi = 230°$

Compare your results in (a) and (b) and explain any differences. What is v_* in parts (a) and (b)?

12.2-4 Find $v_B(\rho, \phi^\pm)$ for a 90° interior wedge angle (both polarizations) when:

a. $\phi' = 45°$, $\rho = 10\lambda$, $\phi = 90°$

b. $\phi' = 45°$, $\rho = 10\lambda$, $\phi = 138°$

c. $\phi' = 45°$, $\rho = 10\lambda$, $\phi = 180°$

Comment on your results and justify the formulas you used to evaluate the diffracted field in each case.

12.2-5 A vertically polarized cellular antenna transmits 20 W at 860 MHz. A receiving antenna is shadowed by a 0.3-km-high ridge normal to a line drawn between the two antennas as shown. How much power is available at the terminals of the receiving antenna if the gain of the receiving antenna in the direction of the ridge is 4 dB and that of the transmitting antenna is 15 dB toward the ridge? As a rough approximation, assume the ridge is perfectly conducting.

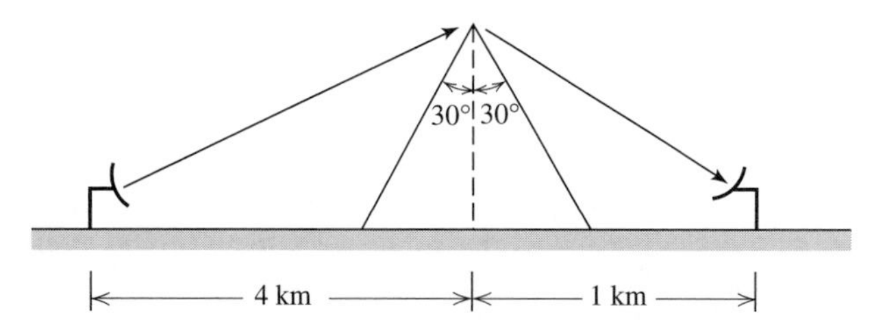

12.2-6 Substitute (12-34) and (12-35) into (12-33) and explain the physical significance of each of the four terms you obtain.

12.2-7 Draw a sketch that illustrates the first postulate of Keller's theory. Include both a direct ray and a diffracted ray in your sketch.

12.3-1 Consider a magnetic line source parallel to the edge of a half-plane as shown. In this situation, the diffracted field appears to originate from a magnetic line source located at the edge. Using the flux tube concept of Fig. 12-3, show that the diffracted field may be written as

$$E_{\perp}^{d}(\rho) = -D_{\perp}E_{\perp}^{i}(Q)\frac{e^{-j\beta\rho}}{\sqrt{\rho}}$$

where $E_{\perp}^{i}(Q)$ is the value of the incident field at the edge.

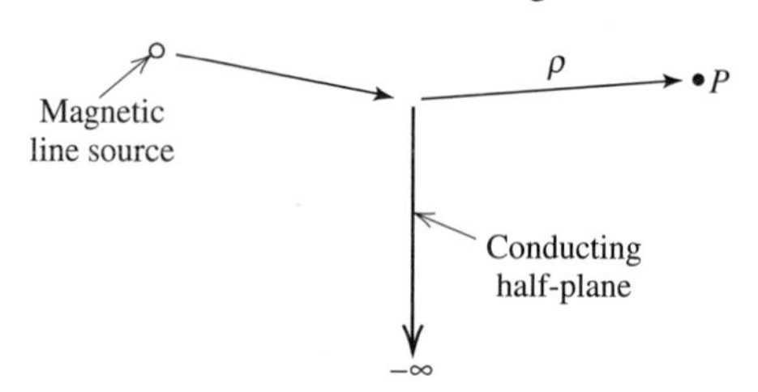

12.3-2 Repeat Prob. 12.3-1 when the magnetic line source is replaced by an electric line source and show that

$$E_{\parallel}^{d}(\rho) = -D_{\parallel}E_{\parallel}^{i}(Q)\frac{e^{-j\beta\rho}}{\sqrt{\rho}}$$

12.3-3 Consider the situation where a point source illuminates the edge of a half-plane at normal incidence. Unlike the previous two problems, in this case there will be spreading in both principal planes. Using the flux tube concept of Fig. 12-3, show that the diffracted field may be written as either

$$E_{\parallel}^{d}(s) = -D_{\parallel}E_{\parallel}^{i}(Q)\sqrt{\frac{s'}{s'+s}}\frac{e^{-j\beta s}}{\sqrt{s}}$$

or

$$E_{\perp}^{d}(s) = -D_{\perp}E_{\perp}^{i}(Q)\sqrt{\frac{s'}{s'+s}}\frac{e^{-j\beta s}}{\sqrt{s}}$$

12.3-4 Show that the diffraction coefficient matrix $[D]$ in (12-46) will generally have seven nonvanishing coefficients if an edge-fixed coordinate system is used rather than a ray-fixed system.

12.4-1 Derive (12-59) from (12-52) and show that (12-59) is the same as (12-42).

12.4-2 Consider the case where a half-plane is illuminated by a plane wave and the observation point is near the edge of the wedge.

a. Show that UTD reduces to the Sommerfeld–Pauli result in (12-44) and hence the UTD is exact.

b. Is UTD an exact solution if the source is near the wedge edge and the observation point is at a very large distance? Why?

c. If both the source and observation points are near the wedge edge, the UTD solution will not be exact. Why? (Although the solution may not be exact, the results may still be useful—see Sec. 12.12.)

12.4-3 Show that an alternative to (12-55) would be to define N^{+} as the value of $[(\phi \pm \phi') + \pi]/2\pi n$ rounded to the nearest integer. Define a similar alternative to (12-56).

12.4-4 Consider a wedge illuminated by either an electric or magnetic line source parallel to the edge and at some distance from it ($\rho' \gg \lambda$).

a. At the reflection boundary (or incident boundary), show that the diffraction coefficient must have a discontinuity of magnitude $\sqrt{\rho'\rho/(\rho + \rho')}$.

b. Show that at the reflection boundary (or incident boundary), the UTD diffraction coefficient is discontinuous by an amount $\pm\sqrt{L}$. What determines the sign of the discontinuity? The following approximation is useful:

$$F(X) \simeq \left[\sqrt{\pi X} - 2Xe^{j(\pi/4)} - \frac{2}{3}X^2e^{-j(\pi/4)}\right]e^{j(\pi/4+X)}$$

which is valid when X is small.

c. From the results of (a) and (b), show that the total field is continuous across the reflection (or incident) shadow boundary.

12.4-5 A plane wave is incident at an angle of $\gamma_o' = 45°$, $\phi' = 30°$ on the edge of a 90° ($n = \frac{3}{2}$) conducting wedge.

a. Use (12-52) and Fig. 12-17 to calculate $E_\perp^d$ at a distance $s = 2\lambda$ when $\phi = 120°$, 132°, 138°, 180°, 222°, 228°, and 260° when $E_\perp^i = 1$ V/m.

b. Repeat (a) for $E_\parallel^d$ when $E_\parallel^i = 1$ V/m.

12.4-6 a. Use (12-36), (12-37), and (12-42) to compute the total field in Fig. 12-14a. Your result will differ from that in Fig. 12-14a at the reflection and shadow boundaries. Why?

b. Recompute (a) but use (12-52) instead of (12-42).

c. Comment on the difference between the results in (a) and (b) above.

12.4.7 a. Using trig identities, put (12-42) into a form similar to (12-52) but without the transition functions, F.

b. Comment on the purpose of the transition functions.

12.5-1 Use the E-plane model in Fig. 12-18b and a computer program for wedge diffraction to verify the curves in Fig. 7-16 that were obtained by aperture integration.

12.5-2 Explain why the rays in Fig. 12-18c make a negligible contribution to the radiation pattern except when $\zeta \approx 90°$.

12.5-3 Show that the doubly diffracted field from Q_1 in Fig. 12-18c can be written as

$$E_{1,2}^d(P) = \frac{1}{2}\frac{e^{-j\beta\rho_E}}{\sqrt{\rho_E}}D_\perp(Q_2)D_\perp(Q_1)\frac{e^{-j\beta 2a}}{\sqrt{2a}}\frac{e^{-j\beta r}}{\sqrt{r}}e^{j\beta a \sin\zeta}$$

12.5-4 Review Prob. 11.8-5.

12.5-5 Calculate the diffracted field from the 194° interior wedge angle edge on the outside of the horn antenna in Fig. 11-18 that is formed by the join of the horn wall with the waveguide. Use $\rho = \lambda$. Assume that the electric field incident on the join is 1V/m. Compare your result to what you observe in Fig. 11-21d.

12.6-1 Use a computer program for wedge diffraction to calculate the total diffracted field for $0 \le \zeta \le 2\pi$ for the antenna of Fig. 12-20. Compare your results with Fig. 12-21. Why is there a difference?

12.6-2 Draw a sketch of the "creeping wave" rays (see Sec. 12.11) on the back side of the parabolic reflector of Fig. 12-20. Now draw rays that originate at Q_1 or Q_2 and reflect several times along the *inside* surface parabolic reflector. These rays are called whispering gallery rays.

12.6-3 Show that the doubly diffracted ray from Q_1 in Fig. 12-20 can be written as

$$E_{1,2}^d(P) = \frac{e^{-j\beta\rho_o}}{\sqrt{\rho_o}}f(\theta_o)D_\parallel(Q_2)D_\parallel(Q_1) \cdot \frac{e^{-j\beta 2a}}{\sqrt{2a}}\frac{e^{-j\beta r}}{\sqrt{r}}e^{j\beta a \sin\zeta}$$

12.6-4 Derive (12-71).

12.6-5 If the line source in Fig. 12-20 is a magnetic line source, calculate the far-field pattern. Your result will be similar to that in Fig. 12-21, except that the discontinuity at $\zeta = 90°$ will be greater and the back lobes will be about 8 dB higher. Why?

12.6-6 Use the UTD to calculate the H-plane pattern of a 90° corner reflector antenna with a dipole feed. The dipole feed is 0.5λ from the apex of the reflector, the reflector sides are 1.0λ long, and the aperture of the corner reflector is 1.414λ across.

12.7-1 The diffracted field that is neglected in (12-86) may be written generally as [14]

$$E_{SD}^{d}(P) = \frac{1}{2j\beta}\frac{\partial E^{i}(Q)}{\partial n}\frac{\partial}{\partial \phi'}D_{\parallel}\Bigg|_{\phi'=0}\sqrt{\frac{\rho}{s(\rho+s)}}\,e^{-j\beta s}$$

Compare the value of this slope diffracted field with the direct field in (12-86) when $\theta = 90°$.

12.7-2 a. Using (12-36) and (12-37) in (12-33), show how a factor of 2 arises in $E(\rho, \phi)$ for the perpendicular polarization in the infinite ground plane case ($v_B = 0$) when the plane wave has grazing incidence to the ground plane ($\phi' = 0$).

b. Then verify, in general, that at grazing incidence the diffracted field must be multiplied by $\frac{1}{2}$, as in (12-80) and (12-82), to obtain the correct value of the diffracted field. To do this, use either the asymptotic form in (12-42) or (12-59) to show that $D_{\parallel} \to 0$ and a factor of 2 naturally arises in $D_{\perp}$.

12.8-1 A short monopole (stub antenna) is mounted at the center of a square ground plane 6λ on a side as shown in Fig. 12-27*a*.

a. Using the two-point approximation, show that the relative diffracted field in the region $200° < \phi_1 < 340°$ can be expressed by

$$E^{d} = \frac{e^{-j(\beta r+\pi/4)}}{\sqrt{2\pi\beta r}}\left[\frac{1}{\cos\frac{\phi_1}{2}} - \frac{e^{-j12\pi\cos\phi_2}}{\cos\frac{\phi_2}{2}}\right]$$

where $\phi_2 = 2\,\pi - (\phi_1 - \pi)$

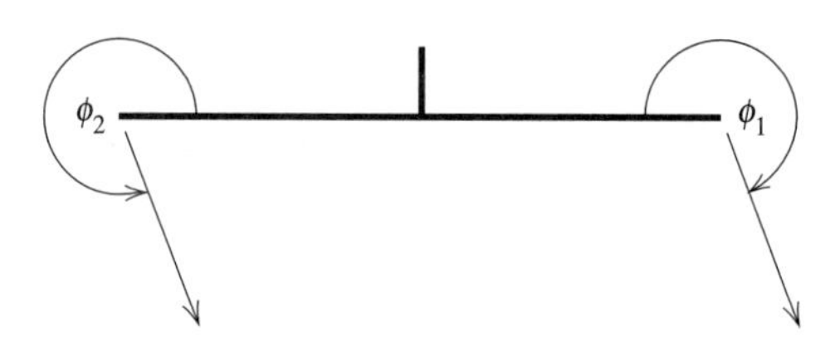

b. Why must the diffracted field be zero when $\phi_1 = 270°$ for this problem? Use a sketch and physical reasoning to explain why.

c. Calculate and plot a graph of the diffracted field for $200° < \phi_1 < 340°$. Compare your results with Fig. 12-28.

12.9-1 Derive (12-97) and (12-98).

12.9-2 Derive (12-101) and (12-102).

12.9-3 A short monopole (stub antenna) is mounted at the center of a circular ground plane 6λ in diameter as shown in Fig. 12-27*b*.

a. Using the equivalent concept, show that the relative diffracted field in the region $90° \le \theta \le 180°$ can be expressed by

$$E^{d} = -\frac{e^{-j(\beta r+\pi/4)}}{\sqrt{2\pi\beta r}}\frac{1}{\cos(\phi/2)}2\pi j J_1(6\pi\sin\theta)$$

where J_1 is the first-order Bessel function. Note that

$$\int_0^{2\pi}\cos(\xi-\xi')e^{jx\cos(\xi-\xi')}\,d\xi' = 2\pi j J_1(x)$$

b. Calculate the diffracted field and compare with that calculated in Prob. 12.8-1.

12.10-1 A triangular cylinder is illuminated by a line source as shown. Apply the self-consistent method to this problem by setting up (12-110) in a form similar to (12-109). Note that some of the matrix elements will be zero. Check your solution with that in [18].

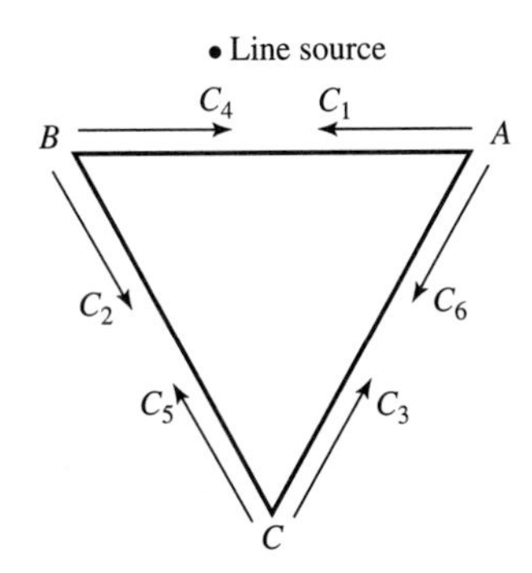

12.11-1 a. Use the information in Table 12-1 to compute σ_w in (12-127). Plot σ_w vs. βa for $0.1 \leq \beta a \leq 10$ for both polarizations on the same graph.

b. Repeat part (a) when the geometrical optics contribution (only) is multiplied by $\left[1 - j\dfrac{5}{16(\beta a)} + \dfrac{127}{512(\beta a)^2}\right]$ in the parallel case and by $\left[1 + j\dfrac{11}{16(\beta a)} - \dfrac{353}{512(\beta a)^2}\right]$ in the perpendicular case [22]. Plot both results on the same graph. Compare with Fig. 12-34.

c. Plot the parallel polarization results from (a) and (b) on one graph and the perpendicular polarization results on another. Comment on your results.

12.12-1 Consider a monopole at the center of a square ground plane whose sides are $\lambda/2$ long. The monopole is to be represented using pulse basis functions and delta weighting functions. The four sides of the ground plane are to be accounted for using wedge diffraction. Diffraction by the four corners is to be ignored. Derive the necessary equations that would enable you to calculate Z_{mn}^g in (12-131).

12.12-2 Derive (12-137).

12.12-3 For the problem in Fig. 12-42, show all possible ray paths that do not involve double (or higher-order) diffractions.

12.12-4 A dipole of length ℓ is located a distance d from the surface of an infinitely long circular cylinder of radius a. The dipole is parallel to the axis of the cylinder. Show how you would account for the presence of the cylinder if only the dipole is represented by the method of moments.

12.13-1 Using physical optics, show that the radar cross section of a flat rectangular plate at normal incidence is $\sigma = 4\pi\,(A^2/\lambda^2)$ where A is the area of the plate.

12.13-2 Equation (12-155) can be converted to a different and often useful form by noting that $(\hat{\mathbf{z}} \cdot \hat{\mathbf{n}})\, ds$ is the projection of the element of surface area ds onto the xy-plane. Thus, $(\hat{\mathbf{z}} \cdot \hat{\mathbf{n}})\, ds = ds_z = (ds_z/d\ell)\, d\ell$ where ds_z is the projection of ds onto the xy-plane. Then (12-155) becomes

$$\sigma = \frac{4\pi}{\lambda^2}\left(\int_0^L e^{-j2\beta\ell}\,\frac{ds_z}{d\ell}\,d\ell\right)$$

where ℓ is the distance from the reference plane to the surface. Use the above expression for the radar cross section to derive the physical optics expression for the RCS of the sphere.

12.13-3 Show that the RCS of an infinite cone (as shown) is $\sigma = (\lambda^2 \tan^4 \alpha)/16\pi$

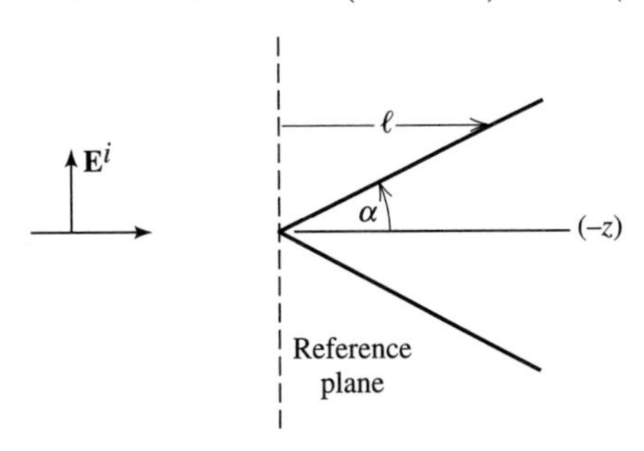

12.13-4 Show that the RCS of a square flat plate with edges parallel to the x- and y-axes, and direction of incidence in the xz-plane, is

$$\sigma = \frac{4\pi a^4}{\lambda^2}\left[\frac{\sin(\beta a \sin\theta)}{\beta a \sin\theta}\right]^2 \cos^2\theta$$

where a is the length of one side. Compare the angular variation of this result with that of the uniformly illuminated line source in Chap. 4.

12.13-5 Show that the RCS of a circular flat plate, or disk, in the xy-plane is

$$\sigma = \frac{\pi a^2}{\tan^2\theta}\left[J_1\left(\frac{4\pi a \sin\theta}{\lambda}\right)\right]^2$$

where a in the radius of the disk and $J_1(x)$ the Bessel function of order one. Also show that at $\theta = 0°$, the above expression reduces to $\sigma = (4\pi/\lambda^2)\, A^2$ where A is the area of the disk [4].

12.14-1 After writing

$$I_b = \int_b^\infty f(x)e^{j\beta\gamma(x)}\, dx = \frac{1}{j\beta}\int_b^\infty \frac{f(x)}{\gamma'(x)}\, j\beta\gamma'(x)e^{j\beta\gamma(x)}\, dx$$

integrate by parts to obtain (12-170).

12.14-2 Show that

$$I_{-a} \cong -\frac{1}{j\beta}\frac{f(-a)}{\gamma'(-a)}\, e^{j\beta\gamma(-a)}$$

12.14-3 Show that (12-167) follows from (12-165) and (12-166).

12.14-4 Interpret the discussion between (12-29) and (12-30) in terms of the concept of stationary phase.

12.15-1 Show that $\begin{Bmatrix} f \\ g \end{Bmatrix}$ in (12-187) can be expressed as

$$\begin{Bmatrix} f \\ g \end{Bmatrix} = \frac{-1}{2n}\left[\cot\left(\frac{\pi + \phi^-}{2n}\right) + \cot\left(\frac{\pi - \phi^-}{2n}\right)\right] \pm \frac{1}{2n}\left[\cot\left(\frac{\pi + \phi^+}{2n}\right) + \cot\left(\frac{\pi - \phi^+}{2n}\right)\right]$$

12.15-2 For the half-plane case in Fig. 12-14, analytically show that at the reflection and shadow boundaries, the singularity in f is cancelled by the singularity in f_o making f' in (12-185) continuous.

HINT: Identify which cotangent term in Problem 12.5-1 is singular at the reflection boundary and which term is singular at the shadow boundary. Cancel these terms with the singularity in f_o by letting $\phi^+ = \pi \pm \delta$ at the GO reflection boundary and by letting $\phi^- = \pi \pm \delta$ at the GO shadow boundary.

12.15-3 For the half-plane case in Fig. 12-14a show by numerical computation that f' in (12-185) is continuous across the reflection and shadow boundaries. Also compute separately f and f_o at these boundaries. Use the geometry in Fig. 12-14.

12.15-4 Evaluate I_o in (12-197).

12.15-5 Show that (12-181) substituted into (12-182) leads to (12-179).

Appendix A

Tables of Commonly Used Frequencies

(in U.S.)

A.1 RADIO FREQUENCY BANDS

Frequency ⟶

3 kHz		30 kHz		300 kHz		3 MHz		30 MHz		300 MHz		3 GHz		30 GHz		300 GHz
	VLF		LF		MF		HF		VHF		UHF		SHF		EHF	
100 km		10 km		1 km		100 m		10 m		1 m		10 cm		1 cm		1 mm

⟵ Wavelength

A.2 TELEVISION CHANNEL FREQUENCIES

VHF

Channel No.	Frequency Range (MHz)	Channel No.	Frequency Range (MHz)
2	54–60	8	180–186
3	60–66	9	186–192
4	66–72	10	192–198
5	76–82	11	198–204
6	82–88	12	204–210
7	174–180	13	210–216

UHF

Channel No.	Frequency Range (MHz)	Channel No.	Frequency Range (MHz)	Channel No.	Frequency Range (MHz)
14	470–476	30	566–572	46	662–668
15	476–482	31	572–578	47	668–674
16	482–488	32	578–584	48	674–680
17	488–494	33	584–590	49	680–686
18	494–500	34	590–596	50	686–692
19	500–506	35	596–602	51	692–698
20	506–512	36	602–608	52	698–704
21	512–518	37	608–614	53	704–710
22	518–524	38	614–620	54	710–716
23	524–530	39	620–626	55	716–722
24	530–536	40	626–632	56	722–728
25	536–542	41	632–638	57	728–734
26	542–548	42	638–644	58	734–740
27	548–554	43	644–650	59	740–746
28	554–560	44	650–656	⋮	
29	560–566	45	656–662	69	800–806

Note: The carrier frequency for the video portion is the lower frequency plus 1.25 MHz. The audio carrier frequency is the upper frequency minus 0.25 MHz. All channels have a 6-MHz bandwidth. For example, the Channel 2 video carrier is at 55.25 MHz and the audio carrier is at 59.75 MHz.

A.3 MOBILE TELEPHONE BANDS

Cellular	824–894 MHz
PCS	1850–1990 MHz

A.4 RADAR BANDS

World War II Band Designations		IEEE Band Designations	
		HF	3–30 MHz
		VHF	30–300 MHz
		UHF	300–1000 MHz
L	390–1550 MHz	*L*-band	1–2 GHz
S	1550–3900 MHz	*S*-band	2–4 GHz
C	3.9–6.2 GHz	*C*-band	4–8 GHz
X	6.2–12.9 GHz	*X*-band	8–12 GHz
Ku	12.9–18 GHz	*Ku*-band	12–18 GHz
K	18–26.5 GHz	*K*-band	18–27 GHz
Ka	26.5–40 GHz	*Ka*-band	27–40 GHz
		V-band	40–75 GHz
		W-band	75–110 GHz
		Millimeter	110–300 GHz

Appendix B

Material Data and Other Constants

B.1 CONDUCTIVITIES OF GOOD CONDUCTORS

Conductor	Conductivity (S/m)
Silicon steel	2×10^6
Brass	1.1×10^7
Aluminum	3.5×10^7
Gold	4.1×10^7
Copper	5.7×10^7
Silver	6.1×10^7

B.2 WIRE DATA

Wire Size AWG	Diameter in mm (in.)	Single Copper Wire Continuous Duty Current Capacity (A)	Copper Wire dc Resistance per Unit Length (Ω/100 m)
8	3.264 (0.1285)	73	0.1952
9	2.906 (0.1144)	—	0.2462
10	2.588 (0.1019)	55	0.3103
11	2.305 (0.0907)	—	0.3914
12	2.053 (0.0808)	41	0.4935
13	1.828 (0.0720)	—	0.6224
14	1.628 (0.0641)	32	0.7849
16	1.291 (0.0508)	22	1.248
18	1.024 (0.0403)	16	1.984
20	0.812 (0.0320)	11	3.155
22	0.644 (0.0253)	—	5.017
24	0.511 (0.0201)	—	7.98
26	0.405 (0.0159)	—	12.69
28	0.321 (0.0129)	—	20.17
30	0.255 (0.0100)	—	32.06

B.3 DIELECTRIC CONSTANT: PERMITTIVITY

$$\varepsilon_o = 8.854 \times 10^{-12} \text{ F/m}$$
$$\approx 10^{-9}/36\pi \quad \text{F/m}$$

B.4 PERMEABILITY

$$\mu_o = 1.26 \times 10^{-6} \text{ H/m}$$
$$\approx 4\pi \times 10^{-7} \text{ H/m}$$

B.5 VELOCITY OF LIGHT

$$c = 1 \text{ layer}/\sqrt{\mu_o \varepsilon_o} = 2.997925 \times 10^8 \text{ m/s}$$

B.6 INTRINSIC IMPEDANCE OF FREE SPACE

$$\eta_o = \sqrt{\frac{\mu_o}{\varepsilon_o}} = 376.73\ \Omega \approx 120\pi\ \Omega$$

Appendix C

Vectors

C.1 UNIT VECTOR REPRESENTATIONS

$$\hat{\mathbf{x}} = \hat{\mathbf{r}} \sin\theta\cos\phi + \hat{\boldsymbol{\theta}} \cos\theta\cos\phi - \hat{\boldsymbol{\phi}} \sin\phi \tag{C-1}$$

$$\hat{\mathbf{y}} = \hat{\mathbf{r}} \sin\theta\sin\phi + \hat{\boldsymbol{\theta}} \cos\theta\sin\phi + \hat{\boldsymbol{\phi}} \cos\phi \tag{C-2}$$

$$\hat{\mathbf{z}} = \hat{\mathbf{r}} \cos\theta - \hat{\boldsymbol{\theta}} \sin\theta \tag{C-3}$$

$$\hat{\mathbf{r}} = \hat{\mathbf{x}} \sin\theta\cos\phi + \hat{\mathbf{y}} \sin\theta\sin\phi + \hat{\mathbf{z}} \cos\theta \tag{C-4}$$

$$\hat{\boldsymbol{\theta}} = \hat{\mathbf{x}} \cos\theta\cos\phi + \hat{\mathbf{y}} \cos\theta\sin\phi - \hat{\mathbf{z}} \sin\theta \tag{C-5}$$

$$\hat{\boldsymbol{\phi}} = -\hat{\mathbf{x}} \sin\phi + \hat{\mathbf{y}} \cos\phi \tag{C-6}$$

C.2 VECTOR IDENTITIES

$$\mathbf{A} \times (\mathbf{B} \times \mathbf{C}) = (\mathbf{A} \cdot \mathbf{C})\mathbf{B} - (\mathbf{A} \cdot \mathbf{B})\mathbf{C} \tag{C-7}$$

$$(\mathbf{A} \times \mathbf{B}) \times \mathbf{C} = (\mathbf{C} \cdot \mathbf{A})\mathbf{B} - (\mathbf{C} \cdot \mathbf{B})\mathbf{A} \tag{C-8}$$

$$\nabla \cdot (\nabla \times \mathbf{G}) = 0 \tag{C-9}$$

$$\nabla \times \nabla g = 0 \tag{C-10}$$

$$\nabla \cdot \nabla g = \nabla^2 g \tag{C-11}$$

$$\nabla(f + g) = \nabla f + \nabla g \tag{C-12}$$

$$\nabla \cdot (\mathbf{F} + \mathbf{G}) = \nabla \cdot \mathbf{F} + \nabla \cdot \mathbf{G} \tag{C-13}$$

$$\nabla(fg) = g\nabla f + f\nabla g \tag{C-14}$$

$$\nabla \cdot (f\mathbf{G}) = \mathbf{G} \cdot (\nabla f) + f(\nabla \cdot \mathbf{G}) \tag{C-15}$$

$$\nabla \times (f\mathbf{G}) = (\nabla f) \times \mathbf{G} + f(\nabla \times \mathbf{G}) \tag{C-16}$$

$$\nabla \times (\nabla \times \mathbf{G}) = \nabla(\nabla \cdot \mathbf{G}) - \nabla^2 \mathbf{G} \tag{C-17}$$

$$\nabla^2 \mathbf{G} = \hat{\mathbf{x}}\nabla^2 G_x + \hat{\mathbf{y}}\nabla^2 G_y + \hat{\mathbf{z}}\nabla^2 G_z \tag{C-18}$$

$$\nabla \cdot (\mathbf{F} \times \mathbf{G}) = \mathbf{G} \cdot (\nabla \times \mathbf{F}) - \mathbf{F} \cdot (\nabla \times \mathbf{G}) \tag{C-19}$$

$$\mathbf{F} \cdot (\mathbf{G} \times \mathbf{H}) = \mathbf{G} \cdot (\mathbf{H} \times \mathbf{F}) = \mathbf{H} \cdot (\mathbf{F} \times \mathbf{G}) \tag{C-20}$$

$$\nabla \times (\mathbf{F} \times \mathbf{G}) = \mathbf{F}(\nabla \cdot \mathbf{G}) - \mathbf{G}(\nabla \cdot \mathbf{F}) + (\mathbf{G} \cdot \nabla)\mathbf{F} - (\mathbf{F} \cdot \nabla)\mathbf{G} \tag{C-21}$$

$$\nabla(\mathbf{F} \cdot \mathbf{G}) = (\mathbf{F} \cdot \nabla)\mathbf{G} + (\mathbf{G} \cdot \nabla)\mathbf{F} + \mathbf{F} \times (\nabla \times \mathbf{G}) + \mathbf{G} \times (\nabla \times \mathbf{F}) \tag{C-22}$$

$$\iiint_v \nabla \cdot \mathbf{G}\, dv = \oiint_s \mathbf{G} \cdot d\mathbf{s} \qquad \text{divergence theorem} \tag{C-23}$$

$$\iint_s (\nabla \times \mathbf{G}) \cdot d\mathbf{s} = \oint_l \mathbf{G} \cdot d\boldsymbol{\ell} \qquad \text{Stokes' theorem} \tag{C-24}$$

C.3 VECTOR DIFFERENTIAL OPERATORS

Rectangular Coordinates

$$\nabla g = \hat{\mathbf{x}} \frac{\partial g}{\partial x} + \hat{\mathbf{y}} \frac{\partial g}{\partial y} + \hat{\mathbf{z}} \frac{\partial g}{\partial z} \tag{C-25}$$

$$\nabla \cdot \mathbf{G} = \frac{\partial G_x}{\partial x} + \frac{\partial G_y}{\partial y} + \frac{\partial G_z}{\partial z} \tag{C-26}$$

$$\nabla \times \mathbf{G} = \hat{\mathbf{x}}\left(\frac{\partial G_z}{\partial y} - \frac{\partial G_y}{\partial z}\right) + \hat{\mathbf{y}}\left(\frac{\partial G_x}{\partial z} - \frac{\partial G_z}{\partial x}\right) + \hat{\mathbf{z}}\left(\frac{\partial G_y}{\partial x} - \frac{\partial G_x}{\partial y}\right) \tag{C-27}$$

$$\nabla^2 g = \frac{\partial^2 g}{\partial x^2} + \frac{\partial^2 g}{\partial y^2} + \frac{\partial^2 g}{\partial z^2} \tag{C-28}$$

Cylindrical Coordinates

$$\nabla g = \hat{\mathbf{r}} \frac{\partial g}{\partial r} + \hat{\boldsymbol{\phi}} \frac{1}{r} \frac{\partial g}{\partial \phi} + \hat{\mathbf{z}} \frac{\partial g}{\partial z} \tag{C-29}$$

$$\nabla \cdot \mathbf{G} = \frac{1}{r} \frac{\partial}{\partial r}(rG_r) + \frac{1}{r} \frac{\partial G_\phi}{\partial \phi} + \frac{\partial G_z}{\partial z} \tag{C-30}$$

$$\nabla \times \mathbf{G} = \hat{\mathbf{r}}\left(\frac{1}{r} \frac{\partial G_z}{\partial \phi} - \frac{\partial G_\phi}{\partial z}\right) + \hat{\boldsymbol{\phi}}\left(\frac{\partial G_r}{\partial z} - \frac{\partial G_z}{\partial r}\right) + \hat{\mathbf{z}} \frac{1}{r}\left[\frac{\partial}{\partial r}(rG_\phi) - \frac{\partial G_r}{\partial \phi}\right] \tag{C-31}$$

$$\nabla^2 g = \frac{1}{r} \frac{\partial}{\partial r}\left(r \frac{\partial g}{\partial r}\right) + \frac{1}{r^2} \frac{\partial^2 g}{\partial \phi^2} + \frac{\partial^2 g}{\partial z^2} \tag{C-32}$$

Spherical Coordinates

$$\nabla g = \hat{\mathbf{r}} \frac{\partial g}{\partial r} + \hat{\boldsymbol{\theta}} \frac{1}{r} \frac{\partial g}{\partial \theta} + \hat{\boldsymbol{\phi}} \frac{1}{r \sin\theta} \frac{\partial g}{\partial \phi} \tag{C-33}$$

$$\nabla \cdot \mathbf{G} = \frac{1}{r^2} \frac{\partial}{\partial r}(r^2 G_r) + \frac{1}{r \sin\theta} \frac{\partial}{\partial \theta}(G_\theta \sin\theta) + \frac{1}{r \sin\theta} \frac{\partial G_\phi}{\partial \phi} \tag{C-34}$$

$$\nabla \times \mathbf{G} = \hat{\mathbf{r}} \frac{1}{r \sin\theta} \left[\frac{\partial}{\partial \theta} (G_\phi \sin\theta) - \frac{\partial G_\theta}{\partial \phi} \right] + \hat{\boldsymbol{\theta}} \frac{1}{r} \left[\frac{1}{\sin\theta} \frac{\partial G_r}{\partial \phi} - \frac{\partial}{\partial r} (r G_\phi) \right] + \hat{\boldsymbol{\phi}} \frac{1}{r} \left[\frac{\partial}{\partial r} (r G_\theta) - \frac{\partial G_r}{\partial \theta} \right] \tag{C-35}$$

$$\nabla^2 g = \frac{1}{r^2} \frac{\partial}{\partial r} \left(r^2 \frac{\partial g}{\partial r} \right) + \frac{1}{r^2 \sin\theta} \frac{\partial}{\partial \theta} \left(\sin\theta \frac{\partial g}{\partial \theta} \right) + \frac{1}{r^2 \sin^2\theta} \frac{\partial^2 g}{\partial \phi^2} \tag{C-36}$$

Appendix D

Trigonometric Relations

$$\sin(\alpha \pm \beta) = \sin\alpha\cos\beta \pm \cos\alpha\sin\beta \tag{D-1}$$

$$\cos(\alpha \pm \beta) = \cos\alpha\cos\beta \mp \sin\alpha\sin\beta \tag{D-2}$$

$$\sin\left(\frac{\pi}{2} \pm \alpha\right) = \cos\alpha \tag{D-3}$$

$$\cos\left(\frac{\pi}{2} \pm \alpha\right) = \mp\sin\alpha \tag{D-4}$$

$$\sin\alpha\cos\beta = \tfrac{1}{2}[\sin(\alpha+\beta) + \sin(\alpha-\beta)] \tag{D-5}$$

$$\cos\alpha\sin\beta = \tfrac{1}{2}[\sin(\alpha+\beta) - \sin(\alpha-\beta)] \tag{D-6}$$

$$\cos\alpha\cos\beta = \tfrac{1}{2}[\cos(\alpha+\beta) + \cos(\alpha-\beta)] \tag{D-7}$$

$$\sin\alpha\sin\beta = -\tfrac{1}{2}[\cos(\alpha+\beta) - \cos(\alpha-\beta)] \tag{D-8}$$

$$\sin\alpha = 2\sin\frac{\alpha}{2}\cos\frac{\alpha}{2} \tag{D-9}$$

$$\sin 2\alpha = 2\sin\alpha\cos\alpha \tag{D-10}$$

$$\cos\alpha = 2\cos^2\frac{\alpha}{2} - 1 = 1 - 2\sin^2\frac{\alpha}{2} \tag{D-11}$$

$$\cos 2\alpha = 2\cos^2\alpha - 1 = \cos^2\alpha - \sin^2\alpha = 1 - 2\sin^2\alpha \tag{D-12}$$

$$\cos 3\alpha = 4\cos^3\alpha - 3\cos\alpha \tag{D-13}$$

$$\cos 4\alpha = 8\cos^4\alpha - 8\cos^2\alpha + 1 \tag{D-14}$$

$$\cos m\alpha = 2^{m-1}\cos^{m}\alpha - \frac{m}{1!}2^{m-3}\cos^{m-2}\alpha + \frac{m(m-3)}{2!}2^{m-5}\cos^{m-4}\alpha + \cdots \tag{D-15}$$

$$1 = \sin^2\alpha + \cos^2\alpha \tag{D-16}$$

$$\sec^2\alpha = \frac{1}{\cos^2\alpha} = 1 + \tan^2\alpha \tag{D-17}$$

$$\sin \alpha = \alpha - \frac{\alpha^3}{3!} + \frac{\alpha^5}{5!} - \frac{\alpha^7}{7!} + \cdots \tag{D-18}$$

$$\cos \alpha = 1 - \frac{\alpha^2}{2!} + \frac{\alpha^4}{4!} - \frac{\alpha^6}{6!} + \cdots \tag{D-19}$$

$$e^{\pm j\alpha} = \cos \alpha \pm j \sin \alpha \tag{D-20}$$

$$\tan \alpha = \frac{\sin \alpha}{\cos \alpha} \tag{D-21}$$

Appendix E

Hyperbolic Relations

$$\sinh \alpha = \frac{e^{\alpha} - e^{-\alpha}}{2} = \alpha + \frac{\alpha^3}{3!} + \frac{\alpha^5}{5!} + \frac{\alpha^7}{7!} + \cdots \tag{E-1}$$

$$\cosh \alpha = \frac{e^{\alpha} + e^{-\alpha}}{2} = 1 + \frac{\alpha^2}{2!} + \frac{\alpha^4}{4!} + \frac{\alpha^6}{6!} + \cdots \tag{E-2}$$

$$\tanh \alpha = \frac{\sinh \alpha}{\cosh \alpha} = \frac{1}{\coth \alpha} \tag{E-3}$$

$$\sinh(\alpha \pm j\beta) = \sinh \alpha \cos \beta \pm j \cosh \alpha \sin \beta \tag{E-4}$$

$$\cosh(\alpha \pm j\beta) = \cosh \alpha \cos \beta \pm j \sinh \alpha \sin \beta \tag{E-5}$$

$$\sinh(j\alpha) = j \sin \alpha = \frac{e^{j\alpha} - e^{-j\alpha}}{2} \tag{E-6}$$

$$\cosh(j\alpha) = \cos \alpha = \frac{e^{j\alpha} + e^{-j\alpha}}{2} \tag{E-7}$$

Appendix F

Mathematical Relations

F.1 DIRAC DELTA FUNCTION

The Dirac delta function (or impulse function) is zero everywhere except when the argument is zero.

$$\delta(x - x_o) = 0 \qquad \text{for} \qquad x \neq x_o \tag{F-1}$$

For the zero argument case, the function is singular but in a special way: The area is unity, that is,

$$\int_{-\infty}^{\infty} \delta(x - x_o)\, dx = 1 \tag{F-2}$$

Another useful property of the Dirac delta function follows:

$$\int_{-\infty}^{\infty} g(x)\, \delta(x - x_o)\, dx = g(x_o) \tag{F-3}$$

F.2 BINOMIAL THEOREM

$$\begin{aligned}(a + b)^n = a^n + na^{n-1}b &+ \frac{n(n-1)}{2!} a^{n-2}b^2 \\ &+ \frac{n(n-1)(n-2)}{3!} a^{n-3}b^3 + \cdots\end{aligned} \tag{F-4}$$

$$(1 \pm x)^n \approx 1 \pm nx \qquad \text{for} \qquad x \ll 1 \tag{F-5}$$

F.3 BESSEL FUNCTIONS

$$J_0(x) = \frac{1}{2\pi}\int_0^{2\pi} e^{jx\cos\alpha}\, d\alpha \tag{F-6}$$

$$\begin{aligned}J_n(x) &= \frac{j^{-n}}{2\pi}\int_0^{2\pi} e^{jx\cos\alpha}\cos(n\alpha)\, d\alpha \\ &= \sum_{m=0}^{\infty} \frac{(-1)^m x^{2m+n}}{m!(m+n)!2^{2m+n}}\end{aligned} \tag{F-7}$$

$$J_n(x) = \frac{2(n-1)}{x} J_{n-1}(x) - J_{n-2}(x) \tag{F-8}$$

$$\int x^{n+1} J_n(x)\, dz = x^{n+1} J_{n+1}(x) \tag{F-9}$$

$$\int_0^1 (1-x^2)^n x J_0(bx)\, dx = \frac{2^n n!}{b^{n+1}} J_{n+1}(b) \tag{F-10}$$

F.4 SOME USEFUL INTEGRALS

$$\int \sin(a+bx) e^{cx}\, dx = \frac{e^{cx}}{b^2+c^2} [c \sin(a+bx) - b\cos(a+bx)] \tag{F-11}$$

$$\int_{-\infty}^{\infty} \frac{\sin^2 x}{x^2}\, dx = \pi \tag{F-12}$$

$$\mathrm{Si}(x) = \int_0^x \frac{\sin \tau}{\tau}\, d\tau \qquad \text{sine integral} \tag{F-13}$$

$$\mathrm{Ci}(x) = -\int_x^{\infty} \frac{\cos \tau}{\tau}\, d\tau \qquad \text{cosine integral} \tag{F-14}$$

$$\mathrm{Cin}(x) = \int_0^x \frac{1-\cos \tau}{\tau}\, d\tau \tag{F-15}$$

$$\mathrm{Cin}(x) = 0.5772 + \ln(x) - \mathrm{Ci}(x) \tag{F-16}$$

Fresnel integrals:

$$C(x) = \int_0^x \cos\left(\frac{\pi}{2}\tau^2\right) d\tau; \qquad C(-x) = -C(x) \tag{F-17a}$$

$$S(x) = \int_0^x \sin\left(\frac{\pi}{2}\tau^2\right) d\tau; \qquad S(-x) = -S(x) \tag{F-17b}$$

Appendix G

Computing Packages

Most of the problems in this book can be coded easily using a commercial mathematics applications package. There are many of these in use and supplying files for them is not useful. Instead, we provide computing modules for a few important antenna topics. Most of them allow the student to change parameter values and immediately see the effect. Directions for accessing these packages are posted on the World Wide Web at the following address:

www.wiley.com/college/stutzman

G.1 GENERAL ANTENNA PACKAGE: ANTENNA PATTERN VISUALIZATION (APV)

This package is made available by A. Z. Elsherbeni and C. D. Taylor of the University of Mississippi. It is a user-friendly package that presents radiation patterns in two- or three-dimensional views under user control. The following antennas are included:

Dipoles
Arrays
Loops
Corner reflectors

The arrays portion of the package is especially valuable for investigating the influence of the array geometry on the antenna pattern. The array variables are easy to change and the pattern is displayed immediately.

G.2 ARRAY PLOTTING PACKAGE: PCARRPAT

This package provides polar pattern plots in all three principal planes for an arbitrary array that can have elements at any locations in three dimensions and can have any excitations. An input file must be created in the following format:

```
N       NETYPE      NPOINT
THETA0      PHI0        (use only if NPOINT = 1)
X      Y      Z      A      ALPHA      (one line for each
                                        element)
```

```
      X    Y    Z    A    ALPHA
      .
      .
      .

where
      N = total number of elements

      NETYPE = 0 for isotropic elements
               1 for collinear half-wave dipoles parallel
                 to the z-axis
               2 for parallel half-wave dipoles parallel
                 to the x-axis
               3 for collinear short dipoles parallel to
                 the z-axis
               4 for parallel short dipoles parallel to
                 the x-axis

      NPOINT = 1 if element phases are to be adjusted in
                 the program to steer the main beam to
                 direction (THETA0, PHI0)
               0 if not
      X, Y, Z,  = element center locations
      A =         amplitude of current excitation
      ALPHA =     phase of current excitation
```

G.3 WIRE, A GENERAL WIRE ANTENNA PROGRAM

WIRE permits the user to specify arbitrary arrangements of straight wires of finite size, with or without lumped loads, and with arbitrary connectivity. Both antenna and scattering problems can be solved. Many antenna configurations can be modeled, including arrays. The method of moments solution approach is used, so full mutual coupling is accounted for. The following are available outputs: values for current distribution on the wires, input impedance, radiation patterns, and gain; and plots of current distributions and patterns.

G.4 PARABOLIC REFLECTOR ANTENNA CODE: "PRAC"

PRAC is a user-friendly program for analysis of reflector antennas. The main reflector geometry as well as the desired illumination are specified by the user. The program returns the required feed pattern, gain, and radiation patterns, including cross polarization patterns.

G.5 DIFFRACTION CODES

The subroutine DW computes the diffraction coefficients $D_\perp$ and $D_\parallel$ presented in Section 12.4 for the wedge of interior angle $(2 - n)\pi$. The subroutine will also compute the slope diffraction coefficients associated with the perpendicular and parallel cases. The latter slope diffraction coefficient is discussed in Section 12.7.

To use the subroutine, it is only necessary to know the calling parameters in line 1, which are

DS = diffraction coefficient $D_{\parallel}(L, \phi, \phi')$
DH = diffraction coefficient $D_{\perp}(L, \phi, \phi')$
DPS = slope diffraction coefficient for the parallel case
DPH = slope diffraction coefficient for the perpendicular case
R = distance parameter L
PH = angle ϕ
PHP = angle ϕ'
BO = angle γ_o
FN = n of the interior wedge angle $(2 - n)\pi$

As an example of the use of subroutine DW, consider the *E*-plane analysis of the horn antenna in Section 12.5. In writing a "main program" to analyze the horn antenna we would call, for example, DW (X, DPER, X, X, RL, PHI, 0.0, 90.0, 2.0) where X is a variable not used in the program. We must supply the subroutine with RL and PHI, and it will return DPER.

The user of subroutine DW may verify the statement listing by calculating the diffracted field in Fig. 12-14.

Appendix H

Bibliography

H.1 DEFINITIONS

1. *IEEE Standard Definitions of Terms for Antennas*, IEEE Standard 145-1993, IEEE: 445 Hoes Lane, Piscataway, NJ, 1993.

H.2 FUNDAMENTAL BOOKS

1. S. Silver, Editor, *Microwave Antenna Theory and Design*, MIT Radiation Laboratory Series Vol. 12, McGraw-Hill Book Co.: NY, 1949. Available from PPL Dept. IEEE Service Center, Piscataway, NJ 08855-1331.
2. W. L. Stutzman and G. A. Thiele, *Antenna Theory and Design*, John Wiley & Sons: NY, 1981. Second Edition, 1997.
3. C. A. Balanis, *Antenna Theory*, John Wiley & Sons: NY, 1982. Second Edition, 1997.
4. J. D. Kraus, *Antennas*, Second Edition, McGraw-Hill Book Co.: NY, 1988.
5. R. S. Elliott, *Antenna Theory and Design*, Prentice-Hall: Englewood Cliffs, NJ, 1981.
6. Thomas A. Milligan, *Modern Antenna Design*, McGraw-Hill Book Co.: NY, 1985.
7. R. E. Collin and F. J. Zucker, Editors, *Antenna Theory* Parts 1 and 2, McGraw-Hill Book Co.: NY, 1969.
8. E. Wolff, *Antenna Analysis*, John Wiley & Sons: New York, 1966, Artech House Inc.: 625 Canton St., Norwood, MA 02062, 1988.
9. S. A. Schelkunoff, *Advanced Antenna Theory*, John Wiley & Sons: NY, 1952.
10. S. A. Schelkunoff and H. T. Friis, *Antenna Theory and Practice*, John Wiley & Sons: NY, 1952.
11. R. Chatterjee, *Antenna Theory and Practice*, John Wiley & Sons: NY, 1988.
12. Kai Fong Lee, *Principles of Antenna Theory*, John Wiley & Sons: NY, 1984.
13. Lamont V. Blake, *Antennas*, First Edition, 1966; Artech House Inc.: 625 Canton St., Norwood, MA 02062, 1987.
14. T. S. M. Maclean, *Principles of Antennas—Wire and Aperture*, Cambridge Press: Cambridge, 1986.
15. W. L. Weeks, *Antenna Engineering*, McGraw-Hill Book Co.: NY, 1968.
16. E. Jordan and K. Balmain, *Electromagnetic Waves and Radiating Systems*, Prentice-Hall: Englewood Cliffs, NJ, 1950. Second edition, 1968.
17. R. E. Collin, *Antennas and Radiowave Propagation*, McGraw-Hill Book Co.: NY, 1985.
18. John Griffiths, *Radio Wave Propagation and Antennas: An Introduction*, Prentice-Hall International: Englewood Cliffs, NJ, 1987.
19. George Monser, *Antenna Design: A Practical Guide*, McGraw-Hill Book Co.: NY, 1996.
20. J. A. Kuecken, *Antennas and Transmission Lines*, Howard Sams: Indianapolis, 1969.
21. B. Rulf and G. A. Robertshaw, *Understanding Antennas for Radar, Communications, and Avionics*, Van Nostrand Reinhold Co.: NY, 1987.
22. B. D. Steinberg, *Principles of Aperture & Array System Design*, John Wiley & Sons: NY, 1976.
23. F. R. Connor, *Antennas*, Edward Arnold: London, 1989.
24. Martin S. Smith, *Introduction to Antennas*, MacMillan Education Ltd: London, 1988.
25. E. A. Laport, *Radio Antenna Engineering*, McGraw-Hill Book Co.: NY, 1952.

26. D. W. Fry and F. K. Goward, *Aerials for Centimeter Wave-Lengths*, Cambridge University Press: Cambridge, 1950.
27. R. W. P. King and C. W. Harrison, *Antennas and Waves: A modern approach*, MIT Press: Cambridge, MA, 1969.
28. R. W. P. King, H. R. Mimno, and A. H. Wing, *Transmission Lines, Antennas and Waveguides*, McGraw-Hill: NY, 1945.

H.3 HANDBOOKS AND GENERAL REFERENCE BOOKS

1. R. C. Johnson, *Antenna Engineering Handbook*, Third Edition, McGraw-Hill Book Company: NY, 1993.
2. Y. T. Lo and S. W. Lee, Editors, *Antenna Handbook*, Van Nostrand Reinhold: NY, 1988.
3. A. W. Rudge, K. Milne, A. D. Olver, P. Knight, editors, *The Handbook of Antenna Design*, Vols. I and II, Peregrinus: London, 1982.
4. Richard C. Johnson, *Designer Notes for Microwave Antennas*, Artech House: Norwood, MA 1991.
5. R. A. Burberry, *VHF and UHF Antennas*, IEE Electromagnetic Waves Series No. 35, Peter Peregrinus Ltd.: London, 1992.
6. R. C. Hansen, Editor, *Microwave Scanning Antennas*, Vol. I–*Apertures* Vol. II–*Arrays* and Vol. III–*Frequency Scanning Arrays*, Academic Press: NY 1964. Reprinted in one volume by Penninsula Publishing, P.O. Box 867, Los Altos, CA, 94022.
7. P. J. B. Clarricoats, editor, *Advanced Antenna Technology*, Microwave Exhibitions and Publications, Ltd., UK, 1981.
8. Kai Chang, editor, *Handbook of Microwave and Optical Components, Vol. 1: Microwave Passive and Antenna Components*, John Wiley & Sons, 1989.

H.4 MEASUREMENTS BOOKS

1. *IEEE Standard Test Procedures for Antennas*, IEEE Standard 149-1979, IEEE: 445 Hoes Lane, Piscataway, NJ 08854, 1979.
2. Gary E. Evans, *Antenna Measurement Techniques*, Artech House: Norwood, MA, 1990.
3. Dan Slater, *Near-Field Antenna Measurements*, Artech House: Norwood, MA, 1991.
4. J. E. Hansen, editor, *Spherical Near-Field Antenna Measurements*, IEE Electromagnetic Wave Series, PPL Dept., IEEE Service Center Piscataway, NJ 08855-1331, 1988.

H.5 SPECIALIZED ANTENNA TOPICS BOOKS

H.5.1 Wire Antennas

1. J. Rockway, J. Logan, D. Tam, and S. Li, *The MININEC SYSTEM: Microcomputer Analysis of Wire Antennas*, Artech House: Norwood, MA, 1988.
2. S. T. Li, J. W. Rockway, J. C. Logan, and D. W. S. Tam, *Microcomputer Tools for Communications Engineering*, Artech House: Norwood, MA, 1983.
3. B. K. Kolundzija, J. S. Ognjanovic, T. K. Sarkar, and R. F. Harrington, *WIPL: Electromagnetic Modeling of Composite Wire and Plate Structures, Software and User's Manual*, Artech House: Norwood, MA, 1995.
4. B. D. Popovic, *CAD of Wire Antennas and Related Radiating Structures*, J. Wiley Research Studies Press Ltd.: NY, 1991.
5. R. W. P. King, *The Theory of Linear Antennas*, Harvard University Press: Cambridge, MA, 1956.
6. R. W. P. King, *Tables of Antenna Characteristics*, IFI/Plenum: NY, 1971.
7. R. W. P. King and G. S. Smith, *Antennas in Matter*, MIT Press: Cambridge, 1981.
8. M. L. Burrows, *ELF Communications Antennas*, Peregrinus, London, 1978.
9. S. Uda and Y. Mushiake, *Yagi-Uda Antenna*, Saski Printing and Publishing Co., Sendai, Japan, 1954.
10. A. E. Harper, *Rhombic Antenna Design*, Van Nostrand: NY, 1941.
11. R. M. Bevensee, *Handbook of Conical Antennas and Scatterers*, Gordon and Breach Science: NY, 1973.
12. M. M. Weiner, S. P. Cruze, C. C. Li, and W. J. Wilson, *Monopole Elements on Circular Ground Planes*, Artech House: Norwood, MA, 1987.
13. F. M. Landstorfer and R. R. Sacher, *Optimization of Wire Antennas*, John Wiley & Sons: NY, 1985.

14. J. R. Wait, *Electromagnetic Radiation from Cylindrical Structures*, IEE Electromagnetic Wave Series, PPL Dept., IEEE Service Center, Piscataway, NJ 08855-1331, 1988.
15. W. I. Orr, *Simple, Low-Cost Wire Antennas for Radio Amateurs*, Radio Publications: Wilton, CT, 1972.

H.5.2 Arrays

1. N. Amitay, V. Galindo, and C. P. Wu, *Theory and Analysis of Phased Array Antennas*, John Wiley & Sons: NY, 1972.
2. M. T. Ma, *Theory and Application of Antenna Arrays*, John Wiley & Sons: NY, 1974.
3. A. Kumar, *Antenna Design with Fiber Optics*, Artech House: Norwood, MA, 1996.
4. E. Brookner, *Practical Phased Array Systems*, Artech House: Norwood, MA, 1991.
5. R. C. Hansen, editor, *Significant Phased Array Papers*, Artech House: Norwood, MA, 1964.
6. A. A. Oliner and G. H. Knittel, Editors, *Phased Array Antennas*, Artech House: Norwood, MA, 1972.
7. M. Mikavica and A. Nesic, *CAD for Linear and Planar Antenna Arrays of Various Radiating Elements*, Disks and Users Manual, Artech House: Norwood, MA, 1991.
8. J. P. Scherer, *LAARAN: Linear Antenna Array Analysis Software and User's Manual*, Artech House: Norwood, MA, 1989.
9. B. D. Steinberg, *Microwave Imaging with Large Antenna Arrays*, John Wiley & Sons: NY, 1983.
10. S. Haykin, Editor, *Array Signal Processing*, Prentice-Hall: Englewood Cliffs, NJ, 1985.
11. R. W. P. King, R. B. Mack, and S. S. Sandler, *Arrays of Cylindrical Dipoles*, Cambridge: London, 1968.
12. M. T. Ma and D. C. Hyovalti, *A Table of Radiation Characteristics of Uniformly Spaced Optimum Endfire Arrays with Equal Sidelobes*, National Bureau of Standards, 1965.
13. R. C. Hansen, *Phased Array Antennas*, John Wiley & Sons: NY, 1997.

H.5.3 Broadband Antennas

1. V. Rumsey, *Frequency Independent Antennas*, Academic Press: NY, 1966.
2. Y. Mushiake, *Self-Complementary Antennas*, Springer-Verlag, Berlin, 1996.
3. Carl E. Smith, *Log Periodic Antenna Design Handbook*, Smith Electronics, Inc.: Cleveland, OH, 1966.
4. H. Nakano, *Helical and Spiral Antennas—A Numerical Approach*, John Wiley and Sons: NY, 1987.
5. R. G. Corzine and J. A. Mosko, *Four-Arm Spiral Antennas*, Artech House: Norwood, MA, 1989.

H.5.4 Traveling Wave Antennas

1. C. H. Walter, *Traveling Wave Antennas*, McGraw Hill, NY, 1965; Peninsula Pub, 1990.

H.5.5 Microstrip Antennas and Printed Antennas

1. J. R. James and P. S. Hall, editors, *Handbook of Microstrip Antennas*, Vols. I and II, Peter Peregrinis: London, 1989.
2. I. J. Bahl and P. Bhartia, *Microstrip Antennas*, Artech House, Inc.: Norwood, MA, 1980.
3. J. R. James, P. S. Hall, and C. Wood, *Microstrip Antenna Theory and Design*, IEE Electromagnetic Waves Series 12, IEE/PPL, IEEE Service Center, 445 Hoes Lane, Piscataway, NJ 08854, 1981.
4. K. C. Gupta and A. Benella, editors, *Microstrip Antenna Design*, Artech House: Norwood, MA, 1988.
5. D. M. Pozar and D. H. Schaubert, editors, *Microstrip Antennas: The Analysis and Design of Microstrip Antennas and Arrays*, IEEE Press, 1995.
6. N. Herscovici, *CAD of Aperture—Fed Microstrip Transmission Lines and Antennas*, Artech House: Norwood, MA, 1996.
7. R. A. Sainati, *CAD of Microstrip Antennas for Wireless Applications*, Artech House: Norwood, MA, 1996.
8. J.-F. Zurcher and F. Gardiol, *Broadband Patch Antennas*, Artech House: Norwood, MA, 1995.

9. P. Bhartia, K. V. S. Rao, and R. S. Tomar, *Millimeter-Wave Microstrip and Printed Circuit Antennas*, Artech House: Norwood, MA, 1988.
10. T. C. Edwards, *Foundations of Microstrip Antennas*, John Wiley & Sons: NY, 1981.
11. G. Dubost, *Flat Radiating Dipoles and Applications to Arrays*, Research Studies Press, John Wiley: Chichester, 1981.
12. A. K, Bhattacharyya, *Electromagnetic Fields in Multilayered Structures: Theory and Applications*, Artech House: Norwood, 1994.

H.5.6 Reflector and Lens Antennas

1. B. S. Wescott, *Shaped Antenna Reflector Design*, John Wiley & Sons: NY, 1983.
2. A. W. Love, Editor, *Reflector Antennas*, IEEE Press: NY, 1978.
3. W. V. T. Rusch and P. D. Potter, *Analysis of Reflector Antennas*, Academic Press: NY, 1970.
4. R. Mittra, et al., eds., *Satellite Communications Antenna Technology*, Elsevier, 1983.
5. P. J. Wood, *Reflector Antenna Analysis and Design*, IEE/PPL, IEEE Service Center, 445 Hoes Lane, Piscataway, NJ 08854, 1986.
6. C. J. Sletten, editor, *Reflector and Lens Antennas*, Artech House Inc.: Norwood, MA, 1988.
   Reflector and Lens Antennas: Analysis and Design Using PCs, Software and Users Manual, Version 2.0, 1991.
7. Craig R. Scott, *Modern Methods of Reflector Antenna Analysis and Design*, Artech House: Norwood, MA, 1989.
8. Roy Levy, *Structural Engineering of Microwave Antennas for Electrical, Mechanical, and Structural Engineers*, IEEE Press, 1996.
9. J. Brown, *Microwave Lenses*, John Wiley: London, 1953.

H.5.7 Horns/Feeds

1. P. J. B. Clarricoats and A. D. Olver, *Corrugated Horns for Microwave Antennas*, IEEE Service Center, PPL Dept.: 445 Hoes Lane, Piscataway NJ 08854, 1984.
2. A. O. Olver, P. J. B. Clarricoats, A. A. Kisk, and L. Shafai, *Microwave Horns and Feeds*, IEEE Press, 1994.
3. J. Uher, J. Bornemann, and U. Rosenberg, *Waveguide Components for Antenna Feed Systems: Theory and CAD*, Artech House: Norwood, MA, 1993.
4. A. W. Love, Editor, *Electromagnetic Horn Antennas*, IEEE Press: NY, 1976.

H.5.8 Moment Methods

1. R. F. Harrington, *Field Computation by Moment Methods*, Macmillan: NY, 1968.
2. M. N. O. Sadiku, *Numerical Techniques in Electromagnetics*, CRC Press: Boca Raton, FL, 1992.
3. Richard C. Booton, Jr., *Computational Methods for Electromagnetics and Microwaves*, John Wiley: NY, 1992.
4. B. D. Popovic, M. B. Dragovic, and A. R. Djordjevic, *Analysis and Synthesis of Wire Antennas*, J. Wiley Research Studies Press, 1982.
5. J. Moore and R. Pizer, editors, *Moment Methods in Electromagnetics*, Research Studies Press, John Wiley: Letchworth, England, 1984.
6. C. Hafner, *The Generalized Multipole Technique for Computational Electromagnetics*, Artech House: Norwood, MA, 1990.
7. C. Hafner, *2-DMMP: Two-Dimensional Multiple Multipole Software and User's Manual*, Artech House: Norwood, MA, 1990.
8. A. R. Djordjevic, M. B. Bazdar, G. M. Vitosevic, T. K. Sarkar, and R. F. Harrington, *Analysis of Wire Antennas and Scatterers: Software and User's Manual*, Artech House: Norwood, MA, 1990.
9. E. K. Miller, L. Medgyesi-Mitschang, and E. H. Newman, Editors, *Computational Electromagnetics: Frequency-Domain Method of Moments*, IEEE Press: IEEE Service Center, Piscataway, NJ, 1991.
10. Johnson J. H. Wang, *Generalized Moment Methods in Electromagnetics*, John Wiley & Sons: NY, 1991.
11. Robert C. Hansen, *Moment Methods in Antennas and Scattering*, Artech House: Norwood, MA, 1990.
12. R. Mittra, Editor, *Computer Techniques for Electromagnetics*, Pergamon Press: Oxford, 1973.

H.5.9 FD-TD

1. K. S. Kunz and R. J. Luebbers, *Finite Difference Time Domain Method for Electromagnetics*, CRC Press: Boca Raton, FL, 1993.
2. A. Taflove, *Computational Electromagnetics: The Finite-Difference Time-Domain Method*, Artech House: Norwood, 1995.

H.5.10 High Frequency Methods

1. D. A. McNamara, J. A. G. Malherbe, and C. W. Pistorius, *Introduction to the Uniform Geometrical Theory of Diffraction*, Artech House: Norwood, MA, 1989.
2. G. L. James, *Geometrical Theory of Diffraction for Electromagnetic Waves*, revised edition, PPL Dept. IEEE Service Center, Piscataway, NJ 08855-1331, 1986.
3. L. B. Felsen and N. Marcuvitz, *Radiation and Scattering of Waves*, Prentice-Hall, Englewood Cliffs, NJ, 1973.
4. B. S. Cornbleet, *Microwave and Optical Ray Geometry*, John Wiley & Sons, NY, 1984.
5. E. V. Jull, *Aperture Antennas and Diffraction Theory*, IEE/PPL, IEEE Service Center, 445 Hoes Lane, Piscataway, NJ 08854, 1981.
6. R. C. Hansen, Editor, *Geometric Theory of Diffraction*, IEEE Press, 1981.
7. B. S. Cornbleet, *Microwave Optics: The Optics of Microwave Antenna Design*, Academic Press: London, 1976.
8. R. H. Clarke and J. Brown, *Diffraction Theory and Antennas*, John Wiley & Sons: NY, 1980.

H.5.11 Adaptive Antennas

1. J. E. Hudson, *Adaptive Array Principles*, Peter Peregrinus: Stevenage UK, 1981.
2. R. T. Compton, Jr., *Adaptive Antennas: Concepts and Performance*, Prentice-Hall: Englewood Cliffs, NJ, 1988.
3. R. A. Monzingo and T. W. Miller, *Introduction to Adaptive Arrays*, John Wiley & Sons: New York, 1980.
4. A. Farina, *Antenna-Based Signal Processing Techniques for Radar Systems*, Artech House: Norwood, MA, 1991.
5. E. Nicolau and D. Zaharia, *Adaptive Arrays*, Elsevier, 1989.

H.5.12 Mobile, Personal, and Satellite Communications Antennas

1. K. Fujimoto and J. R. James, editors, *Mobile Antenna Systems Handbook*, Artech House: Norwood, MA, 1989.
2. K. Siwiak, *Radiowave Propagation and Antennas for Personal Communications*, Artech House, Norwood, 1995.
3. Preston E. Law, *Shipboard Antennas*, Artech House, Norwood, MA, 1983.
4. T. Kitsuregawa, *Satellite Communication Antennas: Electrical and Mechanical Design*, Artech House: Norwood, MA, 1989.

H.5.13 Polarization Topics

1. W. L. Stutzman, *Polarization in Electromagnetic Systems*, Artech House: Norwood, MA, 1993.
2. H. Mott, *Polarization in Antennas and Radar*, John Wiley & Sons: NY, 1986.

H.5.14 Radomes

1. D. J. Kazakoff, *Analysis of Radome Enclosed Antennas*, Artech House: Norwood, MA, 1997.
2. J. D. Walton, Jr., editor, *Radome Engineering Handbook*, Marcel Dekker: NY, 1981.
3. H. L. Hirsch and D. C. Grove, *Practical Simulation of Radar Antennas and Radomes*, Artech House: Norwood, MA, 1988.

H.5.15 Software for General Antenna Applications

1. L. Diaz and T. Milligan, *Antenna Engineering Using Physical Optics: Practical CAD Techniques and Software*, Artech House: Norwood, MA, 1996.
2. D. M. Pozar, *PCAAD–Personal Computer Aided Antenna Design*, Antenna Design Associates: 55 Teawaddle Hill Road, Leverett, MA 01002, 1992.
3. D. Pozar, *Antenna Design Using Personal Computers*, Artech House, Norwood, MA, 1985.
4. J. A. Kuecken, *Exploring Antennas and Transmission Lines by Personal Computer*, Van Nostrand Reinhold, 1986.

H.5.16 Small Antennas

1. K. Hirasawa and M. Haneishi, *Analysis, Design, and Measurement of Small and Low-Profile Antennas*, Artech House: Norwood, MA, 1992.
2. K. Fujimoto, A. Henderson, K. Kirasawa, and J. James, *Small Antennas*, John Wiley & Sons: NY, 1987.

H.5.17 Other Topics

1. A. Kumar and H. D. Hristov, *Microwave Cavity Antennas*, Artech House: Norwood, MA, 1989.
2. Rajeswari Chatterjee, *Dielectric and Dielectric-Loaded Antennas*, John Wiley & Sons: NY, 1985.
3. D. G. Kiely, *Dielectric Aerials*, Methuen, 1952.
4. J. Bach Anderson, *Metallic and Dielectric Antennas*, Polyteknisk Forlag: Denmark, 1971.
5. T. Macnamara, *Handbook of Antennas for EMC*, Artech House: Boston, 1995.
6. D. R. Rhodes, *Synthesis of Planar Antenna Sources*, Clarendon Press: Oxford, 1974.
7. F. Sporleder and H-G Unger, *Waveguide Tapers, Transitions, and Couplers*, IEE Electromagnetic Waves Series 6, PPL Dept. IEEE Service Center, Piscataway, NJ 08855-1331, 1979.
8. G. A. Savitskii, *Calculations for Antenna Installations*, Amerind Pub. Co.: New Delhi, 1982.
9. G. W. Wiskin, R. Manton, and J. Causebrook, *Masts, Antennas, and Service Planning*, Focal Press: Oxford, England, 1992.
10. Kai Chang, *Microwave Ring Circuits and Antennas*, John Wiley & Sons: NY, 1996.

Index